Linear Algebra

John de Pillis
UNIVERSITY OF CALIFORNIA, RIVERSIDE

Linear Algebra

Holt, Rinehart and Winston, Inc.
NEW YORK / CHICAGO / SAN FRANCISCO / ATLANTA
DALLAS / MONTREAL / TORONTO / LONDON / SYDNEY

Library of Congress Catalog Card Number: 69-14251
SBN: 03-073595-5
Printed in the United States of America
1 2 3 4 5 6 7 8 9

In memory of my mother
Lili de Pillis

Preface

Description. Suitable for freshman or sophomore students, this introduction to linear algebra begins with a discussion of concrete vector spaces after which systems of linear equations are encountered. Vector-space notions are immediately applied in presenting a theory of linear programming in the plane. We then proceed to linear transformations and their matrices, followed by a deeper study of linear equations. Next, determinants are studied from both a classical and nonclassical approach. The final chapter deals with the structure of operators and introduces ideas of inner product, diagonalization, and the spectral decomposition theorems for self-adjoint operators and Hermitian matrices. Each chapter contains examples, exercises, and a summary of the material discussed.

A complete chart of all theorems and definitions of the book is followed by a self-contained appendix on mathematical induction and an appendix on complex numbers.

Features. We are guided by two general principles: First, new ideas should be offered in a concrete (or particular) setting and should then evolve into the more abstract (or general) situation. Second, sufficient space should be given to establishing motivation and anticipating the student's questions. (Why is this definition necessary? Where did it come from? Why is this theorem given in this form?)

Consistent with these general aims, the following particular features are incorporated in this text:

- Concrete n-vectors are introduced at the outset and vector-space notions are immediately used to develop a theory of linear programming in the plane. As the first chapter unfolds, we are systematically led to the definition of an *abstract* vector space.
- Systems of simultaneous linear equations are briefly explored in Chapter I

where the technique of Gaussian Elimination is described. The justification of this technique, along with a deeper study of linear systems, appears later in Chapter III.

- Each chapter contains an abundance of exercises and illustrative examples.
- By use of the notation $mx(\mathbf{T})_{\mathscr{X}\mathscr{Y}}$ for the matrix of a linear transformation **T**, we have tried to underscore the fact that the matrix depends on the *three* entities which are: the linear transformation **T**, the ordered basis $\mathscr{X}$ of the domain space, and the ordered basis $\mathscr{Y}$ of the range space. Many examples and exercises dwell on this dependence, as well.
- Matrices are discussed for the 2×2 case first. This special case may be omitted, if desired, in favor of the general $m \times n$ case which follows.
- Determinants are presented in two ways. First, in Chapter IV, we have the classical development depending on properties of permutation functions. Alternately, Chapter V offers the nonclassical multilinear viewpoint which does not require permutations.
- The spectral theorem for self-adjoint operators and hermitian matrices is not presented as a dead-end finale. This important decomposition finds extensive use at the end of Chapter VI.
- Important computational techniques of the text are immediately retrievable via the unique "computations" entry in the index. For example, five references are given for techniques of finding the inverse of a matrix under the "matrix inverse" subheading of the "Computations" entry.
- For easy location of theorems and definitions, we use a consecutive triple entry numbering system which appears in the running head of every page. As an example, item (theorem or definition) II.3.9 appears in Chapter II, Section 3, and is the ninth item of that section.
- Each chapter concludes with a summary.
- A bird's-eye view of the entire text appears for the student's convenience in the form of a chart of all theorems and definitions (following Chapter VI).

Acknowledgments. Most importantly, it was my wife Susie who encouraged me in the writing of this book. I am grateful for the helpful comments and suggestions of Professors R. M. Thrall and J. Murtha, who read the entire manuscript in raw form. Also, much credit is due Mrs. Ollie Cullers, Mrs. Jane Scully, and Mrs. Sharon Unangst, who translated scribbled notes into a polished typed form. For the use of their facilities, I wish to thank the University of California at Riverside, and Brookhaven National Laboratory, Upton, New York.

Riverside, California JOHN DE PILLIS
January 1969

Contents

Linear Algebra

Introduction

The purpose of this introduction is to set down the language of set theory. The definitions presented are somewhat inexact in that the undefined terms are not explicitly cited. The reason for this omission is that a rigorous development of the theory of sets would require techniques that are quite advanced. However, it is hoped that the concepts will sit well with the readers' intuition.

Definition 1 A **set** will denote a collection (family, aggregate) of distinguishable objects, which will be called **elements** or **members** of the set.

Notation 2 If A represents a set, and x represents one of its elements (members), then we use the notation

$$x \in A, \quad \text{read} \quad \text{"}x \text{ belongs to the set } A\text{,"}$$

or "x is an element of A," or "x is a member of A." Not too surprisingly, then, we use the notation $x \notin A$ to stand for the negative of the previous statement(s), that is, "x is not an element of A," and so forth.

Notation 3 Suppose the symbols $x_1, x_2, \ldots, x_n, \ldots$ represent a "listing" of the elements of a set A. (Here, the dots ... are to stand for the phrase "and so on" which indicates that we really know what all the elements of A are (somehow), but use this shorthand as a convenience). In this case, we could simply write

$$A = \{x_1, x_2, \ldots, x_n, \ldots\}.$$

Example 1 Let A be the 4-element set of letters y, a, r, g. That is (Notation 3), $A = \{y,a,r,g\}$. Alternatively, $A = \{g,r,a,y\}$ or $A = \{r,a,y,g\}$. If you like to think of A as the word "yarg," then you are artificially imposing an *ordering* on the elements of A. Heretofore, we have mentioned nothing of ordered sets, although they will prove useful to us later.

Example 2 Let A represent the 3-element set whose elements are $\{y,a,r,g\}$, $\{p,o,t\}$, and $\{n,u,t,s\}$. Notice that the elements are themselves *sets*. Thus, $A = \{\{y,a,r,g\}, \{p,o,t\}, \{n,u,t,s\}\}$. According to Notation 2, we make the statement,

$$\{p,o,t\} \in A,$$

or

$$\{p,o,t\} \in \{\{y,a,r,g\}, \{p,o,t\}, \{n,u,t,s\}\}.$$

Now, each *letter* is *not* an element of A. That is,

$$p \notin A;$$

rather, $p \in \{p,o,t\}$, and the set $\{p,o,t\}$ is an element of A.

EXERCISES

1 Let A be the set whose elements are the integers from 3 to 7 inclusive.
(a) Is it true that $4 \in A$?
(b) Is it true that $\{3,4\} \in A$?
(c) Is it true that $2 \in A$?

2 Let A be the following set of *words* (here, a *word* is to be interpreted as an *ordered set* in itself, for example $\{c,a,t\}$ is a word, but is not the same word as $\{t,a,c\}$ since the *orderings* of the letters differ):

$$A = \{\{c,a,t\}, \{t,a,c\}, \{f,r,o,g\}, \{g,a,r,f\}\}.$$

Which of the following statements is true? Explain.
(a) $\{f,r,o,g\} \in A$
(b) $f \in \{f,r,o,g\}$
(c) $f \in A$
(d) $\{f,r,o,g,a,r,f\} \in A$
(e) $\{t,a\} \in A$.

Definition 4 Let A and B be sets. Then the symbols $A \subset B$ (equivalently, $B \supset A$), read **A is contained in B**, **B contains A** or **A is a subset of B**, is to mean that each and every element of A is also an element of B. That is to say, $A \subset B$ means:

$$\text{whenever} \quad x \in A, \quad \text{then} \quad x \in B.$$

Example 3 Let $A = \{1,2\ 3\ 4\}$ and $B = \{1,3\}$. We assert that $B \subset A$, since (as Definition 4 requires), every element of B is also an element of A. The elements of B are exactly the integers 1 and 3 and surely, $1 \in \{1,2,3,4\}$ and $3 \in \{1,2,3,4\} = A$.

With the notion of inclusion (Definition 4) in hand, we are able to say precisely what is meant in saying that one set A is "equal to" another set B. Informally, we shall write $A = B$ whenever elements of A are always elements of B, and *vice versa*. Formally, we have Definition 5.

Definition 5 Let A and B be sets. The symbol $A = B$, read **A equals B** or **B equals A**, is to mean that $A \subset B$ *and* $B \subset A$.

Example 4 Let A be the set of all even integers, that is, integers which are of the form $2 \cdot q$ where q is an arbitrary integer. We construct the set B as follows: Select the integer 0. Then leave the adjacent integers alone (in this case leave 1 and -1 alone) and select the "next" integers (in this case, select 2 and -2). Repeat this selection process of selecting "every other" integer. The set of all selected integers (elements) will be our set B.

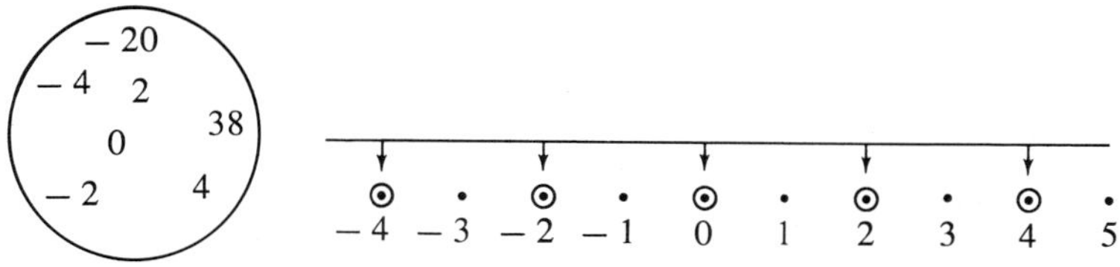

$A =$ All integers $2 \cdot q$ $\qquad$ $B =$ Every "other" integer, $0 \in B$

(a) $\qquad$ (b)

Figure 1

Note that we've described elements of 2 (apparently different) sets A and B. Certainly, these descriptions are not the same, for elements of A were selected by their *algebraic* properties, namely, divisibility by 2, while elements of B were selected by their geometric arrangement on a line, namely, the arrangement dictated by their order (see Figure 1). However, every integer (in A) of the form $2 \cdot q$, q an integer, is an integer in B, that is, $x \in A$ implies $x \in B$, or $A \subset B$. Every integer of B is also of the form $2 \cdot q$ for some integer q, that is, $x \in B$ implies $x \in A$, or $B \subset A$. That is, $A = B$.

Definition 6 The **empty set** (void set, null set), symbolized by the letter $\varnothing$, is the set that has no elements.

Note that Definition 4 concerned itself with a certain relationship (inclusion) between pairs of sets. In what follows, we will define 2 operations on pairs of sets (union and intersection) that will allow us to generate a third set.

Definition 7 Suppose A and B are sets. The (third) set, ***A* union *B***, symbolized $\boldsymbol{A \cup B}$, is the set of elements that belong either to A or to B (or possibly to both A and B). In other words an element qualifies as a member of $A \cup B$ if and only if that element is already an element of (*at least*) one (but not necessarily both) of the sets, A or B. Thus,

$$x \in A \cup B \quad \text{if and only if} \quad x \in A \quad \text{or} \quad x \in B.$$

Example 5 Let $A = \{4,a,D\}$ and $B = \{a,b,c\}$. Then $A \cup B = \{4,a,D,b,c\}$. Note that $a \in A$ and $a \in B$, but it is not "counted twice" in $A \cup B$. The element a meets the test of Definition 7 in that it is an element of at least one of the sets, A or B, and that's all we need to know.

Example 6 Let $A = \{1,2,3,4\}$, and $B = \{a,b,c\}$. Then $A \cup B = \{1,2,3,4,a,b,c\}$.

Definition 8 Suppose A and B are sets. The (third) set, ***A* intersect *B***, symbolized $\boldsymbol{A \cap B}$, is the set of elements that belong to both A and B (at the same time). Thus,

$$x \in A \cap B \quad \text{if and only if} \quad x \in A \quad \text{and} \quad x \in B.$$

Example 7 Let A and B be as in Example 5. Looking for the elements which are common to both sets, we are able to unearth only the element a. Thus, $A \cap B = \{a\}$.

Example 8 Let A and B be as in Example 6. There are absolutely no elements common to *both* A and B in this case. Thus $A \cap B = \varnothing$, the set with no elements (Definition 6).

Sets of the form {3,chair, #} are not frequent, and, indeed, may raise the question, "Is any and every object eligible to be a member of a set?" Were we to answer in the affirmative, we would soon arrive at some unhappy paradoxes. For example, admitting "too many" sets leads to such puzzles as deciding whether the following sentence is true or false:

"This sentence is false."

If the sentence is true, then we are obliged to believe its statement, namely, it is a false sentence. If, on the other hand, we decide to judge the sentence as false, then we must believe the opposite of its message, namely, that the sentence is not false. Such a dilemma might impel one to abandon mathematics altogether, but a judicious limiting of eligible sets does put our house in order. It is for this reason that we prescribe all of our "eligible" elements in advance. The set of elements under consideration will be called the **universal set** (universe) and is denoted U. For example, U may stand for all integers. We then are permitted to speak only of subsets whose elements are integers.

Example 9 Let $U = \{-2,-1,0,1,2\}$. In *this* context, now, the set $\{1,5\}$ has no meaning, since the element 5 is not drawn from our universe U.

Definition 9 Suppose A is a set. The set ***A* complement** or **complement of *A*** (symbolized $\sim A$), is exactly the set of elements which are in U, but not in A. Thus, $x \in \sim A$ if and only if $x \in U$ and $x \notin A$.

Example 10 Let $U = \{s,u,v,w,3,4\}$, and let $A = \{u,v,3\}$. Then $\sim A = \{s,w,4\}$, the set of elements of U, which are not in A.

The so-called "Venn diagram" is a useful device for illustrating the previously defined operations and set relations. The idea of a Venn diagram is to present a set by a certain set of points in the plane (a 2-dimensional "surface"). Observations (however obvious) from Venn diagrams do not *prove* anything about general sets. For one thing, not every set is, in fact, a subset of the plane, and secondly, all proofs must flow from the explicit use of the *definitions*.

We present examples of Venn diagrams in Figure 2. The shaded regions represent the sets indicated under the set U. Notice that in (b) and (d) the set B is the set of points (in U) which are outside the rectangle.

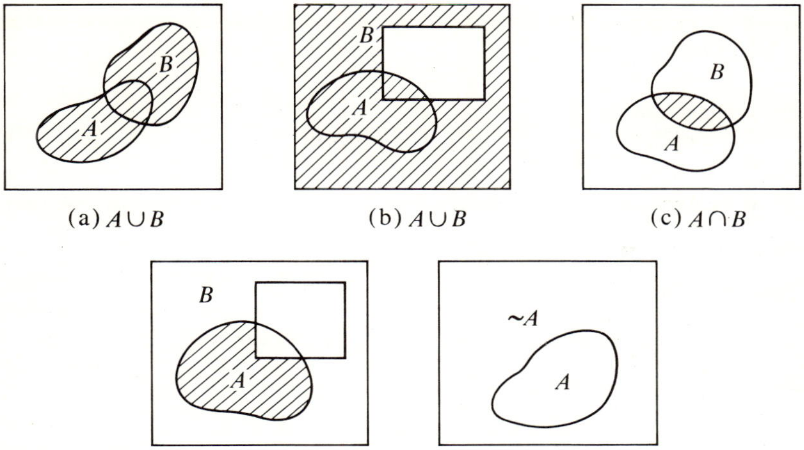

(a) $A \cup B$ (b) $A \cup B$ (c) $A \cap B$

(d) $A \cap B$ (e)

Figure 2

EXERCISES

Verify the following statements (Exercises 3 through 13) for sets A, B, and C in the universe U. Use the definitions and illustrate with a Venn diagram.

3 $A \subset A$

4 $A \cup A = A$

5 $A \cap A = A$

6 $A \cap \varnothing = \varnothing$

7 $A \cup \varnothing = \varnothing$

8 $\varnothing \subset A$

9 $A \cap (B \cup C) = (A \cap B) \cup (A \cap C)$

10 $A \cup (B \cap C) = (A \cup B) \cap (A \cup B)$

11 $\sim(\sim A) = A$

12 $\sim\varnothing = U$

13 $\sim U = \varnothing$

14 Let $U = \{1,2,3,4,5,6\}$, $A = \{1,3,5\}$, $B = \{2\}$, and $C = \{1,2,3\}$. Describe the following sets:

(a) $\sim A$, $\sim B$, $\sim C$.

(b) $A \cap B$, $A \cap \sim B$, $\sim(\sim A \cup B)$

(c) $A \cap C$, $A \cap \sim C$, $\sim(\sim A \cup C)$

(d) $A \cup B$, $A \cup \sim B$, $\sim(A \cap B)$

(e) $A \cup \sim B$, $\sim(\sim A \cap B)$

(f) $(A \cap B) \cap C$

(g) $A \cap (B \cap C)$

(h) $(A \cap \sim B) \cap \sim C$

(i) $A \cap (B \cup C)$

(j) $(A \cap B) \cup (A \cap C)$

15 Show by example, that for certain sets A and B, it is not always true that

$$\sim(A \cup B) = \sim A \cup \sim B.$$

Show that, however,

$$\sim(A \cup B) = \sim A \cap \sim B \quad \textit{is} \text{ always true.}$$

16 Show by example, that for certain sets A and B, it is not always true that

$$\sim(A \cap B) = \sim A \cap \sim B.$$

Show that, however,

$$\sim(A \cap B) = \sim A \cup \sim B \quad \textit{is} \text{ always true.}$$

17 Prove that, for any sets A and B,

$$(A \cap B) \subset B \quad \text{and} \quad (A \cap B) \subset A$$

18 Prove that, for any sets A and B,

$$A \subset (A \cup B) \quad \text{and} \quad B \subset (A \cup B).$$

CHAPTER I

Vector Spaces and Applications

1. Addition and Scalar Multiplication of Vectors

Definition I.1.1 A **concrete column vector** is an ordered set of n numbers (or **scalars**)

$$\begin{pmatrix} x_1 \\ x_2 \\ \vdots \\ x_n \end{pmatrix}$$

written in vertical fashion, where n is some positive integer. A **concrete row vector** is an ordered set of numbers $(x_1,x_2,\ldots,x_n)$ written in horizontal fashion. In both cases, the numbers $x_1, x_2,\ldots, x_n$, are called **coordinates**, **entries**, or **components**, of the respective vectors.

The set of all column vectors is denoted by the symbol V_n. V_n is also denoted by $V_n(\mathbf{R})$ or $V_n(\mathbf{C})$ according to whether the entries are elements of the real numbers $\mathbf{R}$, or complex numbers $\mathbf{C}$, respectively.

Example I.1

$$\begin{pmatrix} 2 \\ 0 \\ 1 \end{pmatrix}, \begin{pmatrix} -3 \\ 1 \end{pmatrix}, \begin{pmatrix} \frac{1}{2} \\ 0 \\ -1 \\ 1 \end{pmatrix}, (1,3,0,2),\ (1,1),\ (0,\tfrac{1}{2},-\tfrac{1}{2}).$$

The first 3 vectors are column vectors (in V_3) with 3, 2, and 4 components (coordinates), respectively. The last 3 vectors are row vectors with 4, 2, and 3 components, respectively.

Example I.2 For each family on the block, we may construct a column (or row) vector with 3 coordinates. Let the first coordinate represent the pennies spent each week on milk, the second coordinate represent the weekly expenditure for garlic, and the third coordinate agree with the weekly expenditure for mouthwash. Thus, if there are only 2 families on the block, family A and family B, we might represent their weekly expenditures:

$$X_A = \begin{pmatrix} 250 \\ 10 \\ 60 \end{pmatrix} \qquad X_B = \begin{pmatrix} 125 \\ 40 \\ 730 \end{pmatrix} \begin{matrix} \leftarrow \text{milk} \\ \leftarrow \text{garlic} \\ \leftarrow \text{mouthwash.} \end{matrix}$$

Suppose we raise the question: How much money is spent by the families on the block in a week? It seems clear that to find the *combined* weekly expenditure profile, we add componentwise, that is, if expenses of family $A = X_A$, expenses of family $B = X_B$, then

$$X_A \text{ plus } X_B = \begin{pmatrix} 250 \\ 10 \\ 60 \end{pmatrix} \text{ plus } \begin{pmatrix} 125 \\ 40 \\ 730 \end{pmatrix} = \begin{pmatrix} 250 + 125 \\ 10 + 40 \\ 60 + 730 \end{pmatrix} = \begin{pmatrix} 375 \\ 50 \\ 790 \end{pmatrix} \begin{matrix} \leftarrow \text{milk} \\ \leftarrow \text{garlic} \\ \leftarrow \text{mouthwash.} \end{matrix} \qquad \text{(I.1.1)}$$

Actually, Equation (I.1.1) sets the pattern for the notion of *addition of one vector to another*. We set down a formal definition of vector addition with

Definition I.1.2 Suppose X and Y are column vectors, each with n coordinates. Then the **sum of *X* and *Y*, *X* ∔ *Y***, is defined by

$$X \dotplus Y = \begin{pmatrix} x_1 \\ x_2 \\ \vdots \\ x_n \end{pmatrix} \dotplus \begin{pmatrix} y_1 \\ y_2 \\ \vdots \\ y_n \end{pmatrix} = \begin{pmatrix} x_1 + y_1 \\ x_2 + y_2 \\ \vdots \\ x_n + y_n \end{pmatrix}$$

Similarly, if X and Y are each row vectors with n coordinates, then ***X* plus *Y*, *X* ∔ *Y***, is defined by

$$(x_1, x_2, \ldots, x_n) \dotplus (y_1, y_2, \ldots, y_n) = (x_1 + y_1, x_2 + y_2, \ldots, x_n + y_n).$$

REMARK. The sum of a pair of concrete vectors is meaningful only when both vectors have the *same* number of components (coordinates). Moreover, we observe that the sum of 2 vectors, each having n components, yields a vector of the *same type*, that is, one with n components, Also, we have used the symbol $\dotplus$ to underscore the fact that we are adding something *other* than numbers; addition of numbers is denoted by the usual symbol $+$. The following example will motivate our definition of what is meant by the multiplication of a vector by a number.

Example I.3 Suppose we construct a "happiness profile" for a certain child over a year's time which lists the following: (1) The number of pounds of popcorn received, and (2) the number of hours spent juggling oranges. If, in the year 1878, our child received 21 pounds of popcorn, and spent 48 hours juggling oranges, then his happiness profile would be represented by the column vector

$$X = \begin{pmatrix} 21 \\ 48 \end{pmatrix} \begin{matrix} \leftarrow \text{pounds of popcorn received during 1878} \\ \leftarrow \text{hours juggling oranges in 1878.} \end{matrix}$$

Now if, in the following year of 1879, we desired to "increase the happiness" of the child $2\frac{1}{2}$ times, we could reasonably expect to achieve this goal by multiplying each component of the happiness profile vector, X, by the number $2\frac{1}{2}$. That is,

$$2\tfrac{1}{2} \text{ times} \begin{pmatrix} 21 \\ 48 \end{pmatrix} = \begin{pmatrix} 2\frac{1}{2} \cdot 21 \\ 2\frac{1}{2} \cdot 48 \end{pmatrix} = \begin{pmatrix} 52\frac{1}{2} \\ 120 \end{pmatrix},$$

which tells us that in the year 1879, this child will have been $2\frac{1}{2}$ times as happy as he was the previous year, after receiving a total of $52\frac{1}{2}$ pounds of popcorn, and after juggling oranges for 120 hours. To each his own. Nevertheless, we now have a model from which to pattern a *general* definition for the product of a number against a vector.

Definition I.1.3 Let the symbol α (alpha) represent a number, and let X be a row (respectively, column) vector with n coordinates $x_1, x_2, \ldots, x_n$. The vector $\alpha \cdot X$, **α times X**, is the row (respectively, column) vector whose n coordinates are the numbers $\alpha x_1, \alpha x_2, \ldots, \alpha x_n$. That is, for column vectors and for row vectors, we have

$$\alpha \cdot \begin{pmatrix} x_1 \\ x_2 \\ \vdots \\ x_n \end{pmatrix} = \begin{pmatrix} \alpha x_1 \\ \alpha x_2 \\ \vdots \\ \alpha x_n \end{pmatrix},$$

and

$$\alpha \cdot (x_1, x_2, \ldots, x_n) = (\alpha x_1, \alpha x_2, \ldots, \alpha x_n).$$

As an extension of the notion of equality of numbers, we define what is meant by the equality of two *vectors*.

Definition I.1.4 Let X and Y both be (row or column) vectors with components $x_1, x_2, \ldots, x_n$, and $y_1, y_2, \ldots, y_n$, respectively. Then the vectors X and Y are said to be **equal**, symbolized $\mathbf{X = Y}$, or $\mathbf{Y = X}$, if and only if corresponding components agree, that is, if and only if $x_1 = y_1$, $x_2 = y_2, \ldots,$ and $x_n = y_n$.

EXERCISES

I.1.1 Given the following column vectors

$$X = \begin{pmatrix} 1 \\ 0 \\ 2 \end{pmatrix}, \qquad Y = \begin{pmatrix} 2 \\ 0 \\ 4 \end{pmatrix}, \qquad Z = \begin{pmatrix} 1 \\ -1 \\ 1 \end{pmatrix},$$

compute

(a) $X + Y$
(b) $Y + Z$
(c) $2 \cdot X + (-1) \cdot Y$
(d) $(2 \cdot X + (-1) \cdot Y) + Z$
(e) $4 \cdot Z$
(f) $(-4) \cdot Z$
(g) $4 \cdot Z + (-4) \cdot Z$
(h) $(X + Y) + Z$
(i) $X + (Y + Z)$
(j) $3 \cdot X + 2 \cdot X$
(k) $5 \cdot X$
(l) $(Z + 3 \cdot Y) + (-1) \cdot X$

I.1.2 With the following row vectors,

$$X = (2, -1, 0, 1), \qquad Y = (4, 0, 0, 2), \qquad Z = (-1, -1, 1, 1),$$

compute parts (a) through (l) of Exercise I.1.1.

I.1.3 Find the unknown coordinates if

(a) $\begin{pmatrix}1\\2\\1\end{pmatrix} \dotplus \begin{pmatrix}x_1\\x_2\\x_3\end{pmatrix} = \begin{pmatrix}4\\1\\5\end{pmatrix}.$

(b) $\begin{pmatrix}x_1\\2\\1\end{pmatrix} \dotplus \begin{pmatrix}3\\y_2\\y_3\end{pmatrix} = \begin{pmatrix}0\\1\\2\end{pmatrix}.$

(c) $2 \cdot (x_1,3,x_3) \dotplus 3 \cdot (1,y_2,1) = (0,0,0).$
(d) $(-2) \cdot (x_1,3,x_3) \dotplus 3 \cdot (1,y_2,1) = (0,0,0).$
(e) $(x_1,x_2,x_3,x_4) \dotplus (1,2,3,4) = (1,2,3,4).$

I.1.4 Find the scalars (numbers) α and β, if

(a) $\alpha \cdot \begin{pmatrix}1\\2\\0\\1\end{pmatrix} \dotplus \beta \cdot \begin{pmatrix}1\\1\\0\\0\end{pmatrix} = \begin{pmatrix}5\\8\\0\\3\end{pmatrix}.$

(b) $\alpha \cdot (1,2) \dotplus \beta \cdot (3,5) = (-2,-3).$
(c) $\alpha \cdot (x_1,x_2) \dotplus \beta \cdot (0,y_2) = (-x_1,-x_2).$

(d) $\alpha \cdot \begin{pmatrix}1\\0\\1\end{pmatrix} \dotplus \beta \cdot \begin{pmatrix}0\\2\\0\end{pmatrix} = \begin{pmatrix}-x_1\\x_2\\-x_1\end{pmatrix}.$

(e) $(2\alpha) \cdot \begin{pmatrix}1\\0\end{pmatrix} \dotplus (3\beta) \cdot \begin{pmatrix}1\\1\end{pmatrix} = \begin{pmatrix}1\\6\end{pmatrix}.$

I.1.5 Show that no numbers α and β can exist such that

$$\alpha \cdot \begin{pmatrix}1\\1\\0\end{pmatrix} \dotplus \beta \cdot \begin{pmatrix}-1\\1\\0\end{pmatrix} = \begin{pmatrix}2\\3\\1\end{pmatrix}.$$

I.1.6 Prove that for any row or column vector X, with 3 components x_1, x_2, x_3, and for any pair of numbers α, β,

(a) $(\alpha + \beta) \cdot X = \alpha \cdot X \dotplus \beta \cdot X,$

and

(b) $\alpha \cdot (\beta \cdot X) = (\alpha\beta) \cdot X.$

Are these equations true if X has n (n not necessarily equal to 3) components, for n a positive integer?

Note: The left-hand side and the right-hand side of Exercises I.1.6a and I.1.6b represent *different* sequences of operations. The left-hand side of Exercise I.1.6a requires that the vector X be multiplied by the number $(\alpha + \beta)$; the right-hand side, however, is a sum (not a product like the left-hand side) of the 2 vectors $\alpha \cdot X$ and $\beta \cdot X$. Similarly, the left-hand side of Exercise I.1.6b represents the product of the number α against the vector $\beta \cdot X$; the right-hand side, however, represents the product of the (different) scalar $\alpha\beta$ against the (different) vector X.

I.1.7 If X and Y are each (row or column) vectors with 3 components, is it true that for any and all numbers α, that the vector $\alpha \cdot (X + Y)$ is the same as the sum of the 2 vectors $\alpha \cdot X$ and $\alpha \cdot Y$? That is, do we always have

$$\alpha \cdot (X + Y) = \alpha \cdot X + \alpha \cdot Y \text{ ?}$$

2. *Products of Vectors*

In this instance, we consider 2 profile vectors; first, a profile vector which lists quantities of items purchased, and second, a profile vector which lists the cost per item. To fix ideas, let us consider the items be bread, eggs, apples, and hubcaps. Our mythical housewife defines a quantity profile vector by purchasing 2 loaves of bread, 1 dozen eggs, 7 apples, and 10 hubcaps. We represent this information with the row vector

$$X = (2, 1, 7, 10)$$

loaves of bread
dozens of eggs
number of apples
number of hubcaps

Quantity Profile

In a similar fashion, a cost-per-unit profile vector is constructed once we know the cost per unit of each of our 4 commodities. Let us say, then, that each loaf of bread is \$0.40, each dozen eggs is \$0.82, each apple is \$0.10, and each hubcap is \$4.98. This information may be represented by the column vector

$$Y = \begin{pmatrix} 0.40 \\ 0.82 \\ 0.10 \\ 4.98 \end{pmatrix} \begin{matrix} \leftarrow \text{per loaf of bread} \\ \leftarrow \text{per dozen eggs} \\ \leftarrow \text{per apple} \\ \leftarrow \text{per hubcap} \end{matrix}$$

Cost-per-unit Profile

Now what was the total cost of this transaction to the housewife? As our experience in the real world would lead us to believe, we multiply the number of units of each item by the cost per unit to find the cost of each item. Sum the cost of each of the four products, and we have our total cost, that is,

$$\text{Total cost} = 2 \times (0.40) + 1 \times (0.82) + 7 \times (0.10) + 10 \times (4.98), \quad \text{(I.2.1)}$$

which is in accordance with standard grocery-store procedure around the world. Now we may abbreviate the information of (I.2.1) by representing the total cost as a "combination" or "product" of the quantity vector X, and the cost-per-unit vector Y. Thus, we may *define* the *number* of (I.2.1) as a product or combination of the 2 *vectors* X and Y. Thus, we agree that the symbol

$$(2,1,7,10) \cdot \begin{pmatrix} 0.40 \\ 0.82 \\ 0.10 \\ 4.98 \end{pmatrix}$$

will mean the same thing as $2 \times (0.40) + 1 \times (0,82) + 7 \times (0.10) + 10 \times (4.98)$. That is to say,

$$(2,1,7,10) \cdot \begin{pmatrix} 0.40 \\ 0.82 \\ 0.10 \\ 4.98 \end{pmatrix} = 2 \times (0.40) + 1 \times (0.82) + 7 \times (0.10) + 10 \times (4.98)$$

by definition. In this example, $X \cdot Y$ happens to equal the number 51.12, the total cost (in dollars) to the housewife.

This last equation sets a pattern for a *general* definition of the number produced by "multiplying" a concrete row vector by a concrete column vector. Here it is.

Definition I.2.1 Given the concrete vector X and the concrete column vector Y, with real number components $x_1, x_2, \ldots, x_n$, and $y_1, y_2, \ldots, y_n$, respectively, then the **product of X with Y** (X *and* Y, X *against* Y) is the number defined by

$$X \cdot Y = (x_1, x_2, \ldots, x_n) \cdot \begin{pmatrix} y_1 \\ y_2 \\ \vdots \\ y_n \end{pmatrix} = x_1 y_1 + x_2 y_2 + \cdots + x_n y_n.$$

Example I.4 If

$$Y = \begin{pmatrix} 1 \\ 0 \\ -1 \end{pmatrix}$$

and $X = (2,3,4)$, then

$$\begin{aligned} X \cdot Y = (2,3,4) \cdot \begin{pmatrix} 1 \\ 0 \\ -1 \end{pmatrix} &= 2 \cdot 1 + 3 \cdot 0 + 4 \cdot (-1) \\ &= 2 + 0 - 4 \\ &= -2. \end{aligned}$$

We shall present a situation which will allow us to extend Definition I.2.1 (which defines the product of a row vector and a column vector) to the case where we shall be multiplying certain rectangular arrays (not necessarily rows or columns) of numbers together. In order to do this, we define these arrays now.

Definition I.2.2 A rectangular array of numbers of the form

$$\begin{pmatrix} x_{11} & x_{12} & \cdots & x_{1n} \\ x_{21} & x_{22} & \cdots & x_{2n} \\ \vdots & \vdots & \ddots & \vdots \\ x_{m1} & x_{m2} & \cdots & x_{mn} \end{pmatrix}, \text{ denoted } (x_{ij})_{ij},$$

which consists of m rows and n columns, is called an $\boldsymbol{m \times n}$ (*read* "m *by* n") **matrix**. Here, the symbol x_{ij} denotes the number or entry which lies in the ith row and jth column, $i = 1, 2, \ldots, m$, $j = 1, 2 \ldots, n$.

Actually, our row vectors are merely $1 \times n$ matrices (plural of matrix), and column vectors are $n \times 1$ matrices (see Definition I.1.1). A matrix offers a convenient means of representing several profiles simultaneously. The following example illustrates this.

Example I.5 Suppose we observe a team of elves that bottles 4 different beverages —cherry punch, apricot ambrosia, strawberry malts, and whiskey. Now suppose that on Monday, we observe the elves bottling these fluids at the following rates:

Cherry punch: 3 quarts per hour.
Apricot ambrosia: $4\frac{1}{2}$ quarts per hour.
Strawberry malts: 2 quarts per hour.
Whiskey: 6 quarts per hour.

For Monday, then, we could construct an output-rate vector profile for this team of elves.

Cherry punch, apricot ambrosia, strawberry malt, whiskey

Monday Output-rate Vector = (3 , 4.5 , 2 , 6) in quarts/hour

So much for Monday. But as everyone knows, elves also work on Wednesdays and Fridays (unless Snow White's birthday falls on one of these days). Accordingly, we construct a row vector with 4 components (similar to the one above) for *each* of these days. Placing one row below the other, we obtain a 3×4 matrix, which is an output-rate matrix profile for the elves. To wit:

Output Rate Matrix Profile

	cherry punch, ↓	apricot ambrosia, ↓	strawberry malt, ↓	whiskey ↓	(quarts/hour)
Monday →	3	4.5	2	6	
Wednesday →	2.5	4	1.6	8	(I.2.2)
Friday →	0	0	0	17	

We continue our study of the elves' work habits by constructing a 4×2 matrix to represent their time profile. Thus, let us say that of the two daily shifts, various elves budget their time as follows:

Morning shift: Two hours on the cherry punch, 3 hours on the apricot ambrosia, $1\frac{1}{2}$ hours on the strawberry malt, and 4 hours on the whiskey.

Afternoon shift: One hour on the cherry punch, 2 hours on the apricot ambrosia, 2 hours on the strawberry malt, and 4 hours on the whiskey.

To diagram this information, we set down the matrix

Time Matrix Profile

$$\begin{array}{l} \\ \\ \text{Hours on cherry punch} \rightarrow \\ \text{Hours on apricot ambrosia} \rightarrow \\ \text{Hours on strawberry malt} \rightarrow \\ \text{Hours on whiskey} \rightarrow \end{array} \begin{array}{cc} \text{morning shift} & \text{afternoon shift} \\ \downarrow & \downarrow \\ \left(\begin{array}{c} 2 \\ 3 \\ 1.5 \\ 4 \end{array}\right. & \left.\begin{array}{c} 1 \\ 2 \\ 2 \\ 4 \end{array}\right) \end{array} \tag{I.2.3}$$

We pose a question that will combine the information of our Output Rate Matrix (I.2.2), and our Time Matrix (I.2.3). The question is: How many quarts of all the beverages combined are produced by the elves each morning and each afternoon? Naturally, we are seeking 6 numbers, namely, the quarts produced on the mornings of Monday, Wednesday, and Friday, and the quarts produced on the afternoons of Monday, Wednesday, and Friday.

But observe! Each of these numbers may be obtained by multiplying a *row* of matrix (I.2.2), by a *column* of matrix (I.2.3) (see Definition I.2.1). For example, should we desire the totality of quarts produced by the elves on Wednesday's morning shift, we multiply the second row of (I.2.2) (the output rate vector for Wednesday) by the first column of (I.2.3) (the time profile of the morning shift). Thus,

$$\begin{aligned} \begin{array}{c}\text{Total quarts produced} \\ \text{Wednesday morning}\end{array} &= \underbrace{(2.5, 4, 1.6, 8)}_{\text{quarts per hour}} \cdot \left.\begin{pmatrix} 2 \\ 3 \\ 1.5 \\ 4 \end{pmatrix}\right\}\text{hours} \\ &= 2.5 \times (2) + 4 \times (3) + 1.6 \times (1.5) + 8 \times (4) \\ &= 5.0 + 12 + 2.4 + 32 = 51.4 \text{ quarts.} \end{aligned}$$

In obtaining the other 5 numbers, we will say that by multiplying rows of (I.2.2) by columns of (I.2.3), we are *multiplying the two matrices together*. That is to say, we may write the product of matrix (I.2.2) and matrix (I.2.3) as

$$\begin{array}{l} \\ \\ \text{Mon} \rightarrow \\ \text{Wed} \rightarrow \\ \text{Fri} \rightarrow \end{array} \underbrace{\begin{array}{c} \begin{array}{cccc} \text{CP} & \text{AA} & \text{SM} & \text{W} \\ \downarrow & \downarrow & \downarrow & \downarrow \end{array} \\ \begin{pmatrix} 3 & 4.5 & 2 & 6 \\ 2.5 & 4 & 1.6 & 8 \\ 0 & 0 & 0 & 17 \end{pmatrix} \end{array}}_{\text{quarts per hour}} \times \cdot \underbrace{\begin{array}{c} \begin{array}{cc} \text{Morning} & \text{Afternoon} \\ \downarrow & \downarrow \end{array} \\ \begin{pmatrix} 2 & 1 \\ 3 & 2 \\ 1.5 & 2 \\ 4 & 4 \end{pmatrix} \end{array}}_{\text{hours}} \begin{array}{l} \\ \\ \leftarrow \text{CP} \\ \leftarrow \text{AA} \\ \leftarrow \text{SM} \\ \leftarrow \text{W} \end{array} = \begin{array}{l} \\ \\ \text{Mon} \rightarrow \\ \text{Wed} \rightarrow \\ \text{Fri} \rightarrow \end{array} \underbrace{\begin{array}{c} \begin{array}{cc} \text{Morning} & \text{Afternoon} \\ \downarrow & \downarrow \end{array} \\ \begin{pmatrix} 16.5 & 40 \\ 51.4 & 45.7 \\ 68 & 68 \end{pmatrix} \end{array}}_{\text{quarts}}, \tag{I.2.4}$$

quarts per hour × hours ⟶ quarts

which, at a glance, tells the story of the elves. In any case, (I.2.4) sets the pattern for the multiplication of a pair of matrices, provided the left-hand matrix has exactly as many *columns* as the right-hand matrix has *rows*. We define this now along with scalar multiplication of matrices, which will generalize Definition I.2.1. Recall that Definition I.2.1 concerns only *single*-row matrices and *single*-column matrices.

Definition I.2.3 Let A be an $m \times k$ matrix (m rows and k columns) and let B be a $k \times n$ matrix (k rows and n columns). Then the $m \times n$ matrix $A \cdot B$, **A multiplied by B**, is defined as follows: The number which is entered into the ith row and jth column (the ijth entry) of $A \cdot B$ is obtained by multiplying the ith row of A by the jth column of B, as per Definition I.2.1.

If A is any rectangular matrix, then for any number α, the matrix **$\alpha \cdot A$** or just **αA, α times A**, is obtained from A by multiplying each and every entry of A by α.

If A and B are matrices, each with m rows and n columns, then the **sum, $A + B$**, is also an $m \times n$ matrix which is obtained by adding the entries of A to corresponding entries of B.

Example I.6 Set

$$A = \begin{pmatrix} 1 & 3 & 5 \\ 2 & 0 & -1 \end{pmatrix}, \qquad B = \begin{pmatrix} 1 & 2 & 0 \\ 0 & 0 & 1 \\ -1 & 0 & 1 \end{pmatrix}$$

$A \cdot B$ is defined, while $B \cdot A$ has no meaning, according to Definition I.2.3. Now

$$A \cdot B = \begin{pmatrix} 1 & 3 & 5 \\ 2 & 0 & -1 \end{pmatrix} \cdot \begin{pmatrix} 1 & 2 & 0 \\ 0 & 0 & 1 \\ -1 & 0 & 1 \end{pmatrix}$$

$$= \begin{pmatrix} 1 \cdot 1 + 3 \cdot 0 + 5 \cdot (-1) & 1 \cdot 2 + 3 \cdot 0 + 5 \cdot 0 & 1 \cdot 0 + 3 \cdot 1 + 5 \cdot 1 \\ 2 \cdot 1 + 0 \cdot 0 + (-1) \cdot (-1) & 2 \cdot 2 + 0 \cdot 0 + (-1) \cdot 0 & 2 \cdot 0 + 0 \cdot 1 + (-1) \cdot 1 \end{pmatrix}$$

$$= \begin{pmatrix} -4 & 2 & 8 \\ 3 & 4 & -1 \end{pmatrix}.$$

For $\alpha = -3$, we have

$$\alpha A = (-3) \cdot \begin{pmatrix} 1 & 3 & 5 \\ 2 & 0 & -1 \end{pmatrix}$$

$$= \begin{pmatrix} (-3) \cdot 1 & (-3) \cdot 3 & (-3) \cdot 5 \\ (-3) \cdot 2 & (-3) \cdot 0 & (-3) \cdot (-1) \end{pmatrix}$$

$$= \begin{pmatrix} -3 & -9 & -15 \\ -6 & 0 & 3 \end{pmatrix}.$$

For

$$C = \begin{pmatrix} 3 & -1 & -2 \\ 0 & 1 & 0 \end{pmatrix},$$

$$A + C = \begin{pmatrix} 1 & 3 & 5 \\ 2 & 0 & -1 \end{pmatrix} + \begin{pmatrix} 3 & -1 & -2 \\ 0 & 1 & 0 \end{pmatrix}$$

$$= \begin{pmatrix} 1+3 & 3+(-1) & 5+(-2) \\ 2+0 & 0+1 & -1+0 \end{pmatrix} = \begin{pmatrix} 4 & 2 & 3 \\ 2 & 1 & -1 \end{pmatrix}.$$

EXERCISES

I.2.1 (a) In 1860, Santa Claus decided to distribute 2 dolls per child, 1 bugle per child, and 3 skates per child. Construct a row vector which represents the toy-per-child profile for Santa in 1860.

(b) In the following years of 1861, 1862, and 1863, Santa distributed the same types of toys as follows:

1861: 3 dolls/child, 2 bugles/child, 1 skate/child
1862: 1 doll/child, 2 bugles/child, no skates/child
1863: 6 dolls/child, no bugles/child, 1 skate/child.

Construct a 4×3 matrix which represents Santa's toy-per-child profile over the 4 years 1860, 1861, 1862, and 1863.

(c) Each year, at a certain white house, there are always 3 children receiving dolls, 6 children receiving bugles, and 3 other children receiving skates (assume no child receives 2 kinds of toys). Construct a column vector representing the child profile of toys received.

(d) Multiply the toy-per-child matrix of part (b) against the column matrix of part (c) to obtain a profile for the total number of toys distributed at the white house in each of the four years 1860, 1861, 1862, and 1863.

(e) Each year, in a certain yellow house, there are always 4 children receiving dolls, 7 children receiving bugles, and no children receiving skates. Construct a 3×2 matrix profile of the toy receivers of the white house and of the yellow house. Let each column represent one of the houses.

(f) Multiply the toy-per-child matrix of part (b) against the matrix of part (e) to obtain the 4×2 matrix whose 8 entries relate the totality of toys distributed at each of the 2 houses, during each of the 4 years 1860, 1861, 1862, and 1863.

I.2.2 Given the matrices A and B, compute $A \cdot B$ and $B \cdot A$, whenever the product is defined.

(a) $A = \begin{pmatrix} 2 & 0 & -1 \\ 3 & 1 & 1 \end{pmatrix}$, $\quad B = \begin{pmatrix} 1 & 0 & 0 \\ 0 & 1 & 0 \\ 0 & 0 & 1 \end{pmatrix}$.

(b) $A = \begin{pmatrix} 1 & 0 \\ 0 & 1 \end{pmatrix}$, $\quad B = \begin{pmatrix} 1 & 3 \\ 4 & 2 \end{pmatrix}$.

(c) $A = \begin{pmatrix} 1 \\ 2 \\ 3 \end{pmatrix}$, $\quad B = (4,1,3)$.

(d) $A = \begin{pmatrix} 0 & 1 & 0 & -1 & 3 \\ 1 & 1 & 2 & 1 & 1 \end{pmatrix}$, $B = \begin{pmatrix} 0 & 5 \\ 1 & -1 \\ 1 & 0 \\ 6 & 1 \end{pmatrix}$.

(e) $A = \begin{pmatrix} 0 & 0 \\ 0 & 0 \\ 0 & 0 \end{pmatrix}$, $B = \begin{pmatrix} 1 & 5 & 7 & 9 \\ 1 & 5 & 9 & 7 \end{pmatrix}$.

(f) $A = \begin{pmatrix} 1 & 1 & 1 \\ 0 & 0 & 0 \\ -1 & -1 & -1 \end{pmatrix}$, $B = \begin{pmatrix} 2 & 1 & 0 \\ 2 & 1 & 0 \\ 2 & 1 & 0 \end{pmatrix}$.

I.2.3 What constraints must be put on the rectangular matrices A and B, in order that *both* products, $A \cdot B$ and $B \cdot A$, are defined?

I.2.4 Given the matrices A, B, and the number α. In each case, compute $A \cdot B$, αA, αB, $A \cdot (\alpha B)$, $(\alpha A) \cdot B$, $\alpha(A \cdot B)$.

(a) $A = \begin{pmatrix} 1 & 0 \\ 0 & 1 \end{pmatrix}$, $B = \begin{pmatrix} 3 & 2 & 1 \\ 1 & 1 & 2 \end{pmatrix}$, $\alpha = -2$

(b) $A = (3,1,2)$, $B = \begin{pmatrix} 1 \\ 5 \\ 6 \end{pmatrix}$, $\alpha = 3$

(c) $A = \begin{pmatrix} 4 & 0 & 1 \\ 1 & 1 & -1 \\ 3 & 2 & -1 \\ 0 & 1 & -2 \end{pmatrix}$, $B = \begin{pmatrix} 5 & 5 \\ 1 & 1 \\ 4 & -4 \end{pmatrix}$, $\alpha = -4$

(d) $A = \begin{pmatrix} 1 & 0 & 0 \\ 0 & 1 & 0 \\ 0 & 0 & 1 \end{pmatrix}$, $B = \begin{pmatrix} a & d \\ b & e \\ c & f \end{pmatrix}$, $\alpha = 7$

(e) $A = \begin{pmatrix} 0 & 1 & 0 \\ 1 & 0 & 0 \\ 0 & 0 & 1 \end{pmatrix}$, $B = \begin{pmatrix} a & d \\ b & e \\ c & f \end{pmatrix}$, $\alpha = -3$.

I.2.5 Given matrices A, B, and C, compute $A + B$, $(A + B) \cdot C$, $A \cdot C$, $B \cdot C$, and $A \cdot C + B \cdot C$.

(a) $A = (1,2)$, $B = (2,3)$, $C = \begin{pmatrix} 1 & 2 & 4 \\ 0 & 1 & -3 \end{pmatrix}$

(b) $A = \begin{pmatrix} 1 & 0 & 0 \\ 0 & 1 & 0 \\ 0 & 0 & 1 \end{pmatrix}$, $B = \begin{pmatrix} 1 & -1 & 0 \\ 0 & 3 & 1 \\ 2 & 1 & 0 \end{pmatrix}$, $C = \begin{pmatrix} -1 \\ 0 \\ 1 \end{pmatrix}$

(c) $A = \begin{pmatrix} 5 & 3 \\ 1 & 0 \\ 1 & 1 \end{pmatrix}$, $B = \begin{pmatrix} -5 & -3 \\ -1 & 0 \\ -1 & -1 \end{pmatrix}$, $C = \begin{pmatrix} 4 & 2 \\ 4 & 2 \end{pmatrix}$

(d) $A = \begin{pmatrix} 6 & 8 & 1 \\ 2 & 1 & 1 \end{pmatrix}$, $B = \begin{pmatrix} 6 & 1 & 0 \\ 1 & 1 & 6 \end{pmatrix}$, $C = \begin{pmatrix} 1 & 0 \\ 0 & 0 \\ 0 & 0 \end{pmatrix}$

(e) $A = \begin{pmatrix} a_{11} & a_{12} \\ a_{21} & a_{22} \end{pmatrix}$, $B = \begin{pmatrix} b_{11} & b_{12} \\ b_{21} & b_{22} \end{pmatrix}$, $C = \begin{pmatrix} c_{11} & c_{12} \\ c_{21} & c_{22} \end{pmatrix}$.

3. Simultaneous Linear Equations

In the first instance, we shall consider the following 3 equations in the 3 unknowns x_1, x_2, and x_3:

$$
\begin{aligned}
&\text{(a)} \qquad & x_1 - 2x_2 + 3x_3 &= 7 \\
&\text{(b)} \qquad & 2x_1 + x_2 - x_3 &= 1 \\
&\text{(c)} \qquad & x_1 - x_2 - x_3 &= -6.
\end{aligned}
\tag{I.3.1}
$$

We observe in passing, that the 3 parts of Equations (I.3.1) may be regarded as equivalent to the *matrix* equation

$$\begin{pmatrix} 1 & -2 & 3 \\ 2 & 1 & -1 \\ 1 & -1 & -1 \end{pmatrix} \cdot \begin{pmatrix} x_1 \\ x_2 \\ x_3 \end{pmatrix} = \begin{pmatrix} 7 \\ 1 \\ -6 \end{pmatrix}.$$

In fact, if the left-hand side of this matrix equation is multiplied out, then the 3 equations of (I.3.1) reappear when corresponding coordinates of the two 3×1 column vectors (namely, the left-hand and right-hand sides of the matrix equation) are set equal to each other.

A straightforward method of solution, known as **Gaussian Elimination,** involves a successive "whittling away" of variables in order to isolate their values. This method is based on the fact that there are 3 operations which alter the *form* of the equations, but not the *solutions*. These operations are:

(a) Interchanging a pair of equations
(b) Multiplying an equation through by a nonzero number
(c) Adding a multiple of one equation to another equation.

Let us apply these operations to (I.3.1). Our first step involves the x_1 variable in the top equation of (I.3.1), to annihilate the x_1 variables in the subsequent equations. The annihilation is effected by two uses of operation (c), namely,

add -2 times Equation (a) to Equation (b);
add -1 times Equation (a) to Equation (c).

We thus obtain the new system [with the same solutions as the system (I.3.1)],

$$
\begin{aligned}
&\text{(a)} \qquad & x_1 - 2x_2 + 3x_3 &= 7 \\
&\text{(b)} \qquad & 5x_2 - 7x_3 &= -13 \\
&\text{(c)} \qquad & x_2 - 4x_2 &= -13.
\end{aligned}
\tag{I.3.2}
$$

We could proceed to eliminate the x_2 variable starting with equation (b) of (I.3.2). That is, by adding $-\frac{1}{5}$ times Equation (b) of (I.3.2), to Equation (c) of (I.3.2), we would eliminate the x_2 from the third equation. But this is not the only way. Since it would be pleasant to avoid fractions, we use operation (a) to interchange Equations (I.3.2b) and (I.3.2c). Thus we have

$$\begin{aligned} &\text{(a)} & x_1 - 2x_2 + 3x_3 &= 7 \\ &\text{(b)} & x_2 - 4x_3 &= -13 \\ &\text{(c)} & 5x_2 - 7x_3 &= -13. \end{aligned} \qquad \text{(I.3.3)}$$

Using the operation (c), where -5 times Equation (b) of (I.3.3) is added to Equation (c) of (I.3.3), we obtain

$$\begin{aligned} &\text{(a)} & x_1 - 2x_2 + 3x_3 &= 7 \\ &\text{(b)} & x_2 - 4x_3 &= -13 \\ &\text{(c)} & 13x_3 &= 52. \end{aligned} \qquad \text{(I.3.4)}$$

To tidy Equation (c) of (I.3.4) a bit, we may use operation (b), page 18, and multiply through by the number 1/13. Thus, we have the system

$$\begin{aligned} &\text{(a)} & x_1 - 2x_2 + 3x_3 &= 7 \\ &\text{(b)} & x_2 - 4x_3 &= -13 \\ &\text{(c)} & x_3 &= 4, \end{aligned} \qquad \text{(I.3.5)}$$

in which all but one of the x_1 variables, have been eliminated and one x_2 variable has been eliminated. Equation (c) of our newest system (I.3.5) already yields a partial solution, namely, $x_3 = 4$. We could, if we choose, work with (I.3.5) as its stands, substituting $x_3 = 4$ in Equation (b), and solving for x_2. Thereafter, we could substitute these values of x_2 and x_3 into Equation (a) and solve for the final variable x_1. However, we choose to continue to alter the "triangular" system (I.3.5) and eliminate variables. Working "in reverse," we eliminate the x_3 variables from Equations (a) and (b) by the following operations of type (c) (page 18):

> Add 4 times Equation (c) of (I.3.5), to Equation (b) of (I.3.5),
> Add -3 times Equation (c) of (I.3.5), to Equation (a) of (I.3.5).

In so doing, we obtain the system

$$\begin{aligned} &\text{(a)} & x_1 - 2x_2 &= -5 \\ &\text{(b)} & x_2 &= 3 \\ &\text{(c)} & x_3 &= 4. \end{aligned} \qquad \text{(I.3.6)}$$

Add 2 times Equation (b) of (I.3.6) to Equation (a) of (I.3.6), to obtain

$$\begin{aligned} x_1 \qquad\qquad &= 1 \\ x_2 \qquad &= 3 \\ x_3 &= 4, \end{aligned} \tag{I.3.7}$$

which is a system in which the values for the unknown x_1, x_2, and x_3 are perhaps more apparent than they were in the original form (I.3.1).

EXERCISES

Solve for the unknown variables of the following systems of equations by means of Gaussian Elimination.

I.3.1

$$\begin{aligned} 3x_1 + x_2 &= 2 \\ x_2 &= -1. \end{aligned}$$

I.3.2

$$\begin{aligned} 3x_1 + x_2 &= 2 \\ -x_1 + 2x_2 &= 4. \end{aligned}$$

I.3.3

$$\begin{aligned} ax_1 + bx_2 &= 0 \\ cx_1 + dx_2 &= 1, \quad ad \neq bc. \end{aligned}$$

I.3.4

$$\begin{aligned} ax_1 + bx_2 &= p \\ cx_1 + dx_2 &= q, \quad ad \neq bc. \end{aligned}$$

I.3.5

$$\begin{aligned} 3x_1 + 6x_2 \qquad &= 3 \\ x_1 \qquad\qquad &= 1 \\ 2x_1 - 2x_2 + x_3 &= 1. \end{aligned}$$

I.3.6

$$\begin{aligned} x_1 - 2x_2 - 3x_3 &= -5 \\ 2x_1 + x_2 + x_3 &= 5 \\ 3x_1 - x_2 + 2x_3 &= 12. \end{aligned}$$

I.3.7

$$\begin{aligned} x_1 + x_2 + x_3 + x_4 &= 3 \\ 2x_1 - x_2 - x_3 + x_4 &= 4 \\ -x_1 + 3x_2 - x_3 + 2x_4 &= 12 \\ 2x_1 + 2x_2 + x_3 + x_4 &= 8. \end{aligned}$$

I.3.8

$$\begin{aligned} x_1 - x_2 + 2x_3 - x_4 &= -6 \\ x_2 \qquad + x_4 &= 5 \\ 3x_1 \qquad + x_3 \qquad &= 2 \\ x_2 - x_3 + 3x_4 &= 13. \end{aligned}$$

I.3.9

$$\begin{aligned} x_1 + 2x_2 + x_3 &= a \\ x_1 - 2x_2 + x_3 &= a \\ 2x_1 + x_2 - x_3 &= 0. \end{aligned}$$

I.3.10

$$\begin{aligned} 4x_1 - x_2 - x_3 + 4x_4 &= 0 \\ 2x_1 - x_2 + x_3 + x_4 &= 0 \\ 3x_1 + x_2 + x_3 + 3x_4 &= 0 \\ 6x_1 + x_2 + 3x_3 - x_4 &= 0. \end{aligned}$$

So far, we have seen examples of systems of equations where at *most 1*, and *at least 1* (that is, a *unique*) solution set $x_1, x_2, \ldots, x_n$ could be found. Such is not always the case, however, as there are systems where *no solution exists*, and there are systems where *infinitely many* solution sets exist. For example, the system of 2 equations,

$$\begin{aligned} x_1 + x_2 &= 1 \\ x_1 + x_2 &= 2 \end{aligned}$$

cannot have *any* solutions. Intuitively, the left-hand quantity $(x_1 + x_2)$, is supposed to equal both the number 1 and the number 2, at "the same time." In our hearts, then, we feel that this system has no solution. More formally, Gaussian Elimination may be used: Add -1 times the top equation to the bottom equation to obtain the equivalent system

$$\begin{aligned} x_1 + x_2 &= 1 \\ 0 &= 1. \end{aligned}$$

Almost anybody would agree, now, that the second equation is absurd. No solutions x_1, x_2 exist for this system.

On the other hand, consider the simple system of 1 equation in 2 unknowns:

$$x_1 + 2x_2 = 2.$$

Clearly, there are *infinitely many* solution pairs x_1, x_2, as one of the x_i's may be chosen at will, thereby fixing the other. For example, choose $x_1 = 0$. Then $x_2 = 1$; choose $x_1 = 2$, then, $x_2 = 0$; choose $x_1 = -1$, then $x_2 = \frac{3}{2}$, and so on. We present an example of a system of 3 equations in 3 unknowns, having infinitely many solution triples x_1, x_2, x_3.

Example I.7 Consider the system of equations

$$\begin{aligned} &\text{(a)} & x_1 + 2x_2 - x_3 &= 2 \\ &\text{(b)} & 2x_1 + x_2 + x_3 &= 1 \\ &\text{(c)} & x_1 + 5x_2 - 4x_3 &= 5. \end{aligned} \qquad \text{(I.3.8)}$$

As before, we proceed by Gaussian Elimination to eliminate the variable x_1 from Equations (b) and (c) by adding -2 times Equation (a) to Equation (b), and by adding -1 times Equation (a) to Equation (c). We thus obtain the equivalent system

$$\begin{aligned} &\text{(a)} & x_1 + 2x_2 - x_3 &= 2 \\ &\text{(b)} & -3x_2 + 3x_3 &= -3 \\ &\text{(c)} & 3x_2 - 3x_3 &= 3. \end{aligned} \qquad \text{(I.3.9)}$$

But observe! We have in (I.3.9) only 2 "different" equations, not 3, since Equation (c) is just the negative of Equation (b). To underscore this point, add Equation (b) to Equation (c), and multiply Equation (b) by $-\frac{1}{3}$, to obtain

$$\begin{aligned} &(a) \quad & x_1 + 2x_2 - x_3 &= 2 \\ &(b) \quad & x_2 - x_3 &= 1 \\ &(c) \quad & 0 &= 0. \end{aligned} \tag{I.3.10}$$

From this form, we work toward eliminating the variable x_2 from Equation (a); to do so, we add -2 times Equation (b) to Equation (a), which yields our final form for the system, namely,

$$\begin{aligned} &(a) \quad & x_1 + 0 + x_3 &= 0 \\ &(b) \quad & x_2 - x_3 &= 1. \end{aligned} \tag{I.3.11}$$

From (I.3.11), we see that x_3 may take on *any* value desired, but once it does so, the values for x_1 and x_2 are *fixed*. For example, let us tabulate the results in each instance, when we arbitrarily assign to x_3, the 7 values -2, -1, 0, 1, $\sqrt{2}$, 2, and π, respectively.

Computed from Equation (I.3.11a), namely, $x_1 = -x_3$ ↓

computed from Equation (I.3.11b), namely, $x_2 = 1 + x_3$ ↓

x_3	x_1	x_2
-2	2	-1
-1	1	0
0	0	1
1	-1	2
$\sqrt{2}$	$-\sqrt{2}$	$\sqrt{2}+1$
2	-2	3
π	$-\pi$	$\pi+1$
⋮	⋮	⋮

Example I.8

$$\begin{aligned} &(a) \quad & x_1 + 2x_2 - x_3 + 2x_4 &= 0 \\ &(b) \quad & x_1 + x_2 + x_3 + 2x_4 &= 1 \\ &(c) \quad & -x_1 + x_2 - 5x_3 - 2x_4 &= -3. \end{aligned} \tag{I.3.12}$$

By adding -1 times Equation (a) to Equation (b), and the adding Equation (a) to Equation (c), we obtain the following equivalent system with the x_1 variable eliminated from Equations (b) and (c):

$$\begin{aligned} &(a) \quad & x_1 + 2x_2 - x_3 + 2x_4 &= 0 \\ &(b) \quad & -x_2 + 2x_3 + 0 &= 1 \\ &(c) \quad & 3x_2 - 6x_3 + 0 &= -3. \end{aligned} \tag{I.3.13}$$

To eliminate x_2 from Equation (c) of (I.3.13), add 3 times Equation (b) to Equation (c) to obtain

$$\begin{array}{ll} \text{(a)} & x_1 + 2x_2 - x_3 + 2x_4 = 0 \\ \text{(b)} & -x_2 + 2x_3 + 0 = 1 \\ \hline \text{(c)} & 0 = 0. \end{array} \tag{I.3.14}$$

Finally, we eliminate x_2 in our 2-equation system, from Equation (a) of (I.3.14), by adding 2 times Equation (b) to equation (a). This yields the system

$$\begin{array}{ll} & \overbrace{\hphantom{x_1 + 0 + 3x_3 + 2x_4 = 2}}^{\text{2 degrees of freedom: } x_3,\ x_4.} \\ \text{(a)} & x_1 + 0 + [3x_3 + 2x_4] = 2 \\ \text{(b)} & -x_2 + [2x_3 + 0] = 1. \end{array} \tag{I.3.15}$$

From (I.3.15), we see that x_3 *and* x_4 may be chosen arbitrarily, and once this is done, the values of x_1 [from Equation (a) of (I.3.15)] and of x_2 [from Equation (b) of (I.3.15)] are fixed. For example, let us assign the following, randomly selected values for x_3 and x_4:

$$x_3 = 1 \quad \text{and} \quad x_4 = -2,$$

or else

$$x_3 = \sqrt{2} \quad \text{and} \quad x_4 = 0,$$

or else

$$x_3 = -\pi \quad \text{and} \quad x_4 = 100.$$

From Equation (a) of (I.3.15), which says

$$x_1 = 2 - 3x_3 - 2x_4,$$

and from Equation (b) of (I.3.15), which says

$$x_2 = 2x_3 - 1,$$

we tabulate the effect of our 3 pairs of arbitrary selections for values of x_3 and x_4 as follows:

x_3	x_4	x_1 $(= 2 - 3x_3 - 2x_4)$	x_2 $(= 2x_3 - 1.)$
1	-2	3	1
$\sqrt{2}$	0	$2 - 3\sqrt{2}$	$2\sqrt{2} - 1$
$-\pi$	100	$3\pi - 198$	$-2\pi - 1$

If these rows, indicating values for x_1, x_2, x_3, and x_4, were substituted into the original system (I.3.12), one would see that they represent solutions.

CASE OF n EQUATIONS, $n+k$ UNKNOWNS. We suppose that the process of Gaussian Elimination, in this case, leads us to an equivalent system of equations of the following form

$$
\begin{array}{lrl}
\text{(a)} & x_1 + 0 + 0 + \cdots + 0 & \Big| + a_{11}x_{n+1} + \cdots + a_{1k}x_{n+k} = b_1 \\
\text{(b)} & x_2 + 0 + \cdots + 0 & \Big| + a_{21}x_{n+1} + \cdots + a_{2k}x_{n+k} = b_2 \\
\text{(c)} & x_3 + \cdots + 0 & \Big| + a_{31}x_{n+1} + \cdots + a_{3k}x_{n+k} = b_3 \\
\vdots & \ddots \quad \vdots & \Big| \quad \vdots \qquad\qquad\qquad \vdots \qquad \vdots \\
(n) & x_n & \Big| + a_{n1}x_{n+1} + \cdots + a_{nk}x_{n+k} = b_n
\end{array}
\qquad \text{(I.3.16)}
$$

$\leftarrow k$ degrees of freedom in choice $\rightarrow$ of $x_{n+1}, x_{n+2}, \ldots, x_{n+k}$

which is a system having *at least as many unknowns as there are equations.* The number k, the excess of the number of unknowns over the number of equations *always* represents the "degrees of freedom" one has, or the number of variables which may be assigned arbitrary values. Consequently, if the Gaussian elimination leads to the *same* number of equations as unknowns, then $k = 0$, that is, the solution of (I.3.16) will prove to be necessarily unique.

CASE OF FEWER UNKNOWNS THAN EQUATIONS. If we commence with a system having *fewer* equations than unknowns, then the process of Gaussian Elimination will unearth the true story in the sense that the system is really of the form discussed above; or else *no* solution exists. For example, consider the system

$$
\begin{aligned}
x_1 + 2x_2 + x_3 &= 1 \\
2x_1 + 4x_2 + 2x_3 &= 2 \\
3x_1 + 6x_2 + 3x_3 &= 3 \\
4x_1 + 8x_2 + 4x_3 &= 4 \\
5x_1 + 10x_2 + 5x_3 &= 5,
\end{aligned}
$$

which *apparently* has more equations than unknowns. However, once Gaussian Elimination is applied to these "5" equations, we obtain the *single*-equation system

$$x_1 + \mid 2x_2 + x_3 = 1,$$

which is of the form (I.3.16).

As another example, consider

$$\begin{aligned} &\text{(a)} & x_1 + x_2 &= 1 \\ &\text{(b)} & x_1 + 2x_2 &= 3 \\ &\text{(c)} & x_1 - x_2 &= 2, \end{aligned} \tag{I.3.17}$$

which also has, apparently, more equations (3) than unknowns (2). Appealing to Gaussian Elimination to reveal the *true* nature of this system, we effect the following operations: To clear x_1 from Equations (I.3.17b) and (I.3.17c), subtract Equation (I.3.17a) from each. Thus, we have the equivalent system

$$\begin{aligned} &\text{(a)} & x_1 + x_2 &= 1 \\ &\text{(b)} & x_2 &= 2 \\ &\text{(c)} & -2x_2 &= 1. \end{aligned} \tag{I.3.18}$$

Clearly, Equations (b) and (c) of (I.3.18) are in conflict, for (b) suggests that $x_2 = 2$, while (c) requires $x_2 = -\frac{1}{2}$. Another means of revealing this contradiction lies in continuation of the Gaussian Elimination process, namely, clear x_2 from Equations (c) and (a) of (I.3.18) by adding 2 times Equation (b) to Equation (c), and by adding -1 times Equation (b) to Equation (a). This yields the absurd system

$$\begin{aligned} &\text{(a)} & x_1 + 0 &= -1 \\ &\text{(b)} & x_2 &= 2 \\ &\text{(c)} & 0 &= 5, \end{aligned} \tag{I.3.19}$$

which is somewhat marred by Equation (I.3.19c). The system (I.3.17), which is equivalent to (I.3.19), has *no* solutions whatever.

Thus, confronted with any system of simultaneous linear equations, Gaussian Elimination leads to solution (unique or not) or to a stark realization that solutions do not exist. This is all that any reasonable person would expect from a method of solution.

EXERCISES

Find solutions for the following systems of linear equations:

I.3.11 $$x_1 + x_2 = 1.$$

I.3.12 $$\begin{aligned} x_1 + x_2 &= 1 \\ 2x_1 + 2x_2 &= 2. \end{aligned}$$

I.3.13

$$\begin{aligned} x_1 + x_2 &= 1 \\ 2x_2 + 2x_2 &= 1. \end{aligned}$$

I.3.14

$$\begin{aligned} x_1 + x_2 - x_3 &= 3 \\ x_1 - x_2 + x_3 &= 1. \end{aligned}$$

I.3.15

$$\begin{aligned} 2x_1 - 2x_2 + x_3 &= 0 \\ 3x_1 + x_2 + x_3 &= 0. \end{aligned}$$

I.3.16

$$\begin{aligned} x_1 - 2x_2 + 3x_3 + x_4 &= 0 \\ x_1 + x_2 - 3x_3 - x_4 &= 0 \\ 2x_1 - x_2 + x_3 - x_4 &= 0 \\ 2x_1 + 2x_2 - 5x_3 - 3x_4 &= 0. \end{aligned}$$

I.3.17

$$\begin{aligned} x_1 + 2x_2 &= 0 \\ 2x_1 + x_2 &= 0 \\ x_1 + x_2 &= 1. \end{aligned}$$

I.3.18

$$\begin{aligned} x_1 + 2x_2 &= 3 \\ x_1 - x_2 &= 0 \\ x_1 + 8x_2 &= 9. \end{aligned}$$

I.3.19

$$\begin{aligned} x_1 + 2x_2 + x_3 + x_4 &= 0 \\ x_1 - x_2 + x_3 - x_4 &= 0 \\ x_1 + 8x_2 + x_3 + 5x_4 &= 0 \\ 2x_1 + 7x_2 + 2x_3 + 4x_4 &= 0. \end{aligned}$$

I.3.20 (Exercise I.3.19).

$$\begin{aligned} x_1 + 2x_2 + x_3 + x_4 &= 6 \\ x_1 - x_2 + x_3 - x_4 &= -2 \\ x_1 + 8x_2 + x_3 + 5x_4 &= 22 \\ 2x_1 + 7x_2 + 2x_3 + 4x_4 &= 20. \end{aligned}$$

I.3.21 (Exercises I.3.19 and I.3.20).

$$\begin{aligned} x_1 + 2x_2 + x_3 + x_4 &= 6 \\ x_1 - x_2 + x_3 - x_4 &= -2 \\ x_1 + 8x_2 + x_3 + 5x_4 &= 22 \\ 2x_1 + 7x_2 + 2x_3 + 4x_4 &= 0. \end{aligned}$$

I.3.22 For which numbers a, b, and c, does the following system of equations have a solution?

$$\begin{aligned} x_1 + 2x_2 - x_3 &= a \\ 2x_1 - x_2 + x_3 &= b \\ x_1 + 7x_2 - 2x_3 &= c. \end{aligned}$$

I.3.23 For which column vectors

$$X = \begin{pmatrix} x_1 \\ x_2 \\ x_3 \end{pmatrix}$$

is it true that

$$\overbrace{\begin{pmatrix} 3 & 2 & 3 \\ -12 & 3 & -6 \\ 0 & 4 & 9 \end{pmatrix}}^{A} \cdot \begin{pmatrix} x_1 \\ x_2 \\ x_3 \end{pmatrix} = 3 \begin{pmatrix} x_1 \\ x_2 \\ x_3 \end{pmatrix}?$$

Note: X is not unique, for if $A \cdot X = 3X$ then $A \cdot (\lambda X) = 3 \cdot (\lambda X)$ for any number $\lambda \neq 0$. Thus λX works as well as X, relative to the matrix A.

I.3.24 For which column vectors is it true that

$$\begin{pmatrix} \frac{1}{4} & \frac{1}{4} & \frac{\sqrt{2}}{4} \\ \frac{1}{4} & \frac{1}{4} & \frac{\sqrt{2}}{4} \\ \frac{\sqrt{2}}{4} & \frac{\sqrt{2}}{4} & \frac{1}{2} \end{pmatrix} \cdot \begin{pmatrix} x_1 \\ x_2 \\ x_3 \end{pmatrix} = \begin{pmatrix} x_1 \\ x_2 \\ x_3 \end{pmatrix}?$$

4. *Linear Programming*

We shall present a problem whose solution will involve the use of some of the notions we've introduced regarding vectors, products of vectors, and Gaussian Elimination. Here is the problem:

A toymaker produces dolls for little girls, and toy cars for little boys. In the manufacturing process, each toy, be it doll or car, must pass through each of 3 machines in succession. Now, these machines have certain built-in limitations. The times required to process each toy are given in the following tabulation:

(*painting*) *machine A*: works 1 hour per doll, or 4 hours per car.
(*drying*) *machine B*: works 1 hour per doll, or 1 hour per car. (I.4.1)
(*wrapping*) *machine C*: works 2 hours per doll, or 1 hour per car.

Furthermore, to avoid mechanical failure, each machine has a maximum number of hours per week during which it may be operated. Specifically,

(*painting*) *machine A*: works at most 40 hours per week.
(*drying*) *machine B*: works at most 16 hours per week. (I.4.2)
(*wrapping*) *machine C*: works at most 28 hours per week.

Of course the toymaker has all sorts of choices; he may produce only dolls, or only cars, or else he can *divide* the available times on each of his machines between the dolls and the cars. However, the toymaker wishes to appropriate just the right amount of time on each machine, for each toy, so that the *maximum number of children* will enjoy his toys. Depending on the number of children who share any of the toys, the answer to the toymaker's time-budgeting problem varies. To illustrate this, we shall consider 2 sharing patterns, namely,

CASE I. Only 1 girl is to play with a doll, and only 1 boy is to play with a toy car. (I.4.3)

CASE II. 3 girls share each doll, and 2 boys share each car.

Naturally, in Case II, where there are more children per toy than in Case I, we can expect that more children will be accommodated by the toymaker. In fact, the following answers shall emerge:

CASE I. Sixteen children per week at most may be accommodated—and this may be achieved in *more than one way*. For example, the toymaker could produce 8 dolls and 8 cars per week by running machines A and B for their maximum times [40 hours and 16 hours, respectively; see (I.4.1) and (I.4.2)], and by working machine C, therefore, for the necessary 24 hours (out of a possible weekly maximum of 28 hours). Sixteen toys can be produced in yet another way. By working machines B and C for their maximum times (16 hours and 28 hours, respectively), the toymaker can produce 12 dolls and 4 cars each week. This means that machine A needs to work only 28 hours [see (I.4.1)] out of a weekly limit of 40 hours [see (I.4.2)].

CASE II. If 3 girls share a doll, and 2 boys share each car, then a maximum of 44 children per week can be accommodated with the weekly production of 12 dolls and 4 cars. The solution here, unlike that of Case I, is *unique*, in that any other combination of dolls and cars will accommodate *fewer* children. (Try a combination of your own; see whether the machines can fill your order from (I.4.1) and (I.4.2), and from (I.4.3); figure the number of children sharing this total of toys. The result will always be a number *less* than 44.) Observe that, as before, 12 dolls and 4 cars can be produced by running machines B and C for their maximum weekly times, that is, 16 hours and 28 hours, respectively, which requires only 36 hours of work from machine A. Moreover, this is the *only* way in which 12 dolls and 4 cars can be produced.

How are these answers arrived at? The development of the appropriate techniques follows; the solution materializes on page 49.

Relative to the toymaker's problem, we were seeking a certain *pair* of numbers (u,v), where u represents the weekly output of dolls, and v represents the weekly output of toy cars. As we've stated, without proof, the maximum number of children can be accommodated weekly (when one girl is assigned to each doll and one boy is assigned to each car) with a production of 8 dolls and 8 cars, or 12 dolls and 4 cars. Thus, at least 2 maximizing combinations can be represented efficiently by the ordered pairs, or row vectors

$$(8, \ 8) \quad \text{and} \quad (12, \ 4)$$

number of dolls   number of cars

Now, row vectors also have a physical model in the plane, where the u-axis or doll-axis, is constructed horizontally, and the v-axis, or car-axis is constructed vertically. Thus, our row vectors have a portrait or picture-model, as given by Figure 1.1. As is usual, the portrait of the row vector (12,4), say, is the point arrived at by traveling 12 units to the right along the u-axis (starting from the origin 0), and then traveling 4 units vertically. The point arrived at is the picture-model of the row vector (12,4). To reinforce this picture, sometimes our arrow is drawn emanating from the origin, with its head at the point; in Figure 1.1, the points at which the heads of our arrows rest are (8,8) and (12,4).

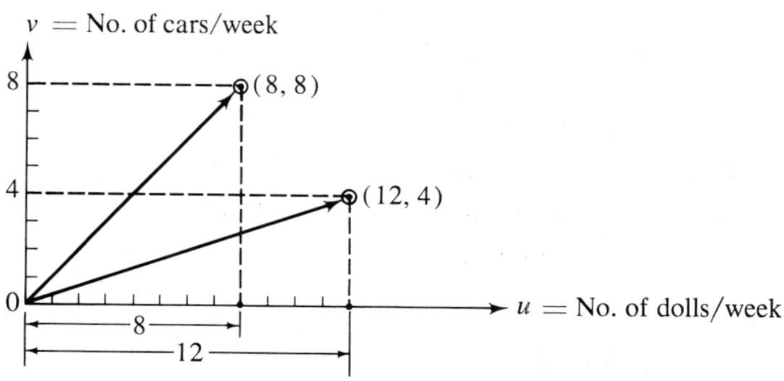

Figure 1.1

At the core of understanding the toymaker's problem, and problems like it, is an adequate description of points in the plane, or row vectors, which lie along a line segment connecting 2 fixed points in the plane. For example, how can we describe *all* points which lie on the straight line connecting the couple of points (8,8) and (12,4)? In order to do this, we shall use the geometric interpretation of vector addition in the plane, namely, the parallelogram method, and we shall envision the multiplication of a vector $X = (x_2,x_2)$ by a number α, as a "stretching" or "contraction" of the original vector X, where the sense of the vectors X and $\alpha \cdot X$ is the same. The diagrams of Figure 1.2 illustrate this where we see the relative positions of vectors X, Y, and $X + Y$. Line A is shown as parallel to the vector (arrow) X, and line A is also the same *length* as the vector X.

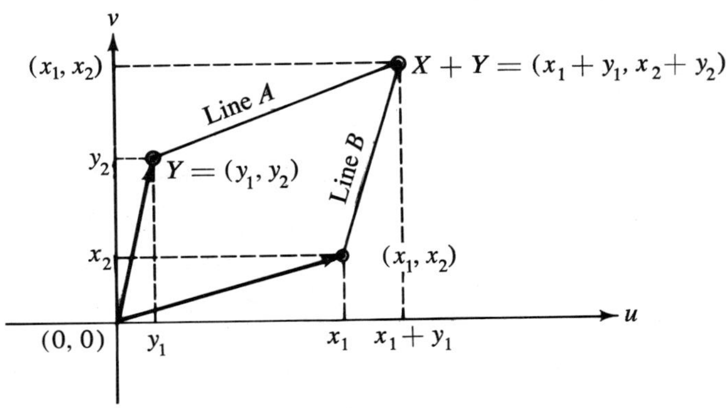

***Figure 1.2** Vector Addition in the Plane*

$$X = (x_1, x_2) \qquad Y = (y_1, y_2)$$
$$X + Y = (x_1 + y_1, x_2 + y_2)$$

Similarly, line B is the same length as, and is parallel to, the arrow which represents the vector Y. A geometric interpretation of vector addition follows from this, for if line A is regarded as a translate of the arrow vector X, then the two vectors, X and Y, are in a head-to-tail relationship where the "final" head indicates the point in the plane corresponding to the sum. This is illustrated in Figure 1.3. This geometric head-to-tail procedure may be extended for any finite number of vectors in the plane (and even in 3-dimensional space). Figure 1.4 illustrates the sum of 3

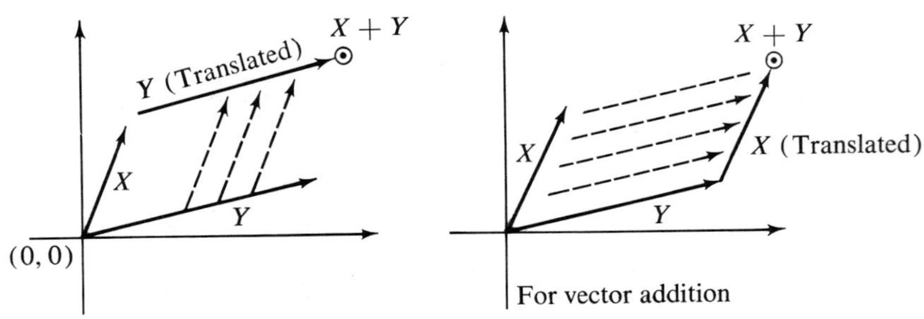

(a) (b)

Figure 1.3

vectors X, Y, and Z (indicated by arrows emanating from the origin). As there are 6 different ways in which we may attach heads-to-tails, we illustrate only 3. In each case, however, the *last* arrow head, the one which has no other arrow attached to it, rests at the point representing the sum $X + Y + Z$.

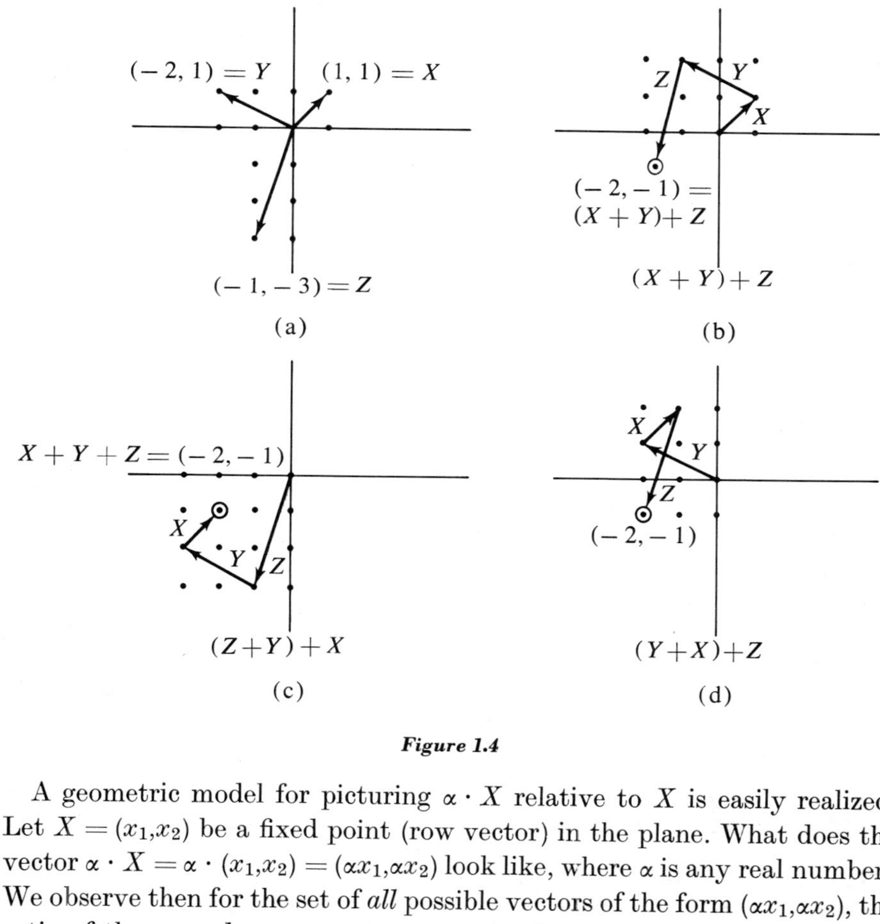

Figure 1.4

A geometric model for picturing $\alpha \cdot X$ relative to X is easily realized. Let $X = (x_1,x_2)$ be a fixed point (row vector) in the plane. What does the vector $\alpha \cdot X = \alpha \cdot (x_1,x_2) = (\alpha x_1,\alpha x_2)$ look like, where α is any real number? We observe then for the set of *all* possible vectors of the form $(\alpha x_1,\alpha x_2)$, the ratio of the second component αx_2, to the first component αx_1 is always the *same*, provided $\alpha x_1 \neq 0$. That is, for any nonzero α,

$$\frac{\text{second component}}{\text{first component}} = \frac{\alpha x_2}{\alpha x_1} = \frac{x_2}{x_1} = \text{constant},$$

since the numbers x_1 and x_2 were fixed in advance when we fixed $X = (x_1,x_2)$. The set of all such vectors (u,v) whose components are so related (that is $v/u = \text{constant}$ or $v = \text{constant} \times u$), lies on the *straight line* which passes through the 2 points (0,0) and (x_1,x_2). Therefore, the set of all vectors of the form $(\alpha x_1,\alpha x_2)=\alpha\cdot(x_1,x_2)=\alpha\cdot X$, lies on the line comprised of *all* (u,v) such that $(v/u)=(x_2/x_1)$ or $v=(x_2/x_1)\cdot u$: See Figure 1.5. Typical (u,v) of this type are $(u,v)=(\alpha x_1,\alpha x_2)$, where α is a real number.

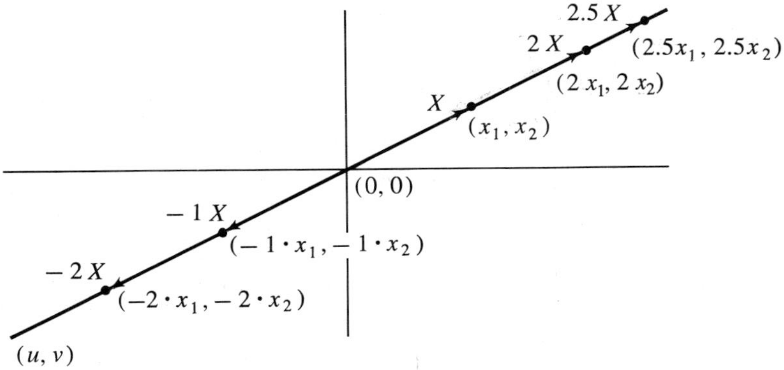

Figure 1.5

EXERCISES

I.4.1 Plot the following row vectors in the plane:

$$X=(0,1),\ Y=(-1,1),\ Z=(-2,-3).$$

Draw the 6 head-to-tail diagrams which represent the vector sum

$$X+Y+Z=(-3,-1),$$

that is, graph the vectors

(a) X – to – Y – to – Z
(b) X – to – Z – to – Y
(c) Y – to – X – to – Z
(d) Y – to – Z – to – X
(e) Z – to – X – to – Y
(f) Z – to – Y – to – X [see (I.4.4)].

I.4.2 Repeat Exercise I.4.1 with the row vectors

$$X=(0,4),\ Y=(2,2),\ Z=(-2,2).$$

I.4.3 Repeat Exercise I.4.1 with the row vectors

$$X=(2,1),\ Y=(1,3),\ Z=(1,1).$$

I.4.4 Plot the row vectors $\alpha\cdot X$, $\alpha\cdot Y$, and $\alpha\cdot Z$ in the plane where $X=(0,1)$, $Y=(-1,2)$, $Z=(3,-2)$, for the cases

(a) $\alpha=1$
(b) $\alpha=-1$
(c) $\alpha=2$
(d) $\alpha=-2\frac{1}{2}$.

I.4.5 Given the row vector $X = (2,1)$ and $Y = (1,3)$ in the plane, graph the vector $\alpha \cdot X + \beta \cdot Y$, where

(a) $\alpha = 2, \beta = -1$
(b) $\alpha = 4, \beta = -2$
(c) $\alpha = 1, \beta = 1$
(d) $\alpha = \frac{1}{2}, \beta = \frac{1}{2}$
(e) $\alpha = \frac{1}{4}, \beta = \frac{3}{4}$.

I.4.6 Repeat Exercise I.4.5 with the row vectors

$$X = (3,0),\ Y = (-1,0).$$

I.4.7 Repeat Exercise I.4.5 with the row vectors

$$X = (-2,3),\ Y = (3,-2).$$

I.4.8 Show that the arrow denoted by "?" in Figure 1.6 is necessarily a translate of the row vector $Y - X$.

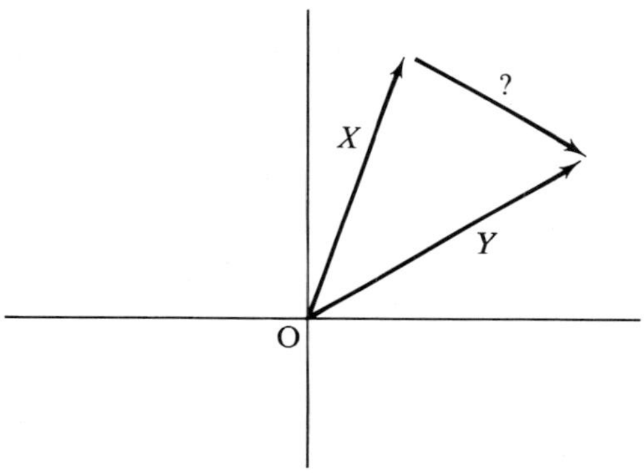

Figure 1.6

I.4.9 Given the vectors X and Y, graph the vectors X, Y, $X + Y$, and $Y - X$.

(a) $X = (2,1)$, $Y = (-3,1)$
(b) $X = (0,2)$, $Y = (3,0)$
(c) $X = (-1,-2)$, $Y = (-1,-3)$
(d) $X = (2,3)$, $Y = (4,6)$
(e) $X = (-1,-1)$, $Y = (1,1)$.

I.4.10 Graph the vectors X, Y, $\alpha \cdot X$, $\beta \cdot Y$, and $\alpha \cdot X - \beta\, Y$ for $\alpha = \frac{1}{2}, \beta = 2$, when

(a) $X = (2,0)$, $Y = (0,\frac{1}{2})$,
(b) $X = (0,0)$, $Y = (-1,4)$,
(c) $X = (-4,1)$, $Y = (1,1)$.

The stage is set so that we can characterize certain subsets of the plane. In particular, relative to points (vectors) X and Y in the plane, we shall be concerned with a characterization of (see Figure 1.7):

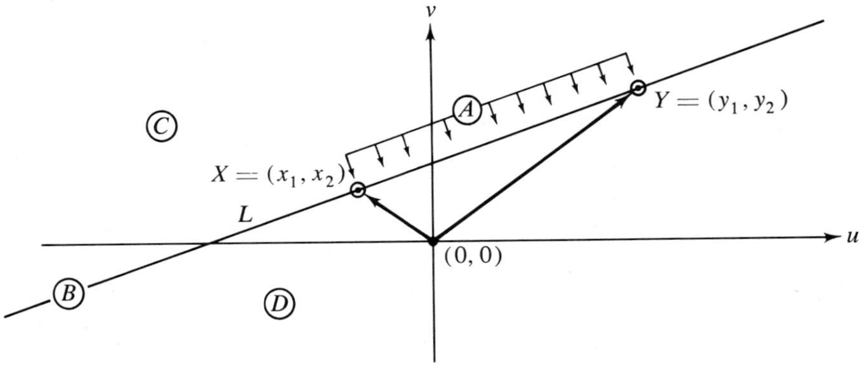

Figure 1.7

(A) All points (u,v) lying on the line *between* $X=(x_1,x_2)$ and $Y=(y_1,y_2)$.

As an extension of (A), we shall characterize:

(B) All points (u,v) lying on the *entire* line L which passes through points $X=(x_1,x_2)$ and $Y=(y_1,y_2)$.

(C) All points (u,v) which lie *above* the line L.

(D) All points (u,v) which lie *below* the line L.

Let us consider the question: If we start at the point $X=(x_1,x_2)$, how can we proceed to the point Y? Naturally, if we add the vector $Y-X$ to X, we obtain $X+Y(-X)=Y$, which puts us at Y. But what if we wanted only to proceed to the *midpoint* between the vectors X and Y? The answer is simple. Instead of adding all of $(Y-X)$ to X, just add $\frac{1}{2}(Y-X)$ to X; this puts us halfway between X and Y. Similarly, add $\frac{2}{3}(Y-X)$ to X to put yourself $\frac{2}{3}$ of the way between X and Y (see Figure 1.8). In this manner, we see geometrically, that the points between X and Y are all points of the form $X+\lambda(Y-X)$, where λ varies between $\lambda=0$ (which puts you at X) and $\lambda=1$ (which puts you at Y).

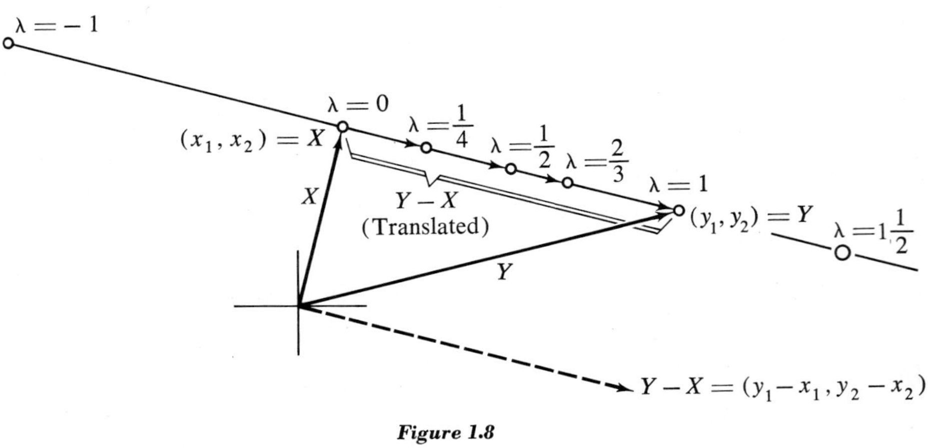

Figure 1.8

This line between X and Y can now be extended infinitely in both directions by allowing the number λ to vary without the restriction above, namely, that λ is between zero and 1. As Figure 1.8 illustrates, $\lambda = 1\frac{1}{2}$ corresponds to an "overshooting" of Y, starting from X, by a factor of $\frac{1}{2}$. When $\lambda = -1$, then we travel in a "negative direction" or "backward" from Y by one multiple of the distance from X to Y. Thus, we come to Proposition I.4.1.

Proposition I.4.1 Let $X = (x_1,x_2)$ and $Y = (y_1,y_2)$ be vectors in the plane. Then the straight line L, in the plane which passes through the points X and Y, is exactly the set of points (u,v) in the plane of the form

$$(u,v) = X + \lambda(Y - X) = (1 - \lambda)X + \lambda Y, \qquad \text{(I.4.4)}$$

where λ varies over all possible real numbers.

[Equation (I.4.4) is called a **parametric** form since a parameter λ is involved.] Moreover, that portion of the line which forms the segment *between* the vectors X and Y, and includes X and Y as endpoints is characterized by (I.4.4) where the parameter (variable) λ is constrained by the condition

$$0 \leq \lambda \leq 1. \qquad \text{(I.4.5)}$$

(As is standard, the symbol "$\leq$" is read "is less than or equal to"; "$<$" is read "is less than"; "$\geq$" is read "is greater than or equal to", and "$>$" is read "is greater than.")

With Equations (I.4.4) and (I.4.5) we have characterized regions A and B described in Figure 1.7. We offer another word on region A, the line segment between X and Y, by rewriting (I.4.4) with the constraint of (I.4.5) as follows:

(u,v) is on the line segment between X and Y if and only if

$$(u,v) = X + \lambda(Y - X), \quad \text{where} \quad 0 \leq \lambda \leq 1,$$

that is,

$$(u,v) = X + \lambda Y - \lambda X, \quad \text{where} \quad 0 \leq \lambda \leq 1,$$

or

$$(u,v) = \underset{\substack{\uparrow \\ \mu_1}}{(1 - \lambda)}X + \underset{\substack{\uparrow \\ \mu_2}}{\lambda}Y, \quad \text{where} \quad 0 \leq \lambda \leq 1.$$

This last equation says that the points on the line segment between X and Y are exactly those of the form $\mu_1 X + \mu_2 Y$, where the numbers μ_1 and μ_2 are between zero and 1, inclusive, and their sum is 1; that is to say,

> The points on the line segment between the vectors X and Y are all vectors of the form
>
> $$(u,v) = \mu_1 X + \mu_2 Y, \tag{I.4.6}$$
>
> where $0 \leq \mu_1 \leq 1$, $0 \leq \mu_2 \leq 1$, and $\mu_1 + \mu_2 = 1$.

Any vector of the form (I.4.6) is called a **convex combination of the vectors X and Y.**

We shall see that our Equation (I.4.4) describing the points (u,v) which comprise the line passing through $X = (x_1,x_2)$ and $Y = (y_1,y_2)$, is equivalent to the "usual" linear equations

(a) $au + bv = c$ for certain constants a, b, c, or

(b) $v = mu + b$, where m is the slope of the line, and b is the vertical-axis intercept. (I.4.7)

In fact, for $X = (x_1,x_2)$, $Y = (y_1,y_2)$, we have, from (I.4.4), that (u,v) is a row vector (point in the plane) on the line through X and Y, if and only if, for all λ,

$$\begin{aligned}(u,v) &= (x_1,x_2) + \lambda((y_1,y_2) - (x_1,x_2)) \\ &= (x_1,x_2) + \lambda \cdot (y_1 - x_1, y_2 - x_2) && \text{(see Definition I.1.2)} \\ &= (x_1,x_2) + (\lambda(y_1 - x_1), \lambda(y_2 - x_2)) && \text{(see Definition I.1.3)} \\ &= (x_1 + \lambda(y_1 - x_1),\ x_2 + \lambda(y_2 - x_2)).\end{aligned} \tag{I.4.8}$$

For convenience, put

$$z_1 = y_1 - x_1 \quad \text{and} \quad z_2 = y_2 - x_2. \tag{I.4.9}$$

Thus, (I.4.8) obtains, if and only if

$$(u,v) = (x_1 + \lambda z_1,\ x_2 + \lambda z_2), \tag{I.4.10}$$

which occurs if and only if corresponding components are equal (Definition I.1.4). Thus, (I.4.10) leads to the 2 equations

$$\begin{aligned}&\text{(a)} \quad u = x_1 + \lambda z_1 \\ &\text{(b)} \quad v = x_2 + \lambda z_2, \text{ for all numbers } \lambda.\end{aligned} \tag{I.4.11}$$

To eliminate the number λ from the equations of (I.4.11), let us multiply Equation (a) by z_2, and Equation (b) by $-z_1$, and then add the 2 resulting equations. This process leads us to the single equation involving the variables u and v, and the *constants* x_1, x_2, z_1 (equals $y_1 - x_1$) and z_2 (equals $y_2 - x_2$),

$$\underset{\substack{\uparrow\\ a}}{z_2} \cdot u + \underset{\substack{\uparrow\\ b}}{(-z_1)} \cdot v = \underset{\substack{\uparrow\\ c}}{z_2 x_1 - z_1 x_2}, \tag{I.4.12}$$

which is an equation of the form

$$a \cdot u + b \cdot v = c \quad \text{for certain constants } a,\, b, \text{ and } c.$$

Thus, we have established that Equation (I.4.4) for the vector (u,v) leads to Equation (I.4.7a), which relates the coordinates, u and v, to each other.

Equation (I.4.12) can be refined further if $z_1 = y_1 - x_1 \neq 0$. If $z_1 = y_1 - x_1 = 0$, which says that X and Y lie on a vertical line, then (I.4.12) tells us that $u = x_1$, a constant. Thus, all points on the (necessarily vertical) line (I.4.4), are of the form (x_1,v) where v is unrestricted. But if $z_1 = y_1 - x_1 \neq 0$, then we solve (I.4.12) for v to obtain

$$v = \underbrace{\frac{z_2}{z_1}}_{m} \cdot u + \underbrace{\frac{z_1 x_2 - z_2 x_1}{z_1}}_{b}$$

which is in the slope-intercept form of (I.4.7b); the number

$$m = \frac{z_2}{z_1} = \frac{y_2 - x_2}{y_1 - x_1}$$

is the slope of the line, and the number

$$b = \frac{z_1 x_2 - z_2 x_1}{z_1}$$

is the vertical-axis intercept.

A fact that will prove to be of great importance is that although parametric equations of form (I.4.4) lead to form (I.4.7a), the *converse* situation also obtains, that is, each equation of form (I.4.7a), defines a parametric equation of form (I.4.4). Thus, there are 2 forms by which we describe points (u,v) on a straight line in the plane. It might be mentioned, however, that the parametric form (I.4.4) has one advantage over (I.4.7a) or (I.4.7b); due to its geometric development, (I.4.4) can be used to describe straight lines in n-space for arbitrary n. For example, if we interpret ordered triples, or row vectors $(x_1,x_2,x_3,)$, as points in 3-dimensional space, then the set of points $X + \lambda(Y - X)$, λ varies over all numbers, describes the line in 3-space passing through $X = (x_1,x_2,x_3)$ and $Y = (y_1,y_2,y_3)$.

EXERCISES

In Exercises I.4.11 through I.4.20, you are given a pair of points $X = (x_1, x_2)$ and $Y = (y_1, y_2)$ in the plane. In each exercise, you are asked the following:

(1) For the parametric equation $(u,v) = X + \lambda(Y - X) = (1 - \lambda)X + \lambda Y$, plot points in the plane for the values $\lambda = 0$, $\lambda = \pm 1$, $\lambda = \pm 2$ $\lambda = \pm 3$, $\lambda = \frac{1}{4}$, $\lambda = \frac{1}{2}$, $\lambda = \frac{3}{4}$.

(2) Find constants a, b, and c such such that for all (u,v) on the line L, we have the relation $au + bv = c$, that is, find a nonparametric equation which relates u to v where (u,v) is on the line L.

(3) Find the vertical-axis and horizontal-axis intercepts of the line L, that is, for which numbers v and u are the points $(0,v)$ and $(u,0)$ on L? Graph the line L.

I.4.11 $X = (1,1)$, $Y = (0,3)$.
I.4.12 $X = (0,3)$, $Y = (1,1)$.
I.4.13 $X = (-1,-1)$, $Y = (2,-3)$.
I.4.14 $X = (2,-1)$, $Y = (2,3)$.
I.4.15 $X = (2,-1)$, $Y = (5,-1)$.
I.4.16 $X = (3,1)$, $Y = (7,5)$.
I.4.17 $X = (3,1)$, $Y = (5,3)$.
I.4.18 $X = (5,3)$, $Y = (8,6)$.
I.4.19 $X = (0,-2)$, $Y = (9,7)$.
I.4.20 $X = (0,0)$, $Y = (0,8)$.
I.4.21 Consider the equation $au + bv = c$ (a and b are nonzero numbers) which relates the number u to the number v.

(a) Find the value of v when $u = 0$, and find the value of u when $v = 0$, that is, find $X = (0,v)$ and $Y = (u,0)$, which are solution vectors for the equation.

(b) Show that u and v are solutions to the equation $au + bv = c$ if and only if the vector equation $(1 - \lambda)X + \lambda Y = (u,v)$ holds true for all λ [X and Y are given in (a)].

Where do we stand at this moment? We have seen that any expression of the form $au + bv = c$ defines an infinity of row vectors (points in the plane) (u,v) which lie on a straight line. In case $b = 0$, we see that only u, the *first* coordinate u of the solution vector (u,v), is constrained; in fact $u = c/a$, so that the (vertical) line in question is the set of all vectors $(c/a,v)$, where v is unconstrained. In case $b \neq 0$, $a \neq 0$, we see that when $u = 0$, $v = c/b$, or equivalently, the vector $(u,v) = (0, c/b)$ lies on the line. In plotting the line, we can plot these axis intercepts $(0,c/b)$ (on the vertical v-axis) and $(c/a,0)$ (on the horizontal u-axis), and connect these 2 points. We raise the question:

Which points (u,v) in the plane correspond to the *inequality*

$$au + bv > c?$$

For the sake of illustration, let us assume $b > 0$. We alter the inequality as follows:

$$au + bv > c$$

$bv > -au + c$	add $-au$ to both sides
$v > \dfrac{-a}{b}u + \dfrac{c}{b}$	divide through by *positive* number b, thus preserving inequality.

This last line says that v is *greater than* the number $-(a/b)u + c/b$, or on the plane, v is *above* the number $-(a/b)u + c/b$. In other words, the coordinates u and v of the vector (u,v) satisfy the inequality $au + bv > c$ if and only if (u,v) is *above* the line defined by the corresponding *equality* $au + bv = c$, provided $b > 0$. As Figure 1.9 indicates, the region *above* the line is the set of vectors (u,v), where u and v are constrained by the inequality $au + bv > c$. Similarly, the region below the line L consists of those (u,v) where $au + bv < c$, provided $b > 0$.

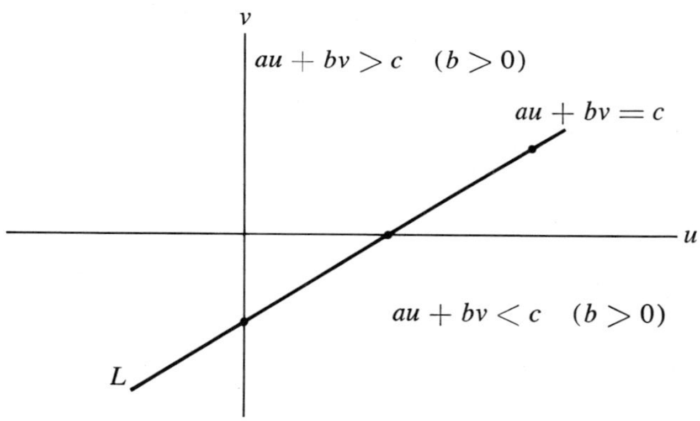

Figure 1.9

Example Find all (u,v) in the plane such that $3u - 4v > 12$.

We shall present 2 methods, and in both cases, it will be useful to graph the boundary line defined by the corresponding equality $3u - 4v = 12$. Observe that when $u = 0$, $v = -3$, so that $(u,v) = (0, -3)$ is on the line. When $v = 0$, u automatically equals 4, so that $(u,v) = (4,0)$ is also on the line. A graph of the line is shown in Figure 1.10.

Method 1: Isolate the variable v. Accordingly,

(a)	$3u - 4v > 12$	given
(b)	$-4v > -3u + 12$	add $-3u$ to both sides of (a)
(c)	$v < \frac{3}{4}u - 3$	multiply both sides of (b) through by $-\frac{1}{4}$, thereby reversing inequality.

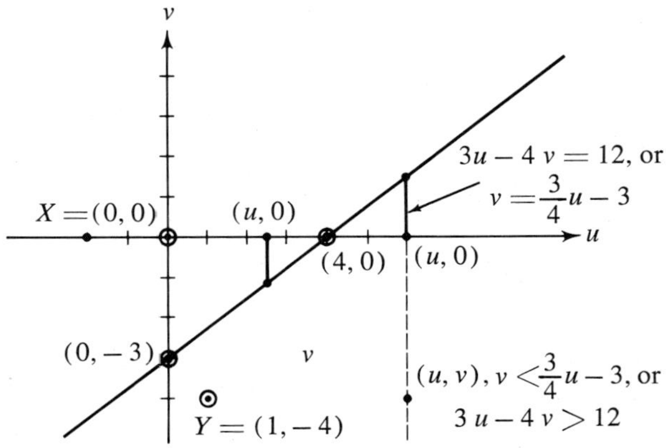

Figure 1.10

Line (c) tells us explicitly, that v is *less* than, or *below* the line $\frac{3}{4}u - 3$. Our desired region, described by the inequality $3u - 4v > 12$, is that region *below* the line $3u - 4v = 12$.

Method 2: Since we know in advance that we shall choose 1 side of the line $3u - 4v = 12$, or the other, simply check the inequality $3u - 4v > 12$ for 2 vectors $X = (u_1, v_1)$, and $Y = (u_2, v_2)$, 1 vector from each side. Actually, 1 vector would suffice, but in choosing 2, we check our results. For a point above the line, choose $X = (0,0) = (u_1, v_1)$. Clearly, the inequality $3u_1 - 4v_1 > 12$, or $3 \cdot 0 - 4 \cdot 0 > 12$ is not valid for $u_1 = 0$, $v_1 = 0$. For a point *below* the line, choose $Y = (1, -4) = (u_2, v_2)$. Sure enough,

$$3u_2 - 4v_1 = 3 \cdot 1 - 4 \cdot (-4) = 3 + 16 > 12 \text{ is valid.}$$

The region *below* the line is our desired region.

This example illustrates a characterization of regions C and D of Figure 1.7 which we formulate in Proposition I.4.2.

Proposition I.4.2 The inequalities $au + bv > c$ and $au + bv < c$, respectively, define the sets of vectors (u,v) in the plane which lie on each side of the line defined by the corresponding equation $au + bv = c$. Such a region (excluding the points *on* the line) is called an **open half-plane**. The set of points on 1 side of the line, along with (union) the boundary line $au + bv = c$, is called a **closed half-plane**.

Of interest to us will be the *intersection* of a number of half-planes which we illustrate by example.

Example Find all vectors (u,v) in the plane such that the following 3 inequalities hold simultaneously:

(i) $u \geq 1$

(ii) $u + 2v \leq 8$

(iii) $v - u \geq -2.$

We begin by graphing the lines defined by the corresponding *equalities* $u = 1$, $u + 2v = 8$, and $v - u = -2$ (Figure 1.11). As the arrows indicate, the region which falls on the appropriate sides of lines (i), (ii), and (iii), is the triangle whose vertices are formed by the intersections of these lines, namely, the vectors (points) $(1,-1)$, $(4,2)$, and $(1,\frac{7}{2})$.

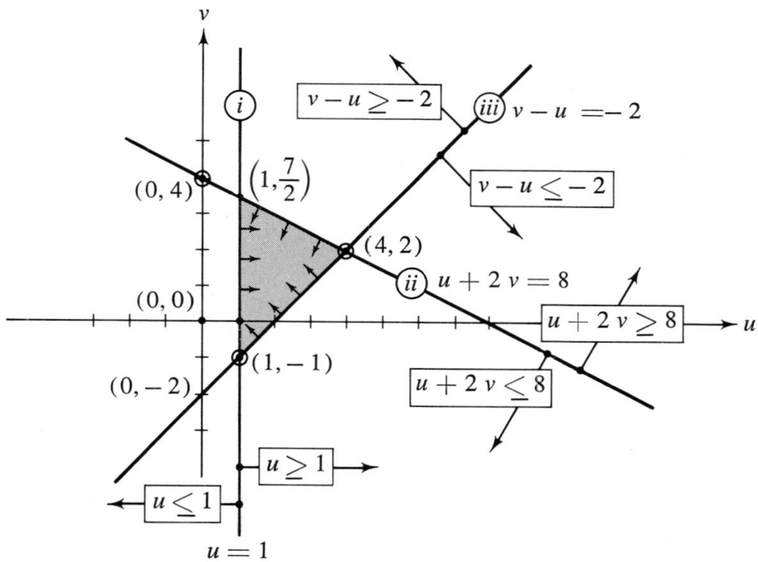

Figure 1.11

It must be pointed out that a region which satisfies all inequalities *simultaneously* may, on occasion, not exist. For example, if we were to reverse the inequalities for (i), (ii), and (iii), we would obtain

(i)′ $u \leq 1$

(ii)′ $u + 2v \geq 8$

(iii)′ $v - v \leq -2.$

We diagram these 3 closed half-planes, as shown in Figure 1.12. The wedge-shaped shaded region is where inequalities (i) and (iii) obtain simultaneously. The half-plane defined by (ii) does not interest this wedge-shaped region, and so there is *no* (u,v) in the plane, which lies in all 3 regions, that is, there is *no* (u,v) for which *all* 3 inequalities, (i), (ii), *and* (iii), obtain.

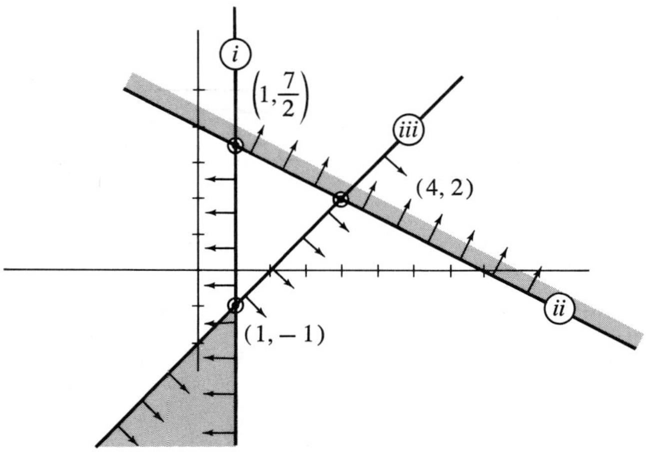

Figure 1.12

Example Graph the region [the set of all (u,v)] in the u - v plane where

(i) $$v + u \geq 0$$

(ii) $$v - u \geq -2$$

(iii) $$v + 4u < 13$$

(iv) $$3v - 4u < 7.$$

We proceed to sketch the boundary lines of these 4 half-planes which are given by the corresponding equalities (a) $v + u = 0$, (b) $v - u = 2$, (c) $v + 4u = 13$, and (d) $3v - 4u = 7$ (Figure 1.13). The shaded region indicates the set of (u,v) in the

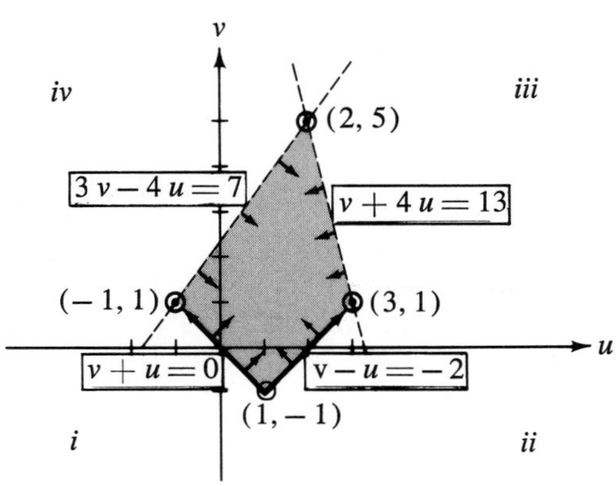

Figure 1.13

plane, which satisfy *all* 4 inequalities, that is, the intersection of the 4 half-planes (i), (ii), (iii), and (iv). Note that 2 of our half-planes are closed [namely, (i) and (ii)], and the other 2 are open. For this reason, we use dotted lines to denote the fact that the boundary lines are *not* included in the open half-planes. Also the "corner" points $(-1,1)$ and $(3,1)$ are not included since they fail to satisfy the inequalities (iv) and (iii), respectively. Of the 4 vertices, only the vertex $(1,-1)$ is included in the intersection of the 4 half-planes.

EXERCISES

In Exercises I.4.22 through I.4.28, graph the intersection of the half-planes defined by the given inequalities.

I.4.22 (i) $u > 0$

(ii) $v \geq 0$

(iii) $2u + 3v \geq 6.$

I.4.23 (i) $u > 0$

(ii) $v \geq 0$

(iii) $2u + 3v \leq 6.$

I.4.24 (i) $u > 0$

(ii) $v > 1$

(iii) $u < 3$

(iv) $v \leq 2.$

I.4.25 (i) $u > 0$

(ii) $v > 1$

(iii) $u < 3$

(iv) $v \geq 2.$

I.4.26 (i) $8v + 3u > 23$

(ii) $v + 4u < 21$

(iii) $7v - u \leq 31.$

I.4.27 (see Exercise I.4.26)

(i) $8v + 3u > 23$

(ii) $v + 4u < 21$

(iii) $7v - u \geq 31.$

I.4.28 (i) $8v + 3u > 23$

(ii) $v + 4u > 21$

(iii) $7v - u \leq 31.$

I.4.29 In this exercise, let $u =$ number of apples, and $v =$ number of oranges. In a grocery store which sells apples and oranges, are 3 workers. Worker A unpacks the fruit, then Worker B displays the fruit, and finally, Worker C sells the fruit. The time each worker spends per apple or per orange is given in the following table.

	Apples	Oranges
Worker A (unpacking)	1 min/apple	3 min/orange
Worker B (displaying)	2 min/apple	2 min/orange
Worker C (selling)	4 min/apple	1 min/orange

Thus, for each vector (u,v) in the apple-orange plane, we could find the *time* spent by each worker. For example, the *time* t_A spent by worker A is expressed as a function of u, the number of apples, and v, the number of oranges, as follows:

$$\begin{aligned} t_A(u,v) &= 1\,\frac{\text{min}}{\text{apple}} \times u \text{ apples} + 3\,\frac{\text{min}}{\text{orange}} \times v \text{ oranges} \\ &= 1 \cdot u + 3 \cdot v \text{ (minutes).} \end{aligned}$$

One interpretation of this is the following: We have taken Worker A's rate profile, considered as a row vector $(1,3)$, and multiplied it on the right by a column vector $\begin{pmatrix} u \\ v \end{pmatrix}$, representing the apple-orange output.

Now each worker has a time *limitation* due to poor stamina. Worker A can spend no more than 27 minutes per day working, Worker B can spend no more than 22 minutes per day, and Worker C can spend no more than 32 minutes per day. That is,

$$\begin{aligned} t_A(u,v) &\leq 27 \text{ minutes} \\ t_B(u,v) &\leq 22 \text{ minutes} \\ t_C(u,v) &\leq 32 \text{ minutes.} \end{aligned} \tag{1}$$

In this problem, you are asked the following:

(a) Find specific expressions for the times spent by each of the workers as a function of the quantities u and v. That is, find $t_A(u,v)$ (given above), $t_B(u, v)$, and $t_C(u,v)$.

(b) In the u-v (apple-orange) plane, graph each of the 3 half-planes defined by the inequalities of (1).

(c) Although (1) presents only 3 inequalities (or half-planes), 2 *more* inequalities are implicit by virtue of the physical nature of the problem. That is, we also have the number of apples, $u \geq 0$, and the number of oranges handled, $v \geq 0$. Thus, graph the intersection of these 5 half-planes, that is, the 3 given by (1), and the 2 just described. Find the 5 vertices or "corner points" of this region.

(d) Show that the vertices, or corner points, are extreme in the following sense (see Figure 1.14): For 1 of the vertices (u,v), each worker works a minimum length of time, namely, zero minutes, or $t_A = t_B = t_C = 0$. For 2 of the vertices, exactly 1 worker is working for his maximum length of time [see (1)], and for the other 2 vertices, exactly 2 workers are working

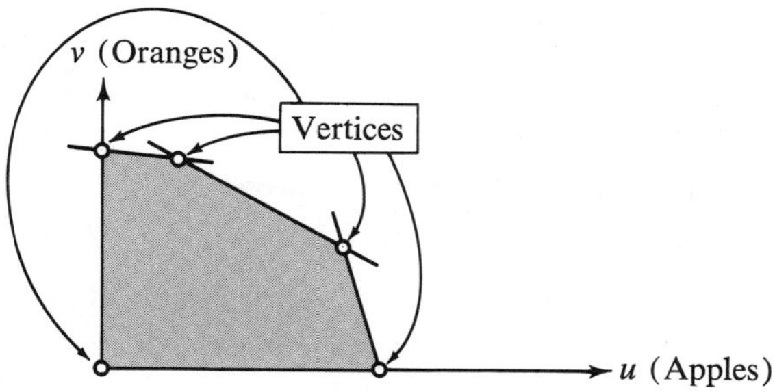

Figure 1.14

their maximum times. In other words, compute $t_A(u,v)$, $t_B(u,v)$, $t_C(u,v)$, in each of the 5 instances where (u,v) is a vertex, and compare with (1).

(e) Is it possible to handle (that is, unpack, display, and sell) 3 apples and 8 oranges with *all* workers working less than their maximum times?

(f) Is it possible to handle 3 apples and 10 oranges? Why or why not?

(g) If, moreover, 1¢ profit is made on each apple, and 2¢ profit is made on each orange, then find an expression for $p(u,v)$, the profit made on u apples and v oranges.

(h) Evaluate the profit function $p(u,v)$ from part (g), at each of the 5 vertices of the region described in part (c).

(i) Graph each of the lines in the u-v plane corresponding to the respective equations $p(u,v) = 20¢$, $p(u,v) = 19¢$, $p(u,v) = 15¢$, $p(u,v) = 8¢$, $p(u,v) = 0¢$, and show that each of these (parallel) lines passes through a vertex [see part (h)]. Is the profit a maximum for the region at a vertex? Is it minimal at a vertex?

Our last Exercise I.4.29 brings us to the kind of question we wish to consider. Roughly speaking, it is this: For which points (u,v), is a linear expression of the form $p(u,v) = au + bv$ a maximum, or a minimum, where the points (u,v) belong to an intersection of closed half-planes? In Exercise I.4.29, $p(u,v) = u + 2v$ represents the profit due to u apples and v oranges; in our toymaker's problem, $p(u,v) = u + v$ (for Case I), or $p(u,v) = 3u + 2v$ (for Case II) represents the number of children serviced by u dolls and v toy cars.

To answer this question, we prove that when linear, numerical-valued functions (or more exactly, functionals) of the form $p(u,v) = au + bv$, when restricted to a finite, nonempty intersection of closed half-planes, always take on their maximum and their minimum values, at *vertices* (but not necessarily *only* at vertices). We first describe the behavior of the values of $p(u,v) = au + bv$ when the set of points (u,v) are restricted to a line segment in the plane. Accordingly, we present:

Proposition I.4.3 Let the function $p(u,v)$ be defined by the equation $p(u,v) = au + bv + c$ for certain fixed numbers a, b, and c. Let L be the straight-line segment from the point $X = (x_1,x_2)$ to the point $Y = (y_1,y_2)$, which includes the endpoints X and Y. Then

(1) The values of $p(u,v)$, for (u,v) on the line segment L, are such that $p(u,v)$ is a maximum at 1 of the endpoints, and a minimum at the other endpoint.

(2) If $p(u,v)$ has the same value at the endpoints X and Y [that is, if $p(X) = p(x_1,x_2) = p(Y) = p(y_1,y_2)$], then $p(u,v)$ is constant on *all* of L.

Proof: Let us assume, without loss of generality, that $p(X) < p(Y)$, that is, $p(x_1,x_2) < p(y_1,y_2)$, so that the function p assumes *different* values at the endpoints X and Y. We are to show that the value of $p(u,v)$, for any and all points (u,v) between X and Y, falls *between* $p(X)$ and $p(Y)$. Thus, $p(X)$ will be a minimum value, and $p(Y)$ will be a maximum value, and Proposition I.4.3 will be proved.

Now, from Proposition I.4.1, any point on the line segment between $X = (x_1,x_2)$ and $Y = (y_1,y_2)$ is of the form

$$(u,v) = (1-\lambda)X + \lambda Y \qquad \text{where } 0 < \lambda < 1.$$

That is, (u,v) is on L, between X and Y if and only if

$$\begin{aligned}(u,v) &= (1-\lambda)(x_1,x_2) + \lambda(y_1,y_2) && \text{(where } 0<\lambda<1) \\ &= (\underbrace{(1-\lambda)x_1 + \lambda y_1}_{u},\ \underbrace{(1-\lambda)x_2 + \lambda y_2}_{v}) && \text{Definitions I.1.2, I.1.3.}\end{aligned}$$

This implies, from Definition I.1.4, that

$$u = (1-\lambda)x_1 + \lambda y_1$$

and

$$v = (1-\lambda)x_2 + \mu y_2 \quad \text{for } 0 < \lambda < 1. \qquad \text{(I.4.13)}$$

Now we are in a position to evaluate $p(u,v)$ for this "in-between point" (u,v).

$$\begin{aligned}p(u,v) &= au + bv + c && \text{hypothesis (definition of } p) \\ &= a((1-\lambda)x_1 + \lambda y_1) + b((1-\lambda)x_2 + \lambda y_2) + c && \text{from (1.4.13)} \\ &= (1-\lambda)ax_1 - \lambda ay_1 + (1-\lambda)bx_2 + \lambda by_2 + c \\ &= \underbrace{(1-\lambda)ax_1 + (1-\lambda)bx_2} + \underbrace{\lambda ay_1 + \lambda by_2} + c && \text{(I.4.14)} \\ &= (1-\lambda)\underbrace{(ax_1 + bx_2 + c)} + \lambda\underbrace{(ay_1 + by_2 + c)} \\ &= (1-\lambda)p(x_1,x_2) + \lambda p(y_1,y_2) && \text{Hypothesis} \\ &= (1-\lambda)p(X) + \lambda p(Y).\end{aligned}$$

The proposition follows from (I.4.14). Let us denote the difference of $p(X)$ and $p(Y)$ by the symbol Δ. That is, we set

$$p(Y) - p(X) = \Delta. \tag{I.4.15}$$

We have assumed that $\Delta \geq 0$. We continue to write (I.4.14) as

$$\begin{aligned} p(u,v) &= (1-\lambda)p(X) + \lambda \underbrace{p(Y)}_{\searrow} \\ &= (1-\lambda)p(X) + \lambda\overbrace{(\Delta + p(X))} && \text{from (I.4.15)} \\ &= (1-\lambda)p(X) + \lambda\Delta + \lambda p(X) && \text{(1.4.16)} \\ &= p(X) + \underbrace{\lambda\Delta}_{\text{non-negative}} \\ &\geq p(X), \quad \text{since} \quad \lambda\Delta \quad \text{is non-negative.} \end{aligned}$$

Similarly,

$$\begin{aligned} p(u,v) &= (1-\lambda)\underbrace{p(X)}_{\downarrow} + \lambda p(Y) && \text{from (I.4.14)} \\ &= \overbrace{(1-\lambda)(p(Y) - \Delta)} + \lambda p(Y) && \text{from (I.4.15)} \\ &= (1-\lambda)p(Y) - \Delta(1-\lambda) + \lambda p(Y) && \text{(I.4.17)} \\ &= p(Y) - \underbrace{\Delta(1-\lambda)}_{\text{non-negative}} \\ &\leq p(Y) \quad \text{since} \quad -\Delta(1-\lambda) \quad \text{is negative.} \end{aligned}$$

Thus, (I.4.16) and (I.4.17) reveal that if (u,v) falls on the line between X and Y, then for $p(X) < p(Y)$,

$$p(X) < p(u,v) < p(Y),$$

which proves part (1) of Proposition I.4.3.

From Equations (I.4.16) and (I.4.17), if $p(X) = p(Y)$, so that $\Delta = 0$, we have for *all* (u,v) between X and Y,

$$p(X) = p(u,v) = p(Y),$$

which says that p is constant on the *entire straight-line* segment from X to Y.

As a corollary to this result, we shall be able to say that linear functionals of the form $p(u,v) = au + bv$ attain their maximum values and their minimum values at the vertices of certain (convex) regions which we describe with:

Definition I.4.4 A **finite covex polygon** denotes a region of finite area which is the intersection of a finite number of closed half-planes.

REMARK. Although we shall not prove it formally, the finite intersection of half-planes is **convex** in the sense that whenever 2 points X and Y are in the region, then the line segment joining them lies with the region as well. This "convex" property obtains for half-planes, and since convexity (in the sense described above) is preserved after intersection, finite convex polygons are also convex. The following 3 illustrations of Figure 1.15 exhibit regions that are and are not finite convex polygons.

(a) (b) (c)

Figure 1.15

Region (*a*) is *not* a finite convex polygon. Although (a) is the intersection of three half-planes, its area is not finite.

Region (*b*) is *not* a finite convex polygon. Although its area is finite, it is not the intersection of half-planes. In fact, as the diagram indicates, it is not convex since there are 2 points, both *in* the region, but the (dotted) line segment joining them is not *wholly* within the region itself.

Region (*c*) is a finite, convex polygon with $n = 5$ sides and $n = 5$ vertices.

Theorem I.4.5 Any linear functional $p(u,v) = au + bv + c$ defined for points (u,v) in the plane, when restricted to a finite convex polygon, attains its maximum value at a vertex and its minimum value at a vertex. If the maximum value (respectively, the minimum value) is attained at its *adjacent* vertices (vertices connected by a boundary side), then $p(u,v)$ attains its maximum (respectively, minimum) at all points along the connecting side.

Informal Proof: The result is a direct consequence of the previous proposition. Let us consider a typical finite, convex polygon to which the functional p is restricted. (See Figure 1.16.) Construct a line L through the

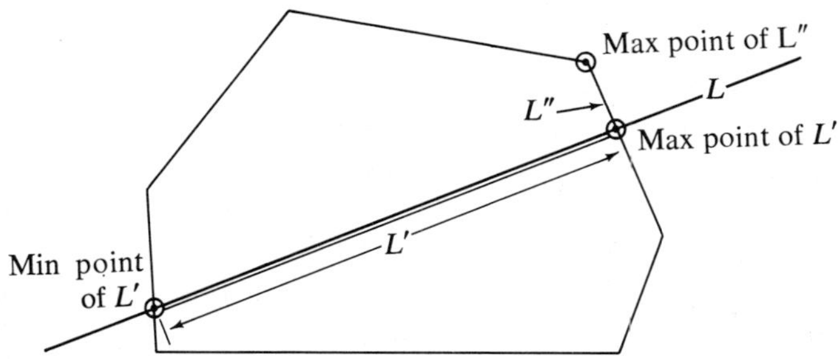

Figure 1.16

polygon so that it cuts at 2 points. From Proposition I.4.3, p assumes its maximum (respectively, minimum) over the line segment L' at one of the endpoints, that is, at a point on the *boundary* of the polygon. As long as we are on a boundary segment, we again apply Proposition I.4.3 to *that* straight line segment L'' to conclude that p assumes its maximum (respectively, minimum) at the endpoints; the endpoints of a boundary side are *vertices* of the polygon. Now that we are at the vertex, where a maximum exists, we may proceed counter-clockwise, say, to compare the value of p for *adjacent* vertices. After a finite number of comparisons, we shall have our "best" vertex where p assumes its maximum (respectively, minimum).

If the maximum happens to occur at adjacent *vertices*, then p is constant on the boundary side connecting them (see Proposition I.4.3). The proof is done.

REMARK. Theorem I.4.5 says that if you have a function $p(u,v) = au + bv + c$, where the argument vectors (u,v) are restricted to a finite convex polygon, then in order to find the point or points for which $p(u,v)$ is a maximum or a minimum over that region, all you need to do is to evaluate the function at the n vertices, and compare accordingly. This finite process of search and comparison is relatively easy to program on a high-speed computer.

Example We solve the **toymaker's problem** which was presented at the beginning of this section. We are given the following time-rate profiles of each of the doll-producing machines [see (I.4.1)]:

		Dolls (u)	Toy cars (v)
	Machine A (painting)	1 hr/doll	4 hr/car
(1)	Machine B (drying)	1 hr/doll	1 hr/car
	Machine C (wrapping)	2 hr/doll	1 hr/car

We are given certain time limitations, as well [see (I.4.2)]:

(2)
$$t_A \leq 40 \text{ hr/week}$$
$$t_B \leq 16 \text{ hr/week}$$
$$t_C \leq 28 \text{ hr/week}$$

t_A = time worked by machine A, and so on.

So it is, that (1) and (2) define a finite convex polygon in the u-v (doll-car) plane; for if we calculate the time required of each machine as a function of u and v, the number of dolls and cars, respectively, processed by each machine, we have

(a) $$t_A(u,v) = 1 \cdot u + 4 \cdot v \leq 40$$

(b) $$t_B(u,v) = 1 \cdot u + 1 \cdot v \leq 16$$

(c) $$t_C(u,v) = 2 \cdot u + 1 \cdot v \leq 28.$$

from (1) from (2)

Also, from the physical nature of the problem, we have that the number of toys is not negative, that is,

(d) $$u \geq 0$$

(e) $$v \geq 0.$$

The finite convex polygon in the u-v plane, which depicts all doll-car combinations (u,v) that satisfy *all* 5 inequalities above, is given in Figure 1.17.

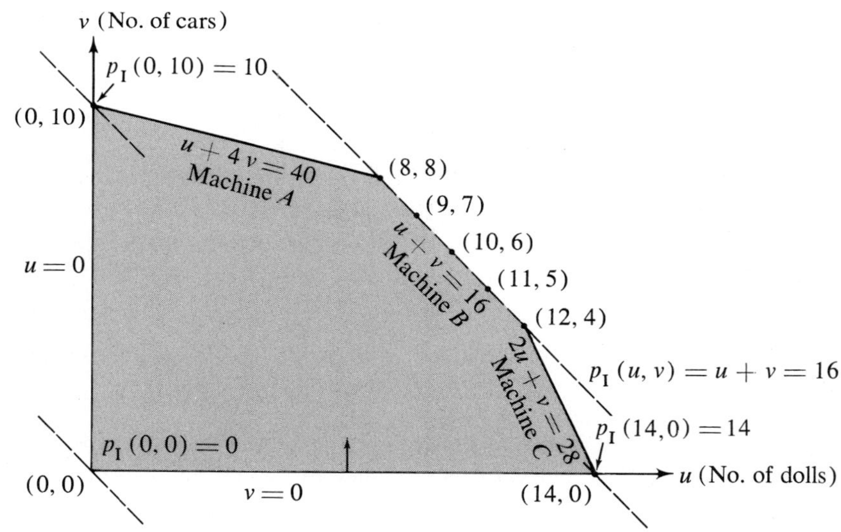

Figure 1.17

The shaded region of the polygon (Figure 1.17) represents all feasible solutions, that is, all the possible doll-car combinations (u,v) that the toymaker can produce subject to the constraints of (1) and (2).

Now, what are we to maximize?

CASE I. At most 1 girl is to have 1 doll, and at most 1 boy is to have 1 toy car. Thus, the number of children served in this case, as a function of u (the number of dolls) and v (the number of cars) is

$$p_{\mathrm{I}}(u,v) = 1\,\frac{\text{girl}}{\text{doll}} \times u \text{ dolls} + 1\,\frac{\text{boys}}{\text{car}} \times v \text{ cars}$$

$$= 1 \cdot u + 1 \cdot v \quad (\text{boys} + \text{girls} = \text{children}).$$

What good luck! The child function $p_{\mathrm{I}}(u,v) = 1 \cdot u + 1 \cdot v = au + bv + c$ ($a = b = 1$, $c = 0$) is of the desired form of Theorem I.4.5. Its maximum occurs, then, at one of the 5 vertices (0,0), (0,10), (8,8), (12,4), or (0,14). Only these 5 computations of $p_{\mathrm{I}}(u,v)$ need be made in order to find the maximum (and minimum) number of children served. Now

$$\begin{aligned} p_{\mathrm{I}}(0,0) &= 0 + 0 = 0 \\ p_{\mathrm{I}}(0,10) &= 0 + 10 = 10 \\ p_{\mathrm{I}}(8,8) &= 8 + 8 = 16 \\ p_{\mathrm{I}}(12,4) &= 12 + 4 = 16 \\ p_{\mathrm{I}}(0,14) &= 0 + 14 = 14. \end{aligned}$$

We conclude that $p_{\mathrm{I}}(u,v)$ is at its maximum value [namely, $p(u,v) = 16$] at the 2 vertices (8,8) and (12,4). According to Theorem I.4.5, $p_{\mathrm{I}}(u,v)$ is at its maximum value 16 for *all* (u,v) along the boundary side which connects (8,8) and (12,4). If we assume only whole numbers of toys are produced, then the allowable points at which $p_{\mathrm{I}}(u,v)$ is at its maximum value are the *five* points (8,8) (vertex), (9,7), (10,6), (11,5), and (12,4) (vertex).

As intuition would have it, the maximum number of children served is attained when at least 1 of the machines is working for its maximum time (although this is not always to be expected). The following table illustrates this.

max point	*Machine A*	*Machine B*	*Machine C*
(8,8)	Max time	Max time	
(9,7)		Max time	
(10,6)		Max time	
(11,5)		Max time	
(12,4)		Max time	Max time
↑ ↑ dolls cars			

In *any* case, we see that machine B, the drying machine, will always have to be used for its maximum time of 16 hours.

CASE II. At most 3 girls share a doll, at most 2 boys share a car. The only difference introduced here (see Figure 1.18) is the child function, which now takes the form

$$\text{Number of children} = p_{\text{II}}(u,v) = 3\,\frac{\text{girls}}{\text{doll}} \times u \text{ dolls} + 2\,\frac{\text{boys}}{\text{car}} \times v \text{ cars}$$
$$= 3 \cdot u + 2 \cdot v.$$

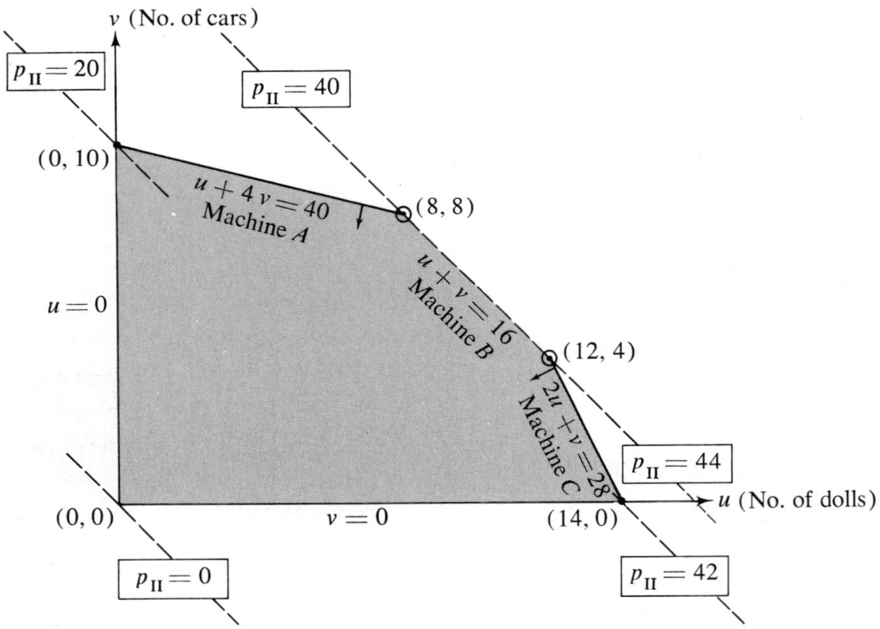

Figure 1.18

Again, all we need do is evaluate p_{II} at the five vertices.

$$\begin{aligned}
p_{\text{II}}(0,0) &= 3 \cdot 0 + 2 \cdot 0 = 0 \text{ children}\\
p_{\text{II}}(0,10) &= 3 \cdot 0 + 2 \cdot 10 = 20 \text{ children}\\
p_{\text{II}}(8,8) &= 3 \cdot 8 + 2 \cdot 8 = 40 \text{ children}\\
p_{\text{II}}(12,4) &= 3 \cdot 12 + 2 \cdot 4 = 44 \text{ children}\\
p_{\text{II}}(14,0) &= 3 \cdot 14 + 2 \cdot 0 = 42 \text{ children.}
\end{aligned}$$

In this case, the maximal vertex (12,4) is *unique*; 44 children may be served by the production of 12 dolls and 4 toy cars (which is accomplished by working machines B and C for their maximum times of 16 and 28 hours, respectively).

EXERCISES

I.4.30 For Exercise I.4.29, find the maximum profit and the number of apples and oranges which must be sold to achieve this profit.

I.4.31 A manufacturer of sweatshirts decides to give out free sweatshirts to students of a certain school. His constraints are these:

(1) He will give 1 sweatshirt to each boy, and 3 sweatshirts to each girl.

(2) He will give away no more than 42 sweatshirts in all.

(a) Graph the finite convex polygon on the boy-girl axis (let $u =$ number of boys, $v =$ number of girls), which consists of all boy-girl combinations (u,v) who may receive sweatshirts subject to constraints (1) and (2) along with the 2 implicit constraints that u, the number of boys, and v, the number of girls, are non-negative.

(b) Suppose the manufacturer knows that each boy receiving 1 free sweatshirt will tell exactly 1 other person of his generous deed, while each girl receiving 3 free sweatshirts will tell 9 other people of the giveaway. Find an expression for $p(u,v)$, the number of people who have been told, by u boys and v girls, about the sweatshirt gifts.

(c) Find the number of boys u, and the number of girls v, who receive free sweatshirts in order that the maximum number of people are told of the sweatshirt giveaway. That is, for which (u,v) is $p(u,v)$ a maximum?

I.4.32 (See Exercise I.4.31.) Suppose, after seeing the results of Exercise I.4.31, the manufacturer takes pity on the boys, and decides to constrain himself by the further condition that at least 9 boys shall each be given 1 free sweatshirt. What is the maximum number of people *now*, who will hear of the sweatshirt giveaway from the boys and girls receiving them? In other words, how many fewer people will hear of the manufacturer's gifts if at least 9 boys are to receive free sweatshirts?

I.4.33 (See Exercise I.4.31.) To increase favorable publicity, the manufacturer decides to tell 10 other people himself.

(a) Find an expression for $q(u,v)$, the number of people who will hear of the manufacturer's generosity from those receiving free sweatshirts, *and* from the manufacturer himself.

(b) What is the maximum of $q(u,v)$, the number of people who hear of the sweatshirt giveaway, and how many boys and how many girls receive sweatshirts?

I.4.34 A candy man manufacturers cherry candies and rum candies. He decides that each cherry candy shall weigh at least 1 ounce, and each rum goodie shall weigh at least 2 ounces. Accordingly, each box of mixed cherry and rum candies, shall weigh at least 1 pound (16 ounces). On the other hand, he doesn't want to put more than 6 pieces of (the 2-ounce) rum candy or 10 pieces of (the 1-ounce) cherry candy in any 1 box.

(a) Is it possible or feasible for the manufacturer to put the maximum number of candy pieces (6 rum and 10 cherry pieces) in a box and still be consistent with his minimum-weight-per-box constraint?

(b) On the cherry-rum plane (let $u =$ number of cherry candies, $v =$ number of rum candies), plot the finite convex polygon consisting of all possible combinations (u,v) of cherry and rum pieces, which are consistent with the constraints of minimum weight per box, and maximum number of each type of candy pieces per box.

(c) If the cost of packing these candies averages to 3¢ for each cherry piece, and 4¢ for each rum piece, plus a fixed cost of 10¢ per box, then find an expression for $p(u,v)$, the cost in pennies of packing a box of u cherry pieces and v rum pieces.

(d) For what combination of cherry pieces and rum pieces in each box will the candy maker incur the least cost? That is, for which (u,v) is $p(u,v)$ a minimum? What *is* the least cost per box? (Does it seem reasonable to suppose that his least cost occurs when he packs the minimum *weight* of candy, namely, 16 ounces?)

(e) For what combination of cherry pieces and rum pieces in each box will the cost be the largest possible? That is, for which (u,v) is $p(u,v)$ a maximum? What *is* the maximum cost per box?

5. *Abstract Vector Spaces*

Let us return, momentarily, to the past—to section 1 in particular, where we defined the so-called "concrete" n-vectors (Definition I.1.1). Let us oversimplify the situation somewhat by saying that the concrete n-vectors represent a collection of objects that we could

(1) add pairwise (Definition I.1.2), and

(2) multiply on the outside by scalars (Definition I.1.3).

A natural question arises in this context, namely: Are there other kinds of objects (besides columns or rows of scalars) which can be added pairwise and multiplied by scalars? Well, once we understand fully what will then be meant by "pairwise addition," and "scalar multiplication," we shall see that the answer is a resounding *yes*. In short, we shall indeed be able to exhibit sets of objects which, like sets of column vectors, accept a pairwise addition (whatever that means) along with a scalar multiplication (whatever that means). That is, these "other" sets will behave very much like our concrete n-vectors in that a Definition I.1.2 (addition) and a Definition I.1.3 (scalar multiplication) will be defined for them. Rather than call these elements "other vectors," or "kind-of-like-concrete-vectors," or something equally gross, we shall call these elements "abstract vectors"; certain collections of "abstract vectors" will then be called "abstract vector spaces" (detailed definitions to follow).

We present a discuss on of real concrete 2-vectors, which shows that each 2-vector may be thought of in an abstract sense, that is, as a point, or as a physical arrow in the plane which emanates from some fixed point in the plane. The student who has absorbed Section 4 dealing with linear programming, may pass directly now to page 59.

REMARKS ON CONCRETE 2-VECTORS. Consider the plane. Suppose we have 2 perpendicular axes (lines) in the plane which cross at a point, which we call the origin. A problem now arises: If we were to select a point in the plane, could we give it a "name"? Conversely, if we were told the "name" of a certain point, could we go to the plane and find that point? In other words, is there some way in which we can assign "names" to points in the plane such that (1) for every point, there is one and only one "name," and such that (2) for every "name," there is one and only one point? The question may be stated more formally: Is there a 1-1 correspondence between the set of points in the plane and a certain set of "names"?

If there is a measurement of *length* along the axes, the answer is yes. In this case, a pair of real numbers (a,b) can be assigned to each point in the plane. In Figure 1.19, we have named 3 points with pairs of numbers.

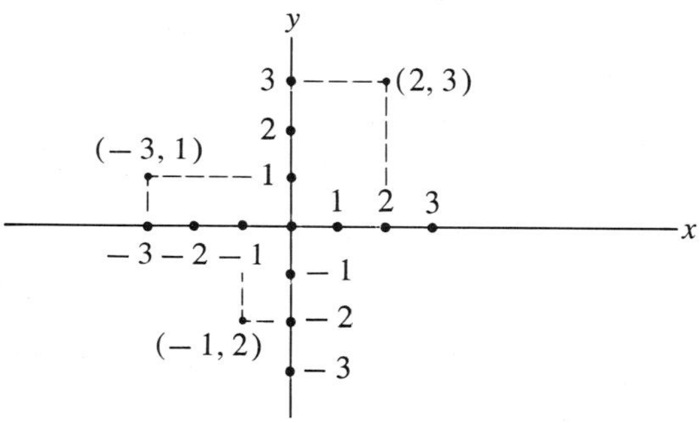

Figure 1.19

These pairs also tell us how to find the point. In fact, if we read the first coordinate (entry) of the pair, we know how many units along the x-axis we must travel (positive numbers indicate a travel to the right, while negative numbers indicate a travel to the left). Once we have moved the required distance along the x axis, upon reading the second coordinate (entry), we see how far parallel to the y axis we must continue to travel (positive second coordinates indicate upward travel, and negative numbers indicate downward travel). These remarks tell use that for every name there is 1 and only 1 point, and question (b) is answered in the affirmative. Question (a) also carries an affirmative answer, since we can name every point simply by "going to it" as follows: From the origin, travel along the x axis until you are directly below, or above the point. The distance you have travelled is placed in the first entry of your pair (remember that distance can be negative or non-negative according to whether you travel left or right from the origin). The second entry is made once you discover your distance travelled up or down from the x axis in order to reach the given point. Hence, there *is* a 1-1 correspondence between points on the plane and "names" which we now see are ordered pairs of real numbers.

Note that not only is the pair (3,1), for example, important, but the *ordering* is important as well. By these rules of location, (3,1) is not the same as (1,3), that is, they represent (or name, or locate) 2 *different* points.

From now on, we will think of ordered pairs of real numbers as being the *same as* the point it locates, since there is a 1-1 correspondence. This deliberate confusion occurs often in real life, after all, with no evidence of discomfort. Someone may say, "I am shaking hands with Willy Schniggleburger," when, in fact, he means that he shaking hands with a person (point in the earth-plane) who is *named* "Willy Schniggleburger." "Willy Schniggleburger" is only a name (hence impossible to shake hands with) but this deliberate confusion between the name and the thing being named offers no difficulties.

Referring to Definition I.1.1, we see that these ordered pairs of numbers are 2-vectors, and that 2-vectors are (the same as, names for) points in the plane.

For convenience in diagramming, a point in the plane is indicated by an arrow, with its tail at the origin and its head at the point in question. Figure 1.20 illustrates this device.

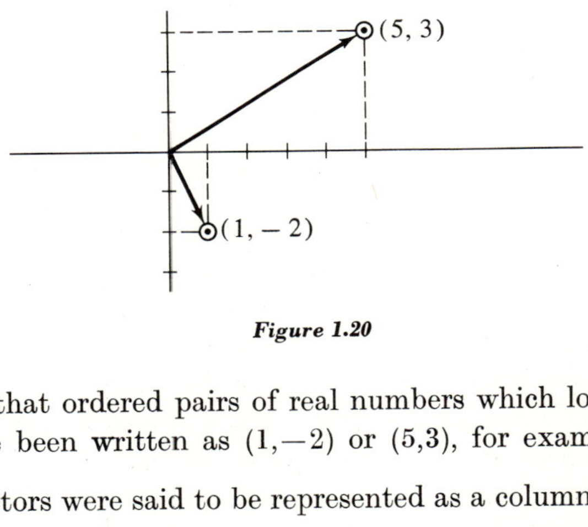

Figure 1.20

Observe that ordered pairs of real numbers which locate points in the plane, have been written as (1,—2) or (5,3), for example; on the other hand, 2-vectors were said to be represented as a column, such as $\begin{pmatrix} 1 \\ -2 \end{pmatrix}$ or $\begin{pmatrix} 5 \\ 3 \end{pmatrix}$. No matter whether the numbers are written horizontally or vertically, in both cases we are talking about an *ordered pair of numbers*, and it makes no difference, conceptually, whether we order this pair horizontally or vertically. In each case, we understand which is the *first* coordinate and which is the *second* coordinate.

We note that it is not necessary to have chosen the axes to be perpendicular; moreover, the scales of measurement on each of the axes do not have to be the same.

Unless otherwise stated, we shall deal only with *real* numbers, referring to them as scalars. The ideas to be presented are consistent, however, for complex numbers (scalars).

EXERCISES

I.5.1 Diagram the following concrete n-vectors as (abstract) points in the plane.

(a) $\begin{pmatrix} 1 \\ 0 \end{pmatrix}$ (e) $\begin{pmatrix} 3 \\ -2 \end{pmatrix}$

(b) $\begin{pmatrix} 0 \\ 1 \end{pmatrix}$ (f) (-5)

(c) (3) (g) $\begin{pmatrix} 0 \\ 0 \end{pmatrix}$

(d) $\begin{pmatrix} -2 \\ 3 \end{pmatrix}$ (h) $\begin{pmatrix} -7 \\ 0 \end{pmatrix}$.

I.5.2 Identify the vectors diagrammed below (Figure 1.21).

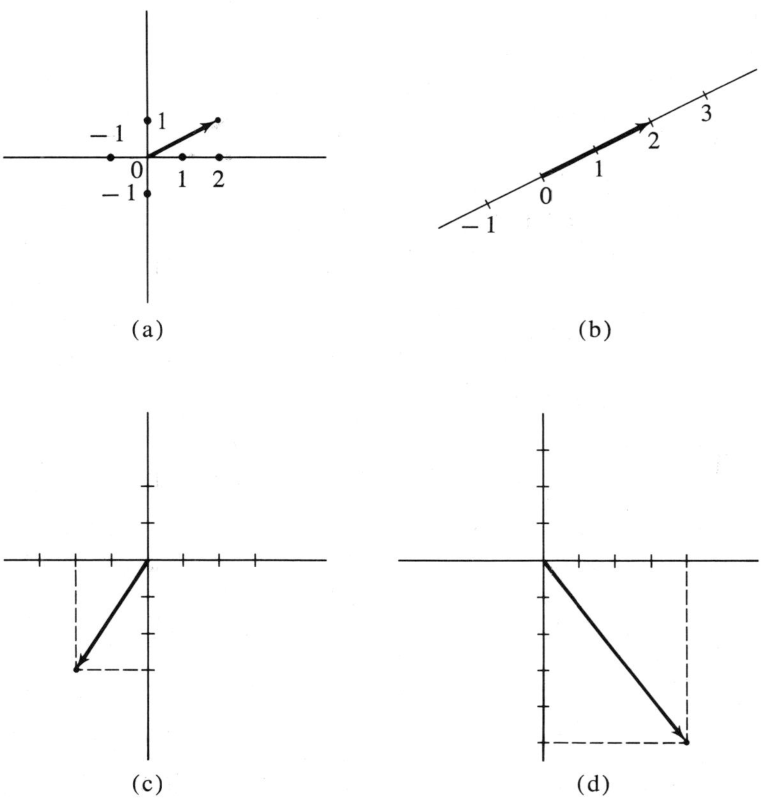

Figure 1.21

I.5.3 Give a geometric interpretation relative to 3-dimensional space for V_3, the set of all 3-vectors $\begin{pmatrix} x \\ y \\ z \end{pmatrix}$, where x, y, and z are real numbers.

I.5.4 Find the vectors indicated:

(a) $\begin{pmatrix} 3 \\ -1 \\ 2 \end{pmatrix} \dotplus \begin{pmatrix} -1 \\ 3 \\ 0 \end{pmatrix} = ?$

(b) $\begin{pmatrix} 2 \\ 0 \end{pmatrix} \dotplus \begin{pmatrix} 0 \\ -2 \end{pmatrix} = ?$

(c) $\begin{pmatrix} 7 \\ 9 \\ 1 \\ 3 \\ 0 \end{pmatrix} \dotplus \begin{pmatrix} 0 \\ 0 \\ 0 \\ 0 \\ 0 \end{pmatrix} = ?$

(d) $(5) \dotplus (2) = ?$

(e) $\begin{pmatrix} -3 \\ -4 \\ -\frac{1}{2} \end{pmatrix} \dotplus \begin{pmatrix} -\frac{3}{4} \\ -1 \\ -\frac{1}{2} \end{pmatrix} = ?$

(f) $\begin{pmatrix} 9.2 \\ -8.3 \\ 6.0 \end{pmatrix} \dotplus \begin{pmatrix} 1.1 \\ 1.1 \\ -1.1 \end{pmatrix} = ?$

I.5.5 Find the vectors indicated.

(a) $\left(\begin{pmatrix} 3 \\ 2 \\ -1 \end{pmatrix} \dotplus \begin{pmatrix} 5 \\ 0 \\ 6 \end{pmatrix}\right) \dotplus \begin{pmatrix} -2 \\ 6 \\ 8 \end{pmatrix}$

(b) $\begin{pmatrix} 3 \\ 2 \\ -1 \end{pmatrix} \dotplus \left(\begin{pmatrix} 5 \\ 0 \\ 6 \end{pmatrix} \dotplus \begin{pmatrix} -2 \\ 6 \\ 8 \end{pmatrix}\right)$

(c) $\left(\begin{pmatrix} 9 \\ 5 \\ 4 \end{pmatrix} \dotplus \begin{pmatrix} 3 \\ -2 \\ -7 \end{pmatrix}\right) \dotplus \begin{pmatrix} -3 \\ 2 \\ 7 \end{pmatrix}$

(d) $\begin{pmatrix} 9 \\ 5 \\ 4 \end{pmatrix} \dotplus \left(\begin{pmatrix} 3 \\ -2 \\ -7 \end{pmatrix} \dotplus \begin{pmatrix} -3 \\ 2 \\ 7 \end{pmatrix}\right)$

I.5.6 Solve for the concrete vector X; how many such X's can be found?

(a) $\begin{pmatrix} 8 \\ 0 \\ 1 \\ -1 \end{pmatrix} \dotplus X = \begin{pmatrix} 4 \\ 0 \\ 3 \\ 2 \end{pmatrix}$

(b) $\begin{pmatrix} -4 \\ -6 \\ 0 \end{pmatrix} \dotplus X = \begin{pmatrix} 1 \\ 2 \\ 3 \end{pmatrix}$

(c) $X \dotplus X \dotplus X = \begin{pmatrix} \frac{1}{2} \\ 0 \\ 3 \end{pmatrix}$

(d) $\begin{pmatrix} 9 \\ 5 \\ 4 \end{pmatrix} \dotplus X \dotplus X = \begin{pmatrix} 4 \\ 3 \\ -8 \end{pmatrix}$

I.5.7 How many *pairs* of vector solutions X, Y can be found for the following? Give at *least* one specific solution if possible.

(a) $X \dotplus Y = \begin{pmatrix} 2 \\ 1 \\ 0 \end{pmatrix}$ (c) $X \dotplus Y = \begin{pmatrix} 0 \\ 0 \\ 0 \end{pmatrix}$

(b) $X \dotplus X \dotplus Y = \begin{pmatrix} -1 \\ 0 \\ \frac{1}{2} \end{pmatrix}$

I.5.8 If we identify ordered pairs of real numbers $\begin{pmatrix} x \\ y \end{pmatrix}$ with points in the plane, and if we identify each point in the plane with an arrow in the plane which has its tail at the origin and its head at that point, show that Definition **I.1.2** can be represented, geometrically, by the parallelogram law (see Figure 1.22). That is, if a parallelogram is completed by drawing (dotted) lines parallel to the vectors X and Y, then the point (vector) indicated by the intersection of the dotted lines is equal to $X \dotplus Y$.

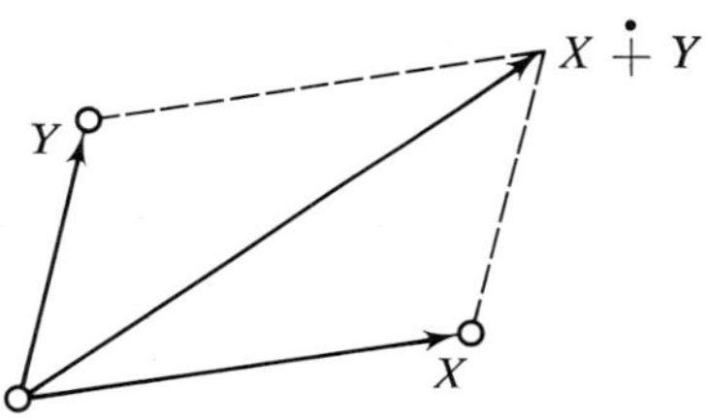

Figure 1.22

It is high time to at least partially describe our objectives concerning so-called "abstract addition" and "scalar multiplication."

ADDITION. Suppose we have a nonempty set U, the elements of which we shall call "vectors." Roughly speaking, an *addition* of a pair of vectors X and Y in U, is a rule by which a *third* vector is constructed when vectors X and Y are combined (that is, added). This third vector, the sum of X and Y, is denoted by a symbol which carries the symbols X and Y, that is, X-added-to-Y is denoted $X \dotplus Y$. Of course, we might have used any number of symbols to represent the sum, that is, that vector resulting from X and Y, such as $\frac{X}{Y}$, $X * Y$, $X \Delta Y$, and so on. (See Figure 1.23.) The property we have described above is **closure under addition.** That is, U is said to be closed under the operation of addition if pairs of vectors (elements of U) combine or add to yield not a new kind of animal, but rather a familiar vector (element) in U. We observe that Definition I.1.2 describes an addition between pairs of concrete n-vectors, resulting in a concrete n-vector.

(a) (b) (c)

Figure 1.23

Note that the operation $\dotplus$ is defined between *pairs* of vectors only. What, then, is the meaning of the expression

$$X \dotplus Y \dotplus Z?$$

As it stands, this expression has *no* meaning, since $\dotplus$ acts *only* between pairs of vectors; but we may impose 2 interpretations as follows:

(a) $\quad X \dotplus Y \dotplus Z$ is defined as $(X \dotplus Y) \dotplus Z$ or

(b) $\quad X \dotplus Y \dotplus Z$ is defined as $X \dotplus (Y \dotplus Z)$.

(a) and (b) are each consistent with our definition of $\dotplus$ for concrete vectors (Definition I.1.2) that is, (a) tells us to combine (add) the *pair* X and Y, and then to combine the *pair* $(X \dotplus Y)$ (which is a vector) and Z. (b) tells us, on the other hand, that we combine the *pair* X and $(Y \dotplus Z)$. In both cases, the final results are equal, as we will now show.

Now let

$$X = \begin{pmatrix} x_1 \\ x_2 \\ \vdots \\ x_n \end{pmatrix}, \quad Y = \begin{pmatrix} y_1 \\ y_2 \\ \vdots \\ y_n \end{pmatrix} \quad \text{and} \quad Z = \begin{pmatrix} z_1 \\ z_2 \\ \vdots \\ z_n \end{pmatrix}.$$

Then, following Definition I.1.2,

$$\text{(a)}' \qquad (X \dotplus Y) \dotplus Z = \begin{pmatrix} x_1 + y_1 \\ x_2 + y_2 \\ \vdots \\ x_n + y_n \end{pmatrix} \dotplus \begin{pmatrix} z_1 \\ z_2 \\ \vdots \\ z_n \end{pmatrix} = \begin{pmatrix} (x_1 + y_1) + z_1 \\ (x_2 + y_2) + z_2 \\ \vdots \\ (x_n + y_n) + z_n \end{pmatrix},$$

and

$$\text{(b)}' \qquad X \dotplus (Y \dotplus Z) = \begin{pmatrix} x_1 \\ x_2 \\ \vdots \\ x_n \end{pmatrix} \dotplus \begin{pmatrix} y_1 + z_1 \\ y_2 + z_2 \\ \vdots \\ y_n + z_n \end{pmatrix} = \begin{pmatrix} x_1 + (y_1 + z_1) \\ x_2 + (y_2 + z_2) \\ \vdots \\ x_n + (y_n + z_n) \end{pmatrix}$$

The final concrete vectors in (a) and in (b) (which result by applying the rule of addition twice in each case) are equal if and only if corresponding coordinates (entries) are equal (Definition I.1.4). That is, we need $(x_i + y_i) + z_i = x_i + (y_i + z_i)$ for *all* $i = 1, 2, 3, \ldots, n$. But x_i, y_i and z_i are numbers, after all, and such an equation holds for *all* triples of numbers, and in particular for x_i, y_i and z_i. This proves the equivalence desired; that is, the problem of equating $(X \dotplus Y) \dotplus Z$ with $X \dotplus (Y \dotplus Z)$ (which is a vector property), reduces to a similar problem with *numbers*, something with which we are familiar.

Only after having proved that $(X \dotplus Y) \dotplus Z = X \dotplus (Y \dotplus Z)$, does the symbol $X \dotplus Y \dotplus Z$ (without parentheses) have meaning, at least for triples of concrete n-vectors; no matter which interpretation (a) or (b) is given, the result is the same. Hence, we may extend Definition I.1.2 with:

Definition I.5.1 $\boldsymbol{X \dotplus Y \dotplus Z}$ is defined to be the vector $(X \dotplus Y) \dotplus Z$ [or, equivalently, $X \dotplus (Y \dotplus Z)$] for all concrete vectors X, Y, and Z in V_n.

This feature of being able to associate (relative to addition) X with Y, and *then* with Z, or to first associate (add) Y with Z, and then to associate that combination $(Y \dotplus Z)$ with X, and in both cases get the same result, is highly regarded by mathematicians. It is said that the *vector-space* V_n is **associative** *under the operation* $\dotplus$, *of addition*. We have observed that V_n, a set of concrete vectors is associative under $\dotplus$, only because the set of numbers is associative under its *own* addition, $+$.

Example If

$$X = \begin{pmatrix} 1 \\ -1 \\ 3 \end{pmatrix}, \quad Y = \begin{pmatrix} \frac{1}{2} \\ 0 \\ -\frac{1}{2} \end{pmatrix} \quad \text{and} \quad Z = \begin{pmatrix} -\frac{3}{2} \\ 2 \\ -5 \end{pmatrix}$$

compute $(X \dotplus Y) \dotplus Z$ and $X \dotplus (Y \dotplus Z)$.

So far, we have stumbled into 2 properties of the addition $\dotplus$ (Definition I.1.2) between pairs of concrete vectors; namely, *closure* ($X \dotplus Y$ is always (another) n-vector) and *associativity* $(X \dotplus (Y \dotplus Z) = (X \dotplus Y) \dotplus Z)$, We shall presently require an arbitrary (abstract) vector-space to be a set where at least such a closed, associative-addition operation prevails.

SCALAR MULTIPLICATION. If we use our concrete n-vectors as a model, we see that for every vector X, and for every scalar α, a "new" vector, αX is defined (Definition I.1.3). That is, we have a means of hitting an n-vector with a scalar, producing an n-vector in the end. Thus, we may say that a *scalar multiplication* exists on concrete n-vectors and, moreover, this multiplication is *closed* (nothing new is produced by this V_n multiplication; just another n-vector); more exactly, the set V_n which contains the vectors is **closed under scalar multiplication.**

At this point, then, we know how to add 2 concrete vectors (thus producing a third concrete vector), and we know how to multiply any vector by a scalar (thus producing a second vector).

We raise the question: Are there any other properties of the numbers which are reflected in the structure of concrete vectors? For example, we have seen that the concrete vector-space V_n is *associative* under the operation of $+$ because the scalars are associative under the operation of $+$, their own addition.

Some properties of numbers are expressed in the following axioms.

(a) $a + b = b + a$ for all a, b, c. (**commutivity**)

(b) $a + (b + c) = (a + b) + c$ for all a, b, c. (**associativity**)

(c) There is an element, denoted 0, such that $0 + a = a$ for all a; 0 is called an **identity** under addition.

(d) Given any number a, there is a number, denoted a', such that $a + a' = 0$. a' is called the **additive inverse** of a.

Property (a) says that the order of addition does not effect the result. We have already discussed property (b). Property (c) says that there is a "feeble" element with the property that when it is combined with (added to) *any* other element, it "leaves that element intact." There is a similar "feeble" element (identity) for numbers under the other operation of multiplication; that is, there is a number, denoted e, such that for *any* number a, $e \cdot a = a$. Usually, this identity e is symbolized by 1, the number 1. Property (d) says that for any number a, there is some other number a' that "returns us to the identity." This additive inverse a', which depends on a, is often denoted $-a$. Hence $a + (-a) = 0$. Once again we refer to the operation of multiplication on the numbers. Is there a multiplicative inverse for each number a; that is, is there a number $\bar{a}$ that "brings us back to the identity e"? More exactly, for every number a, is there a number $\bar{a}$ such that $a \cdot \bar{a} = e$, the identity? It turns out that such an $\bar{a}$ exists for all a if and only if $a \neq 0$, the additive identity. This inverse $\bar{a}$ is often denoted by the symbol a^{-1}, or the symbol $1/a$.

We have not mentioned the powerful and important property of numbers that relates the 2 operations of addition and multiplication, namely, the **distributive law** which says $a \cdot (b + c) = a \cdot b + a \cdot c$ for all

a, b, c. We have no closed multiplication defined on pairs of vectors, however, so our attention will turn to showing only that the *additive* properties (a), (b), (c), and (d) hold for concrete vectors, as well.

Property (a) is an exercise (Exercise I.5.9) for the reader. Property (b) has already been established for concrete vectors, so we pass to (c). The problem is the following: Is there a "feeble" concrete *vector*, call it F, such that $F \dotplus X = X$ for any and all concrete vectors X? To be more specific, let

$$F = \begin{pmatrix} f_1 \\ f_2 \\ \vdots \\ f_n \end{pmatrix} \quad \text{and} \quad X = \begin{pmatrix} x_1 \\ x_2 \\ \vdots \\ x_n \end{pmatrix}$$

We desire

$$\begin{pmatrix} f_1 \\ f_2 \\ \vdots \\ f_n \end{pmatrix} \dotplus \begin{pmatrix} x_1 \\ x_2 \\ \vdots \\ x_n \end{pmatrix} = \begin{pmatrix} x_1 \\ x_2 \\ \vdots \\ x_n \end{pmatrix} \quad \text{for } all \ X = \begin{pmatrix} x_1 \\ x_2 \\ \vdots \\ x_n \end{pmatrix}$$

or,

$$\begin{pmatrix} f_1 + x_1 \\ f_2 + x_2 \\ \vdots \\ f_n + x_n \end{pmatrix} = \begin{pmatrix} x_1 \\ x_2 \\ \vdots \\ x_n \end{pmatrix} \quad \text{for all } X = \begin{pmatrix} x_1 \\ x_2 \\ \vdots \\ x_n \end{pmatrix}.$$

Now 2 vectors are *equal* (Definition I.1.4) if and only if their corresponding coordinates are equal. The above equality holds, therefore, if and only if $f_1 + x_1 = x_1$, and $f_2 + x_2 = x_2$, and so on until $f_n + x_n = x_n$, for all *numbers* $x_1, x_2, \ldots, x_n$. But $f_1 + x_1 = x_1$ for all numbers x_1 if and only if f_1 equals the *number* 0. Similar argument yields the fact that f_2 must be 0, $f_3 = 0$, and so on, until we see that *every* coordinate of F, is the number 0. That is,

$$F = \begin{pmatrix} 0 \\ 0 \\ \vdots \\ 0 \end{pmatrix}$$

is a "feeble" *vector* (identity) under addition $\dotplus$. From now on, F will be denoted by O, which suggests the nature of its coordinates.

REMARK. Regarding the identity 0 for numbers, or O for vectors, we have not spoken of the point of uniqueness. That is, could there be *more* than 1 number or concrete vector with the "feeble" property under addition? The answer is that if such an identity exists at all, then there is only 1 identity. To see this, suppose we have 2 identities, O_1 and O_2, say. That is,

(a) $O_1 + X = X$ for every X and

(b) $O_2 + X = X$ for every X. Now consider the expression $O_1 + O_2$. But O_1 leaves every element, including O_2, alone so that:

(c) $O_1 + O_2 = O_2$. by (a).

But O_2 leaves O_1 alone, so that

(d) $O_1 + O_2 = O_1$ by (b). We have used commutivity here. (How?) From (c) and (d) we have

(e) $O_2 = O_1 + O_2 = O_1$.

That is, $O_2 = O_1$, which says that any 2 identities (if they exist) are, in fact, the same element. The above argument, incidentally, holds for more general situations; any set of objects with a commutative operation (such as addition on the numbers) has only 1 identity relative to that operation, provided an identity exists at all.

We now list 10 properties (see Definition I.5.2 following) which we shall use to define an abstract vector-space. We suppress the adjective "concrete" that was used in connection with columns or rows of numbers. That is, there are *other* collections of objects (besides columns of numbers) which may be multiplied by scalars and pairwise added very much like our *concrete* vectors (see Exercise I.1.4). These sets will be called *abstract* or *general* vector-spaces. To state precisely what we mean by a set of objects (a vector-space) which behaves "very much like" our concrete vector-space of n-tuples, we offer Definition I.5.2.

Definition I.5.2 Let V be a nonempty set of objects. V will be called a **vector-space over the scalar number-field** (*or*, **over the scalar numbers**), or a **scalar vector-space,** if and only if:

There is an operation $+$ defined between pairs of elements of V such that

A.1 $X + Y \in V$ whenever $X \in V$ and $Y \in V$ (closure under the operation $+$)

A.2 $X + Y = Y + X$ for all $X \in V$ and all $Y \in V$ (commutivity under $+$)

A.3 There exist an element, call it O, with the property that $X + O = X$ for all $X \in V$. (existence of an additive identity)

A.4 For every $X \in V$, there is an element in V (depending on X), call it $(-X)$, such that $X + (-X) = O$ (existence of an additive inverse for every element X)

A.5 $(X \dotplus Y) \dotplus Z = X \dotplus (Y \dotplus Z)$ for all $X \in V$, $Y \in V$, and $Z \in V$ (associative law)

There is an operation $\circ$ defined between a scalar (number) and an element (vector) of V such that

B.1 $\alpha \circ X \in V$ for all $X \in V$, and all real numbers α (closure under scalar multiplication).

B.2 $\alpha \circ (\beta \circ X) = (\alpha\beta) \circ X$ for all $X \in V$ and all numbers α and β.

B.3 $1 \circ X = X$ for all $X \in V$ where 1 is the multiplicative identity in the field of numbers. Moreover, the operations $\circ$ and $\dotplus$ are related by the following properties:

B.4 $(\alpha + \beta) \circ X = \alpha \circ X \dotplus \beta \circ X$ for all $X \in V$ and all numbers α and β.

B.5 $\alpha \circ (X \dotplus Y) = (\alpha \circ X) \dotplus (\alpha \circ Y)$ for all $X \in V$, all $Y \in V$ and all numbers α.

Some of the following exercises involve only the checking of certain properties on this list relative to certain sets of mathematical objects.

We point out that there is nothing sacred about the particular symbols $\dotplus$ and $\circ$; in fact, any symbols may be used so long as their meanings are clearly understood. See, for example, Exercises I.1.14, I.1.15, I.1.16, and I.1.17.

EXERCISES

I.5.9 Show that for any pair of concrete vectors $X, Y \in V_n$, $X \dotplus Y = Y \dotplus X$ (see Definition I.1.2).

I.5.10 Show that for any scalars α and β, and for any concrete vector $X \in V_n$, $\alpha(\beta X) = (\alpha\beta) \circ X$. Note the apparent difference between the 2 sides of this equation. The left-hand side tells us to multiply X first by the scalar β, and then to multiply that result by the scalar α. The right-hand side tells us to multiply the vector X by the scalar $\alpha\beta$.

I.5.11 (a) Show that for any scalars α and β, and any vector $X \in V_n$, $(\alpha + \beta) \circ X = \alpha \circ X \dotplus \beta \circ X$. Again, note the apparent difference between the 2 sides.
(b) Show that for every concrete $X \in V_n$ an additive inverse, $(-X)$ exists in V_n according to A.4 of Definition I.5.2 of this equation.

I.5.12 Show that for any scalar α, and any concrete vectors $X \in V_n$ and $Y \in V_n$,

$$\alpha \circ (X \dotplus Y) = \alpha \circ X \dotplus \alpha \circ Y.$$

I.5.13 Solve (if possible) for the scalars α and β:

(a) $\alpha \circ \begin{pmatrix} 2 \\ -1 \end{pmatrix} \dotplus \beta \circ \begin{pmatrix} 0 \\ 3 \end{pmatrix} = \begin{pmatrix} -1 \\ -1 \end{pmatrix}$

(b) $\alpha \circ \begin{pmatrix} 1 \\ 8 \end{pmatrix} \dotplus \beta \circ \begin{pmatrix} -\frac{1}{2} \\ 7 \end{pmatrix} = \begin{pmatrix} 1 \\ 2 \end{pmatrix}$

(c) $\alpha \circ \begin{pmatrix} 1 \\ 2 \\ 3 \end{pmatrix} \dotplus \beta \circ \begin{pmatrix} 1 \\ 2 \\ 3 \end{pmatrix} = \begin{pmatrix} 2 \\ 4 \\ 6 \end{pmatrix}$

(d) $\alpha \circ \begin{pmatrix} 1 \\ 2 \\ 3 \end{pmatrix} \dotplus \beta \circ \begin{pmatrix} 1 \\ 2 \\ 3 \end{pmatrix} = \begin{pmatrix} 1 \\ -1 \\ 0 \end{pmatrix}$

(e) $\alpha \circ \begin{pmatrix} -2 \\ 1 \end{pmatrix} \dotplus \beta \circ \begin{pmatrix} 0 \\ 1 \end{pmatrix} = \begin{pmatrix} 0 \\ 0 \end{pmatrix}$

(f) $\alpha \circ \begin{pmatrix} 3 \\ 0 \\ 1 \end{pmatrix} \dotplus \beta \circ \begin{pmatrix} 6 \\ 0 \\ 2 \end{pmatrix} = \begin{pmatrix} 0 \\ 0 \\ 0 \end{pmatrix}$

I.5.14 Consider an expression of the form $p(x) = a_0 + a_1x$, where a_0, a_1 are arbitrary real numbers, and the symbol x is to be thought of as just that—a symbol (usually called an *indeterminant*). Such an expression is called a *polynomial of formal degree one* over the real field. Let $p(x)$ and $q(x)$ be 2 such polynomials of degree 1, that is, $p(x) = a_0 + a_1x$, and $q(x) = b_0 + b_1x$ for some real numbers a_0, a_1, b_0, b_1. Suppose an operation $\perp$ is *defined* between pairs of polynomials by

$\perp$ – *Def*: $$p(x) \perp q(x) = (a_0 + a_1x) \perp (b_0 + b_1x) = (a_0 + b_0) + (a_1 + b_1)x$$

(which is, itself, a new polynomial of degree 1). Suppose the scalar-multiplication operator $\times$ is *defined* by

$\times$ – *Def*: $$c \times p(x) = c \times (a_0 + a_1x) = (ca_0) + (ca_1)x$$

for all real numbers c.

Show that V, the set of *all* polynomials of degree 1, forms an abstract vector-space (satisfies *all* the properties on the list of Definitions I.5.2). Note that we are using $\perp$ for $\dotplus$ and $\times$ for $\circ$.

I.5.15 (See Exercise I.5.14.) Suppose the polynomial $p(x) = a_0 + a_1x$ is identified with the column $\begin{pmatrix} a_0 \\ a_1 \end{pmatrix}$. That is, we will say, "The column of $p(x)$" denoted col $(p(x))$, or col $(a_0 + a_1x)$, is equal to $\begin{pmatrix} a_0 \\ a_1 \end{pmatrix}$. For example, if $p(x) = 3 + 5x$, then col $(p(x)) = \begin{pmatrix} 3 \\ 5 \end{pmatrix}$.

Suppose, further, that these columns are thought of as concrete 2-vectors, endowed with their usual (coordinate-wise) addition $\dotplus$, and scalar multiplication. Verify that the following 2 equations hold:

(a) $$\text{col}\,(p(x)) \dotplus \text{col}\,(q(x)) = \text{col}\,(p(x) \perp q(x))$$

(b) $$c \circ \text{col}\,(p(x)) = \text{col}\,(c \times p(x))$$

for all polynomials $p(x)$ and $q(x)$, and all real numbers c. Notice that the use of the 2 "addition" symbols $\dotplus$ and $\perp$ are used in a consistent and meaningful manner, that is, $\dotplus$ is defined between *columns* and $\perp$ is defined between *polynomials*. This exercise asks you to verify the 2 equations by computing separately, the left-hand sides and the right-hand sides for arbitrary polynomials and their columns; then, compare the 2 sides to check whether or not they agree.

I.5.16 Suppose $p(x) = a_0 + a_1x + a_2x^2 + \cdots + a_nx^n$. Such an expression is called a *polynomial of degree n over the real field*, if the $n+1$ elements a_0, a_1, a_2, $\ldots$, a_n are real numbers. An operation $\perp$ is *defined* between pairs of such polynomials by

$\perp - Def$: $p(x)q(x) = (a_0 + b_0) + (a_1 + b_1)x + (a_2 + b_2)x^2 + \cdots + (a_n + b_n)x^n$

where

$$p(x) = a_0 + a_1x + a_2x^2 + \cdots + a_nx^n$$

and

$$q(x) = b_0 + b_1x + b_2x^2 + \cdots + b_nx^n.$$

The scalar-multiplication operator $\times$ will be *defined* by

$\times - Def$: $c \times p(x) = ca_0 + (ca_1)x + (ca_2)x^2 + \cdots + (ca_n)x^n$

for all real numbers c, where

$$p(x) = a_0 + a_1x + a_2x^2 + \cdots + a_nx^n.$$

Show that V, the set of *all* polynomials of degree n, forms a real (abstract) vector-space relative to the operations of addition $\perp$ and scalar multiplication $\times$.

I.5.17 (See Exercise I.5.16.) Let us identify the polynomial $p(x) = a_0 + a_1x + a_2x^2 + \cdots + a_nx^n$ with the $(n+1)$-entried (concrete) column

$$\begin{pmatrix} a_0 \\ a_1 \\ a_2 \\ \vdots \\ a_n \end{pmatrix},$$

which will be symbolized by *col*$(p(x))$.

As in Exercises I.5.15 and I.5.16, we have established a 1-1 correspondence between polynomials of degree n and (concrete) column vectors with $n+1$ real-number coordinates (entries); that is, these columns are thought of as real $(n+1)$-vectors endowed with their usual (coordinate-wise) addition $\dot{+}$ and scalar multiplication. Verify that the following 2 equations hold.

(a) $\text{col}\ (p(x)) \dot{+} \text{col}\ (q(x)) = \text{col}\ (p(x) \perp q(x))$

(b) $c \circ \text{col}\ (p(x)) = \text{col}\ (c \times p(x))$

for all polynomials $p(x)$ and $q(x)$, and all real numbers c.

I.5.18 Let Z be the set of all integers, that is $Z = \{0, \pm 1, \pm 2, \ldots, \pm n, \ldots\}$, with the usual addition $+$ and multiplication $\times$ defined on it.

(a) Which properties of Definition I.5.2 does V satisfy?

(b) Is V a real (abstract) vector-space?

(c) Suppose the numbers α and β in Definition I.5.2 were *restricted* to the set of integers Z. Show that *all* properties of Definition I.5.2 hold true (relative to Z) with this restriction. Note: since all the properties *do* hold (with this restriction on the α's and β's), this does *not* make Z a *vector-space*, since the definition (which is the Word of Truth, as far as we are concerned) allows the α's and β's to be *any* real numbers, *without* restriction. Nonetheless, the fact that we still have structure on Z, even though the nature of the α's and β's is modified somewhat, is a still pleasant situation. What we have here is a simple example of what is called a *module*, which is a generalization of a vector-space.

I.5.19 Let V be the set of all columns of the form $\begin{pmatrix} x \\ y \end{pmatrix}$ where x and y are non-negative numbers. Suppose addition $\dotplus$ between columns and scalar multiplication are defined according to Definitions I.1.2 and I.1.3. Is V a real vector-space? (Use Definition I.5.2 as a test.)

I.5.20 Suppose V is the set of *all* columns $\begin{pmatrix} x \\ y \end{pmatrix}$ where x and y are arbitrary real numbers. Suppose addition $\dotplus$ between columns is defined according to Definition I.1.2. Suppose we *re*define scalar multiplication $\circ$ according to

$\circ$ — *Def*:
$$\alpha \circ \begin{pmatrix} x \\ y \end{pmatrix} = \begin{pmatrix} 3\alpha x \\ 3\alpha y \end{pmatrix},$$

for all real numbers α, x, and y. Show that *all* properties of Definition I.5.2 are satisfied here, except for B.2 and B.3.

I.5.21 Let V_n be the set of all real (concrete) n-vectors with the addition operator $\dotplus$ as defined according to Definition I.1.2. Suppose we redefine scalar multiplication $\circ$ according to

$\circ$ — *Def:*
$$\alpha \circ X = 0$$

for all real numbers α and all $X \in V_n$. Show that all properties of Definition I.5.2 obtain in this case, except for B.3.

I.5.22 Let V be the set containing only 1 element, namely the zero vector (which may be thought of as the n-entry column, with every coordinate equal to the number 0). Show that with the usual definitions of $\dotplus$ and $\circ$ (Definitions I.1.2 and I.1.3), V forms a real abstract vector space (satisfies all properties listed in Definition I.5.2). For obvious reasons, this space is known as the *trivial vector-space.*

I.5.23 (See Exercise I.5.22.) Suppose V is a bonafide real vector space, and that the set V has only 1 element. Is V the trivial vector-space?

I.5.24 Is there any real vector-space V such that V has
(a) only 2 elements?
(b) only 3 elements?
(c) only a finite number of elements?

I.5.25 Let V be the set of all n-entry columns with real numbers as coordinates. Suppose we retain the usual scalar multiplication on V (Definition I.1.3) but *redefine* addition $\dotplus$ between columns according to

$\dotplus$ — *Def*:
$$X \dotplus Y = O$$

for all $X \in V$ and all $Y \in V$. Show that V is not a real vector-space (according to Definition I.5.2). Which properties on the list fail in this case?

6. Bases and Dimension

From this point on, we will abandon the $\circ$ notation for scalar multiplication and simply write αX for $\alpha \circ X$, since no confusion will result. Also $+$ will replace $\dotplus$.

Consider the concrete vector-space V_3. The vector

$$\begin{pmatrix} 15 \\ -3 \\ 6 \end{pmatrix}$$

in V_3 may be represented by the following sum:

$$15\begin{pmatrix} 1 \\ 0 \\ 0 \end{pmatrix} + (-3)\begin{pmatrix} 0 \\ 1 \\ 0 \end{pmatrix} + 6\begin{pmatrix} 0 \\ 0 \\ 1 \end{pmatrix}$$

In fact, if

$$\begin{pmatrix} x_1 \\ x_2 \\ x_3 \end{pmatrix}$$

is any vector in V_3, we have

$$\begin{pmatrix} x_1 \\ x_2 \\ x_3 \end{pmatrix} = x_1\begin{pmatrix} 1 \\ 0 \\ 0 \end{pmatrix} + x_2\begin{pmatrix} 0 \\ 1 \\ 0 \end{pmatrix} + x_3\begin{pmatrix} 0 \\ 0 \\ 1 \end{pmatrix}$$

The student should verify the equality by applying the definitions of scalar multiplication and vector addition to the right-hand side. We notice that *any* vector in V_3 can be expressed in terms of the 3 fixed vectors

$$\begin{pmatrix} 1 \\ 0 \\ 0 \end{pmatrix}, \quad \begin{pmatrix} 0 \\ 1 \\ 0 \end{pmatrix}, \quad \begin{pmatrix} 0 \\ 0 \\ 1 \end{pmatrix}$$

It is our purpose to study this "generating" phenomenon more closely. We are led to Definition I.6.1.

Definition I.6.1 Given the ordered set of vectors $\{X_1, X_2, \ldots, X_k\}$ in some vector-space V_n, then any vector of the form

$$\alpha_1 X_1 + \alpha_2 X_2 + \alpha_3 X_3 + \cdots + \alpha_k X_k$$

is called a **linear combination** of the vectors $X_1, X_2, \ldots, X_k$, where $\alpha_1, \alpha_2, \ldots, \alpha_k$ are scalars (numbers). We remark that there is no relation between k (the numbers of ordered vectors given) and n, the number of coordinates (entries) per vector.

We have already seen that the vector

$$\begin{pmatrix} 15 \\ -3 \\ 6 \end{pmatrix}$$

is equal to

$$15\begin{pmatrix} 1 \\ 0 \\ 0 \end{pmatrix} + (-3)\begin{pmatrix} 0 \\ 1 \\ 0 \end{pmatrix} + 6\begin{pmatrix} 0 \\ 0 \\ 1 \end{pmatrix},$$

and so it can be said that the vector

$$\begin{pmatrix} 15 \\ -3 \\ 6 \end{pmatrix}$$

is a *linear combination* of the ordered vectors

$$\left\{\begin{pmatrix} 1 \\ 0 \\ 0 \end{pmatrix}, \begin{pmatrix} 0 \\ 1 \\ 0 \end{pmatrix}, \begin{pmatrix} 0 \\ 0 \\ 1 \end{pmatrix}\right\}.$$

The scalars which are "attached" to these vectors are, respective to their order, 15, -3, and 6. We have also seen a stronger property; namely, that *any* vector

$$\begin{pmatrix} x_1 \\ x_2 \\ x_3 \end{pmatrix}$$

is a linear combination of the vectors in the ordered set

$$\left\{\begin{pmatrix} 1 \\ 0 \\ 0 \end{pmatrix}, \begin{pmatrix} 0 \\ 1 \\ 0 \end{pmatrix}, \begin{pmatrix} 0 \\ 0 \\ 1 \end{pmatrix}\right\}.$$

In this general case, the appropriate scalars which are used, are (respective to the order of the vectors), x_1, x_2, and x_3. That is,

$$\begin{pmatrix} x_1 \\ x_2 \\ x_3 \end{pmatrix} = x_1\begin{pmatrix} 1 \\ 0 \\ 0 \end{pmatrix} + x_2\begin{pmatrix} 0 \\ 0 \\ 1 \end{pmatrix} + x_3\begin{pmatrix} 0 \\ 0 \\ 1 \end{pmatrix},$$

which agrees with our Definition I.6.1 of linear combination.

We have encountered a situation where *any* vector (in V_3) can be obtained as a certain linear combination of the ordered vectors

$$\left\{\begin{pmatrix} 1 \\ 0 \\ 0 \end{pmatrix}, \begin{pmatrix} 0 \\ 1 \\ 0 \end{pmatrix}, \begin{pmatrix} 0 \\ 0 \\ 1 \end{pmatrix}\right\}.$$

When we have a certain ordered set of "building block" vectors (or, "generating" vectors) for the entire space V_n we say that this "building-block" set *spans* the space. More exactly,

Definition 1.6.2 Let $\{X_1,X_2,\ldots,X_k\}$ be an ordered set of vectors in the space V_n. If *each* (and every) vector Y, say, can be represented as a linear combination of the vectors $X_1, X_2,\ldots, X_k$ (that is, if $Y=\alpha_1X_1+\alpha_2X_2+\cdots+\alpha_kX_k$ for certain scalars $\alpha_1, \alpha_2,\ldots, \alpha_k$) then the ordered set of vectors $\{X_1,X_2,\ldots,X_k\}$ is said to **span** the space V_n.

This question arises: Given a vector-space, is there a "smallest" set of vectors that spans the whole space? When do we have "too many" spanning vectors to do the job (of generating all vectors of the vector-space V_n)? The answer is closely connected with the notion of *linear independence* and *linear dependence*. We develop these notions now, and state the definitions formally.

Definition I.6.3 An ordered set of vectors $\{X_1,X_2,\ldots,X_k\}$, of a vector-space V_n, is called **linearly dependent** (abbreviated **LD**) if and only if (at least) one of the X_i's is a linear combination of the other X_i's; that is, if and only if for some X_i, we can find suitable scalars (α_i's) such that

$$X_i=\alpha_1X_1+\alpha_2X_2+\cdots+\alpha_{i-1}X_{i-1}+\alpha_{i+1}X_{i+1}+\cdots+\alpha_kX_k.$$

The choice of the term "dependent" is suggestive. A set of vectors is linearly dependent, means that at least 1 vector is "dependent" on the others in the sense that it is a linear combination of the others. The following definition should come as no surprise.

Definition I.6.4 An ordered set of (nonzero) vectors $\{X_1,X_2,\ldots,X_k\}$ of a vector-space is called **linearly independent** (abbreviated **LI**) if and only if the set is not linearly dependent.

Thus, a set of vectors is linearly independent if *no* vector in the set is a linear combination of the remaining vectors. We note that the adjective "nonzero" is a redundancy. It is automatic that any linearly independent set of vectors consists of nonzero vectors only. We cite some examples.

Example Consider the following ordered set in V_2:

$$\left\{\begin{pmatrix}1\\2\end{pmatrix}, \begin{pmatrix}2\\3\end{pmatrix}, \begin{pmatrix}-1\\-1\end{pmatrix}\right\}.$$

This set is linearly dependent since $\begin{pmatrix}-1\\-1\end{pmatrix}=(1)\begin{pmatrix}1\\2\end{pmatrix}+(-1)\begin{pmatrix}2\\3\end{pmatrix}$; that is, the third (concrete) vector listed in the set, $\begin{pmatrix}-1\\-1\end{pmatrix}$, is a linear combination of the remaining vectors $\begin{pmatrix}1\\2\end{pmatrix}$ and $\begin{pmatrix}2\\3\end{pmatrix}$. (Note that since the set is ordered, the "third vector" is a well-defined notion.)

Example Consider the following ordered set in V_3:

$$\left\{\begin{pmatrix}-1\\0\\1\end{pmatrix}, \begin{pmatrix}2\\0\\0\end{pmatrix}\right\}.$$

This set is linearly independent since neither vector is a linear combination of the other vector. In fact, since there is only one "other" vector, a "linear combination of the other vector" now means "a scalar multiple of that other vector."

To state the fact of linear independence with a supporting argument, we will try to represent

$$\begin{pmatrix}-1\\0\\1\end{pmatrix}$$

as a linear combination of the remaining vector

$$\begin{pmatrix}2\\0\\0\end{pmatrix},$$

and we will see that this cannot be done, after all. This impossible situation has already been labelled as linear independence in Definition I.6.4. Suppose for some scalar α,

$$\begin{pmatrix}-1\\0\\1\end{pmatrix} = \alpha\begin{pmatrix}2\\0\\0\end{pmatrix}.$$

From the definition of scalar multiplication (Definition I.1.3), the right-hand vector equals

$$\begin{pmatrix}2\alpha\\0\\0\end{pmatrix}$$

Thus,

$$\begin{pmatrix}-1\\0\\1\end{pmatrix} = \begin{pmatrix}2\alpha\\0\\0\end{pmatrix}.$$

Definition I.1.4 tells us that 2 concrete vectors are equal if and only if their corresponding coordinates agree. In our particular case, we must have

(a) $$-1 = 2\alpha, \quad 0 = 0, \quad \text{and} \quad 1 = 0.$$

Can an α be found to make the above equations hold? Certainly, if $\alpha = -\frac{1}{2}$, then the first equation of (a) holds true. The second equation holds true in any case, but no α on earth can possibly rescue us from the nonvalidity of the third equation, $1 = 0$.

We can*not* write

$$\begin{pmatrix} -1 \\ 0 \\ 1 \end{pmatrix}$$

as a linear combination of

$$\begin{pmatrix} 2 \\ 0 \\ 0 \end{pmatrix},$$

and so the set

$$\left\{ \begin{pmatrix} -1 \\ 0 \\ 1 \end{pmatrix}, \begin{pmatrix} 2 \\ 0 \\ 0 \end{pmatrix} \right\}$$

is linearly independent.

We will now present the notions of linear independence and dependence from a slightly different point of view. Recall that the ordered set of vectors $\{X_1,X_2,\ldots,X_k\}$ is linearly dependent if one of the vectors X_1, say can be written

$$X_1 = \alpha_2 X_2 + \alpha_3 X_3 + \cdots + \alpha_k X_k$$

for some scalars $\alpha_2, \alpha_3, \ldots, \alpha_k$. If we add (the additive inverse) $-X_1$ to both sides of the equation, we obtain

$$O = -X_1 + \alpha_2 X_2 + \alpha_3 X_3 + \cdots + \alpha_k X_k.$$

That is, *if* the set is linearly dependent, *then* the O vector can be written as a *nontrivial* linear combination of the vectors $X_1, X_2, \ldots, X_k$. The term nontrivial means that *not all* of the α_i's are zero (although *some* of the α_i's may be zero). In particular, α_1, which is the scalar multiplying X_1, turns out to be equal to -1, which is not zero.

We establish the converse. We will show that *if* the O vector is a nontrivial linear combination of the vectors $X_1, X_2, \ldots, X_k$, *then* the ordered set $\{X_1,X_2,\ldots,X_k\}$ is linearly dependent. To see this, set

$$O = \alpha_1 X_1 + \alpha_2 X_2 + \cdots + \alpha_k X_k.$$

where (by the nontriviality assumption) not *all* the α_i's are zero, although *some* α_i's may be zero. By renumbering (that is, reordering) the X_i's if necessary, we may suppose that $\alpha_1 \neq 0$. After multiplying the above equation through by $1/\alpha_1$ (the multiplicative inverse of α_1), we obtain, by properties B.5, B.2, and A.5 of Definition I.5.2,

$$O = X_1 + \frac{\alpha_2}{\alpha_1} X_2 + \cdots + \frac{\alpha_k}{\alpha_1} X_k,$$

and adding

$$\left(-\frac{\alpha_2}{\alpha_1} X_2\right) + \left(-\frac{\alpha_3}{\alpha_1} X_3\right) + \cdots + \left(-\frac{\alpha_k}{\alpha_1} X_k\right)$$

to both sides, we have

$$\left(-\frac{\alpha_2}{\alpha_1} X_2\right) + \left(-\frac{\alpha_3}{\alpha_1} X_3\right) + \cdots + \left(-\frac{\alpha_k}{\alpha_1} X_k\right) = X_1$$

which says that X_1 is a linear combination of the remaining vectors X_2, $X_3, \ldots, X_k$. In this case, the strategic scalars we have used (relative to the ordering) are the numbers $-\alpha_2/\alpha_1$, $-\alpha_3/\alpha_1, \ldots, -\alpha_k/\alpha_1$, respectively. We have proven the following theorem.

Theorem I.6.5 An ordered set of vectors $\{X_1, X_2, \ldots, X_k\}$ is linearly dependent if and only if

$$O = \alpha_1 X_1 + \alpha_2 X_2 + \cdots + \alpha_k X_k$$

for some appropriate (ordered) set of scalars $\{\alpha_1, \alpha_2, \ldots, \alpha_k\}$ which are *not* all zero.

Of course, setting each α_i equal to zero would *suffice* in generating the zero vector O, but linear dependence says that such an extreme action is *not* necessary; O can be generated by using some nonzero α_i's.

REMARK. Note that we have expressed the notion of linear dependence in 2, apparently different ways. First, to say an ordered set was linearly independent was to say that at least 1 vector in that set was a linear combination of the others (Definition I.6.3). Secondly, we considered a set of vectors with the property that we could generate (by means of linear combination) the zero vector without *having* to choose each α_i equal to zero. We then showed that if we assumed an ordered set was linearly dependent, then the second property was an inevitable consequence. We also saw that if the second property was assumed, linear dependence was a necessary consequence. That is, one property implied the other, which is to say, these properties are *equivalent*; they are, in fact, the same property. This is exactly the meaning of the phrase "property A holds if and only if property B holds."

One reason for establishing such equivalence is that one form may be more convenient than the other for developing further properties. It will be more convenient, for example, to think of Theorem I.6.5 as an *alternate definition* for linear dependence since this is the viewpoint we shall use from now on. There is no harm in this, since its equivalence to the original definition is now a matter of record. Suppose $\{X_1,X_2,\ldots,X_k\}$ is a linearly independent ordered set of (necessarily nonzero) vectors. Since a linearly independent set of vectors is just a set which is not linearly dependent, we see that the *only* way we can generate the O vector (by means of linear combination) is by setting *each* of the α_1's equal to zero. More exactly, we have proved a new theorem.

Theorem I.6.6 An ordered set of (nonzero) vectors $\{X_1,X_2,\ldots,X_k\}$ is *linearly independent* if and only if whenever we set

$$O = \alpha_1 X_1 + \alpha_2 X_2 + \cdots + \alpha_k X_k,$$

then, *necessarily*, $\alpha_1 = \alpha_2 = \cdots = \alpha_k = 0$ (each $\alpha_i = 0$).

The ideas of linear dependence and independence (Theorems I.6.5 and I.6.6) may be summarized as follows: Consider the ordered set of vectors $\{X_1,X_2,\ldots,X_k\}$. Now by means of linear combination of these vectors, try to generate the zero vector. That is, try to find scalars $\alpha_1, \alpha_2,\ldots, \alpha_k$ such that

$$O = \alpha_1 X_1 + \alpha_2 X_2 + \cdots + \alpha_k X_k.$$

If the trivial scalars $\alpha_1 = 0$, $\alpha_2 = 0,\ldots, \alpha_k = 0$ are the *only* ones possible to produce O, then the ordered set $\{X_1,X_2,\ldots,X_k\}$ is *linearly independent.* If, on the other hand, you don't have to take *all* the α_i's equal to zero (some $\alpha_i \neq 0$, and O can be generated according to the above equation), then the ordered set $\{X_1,X_2,\ldots,X_k\}$ is *linearly dependent.*

Example Is the ordered set of vectors

$$\left\{\begin{pmatrix} 7 \\ \frac{1}{2} \\ -2 \end{pmatrix}, \begin{pmatrix} 0 \\ 2 \\ 1 \end{pmatrix}, \begin{pmatrix} 0 \\ 2 \\ 1 \end{pmatrix}\right\}$$

linearly independent? Let us (by means of linear combination) attempt to generate the O vector

$$\begin{pmatrix} 0 \\ 0 \\ 0 \end{pmatrix}.$$

That is, for which α_1, α_2, α_3 (scalars) is it true that

$$\alpha_1 \begin{pmatrix} 7 \\ \frac{1}{2} \\ -2 \end{pmatrix} + \alpha_2 \begin{pmatrix} 0 \\ 2 \\ 1 \end{pmatrix} + \alpha_3 \begin{pmatrix} 0 \\ 2 \\ 0 \end{pmatrix} = \begin{pmatrix} 0 \\ 0 \\ 0 \end{pmatrix}?$$

Certainly, if $\alpha_1 = \alpha_2 = \alpha_3 = 0$, then we do get O, but do we *have* to resort to this trivial case? If we do, the set of vectors is linearly independent; if we do not have to take all the α_1's equal to zero, then the set is linearly dependent.

Now by definition of scalar multiplication (Definition I.1.3), the above equation becomes

$$\begin{pmatrix} 7\alpha_1 \\ \frac{1}{2}\alpha_1 \\ -2\alpha_1 \end{pmatrix} + \begin{pmatrix} 0 \\ 2\alpha_2 \\ 1\alpha_2 \end{pmatrix} + \begin{pmatrix} 0 \\ 2\alpha_3 \\ 0 \end{pmatrix} = \begin{pmatrix} 0 \\ 0 \\ 0 \end{pmatrix}.$$

From the definition of vector addition (Definition I.1.2) we add the vectors on the left coordinate-wise to obtain the vector equation

$$\begin{pmatrix} 7\alpha_1 \quad 0 \quad 0 \\ \frac{1}{2}\alpha_1 + 2\alpha_2 + 2\alpha_3 \\ -2\alpha_1 + 2\alpha_2 + \ 0 \end{pmatrix} = \begin{pmatrix} 0 \\ 0 \\ 0 \end{pmatrix}.$$

Now, Definition I.1.4 tells us when 2 concrete vectors are equal; namely, if and only if the corresponding coordinates (which are numbers) are equal. This, in a sense, "transforms" the above vector equation to the problem of 3 separate scalar equations, namely:

(a) $$7\alpha_1 + \ 0 \ + \ 0 \ = 0$$
(b) $$\tfrac{1}{2}\alpha_1 + 2\alpha_2 + 2\alpha_3 = 0$$
(c) $$-2\alpha_1 + \ \alpha_2 \ + \ 0 \ = 0.$$

A most unavoidable prospect will soon become apparent. We will actually be *forced* to choose $\alpha_1 = \alpha_2 = \alpha_3 = 0$. That is, whether we like it or not, in order that Equations (a), (b), and (c) be simultaneously satisfied, all the α_i's will *have* to be zero. This, according to Theorem I.6.6, means that the set we are testing is linearly independent.

To ask that Equation (a) hold true, immediately forces α_1 to be zero. With this fact in mind, observe Equation (c), which formally written, becomes

$$\begin{array}{c} \quad\ \alpha_1 \\ \quad\ \downarrow \\ -2 \cdot 0 + \alpha_2 = 0. \end{array}$$

Inescapably, then, α_2 is also equal to zero. Since we must set α_1 and α_2 equal to zero, Equation (*b*) becomes

$$\tfrac{1}{2} \cdot \underset{\uparrow \alpha_1}{0} + 2 \cdot \underset{\uparrow \alpha_2}{0} + 2 \cdot \alpha_3 = 0.$$

But $2\alpha_3 = 0$ if and only if $\alpha_3 = 0$ and we are done. We have shown that

$$\alpha_1 \begin{pmatrix} 7 \\ \frac{1}{2} \\ -2 \end{pmatrix} + \alpha_2 \begin{pmatrix} 0 \\ 2 \\ 1 \end{pmatrix} + \alpha_3 \begin{pmatrix} 0 \\ 1 \\ 0 \end{pmatrix} = \begin{pmatrix} 0 \\ 0 \\ 0 \end{pmatrix}$$

only when all the α_i's equal zero.

Example Show that the vectors $\begin{pmatrix} 2 \\ 1 \end{pmatrix}, \begin{pmatrix} -1 \\ 0 \end{pmatrix}, \begin{pmatrix} 0 \\ 1 \end{pmatrix}$ are linearly dependent. We try to find $\alpha_1, \alpha_2, \alpha_3$ such that

$$\alpha_1 \begin{pmatrix} 2 \\ 1 \end{pmatrix} + \alpha_2 \begin{pmatrix} -1 \\ 0 \end{pmatrix} + \alpha_3 \begin{pmatrix} 0 \\ 1 \end{pmatrix} = \begin{pmatrix} 0 \\ 0 \end{pmatrix},$$

and hope that we can do it *without* having to set each α_i equal to zero. As in the previous example, we use Definitions I.1.2, I.1.3, and I.1.4 to obtain the 2 scalar equations

(a) $$2 \cdot \alpha_1 - \alpha_2 + 0 = 0$$

(b) $$\alpha_1 + 0 + \alpha_3 = 0.$$

By adding α_2 to both sides of (a), we have our first constraint for the α_i's, namely

(c) $$2\alpha_1 = \alpha_2.$$

Adding the additive inverse of α_1, $(-\alpha_1)$, to both sides of (b) yields the second constraint

(d) $$\alpha_3 = -\alpha_1.$$

In other words, we have shown that $\alpha_1\begin{pmatrix}2\\1\end{pmatrix} + \alpha_2\begin{pmatrix}-1\\0\end{pmatrix} + \alpha_3\begin{pmatrix}0\\1\end{pmatrix} = \begin{pmatrix}0\\0\end{pmatrix}$ if and only if the second alpha, α_2, is twice the first alpha, α_1 [Equation (c)], and the third alpha, α_3, is equal to the negative of the first alpha, $-\alpha_1$ [Equation (d)]. There is nothing here that *forces* us to choose *all* the alphas equal to zero. In fact, choose α_1 to be *any* nonzero scalar, say 1. Then if $\alpha_1 = 1$, Equation (c) tells us that $\alpha_2 = 2$, and Equation (d) says that $\alpha_3 = 1$. Indeed,

$$\overset{\alpha_1}{\underset{\downarrow}{}}\;(1)\begin{pmatrix}2\\1\end{pmatrix} + \overset{\alpha_2}{\underset{\downarrow}{}}\;(2)\begin{pmatrix}-1\\0\end{pmatrix} + \overset{\alpha_3}{\underset{\downarrow}{}}\;(-1)\begin{pmatrix}0\\1\end{pmatrix} = \begin{pmatrix}0\\0\end{pmatrix} = O.$$

Since we were *not* obliged to choose $\alpha_1 = \alpha_2 = \alpha_3 = 0$, in order to generate the zero vector, the ordered set of vectors $\left\{\begin{pmatrix}2\\1\end{pmatrix}, \begin{pmatrix}-1\\0\end{pmatrix}, \begin{pmatrix}0\\1\end{pmatrix}\right\}$ is linearly dependent.

EXERCISES

Determine whether the ordered sets of concrete vectors in Exercises I.6.1 through I.6.8 are linearly independent or linearly dependent.

I.6.1 $\left\{\begin{pmatrix}1\\0\end{pmatrix}, \begin{pmatrix}-1\\0\end{pmatrix}\right\}$

I.6.2 $\left\{\begin{pmatrix}1\\0\end{pmatrix}, \begin{pmatrix}1\\1\end{pmatrix}\right\}$

I.6.3 $\left\{\begin{pmatrix}1\\0\end{pmatrix}, \begin{pmatrix}1\\0\end{pmatrix}, \begin{pmatrix}-1\\0\end{pmatrix}\right\}$

I.6.4 $\left\{\begin{pmatrix}3\\0\\1\end{pmatrix}\right\}$

I.6.5 $\left\{\begin{pmatrix}3\\0\\1\end{pmatrix}, \begin{pmatrix}0\\0\\1\end{pmatrix}\right\}$

I.6.6 $\left\{\begin{pmatrix}3\\0\\1\end{pmatrix},\begin{pmatrix}0\\0\\1\end{pmatrix},\begin{pmatrix}3\\0\\2\end{pmatrix}\right\}$

I.6.7 $\left\{\begin{pmatrix}3\\0\\1\end{pmatrix},\begin{pmatrix}3\\0\\2\end{pmatrix}\right\}$

I.6.8 $\left\{\begin{pmatrix}1\\1\end{pmatrix},\begin{pmatrix}0\\1\end{pmatrix},\begin{pmatrix}-1\\-1\end{pmatrix}\right\}$

For which values of the scalar, r, are the following ordered sets of concrete vectors (Exercises I.6.9 through I.6.15) linearly independent; linearly dependent?

I.6.9 $\left\{\begin{pmatrix}1\\0\end{pmatrix},\begin{pmatrix}r\\1\end{pmatrix}\right\}$

I.6.10 $\left\{\begin{pmatrix}1\\0\end{pmatrix},\begin{pmatrix}1\\r\end{pmatrix}\right\}$

I.6.11 $\left\{\begin{pmatrix}1\\0\\0\end{pmatrix},\begin{pmatrix}0\\1\\0\end{pmatrix},\begin{pmatrix}0\\r\\1\end{pmatrix}\right\}$

I.6.12 $\left\{\begin{pmatrix}1\\0\\0\end{pmatrix},\begin{pmatrix}0\\1\\0\end{pmatrix},\begin{pmatrix}0\\1\\r\end{pmatrix}\right\}$

I.6.13 $\left\{\begin{pmatrix}1\\0\end{pmatrix},\begin{pmatrix}0\\1\end{pmatrix},\begin{pmatrix}r\\2\end{pmatrix}\right\}$

I.6.14 $\left\{\begin{pmatrix}r\\0\\1\end{pmatrix},\begin{pmatrix}0\\1\\0\end{pmatrix}\right\}$

I.6.15 $\left\{\begin{pmatrix}r\\0\\1\end{pmatrix}\right\}$

I.6.16 (a) Suppose the ordered set of nonzero vectors $\mathscr{X} = \{X_1,X_2,\ldots,X_k\}$ is linearly independent in the vector space V. Prove that any nonempty subset of $\mathscr{X}$ is also linearly independent in V.
(b) Prove that if the ordered set $\{X_1,X_2,\ldots,X_k\}$ is linearly independent, then no $X_i = O$ for $i = 1, 2,\ldots, k$.

I.6.17 Suppose the ordered set of nonzero vectors $\mathscr{X} = \{X_1,X_2,\ldots,X_k\}$ is linearly independent in the vector space V. Prove that for any finite, nonempty ordered set $\mathscr{Y} = \{Y_1,Y_2,\ldots,Y_l\}$, the ordered set $\mathscr{X} \cup \mathscr{Y} = \{X_1,\ldots,X_k,Y_1, \ldots Y_k\}$ is also linearly dependent in V.

I.6.18 Show that V_2 is spanned by the ordered set $\left\{\begin{pmatrix}1\\0\end{pmatrix},\begin{pmatrix}0\\1\end{pmatrix}\right\}$, that is, every vector in V_2 can be represented as a linear combination of the vectors given.

I.6.19 Show that V_2 is spanned by the ordered set $\left\{\begin{pmatrix}1\\0\end{pmatrix},\begin{pmatrix}0\\1\end{pmatrix},\begin{pmatrix}1\\-1\end{pmatrix}\right\}$. Is this set LI?

I.6.20 Is V_2 spanned by the ordered set $\left\{\begin{pmatrix}\frac{1}{2}\\0\end{pmatrix},\begin{pmatrix}0\\\frac{1}{2}\end{pmatrix}\right\}$?

I.6.21 Is V_2 spanned by the ordered set $\left\{\begin{pmatrix}\frac{1}{2}\\0\end{pmatrix},\begin{pmatrix}1\\0\end{pmatrix}\right\}$?

I.6.22 Show that V_3 is spanned by the ordered set

$$\left\{\begin{pmatrix}1\\0\\0\end{pmatrix},\begin{pmatrix}0\\1\\0\end{pmatrix},\begin{pmatrix}0\\1\\1\end{pmatrix}\right\}.$$

I.6.23 Show that the set of all vectors of the form

$$\begin{pmatrix}x\\x\\0\end{pmatrix}$$

is a real vector-space, where x is an arbitrary real number. Find an ordered set of vectors which spans this vector-space.

I.6.24 Suppose $Y = \alpha_1X_1 + \alpha_2X_2 + \cdots + \alpha_kX_k$, where $Y, X_1, X_2,\ldots, X_k$ are nonzero vectors in V_n, and $\alpha_1,\alpha_2,\ldots, \alpha_k$ are scalars. Assume, moreover, that $\alpha_1 \neq 0$.

(a) Prove that X_1 can be represented as a linear combination of elements of the ordered set $\{Y,X_2,X_3,\ldots,X_k\}$.

(b) Prove that any vector in V_n that is a linear combination of the elements of $\{X_1,X_2,\ldots,X_k\}$ is also a linear combination of the elements of the set $\{Y,X_2,\ldots,X_k\}$ and conversely, that the sets of vectors spanned by the 2 ordered sets $\{Y,X_2,X_3,\ldots,X_k\}$, $\{X_1,X_2,\ldots,X_k\}$ respectively, are identical.

I.6.25 Suppose the ordered set $\{X_1,X_2,\ldots,X_k\}$ spans the vector-space V_n. Prove that if Y is any nonzero vector of V_n, then the ordered set $\{Y,X_1,X_2,\ldots,X_k\}$ is linearly dependent.

We have seen (following Definition I.6.1) that the ordered set of vectors

$$\left\{\begin{pmatrix}1\\0\\0\end{pmatrix},\begin{pmatrix}0\\1\\0\end{pmatrix},\begin{pmatrix}0\\0\\1\end{pmatrix}\right\}$$

spans (by means of linear combination) the whole space V_3. It is easily verified that this same set of vectors is linearly independent, as well (see Exercise I.6.11). For an ordered set of vectors to have both these properties (spanning the whole space, and linear independent) is so important that such sets are given a special name. Toward this end, we set down:

Definition I.6.7 A nonempty, ordered set of vectors $\{X_1,X_2,\ldots,X_k\}$ of a vector space V_n, is called a **basis** of V_n if and only if

(a) $\{X_1,X_2,\ldots,X_k\}$ is a linearly independent set, and

(b) $\{X_1,X_2,\ldots,X_k\}$ spans the vector-space V_n.

For the trivial vector-space $V = \{O\}$, we will say that $\varnothing$, the empty set is a basis for $\{O\}$.

In other words, an ordered set of vectors is called a *basis* for a nonzero vector-space if and only if it possesses the properties of Definition I.6.2 and Theorem I.6.6, *simultaneously*. (The plural of "basis" is "bases.") Thus the ordered set of vectors

$$\left\{\begin{pmatrix}1\\0\\0\end{pmatrix}, \begin{pmatrix}0\\1\\0\end{pmatrix}, \begin{pmatrix}0\\0\\1\end{pmatrix}\right\}$$

in V_3 is a basis of V_3. For all we know at this point, this set may be the *only* basis for V_3.

Discussion. At least this example establishes *existence* of such an animal as a "basis." We are assured, then, that properties (a) and (b) in Definition I.6.7 do not conflict with each other. A set *can* possess both these properties simultaneously. This observation raises the question of *consistency* for mathematical definitions in general. Whenever a mathematician asks for more than 1 property to obtain (hold true) at once (on a vector-space, say), he must assure himself of the fact that these properties do not contradict each other. For example, suppose you are a sociologist and state the following "definition:" A person is called a **shmoggle** if and only if

(a) that person is a male,

and

(b) that person is a female.

Since no person is both a male *and* a female (at least to the author's knowledge), this "definition" is inconsistent. Properties (a) and (b) conflict. Formally, it might be stated that the set of people who are both male and female, is *empty* (there is no person qualified to be a shmoggle), or the set of shmoggles is *empty*.

By means of the following example, we will exhibit a basis *other* than the one we have already discussed for V_3.

Example Consider the ordered set of vectors

$$\left\{\begin{pmatrix}3\\-1\\0\end{pmatrix}, \begin{pmatrix}0\\2\\1\end{pmatrix}, \begin{pmatrix}0\\1\\0\end{pmatrix}\right\}.$$

We first establish property (a) of Definition I.6.7 (linear independence). Let

$$\alpha_1\begin{pmatrix}3\\-1\\0\end{pmatrix}+\alpha_2\begin{pmatrix}0\\2\\1\end{pmatrix}+\alpha_3\begin{pmatrix}0\\1\\0\end{pmatrix}=\begin{pmatrix}0\\0\\0\end{pmatrix}=O.$$

Using Definitions I.1.2, I.1.3, and I.1.4, we obtain the 3 scalar equations

(1) $$3\alpha_1+0\quad+0\ =0$$

(2) $$-\alpha_1+2\alpha_2+\alpha_3=0$$

(3) $$0\quad+\ \alpha_2+0\ =0.$$

Inspection of Equation (1) tells us that α_1 must equal zero, and Equation (3) says that $\alpha_2=0$ as well. Substituting these values into Equation (3), we see that $\alpha_3=0$. All the alphas, α_1, α_2, α_3 *must* equal zero, and hence the set is linearly independent, thereby establishing property (a) of Definition I.6.7.

As for property (b), let

$$\begin{pmatrix}x_1\\x_2\\x_3\end{pmatrix}$$

be an *arbitrary* vector of V_3. That is, the coordinates x_1, x_2, and x_3 are not explicitly restricted (but are fixed throughout). It is hoped that scalars (real numbers, here) α_1, α_2, and α_3 can be found so that

$$\alpha_1\begin{pmatrix}3\\-1\\0\end{pmatrix}+\alpha_2\begin{pmatrix}0\\2\\1\end{pmatrix}+\alpha_3\begin{pmatrix}0\\1\\0\end{pmatrix}=\begin{pmatrix}x_1\\x_2\\x_3\end{pmatrix}.$$

In other words, it is hoped that the 3 vectors under consideration *span* the space V_3. Again, applying Definitions I.1.2, I.1.3, and I.1.4, we arrive at the 3 scalar equations

(4) $$3\alpha_1+\ 0\ +0\ =x_1$$

(5) $$-\alpha_1+2\alpha_2+\alpha_3=x_2$$

(6) $$0\quad+\ \alpha_2+0\ =x_3.$$

Observing Equation (4), we see that it holds true if and only if $\alpha_1=x_1/3$, and Equation (6) obtains if and only if $\alpha_2=x_3$. So far, so good. We have found 2 alphas, which, naturally, depend on the vector

$$\begin{pmatrix}x_1\\x_2\\x_3\end{pmatrix}$$

that we are trying to generate. Solving for α_3 in Equation (5), and substituting the number $x_1/3$ for α_1, and the number x_3 for α_2, we see that

$$\alpha_3=x_2-2\cdot x_3+\frac{x_1}{3}.$$

Thus, all 3 alphas, a_1, α_2, and α_3, *can* be found. In fact it is easily verified that for *any* concrete vector

$$\begin{pmatrix} x_1 \\ x_2 \\ x_3 \end{pmatrix},$$

$$\underset{\displaystyle \frac{x_1}{3}}{\overset{\alpha_1}{\downarrow}} \begin{pmatrix} 3 \\ -1 \\ 0 \end{pmatrix} + \underset{\displaystyle x_3}{\overset{\alpha_2}{\downarrow}} \begin{pmatrix} 0 \\ 2 \\ 1 \end{pmatrix} + \underset{\displaystyle \left(x_2 - 2x_3 + \frac{x_1}{3}\right)}{\overset{\alpha_3}{\downarrow}} \begin{pmatrix} 0 \\ 1 \\ 0 \end{pmatrix} = \begin{pmatrix} x_1 \\ x_2 \\ x_3 \end{pmatrix}$$

which establishes property (b) of Definition I.6.7. We have shown that the set of vectors

$$\left\{ \begin{pmatrix} 3 \\ -1 \\ 0 \end{pmatrix}, \begin{pmatrix} 0 \\ 2 \\ 1 \end{pmatrix}, \begin{pmatrix} 0 \\ 1 \\ 0 \end{pmatrix} \right\}$$

is a basis for V_3.

We will present another property of sets of vectors which is equivalent to the basis property of Definition I.6.7. Instead of reverting to Definition I.6.3 in dealing with the linear independence property, we shall see that its equivalent, Theorem I.6.6, presents a more convenient form.

Theorem I.6.8 The following properties are equivalent for an ordered set of (nonzero) vectors $\{X_1, X_2, \ldots, X_k\}$ in V_n.

(a) The set is a basis for V_n.

(b) Each and every vector Y in V_n can be represented as a linear combination of the vectors $X_1, X_2, \ldots, X_k$, that is, $Y = \alpha_1 X_1 + \alpha_2 X_2 + \cdots + \alpha_k X_k$ for some set of scalars $\alpha_1, \alpha_2, \ldots, \alpha_k$, and, moreover, this representation is *unique*. That is, there is 1 and only 1 way that the alphas $\alpha_1, \alpha_2, \ldots, \alpha_k$ may be chosen so as to generate the vector Y.

Proof: We will show that (a) implies (b). Assume, then, that the ordered set of nonzero vectors $\{X_1, X_2, \ldots, X_k\}$ forms a basis for V_n. But a basis, by definition (Definition I.6.7) spans the space V_n, so that any Y in V_n *can* be represented by $Y = \alpha_1 X_1 + \alpha_2 X_2 + \cdots + \alpha_k X_k$, for at least one set of scalars $\alpha_1, \alpha_2, \ldots, \alpha_k$. Only the question of uniqueness remains; we must show that there is *only* 1 set of alphas that suffices to generate Y. In fact, suppose

$$Y = \alpha_1 X_1 + \alpha_2 X_2 + \cdots + \alpha_k X_k. \tag{1}$$

$$Y = \alpha_1' X_1 + \alpha_2' X_2 + \cdots + \alpha_k' X_k. \tag{2}$$

Subtracting (2) from (1) yields the equation

$$O = (\alpha_1 - \alpha_1')X_1 + (\alpha_2 - \alpha_2')X_2 + \cdots + (\alpha_k - \alpha_k')X_k.$$

Not forgetting the important assumption that the set $\{X_1,X_2,\ldots,X_k\}$ is linearly independent, we utilize Theorem I.6.6; O can be generated by this set *only* if all the scalars are zero. In this case, this means that $(\alpha_1 - \alpha_1') = 0$, $(\alpha_2 - \alpha_2') = 0,\ldots,$ $(\alpha_k - \alpha_k') = 0$. That is, $\alpha_1 = \alpha_1'$, $\alpha_2 = \alpha_2',\ldots,$ $\alpha_k = \alpha_k'$. Thus, try as we may, the assumption of linear independence forces us to admit that Equations (1) and (2) above, are one and the same equation. Since every vector can be generated by this set, and this can be done with only 1 set of scalars $\alpha_1, \alpha_2,\ldots, \alpha_k$, we have shown that (a) implies (b).

We will now show the converse, that is, that (b) implies (a).

Assume that the set of vectors has the property that any vector, Y, say, can be written in 1 and only 1 way is a linear combination of the vectors $X_1, X_2,\ldots, X_k$. Now assuming that *any* vector can be written as a linear combination at all, says that $X_1, X_2,\ldots, X_k$ *spans* the space V_n. This establishes "half" (the spanning half) of the basis property.

Now, every vector can be written in only 1 way as a linear combination of the vectors $X_1, X_2,\ldots, X_k$ by hypothesis (b). In particular, O has only 1 representation, namely,

$$O = a_1X_1 + a_2X_2 + \cdots + a_kX_k.$$

In any event, setting $a_1 = a_2 = \cdots = a_k = 0$, *always* produces the zero vector O. (Why?) This very representation, then, must be that *unique* representation that we have assumed exists. That is, the *only* way to generate the zero vector O, is by means of the trivial linear combination where $a_1 = a_2 = \cdots = a_k = 0$. This proves that (b) implies (a), which completes the proof.

Admittedly, we have made quite a to-do over bases. We defined a basis as a set of linearly independent (nonzero) vectors which spans the whole vector-space V_n, and then showed that an arbitrary vector in V_n can be represented in only one way as a linear combination of the basis vectors. We should pause a moment to assure ourselves that we are not speaking in a void. That is, for any vector-space, can sets always be found that are both linearly independent *and* generating? We have seen by example that at least *one* set of vectors is a basis for the concrete vector space V_3, namely,

$$\left\{\begin{pmatrix}1\\0\\0\end{pmatrix},\begin{pmatrix}0\\1\\0\end{pmatrix},\begin{pmatrix}0\\0\\1\end{pmatrix}\right\}.$$

(It is an exercise to show that "this kind" of a set always forms a basis for a concrete vector-space.) It is no coincidence that there are 3 vectors in this basis for V_3, the space of all 3-vectors (columns with 3 entries). Heretofore, when we indicated the ordered set $\{X_1,X_2,\ldots,X_k\}$ as a basis for V_n, we made no mention of the *number of (nonzero) vectors* in the basis, as compared with n, the *number of entries* in each vector. This question will be settled after we prove Theorem I.6.9, which says that any spanning set of V_n contains a greater or equal number of vectors than does any linearly independent set of V_n. We state and prove this now.

Theorem I.6.9 Suppose the ordered set of vectors $\{X_1,X_2,\ldots,X_k\}$ spans V_n, and the set of vectors $\mathcal{Y}=\{Y_1,Y_2,\ldots,Y_j\ldots\}$ is linearly independent in V_n. Then, necessarily, $\mathcal{Y}$ has only finitely many vectors and, moreover, this finite number j is less than or equal to k ($j\leq k$) (symbolically, LI $\leq$ span).

Proof: Consider the amended ordered set $\{Y_1,X_1,X_2,\ldots,X_k\}$ which has $k+1$ vectors. This set is now linearly dependent since Y_1 can be expressed as a linear combination of the X_i's (we are using the fact that the set $\{X_1,X_2,\ldots,X_k\}$ *spans* the space V_n). Thus,

$$\text{(a)}\qquad \alpha_0 Y_1+\alpha_1 X_1+\alpha_2 X_2+\cdots+\alpha_k X_k=O$$

for some set of scalars α_0, α_1, $\alpha_2,\ldots,$ α_k which are *not all* zero. We can be more specific. Not all of the alphas "attached" to the X_i's are equal to zero; that is, not all of the scalars α_1, $\alpha_2,\ldots,$ α_k equal zero. If they *were all* equal to zero, we would have

$$\alpha_0 Y_1+0X_1+0X_2+\cdots+0X_k=O,$$

which says that $\alpha_0 Y_1=O$ which is impossible. (Why?)

Since at least 1 of the numbers α_1, $\alpha_2,\ldots,$ α_k is not zero, let us assume (by renumbering the X_i's if necessary) that $\alpha_1\neq 0$. This means that X_1 can be expressed as a linear combination of the vectors Y_1, X_2, X_3, $X_4,\ldots,$ X_k. This follows from observation of Equation (a). But then X_1 may be eliminated and the set $\{Y_1,X_2,X_3,\ldots,X_k\}$ with k vectors (where Y_1 replaces X_1) spans V_n. (See Exercise I.6.24.)

We argue similarly with respect to the set $\{Y_1,Y_2,X_2,\ldots,X_k\}$, which has $k+1$ vectors. First, as before, the set is linearly dependent. Since the set $\{Y_1,X_2,\ldots,X_k\}$ spans the space, certainly Y_2 can be expressed as a linear combination of these vectors (see Definition I.6.2). Thus,

$$\text{(b)}\qquad \alpha_0 Y_1+\alpha_1 Y_2+\alpha_2 X_2+\alpha_3 X_3+\cdots+\alpha_k X_k=O$$

where not *all* of the $k+1$ scalars α_0, $\alpha_1,\ldots,$ α_k are equal to zero. Again, we will be more specific by showing that at least 1 of the alphas attached to the X's (in this case, at least one of the numbers $\alpha_2, \alpha_3, \ldots, \alpha_k$) is not zero. In fact, suppose they were all zero. Then we would have

$$\alpha_0 Y_1 + \alpha_1 Y_2 + 0X_2 + 0X_3 + \cdots + 0X_k = O,$$

which is Equation (b) rewritten. But this would say that $\alpha_0 Y_1 + \alpha_1 Y_2 = O$. Recall that the set $\{Y_1, Y_2, \ldots, Y_j, \ldots\}$ is linearly independent (see Exercise I.6.16). Thus, $\alpha_0 Y_1 + \alpha_1 Y_2 = O$ *only* for $\alpha_0 = \alpha_1 = 0$. But this would mean that *all* the $k+1$ alphas $\alpha_0, \alpha_1, \alpha_2, \alpha_3, \ldots, \alpha_k$ are equal to zero. This contradicts Equation (b). Our initial supposition that *all* the alphas attached to the X's were zero led us to this unhappy contradiction. We must logically abandon it and accept the truth of the statement that *not all* of the scalars $\alpha_2, \alpha_3, \ldots, \alpha_k$ equal zero.

Again, by relabelling the X's, if necessary, suppose $\alpha_2 \neq 0$. Appealing to Equation (b) once more, we see that X_2 can be expressed in terms of the k remaining vectors Y_1, Y_2, X_3, $X_4, \ldots, X_k$. We claim that this set spans V_n. As we have just shown, this set can generate X_2. This set can also generate *any* vector since the set $\{Y_1, X_2, X_3, \ldots, X_k\}$ generates any vector, and X_2, being a linear combination of the vectors Y_1, Y_2, $X_3, \ldots, X_k$, allows an *arbitrary* vector to be generated by these vectors (Exercise I.6.24).

We continue with this replacement procedure. As argued before, it can be shown that the set $\{Y_1, Y_2, Y_3, X_3, \ldots, X_k\}$ is linearly dependent, and that *not all* of the alphas attached to the X's equal zero, where

$$\alpha_0 Y_1 + \alpha_1 Y_2 + \alpha_2 Y_3 + \alpha_3 X_3 + \cdots + \alpha_k X_k = O.$$

(This is a consequence of the fact that the set $\{Y_1, Y_2, Y_3\}$ is linearly independent.) Hence, one of the X's, X_3 say, can be eliminated, leaving the set of k vectors $\{Y_1, Y_2, Y_3, X_4, X_5, \ldots, X_k\}$ as a spanning set for V_n.

How many times can we repeat this replacement process, so as to have a spanning set? We will now see (since the Y's are linearly independent) that there can be no *more* Y's than there are X's, that is, there are only a finite number of Y's, and this finite number is less than or equal to k. Suppose the opposite. That is, we will assume that the number of Y's, is strictly *greater* than k, the number of X's, and we will arrive at a contradiction. Since there are more Y's than X's, assume that *all* of the X's have been replaced so that the set $\{Y_1, Y_2, \ldots, Y_k\}$ is, as in each previous step, a spanning set for V_n. But there are some Y_i vectors "left over," namely, Y_{k+1}, $Y_{k+2}, \ldots, Y_j, \ldots$. In particular, Y_{k+1} can be represented as a linear combination of the spanning vectors $\{Y_1, Y_2, \ldots, Y_k\}$, that is,

$$Y_{k+1} = \beta_1 Y_1 + \beta_2 Y_2 + \cdots + \beta_k Y_k,$$

for certain scalars (obviously not all equal to zero) $\beta_1, \beta_2, \ldots, \beta_k$. That is,

$$O = \beta_1 Y_1 + \beta_2 Y_2 + \cdots + \beta_k Y_k + (-1) Y_{k+1}$$

which says that the set $\{Y_1, Y_2, \ldots, Y_{k+1}\}$, and hence the original set $\{Y_1, Y_2, \ldots, Y_{k+1}, \ldots, Y_j, \ldots\}$ is linearly dependent (see Exercise I.6.17). That certainly contradicts the linear *in*dependence assumption on $\{Y_1, Y_2, \ldots, Y_j, \ldots\}$, So we conclude there can be no more (leftover) Y's than X's. This shows that any linearly independent set contains no more vectors than any (other) spanning set, proving the theorem.

One interesting corollary which flows from Theorem I.6.9 is the fact that *any* 2 bases of V_n have the same number of elements. Formally stated, we have

Corollary I.6.10 Suppose the ordered sets $\mathscr{X} = \{X_1, X_2, \ldots, X_k\}$ (finite) and $\mathscr{Y} = \{Y_1, Y_2, \ldots, Y_j, \ldots\}$ (possibly infinite) are bases for V_n. Then the sets $\mathscr{X}$ and $\mathscr{Y}$ have the same (finite) number of elements, namely, k.

Proof: Since $\mathscr{X}$ is a basis, $\mathscr{X}$ is a spanning set for V_n and contains at least as many vectors as the set $\mathscr{Y}$, which is linearly independent (since $\mathscr{Y}$ is a basis), that is k (which is finite) is *greater* than or equal to the number of elements, call it j, in $\mathscr{Y}$. Thus $\mathscr{Y}$ cannot have infinitely many elements, since $j \leq k$.

Reversing the rôles of the sets $\mathscr{X}$ and $\mathscr{Y}$, the spanning set $\mathscr{Y}$ contains at least as many vectors as does the linearly independent set $\mathscr{X}$, that is, j is greater than or equal to k ($j \geq k$).

The only conclusion that can be drawn from the preceding 2 paragraphs is that j equals k, or that, indeed, any 2 bases of V_n have the same number of elements.

Theorem I.6.9 allows for alternate definitions of "basis." We see from this theorem that any linearly independent set is "smaller" than or is "the same size as" any spanning set. A basis can be thought of as a *maximal linearly independent* set, that is, a linearly independent set such that if any vector (zero or not) were amended to it, the augmented set would become linearly dependent. Similarly, a basis can be thought of as a *minimal spanning* set, that is, a spanning set with the property that if any vector were deleted from it, the diminished set would not longer span the vector-space. See Exercises I.6.35b and I.6.35c.

Since the number of vectors in any basis is unique (no 2 bases have different numbers of vectors) exactly how many vectors *does* a basis for V_n contain? To answer this we are led to Definition I.6.11.

Definition I.6.11 The vector E_i of the concrete vector-space V_n, is the vector whose ith coordinate (entry) is the number 1, and all other coordinates (entries) are equal to zero. That is,

$$E_1 = \begin{pmatrix} 1 \\ 0 \\ 0 \\ \vdots \\ 0 \end{pmatrix}, \quad E_2 = \begin{pmatrix} 0 \\ 1 \\ 0 \\ \vdots \\ 0 \end{pmatrix}, \quad E_3 = \begin{pmatrix} 0 \\ 0 \\ 1 \\ 0 \\ \vdots \\ 0 \end{pmatrix},$$

and so on. The ordered collection of all these vectors $\{E_1,E_2,E_3,\ldots,E_n\}$ is called the **standard basis for V_n.**

REMARK. It is easily verified that the set $\{E_1,E_2,\ldots,E_n\}$ is, in fact, a basis. (Exercise I.6.25). Just because the set $\{E_1,E_2,\ldots,E_n\}$ is *labelled* "standard basis," this does not *make* it a basis. Also (assuming verification of the "basis" label) we see that there are n vectors in this set. Hence we have:

Theorem I.6.12 Every basis for the concrete vector-space V_n has precisely n vectors, where n is an integer greater than or equal to 1.

REMARK. It is now meaningful to say that the concrete vector-space V_n is an *n-dimensional* vector-space (see Definition I.1.1). That is, any vector V_n can be generated (by means of linear combination) by the nonzero vectors of a basis.

In fact, we define *dimension* as follows:

Definition I.6.13 Let V be a (concrete or abstract) vector-space which has nonzero members. The **dimension of V** is the number of nonzero vectors in any basis of V, provided a basis exists. In case $V = \{O\}$ (which is a vector-space), then we say that V has **dimension zero**.

More properly, the vector-space which has the zero vector as its only member belongs in the discussion of subspaces, but it does no harm to announce its dimension here.

EXERCISES

I.6.26 Prove that the standard basis (Definition I.6.11) is, in fact, a basis for V_n. Why or why not are the following sets of concrete vectors (Exercises I.6.27 through I.6.32) bases for their respective vector-spaces V_n?

I.6.27 $\begin{pmatrix} 0 \\ 1 \end{pmatrix}, \begin{pmatrix} \frac{1}{2} \\ 1 \end{pmatrix}$

I.6.28 $\begin{pmatrix}1\\0\\0\end{pmatrix}, \begin{pmatrix}0\\0\\1\end{pmatrix}$

I.6.29 $\begin{pmatrix}1\\0\\0\end{pmatrix}, \begin{pmatrix}0\\0\\1\end{pmatrix}, \begin{pmatrix}-1\\1\\0\end{pmatrix}$

I.6.30 $\begin{pmatrix}1\\0\\0\end{pmatrix}, \begin{pmatrix}0\\0\\1\end{pmatrix}, \begin{pmatrix}-1\\1\\0\end{pmatrix}, \begin{pmatrix}0\\1\\0\end{pmatrix}$

I.6.31 $\{(2)\}$

I.6.32 $\begin{pmatrix}3\\5\end{pmatrix}, \begin{pmatrix}-6\\-16\end{pmatrix}$

I.6.33 Consider the sets of concrete vectors in Exercises I.6.1 through I.6.15, and I.6.18 through I.6.22. Which of these sets of vectors are bases for their respective vector-spaces?

I.6.34 Suppose an ordered set of vectors $\mathscr{X} = \{X_1, X_2, \ldots, X_k\}$ is linearly *dependent* in V_n. Show that there is a nonempty linearly independent subset of $\mathscr{X}$, if not all X_i $(i = 1,2,\ldots,k)$ equal O.

I.6.35 (a) Prove that an ordered spanning set of vectors $\mathscr{X} = \{X_1, X_2, \ldots, X_n\}$ in V_n, which has n vectors is necessarily linearly independent.

Hint: Assume that $\mathscr{X} = \{X_1, X_2, \ldots, X_n\}$ in V_n is linearly dependent and that $\mathscr{X}$ spans V_n.

(b) The linearly independent set $\{X_1, X_2, \ldots, X_k\}$ is called **maximal** if and only if for every vector Y, the ordered augmented set $\{X_1, X_2, \ldots, X_k, Y\}$ is linearly dependent. Prove that every basis is a maximal linearly independent set, and conversely, every maximal linearly independent set is a basis.

(c) The spanning set $\{X_1, X_2, \ldots, X_k\}$ is called **minimal** if and only if whenever any X_i $(i = 1,2,\ldots,k)$ is deleted from the set, the resulting set is not a spanning set. Prove that every basis is a minimal spanning set, and conversely, every minimal spanning set is a basis.

Using the replacement idea of Theorem I.6.9 we are able to establish the theorem which states that any linearly independent set of nonzero vectors may be "extended to a basis." More exactly, we state Theorem I.6.14.

Theorem I.6.14 Let $\{X_1, X_2, \ldots, X_k\}$ be an ordered linearly independent set of (nonzero) concrete vectors in V_n, where k is greater than or equal to one, and less than n. We conclude that $n - k$ linearly independent vectors $\{F_{k+1}, F_{k+2}, \ldots, F_n\}$ can be found so that the ordered set $\{X_1, X_2, \ldots, X_k, F_{k+1}, F_{k+2}, \ldots, F_n\}$ is a *basis* for V_n.

Proof: We know that V_n has at least 1 basis $\{F_1, F_2, \ldots, F_n\}$, say. Consider the set of $n+1$ vectors $\{Y_1, F_1, F_2, \ldots, F_n\}$ which is linearly dependent (see Exercise I.6.35). As in the proof of Theorem I.6.9, one of the

F's (F_1, say) can be eliminated, leaving the set of nonzero vectors, $\{Y_1,F_2,F_3,\ldots,F_n\}$ as a spanning set. Since the set spans, and since there are n vectors in the set, the set is necessarily linearly independent (see Exercise I.6.35), that is, the new set is a *basis*.

We repeat this procedure. Consider the augmented set $\{Y_1,Y_2,F_2,F_3,\ldots,F_n\}$ of $n+1$ vectors. This set is linearly dependent. Hence, we may write $O = \alpha_1Y_1 + \alpha_2Y_2 + \alpha_3F_2 + \cdots + \alpha_{n+1}F_{n+1}$ for some $n+1$ scalars $\{\alpha_1,\alpha_2,\alpha_3,\ldots,\alpha_{n+1}\}$ which are *not* all zero. As we argued in the proof of Theorem I.6.9, one of the F's (F_2, say) may be expressed as a linear combination of the other vectors. (Why?) We may cast out the F_2, leaving the set $\{Y_1,Y_2,F_3,F_4,\ldots,F_n\}$. This set is necessarily linearly independent.

By successive repetitions of this procedure (which may involve renumbering of the F_i's), we arrive at the set of n vectors $\{Y_1,Y_2,\ldots,Y_k,F_{k+1},\ldots,F_n\}$ which is a basis for V_n. Thus, we have shown, that every linearly independent set of vectors in V_n (with fewer than n vectors) can always be extended to a basis of V_n.

REMARK. Theorem I.6.14 answers the question: How many distinct bases are there for V_n? Theorem I.6.14 says, in effect, that *if* we have a LI (that is, linearly independent) set of nonzero vectors $\{Y_1,Y_2,\ldots,Y_k\}$ then we can "throw in" just enough linearly independent vectors $\{F_{k+1},\ldots,F_n\}$ so that $\{Y_1,Y_2,\ldots,Y_k,F_{k+1},\ldots,F_n\}$ is a "new" basis. In particular, we may choose the LI, 1-vector set $\{Y\}$, and extend *it* to a basis for some $Y \neq O$. There are infinitely many nonzero Y vectors; hence, infinitely many bases can be constructed.

EXERCISES

Extend the following sets of concrete vectors to bases, where possible.

I.6.36 $\left\{\begin{pmatrix}1\\0\\0\end{pmatrix},\begin{pmatrix}0\\1\\0\end{pmatrix}\right\}$

I.6.37 $\left\{\begin{pmatrix}0\\1\\1\end{pmatrix},\begin{pmatrix}1\\0\\0\end{pmatrix},\begin{pmatrix}0\\1\\0\end{pmatrix}\right\}$

I.6.38 $\left\{\begin{pmatrix}3\\0\end{pmatrix}\right\}$

I.6.39 $\left\{\begin{pmatrix}1\\0\\0\\1\end{pmatrix},\begin{pmatrix}3\\0\\0\\0\end{pmatrix}\right\}$

I.6.40 $\left\{\begin{pmatrix}1\\0\\0\\0\end{pmatrix}, \begin{pmatrix}0\\1\\0\\1\end{pmatrix}, \begin{pmatrix}3\\2\\-1\\0\end{pmatrix}\right\}$

I.6.41 $\left\{\begin{pmatrix}5\\0\\1\end{pmatrix}, \begin{pmatrix}3\\2\\1\end{pmatrix}\right\}$

I.6.42 $\{(3)\}$

I.6.43 $\left\{\begin{pmatrix}0\\2\\1\\0\end{pmatrix}, \begin{pmatrix}1\\3\\0\\0\end{pmatrix}, \begin{pmatrix}1\\1\\1\\1\end{pmatrix}\right\}$

7. *Subspaces*

We will define a subspace of a (concrete or abstract) vector-space U to be a subset of U which is, in itself, a vector-space, or a subvector space of U. Thus, we present Definition I.7.1.

Definition I.7.1 If M is a nonempty set of vectors in vector-space U (that is, $M \subset U$), then M is called a **subspace** (of U) if and only if M is also a vector-space in accordance with the 10 properties of Definition I.5.2, relative to the same operations of U.

Hence, we are to consider the problem of taking a sub*set* M of a vector space U, and deciding whether or not M is also a vector-space in its own right. At first glance, this problem might not seem especially appetizing. After all, whenever we test a set M to see whether it is a vector-space, we are obliged to go through 10 *different tests*, namely, properties A.1 through A.5 and B.1 through B.5 of Definition I.5.2.

But the set M that *we* are testing is not just any set off the street—according to Definition I.7.1, M is imbedded as a subset of a *vector space*! And this makes all the difference. As we shall see, if $M \subset U$, and if U is known to be a vector-space (so that all 10 properties of Definition I.5.2 hold for U), then we shall need to apply only 2 tests (instead of 10!) to M in order to decide whether or not M is a bonafide vector-space. As our next theorem demonstrates, if M is only

(a) closed under addition (A.1 of Definition I.5.2) and

(b) closed under scalar multiplication (B.1 of Definition I.5.2),

then the remaining 8 properties of Definition I.5.2 will automatically hold for M, since M is a subset of a vector-space. Thanks to the following theorem, then, subsets M of vector-spaces which are suspected of being vector-spaces, do not have to be tested against *all* 10 properties of Definition I.5.2 in order to confirm this suspicion; the 2 properties of closure under addition and closure under scalar multiplication will always suffice.

Theorem I.7.2 If M is a nonempty set of vectors in (concrete or abstract) vector-space U (that is, $M \subset U$), then M is a subspace of U if and only if

(a) the vector $X + Y \in M$ whenever $X \in M$ and $Y \in M$, and

(b) for all scalars α, $\alpha X \in M$ whenever $X \in M$.

Proof: ($\Rightarrow$) We assume that the subset $M \subset U$ is a bonafide sub (vector) space, that is, all 10 properties of Definition I.5.2 hold for M. Thus, it is not asking very much to believe that if 10 properties hold, then surely 2 of these properties hold, namely, properties A.1 (closure under addition) and B.1 (closure under scalar multiplication). These are exactly properties (a) and (b) of Theorem I.7.2 which proves the easy part ($\Rightarrow$) of the theorem.

($\Leftarrow$) We show now that if the 2 closure properties (a) and (b) hold on M (properties A.1 and B.1 of Definition I.5.2), then *necessarily*, the other 8 vector-space properties of Definition I.5.2 also hold true on M. Now we observe that the following 6 properties,

(c) A.2 (commutivity), A.5 (distributivity), and
B.2, B.3, B.4, and B.5,

are valid for any and all subsets (not necessarily subvector spaces) or vector-space U. These 6 "inherited" properties are characterized as *equations* which are valid for *any* and *all* vectors of U, and for *all* scalars. Hence, in considering vectors in $M \subset U$, we do not cancel the fact that we are also considering vectors of U.

So far, then, our subset M enjoys 8 of the 10 desired properties, namely, A.1 and B.1 (hypothesis) along with the 6 inherited properties of (c). This leaves only 2 properties of existence, A.3 and A.4, to verify; we effect this verification now.

A.3 Verified: Suppose $X \in M \subset U$. We are to show that $-X$, the additive inverse of X, also belongs to M. Toward this end, use assumption (b) of the hypothesis (choose $\alpha = -1$) which assures us that $(-1) \cdot X \in M$ if $X \in M$. But for any (concrete or abstract) vector-space, $(-1)X = -X$, the additive inverse of X (see Exercise I.7.5). Thus, property A.3, the existence of an additive inverse vector for each vector in M, is established.

A.4 Verified: We are to prove that O, the zero (feeble) vector belongs to M. Now is the time to use assumption (a) which says that M is closed under addition. Thus, choose any $X \in M$. From the previous paragraph (A.3 Verified), we are sure that $-X \in M$ as well. By the closure assumption, Hypothesis (a), the sum $X + (-X)$, which happens to equal O, belongs to M. A.4 is thus verified, implying that nonempty subsets M of a vector-space U which enjoy the 2 closure properties A.1 and B.1, necessarily enjoy all 10 vector-space properties of Definition I.5.2.

We present 2 examples of sub*sets* of V_n, 1 of which is a sub*space*, and the other is not.

Example I Let M be the (sub)set of all vectors of V_3, whose first coordinate (entry) is 2, that is, M is the set of all vectors of the form

$$\begin{pmatrix} 2 \\ x_2 \\ x_3 \end{pmatrix},$$

where x_2 and x_3 are unrestricted. Moreover, it is to be remembered that these are the *only* vectors of M. In this case, M is *not* a subspace since, if we multiply a vector of the form

$$\begin{pmatrix} 2 \\ x_2 \\ x_3 \end{pmatrix}$$

by the scalar 3, say, we do *not* get a vector of the desired form (whose first coordinate is 2). In fact,

$$3 \cdot \begin{pmatrix} 2 \\ x_2 \\ x_3 \end{pmatrix} = \begin{pmatrix} 6 \\ 3x_2 \\ 3x_3 \end{pmatrix}$$

(see Definition I.1.3) and the first coordinate is 6, not 2;

$$3 \cdot \begin{pmatrix} 2 \\ x_2 \\ x_3 \end{pmatrix}$$

is not in M and so M cannot be a subspace since condition (a) of Definition I.7.1 is not satisfied. It is an exercise to show that condition (b) is not satisfied either, so that M is not even close to being a subspace of V_3.

Example II Let M be the (sub)set of all vectors of V_3, whose *second* coordinate is the number 0. Thus, any and all vectors in M are of the form

$$\begin{pmatrix} x_1 \\ 0 \\ x_3 \end{pmatrix},$$

where the coordinates x_1, x_3 are unrestricted. For example, the vectors

$$\begin{pmatrix} \frac{1}{2} \\ 0 \\ -1 \end{pmatrix} \quad \text{and} \quad \begin{pmatrix} 2 \\ 0 \\ 0 \end{pmatrix}$$

are in M, but the vectors

$$\begin{pmatrix} 0 \\ 1 \\ 0 \end{pmatrix} \quad \text{and} \quad \begin{pmatrix} -2 \\ 0.001 \\ 3 \end{pmatrix}$$

are not in M. In order to see that M is a subspace of V_3, we need only verify that conditions (a) and (b) of Definition I.7.1 hold true in this special example. As for checking (a), we must assure ourselves that for *any* scalar α, and for *any* vector in $X \in M$, αX is still in M. But if $X \in M$,

$$X = \begin{pmatrix} x_1 \\ 0 \\ x_3 \end{pmatrix}$$

and hence

$$\alpha \begin{pmatrix} x_1 \\ 0 \\ x_3 \end{pmatrix} = \begin{pmatrix} \alpha x_1 \\ \alpha 0 \\ \alpha x_3 \end{pmatrix} = \begin{pmatrix} \alpha x_1 \\ 0 \\ \alpha x_3 \end{pmatrix}$$

which is, happily, a vector with second coordinate equal to zero. Thus, condition (a) obtains; scalar multiples of vectors of M are still in M. As for condition (b), let us consider 2 vectors of M and let us denote then by

$$\begin{pmatrix} x_1 \\ 0 \\ x_3 \end{pmatrix} \quad \text{and} \quad \begin{pmatrix} y_1 \\ 0 \\ y_3 \end{pmatrix}.$$

We notice that their vector sum

$$\begin{pmatrix} x_1 \\ 0 \\ x_3 \end{pmatrix} + \begin{pmatrix} y_1 \\ 0 \\ y_3 \end{pmatrix} = \begin{pmatrix} x_1 + y_1 \\ 0 + 0 \\ x_3 + y_3 \end{pmatrix} \quad \text{(Definition I.1.2)} = \begin{pmatrix} x_1 + y_1 \\ 0 \\ x_3 + y_3 \end{pmatrix}$$

also belongs to M since the second coordinate is the scalar 0. Thus, sums of vectors in M are still in M and condition (b) holds true. We conclude that M, the set of vectors of V_3 whose second coordinate is zero, is a subspace of V_3 due to Theorem I.7.2.

EXERCISES

I.7.1 Let M be the subset of V_3 whose every element is of the form

$$\begin{pmatrix} 2 \\ x_2 \\ x_3 \end{pmatrix},$$

where x_2 and x_3 are real numbers. Show that M does not satisfy condition (b) of Theorem I.7.2.

I.7.2 Show that the 1-element subset $\{O\}$ in V_n consisting of the zero vector only, is a subspace of V_n (We have already defined the dimension of $\{O\}$ to be zero, but it must be *verified* that $\{O\}$ is a bonafide subspace according to Definition I.7.1 and Theorem I.7.2.)

I.7.3 Let M be the subset of V_n, whose every element has a non-negative first coordinate. Is M a subspace of V_n?

I.7.4 Let M be the subset of V_n (n is greater than or equal to 3) whose every element is a vector with the first and third coordinate equal (the other coordinates are unrestricted). Is M a subspace of V_n?

I.7.5 (Scalars multiplying abstract vectors.) If $X = (x_1, x_2, \ldots, x_n)$ is a *concrete* vector, then there is no difficulty in applying Definition I.1.3 to see that for the scalar $\alpha = 0$,
(1) $0X = O$, the zero (feeble) vector, and for the scalar $\alpha = -1$,
(2) $(-1)X = -X$, the additive inverse of X.
(Note the difference between the 2 minus signs in the last equation; the first denotes the additive inverse of a *scalar*, while the second denotes the additive inverse of a *vector*!) The object of this exercise is to establish (1) and (2) for *any abstract* vector-space, where only the 10 properties of Definition I.5.2 may be used.
(a) Prove that for any and all (abstract) vectors X, $0X = O$, the zero vector of the vector-space. [*Hint:* Consider $0X = (0 + 0)X$. Use the distributive law.]
(b) Prove that for any and every abstract vector X, $(-1)X = -X$, the additive inverse of X. [*Hint:* Recall that $-X$ is that *unique* vector with the property that $X + (-X) = O$. Test $(-1)X$ against this property to see whether it, too, has the property $X + (-1)X = O$.]

We have spoken of basis vectors for the vector-space as certain "generators." We wish to study this generating property relative to subspaces. Accordingly, we present Definition I.7.3.

Definition I.7.3 Given the nonempty set of vectors $\{X_1, X_2, \ldots, X_k\}$ of V_n; the **linear span of the vectors** $X_1, X_2, \ldots, X_k$, denoted by the symbol $\mathbf{sp}[\mathbf{X}_1, \mathbf{X}_2, \ldots, \mathbf{X}_k]$, is precisely the set of *all* possible linear combinations formed by the vectors $X_1, X_2, \ldots, X_k$ (see Definition I.6.1), that is, $Y \in \mathrm{sp}[X_1, X_2, \ldots, X_k]$ if and only if $Y = \alpha_1 X_1 + \alpha_2 X_2 + \cdots + \alpha_k X_k$ for some scalars $\alpha_1, \alpha_2, \ldots, \alpha_k$.

We have defined the symbol $\mathrm{sp}[X_1, X_2, \ldots, X_k]$ to denote the set of vectors spanned by (created from) the vectors $X_1, X_2, \ldots, X_k$. Observe that with this notation, a (nonempty) set of nonzero concrete vectors $\{X_1, X_2, \ldots, X_k\}$ is a *basis* for V_n if and only if the set $\{X_1, X_2, \ldots, X_k\}$ is linearly independent and $\mathrm{sp}\,[X_1, X_2, \ldots, X_k] = V_n$.

The question arises: For any nonempty finite set of (possibly zero) vectors $\{X_1, X_2, \ldots, X_k\}$ is the set $\mathrm{sp}[X_1, X_2, \ldots, X_k]$ also a sup*space*? Motivated by Theorem I.7.2, we ask;

(a) If Y_1 and Y_2 are in $\mathrm{sp}[X_1, X_2, \ldots, X_k]$, is $Y_1 + Y_2$ also in $\mathrm{sp}[X_1, X_2, \ldots, X_k]$?
(b) If Y is in $\mathrm{sp}[X_1, X_2, \ldots, X_k]$ is αY also in $\mathrm{sp}[X_1, X_2, \ldots, X_k]$ for all scalars α? The answer is given in Proposition I.7.4.

Proposition I.7.4 Given the nonempty set of vectors $\{X_1, X_2, \ldots, X_k\}$ in V_n; then $\mathrm{sp}[X_1, X_2, \ldots, X_k]$ is always a subspace of V_n.

Proof: Exercise I.7.6.

Since we have introduced the notation of the linear span of a finite, nonempty ordered set of vectors, we set about to use it. We will show in the next theorem that subspaces of V_n, very like the mother space V_n, also have bases. Thus, from Corollary I.6.10 (which says that any 2 bases have the same number of elements) and Definition I.6.13 (defining "dimension" to be the number of vectors forming a basis), we can speak of the *dimension* of subspaces. We will see that the dimension of any subspace is less than or equal to n, the dimension of the mother space V_n. Formally, we state Theorem I.7.5.

Theorem I.7.5 Any subspace M, of V_n has a basis. Moreover, the number of nonzero vectors in this basis (that is, the dimension of M) is no greater than n, the dimension of V_n.

Proof: There is the possibility that $M = \{O\}$, the trivial subspace which has the vector O as its only element. In this case, the theorem holds, since M is zero-dimensional (see Definition I.6.13).

In the event that M is nontrivial, there exists a nonzero vector X_1, say, which belongs to M. Now sp$[X_1]$ is a 1-dimensional subspace of M. Either sp$[X_1]$ *is* equal to M or it *isn't* equal to M (surely, this is no surprise). If sp$[X_1]$ *is* equal to M, then we are through; that is, M is a 1-dimensional subspace of V_n. However, if sp$[X_1]$ is *not* equal to M, then there exists a nonzero vector, call it X_2, such that X_2 is in M, but is *not* in sp$[X_1]$; that is, $X_2 \in M$ and $X_2 \notin$ sp$[X_1]$. Now observe that $\{X_1,X_2\}$ is a linearly independent set (see Exercise I.7.7). Consider the subspace sp$[X_1,X_2]$, that is, the subspace spanned by the linearly independent set $\{X_1,X_2\}$. Again, as before, sp$[X_1,X_2]$ is a subspace of M. Either sp$[X_1,X_2]$ *is* equal to M or it *is not* equal to M. If sp$[X_1,X_2]$ *is* equal to M, then we are done; that is, M is a 2-dimensional subspace having the set $\{X_1,X_2\}$ as a basis. On the other hand, if sp$[X_1,X_2]$ is *not* equal to M, there exists a third nonzero vector, X_3, such that X_3 is an element of M, yet *not* an element of sp$[X_1,X_2]$. We consider the (3-dimensional) subspace of M, sp$[X_1,X_2,X_3]$. Bear in mind that the set $\{X_1,X_2,X_3\}$ is linearly independent.

We continue this process of successively choosing nonzero vectors X_k which are in M, yet *not* in sp$[X_1,X_2,\ldots,X_{k-1}]$. Throughout, the set $\{X_1,X_2,\ldots,X_{k-1},X_k\}$ is linearly independent, and hence a basis for sp$[X_1,X_2,\ldots,X_k]$. The fact that this selection process *always* leads to a *linearly independent* set $\{X_1,X_2,\ldots,X_k\}$ really saves the day for us. You see, we *already* know that *no* linearly independent subset of V_n (regardless of how the vectors are chosen) can have more than n vectors (see Theorem I.6.9, Corollary I.6.10). This happy fact forces us to halt in our selection of the X_k's *before* $n + 1$ vectors are chosen (since linear independence of the set $\{X_1,X_2,\ldots,X_k\}$ *must* prevail). Finally, we observe that the halting of our selections of the X_k's must be caused by the fact that for some k (no greater than n), $\text{sp}[X_1,X_2,\ldots,X_k]$ is actually *equal* to M. If it *weren't* equal to M, we would then select an X_{k+1}, which would be in M and not in $\text{sp}[X_1,\ldots,X_k]$, leading to the fact that the $k+1$ element set $\{X_1,X_2,\ldots,X_k,X_{k+1}\}$ is linearly independent. This is impossible (Theorem I.6.9, Corollary I.6.10). This proves the theorem that *any* subspace M of V_n has a basis consisting of no more than n nonzero vectors.

EXERCISES

I.7.6 Prove Proposition I.7.4.

I.7.7 Consider the subspace $\text{sp}[X_1,X_2,\ldots,X_k]$ of V_n where $\{X_1,X_2,\ldots,X_k\}$ is a set of nonzero vectors. Suppose there exists a nonzero vector $Y \in V_n$ such that $Y \notin \text{sp}[X_1,X_2,\ldots,X_k]$. Prove that the set $\{X_1,X_2,\ldots,X_k,Y\}$ is linearly independent in V_n.

I.7.8 Let $M \subset V_2$ be the set of all vectors of the form $\begin{pmatrix} x \\ x \end{pmatrix}$. Prove that M is a subspace of V_2. Find a basis (hence the dimension) of M.

I.7.9 Let $M \subset V_n$ be the set of all vectors of the form

$$\begin{pmatrix} x_1 \\ x_1 \\ x_3 \\ \cdot \\ \cdot \\ \cdot \\ x_n \end{pmatrix},$$

that is, the first and second coordinates are equal.
Prove that M is a subspace of V_n and find a basis of M.

I.7.10 Prove that any $n+1$ nonzero vectors $X_1, X_2, \ldots, X_{n+1}$ in V_n form a linearly dependent set. [*Hint:* Consider the subspace $\text{sp}[X_1,X_2,\ldots,X_{n+1}]$ of V_n, and its dimension relation to the dimension of V_n. Use Theorem I.7.5.]

I.7.11 Show that the only n-dimensional subspace of V_n is V_n itself. [*Hint:* Suppose there is an n-dimensional subspace of V_n which is not equal to V_n. See Exercise I.6.35 for a lead to the inevitable contradiction to Theorem I.7.5. See Exercise I.7.7 also.]

The remainder of this section will develop a relationship between the dimensions of pairs of subspaces M_1 and M_2. To this end, we introduce the symbol $M_1 \dotplus M_2$, to indicate a "sum" of subspaces; more exactly, we offer Definition I.7.6.

Definition I.7.6 Given the subspaces M_1, and M_2 in V_n, that is, $M_1 \subset V_n$ and $M_2 \dotplus V_n$. The symbol $\boldsymbol{M_1 \dotplus M_2}$ will denote the set of *all* vectors of the form $X + Y$, where $X \in M_1$, and $Y \in M_2$; $M_1 \dotplus M_2$ is called the **sum of subspaces** M_1 and M_2.

Example In the vector-space V_3, let M_1 be the subspace of vectors whose second and third coordinates (entries) are zero, that is,

$$X \in M_1 \quad \text{if and only if} \quad X = \begin{pmatrix} x \\ 0 \\ 0 \end{pmatrix},$$

where x is a scalar. Let M_2 be the subspace of vectors whose first and third coordinates (entries) are zero, that is,

$$Y \in M_2 \quad \text{if and only if} \quad Y = \begin{pmatrix} 0 \\ y \\ 0 \end{pmatrix},$$

where y is a scalar. $M_1 \dotplus M_2$ is then the set of all vectors of the form

$$\begin{pmatrix} x \\ 0 \\ 0 \end{pmatrix} + \begin{pmatrix} 0 \\ y \\ 0 \end{pmatrix} = \begin{pmatrix} x \\ y \\ 0 \end{pmatrix}.$$

That is, $M_1 \dotplus M_2$ is the sub*space* of vectors whose third coordinate is zero. Of course there are vectors of $M_1 \dotplus M_2$ whose first or second coordinate is zero, but we observe that it is only the third coordinate which must be zero for *every* vector in $M_1 \dotplus M_2$.

Observe, that Definition I.7.6 does not say whether or not the *set* $M_1 \dotplus M_2$, the sum of subspaces, is again a sub*space*. We settle this question now with Theorem I.7.7.

Theorem I.7.7 Suppose M_1 and M_2 are subspaces of V_n. The set $M \dotplus M_2$ is also a subspace of V_n.

Proof: According to Theorem I.7.2, we have only 2 conditions to verify. Condition (b) says that scalar multiples of vectors in $M_1 \dotplus M_2$ must still be in $M_1 \dotplus M_2$ if $M_1 \dotplus M_2$ is to qualify as a subspace. But for any scalar α and any vector $X + Y \in M_1 \dotplus M_2$, $\alpha(X + Y) = (\alpha X) + (\alpha Y)$. But αX and αY are bonafide members of M_1 and M_2, respectively, since M_1 and M_2 are subspaces and satsify condition (a). Hence $\alpha X + \alpha Y$ is an element of $M_1 \dotplus M_2$, since it is the sum of elements from M_1 and M_2.

As for verifying condition (a) of Theorem I.7.2, let us suppose that $(X + Y)$ and $(U + V)$ are 2 typical vectors in $M_1 \dotplus M_2$, where X and U are vectors in M_1, and Y and V are vectors in M_2. Can we say that $(X + Y) + (U + V)$ is a bonafide element of $M_1 \dotplus M_2$, that is, the sum of a vector in M_1 with a vector in M_2? Note that

$$(X + Y) + (U + V) = (X + U) + (Y + V).$$

In establishing this equality we have actually used the properties of associativity and commutivity of addition. On the right-hand side we see that the vector $(X + U) \in M_1$ and the vector $(Y + V) \in M_2$, since M_1 and M_2 are subspaces and hence obey condition (b). Thus the sum of any 2 vectors from $M_1 \dotplus M_2$ is again in $M_1 \dotplus M_2$, establishing (b) and proving the theorem.

We have just shown that given 2 subspaces, M_1 and M_2, we can construct a new subspace, denoted by $M_1 \dotplus M_2$. Another subspace will now be constructed from M_1 and M_2, namely, $M_1 \cap M_2$, the set-theoretic intersection of M_1 and M_2 (see Definition 8). Again, we must *verify* the assertion that $M_1 \cap M_2$ is a *subspace*. Hence, we state Theorem I.7.8.

Theorem I.7.8 If M_1 and M_2 are subspaces of V_n, then $M_1 \cap M_2$ is also a subspace of V_n.

Proof: We simply verify the fact that conditions (a) and (b) of Theorem I.7.2 obtain for the set $M_1 \cap M_2$.

Now suppose a vector $X \in M_1 \cap M_2$. From the very definition of intersection (Definition 8), X belongs to *both* the sets M_1 *and* M_2. Since M_1 and M_2 are each *subspaces*, and so αX must belong to both M_1 and M_2 for any scalar α [property (a) of Theorem I.7.2]. The intersection $M_1 \cap M_2$ is defined to be *exactly* those elements which belong to both M_1 and to M_2; αX belongs to both M_1 and to M_2 and hence $\alpha X \in M_1 \cap M_2$ for any scalar α, and any $X \in M_1 \cap M_2$. This proves condition (a) of Theorem I.7.2.

Now let X and Y be 2 vectors of the set $M_1 \cap M_2$, that is, X and Y each belongs to both of the sets M_1 and M_2 (Definition 8). We wish to show that $X + Y$ also belongs to $M_1 \cap M_2$, thus establishing property (b) of Theorem I.7.2.

We know, however, that the vector $X + Y \in M_1$ since X and $Y \in M_1$, and M_1 is a subspace (closed under $+$). We know that $X + Y \in M_2$, since X and $T \in M_2$, and M_2 is a subspace. The previous 2 sentences say: $X + Y$ belongs to both sets, M_1 and M_2. Since $M_1 \cap M_2$ is precisely the set of elements belonging to both M_1 and to M_2, $X + Y \in M_1 \cap M_2$, verifying condition (b) of Theorem I.7.2. We have shown that for any $X \in M_1 \cap M_2$, αX also belongs to $M_1 \cap M_2$ for any scalar α, and for any 2 vectors, X and Y, in $M_1 \cap M_2$, $X \dotplus Y$ is also in $M_1 \cap M_2$. Thus $M_1 \cap M_2$ is a subspace.

From 2 subspaces, M_1 and M_2, we have constructed the sets $M_1 \dotplus M_2$ and $M_1 \cap M_2$ and have shown them to be sub*spaces*, as well. We relate the dimensions of these sets (now verified to be subspaces) by means of Theorem I.7.9.

Theorem I.7.9 If M_1 and M_2 are subspaces of V_n, then

$$\dim (M_1 \dotplus M_2) = \dim (M_1) + \dim (M_2) - \dim (M_1 \cap M_2)$$

where the symbol "dim (M)" stands for "the dimension of the subspace M."

Proof: The theorem is trivially true if one of the M_i's $= \{O\}$, that is, if $M_1 = \{O\}$, then $M_1 \dotplus M_2 = M_2$, $M_1 \cap M_2 = \{O\}$, and the result follows immediately. Hence we will assume from now on, that $M_1 \neq \{O\}$ and $M_2 \neq \{O\}$.

$M_1 \cap M_2$ is a subspace (Theorem I.7.8) and has a basis (Theorem I.7.5) of p nonzero vectors $\{X_1, X_2, \ldots, X_p\}$, say. Now the subspace $(M_1 \cap M_2)$ is a subset of M_1 and the subspace $(M_1 \cap M_2)$ is a subset of M_2 (Exercise 7). Thus the linearly independent set $\{X_1, X_2, \ldots, X_p\}$ extends to a basis of M_1 and also to a basis of M_2 (Theorem I.6.14). This is true even if $M_1 \cap M_2 = \{O\}$. That is, there is a set of $r - p$, nonzero vectors $\{Y_{p+1}, Y_{p+2}, \ldots, Y_r\}$ such that the r vectors in the set $\{X_1, X_2, \ldots, X_p, Y_{p+1}, \ldots, Y_r\}$ form a *basis* for M_1. Similarly, there is a set of $s - p$, nonzero vectors $\{Z_{p+1}, Z_{p+2}, \ldots, Z_s\}$ such that s vectors in the set $\{X_1, X_2, \ldots, X_p, Z_{p+1}, \ldots, Z_s\}$ form a basis for M_2. Note that $\dim (M_1 \cap M_2) = p$, $\dim (M_1) = r$ and $\dim (M_2) = s$. [It is possible that $p = 0$ $(M_1 \cap M_2 = \{O\})$].

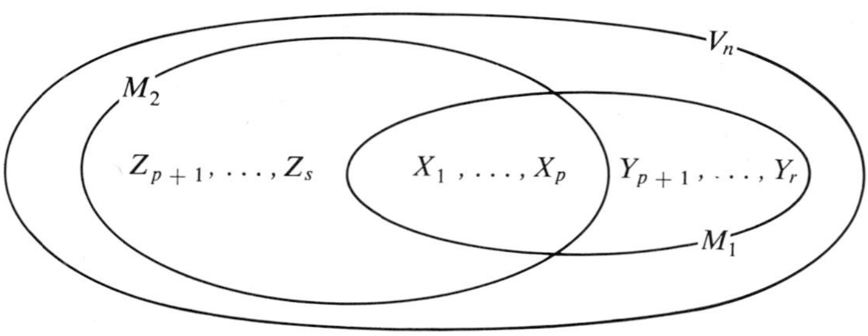

Figure 1.24

Now the set $M_1 \dotplus M_2$ is simply the set of sums of pairs of vectors, one from M_1 and one from M_2 (Definition I.7.6). Since any vector in M_1 can be obtained (by linear combination) from the set $\{X_1, X_2, \ldots, X_p, Y_{p+1}, \ldots, Y_r\}$, and any vector in M_2 can be obtained (by linear combination) from the set $\{X_1, X_2, \ldots, X_p, Z_{p+1}, \ldots, Z_s\}$, we see that any vector in $M_1 \dotplus M_2$ can be obtained (by linear combination) from the set

(a) $$\{X_1, X_2, \ldots, X_p, Y_{p+1}, Y_{p+2}, \ldots, Y_r, Z_{p+1}, Z_{p+2}, \ldots, Z_s\}.$$

In other words, this set (a) *spans* the subspace $M_1 \dotplus M_2$. (If $M_1 \cap M_2 = \{O\}$, no X's would appear in (a).) If we can show this set to be linearly independent as well, we will have a *basis* for $M_1 + M_2$ (Definition I.6.7) on our hands.

We demonstrate the linear independence property. Suppose we have scalars α_i's, β_j's, and γ_k's, such that

$$\text{(b)}\quad (\alpha_1 X_1 + \cdots + \alpha_p X_p) + (\beta_{p+1} Y_{p+1} + \cdots + \beta_r Y_r) + (\gamma_{p+1} Z_{p+1} + \cdots + \gamma_s Z_s) = 0.$$

We will show that *necessarily*, each of the α's, β's, and γ's must be equal to zero (Theorem I.6.6). As Figure 1.24 indicates (but does not prove), the set of vectors $\{Z_{p+1}, Z_{p+2}, \ldots, Z_s\}$ is contained in M_1, but *not* in M_2. This important fact tells us that there is no nonzero vector which belongs to *both* vector spaces $\text{sp}[X_1, X_2, \ldots, X_p, Y_{p+1}, \ldots, Y_r]$ and $\text{sp}[Z_{p+1}, Z_{p+2}, \ldots, Z_s]$. Now we refer to our Equation (b), which is rewritten as follows:

$$\text{(c)}\quad (\alpha_1 X_1 + \cdots + \alpha_p X_p) + (\beta_{p+1} Y_{p+1} + \cdots + \beta_r Y_r) = (-\gamma_{p+1} Z_{p+1} - \cdots - \gamma_s Z_s).$$

The left-hand side is a vector belonging to M_1 (since the X_i's and Y_j's form a basis for M_1), while the right-hand side is a vector in M_2 but *not* in M_1 if the γ_k's are nonzero. The equality of the 2 sides says, therefore, that the right-hand side of (c) must be the zero vector O. But the sets $\{X_1, X_2, \ldots, X_p, Y_{p+1}, \ldots, Y_r\}$ and $\{Z_{p+1}, Z_{p+2}, \ldots, Z_s\}$ are each linearly independent; thus, the *only* way we can generate the zero vector from each of these sets (by linear combination) is to choose each scalar equal to zero, that is,

$$\alpha_1 = \alpha_2 = \cdots = \alpha_p = \beta_{p+1} = \beta_{p+2} = \cdots = \beta_r = \gamma_{p+1} = \gamma_{p+2} = \cdots = \gamma_s = 0.$$

We have shown that the set

$$\{X_1, X_2, \ldots, X_p, Y_{p+1}, Y_{p+2}, \ldots, Y_r, Z_{p+1}, Z_{p+2}, \ldots, Z_s\}$$

is a spanning, linearly independent set in $M \dotplus M_2$, that is, a *basis* of $M_1 + M_2$. In counting the number of vectors in this set, we arrive at

$$\dim (M_1 \dotplus M_2).$$

Now there are r vectors in the set $\{X_1, X_2, \ldots, X_p, Y_{p+1}, \ldots, Y_r\}$ and $s - p$ vectors in the set $\{Z_{p+1}, Z_{p+2}, \ldots, Z_s\}$. Hence,

$$\dim (M_1 + M_2) = r + (s - p) = \dim (M_1) + \dim (M_2) - \dim (M_1 \cap M_2),$$

proving the theorem.

There are some interesting consequences of this equality which we illustrate in the following examples.

Example Consider V_3 to be the vector space of ordered *triples* of real numbers

$$\begin{pmatrix} x \\ y \\ z \end{pmatrix}.$$

As indicated previously, we may consider these triples to be identical with points in 3 space, that is, V_3 "is" 3-space. We will classify all subspaces of V_3 which have dimension 0, 1, 2, and 3 respectively. Since V_3 is itself a 3-dimensional space, any subspace of V_3 can have dimension no *more* than 3 (Theorem I.7.5). The zero-dimensional subspace is unique, by definition, and consists only of the zero vector

$$\{O\} = \left\{ \begin{pmatrix} 0 \\ 0 \\ 0 \end{pmatrix} \right\},$$

the origin. One-dimensional subspaces are of the form sp$[X_0]$, where X_0 is a nonzero vector in 3 space, that is, a typical 1-dimensional space consists of *all* vectors of the form αX_0, where α runs over *all* real numbers. A 1-dimensional subspace, then, is the set of all scalar multiples of a fixed nonzero vector

$$X_0 = \begin{pmatrix} x_0 \\ y_0 \\ z_0 \end{pmatrix}$$

in 3 space. We may geometrically represent this subspace as an infinite line or ray passing through the origin and the point (x_0, y_0, z_0), as indicated by Figure 1.25.

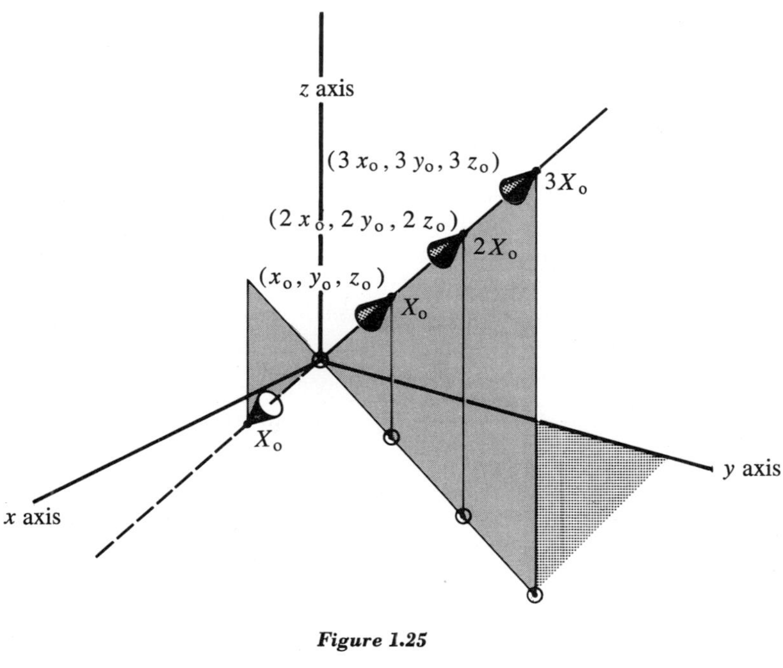

Figure 1.25

What do 2-dimensional subspaces look like? Any 2-dimensional subspace M has a basis in V_3, $\{X_1,X_2\}$, say, that is, $M = \text{sp}[X_1,X_2]$ and $\{X_1, X_2\}$ is a linearly independent set. Now how are 2 vectors related (geometrically) so as to be linearly independent? This is easily answered by observing that 2 nonzero vectors, X_1 and X_2, are linearly dependent if and only if for some nonzero scalars α and β, we have $\alpha X_1 + \beta X_2 = O$. This is equivalent to saying $X_1 = -(\beta/\alpha)X_2$. Thus, 2 nonzero vectors form a linearly *dependent* set if and only if 1 vector is simply a scalar multiple of the other. This means that they lie on the same ray or line. This phenomenon is given a label, namely, 2 nonzero vectors are linearly dependent if and only if they are colinear (lie on the same ray; belong to the same 1-dimensional subspace). Thus, 2 nonzero vectors are linearly *independent* if and only if they are *not* colinear (one is not simply a scalar multiple of the other). Thus, the 2-dimensional spaces $M = \text{sp}[X_1,X_2]$ are represented by the *planes* containing the 3 points (vectors), O, X_1, and X_2. (See Figure 1.26.)

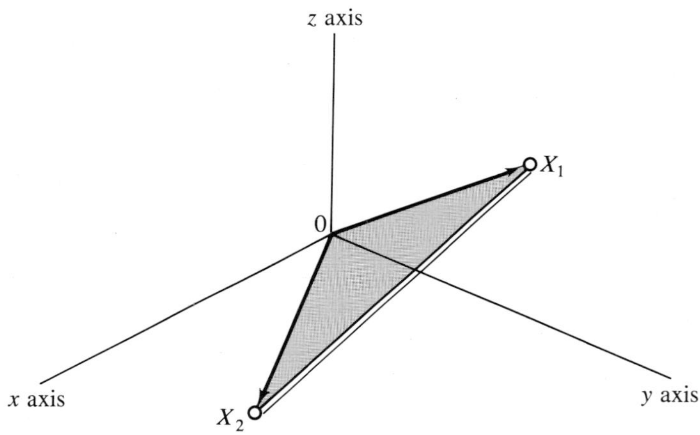

Figure 1.26

There is only one 3-dimensional subspace, V_3 itself. This question has been approached more generally in Exercise I.7.11, but we can see it here as well. If M is a 3-dimensional, then it has for its basis $\{X_1,X_2,X_3\}$. If M were *not* equal to V_3, we would be able to find a vector X_4, say, in V_3 and *not* in M. Thus, $\{X_1,X_2,X_3,X_4\}$ would be a linearly independent set of V_3, which is impossible since any 4 vectors of V_3 are necessarily linearly dependent (Exercise I.7.10). We conclude that there is no such X_4 and that $M = \text{sp}[X_1,X_2,X_3] = V_3$.

We now have a complete geometric interpretation for subspaces of V_3, namely: If $\dim(M) = O$, then $M = O$; if $\dim(M) = 1$, then M is a ray through O; if $\dim(M) = 2$, then M is a plane through O; if $\dim(M) = 3$, then $M = V_3$ itself.

We pose the following problem, and answer it in 2 ways: Suppose M_1 is an infinite ray passing through O (a 1-dimensional subspace of V_3) and M_2 is a plane passing through O (a 2-dimensional subspace of V_3). If no point of the ray M_1 (other than O) lies in the plane M_2, then show that the space $M_1 \dotplus M_2 = V_3$.

Solution I: Let X_1 be a nonzero vector of M_1. Then $M_1 = \text{sp}[X_1]$. Let $\{X_2,X_3\}$ be a basis for M_2. Since X_1 is not in M_2 (not in $\text{sp}[X_2,X_3]$), the set $\{X_1,X_2,X_3\}$ is LI in V_3, and hence $\text{sp}[X_1,X_2,X_3]$, having dimension 3, must be equal to V_3. (See Figure 1.27.)

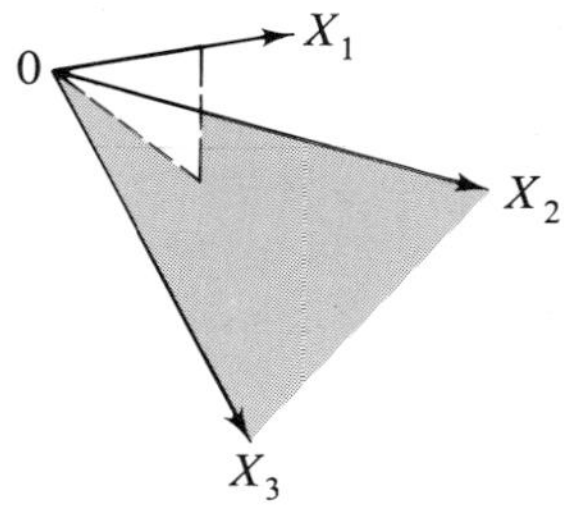

Figure 1.27

Solution II: We are given $\dim(M_1) = 1, \dim(M_2) = 2$, and by hypothesis, $\dim(M_1 \cap M_2) = \dim(O) = 0$. Hence $\dim(M_1 \dotplus M_2) = \dim(M_1) + \dim(M_2) - \dim(M_1 \cap M_2) = 1 + 2 - 0 = 3$. Thus, $M_1 \dotplus M_2 = V_3$, since V_3 is the only 3-dimensional subspace of itself (see Exercise I.7.11).

This example describes the space V_3 as the sum of 2 subspaces (a ray and a plane through O), where the spaces have only the zero vector in common. We define this phenomenon in general in Definition I.7.10.

Definition I.7.10 Suppose M_1 and M_2 are subspaces of V_n. Suppose the subspace $M = M_1 \dotplus M_2$ and $M_1 \cap M_2 = O$. In this event, we say M is the **direct sum of the subspaces** M_1 and M_2; we symbolize the fact (that $M = M_1 \dotplus M_2$ and $M_1 \cap M_2 = O$) by writing $M = M_1 \oplus M_2$.

The previous example illustrated that $V_3 = M_1 \oplus M_2$; where M_1 is a ray not lying in the plane M_2, that is, $M_1 \cap M_2 = O$ and $V_3 = M_1 \dotplus M_2$.

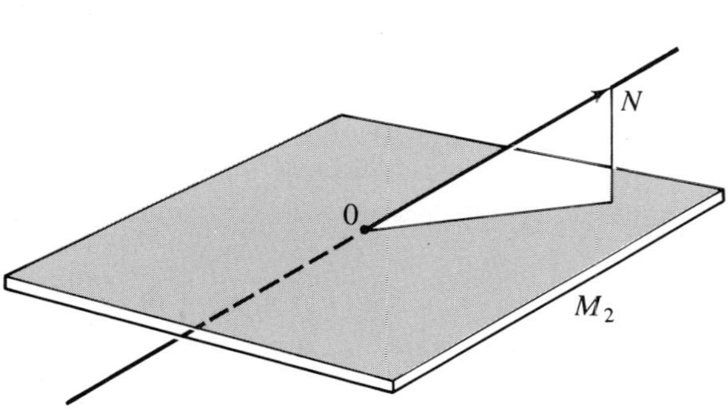

Figure 1.28

We observe from this that given the plane M_2, *any* ray, say N, not lying in M_2, we would have $V_3 = M_2 \oplus N$, and, hence, N is *not unique.* (See Figure 1.28.)

Example Consider the 2-dimensional real plane V_2. Let M_1 and M_2 represent 1-dimensional subspaces such that $M_1 \cap M_2 = \{O\}$. Geometrically, M_1 and M_2 are infinite rays that are distinct, that is, they cross only at the origin (Figure 1.29).

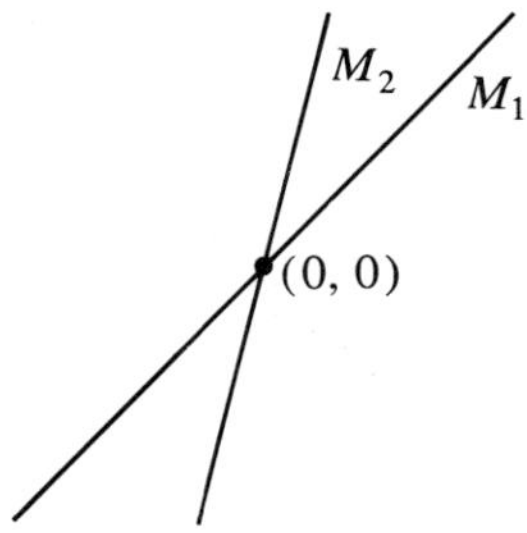

Figure 1.29

Thus, $V_2 = M_1 \dotplus M_2$, since

$$\begin{aligned}\dim(M_1 + M_2) &= \dim(M_1) + \dim(M_2) - \dim(M_1 \cap M_2)\\ &= 1 + 1 - 0\\ &= 2.\end{aligned}$$

The only 2-dimensional subspace of V_2 is V_2 itself. Hence $V_2 = M_1 \dotplus M_2$ and $M_1 \cap M_2 = 0$. In other words, $V_2 = M_1 \oplus M_2$.

As a special case, M_1 could represent the x-axis and M_2 could represent the y-axis. Thus $M_1 = \text{sp}\left[\begin{pmatrix}1\\0\end{pmatrix}\right]$ and $M_2 = \text{sp}\left[\begin{pmatrix}0\\1\end{pmatrix}\right]$. A well-known fact emerges:

$$V_2 = \text{sp}\left[\begin{pmatrix}1\\0\end{pmatrix}, \begin{pmatrix}0\\1\end{pmatrix}\right] = M_1 \dotplus M_2$$

(see Definition I.7.10). That is, the plane is spanned by the standard basis vectors $\begin{pmatrix}1\\0\end{pmatrix}$ and $\begin{pmatrix}0\\1\end{pmatrix}$.

We conclude by stating an extension of Definition I.7.10.

Definition I.7.11 Suppose $M_1, M_2, \ldots, M_k$ are k subspaces of V_n. Let M_i' denote the subspace spanned by all vectors of the form $x_1 + x_2 + \cdots + x_{i-1} + x_{i+1} + \cdots + x_k$, where each $x_j \in M_j$, $j \neq i$. That is, M_i' is the set of all $(k-1)$-length sums whose terms come from each M_j except for $M_j = M_i$; hence, $M_i' = M_1 \dotplus \cdots \dotplus M_{i-1} \dotplus M_{i+1} \dotplus \cdots \dotplus M_k$. If for *each* i, $i = 1, 2, \ldots, k$, we have $M_i \cap M_i' = \{O\}$, then $M = M_1 \dotplus M_2 \dotplus \cdots \dotplus M_k$ will be called the **direct sum** of the subspaces $M_1, M_2, \ldots, M_k$ and will be symbolized by

$$M = M_1 \oplus M_2 \oplus \cdots \oplus M_k.$$

EXERCISES

I.7.12 (a) We are given (Definition I.6.3) the conditions for a set of *nonzero* vectors $\{X_1,X_2,\ldots,X_k\}$ to be linearly independent. Suppose the adjective "nonzero" were eliminated from this definition. What effect would this have on the truth of Theorems I.6.8, I.6.9, and I.6.12?

(b) Suppose the adjective "nonzero" were amended to the vectors in Definition I.7.3. Would Proposition I.7.4 and Theorem I.7.5 still obtain? With this modified definition of a spanning set, would it be possible to find a set whose span is $\{O\}$, the vector-space whose only element is the zero vector O?

Given M_1 and M_2 in Exercises I.7.13 through I.7.18, describe the vectors in the subspaces $M_1 \dotplus M_2$ and $M_1 \cap M_2$.

For which exercises does $M_1 \dotplus M_2 = M_1 \oplus M_2$?

I.7.13 M_1 is the set of all vectors of the form

$$\begin{pmatrix} x \\ x \\ 0 \\ 0 \end{pmatrix}$$

and M_2 is the set of all vectors

$$\begin{pmatrix} y \\ 0 \\ 0 \\ 0 \end{pmatrix};$$

x, y are real numbers.

I.7.14 M_1 is the set of all vectors of the form

$$\begin{pmatrix} x \\ x \\ 0 \\ 0 \end{pmatrix},$$

and M_2 is the set of all vectors of the form

$$\begin{pmatrix} 0 \\ 0 \\ y \\ z \end{pmatrix},$$

where x, y, and z are real numbers.

I.7.15 M_1 is the set of all vectors of the form

$$\begin{pmatrix} x \\ y \\ z \end{pmatrix},$$

and M_2 is the set of all vectors of the form

$$\begin{pmatrix} 0 \\ 0 \\ w \end{pmatrix},$$

where, x, y, z, and w are real numbers.

I.7.16 M_1 is the set of all vectors of the form

$$\begin{pmatrix} x \\ x \\ x \end{pmatrix},$$

and M_2 is the set of all vectors of the form

$$\begin{pmatrix} 0 \\ y \\ z \end{pmatrix},$$

where x, y, and z are real numbers.

I.7.17 M_1 is the set

$$\left\{\begin{pmatrix} 0 \\ 0 \\ 0 \end{pmatrix}\right\},$$

and M_2 is the set of all vectors of the form

$$\begin{pmatrix} x \\ 0 \\ 0 \end{pmatrix},$$

x is a real number.

I.7.18 M_1 is the set of all vectors of the form

$$\begin{pmatrix} x \\ y \\ 0 \end{pmatrix},$$

and M_2 is the set of all vectors of the form

$$\begin{pmatrix} 0 \\ z \\ w \end{pmatrix},$$

where x, y, z, and w are real numbers.

In Exercises I.7.19 through I.7.25, verify Theorem I.7.9 by computing the dimensions of M_1, M_2, $M_1 + M_2$, and $M_1 \cap M_2$.

I.7.19 Choose M_1, M_2 as in Exercise I.7.12.

I.7.20 Choose M_1, M_2 as in Exercise I.7.13.

I.7.21 Choose M_1, M_2 as in Exercise I.7.14.

I.7.22 Choose M_1, M_2 as in Exercise I.7.15.

I.7.23 Choose M_1, M_2 as in Exercise I.7.16.

I.7.24 Choose M_1, M_2 as in Exercise I.7.17.

I.7.25 Choose M_1, M_2 as in Exercise I.7.18.

I.7.26 Consider a set of nonzero vectors $\mathscr{X} = \{X_1, X_2, \ldots, X_k\}$ in V_n. There are 3 properties of such sets that we will relate in this exercise. Just for fun, we will introduce a symbol to represent each of these properties, to wit:

(1) **sp** will denote the spanning property of $\mathscr{X}$, that is, sp obtains if and only if $\text{sp}[X_1, X_2, \ldots, X_k] = V_n$.

(2) **LI** will denote the linear independence of $\mathscr{X}$, that is, LI obtains if and only if $\{X_1, X_2, \ldots, X_k\}$ is linearly independent in V_n.

(3) ***n*-vctrs** will denote the fact that $\mathscr{X}$ has n-vectors, that is, $k = n$.

Now Theorem I.6.9 says, of such sets $\mathscr{X}$, that any set which spans V_n (sp obtains) and is linearly independent (LI obtains), necessarily has n-vectors (n-vctrs obtains). In our terminology, then, Theorem I.6.9 may be expressed by

(A) sp and LI $\Rightarrow$ n-vctrs, where the symbol "$\Rightarrow$" is a shorthand for the word "implies."

Question: (a) Is it true that the statement to be proved in Exercise I.6.35 may be expressed by

(B) sp and n-vctrs $\Rightarrow$ LI?

(b) From Question (a), prove that for any spanning set $\mathscr{X}$, linear independence is equivalent to (implies and is implied by) the fact that $\mathscr{X}$ has n vectors.

(c) Prove the following implication for a set $\mathscr{X}$.

(C) LI and n-vctrs $\Rightarrow$ sp.

[*Hint:* Suppose LI and n-vctrs obtains for X, and that $\mathscr{X} = \{X_1, X_2, \ldots, X_n\}$ does *not* span V_n. Select a nonzero vector, Y say, in V_n which is not an element of $\text{sp}[X_1, X_2, \ldots, X_n]$. Show that the set $\{X_1, X_2, \ldots, X_n, Y\}$ is linearly independent. Consider the dimension of $\text{sp}[X_1, X_2, \ldots, X_n, Y]$ and compare with Theorem I.7.5.]

I.7.27 Exercise I.7.26 shows that if we are given any 2 of the properties sp, LI or n-vctrs for a set X in V_n, then the third property is automatically implied. Use this fact to show the following equivalent statement: Given any 1 of these properties for a set X in V_n, then the other 2 properties are equivalent to each other. For example, if we know that a set X in V_n has n vectors, then it will also be linearly independent if and only if it spans V_n. In our symbolism, given n-vctrs obtains for X. Then LI $\Rightarrow$ sp and sp $\Rightarrow$ LI for X.

I.7.28 Prove that if Y_n is the set of all row vectors $(x_1, x_2, \ldots, x_n)$, then $Y_n = M_1 \oplus \cdots \oplus M_n$, where M_i is the subspace (of dimension 1) spanned by the vector $E_i = (0, \ldots, 0, 1, 0, \ldots, 0)$; E_i is the vector with zeros everywhere except for a 1 in the ith entry.

Chapter I Summary

The objective of Section 1 was to introduce the idea of structure on a set. For example, we considered concrete n vectors, and devised a means (called addition) of combining 2 vectors to generate or produce a third, new vector (Definition I.1.2). Another means for producing 1 vector from another vector and a scalar was also introduced (Definition I.1.3). Section 2 continued this idea with the notion of product. In Definition I.2.1, 2 vectors were multiplied, producing not a vector, but a scalar. In this case, the operation of multiplication between vectors is not *closed*; that is, it does not result in another *vector*. Matrix multiplication was presented next; this time, a matrix times a matrix produced a matrix (Definition I.2.3). Section 3 exhibits a technique for solving systems of equations for an ordered set of scalars, called a solution (more detailed analysis of such systems follows in Chapter III, Section 3). These results, in turn, are applied in the following Section 4 on linear programming, where the notions of linear functionals and convex sets are introduced. Finally, these particular instances of sets with structure (pairwise addition, and so on) inspire a list of 10 test conditions which comprise the very definition of an abstract vector-space over a scalar field (Definition I.5.2). Section 6, therefore, is devoted to an analysis of vector-spaces of finite dimension. This section develops the notions of spanning sets, linear independence, bases, and dimension. In Section 7, we refine our analysis of vector-spaces by considering subsets of vector-spaces, which are vector-spaces in their own right. It is shown that new subspaces can be created from old subspaces, M and N, by means of a super addition, that is, for subspaces M and N, $M + N$ is also a subspace (Definition I.7.6, Theorem I.7.7). Set theoretic intersection generates new subspaces; for subspaces M and N, $M \cap N$ is always a subspace (Theorem I.7.8). Finally, the dimension of the 4 subspaces M, N, $M + N$, $M \cap N$ are all related in Theorem I.7.9 which says that $\dim (M + N) = \dim M + \dim N - \dim (M \cap N)$. After some consequences of this equation, Chapter I concludes.

CHAPTER II

Linear Transformations and Their Matrices

1. *Functions*

Let D and R be any not necessarily distinct nonempty sets. Intuitively, a function f, from the set D to the set R, may be thought of as a rule which associates 1 and only 1 element of the set R with each element of the set D. That is, for every $x \in D$, we associate 1 and only 1 element of R, denoted $f(x)$. Informally, we may suppose the function f to be represented by a machine with an input and an output. Elements of the set D are fed into the machine, transformed, and emitted through the output as elements of the set R. Moreover, for every element $x \in D$ that is fed in, 1 and only 1 element $f(x)$ of R comes out.

The symbol $f(x)$ is read "f of x," "the image of x (under f)," or "f evaluated at x." The input set D is called the *domain* of f, and the output set is called the *range* of f.

Example Suppose f is the function that associates with each real number, its square. That is, for every real number x, we associate with it (or, transform it into) the non-negative number $f(x) = x^2$. The domain D (input set) of f is the set of all real numbers, and the range R of f is the set of all non-negative numbers. Figure 2.1 presents a diagrammatic representation of f.

Moreover, the function transforms every x of the domain, into 1 and only 1 non-negative element of the range. To be sure, an element $f(x)$ of the range may have *come* from more than 1 element of the domain (for example, if $f(x) = x^2$, then 4 is the image of 2 under f, and 4 is the image of -2 under f. That is, $f(2) = 4$ and $f(-2) = 4$). However, the restriction we have placed is that every element of the domain shall have 1 and *only* 1 image in the range, which is not contradicted by the properties of our example.

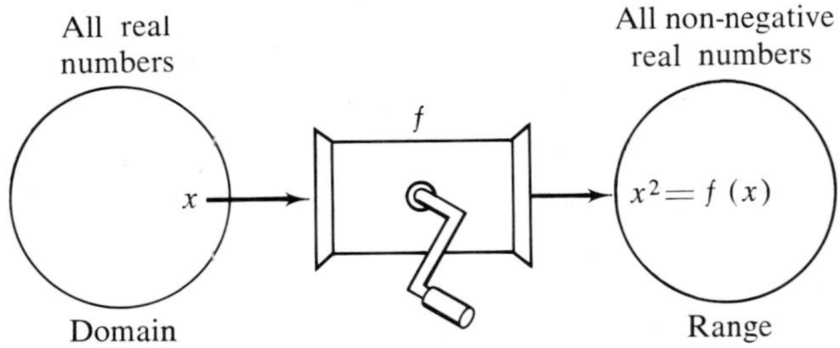

Figure 2.1

Example Let the domain D (input set) be the set of all integers and let the range R (output set) be the set of all odd integers. Suppose f is the function that transforms (maps) each integer into the integer which is 1 more than twice itself; that is, for every $n \in D$, we define $f(n)$ to be $2n + 1$. Figure 2.2 diagrams this function.

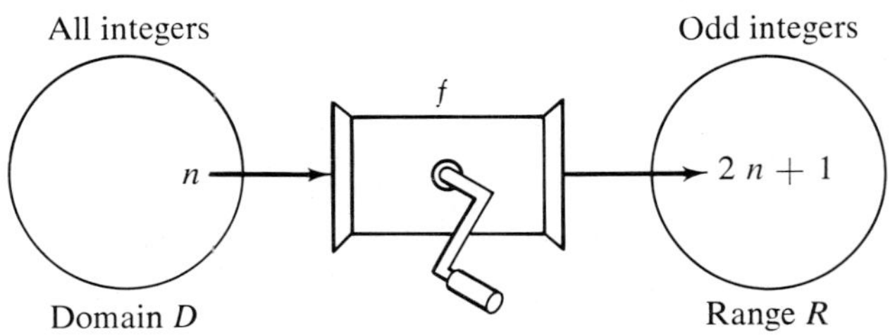

Figure 2.2

From these 2 examples, we are able to discern a certain pattern that will enable us to define a function rigorously. We observe that there is first an "input set" which we call the domain, and secondly there is an "output set" which we call the range. Observing Figures 2.1 and 2.2, we see that the elements are diagrammed in *pairs*; on the left-hand side is an element of the domain, and on the right-hand side is an element of the range, which is the image of the domain element under f. We note that an *ordered pair* of elements is a set of 2 elements, 1 element being identified as the *first* element, and the other element being identified as the *second* element. This leads us to the following definition of function as a certain *set* of *ordered pairs*, with the *first* element of the ordered pair representing the input, and the *second* element of the ordered pair representing the output. More exactly, now, we give Definition II.1.1.

Definition II.1.1 A **function (map, transformation)** f is a set of ordered pairs (x,y) such that if $(x,y) \in f$ and $(x,z) \in f$, then, necessarily, $y = z$. The set **dom** (f), of first elements x, is called the **domain of the function** f, and the set **rg** (f), of second elements y, is called the **range of the function** f. The function f is also denoted by the symbol $f(\cdot)$.

Notation II.1.2 If the ordered pair (x,y) belongs to the function f (which is a certain set of ordered pairs), then y is often denoted $\boldsymbol{f(x)}$, and is read, "f of x," "the image of x under (the function) f," or "f evaluated at x." Thus, the function f is the set of ordered pairs $(x,f(x))$, where $f(x)$ is *uniquely* defined as an element of rg (f) for every element x of dom (f).

Note: There is a difference between the function f, or $f(\cdot)$, (a *set* of ordered pairs) and $f(x)$, a *single element* of rg (f) which is only the image of the particular element x. Crudely put, the dot in the alternate symbol, $f(\cdot)$, for the function f, suggests that it is "not yet evaluated" at the point x; $f(\cdot)$ is not to be confused with $f(x)$.

Example Let $f = \{(1,2),(2,3),(1,4)\}$ be a 3-element set, whose every element is an ordered pair. Is f a function? The answer is "no" since both elements (ordered pairs) $(1,2)$ and $(1,4)$ belong to the set f. This contradicts the definition which states that there is only 1 ordered pair of a function having "1" as its first element. We could diagram this set f (which is not a function) as in Figure 2.3.

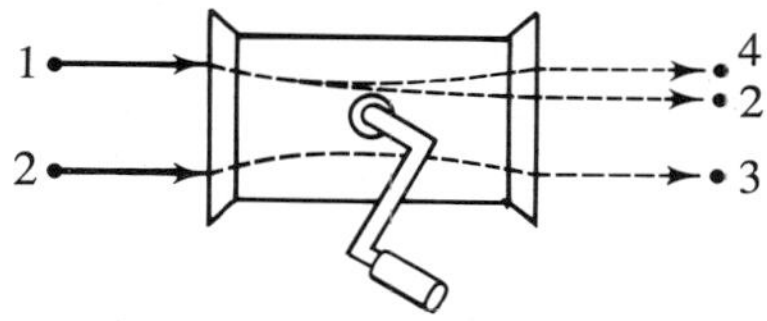

Figure 2.3

Here, we see that the number 1 induces 2 *distinct* outputs, which is a property disallowed by Definition II.1.1.

Example Let $f = \{(1,\frac{1}{2}),(2,\frac{1}{2}),(3,2)\}$. f is a 3-element set which *is* a function. Figure 2.4 shows that for every element x of the domain, there is 1 and only 1 element $f(x)$ in the range. The domain of f is the set of integers $\{1,2,3\}$ and the range of f is the set $\{\frac{1}{2},2\}$.

REMARK. In Chapter I, we described a 2-vector as an ordered pair of numbers (see Definition I.1.1), and the 2-dimensional vector-space as the set of *all* these ordered pairs, whose coordinates (entries) were numbers (see Definition I.1.2). This is not to be confused with a *function*, which could be a set of ordered pairs with numbers as entries. Recall that a function is not just *any* set of ordered pairs; rather, a function is a set of ordered pairs with 1 powerful restriction, namely, that no first element may have more than one second element (image) associated with it. The *vector-space* of ordered pairs includes *every* conceivable ordered pair of numbers *without exception.*

Example Let us construct *all* functions with domain (input set) $\{a,2\}$ and range (output set) $\{M,N\}$.

Now when we say that a set ($\{M,N\}$ in our case) is the range of a function, this

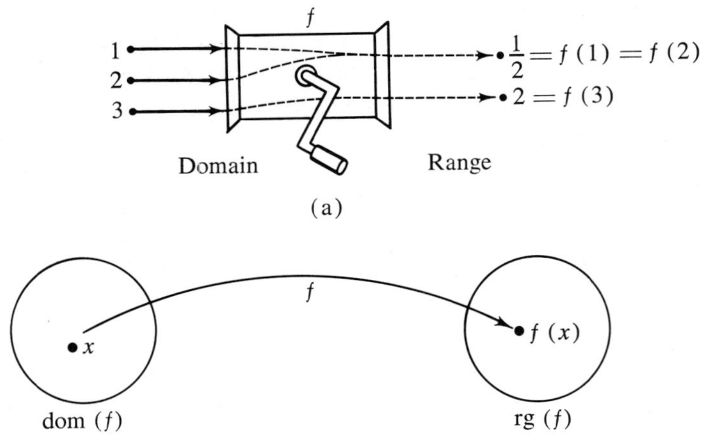

(b)

Figure 2.4

means that each and every element of that set is the image of at least 1 element of the range. Thus, the set

$$F = \{(a,M),(2,M)\}$$

is *not* a function of the desired type; to be sure, F is a function with domain $\{a,2\}$, but the range of this F, $\{M\}$, is "too small" to fulfill the requirements of the first sentence. The set $\{M,N\}$ is what we shall later call an *impact set for the function* F, that is, a set which *contains* the range of F.

To return to the problem at hand, we must assign appropriate output elements to the 2 possible input elements a and 2. We see that if f is a function such that $f(a) = M$, say, then there is only 1 element that we may call $f(2)$, namely, N. That is to say, there happens to be only 1 function f such that $f(a) = M$ (such that the ordered pair $(a,M) \in f$). Similarly, if g is a function such that $g(a) = N$, then g is the only such function, since $g(2)$ *must* be M [if $g(2)$ were set equal to N, then $\{N\}$, the range of g, would be "too small" for our purposes]. Thus, there are only 2 functions that meet the following requirements: The domain is the set $\{a,2\}$ and the range is the *full* set $\{M,N\}$. These functions are just the sets of ordered pairs

$$f = \{(a,M),(2,N)\} \quad (f(a) = M, f(2) = N)$$

$$g = \{(a,N),(2,M)\} \quad (g(a) = N, g(2) = M).$$

EXERCISES

Given the following sets f in Exercises II.1.1 through II.1.5, which of these are functions? For those sets that are functions, describe the domain and range of f.

II.1.1 $f = \{(1,2),(2,3),(3,4)\}$.

II.1.2 f is the set of all ordered pairs of the form $(x,3x - 1)$ where x is an arbitrary real number.

II.1.3 f is as in Exercise II.1.2, except that x is an arbitrary non-negative (possibly zero) real number.

II.1.4 $f = \{(1,1),(2,0),(3,1),(4,0),\ldots\}$. That is, f is the set of ordered pairs of the form (n,ε), where n is a positive integer; $\varepsilon = 0$ if n is even and $\varepsilon = 1$ if n is odd.

II.1.5 $f = \{(3,7)\}$ (f is a set with exactly 1 element).

II.1.6 Suppose $X = \{\text{chair,table,apple}\}$ and $Y = \{3,4\}$. Describe all possible functions which have X as the domain and Y as the range.

II.1.7 Describe all possible functions which have Y as the domain and X as the range, where X and Y are described in Exercise II.1.6.

II.1.8 Describe all possible functions which have X as both the domain and the range, where X is described in Exercise II.1.6.

II.1.9 Construct a set f of ordered pairs (x, y) with the following 2 properties:
(a) the set f is a function
(b) if the first and second entries of each ordered pair (x, y) in the set f are interchanged, the resulting set of (interchanged) ordered pairs is still a function. (This new function is called "f inverse" and is denoted by the symbol f^{-1}.)

II.1.10 Construct a set f of ordered pairs (x, y) with the following properties:
(a) the set f is a function
(b) if the first and second entries of each ordered pair (x, y) in the set f are interchanged, the resulting set is *not* a function.

II.1.11 Let $X = V_3$ and $Y = V_2$. Consider f, the set of ordered pairs (of vectors) of the form

$$\left(\begin{pmatrix} x_1 \\ x_2 \\ x_3 \end{pmatrix}, \begin{pmatrix} x_1 + x_2 \\ x_3 \end{pmatrix}\right), \quad \text{that is, } f\begin{pmatrix} x_1 \\ x_2 \\ x_3 \end{pmatrix} = \begin{pmatrix} x_1 + x_2 \\ x_3 \end{pmatrix},$$

for every vector

$$\begin{pmatrix} x_1 \\ x_2 \\ x_3 \end{pmatrix} \in V_3.$$

Is f a function? If so, what is its domain and range?

We shall soon become interested in sets which are "larger" than just the set of output elements (that is, the range) of certain functions. In naming the sets which accept the range of a function as a subset, we set down Definition II.1.3.

Definition II.1.3 Let f be a function. Then *any* set which contains rg (f), the range of f (set of all "output" elements), is called an **impact set of *f***, or an **impact set for *f***.

Of course, rg $(f) \subset$ rg (f) (see Definition 4) and so rg (f) is an impact set for the function f. In fact, rg (f) is the "smallest" impact set for the function f.

Example Let f be a function whose domain, dom (f), is the set of all integers $Z = 0, \pm 1, \pm 2, \pm 3, \ldots, \pm n, \ldots\}$. The range of f, rg (f), is defined by $f(n) = 3n$ for each integer $n \in Z$. Thus, dom$(f) = Z$, while rg(f) is the set $\{0, \pm 3, \pm 6, \pm 9, \ldots, \pm 3n, \ldots\}$, which we denote by the symbol $3 \cdot Z$. Now rg $(f) = 3 \cdot Z$ is a *subset* of Z, that is, Z is an impact set for f. Also, rg $(f) = 3 \cdot Z$ is a subset of R, the set of *all* real numbers (not just integers); hence, R is another impact set for f.

The notion of impact set leads us, in a natural fashion, to the concept of *into* and *onto* functions; these concepts are always defined relative to certain sets. We present these notions with a bit more precision.

Definition II.1.4 Let f be a function with domain X and impact set Y [that is, $Y \supset \text{rg}(f)$].

Suppose we know that the impact set Y is "bigger" than rg (f) [that is, there is some $y_0 \in Y$ such that no matter which x in X we choose, we always have $f(x) \neq y_0$]. We then use either of the symbols

(a) $$f : X \xrightarrow{\text{into}} Y \quad \text{or} \quad X \xrightarrow[\text{into}]{f} Y,$$

which we read **f sends X into Y**, or **f is an into function from X to Y**.

In the opposite case, suppose we know that the impact set is "no bigger" than rg (f) and is, in fact, equal to rg (f) [that is, for all $y \in Y$, there is always at least one $x \in X$ such that $f(x) = y$]. In this event, we use either of the symbols

(b) $$f : X \xrightarrow{\text{onto}} Y, \quad \text{or} \quad X \xrightarrow[\text{onto}]{f} Y$$

to read **f sends X onto Y**, or **f is an onto function from X to Y**.

In case we only know that f is a function with domain X and impact set Y, but we do not know (or care) whether Y is "bigger" than rg (f) or not, we may simply write

(c) $$f : X \longrightarrow Y, \quad \text{or} \quad X \xrightarrow{f} Y,$$

which we read **f sends X to Y**, or **f is a function from X to Y**.

Notice that (c) is a means of stating (a) or (b) so that a function from X to Y could be either an *onto* function or an *into* function.

Suppose we have a function $f : X \to Y$ [that is, $X = \text{dom}(f)$ and Y is an impact set of f], and another function $g : Y \to Z$. From these 2 functions, we may build a third function; the fact that this third function depends (somehow) on f and g is reflected in the fact that the very symbols "f" and "g" appear in the symbol $g \circ f$ which we use to denote this third function. In order to know exactly what this function $g \circ f$ is, all we need to know is the effect it has upon a typical element x, say, of its domain, that is, we are asking ourselves to define the element $g \circ f(x)$, $g \circ f$ "acting on" x. We do this now.

Definition II.1.5 Let f and g be functions such that $f : X \to Y$ and $g : Y \to Z$ for some fixed sets X, Y, and Z. Then the function $g \circ f : X \to Z$

is defined by the equation

$$g \circ f(x) = g(f(x)) \quad \text{for all} \quad x \in X.$$

That is, the function $g \circ f$ is the set of all ordered pairs $(x, g(f(x)))$, where $x \in X$.

The function $g \circ f$ is called the **composition of g with f**, or the **product of g with f**.

In good conscience, we must verify that the set of ordered pairs $g \circ f$ defined above, is a bonafide function (see Definition II.1.1). To be sure, we have $g \circ f$ defined as the set of *all* ordered pairs $(x, g(f(x)))$, but for each $x \in X$ is there truly only 1 right-hand member of the ordered pair? Thus, we are asking whether for any $x \in X$, there might be 2 answers for the expression $g(f(x))$.

Not a chance.

You see, f and g are, themselves, functions, and therefore enjoy the property of issuing unique images for every element of their respective domains. In fact, for every x, the image (transformation, evaluation) of x, under f, $f(x)$, is unique. Now considering $f(x)$ as an element of the domain (input set) of g, the element $g(f(x))$ is unique in rg (g), the range of g. But the element $g(f(x))$ is, by definition, the image of x under (the function) $g \circ f$, that is, $g(f(x)) = g \circ f(x))$ which we have shown to be a *unique* element in rg (g) for each and every element x in dom (f). Thus, the set $g \circ f$ is always a *function* if the sets f and g are functions.

Figure 2.5 is a diagram of the composition function $g \circ f$. The 2 "machines," f and g in tandem, are equivalent to the composite "machine," $g \circ f$, which is represented by a dotted line.

We note that the function $g \circ f$ can be interpreted as an application (to an element $x \in X$) of the function f followed by an application of the function g, that is, we read $g \circ f$ from right to left.

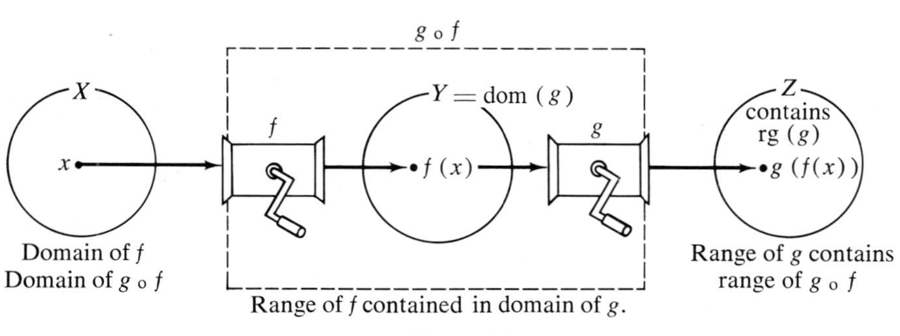

Figure 2.5

EXERCISES

In Exercises II.1.12 through II.1.14, find the composite function $g \circ f$, where possible, and describe its domain and range. State why this is not possible in the other cases.

II.1.12 $f = \{(\frac{1}{2},1),(1,2),(\frac{3}{2},3)\}$ and $g = \{(1,\frac{3}{2}),(2,1),(3,-\frac{1}{2})\}$.

II.1.13 $f = \{(\frac{1}{2},1),(1,2),(\frac{3}{2},3)\}$ and $g = \{(1,2),(2,3),(\frac{1}{2},1)\}$.

II.1.14 f is the (infinite) set of ordered pairs of the form $(n,n+2)$ where $n = 1, 2, 3, \ldots$. g is the (infinite) set of ordered pairs of the form $(n,n-2)$, where $n = 2, 3, 4, 5, \ldots$.

II.1.15 (a) Given the functions f and g as described in Exercise II.1.14, describe the domain and range of each.
(b) What is $f(n)$ for $n = 1, 2, 3, \ldots$?
(c) What is $f(n+3)$, and for which integers n is $f(n+3)$ defined?
(d) What is $g(n+3)$ and for which integers n is $g(n+3)$ defined?
(e) $f(?) = 12$
(f) $g(?) = 12$
(g) $f(?) = 1$
(h) $g \circ f(?) = 9$
(i) $g \circ f(?) = 1$
(j) $g \circ f(?) = n$ for $n = 1, 2, 3, \ldots$.

II.1.16 Given $f = \{(1,\text{chair}),(2,\text{table})\}$ and $g = \{(\text{chair},5),(\text{table},5)\}$.
(a) Show that the sets of ordered pairs, f and g, are functions, and describe their respective domains and ranges.
(b) List the ordered pairs which comprise the function $g \circ f$, and describe the domain and range, if possible.
(c) What is $g(\text{table})$?
(d) What is $g \circ f(2)$?
(e) $g(?) = 5$.
(f) $f(?) = \text{chair}$.
(g) $g \circ f(?) = 1$.
(h) $g \circ f(?) = 5$.

We have noted the important "uniqueness of image" property of a function, that is, for every x in the domain of a function, there is 1 and only 1 element in the range, $y = f(x)$, which is the image of x (under f). In general, the converse does not hold. That is, for some y in the range of f, it is possible that y is the image of *more* than 1 element in the domain of f In other words, for some y in the range of f, it is possible that $y = f(x_1)$. and $y = f(x_2)$ for some x_1 and x_2 in the domain of f, where $x_1 \neq x_2$.

We will now concern ourselves with functions which enjoy the property that every element in the range is the image of 1 and *only* 1 element in the domain, that is, we will study those functions f having the further restriction that for every y in the range of f, $y = f(x)$ for 1 and only 1 x in the domain of f. Such a function is called a one-to-one function and we formally define this notion with Definition II.1.6.

Definition II.1.6 Let f be a function with domain X and range Y (see Definition II.1.1). f is a **one-to-one function** (symbolized **1-1 function**) if and only if, whenever $(v,u) \in f$ and $(w,u) \in f$, then, necessarily, $v = w$ for $v,w \in X$, and $u \in Y$.

Example Let $f = \{(1,2),(2,2),(3,4),(4,3)\}$ which is a function whose domain is, therefore, the set $\{1,2,3,4\}$ and whose range is the set $\{2,3,4\}$. f is *not* 1-1 since the pairs (1,2) and (2,2) both belong to f. That is $f(1) = 2$ and $f(2) = 2$ which says that the number 2 (in the range of f) is the image of more than 1 element in the domain of f. Figure 2.6 illustrates this function.

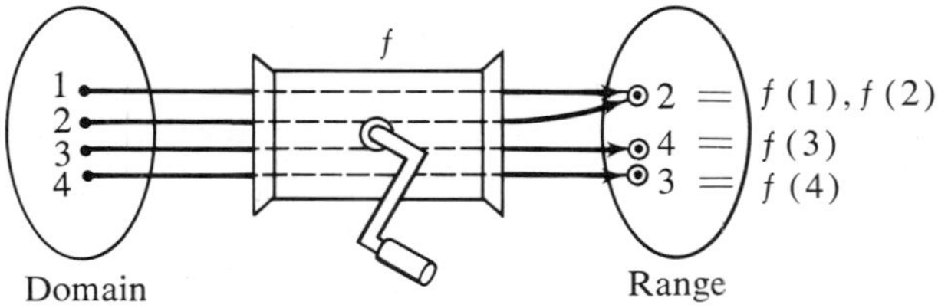

Figure 2.6

A pleasant property of 1-1 functions arises; we may, in a sense, reverse the roles of domain and range. More exactly, if we were to interchange the first and second entries of each ordered pair (x,y) in the function, then the resulting set of ordered pairs would *still* be a function. Not surprisingly, then, this "new" function is called the inverse of f. We establish this fact more formally with Proposition II.1.7.

Proposition II.1.7 Suppose the set f is a 1-1 function with domain X and range Y [elements of f are of the form $(x,f(x))$]. We conclude that the set f^{-1}, of all ordered pairs of the form $(f(x),x)$ is always a *function* with domain Y and range X.

Proof: As given, the set f^{-1} is a set of ordered pairs. In order to qualify as a bonafide function, we must show that if (u,v) and (u,w) belong to the set f^{-1}, then, necessarily, $v = w$. (See Definition II.1.1.) To show that this is the case, suppose that both ordered pairs (u,v) and (u,w) belong to the set f^{-1} (We wish to show that, necessarily, $v = w$.) From the definition of f^{-1} (hypothesis), this means that the ordered pairs (v,u) and (w,u) belong to the set f. That is, $u = f(v)$ and $u = f(w)$. Now is the time to use our hypothesis that f is 1-1. A casual glance at Definition II.1.6 reveals that, necessarily, $v = w$, so that, indeed, the set f^{-1} is a function whose domain is the set Y and range is the set X.

Now that Proposition II.1.7 has established the fact that f^{-1} is a *function* provided f is a 1-1 function, we set down the definition formally.

Definition II.1.8 Let the set f be a 1-1 function whose elements are of the form $(x,f(x))$, for $x \in X$ and $f(x) \in Y$. The function, which is the set of all ordered pairs of the form $(f(x),x)$ is called the **inverse of the function** $\boldsymbol{f}$, and is denoted by the symbol $\boldsymbol{f^{-1}}$, read $\boldsymbol{f}$ **inverse, or the inverse of** $\boldsymbol{f}$.

Example Let f be the infinite set of ordered pairs of the form $(n,2n)$ where $n = 1, 2, 3, \ldots$, that is, $f(n) = 2n$. The domain of f is the set of positive integers $X = \{1,2,3,\ldots\}$ and the range of f is the set of even positive integers $Y = \{2,4,6,\ldots\}$. Interchanging the order of each of the pairs in f yields the set f^{-1}, the set of ordered pairs of the form $(2n,n)$. That is, $f^{-1}(2n) = n$, for $n = 1, 2, 3, \ldots$. The domain of f^{-1} is the set of even positive integers Y, and the range of f^{-1} is the set of positive integers X. It is easily verified that f is, in fact, a 1-1 function, for suppose the ordered pairs (n,p) and (m,p) both belong to f. We must show that, necessarily, $n = m$ (see Definition II.1.5). From the definition of f in this example, the second entry is twice the first entry, that is, since $(n,p) \in f$, we have $p = 2n$, and since $(m,p) \in f$, we have $p = 2m$. Thus $2m = 2n$, which implies $m = n$. Thus, f is a 1-1 function.

EXERCISES

II.1.17 Given $f: X \to Y$ is 1-1 and onto (that is, f is 1-1 and the impact set $Y = \text{rg}(f)$). We know (from Proposition II.1.7) that the function $f^{-1}: Y \xrightarrow{\text{onto}} X$ exists. Show that f^{-1} is also 1-1.

Hence (again by Proposition II.1.7) the inverse of the function f^{-1}, denoted $(f^{-1})^{-1}$, exists. Show that $(f^{-1})^{-1} = f$.

II.1.18 Prove a converse of Proposition II.1.7. That is, suppose $f: X \xrightarrow{\text{onto}} Y$, and suppose $f^{-1}: Y \xrightarrow{\text{onto}} X$ exists as a bonafide function. Show that f is necessarily 1-1. This establishes the fact that any onto function has an inverse if and only if f is 1-1 to begin with.

In Exercises II.1.19 through II.1.23, a function f is given. Find the inverse function f^{-1} and the inverse of the inverse $(f^{-1})^{-1}$. Describe their respective domains and ranges.

II.1.19 $f = \{(1,\text{door}),(\text{door},1)\}$.

II.1.20 $f = \{(\text{happy},5),(\text{auto},\text{club}),(2,3)\}$.

II.1.21 f is the set of *all* ordered pairs of the form $(x,3x + 2)$, where x is an arbitrary real number.

II.1.22 f is the set of *all* ordered pairs of the form $(n,3n + 2)$, where $n = 1, 2, 3, \ldots$.

II.1.23 f is the set of all ordered pairs of the form $\left(\begin{pmatrix} x_1 \\ x_2 \end{pmatrix}, \begin{pmatrix} x_2 \\ 2x_1 \end{pmatrix}\right)$ where $\begin{pmatrix} x_1 \\ x_2 \end{pmatrix}$ is an arbitrary 2-vector in V_2, that is, $f\begin{pmatrix} x_1 \\ x_2 \end{pmatrix} = \begin{pmatrix} x_2 \\ 2x_1 \end{pmatrix}$ for all $\begin{pmatrix} x_1 \\ x_2 \end{pmatrix} \in V_2$.

II.1.24 Given $f: X \xrightarrow{1-1,\ \text{onto}} Y$. Prove that $f^{-1} \circ f: X \to X$ is the identity function on X, that is, $f^{-1} \circ f(x) = x$ for each and every $x \in X$. Prove, also, that $f \circ f^{-1}(y) = y$ for all $y \in Y$.

Suppose we have the composite function $h \circ g \circ f$ (f followed by g followed by h) as defined by $h \circ g \circ f(x) = h(g(f(x)))$, where X is the domain of f, Y is the range of f and the domain of g, Z is the range of g and the domain of h, and W is the range of h.

Diagrammatically, we have $X \xrightarrow[\text{onto}]{} Y \xrightarrow[\text{onto}]{g} Z \xrightarrow[\text{onto}]{h} W$. In Exercises II.1.25 and II.1.26, describe the sets X, Y, Z, and W and describe the elements $h \circ g \circ f(x)$ in terms of the domain elements x.

II.1.25 $f = \{(\text{arm},3),(\otimes,2)\}$
$g = \{(3,\text{six}),(2,\otimes)\}$
$h = \{(\text{six},1),(\otimes,1)\}$
Describe $X \cap Y$, $X \cap Z$.

II.1.26 f is the set of all ordered pairs of the form $(n,n-1)$ for $n = 0, \pm 1, \pm 2, \ldots$.
g is the set of all ordered pairs of the form $(m,2m)$ for $m = 0, \pm 1, \pm 2, \ldots$.
h is the set of all ordered pairs of the form $(p,(p/2)+1)$ for $p = 0, \pm 2, \pm 4, \pm 6, \ldots$.

II.1.27 With the functions f, g, and h as in Exercise II.1.25, describe:
(a) the functions g^{-1} and f^{-1}
(b) the function $f^{-1} \circ g^{-1}$
(c) the function $f \circ f^{-1} \circ g^{-1}$
(d) the elements $g^{-1}(\otimes)$ and $h(\otimes)$
(e) the element $h(f^{-1}(2))$.

II.1.28 With the functions f, g, and h as in Exercise II.1.26, describe:
(a) the element $f \circ h \circ g(n)$ for $n = 0, \pm 1, \pm 2, \ldots$.
(b) the elements $h \circ g \circ f^{-1}(n)$ for $n = 0, \pm 1, \pm 2, \ldots$.
(c) $f^{-1} \circ g^{-1}(n)$, and $h(n)$ for $n = 0, \pm 2, \pm 4, \pm 6, \ldots$.

II.1.29 Let $f = \{(0,2),(\text{cow},3),(1,\text{yes})\}$ and $g = \{(2,S),(3,Q),(\text{yes},5)\}$. Describe:
(a) the domain and range of the functions f and g
(b) the functions f^{-1} and g^{-1}, and their respective domains and ranges
(c) the function $f^{-1} \circ g^{-1}$ and its domain and range
(d) the function $g \circ f$ and its domain and range
(e) the function $(g \circ f)^{-1}$ and its domain and range [compare with part (c)].

II.1.30 Suppose f and g are 1-1 functions, and that the range of f equals the domain of g (so that $g \circ f$ can be defined). Show that the function $(g \circ f)^{-1}$ is the same as the function $f^{-1} \circ g^{-1}$ (see Exercise II.1.29 for a special case; also Definitions II.1.4, II.1.5, and II.1.8).

II.1.31 Let f, g, h be functions $X \xrightarrow{f} Y \xrightarrow{g} Z \xrightarrow{h} W$. Recall that composition of functions is defined only between *pairs* of functions, so that the symbol $f \circ g \circ h$ has no meaning relative to Definition II.1.5. However, the symbols $h \circ (g \circ f)$ (the function h following the function $g \circ f$) and $(h \circ g) \circ f$ (the function $h \circ g$ following the function f) *do* have well-defined meanings. Show that the (apparently different) functions $h \circ (g \circ f) : X \to W$, and $(h \circ g) \circ f : X \to W$, are in fact the *same* function. Hence the symbol $f \circ g \circ h$ may be used to represent $(h \circ g) \circ f = h \circ (g \circ f)$.

II.1.32 Give an example of a composite function $g \circ f$ where f is not onto.

II.1.33 (a) Suppose $g \circ f$ is 1-1. Show that f must then be 1-1.
(b) Suppose $g \circ f$ is onto. Show that g must be onto.

II.1.34 (See Exercise II.1.33.) Give an example of functions $f : X \to Y$, $g : Y \to Z$ such that $g \circ f : X \to Z$ is both 1-1 and onto, yet f is not onto and g is not 1-1.

II.1.35 Prove: For functions $f : X \to Y$, $g : Y \to Z$, we have
(a) f and g 1-1 $\Rightarrow g \circ f$ is 1-1
(b) f and g onto $\Rightarrow g \circ f$ is onto.

2. Linear Transformations

Recall that Chapter I dwelt at some length on the structure of finite-dimensional vector-spaces; "structure" is taken to mean that an *addition*, and *scalar multiplication* are defined (see Definitions I.1.2, I.1.3, and I.5.2). Section 1 of this chapter outlined the notion of a *function*. We will proceed to blend the ideas of *vector-space* and *function*; that is, we will concern ourselves with certain functions whose domains and ranges are finite-dimensional vector-spaces. Moreover, these functions will be obliged to "preserve the vector-space structure." But what is meant by the phrase "a function preserves the vector-space structure?" We must ask this question in two parts:

(a) What is meant when we say that a function f preserves *scalar multiplication*?

(b) What is meant when we say that a function preserves the vector-space *addition*?

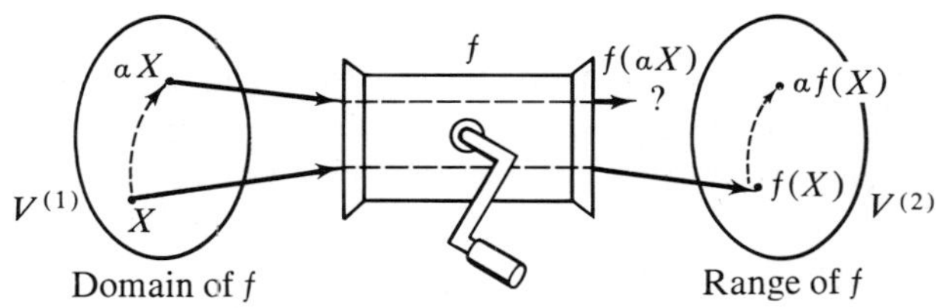

Figure 2.7

In answering question (a), we refer to Figures 2.7 and 2.8. We have denoted by $V^{(1)}$, the vector-space which is the domain of the function f. Similarly, $V^{(2)}$ denotes the vector-space which contains the range of f. $V^{(1)}$ may or may not be equal to $V^{(2)}$. Now let X represent an arbitrary vector of $V^{(1)}$. Since $V^{(1)}$ is a vector-space, scalar multiplication is defined, so that the vector αX is defined in $V^{(1)}$ for any (and every) scalar α. Consider, now, the effect that f has on these 2 vectors X and αX, which are related by scalar multiplication. That is, consider their respective images $f(X)$ and $f(\alpha X)$ which are elements of $V^{(2)}$, which contains the range of f.

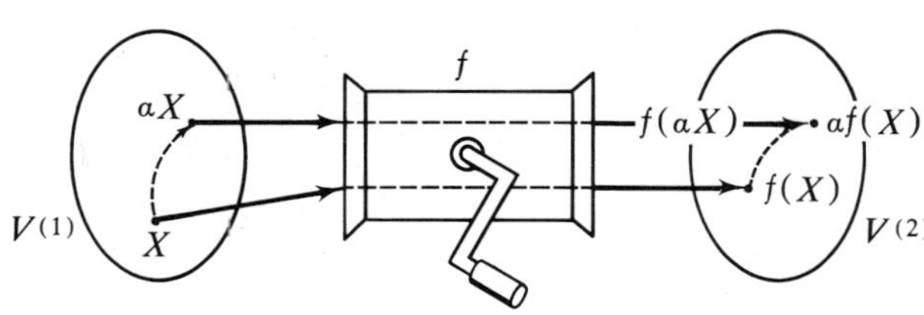

Figure 2.8

In saying that "f preserves scalar multiplication," we are saying that the "same" scalar relation holds between the vectors of the set $\{f(X), f(\alpha X)\}$ as holds between the (original) vectors in the set $\{X, \alpha X\}$, that is, that the second vector is simply equal to the first vector multiplied by α. Thus the phrase *f preserves scalar mutliplication* means

$$f(\alpha X) \text{ (the second vector)} = \alpha f(X) \text{ (\textit{the first vector multiplied by } } \alpha) \tag{II.2.1}$$

for all $X \in V^{(1)}$ and all scalars α. Figure 2.7 diagrams the element $f(\alpha X)$ as undefined, while Figure 2.8 indicates, diagrammatically, the appropriate value for $f(\alpha X)$ if scalar multiplication is to be preserved.

A similar argument establishes a relation which answers question (b).

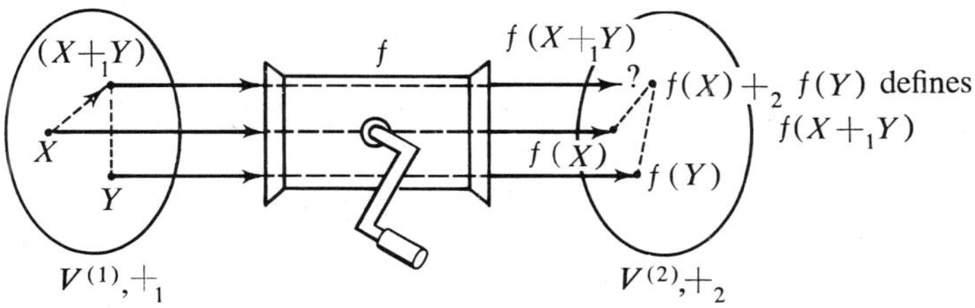

Figure 2.9

Figures 2.9 and 2.10 carry the notations $V^{(1)}$ for the domain of f, and $V^{(2)}$ for the vector-space containing the range of f. The symbols $+_1$ and $+_2$ represent the additions of the vector-spaces $V^{(1)}$ and $V^{(2)}$, respectively. Let X and Y be arbitrary vectors of $V^{(1)}$. Since $V^{(1)}$ is a vector-space endowed with addition $+_1$, there exists a third element $X +_1 Y$ in $V^{(1)}$.

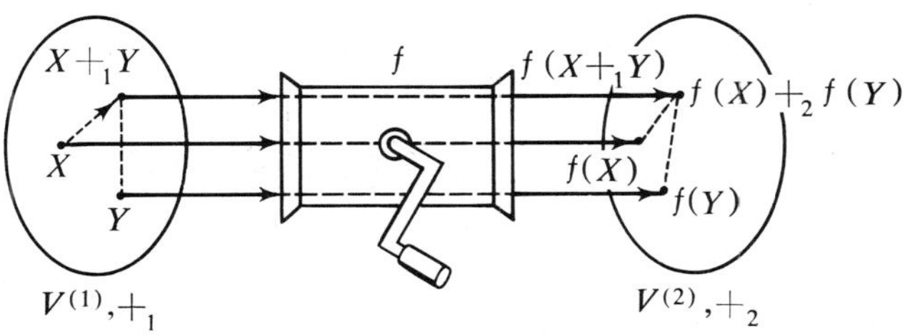

Figure 2.10

Consider, then, the 3 related elements X, Y, and $X +_1 Y$, and their corresponding images $f(X)$, $f(Y)$, and $f(X +_1 Y)$ (which are elements of $V^{(2)}$). In saying that "f preserves addition," we are saying that the "same" additive relation holds among the elements of the sets $\{X, Y, X +_1 Y\}$, and

$\{f(X), f(Y), f(X +_1 Y)\}$, that is, that the third vector is the sum of the first 2 vectors. Thus, the phrase *f preserves addition* means,

$$f(X) +_2 f(Y) = f(X +_1 Y) \quad \text{for all } X \in V^{(1)} \text{ and all } Y \in V^{(1)}. \tag{II.2.2}$$

Figure 2.9 diagrams the element $f(X +_1 Y)$ as undefined, while Figure 2.10 indicates the appropriate value for $f(X +_1 Y)$ if addition is to be preserved.

We note that the symbols $+_1$ and $+_2$ indicate in which space ($V^{(1)}$ or $V^{(2)}$), we are performing our addition. Henceforth, however, we shall *not* distinguish between the 2 (albeit distinct) additions.

We have, now, extracted the properties of functions which preserve the structure (addition and scalar multiplication) of vector-spaces, as indicated in Equations (II.2.1) and (II.2.2). We give such functions a name and define their properties formally with Definition II.2.1. Since these functions are, in a sense, special, we will denote them by the capital boldface letters **T**, **U**, **V**, and so on, and abandon the lower case f and g notation.

Definition II.2.1 Let $\mathbf{T}$ be a function where domain is a vector-space V, and whose range is contained in a vector-space W. $\mathbf{T}$ will be called a **linear transformation, linear map,** or **linear function** on V (from V to W) if and only if

$$\mathbf{T}(\alpha X) = \alpha \mathbf{T}(X) \quad \text{for all } X \in V, \quad \text{all scalars } \alpha, \tag{II.2.1}$$

$$\mathbf{T}(X + Y) = \mathbf{T}(X) + \mathbf{T}(Y) \quad \text{for all } X \in V, \quad \text{and all } Y \in V. \tag{II.2.2}$$

We note that the same symbol $+$ appears on both sides of equation (II.2.2) above, yet it is understood that this symbol could represent 2 different additions if V and W are distinct vector-spaces.

Equations (II.2.1) and (II.2.2) can be combined into 1 equivalent equation. That is, we state Proposition II.2.2.

Proposition II.2.2 Let V and W be vector-spaces. Then $\mathbf{T} : V \to W$ is linear if and only if both the following conditions obtain:

$$\mathbf{T}(\alpha X) = \alpha \mathbf{T}(X) \quad \text{for all } \quad X \in V, \quad \text{all scalars } \alpha \tag{II.2.1}$$

and

$$\mathbf{T}(X + Y) = \mathbf{T}(X) + \mathbf{T}(Y) \quad \text{for all } \quad X \in V, \text{ all } Y \in V. \tag{II.2.2}$$

Equivalently, $\mathbf{T} : V \to W$ is linear if and only if

(a) $$\mathbf{T}(\alpha X + \beta Y) = \alpha \mathbf{T}(X) + \beta \mathbf{T}(Y) \quad \text{for all } X \in V, \; Y \in V \text{ and all scalars } \alpha \text{ and } \beta.$$

Proof: First we will show that if we assume condition (a), then conditions (II.2.1) and (II.2.2) will necessarily follow. Assuming (a), then, simply choose $\beta = 0$. Condition (a) reduces to condition (II.2.1). If α and β are each chosen to equal 1, then condition (a) transforms itself into condition (II.2.2). Hence (a) $\Rightarrow$ (II.2.1) and (II.2.2).

Conversely, assume conditions (II.2.1) and (II.2.2) obtain. We must show condition (a) to be a necessary consequence, that is, (II.2.1) and (II.2.2) $\Rightarrow$ (a). Since condition (II.2.2) is assumed valid, replace X by the vector αX, and Y be the vector βY; condition (II.2.2) then yields

$$\mathbf{T}(\alpha X + \beta Y) = \mathbf{T}(\alpha X) + \mathbf{T}(\beta Y).$$

But condition (II.2.1) obtains as well. Hence we may apply it to the right-hand side of the above equation and write $\alpha\mathbf{T}(X)$ and $\beta\mathbf{T}(Y)$ in place of $\mathbf{T}(\alpha X)$ and $\mathbf{T}(\beta Y)$, respectively. Finally, thanks to conditions (II.2.1) and (II.2.2),

$$\mathbf{T}(\alpha X + \beta Y) = \alpha\mathbf{T}(X) + \beta\mathbf{T}(Y) \quad \text{for all } X \in V,\ Y \in V$$

and all scalars α and β, which is precisely condition (a), that is, (II.2.1) and (II.2.2) $\Rightarrow$(a) and the proof is done.

Example Let $\mathbf{T}$ be defined by the equation

$$\mathbf{T}\begin{pmatrix} x_1 \\ x_2 \\ x_3 \end{pmatrix} = \begin{pmatrix} x_2 \\ x_3 \end{pmatrix}.$$

The domain (input set) of $\mathbf{T}$ is 3-space V_3, and the range (output set) is a subset of 2-space V_2, that is, $\mathbf{T} : V_3 \to V_2$. Figure 2.11 diagrams this function.

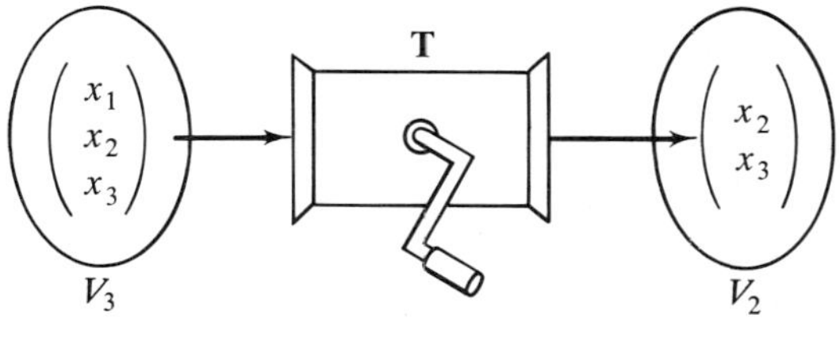

Figure 2.11

This function is, by Definition II.1.1, the set of all ordered pairs of the form

$$\left(\begin{pmatrix} x_1 \\ x_2 \\ x_3 \end{pmatrix}, \begin{pmatrix} x_2 \\ x_3 \end{pmatrix}\right).$$

In this case, the first entry of the ordered pair is a 3-vector (an ordered triple) and the second entry of the ordered pair is a 2-vector (and is, itself, an ordered pair). Crudely speaking, this $\mathbf{T}$ "throws away" the first entry of every 3-vector, transforming it into a 2-vector. We will show that $\mathbf{T}$ is linear by testing it against condition (a) of Proposition II.2.2 [which is equivalent to conditions (II.2.1) and (II.2.2) of Definition II.2.1]. Let

$$X = \begin{pmatrix} x_1 \\ x_2 \\ x_3 \end{pmatrix} \quad \text{and let} \quad Y = \begin{pmatrix} y_3 \\ y_2 \\ y_1 \end{pmatrix}$$

be 2 arbitrary vectors of V_3, the domain of $\mathbf{T}$. Let α and β be arbitrary scalars. Then

$$\begin{aligned}
\mathbf{T}(\alpha X + \beta Y) &= \mathbf{T}\left(\alpha\begin{pmatrix} x_1 \\ x_2 \\ x_3 \end{pmatrix} + \beta\begin{pmatrix} y_1 \\ y_2 \\ y_3 \end{pmatrix}\right) && \text{(notation for } X \text{ and } Y\text{)} \\
&= \mathbf{T}\left(\begin{pmatrix} \alpha x_1 \\ \alpha x_2 \\ \alpha x_3 \end{pmatrix} + \begin{pmatrix} \beta y_1 \\ \beta y_2 \\ \beta y_3 \end{pmatrix}\right) && \text{(definition of scalar multiplication: Definition I.1.3)} \\
&= \mathbf{T}\left(\begin{pmatrix} \alpha x_1 + \beta y_1 \\ \alpha x_2 + \beta y_2 \\ \alpha x_3 + \beta y_3 \end{pmatrix}\right) && \text{(definition of vector addition: Definition I.1.2)} \\
&= \begin{pmatrix} \alpha x_2 + \beta y_2 \\ \alpha x_3 + \beta y_3 \end{pmatrix} && \text{(definition of } \mathbf{T}\text{)} \\
&= \alpha\begin{pmatrix} x_2 \\ x_3 \end{pmatrix} + \beta\begin{pmatrix} y_2 \\ y_3 \end{pmatrix} && \text{(Definitions I.1.2 and I.1.3)} \\
&= \alpha\mathbf{T}\begin{pmatrix} x_1 \\ x_2 \\ x_3 \end{pmatrix} + \beta\mathbf{T}\begin{pmatrix} y_1 \\ y_2 \\ y_3 \end{pmatrix} && \text{(definition of } \mathbf{T}\text{)} \\
&= \alpha\mathbf{T}(X) + \beta\mathbf{T}(Y). && \text{(notation for } X \text{ and } Y\text{).}
\end{aligned}$$

Thus, $\mathbf{T}(\alpha X + \beta Y) = \alpha\mathbf{T}(X) + \beta\mathbf{T}(Y)$ for all $X, Y \in V_3$, all scalars α and β; $\mathbf{T}$ is linear.

Example Let $\mathbf{T}$ be the transformation from V_2 to V_2 which squares each coordinate, that is, $\mathbf{T}\begin{pmatrix} x_1 \\ x_2 \end{pmatrix} = \begin{pmatrix} x_1^2 \\ x_2^2 \end{pmatrix}$ for all $X = \begin{pmatrix} x_1 \\ x_2 \end{pmatrix} \in V_2$. This function is so badly nonlinear, that almost any particular vectors and scalars chosen at random will violate conditions (II.2.1) and (II.2.2) of Definition II.2.1. For example, let $\alpha = 2$ and let $X = \begin{pmatrix} 1 \\ 1 \end{pmatrix}$. On the one hand, we have $\mathbf{T}(\alpha X) = \mathbf{T}\begin{pmatrix} 2 \\ 2 \end{pmatrix} = \begin{pmatrix} 4 \\ 4 \end{pmatrix}$. On the other hand, $\alpha\mathbf{T}(X) = 2\,\mathbf{T}\begin{pmatrix} 1 \\ 1 \end{pmatrix} = 2\begin{pmatrix} 1^2 \\ 1^2 \end{pmatrix} = \begin{pmatrix} 2 \\ 2 \end{pmatrix}$. Thus, $\alpha\mathbf{T}(X)$ and $\mathbf{T}(\alpha X)$ do not agree for our choice of α and X; condition (II.2.2) of Definition II.2.1 is violated. We must conclude that $\mathbf{T}$ is *not* linear.

REMARK. Condition (a) of Proposition II.2.2 describes the property of linearity with respect to the sum of only 2 vectors, αX and βY. Indeed, an apparently stronger property flows from this, namely, $\mathbf{T}$ is a linear operator from V to W if and only if

$$\begin{aligned}
\mathbf{T}(\alpha_1 X_1 + \alpha_2 X_2 + \cdots + \alpha_k X_k)& \\
= \alpha_1\mathbf{T}(X_1) + \alpha_2\mathbf{T}(X_2) + \cdots + \alpha_k\mathbf{T}(X_k)&
\end{aligned} \tag{II.2.3}$$

for all $X_1, X_2, \ldots, X_k \in V$ and all scalars $\alpha_1, \alpha_2, \ldots, \alpha_k$, where k is a positive integer.

We can verify (II.2.3) for 3 vectors ($k = 3$), as we reduce the problem to the original case of 2 vectors. In fact, let $\mathbf{T}$ be linear. Then

$$\begin{aligned}\mathbf{T}(\alpha_1 X_1 + \alpha_2 X_2 + \alpha_3 X_3)& \\ &= \mathbf{T}((\alpha_1 X_1 + \alpha_2 X_2) + (\alpha_3 X_3)) \qquad \text{(associativity, see Definition I.5.1 and Definition I.5.2).}\end{aligned}$$

The right-hand side of this equation interprets $\mathbf{T}$ as acting on the sum of the 2 vectors $(\alpha_1 X_1 + \alpha_2 X_2)$ and $(\alpha_3 X_3)$. Applying condition (II.2.2) of Proposition II.2.2 to the right-hand side, then, yields

$$\mathbf{T}(\alpha_1 X_1 + \alpha_2 X_2 + \alpha_3 X_3) = \mathbf{T}(\alpha_1 X_1 + \alpha_2 X_2) + \mathbf{T}(\alpha_3 X_3).$$

Applying conditions (II.2.2) and (II.2.1) of Proposition II.2.2 to the 2 terms on the right-hand side of this last equation, yields

$$\begin{aligned}\mathbf{T}(\alpha_1 X_1 + \alpha_2 X_2 + \alpha_3 X_3) &= (\alpha_1 \mathbf{T}(X_1) + \alpha_2 \mathbf{T}(X_2)) + \alpha_3 \mathbf{T}(X_3) \\ &= \alpha_1 \mathbf{T}(X_1) + \alpha_2 \mathbf{T}(X_2) + \alpha_3 \mathbf{T}(X_3) \qquad \text{(Definition I.5.1)}\end{aligned}$$

which verifies Equation (II.2.3) at least when $k = 3$. A bonafide proof of this result for general k will be established in the section on mathematical induction.

If does not seem to be a very good idea to discuss linear transformations and their various properties without knowing more *specifically* what they look like. A very concrete characterization arises with Theorem II.2.3.

Theorem II.2.3 The function $\mathbf{T} : V_n \to V_m$ is linear if and only if for certain unique, fixed scalars a_{ji}, $j = 1, 2, \ldots, m$; $i = 1, 2, \ldots, n$,

$$\mathbf{T}(X) = \mathbf{T}\begin{pmatrix} x_1 \\ x_2 \\ \vdots \\ x_i \\ \vdots \\ x_n \end{pmatrix} = \begin{pmatrix} a_{11}x_1 + a_{12}x_2 + \cdots + a_{1i}x_i + \cdots + a_{1n}x_n \\ a_{21}x_1 + a_{22}x_2 + \cdots + a_{2i}x_i + \cdots + a_{2n}x_n \\ \vdots \\ \boxed{a_{j1}x_1 + a_{j2}x_2 + \cdots + a_{ji}x_i + \cdots + a_{jn}x_n} \\ \vdots \\ a_{m1}x_1 + a_{m2}x_2 + \cdots + a_{mi}x_i + \cdots + a_{mn}x_n \end{pmatrix}$$

jth entry of $\mathbf{T}(X) \in V_m$.

(II.2.4)

for each and every n-vector

$$X = \begin{pmatrix} x_1 \\ x_2 \\ \vdots \\ x_n \end{pmatrix} \in V_n.$$

This characterization presents a very graphic picture of what a typical linear transformation does to any n-vector X.

REMARKS. Before us is the rare opportunity to use the lessons of experience (in the form of Definition I.2.3), in an unexpected situation [in the form of (II.2.4)]. Observe that (II.2.4) may be conveniently expressed as a *matrix* equation, that is, Theorem II.2.3 may be rephrased as follows:

▶ The function $\mathbf{T}: V_n \to V_m$ is linear if and only if for a certain $m \times n$ matrix $A = (a_{ij})_{ij}$ (Definition I.2.2),

$$\mathbf{T}(X) = A \cdot X$$

for each and every concrete n-vector $X \in V_n$.

In other words, every linear transformation is actually effected by multiplying each X (considered as an $n \times 1$ matrix) on the left by a certain fixed $m \times n$ matrix A.

Example For all $\begin{pmatrix} x_1 \\ x_2 \end{pmatrix} \in V_2$, define $\mathbf{T}$ by

$$\mathbf{T}\begin{pmatrix} x_1 \\ x_2 \end{pmatrix} = \begin{pmatrix} x_1 x_2 \\ x_2 \end{pmatrix}.$$

Now $\mathbf{T}$ is a function from V_2 to V_2, but $\mathbf{T}$ is easily seen to be not linear. Notice that (II.2.4) forbids products of one entry x_i with itself or with any other entry.

Example $$\mathbf{T}\begin{pmatrix} x_1 \\ x_2 \\ x_3 \end{pmatrix} = \begin{pmatrix} 3x_1 + 2x_2 \\ x_1 \end{pmatrix} \quad \text{for all} \quad \begin{pmatrix} x_1 \\ x_2 \\ x_3 \end{pmatrix} \in V_3$$

defines a function $\mathbf{T} : V_3 \to V_2$. $\mathbf{T}$ is easily seen to be linear from (II.2.4). To be more formal, we could have written

$$\mathbf{T}\begin{pmatrix} x_1 \\ x_2 \\ x_3 \end{pmatrix} = \begin{pmatrix} 3x_1 + 2x_2 + 0x_3 \\ 1x_1 + 0x_2 + 0x_3 \end{pmatrix}$$

to force the form of $\mathbf{T}$ to agree with (II.2.4) more closely.

Roughly speaking then, Theorem II.2.3. says this: The only effect a *linear* function $\mathbf{T}$ can have on a typical n-vector

$$\begin{pmatrix} x_1 \\ x_2 \\ \vdots \\ x_n \end{pmatrix} = X$$

is to transform it to a vector whose typical ith entry is of the form $a_{i1}x_1 + a_{i2}x_2 + \cdots + a_{in}x_n$ for some unique scalars $a_{i1}, a_{i2}, \ldots, a_{in}$. Notice that *no powers or products* (such as $x_2{}^3$ or x_1x_3) of the entries x_i may appear. Only *sums* of numbers of the form $a_{ij}x_j$ may appear in the ith entry of $\mathbf{T}(X)$. Moreover *all* functions of form (II.2.4) are linear.

Proof of Theorem II.2.3: We shall establish only the implication: $\mathbf{T} : V_n \to V_m$ is linear $\Rightarrow \mathbf{T}$ of form in (II.2.4). The converse (and easier) implication: $\mathbf{T}$ of form in (II.2.4) $\Rightarrow \mathbf{T} : V_n \to V_m$ is linear, is left for the student as Exercise II.2.12.

So let us assume that $\mathbf{T} : V_n \to V_m$ is a pleasant linear function in conformity with Definition II.2.1. We aim to show that the form of (II.2.4) is an *inevitable* consequence of this assumption.

Now let $\{E_1, E_2, \ldots, E_n\}$ and $\{F_1, F_2, \ldots, F_m\}$ be the standard bases for V_n and V_m, respectively (Definition I.6.11). Observe that *any* vector X in V_n can be written, then, as

$$X = \begin{pmatrix} x_1 \\ x_2 \\ \vdots \\ x_n \end{pmatrix} = x_1 \begin{pmatrix} 1 \\ 0 \\ \vdots \\ 0 \end{pmatrix} + x_2 \begin{pmatrix} 0 \\ 1 \\ 0 \\ \vdots \\ 0 \end{pmatrix} + \cdots + x_n \begin{pmatrix} 0 \\ 0 \\ \vdots \\ 0 \\ 1 \end{pmatrix} \qquad \text{(II.2.5a)}$$

$$= x_1E_1 + x_2E_2 + \cdots + x_nE_n.$$

Similarly, any vector $Y \in V_m$ can be written

$$Y = \begin{pmatrix} y_1 \\ y_2 \\ \vdots \\ y_m \end{pmatrix} = y_1F_1 + y_2F_2 + \cdots + y_mF_m. \qquad \text{(II.2.5b)}$$

Let us now take this linear $\mathbf{T} : V_n \to V_m$ and apply it to a typical vector $X = x_1E_1 + x_2E_2 + \cdots + x_nE_n$ [see (II.2.5a)].

$$\begin{aligned}\mathbf{T}(X) &= \mathbf{T}\begin{pmatrix} x_1 \\ x_2 \\ \vdots \\ x_m \end{pmatrix} \\ &= \mathbf{T}(x_1E_1 + x_2E_2 + \cdots + x_nE_n) \qquad \text{from (II.2.5a)} \\ &= x_1\mathbf{T}(E_1) + x_2\mathbf{T}(E_2) + \cdots + x_i\mathbf{T}(E_i) + \cdots + x_n\mathbf{T}(E_n). \qquad \text{(II.2.6)}\end{aligned}$$

Now each $\mathbf{T}(E_i)$, $i = 1, 2, \ldots, n$, lies in vector space V_m whose standard basis is the set $\{F_1, F_2, \ldots, F_m\}$. Thus, each of these $\mathbf{T}(E_i)$'s is expressable *uniquely* as a linear combination of the vectors $F_1, F_2, \ldots, F_m$ (Theorem I.6.8). That is to say, for each $\mathbf{T}(E_i)$, there are m unique (remember the "unique" part) scalars $a_{1i}, a_{2i}, \ldots, a_{mi}$ [the "i" appears in the subscripts of these scalars to reflect their dependence on $T(E_i)$], with the property that

$$\mathbf{T}(E_i) = a_{1i}F_1 + a_{2i}F_2 + \cdots + a_{ji}F_j + \cdots + a_{mi}F_m. \qquad \text{(II.2.7)}$$

Thus, from (II.2.7), we obtain

$$\begin{aligned}x_i\mathbf{T}(E_i) &= x_i(a_{1i}F_1 + a_{2i}F_2 + \cdots + a_{mi}F_m) && \text{[from (II.2.7)]} \\ &= x_ia_{1i}F_1 + x_ia_{2i}F_2 + \cdots + x_ia_{mi}F_m && \text{(from axiom B.5, Definition I.5.2).}\end{aligned} \qquad \text{(II.2.8)}$$

In substituting expression (II.2.8) for each and every $x_i\mathbf{T}(E_i)$ in (II.2.6), we obtain

$$\begin{aligned}\mathbf{T}(X) = \; & \\ & (x_1a_{11}F_1 + x_1a_{21}F_2 + \cdots + x_1a_{j1}F_j + \cdots + x_1a_{m1}F_m) \leftarrow x_1\mathbf{T}(E_1) \\ + & (x_2a_{12}F_1 + x_2a_{22}F_2 + \cdots + x_2a_{j2}F_j + \cdots + x_2a_{m2}F_m) \leftarrow x_2\mathbf{T}(E_2) \\ & \quad \vdots \qquad\qquad \vdots \qquad\qquad\qquad \vdots \qquad\qquad\qquad \vdots \\ + & (x_ia_{1i}F_1 + x_ia_{2i}F_2 + \cdots + x_ia_{ji}F_j + \cdots + x_ia_{mi}F_m) \leftarrow x_i\mathbf{T}(E_i) \\ & \quad \vdots \qquad\qquad \vdots \qquad\qquad\qquad \vdots \qquad\qquad\qquad \vdots \\ + & (\underbrace{x_na_{1n}F_1}_{\text{column 1}} + \underbrace{x_na_{2n}F_2}_{\text{column 2}} + \cdots + \underbrace{x_na_{jn}F_j}_{\text{column } j} + \cdots + \underbrace{x_na_{mn}F_m}_{\text{column } m}) \leftarrow x_n\mathbf{T}(E_n).\end{aligned}$$

(II.2.9)

Now by repeated use of axiom B.4 of Definition I.5.2, we may add the "columns" of (II.2.9) so as to combine terms which contain only F_1 vectors, those containing only F_2 vectors, and so on. This procedure has, as its result,

$$\begin{aligned}\mathbf{T}(X) = &(x_1a_{11} + x_2a_{12} + \cdots + x_ia_{1i} + \cdots + x_na_{1n})F_1\\ &+ (x_1a_{21} + x_2a_{22} + \cdots + x_ia_{2i} + \cdots + x_na_{2n})F_2\\ &\vdots\\ &+ (x_1a_{j1} + x_2a_{j2} + \cdots + x_ia_{ji} + \cdots + x_na_{jn})F_j\\ &\vdots\\ &+ (x_1a_{m1} + x_2a_{m2} + \cdots + x_ia_{mi} + \cdots + x_na_{mn})F_m.\end{aligned} \tag{II.2.10}$$

We are almost finished. Let us not forget that the basis $\{F_1,F_2,\ldots,F_m\}$ for V_m is the *standard* basis, so that it becomes an easy matter to reconstruct the particular m-vector which is represented by the unwieldly mass in (II.2.10). In fact, as (II.2.5b) indicates, the jth entry of the m-vector $Y = \mathbf{T}(X)$ is simply the scalar in (II.2.10) which is "attached" to the vector F_j, namely, the scalar

$$a_{ji}x_1 + a_{j2}x_2 + \cdots + a_{ji}x_i + \cdots + a_{jn}x_n$$

is the jth entry of the m-vector $Y = \mathbf{T}(X)$. But this is exactly the statement of the theorem [see (II.2.4)]! Moreover, the scalars a_{ji}, $j = 1, 2,\ldots, m$; $i = 1, 2,\ldots, m$ are necessarily *unique* since they were chosen as coordinates for the vectors $T(E_1)$, $T(E_2),\ldots, T(E_n)$; by Theorem I.6.8, this can be done in only one way [see (II.2.7)]. The implication that $\mathbf{T} : V_n \to V_m$ and $\mathbf{T}$ is linear $\Rightarrow \mathbf{T}$ is a function of the form (II.2.4), has been proven valid. As mentioned previously, the converse implication is left as Exercise II.2.12.

EXERCISES

In Exercises II.2.1 through II.2.10 find out whether the given function $\mathbf{T}$ is linear using only Definition II.2.1 and *not* Theorem II.2.1. In each exercise, describe the domain and the range of $\mathbf{T}$.

II.2.1 $\mathbf{T}\begin{pmatrix}x_1\\x_2\\x_3\end{pmatrix} = \begin{pmatrix}0\\0\\0\end{pmatrix}$ for all $X = \begin{pmatrix}x_1\\x_2\\x_3\end{pmatrix} \in V_3$.

II.2.2 $\mathbf{T}\begin{pmatrix}x_1\\x_2\\x_3\end{pmatrix} = \begin{pmatrix}x_1\\x_3\\x_2\end{pmatrix}$ for all $X = \begin{pmatrix}x_1\\x_2\\x_3\end{pmatrix} \in V_3$.

II.2.3 $\mathbf{T}\begin{pmatrix} x_1 \\ x_2 \\ x_3 \end{pmatrix} = \begin{pmatrix} x_1 - x_2 \\ 0 \\ 0 \end{pmatrix}$ for all $X = \begin{pmatrix} x_1 \\ x_2 \\ x_3 \end{pmatrix} \in V_3$.

II.2.4 $\mathbf{T}\begin{pmatrix} x_1 \\ x_2 \\ x_3 \end{pmatrix} = \begin{pmatrix} x_1 \\ x_2 \end{pmatrix}$ for all $X = \begin{pmatrix} x_1 \\ x_2 \\ x_3 \end{pmatrix} \in V_3$.

II.2.5 $\mathbf{T}\begin{pmatrix} x_1 \\ x_2 \\ 0 \end{pmatrix} = \begin{pmatrix} 0 \\ 0 \end{pmatrix}$ for all $X \in V_3$ whose third coordinate is zero.

II.2.6 $\mathbf{T}\begin{pmatrix} x_1 \\ x_2 \\ 0 \end{pmatrix} = \begin{pmatrix} 0 \\ 0 \end{pmatrix}$ for all $X \in V_3$ whose third coordinate is zero.

$\mathbf{T}\begin{pmatrix} x_1 \\ x_2 \\ x_3 \end{pmatrix} = \begin{pmatrix} 0 \\ x_2 \end{pmatrix}$ for all $X \in V$ whose third coordinate is not zero.

II.2.7 $\mathbf{T}(x) = 3x$ for all (real numbers) x.

II.2.8 $\mathbf{T}(x) = (3x + 2)$ for all $(x) \in V_1$.

II.2.9 $\mathbf{T}\begin{pmatrix} x_1 \\ x_2 \end{pmatrix} = \begin{pmatrix} x_2 \\ x_1 + x_2 \\ 0 \\ x_2 \end{pmatrix}$ for all $X = \begin{pmatrix} x_1 \\ x_2 \end{pmatrix} \in V_2$.

II.2.10 $\mathbf{T}\begin{pmatrix} x_1 \\ x_2 \end{pmatrix} = \begin{pmatrix} 5 \\ x_2 \end{pmatrix}$ for all $X = \begin{pmatrix} x_1 \\ x_2 \end{pmatrix} \in V_2$.

II.2.11 Let $\mathbf{T} : V_1 \to V_1$, where $\mathbf{T}$ sends the vector-space of real numbers, V_1, to itself. We will let the symbol x represent the 1-vector $(x) \in V_1$. Suppose the value of $\mathbf{T}(1)$ is known. Show that if $\mathbf{T}$ is linear, then the value of $\mathbf{T}(x) \in V_1$ is automatically determined for any and every real number $x \in V_1$. (*Hint*: Use condition (II.2.1) of Definition II.2.1. Note that $x = x \cdot 1$.) Show that $\mathbf{T} : V_1 \to V_1$ is linear if and only if for some fixed scalar α_0, $\mathbf{T}(x) = \alpha_0 x$ for all $x \in V_1$.

II.2.12 Suppose that $\mathbf{T} : V_n \to V_m$ is a function of the form (II.2.4) to begin with. Using only Definition II.2.1, prove that $\mathbf{T}$ is necessarily linear.

3. Rank and Nullity of a Linear Transformation

Observe that Definition II.2.1 does not say that the range of $\mathbf{T}$ *is* a vector-space; rather we are told that the range of $\mathbf{T}$ is some *subset* of a vector-space. We will now show that, whether we like it or not, as long as $\mathbf{T}$ is linear, the range of $\mathbf{T}$ is necessarily a bonafide vector-space, and *not* just an unstructured *subset* of a vector-space.

Proposition II.3.1 Let T be a linear transformation whose domain is the vector-space V, and whose range is contained in a vector-space W (see Definition II.2.1). We conclude that the range of $\mathbf{T}$ (the set of all vectors in W of the form $\mathbf{T}(X)$, where $X \in V$; see Definition II.1.1) is a vector-space (see Definition I.5.2).

Proof: First observe that rg ($\mathbf{T}$), the range of $\mathbf{T}$, is a subset of W (rg $(\mathbf{T}) \subset W$), which is a *vector-space*. Thank goodness, then, for Theorem I.7.2, which tells us that we need only test the set rg ($\mathbf{T}$) for 2 vector-space properties of Definition I.5.2, instead of all 10. These 2 properties are closure under addition and closure under scalar multiplication.

THE RANGE OF $\mathbf{T}$ IS CLOSED UNDER $+$. We are to prove that for any 2 elements of rg ($\mathbf{T}$), their sum is always in rg ($\mathbf{T}$). All right, let us stare at 2 typical elements of rg ($\mathbf{T}$). From Definition II.1.1, these elements are images of elements X and Y, say in V, the domain of $\mathbf{T}$, and will necessarily be of the form $\mathbf{T}(X)$ and $\mathbf{T}(Y)$. To say that the sum $\mathbf{T}(X) + \mathbf{T}(Y)$ is also in the range of $\mathbf{T}$, is to say that $\mathbf{T}(X) + \mathbf{T}(Y)$ is the image of some vector in V. But $\mathbf{T}$ is assumed to be linear, and so we are entitled to the benefits of property (II.2.2) in Definition II.2.1. Thus we see that $\mathbf{T}(X) + \mathbf{T}(Y) = \mathbf{T}(X + Y)$; that is, the element $\mathbf{T}(X) + \mathbf{T}(Y)$ is the image under $\mathbf{T}$ of the element $X + Y$, which belongs to V the domain of $\mathbf{T}$.

THE RANGE OF $\mathbf{T}$ IS CLOSED UNDER SCALAR MULTIPLICATION. We are to prove that if an element belongs to the range of $\mathbf{T}$, then this element, multiplied by an arbitrary scalar, still belongs to the range of $\mathbf{T}$. Let $\mathbf{T}(X)$ be such an arbitrary element in the range of $\mathbf{T}$, and let α denote an arbitrary scalar. The element $\alpha\mathbf{T}(X)$ is exactly the element $\mathbf{T}(\alpha X)$ by property (II.2.1) of Definition II.2.1. That is, $\alpha\mathbf{T}(X)$ is the image under $\mathbf{T}$ of the vector αX, and so, indeed, $\alpha\mathbf{T}(X)$ belongs to the range of $\mathbf{T}$ for arbitrary α, and for every $X \in V$. In view of Theorem I.7.2 our proof is done.

Proposition II.3.1 allows us to sharpen Definition II.2.1 by stating that the range of $\mathbf{T}$ is a *subspace* of a certain vector-space W. This leads us in a natural way to the next definition, which has already been stated for general functions in Definition II.1.4.

Definition II.3.2 Let $\mathbf{T}$ be a linear transformation with domain V, whose range is a subspace of the vector-space W. $\mathbf{T}$ will be called **onto** $\boldsymbol{W}$, or simply, **onto** if and only if *every* vector in W is the image of some vector in V, that is, $W =$ the range of $\mathbf{T}$. $\mathbf{T}$ will be called **into** W, or simply, **into** if and only if there is some vector in W which is *not* the image of any vector in V, that is, $W \neq$ the range of $\mathbf{T}$.

Note: Since we are assuming that **T** is linear, and hence, that the range of **T** is a subspace of W (Proposition II.3.1), we may employ Theorem I.7.5 (which says that subspaces have dimension no greater than that of the superspace) to rephrase Definition II.3.2, namely, **T** is *onto* if and only if the dimension of the range of $\mathbf{T} = \dim(W)$, the dimension of W; **T** is *into* if and only if the dimension of the range of **T** is *less* than $\dim(W)$.

There is another subspace (induced by the linearity of **T**) that will be of interest to us. We refer to the set of all vectors in V, the domain of **T**, that are sent to O_2, the zero vector of the range, that is, the set of all $X \in V$ such that $\mathbf{T}(X) = O_2 \in W$. We must verify the fact that this subset of V is, indeed, blessed with the structure of a *subspace* of V. Before we do this, however, we set down some definitions and notations.

Definition II.3.3 Let **T** be a linear transformation from the vector-space V to the vector-space W, that is, V is the domain of **T** and the range of **T** is a subspace of W. The symbol **dom (T)** will denote the domain of **T**; **rg (T)** will denote the range of **T**. The set of all $X \in \operatorname{dom}(\mathbf{T})$ such that $\mathbf{T}(X) = O_2 \in \operatorname{rg}(\mathbf{T})$ is called the **null-space of T**, or the **kernel of T** and is symbolized by either of the symbols **ker (T)** or $\mathbf{T}^{-1}(\boldsymbol{0})$.

Proposition II.3.4 Let **T** be a linear transformation from vector-space V to vector-space W. The subset ker (**T**) is a subspace of $V = \operatorname{dom}(\mathbf{T})$.

Proof: The only properties we must check are the closure-under-addition and closure-under-scalar-multiplication properties (Properties A.1 and B.1 of Definition I.5.2); all other properties are automatic since all elements of $\ker(\mathbf{T}) \subset V$, inherit the remaining vector-space properties (Theorem I.7.2).

▶ ker (**T**) *is closed under addition:* Let X and Y be 2 arbitrary elements of the set ker (**T**), ($\mathbf{T}(X) = O_2$ and $\mathbf{T}(Y) = O_2$). Is $X + Y$ automatically an element of ker (**T**)? That is, does **T** send $X + Y$ to O_2? Observe that

$$\begin{aligned} \mathbf{T}(X+Y) &= T(X) + T(Y) && \text{(linearity of } \mathbf{T}) \\ &\quad\ \downarrow \qquad \swarrow \\ &= O_2 + O_2 && \text{(definition of } X \text{ and } Y) \in \ker(\mathbf{T}). \\ &= O_2 && \text{(definition of zero vector).} \end{aligned}$$

Thus $X + Y$ has met the test ($\mathbf{T}(X + Y) = O_2$) and is a member of ker (**T**).

▶ ker (**T**) *is closed under scalar multiplication:* Let X be an arbitrary element of ker (**T**), and let α be an arbitrary scalar. Is the vector αX

also an element of ker (**T**)? Observe that

$$\begin{aligned} \mathbf{T}(\alpha X) &= \alpha \mathbf{T}(X) && \text{(linearity of } \mathbf{T}) \\ &= \alpha \cdot O_2 && [\text{definition of } X \in \ker(\mathbf{T})] \\ &= O_2. \end{aligned}$$

Thus $\mathbf{T}(\alpha X) = O_2$ and the proposition is proved; ker (**T**) *is* a subspace of $V = \text{dom}(\mathbf{T})$.

REMARK. Proposition II.3.4 shows that *if there are any elements at all,* in the set ker (**T**), then, necessarily, ker (**T**) is a subspace of dom (**T**). How do we know whether the set ker (**T**) is empty or not? Perhaps *no* element of the domain of **T** gets mapped onto O_2, the zero vector of the range of **T**. The following proposition assures us that ker (**T**) is not an empty set. In fact, as long as **T** is linear, O_1, the zero vector of the domain is always sent to O_2, the zero vector of the domain. We present this proposition now.

Proposition II.3.5 Let **T** be a linear transformation from vector-space V to vector-space W, whose zero vectors will be denoted by O_1 and O_2, respectively. Then $\mathbf{T}(O_1) = O_2$. [Equivalently, the ordered pair $(O_1, O_2) = (O_1, \mathbf{T}(O_1))$ is an element of the function **T** provided **T** is linear.]

Proof: We use the fact that any vector, when multiplied by the zero scalar, yields the zero vector of the space. (Why?) Now, if X is an arbitrary vector of $V = \text{dom}(\mathbf{T})$, we have

$$\begin{aligned} \mathbf{T}(O_1) &= \mathbf{T}(0 \cdot X) && (\text{since } O_1 = 0 \cdot X) \\ &= 0 \cdot \mathbf{T}(X) && \text{(Definition II.2.1a)} \\ &= O_2, \end{aligned}$$

which proves Proposition II.3.5.

REMARK. Another path of reasoning leads us to the statement of Proposition II.3.5 as an immediate consequence of Proposition II.3.4, namely, since ker (**T**) is a *subspace*, then $O_1 \in \ker(\mathbf{T})$, since the zero element belongs to *every* vector-space. From the definition of ker (**T**) (Definition II.3.3), $\mathbf{T}(O_1) = O_2$, since *every* element of ker (**T**) is sent to O_2 under the action of **T**.

We have shown that only because a function **T** (from one vector-space to another) is linear, the set ker (**T**) is always a *subspace* of dom (**T**) (Proposition II.3.4), and the set rg (**T**) is *also* a vector-space. (Proposition II.3.1). These 2 vector-spaces are further related by the somewhat surprising (and important) fact that dim (ker (**T**)) + dim (rg (**T**)) = dim (dom (**T**)), where the symbol "dim (V)" stands for "the dimension of the vector space V." Let us develop this fact now.

Theorem II.3.6 Let **T** be a linear transformation whose domain is the vector-space V, and whose range is the vector-space W. We conclude that

$$\dim(\ker(\mathbf{T})) + \dim(\mathrm{rg}(\mathbf{T})) = \dim(\mathrm{dom}(\mathbf{T})). \tag{II.3.1}$$

Note that $V = \mathrm{dom}(\mathbf{T})$, and $W = \mathrm{rg}(\mathbf{T})$.

Proof: Let $V = V_n$, the vector-space of dimension n. We know that ker (**T**) is a (nonempty) subspace of dom $(\mathbf{T}) = V_n$ (Propositions II.3.4, II.3.5), and that its dimension, call it k, is less than or equal to n (Theorem I.7.5). Thus, let the set of nonzero vectors $\{X_1, X_2, \ldots, X_k\}$ be a basis for the subspace ker (**T**). If $\ker(\mathbf{T}) = \{O_1\}$, then the basis of ker (**T**) will be $\varnothing$, the empty set; the basis set with *no* nonzero vectors. Now this basis, a linearly independent set, can be extended to a basis for all V_n (Theorem I.6.14). Hence, let $\{X_1, X_2, \ldots, X_k, Y_1, Y_2, \ldots, Y_l\}$ represent this basis. Since any and every basis of V_n has exactly n elements (Corollary I.6.10), $k + l = n$. This equation can be rewritten as

$$\dim(\ker(\mathbf{T})) + l = \dim(V_n). \tag{II.3.2}$$

As you can see from equation (II.3.2), our proof is complete as soon as we can show that the integer l equals dim (rg (**T**)).

To this end, let us first dispose of the case where T sends every vector of $V = V_n$ to O_2, that is, $\ker(\mathbf{T}) = V_n$, $\mathrm{rg}(\mathbf{T}) = \{O_2\}$. Thus, $\dim(\ker(\mathbf{T})) = n$, $l = \dim(\mathrm{rg}(\mathbf{T})) = 0$ which shows that Equations (II.3.1) and (II.3.2) hold.

For the remainder of the proof, we may suppose that rg (**T**) is *not* the trivial subspace $\{O_2\}$, the case for which we have just shown the theorem to be true.

Now let X be an arbitrary element of V_n, the domain of **T**. X has a (unique) representation as a linear combination of the vectors in our basis (Theorem I.6.8), that is,

$$X = \alpha_1 X_1 + \alpha_2 X_2 + \cdots + \alpha_k X_k + \beta_1 Y_1 + \beta_2 Y_2 + \cdots + \beta_l Y_l,$$

for certain scalars $\alpha_1, \alpha_2, \ldots, \alpha_k, \beta_1, \beta_2, \ldots, \beta_l$.

Now any vector in rg (**T**) is necessarily of the form $\mathbf{T}(X)$, for some $X \in \mathrm{dom}(\mathbf{T})$ (Notation II.1.2). But

$$\begin{aligned}\mathbf{T}(X) &= \mathbf{T}(\beta_1 X_1 + \alpha_2 X_2 + \cdots + \alpha_k X_k + \beta_1 Y_1 + \beta_2 Y_2 + \cdots + \beta_l Y_l) \\ &= \alpha_1 \mathbf{T}(X_1) + \alpha_2 \mathbf{T}(X_2) + \cdots + \alpha_k \mathbf{T}(X_k) + \beta_1 \mathbf{T}(Y_1) + \beta_2 \mathbf{T}(Y_2) + \cdots \\ &\quad + \beta_l \mathbf{T}(Y_l).\end{aligned}$$

[Equation (II.2.3), linearity of **T**]

We observe that each X_i, $i = 1, 2, \ldots, k$ is an element of ker (**T**), so that each $\mathbf{T}(X_i) = O_2$. Thus, any and every $\mathbf{T}(X)$ in the range of $\mathbf{T}$ is of the form

$$\mathbf{T}(X) = \beta_1\mathbf{T}(Y_1) + \beta_2\mathbf{T}(Y_2) + \cdots + \beta_l\mathbf{T}(Y_l).$$

That is, any vector in rg ($\mathbf{T}$) can be expressed as a certain linear combination of the vectors $\mathbf{T}(Y_1)$, $T(Y_2),\ldots,$ $\mathbf{T}(Y_l)$, so that these vectors *span* the vector space rg ($\mathbf{T}$) (see Definition I.6.2). If this set $\{\mathbf{T}(Y_1),\mathbf{T}(Y_2),\ldots,\mathbf{T}(Y_l)\}$ were shown to be linearly independent, then it would, indeed, be a *basis* for rg ($\mathbf{T}$) (Definition I.6.7). In order to show linear independence, we will try to generate the zero vector O_2 as a linear combination of the vectors $\mathbf{T}(Y_1)$, $\mathbf{T}(Y_2),\ldots,$ $\mathbf{T}(Y_l)$, and show that this can be done *only* by setting all the scalars equal to zero, that is, set $O_2 = \beta_1\mathbf{T}(Y_1) + \beta_2\mathbf{T}(Y_2) + \cdots + \beta_l\mathbf{T}(Y_l)$ and show that every β_j, $j = 1, 2,\ldots, l$, equals zero. This will establish the linear independence of the set $\{\mathbf{T}(Y_1),\mathbf{T}(Y_2),\ldots,\mathbf{T}(T_l)\}$ (Theorem I.6.6). Now

$$O_2 = \beta_1\mathbf{T}(Y_1) + \beta_2\mathbf{T}(Y_2) + \cdots + \beta_l\mathbf{T}(Y_l)$$

or

$$O_2 = \mathbf{T}(\beta_1 Y_1 + \beta_2 Y_2 + \cdots + \beta_l Y_l)$$

by linearity of $\mathbf{T}$. This last equation says that the vector $\beta_1 Y_1 + \beta_2 Y_2 + \cdots + \beta_l Y_l \in$ ker ($\mathbf{T}$), so that it must be a linear combination of the nonzero vectors $X_1, X_2,\ldots, X_k$, which form a basis for ker ($\mathbf{T}$). That is, for some scalars $\alpha_1, \alpha_2,\ldots, \alpha_k$, we have

$$\beta_1 Y_1 + \beta_2 Y_2 + \cdots + \beta_l Y_l = \alpha_1 X_1 + \alpha_2 X_2 + \cdots + \alpha_k X_k, \qquad \text{(II.3.3)}$$

or, which is the same thing,

$$\beta_1 Y_1 + \beta_2 Y_2 + \cdots + \beta_l Y_l + (-\alpha)X_1 + (-\alpha_2)X_2 + \cdots + (-\alpha_k)X_k = O_1. \qquad \text{(II.3.4)}$$

Now the n-element ordered set $\{Y_1,Y_2,\ldots,Y_l,X_1,X_2,\ldots,X_k\}$ is a basis for V_n, and is necessarily linearly independent. So it is that each and every scalar in Equation (II.3.4) equals zero (Theorem I.6.6), and this certainly includes the β_i's.

As promised, we have shown the linear independence of the set $\{\mathbf{T}(Y_1),\mathbf{T}(Y_2),\ldots,\mathbf{T}(Y_l)\}$ which spans rg ($\mathbf{T}$). Thus, this set is a basis for rg ($\mathbf{T}$); dim (rg ($\mathbf{T}$)) $= l$. Substituting this value for l in Equation (II.3.2), proves the theorem.

One immediate consequence of this important theorem (although it can be shown directly) is the fact that linear transformations reduce or maintain the dimension of the space; this fact is formally stated as Corollary II.3.7.

Corollary II.3.7 Let **T** be a linear transformation whose domain is the vector-space V, and range is the vector-space W. Then dim (W) is less than or equal to dim (V).

Proof: Inspection of Equation (II.3.1).

Example We define the linear transformation **T** by the condition

$$\mathbf{T}\begin{pmatrix} x_1 \\ x_2 \\ x_3 \end{pmatrix} = \begin{pmatrix} x_1 \\ x_1 \\ x_2 \\ x_1 \end{pmatrix} \quad \text{for all } X = \begin{pmatrix} x_1 \\ x_2 \\ x_3 \end{pmatrix} \in V_3. \tag{II.3.5}$$

It is easily verified that **T** meets the tests required by Definition II.2.1, so that **T** is linear.

T is defined on *all* vectors of V_3, so that V_3 is the domain of **T**. Thus

$$\dim(\text{dom}(\mathbf{T})) = 3.$$

Now which vectors of V_3 get sent to O_2, the zero vector of the range space, that is, what does the subspace ker $(\mathbf{T}) \subset V_3$ look like? According to Equation (II.3.5),

$$\mathbf{T}\begin{pmatrix} x_1 \\ x_2 \\ x_3 \end{pmatrix} = O_2$$

if and only if the coordinates x_1 and x_2 equal zero. Note that the coordinate x_3 does not affect the coordinates of the vector $\mathbf{T}(X)$, as only the coordinates x_1 and x_2 appear. Thus, the vectors of the form (and *only* the vectors of the form)

$$\begin{pmatrix} 0 \\ 0 \\ x_3 \end{pmatrix}$$

are sent to O_2 under the action of **T**, where the coordinate x_3 is arbitrary and unrestricted. This set (subspace) of V_3 is ker $(\mathbf{T})$. Since every vector in this set is of the form

$$\begin{pmatrix} 0 \\ 0 \\ x \end{pmatrix} = x\begin{pmatrix} 0 \\ 0 \\ 1 \end{pmatrix},$$

the set

$$\left\{\begin{pmatrix} 0 \\ 0 \\ 1 \end{pmatrix}\right\} \quad \textit{spans} \quad \ker(\mathbf{T}).$$

This set is (trivially) *linearly independent*, and hence, forms a *basis* (Definition I.6.7) for ker $(\mathbf{T})$. Thus, dim (ker $(\mathbf{T})) = 1$ (Definition I.6.13).

Observe then any vector

$$\begin{pmatrix} x_1 \\ x_1 \\ x_2 \\ x_1 \end{pmatrix}$$

(a 4-vector whose first, second, and fourth coordinates are equal) in the subspace rg(**T**) can be expressed as a linear combination of the vectors in the set

$$\left\{\begin{pmatrix}1\\1\\0\\1\end{pmatrix}, \begin{pmatrix}0\\0\\1\\0\end{pmatrix}\right\},$$

namely

$$\begin{pmatrix}x_1\\x_1\\x_2\\x_1\end{pmatrix} = x_1\begin{pmatrix}1\\1\\0\\1\end{pmatrix} + x_2\begin{pmatrix}0\\0\\1\\0\end{pmatrix}.$$

Thus, this set *spans* rg (**T**). Moreover, this set is *linearly independent*, so that it is, indeed, a basis (Definition I.6.7) of rg (**T**). Thus, dim (rg (**T**)) = 2 (Definition I.6.13).

The previous 3 paragraphs have effected a verification of Theorem II.3.6 for certain special cases.

Observe that **T** sends all of V_3 *into* (not *onto*) the vector-space V_4. Corollary II.3.8 tells us that the dimension of rg (**T**) is less than or equal to the dimension of the domain, which is 3, in this case. Thus we could have seen immediately that **T** could not possibly have sent V_3 *onto* V_4.

At this point, the student must surely be wondering whether the section title "Rank and Nullity" has anything at all to do with the section material. It is the purpose of the following definition to establish this connection.

Definition II.3.8 Let **T** be a linear transformation from the vector space V onto the vector space W, that is, V = dom (**T**) and W = rg (**T**). Then the **rank of T** is defined as dim (W) (that is, the dimension of rg (**T**)), and the **nullity of T** is defined as dim (ker (**T**)), the dimension of the kernel of **T**.

In light of Definition II.3.8, Theorem II.3.6 may be stated, **For any linear transformation, the rank plus the nullity equals the dimension of the domain** [see Equation (II.3.1)]. This well-knit relationship leads us to believe that linearity (as per Definition II.2.1) is a much stronger property than might be supposed from the harmless-looking Equations (II.2.1) and (II.2.2).

EXERCISES

II.3.1 This exercise demonstrates a further property of functions which are *linear*. Let $\{X_1,X_2,\ldots,X_n\}$ be a basis for V_n and let **T** be a linear transformation sending V_n to some vector space W. Show that if only the vectors $\mathbf{T}(X_1)$, $\mathbf{T}(X_2),\ldots,\mathbf{T}(X_n)$ are known, then $\mathbf{T}(X)$ is known for *any* $X \in V_n$. In other words, in order to know what $\mathbf{T}(X)$ is for *any* $X \in V_n$, it is sufficient to know only the images (under $\mathbf{T}$) of some n vectors of V_n which happen to form a basis.

In the following exercises, $\mathbf{T}$ is defined on some basis set of the domain. In each case,

(a) Find $\mathbf{T}(X)$ for *arbitrary* $X \in \text{dom}(\mathbf{T})$, if only basis values are given (see Exercise II.3.1); this may not apply to all exercises.

(b) Find ker $(\mathbf{T})$, rg $(\mathbf{T})$.

(c) Find the dimensions of dom $(\mathbf{T})$, ker $(\mathbf{T})$, and rg $(\mathbf{T})$, and verify Theorem II.3.6.

II.3.2 $\mathbf{T}\begin{pmatrix} x_1 \\ x_2 \end{pmatrix} = \begin{pmatrix} x_1 + x_2 \\ 0 \end{pmatrix}$ for all $X = \begin{pmatrix} x_1 \\ x_2 \end{pmatrix} \in V_2$.

II.3.3 $\mathbf{T}\begin{pmatrix} x_1 \\ x_2 \end{pmatrix} = \begin{pmatrix} x_1 \\ x_2 \end{pmatrix}$ for all $X = \begin{pmatrix} x_1 \\ x_2 \end{pmatrix} \in V_2$.

II.3.4 $\mathbf{T}\begin{pmatrix} 1 \\ 0 \\ 0 \end{pmatrix} = \begin{pmatrix} 0 \\ 1 \\ 0 \end{pmatrix}, \mathbf{T}\begin{pmatrix} 0 \\ 1 \\ 0 \end{pmatrix} = \begin{pmatrix} 0 \\ 0 \\ 0 \end{pmatrix}, \mathbf{T}\begin{pmatrix} 0 \\ 0 \\ 1 \end{pmatrix} = \begin{pmatrix} 1 \\ 0 \\ 0 \end{pmatrix}$.

II.3.5 $\mathbf{T}\begin{pmatrix} 1 \\ 0 \end{pmatrix} = \begin{pmatrix} 2 \\ 0 \end{pmatrix}, \mathbf{T}\begin{pmatrix} 0 \\ 1 \end{pmatrix} = \begin{pmatrix} -1 \\ 0 \end{pmatrix}$.

II.3.6 $\mathbf{T}\begin{pmatrix} 1 \\ 0 \end{pmatrix} = \begin{pmatrix} 1 \\ 1 \end{pmatrix}, \mathbf{T}\begin{pmatrix} 0 \\ 1 \end{pmatrix} = \begin{pmatrix} 1 \\ 1 \end{pmatrix}$.

II.3.7 $\mathbf{T}\begin{pmatrix} x_1 \\ x_2 \\ x_3 \end{pmatrix} = \begin{pmatrix} x_1 \\ 0 \\ 0 \end{pmatrix}$ for all $X = \begin{pmatrix} x_1 \\ x_2 \\ x_3 \end{pmatrix} \in V_3$.

II.3.8 $\mathbf{T}\begin{pmatrix} x_1 \\ x_2 \\ x_3 \end{pmatrix} = \begin{pmatrix} x_1 \\ x_2 \\ 0 \end{pmatrix}$ for all $X = \begin{pmatrix} x_1 \\ x_2 \\ x_3 \end{pmatrix} \in V_3$.

II.3.9 $\mathbf{T}\begin{pmatrix} 1 \\ 0 \\ 0 \end{pmatrix} = \begin{pmatrix} 0 \\ 0 \\ 0 \end{pmatrix}, \mathbf{T}\begin{pmatrix} 0 \\ 1 \\ 1 \end{pmatrix} = \begin{pmatrix} 0 \\ 0 \\ 2 \end{pmatrix}, \mathbf{T}\begin{pmatrix} 0 \\ 0 \\ 1 \end{pmatrix} = \begin{pmatrix} 0 \\ 0 \\ -2 \end{pmatrix}$.

II.3.10 $\mathbf{T}\begin{pmatrix} 1 \\ 0 \\ 0 \end{pmatrix} = \begin{pmatrix} 0 \\ 1 \end{pmatrix}, \mathbf{T}\begin{pmatrix} 0 \\ 1 \\ 0 \end{pmatrix} = \begin{pmatrix} 1 \\ 0 \end{pmatrix}, \mathbf{T}\begin{pmatrix} 0 \\ 0 \\ 1 \end{pmatrix} = \begin{pmatrix} -1 \\ 0 \end{pmatrix}$.

II.3.11 $\mathbf{T}\begin{pmatrix} 1 \\ 0 \end{pmatrix} = \begin{pmatrix} 1 \\ 1 \\ 0 \end{pmatrix}, \mathbf{T}\begin{pmatrix} 1 \\ 1 \end{pmatrix} = \begin{pmatrix} -1 \\ 0 \\ 0 \end{pmatrix}$. Note: verify the fact that $\left\{\begin{pmatrix} 1 \\ 0 \end{pmatrix}, \begin{pmatrix} 1 \\ 1 \end{pmatrix}\right\}$ forms a *basis* for V_2 so that you are sure that you have enough information (see Definition I.6.7 and Exercise II.3.1).

II.3.12 $\mathbf{T}\begin{pmatrix} x_1 \\ x_2 \end{pmatrix} = \begin{pmatrix} x_1 + 2x_2 \\ x_1 - x_2 \\ x_1 + x_2 \end{pmatrix}$ for all $X = \begin{pmatrix} x_1 \\ x_2 \end{pmatrix} \in V_2$.

II.3.13 $\mathbf{T}\begin{pmatrix} x_1 \\ x_2 \end{pmatrix} = \begin{pmatrix} x_2 \\ x_2 \\ x_2 \end{pmatrix}$.

II.3.14 $\mathbf{T}\begin{pmatrix}1\\0\\0\end{pmatrix} = (3), \mathbf{T}\begin{pmatrix}0\\1\\0\end{pmatrix} = (0), \mathbf{T}\begin{pmatrix}0\\0\\1\end{pmatrix} = (-3).$

II.3.15 $\mathbf{T}(2) = \begin{pmatrix}2\\2\\0\end{pmatrix}$ (compare with Exercise II.2.11).

The concluding portion of this section addresses itself to the invertibility of linear transformations. We have already established the fact that any onto function, linear or not, has an inverse if and only if that function is 1-1 (Proposition II.1.7 and Exercise II.1.17). However, we are considering *linear* transformations and in this special case, we are able to isolate additional conditions which are equivalent to invertibility. This is the purpose of the following theorems.

Theorem II.3.9 Let $\mathbf{T}$ be a linear transformation from vector space V to vector-space W. Then $\mathbf{T}$ is 1-1 if and only if $\ker(\mathbf{T}) = \{O_1\}$, the trivial vector-space in V.

Proof: We tackle the easier implication first and show that $\mathbf{T}$ is 1-1 $\Rightarrow \ker(\mathbf{T}) = \{O_1\}$. Now $\ker(\mathbf{T})$ is precisely the set of vectors in V that are sent to O_2, the zero vector of W (Definition II.3.3). In any case, $\mathbf{T}(O_1) = O_2$ since $\mathbf{T}$ is linear (Proposition II.3.5). We are assuming that $\mathbf{T}$ is 1-1; hence, O_1 must be the *only* vector of V that gets sent to O_2. We have shown that if $\mathbf{T}$ is 1-1, then $\ker(\mathbf{T})$ is the 1-element set $\{O_1\}$.

We will establish the converse; that is, $\ker(\mathbf{T}) = \{O_1\} \Rightarrow \mathbf{T}$ is 1-1. Let us pretend for the moment that there are 2 vectors in V, X, and Y, say, that are sent to the same vector under the action of $\mathbf{T}$. It will develop that if $\ker(\mathbf{T})$ is assumed to be just $\{O_1\}$ (which it is), then X and Y must necessarily be one and the same vector. That is, $\mathbf{T}$ cannot send any 2 *distinct* vectors to the same vector, which is to say, $\mathbf{T}$ is 1-1. Watch this development now.

We are supposing that $\mathbf{T}(X) = \mathbf{T}(Y)$ in W. Thus

$$\mathbf{T}(X) - \mathbf{T}(Y) = O_2$$

or

$$\mathbf{T}(X - Y) = O_2 \qquad \text{(Why?)}$$

The previous equation says that the vector $X - Y \in \ker(\mathbf{T})$, which equals $\{O_1\}$ by hypothesis. Since $\ker(\mathbf{T})$ has but 1 element O_1, then, necessarily, the vector $X - Y$ is precisely that element. That is, $X - Y = O_1$, which is equivalent to the statement $X = Y$ (add Y to both sides of the previous equation). Our promise is realized; as long as it is assumed that $\ker(\mathbf{T}) = \{O_1\}$, then $\mathbf{T}$ is necessarily 1-1. The theorem is proved.

We are now in a position to attach the result of Theorem II.3.9 to that of Theorem II.3.6 to obtain another equivalent condition for invertibility of a linear transformation. Accordingly, we set down Corollary II.3.10.

Corollary II.3.10 Let $\mathbf{T}$ be a linear transformation from n-dimensional vector-space V_n to vector-space W. Then $\mathbf{T}$ is 1-1 if and only if the rank of $\mathbf{T}$ $(= \dim(\operatorname{rg}(T))) = n$, the dimension of the domain of $\mathbf{T}$. In other words, $\mathbf{T}$ is 1-1 if and only if $\mathbf{T}$ *preserves* dimension (see Corollary II.3.7).

Proof: According to Theorem II.3.9,

$$\mathbf{T} \quad \text{is} \quad \text{1-1} \rightleftharpoons \ker(\mathbf{T}) = \{O_1\}$$

But

$$\ker(\mathbf{T}) = \{O_1\} \rightleftharpoons \dim(\ker(\mathbf{T})) = 0$$

(see Definition I.6.13). A glance at the result of Theorem II.3.6 [see Equation II.3.1)], tells us that

$$\dim(\ker(\mathbf{T})) = 0 \rightleftharpoons \dim(\operatorname{rg}(\mathbf{T})) = \dim(\operatorname{dom}(\mathbf{T})).$$

There we have it! The 3 displayed equivalences in conjunction tell us that, indeed,

$$\mathbf{T} \quad \text{is} \quad \text{1-1} \rightleftharpoons \dim(\operatorname{rg}(\mathbf{T})) = \dim(\operatorname{dom}(\mathbf{T})),$$

which was to be proved.

We will now list those conditions that we have discovered to be equivalent to invertibility of a linear transformation $\mathbf{T}$ sending n-dimensional space V_n onto (not *into*) vector-space W.

(a) $\mathbf{T}^{-1}$ exists

(b) $\mathbf{T}$ is 1-1

(c) $\ker(\mathbf{T}) = \{O_1\}$

(d) $\operatorname{rank}(\mathbf{T}) = n$

(e) $\mathbf{T}$ preserves dimension.

After our study of matrices and determinants, we shall be able to add other conditions to the list, which are equivalent to invertibility.

It is entirely possible to wonder, at this point, whether it is *worth* studying conditions under which a linear transformation possesses an inverse. A partial justification for the lavish attention spent on the inverse of a linear transformation $\mathbf{T}$, is the fact that if $\mathbf{T}^{-1}$ exists (as a function), then $\mathbf{T}^{-1}$ is automatically a *linear* function (transformation). This is the news that our next theorem brings.

Theorem II.3.11 Let $\mathbf{T}$ be a linear transformation sending n-dimensional space V_n onto n-dimensional space W_n (so that existence of $\mathbf{T}^{-1}$ is a certainty). Then $\mathbf{T}^{-1}$ sending W_n onto V_n, is always a *linear* transformation.

Proof: In order to show that $\mathbf{T}^{-1}$ is linear, we need only to verify that Equations (II.2.1) and (II.2.2) (see Definition II.2.1) obtain.

We will verify (II.2.1). Let X be an arbitrary vector in W_n. Since $\mathbf{T}$ is onto and 1-1, X has 1 and only 1 pre-image in V_n, call it X'. That is, there is a unique $X' \in V_n$ such that

$$\mathbf{T}(X') = X \qquad \text{(see Figure 2.12).} \tag{II.3.6}$$

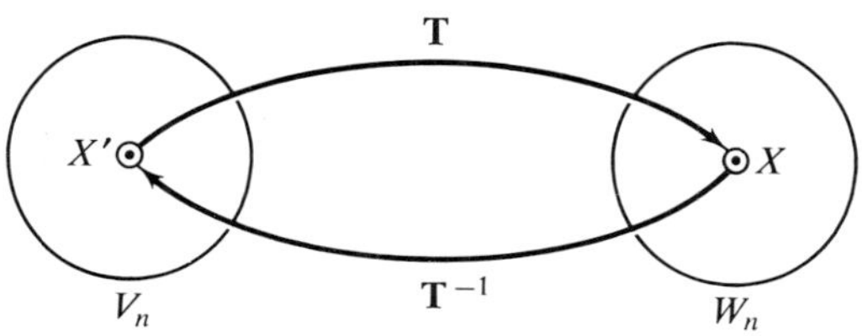

Figure 2.12

Another way of saying the same thing is

$$\mathbf{T}^{-1}(X) = X'. \tag{II.3.7}$$

Multiplying both sides of (II.3.6) by an arbitrary scalar α, yields

$$\alpha\mathbf{T}(X') = \alpha X. \tag{II.3.8}$$

But $\mathbf{T}$ is linear, and is thereby forced to obey condition (II.2.1) of Definition II.2.1, so that (II.3.8) becomes

$$\mathbf{T}(\alpha X') = \alpha X, \tag{II.3.9}$$

which is exactly the same statement as (taking $\mathbf{T}^{-1}$ of both sides)

$$\mathbf{T}^{-1}(\alpha X) = \alpha X', \quad \text{for any scalar } \alpha \text{ and any } X \in W_n. \tag{II.3.10}$$

Now, observe that according to (II.3.7), $X' = \mathbf{T}^{-1}(X)$. If we substitute $\mathbf{T}^{-1}(X)$ for X' in (II.3.10), we obtain

$$\mathbf{T}^{-1}(\alpha X) = \alpha\mathbf{T}^{-1}(X) \quad \text{for all scalars } \alpha \text{ and all } X \in W_n. \tag{II.3.11}$$

That is, $\mathbf{T}^{-1}$ behaves according to Equation (II.2.1) of Definition II.2.1.

The verification of Equation (II.2.2) is left as an exercise for the student (Exercise II.3.16).

EXERCISES

II.3.16 If **T** is a linear transformation from V_n onto W_n, then show that

$$\mathbf{T}^{-1}(X + Y) = \mathbf{T}^{-1}(X) + \mathbf{T}^{-1}(Y) \text{ for all } X, Y \in W_n.$$

In the following exercises, find $\mathbf{T}^{-1}$

II.3.17 $\mathbf{T}\begin{pmatrix} x_1 \\ x_2 \end{pmatrix} = \begin{pmatrix} -x_1 \\ x_2 \end{pmatrix}$ for all $X = \begin{pmatrix} x_1 \\ x_2 \end{pmatrix} \in V_2$.

II.3.18 $\mathbf{T}\begin{pmatrix} x_1 \\ x_2 \end{pmatrix} = \begin{pmatrix} x_2 \\ x_1 \end{pmatrix}$ for all $X = \begin{pmatrix} x_1 \\ x_2 \end{pmatrix} \in V_2$.

II.3.19 $\begin{pmatrix} x_1 \\ x_2 \end{pmatrix} = \begin{pmatrix} 3x_1 \\ 3x_2 \end{pmatrix}$ for all $X = \begin{pmatrix} x_1 \\ x_2 \end{pmatrix} \in V_2$.

II.3.19 $\mathbf{T}\begin{pmatrix} x_1 \\ x_2 \end{pmatrix} = \begin{pmatrix} x_1 + x_2 \\ x_2 \end{pmatrix}$ for all $X = \begin{pmatrix} x_1 \\ x_2 \end{pmatrix} \in V_2$.

4. *Algebraic Properties of Linear Transformations*

We shall return to some of the basic ideas which were introduced in Chapter I. Recall, that after defining a concrete n-vector, we immediately proceeded to the business of combining (adding) pairs of n-vectors (Definition I.1.2) and of multiplying n-vectors by arbitrary numbers (Definition I.1.3). In what follows, it will be shown that we can define operations of addition (between pairs of transformations) and of scalar multiplication on certain sets of linear transformations. Indeed, we shall see that relative to this addition and scalar multiplication, all the axioms given in Definition I.5.2 will be satisfied, that is, this certain set of linear transformations will qualify as a bonafide abstract *vector-space*.

But this is not all. In certain cases, we will be able to effect a multiplication between pairs of linear transformations (a notion we have already seen; Definition II.1.5), yielding a *linear* transformation.

Let us delve into these rules of combination now and see what relations hold among them. Our first definition will describe the sum of 2 transformations. Which function will result from such a sum? A complete description will be had once we know the *image* (output) of any vector in the domain (input) of this function. Accordingly, we set down this definition exactly.

Definition II.4.1 Let $\mathbf{T}_1$ and $\mathbf{T}_2$ be linear transformations, each of which sends vector-space V to vector-space W. The sum of $\mathbf{T}_1$ and $\mathbf{T}_2$, denoted $\mathbf{T}_1 \underset{\cdot}{+} \mathbf{T}_2$, **the sum of** $\mathbf{T}_1$ **and** $\mathbf{T}_2$ is the function defined by the equation

$$(\mathbf{T}_1 \underset{\cdot}{+} \mathbf{T}_2)(X) = \mathbf{T}_1(X) + \mathbf{T}_2(X) \quad \text{for all } X \in V.$$

We make some observation. First, since the linear transformations $\mathbf{T}_1$ and $\mathbf{T}_2$ each sends V and W, so does the function $\mathbf{T}_1 \underset{\cdot}{+} \mathbf{T}_2$. Secondly, the different plus signs $\underset{\cdot}{+}$ and $+$ used in the definition denote additions acting on different types of paired objects; $\underset{\cdot}{+}$ acts between pairs of linear transformations, while $+$ acts between pairs of vectors in W.

It would be hoped that this "new" function generated from the 2 linear transformations $\mathbf{T}_1$ and $\mathbf{T}_2$ is *more* than just a function. Indeed, this is the case, for $(\mathbf{T}_1 \underset{\cdot}{+} \mathbf{T}_2)$ is a *linear* transformation, as we shall show in Theorem II.4.2. Before we prove the linearity property in general, let us consider an example.

Example Let $\mathbf{T}_1$ and $\mathbf{T}_2$ each send V_2 to V_2 according to the following equations

$$\mathbf{T}_1\begin{pmatrix} x_1 \\ x_2 \end{pmatrix} = \begin{pmatrix} x_1 \\ 0 \end{pmatrix} \quad \text{and} \quad \mathbf{T}_2\begin{pmatrix} x_1 \\ x_1 \end{pmatrix} = \begin{pmatrix} -x_1 + x_2 \\ 2x_2 \end{pmatrix}$$

for all $\begin{pmatrix} x_1 \\ x_2 \end{pmatrix} \in V_2$. Then what is the effect of the newly created (linear?) function $(\mathbf{T}_2 \underset{\cdot}{+} \mathbf{T}_2)$ on the arbitrary vector $\begin{pmatrix} x_1 \\ x_2 \end{pmatrix}$? Definition II.4.1 tells us that

$$\begin{aligned}
(\mathbf{T}_1 \underset{\cdot}{+} \mathbf{T}_2)\begin{pmatrix} x_1 \\ x_2 \end{pmatrix} &= \mathbf{T}_1\begin{pmatrix} x_1 \\ x_2 \end{pmatrix} + \mathbf{T}_2\begin{pmatrix} x_1 \\ x_2 \end{pmatrix} \\
&= \begin{pmatrix} x_1 \\ 0 \end{pmatrix} + \begin{pmatrix} -x_1 + x_2 \\ 2x_2 \end{pmatrix} && \text{Definition of } \mathbf{T}_1, \mathbf{T}_2 \\
&= \begin{pmatrix} x_2 \\ 2x_2 \end{pmatrix} && \text{Definition I.1.3.}
\end{aligned}$$

That is, $(\mathbf{T}_1 \underset{\cdot}{+} \mathbf{T}_2)\begin{pmatrix} x_1 \\ x_2 \end{pmatrix} = \begin{pmatrix} x_2 \\ 2x_2 \end{pmatrix}$ for all $\begin{pmatrix} x_1 \\ x_2 \end{pmatrix} \in V_2$. As we see, in *this special case*, $(\mathbf{T}_2 \underset{\cdot}{+} \mathbf{T}_2)$ is linear. One example does not constitute a proof, and so the time is ripe to show that linearity of $(\mathbf{T}_1 \underset{\cdot}{+} \mathbf{T}_2)$ is always an unavoidable consequence of Definition II.4.1. We do this with Theorem II.4.2.

Theorem II.4.2 Given $\mathbf{T}_1$ and $\mathbf{T}_2$, each of which is a linear transformation sending vector-space V to vector-space W. Then the function $(\mathbf{T}_1 \underset{\cdot}{+} \mathbf{T}_2)$ is a linear transformation sending V to W.

Proof: Our ultimate test for linearity is, of course, Definition II.2.1. We establish the validity of Equation (II.2.1) relative to $(\mathbf{T}_1 \underset{\cdot}{+} \mathbf{T}_2)$. Now let α be any scalar and let X be any vector of V. Then

$$\begin{aligned}
(\mathbf{T}_1 \underset{\cdot}{+} \mathbf{T}_2)(\alpha X) &= \mathbf{T}_1(\alpha X) + \mathbf{T}_2(\alpha X) && \text{Definition II.4.1} \\
&= \alpha\mathbf{T}_1(X) + \alpha\mathbf{T}_2(X) && \text{(II.4.1)}
\end{aligned}$$

since $\mathbf{T}_1$ and $\mathbf{T}_2$ are linear, scalars "factor out" according to Equation (II.2.1).

$$= \alpha(\mathbf{T}_1(X) + \mathbf{T}_2(X)) \qquad \text{axiom B.5 of Definition I.5.2.} \qquad (\text{II.4.2})$$

See Exercise I.1.12.

If we refer once more to Definition II.4.1, we see that the expression $\mathbf{T}_1(X) + \mathbf{T}_2(X)$ inside the parentheses of Equation (II.4.2) is defined to be $(\mathbf{T}_1 \dotplus \mathbf{T}_2)(X)$. Substituting, then, allows us to continue our equality as follows

$$= \alpha((\mathbf{T}_1 \dotplus \mathbf{T}_2)(X)). \qquad (\text{II.4.3})$$

Equating (II.4.1) to (II.4.3), yields

$$(\mathbf{T}_1 \dotplus \mathbf{T}_2)(\alpha X) = \alpha((\mathbf{T}_1 \dotplus \mathbf{T}_2)(X)) \qquad (\text{II.4.4})$$

for all scalars α and all $X \in V$. Loosely speaking, scalars "factor out" and the condition of Equation (II.2.1) of Definition II.2.1 are satisfied relative to the function $(\mathbf{T}_1 \dotplus \mathbf{T}_2)$.

With a bit less fanfare, we proceed to establish Equation (II.2.2) of Definition II.2.1 (our final gap toward establishing linearity of $\mathbf{T}_1 \dotplus \mathbf{T}_2$).

Now let X and Y be arbitrary vectors of V. Then

$$(\mathbf{T}_1 \dotplus \mathbf{T}_2)(X + Y) = \mathbf{T}_1(X + Y) + \mathbf{T}_2(X + Y) \qquad \text{Definition II.4.1} \quad (\text{II.4.5})$$

$$= (\mathbf{T}_1(X) + \mathbf{T}_1(Y)) + (\mathbf{T}_2(X) + \mathbf{T}_2(Y)) \qquad \text{linearity of } \mathbf{T}_1, \mathbf{T}_2. \quad (\text{II.4.6})$$

$$= (\mathbf{T}_1(X) + \mathbf{T}_2(X)) + (\mathbf{T}_1(Y) + \mathbf{T}_2(Y)) \qquad \text{commutivity and associativity of vectors in } W. \quad (\text{II.4.7})$$

$$= (T_1 \dotplus T_2(X)) + (T_1 \dotplus T_2)(Y) \qquad \text{Definition II.4.1} \quad (\text{II.4.8})$$

Thus, Equation (II.4.5) coupled with (II.4.8) tells us that

$$(\mathbf{T}_1 \dotplus \mathbf{T}_2)(X + Y) = (\mathbf{T}_1 \dotplus \mathbf{T}_2)(X) + (\mathbf{T}_1 \dotplus \mathbf{T}_2)(Y) \qquad (\text{II.4.9})$$

for all X and $Y \in V$. That is, Equation (II.2.2) of Definition (II.2.1) obtains, and the linearity property of the function $(\mathbf{T}_1 \dotplus \mathbf{T}_2)$ is now an established fact.

As promised, we shall defined a new function, $\alpha \circ \mathbf{T}$, which is obtained by multiplying a linear transformation and a scalar.

Definition II.4.3 Let $\mathbf{T}$ be a linear transformation from vector space V to vector space W. If α is an arbitrary scalar, then the (new) function $\boldsymbol{\alpha}\mathbf{T}$ is defined by the equation

$$\alpha\mathbf{T}(X) = \alpha(\mathbf{T}(X)) \qquad \text{for all } X \in V.$$

Example If

$$\mathbf{T}\begin{pmatrix} x_1 \\ x_2 \\ x_3 \end{pmatrix} = \begin{pmatrix} x_1 + 2x_2 \\ x_3 \end{pmatrix} \quad \text{for all} \quad \begin{pmatrix} x_1 \\ x_2 \\ x_3 \end{pmatrix} \in V_3$$

is a linear transformation (sending V_3 to V_2), and if α is taken to equal 7, then for all

$$\begin{pmatrix} x_1 \\ x_2 \\ x_3 \end{pmatrix} \in V_3,$$

$$7\mathbf{T}\begin{pmatrix} x_1 \\ x_2 \\ x_3 \end{pmatrix} = 7\left(\mathbf{T}\begin{pmatrix} x_1 \\ x_2 \\ x_3 \end{pmatrix}\right) \qquad \text{Definition II.4.3}$$

$$= 7\begin{pmatrix} x_1 + 2x_2 \\ x_3 \end{pmatrix} \qquad \text{definition of } \mathbf{T}\begin{pmatrix} x_1 \\ x_2 \\ x_3 \end{pmatrix}$$

$$= \begin{pmatrix} 7(x_1 + 2x_2) \\ 7x_3 \end{pmatrix} \qquad \text{Definition I.1.3}$$

$$= \begin{pmatrix} 7x_1 + 14x_2 \\ 7x_3 \end{pmatrix}.$$

Were we to take the trouble, we would see that the function (**7T**) is *linear* (Definition II.2.1). Is this accidental? Theorem II.4.4 says quite the contrary.

Theorem II.4.4 Let **T** be a linear transformation sending vector-space V to vector-space W. For any scalar α, the function $\alpha\mathbf{T}$ is a *linear* transformation.

Proof: We first verify Equation (II.2.1) of Definition II.2.1 by showing that scalars "factor out."

Let X be arbitrary in V and let β be any scalar. We have

$(\alpha\mathbf{T})(\beta X) = \alpha(\mathbf{T}(\beta X))$	Definition II.4.3, substitute βX for X.	(II.4.10)
$= \alpha(\beta(\mathbf{T}(X)))$	**T** is linear, hence β factors out, see Equation (II.2.1).	
$= (\alpha\beta)(\mathbf{T}(X))$	axiom B.2 of Definition I.5.2.	
$= (\beta\alpha)(\mathbf{T}(X))$	commutivity, $\alpha\beta = \beta\alpha$.	
$= \beta(\alpha(\mathbf{T}(X)))$	axiom B.2 of Definition I.5.2.	
$= \beta(\alpha\mathbf{T}(X))$	Definition II.4.3.	(II.4.11)

Equations (II.4.10) and (II.4.11) tell us that

$$(\alpha\mathbf{T})(\beta X) = \beta(\alpha\mathbf{T}(X)) \quad \text{for all } X \in V \tag{II.4.12}$$

and all scalars β, that is, β "factors out" and Equation (II.2.1) of Definition II.2.1 holds for $\alpha\mathbf{T}$.

To show that Equation (II.2.2) of Definition II.2.1 obtains for $\alpha\mathbf{T}$, we must show that

$$\alpha\mathbf{T}(X+Y)=\alpha\mathbf{T}(X)+\alpha\mathbf{T}(Y) \quad \text{for all } X,\ Y \in V.$$

Now, for any X, $Y \in V$,

$$\begin{aligned}
\alpha\mathbf{T}(X+Y) &= \alpha(\mathbf{T}(X+Y)) && \text{Definition II.4.3.} && \text{(II.4.13)}\\
&= \alpha(\mathbf{T}(X)+\mathbf{T}(Y)) && \text{linearity of } \mathbf{T}. &&\\
&= \underbrace{\alpha(\mathbf{T}(X))}_{\downarrow} + \underbrace{\alpha(\mathbf{T}(Y))}_{\swarrow} && \text{Axiom B.5 of Definition I.5.2.} &&\\
&= \alpha\mathbf{T}(X) + \alpha\mathbf{T}(Y) && \text{Definition II.4.3.} && \text{(II.4.14)}
\end{aligned}$$

Equations (II.4.13) and (II.4.14) together establish the validity of Equation (II.2.2) of Definition II.2.I relative to the function $\alpha\mathbf{T}$, that is, $\alpha\mathbf{T}$ is *linear* and the proof is done.

REMARK. You may recall that in our discussion of vectors, we were concerned with the mechanics of adding 1 n-vector to another, and of multiplying an n-vector by a scalar. In our present discussion, we have also been concerned with the adding of 1 linear transformation to another, and with the multiplying of a linear transformation by a scalar. Our 2 theorems (Theorems II.4.2 and II.4.4) are closure theorems, that is, they show that the property of *linearity* is *not destroyed* after effecting this addition and scalar multiplication.

So what?

Could it be that we have a "new" *vector-space* on our hands? If, after all, we (very, very loosely) think of a vector-space as a set of objects that can somehow be added together in pairs, and can be multiplied by scalars, in each case, giving us an object of the "same sort," then we are close to this when we speak of the set of all the linear transformations sending 1 vector-space to another. True, enough, we *can* add 2 linear transformations and get from this a linear transformation. We *can* multiply a linear transformation by a scalar and we still have a linear transformation.

This is a good start, but Definition I.5.2 sets out many rules *other* than closure for a vector-space. It so happens that the closure properties (keeping linearity after performing our addition and scalar multiplication) are probably the hardest properties to verify. It will be an exercise for the student to check out the remaining requirements of Definition I.5.2.

We wish to state precisely what our suspected vector-space is. We do this now.

Definition II.4.5 Let $\mathscr{L}(V,W)$ symbolize the set of all linear transformations sending vector-space V to vector-space W.

Theorem II.4.6 Under the operations of addition and scalar multiplication defined according to Definitions II.4.1 and II.4.3, $\mathscr{L}(V,W)$ is a vector-space.

Proof: In consulting our list of vector-space characteristics (Definition I.5.2) we see at once that Theorem II.4.2 is exactly a confirmation of property A.1 $[(\mathbf{T}_1 \dotplus \mathbf{T}_2) \in \mathscr{L}(V,W)$ whenever $\mathbf{T}_1 \in \mathscr{L}(V,W)$ and $\mathbf{T}_2 \in \mathscr{L}(V,W)]$. Also, Theorem II.4.4 tells us that property B.1 obtains $[\alpha\mathbf{T} \in \mathscr{L}(V,W)$ for all $\mathbf{T} \in \mathscr{L}(V,W)$ and all numbers $\alpha]$.

Just for fun, let us show that property B.4 is valid in $\mathscr{L}(V,W)$; the remaining properties are exercises. We wish to show that

$$(\alpha + \beta)\mathbf{T} = \alpha\mathbf{T} \dotplus \beta\mathbf{T} \quad \text{for all } \mathbf{T} \in \mathscr{L}(V,W) \tag{II.4.15}$$

and all scalars α and β. We are to show that the 2 (apparently different) functions of (II.4.15) are, in fact, the same. But 2 functions (linear or not) are the same if and only if each of the functions sends every element of the domain to the same element in the range (equal "inputs" imply equal "outputs"). Thus, the 2 functions of (II.4.15) are the *same* if and only if, for all $X \in V$,

$$(\alpha + \beta)\mathbf{T}(X) = (\alpha\mathbf{T} \dotplus \beta\mathbf{T})(X). \tag{II.4.16}$$

[*Note*: the "$=$" symbol in (II.4.15) indicates equality of functions in $\mathscr{L}(V,W)$, while the "$=$" symbol in (II.4.16) indicates equality of vectors in W.] Beginning with the left-hand side of (II.4.16), we have

$$(\alpha + \beta)\mathbf{T}(X) = (\alpha + \beta)(\mathbf{T}(X)) \qquad \text{Definition II.4.3 where the scalar } (\alpha+\beta) \text{ is substituted for } \alpha. \tag{II.4.17}$$

Now the right-hand side of (II.4.17) tells us that we are multiplying $\mathbf{T}(X)$, a *vector* in W by the scalar $(\alpha + \beta)$. W is a perfectly good vector-space, so we may apply property B.4 of Definition I.5.2 to the right side of (II.4.17), yielding

$$(\alpha + \beta)(\mathbf{T}(X)) = \alpha(\mathbf{T}(X)) + \beta(\mathbf{T}(X)) \tag{II.4.18}$$

$$\begin{aligned} &\quad\;\downarrow \qquad\qquad\quad \downarrow \\ &= \alpha\mathbf{T}(X) \;+\; \beta\mathbf{T}(X) && \text{Definition II.4.3.} \\ &= ((\alpha\mathbf{T}) \dotplus (\beta\mathbf{T}))(X) && \text{Definition II.4.1, replace } \mathbf{T}_1 \text{ with } \alpha\mathbf{T}, \text{ and } \mathbf{T}_2 \text{ with } \beta\mathbf{T}. \end{aligned} \tag{II.4.19}$$

Now the left-hand side of Equation (II.4.17) has been shown to equal (II.4.19), that is, Equation (II.4.16) is valid, which establishes (II.4.15). Therefore property B.4 obtains for the space $\mathscr{L}(V, W)$.

EXERCISES

Exercises II.4.1 through II.4.9 concern the operators $\mathbf{T}_1, \mathbf{T}_2 \in \mathscr{L}(V, W)$ which are defined as follows

$$\mathbf{T}_1\begin{pmatrix} x_1 \\ x_2 \end{pmatrix} = \begin{pmatrix} x_1 \\ 0 \\ 0 \end{pmatrix}, \quad \mathbf{T}_2\begin{pmatrix} x_1 \\ x_2 \end{pmatrix} = \begin{pmatrix} x_1 + x_2 \\ x_1 - x_2 \\ x_2 \end{pmatrix} \quad \text{for all} \quad \begin{pmatrix} x_1 \\ x_2 \end{pmatrix} \in V_2.$$

II.4.1 Find the operator $(\mathbf{T}_1 \dotplus \mathbf{T}_2)$ and verify that it is a genuine *linear* operator in $\mathscr{L}(V_2, V_3)$. (See Definition II.2.1 and axiom A.1 of Definition I.5.2.)

II.4.2 Is it true that $\mathbf{T}_1 \dotplus \mathbf{T}_2 = \mathbf{T}_2 \dotplus \mathbf{T}_1$? (See axiom A.2, Definition I.5.2.)

II.4.3 Is there a linear operator (call it $\mathbf{0}$, the zero operator) in $\mathscr{L}(V_2, V_3)$ with the (feeble) property that $\mathbf{T}_1 \dotplus \mathbf{0} = \mathbf{T}_1$ and $\mathbf{T}_2 \dotplus \mathbf{0} = \mathbf{T}_2$? If so, describe its action on an arbitrary vector in V_2, that is, what is $\mathbf{0}\begin{pmatrix} x_1 \\ x_2 \end{pmatrix}$ (which must be a vector in V_3) for $\begin{pmatrix} x_1 \\ x_2 \end{pmatrix} \in V_2$? (See axiom A.3, Definition I.5.2.)

II.4.4 Does a linear operator exist (call it $\mathbf{T}_2'$) in $\mathscr{L}(V_2, V_3)$ with the property that $\mathbf{T}_2' \dotplus \mathbf{T}_2 = \mathbf{0}$, the zero operator in $\mathscr{L}(V_2, V_3)$? If so, describe $\mathbf{T}_2'\begin{pmatrix} x_1 \\ x_2 \end{pmatrix}$ (a vector in V_3) for every $\begin{pmatrix} x_1 \\ x_2 \end{pmatrix} \in V_2$. We note that $\mathbf{T}_2'$ is usually symbolized $-\mathbf{T}_2$ and is called the *negative of* or the *additive inverse of the operator* $\mathbf{T}_2$. (See axiom A.4, Definition I.5.2.)

II.4.5 If we are given $\mathbf{T}_3 \in \mathscr{L}(V_2, V_3)$ defined by

$$\mathbf{T}_3\begin{pmatrix} x_1 \\ x_2 \end{pmatrix} = \begin{pmatrix} x_1 \\ -x_2 \\ x_1 \end{pmatrix} \quad \text{for all} \quad \begin{pmatrix} x_1 \\ x_2 \end{pmatrix} \in V_2,$$

then verify the following relation (associative law):

$$(\mathbf{T}_1 \dotplus \mathbf{T}_2) \dotplus \mathbf{T}_3 = \mathbf{T}_1 \dotplus (\mathbf{T}_2 \dotplus \mathbf{T}_3).$$

(See axiom A.5, Definition I.5.2.)

II.4.6 Describe the operators $\alpha\mathbf{T}_1$ and $\alpha\mathbf{T}_2$ (see Definition II.4.3) where (a) $\alpha = 2$; (b) $\alpha = -2$; (c) $\alpha = 0$; (d) $\alpha = 1$; (e) $\alpha = \frac{3}{2}$; (f) $\alpha = b$. Verify in each of these cases, that $\alpha\mathbf{T}$ is a bona-fide *linear* operator in $\mathscr{L}(V_2, V_3)$. (See Definition II.2.1, and axioms B.1 and B.3 of Definition I.5.2.)

II.4.7 Is it true that $2\cdot(7\mathbf{T}_1) = (2 \times 7)\mathbf{T}_1$ and that $3\cdot(5\mathbf{T}_2) = (3 \times 5)\mathbf{T}_2$? Prove or disprove this by describing the images of the vector $\begin{pmatrix} x_1 \\ x_2 \end{pmatrix}$ produced by the (apparently) different operators on each side of the equation. (See axiom B.2, Definition I.5.2.)

II.4.8 Verify the relation

$$(3 + 5)\mathbf{T}_i = 3\mathbf{T}_i \dotplus 5\mathbf{T}_i$$

in *each* of the 3 cases $\mathbf{T}_i = \mathbf{T}_1, \mathbf{T}_2, \mathbf{T}_3$. Do this by comparing the images of an arbitrary vector $\begin{pmatrix} x_1 \\ x_2 \end{pmatrix} \in V_2$ produced by the (apparently) different operators $(3+5)\mathbf{T}_i$ and $3\mathbf{T}_i \dotplus 5\mathbf{T}_i$. (See axiom B.4, Definition I.5.2) See also Definitions II.4.1 and II.4.3 and the proof of Theorem II.4.6.)

II.4.9 Show that $8(\mathbf{T}_1 \dotplus \mathbf{T}_2) = 8\mathbf{T}_1 \dotplus 8\mathbf{T}_2$ by comparing the images of $\begin{pmatrix} x_1 \\ x_2 \end{pmatrix}$ produced (in V_3) by the left-hand operator and the right-hand operator of the given equation. (See axiom B.5, Definition I.5.2.)

II.4.10 Exercises II.4.1 through II.4.9 asked for verification of the vector-space axioms (Definition I.5.2) where the linear operators $\mathbf{T}_1$, $\mathbf{T}_2$, and $\mathbf{T}_3$ were far from arbitrary. In this exercise, verify each of the vector-space axioms (except A.1, B.1, and B.4, which have already been verified in the proofs of Theorems II.4.2, II.4.4, and II.4.6, respectively) for the set of operators $\mathscr{L}(V_n, V_m)$. That is, prove that $\mathscr{L}(V_n, V_m)$, the set of all linear functions with domain V_n and range in V_m, is a vector-space over the numbers.

We have just established the fact that $\mathscr{L}(V, W)$ is a vector-space over the numbers. Inspired by Definition II.1.5 (which describes a multiplication or composition of functions), we set our stage so that one linear operator can be followed by another. (See Figure 2.13.)

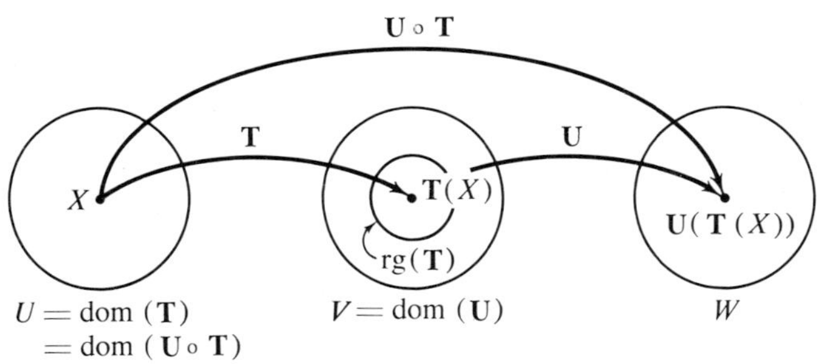

Figure 2.13

Accordingly, let us consider the 2 vector-spaces $\mathscr{L}(U, V)$ and $\mathscr{L}(V, W)$. Let an operator $\mathbf{T}$ belong to $\mathscr{L}(U, V)$ and let $\mathbf{U} \in \mathscr{L}(V, W)$. Then, since rg $(\mathbf{T})$, the range of $\mathbf{T}$, is a subset (in fact, a sub*space*) of V by the very definition of $\mathscr{L}(U, V)$ (Definition II.4.5), the operator $\mathbf{U} \circ \mathbf{T}$ is meaningful in this context (see Definition II.1.5). Recall that $\mathbf{U} \circ \mathbf{T}$ sends any vector $X \in U$ to the vector $\mathbf{U}(\mathbf{T}(X))$, that is, for all $X \in U$, $\mathbf{U} \circ \mathbf{T}(X) = \mathbf{U}(\mathbf{T}(X))$.

Example Let

(a) $$\mathbf{T}\begin{pmatrix} x_1 \\ x_2 \end{pmatrix} = \begin{pmatrix} 2x_1 + x_2 \\ x_2 \\ -x_2 \end{pmatrix} \quad \text{for all} \quad \begin{pmatrix} x_1 \\ x_2 \end{pmatrix} \in V_2,$$

and let

(b) $$\mathbf{U}\begin{pmatrix} x_1 \\ x_2 \\ x_3 \end{pmatrix} = \begin{pmatrix} x_1 + x_2 - x_3 \\ 2x_3 \end{pmatrix} \quad \text{for all} \quad \begin{pmatrix} x_1 \\ x_2 \\ x_3 \end{pmatrix} \in V_3.$$

We note that $\mathbf{T} \in \mathscr{L}(V_2, V_3)$ and $\mathbf{U} \in \mathscr{L}(V_3, V_2)$. Also, $\mathbf{T}$ sends any 2-vector $\begin{pmatrix} x_1 \\ x_2 \end{pmatrix}$ to a 3-vector. Happily, $\mathbf{U}$ accepts 3-vectors for its input (domain of $\mathbf{U} = V_3$) and sends these back to V_2 (range of $\mathbf{U} \subset V_2$). In advance, we can see that the composite (linear?) function $\mathbf{U} \circ \mathbf{T}$ will start with a 2-vector and send it to a 2-vector. We do not know in advance whether or not the function $\mathbf{U} \circ \mathbf{T}$ is a *linear* function. Let us determine $\mathbf{U} \circ \mathbf{T}$. Now

$$\mathbf{U} \circ \mathbf{T}\begin{pmatrix} x_1 \\ x_2 \end{pmatrix} = \mathbf{U}\left(\mathbf{T}\begin{pmatrix} x_1 \\ x_2 \end{pmatrix}\right) \qquad \text{Definition II.1.5}$$

$$= \mathbf{U}\begin{pmatrix} 2x_1 + x_2 \\ x_2 \\ -x_2 \end{pmatrix} \qquad \text{Definition of } \mathbf{T}.$$

From the definition of $\mathbf{U}$ we see that $\mathbf{U}$ "does the following" in order to convert a 3-vector to a 2-vector:

In order to construct the first entry of $\mathbf{U}(X)$ for *any* 3-vector X, we take the sum of the first 2 entries of X and subtract the third entry of X from this sum; in order to construct the second entry of $\mathbf{U}(X)$, simply take the third entry of the 3-vector X and multiply it by 2 [this is exactly a verbalization of the content of (b)]. Abiding by these rules, then,

$$\mathbf{U}\begin{pmatrix} 2x_1 + 2 \\ x_2 \\ -x_2 \end{pmatrix} = \begin{pmatrix} (2x_1 + x_2) + (x_2) - (-x_2) \\ 2(-x_2) \end{pmatrix}$$

$$= \begin{pmatrix} 2x_2 + 3x_2 \\ -2x_2 \end{pmatrix}.$$

That is,

$$\mathbf{U} \circ \mathbf{T}\begin{pmatrix} x_1 \\ x_2 \end{pmatrix} = \begin{pmatrix} 2x_1 + 3x_2 \\ -2x_2 \end{pmatrix}$$

so that $\mathbf{U} \circ \mathbf{T}$ is now given explicitly as a function from V_2 to V_2. A further check against Definition II.2.1 would show that $\mathbf{U} \circ \mathbf{T}$ is even a *linear* function, that is, $\mathbf{U} \circ \mathbf{T} \in \mathscr{L}(V_2, V_2)$. That this is not just dumb luck is the message of Theorem II.4.7.

Theorem II.4.7 Let U, V, and W be vector-spaces. Let $\mathbf{T} \in \mathscr{L}(U, V)$ and $\mathbf{U} \in \mathscr{L}(V, W)$. Then the function $\mathbf{U} \circ \mathbf{T}$ sending U to W is always a linear function (transformation) from U to W, that is, $\mathbf{U} \circ \mathbf{T} \in \mathscr{L}(U, W)$.

Proof: Needless to say, we return to Equations (II.2.1) and (II.2.2) of Definition II.2.1 and check the function $\mathbf{U} \circ \mathbf{T}$ against these 2 conditions. Now for any $X \in U$ and any scalar α, we have $\mathbf{U} \circ \mathbf{T}(\alpha X) = \alpha(\mathbf{U} \circ \mathbf{T}(X))$, that is, the scalar α "factors out" of the function $\mathbf{U} \circ \mathbf{T}$, so that condition (II.2.1) of Definition II.2.1 is satisfied.

In verifying Equation (II.2.2), we let X and Y be arbitrary members of U and consider

$$\begin{aligned} \mathbf{U} \circ \mathbf{T}(X+Y) &= \mathbf{U}(\mathbf{T}(X+Y)) && \text{Definition II.1.5 of } \mathbf{U} \circ \mathbf{T} \\ &= \mathbf{U}(\mathbf{T}(X)+\mathbf{T}(Y)) && \text{since } \mathbf{T} \text{ is linear and obeys} \\ & && \text{(II.2.2) of Definition II.2.1.} \end{aligned} \qquad \text{(II.4.20)}$$

Now the quantities $\mathbf{T}(X)$ and $\mathbf{T}(Y)$ are merely two vectors in V so that in applying the additive property of $\mathbf{U}$ [Equation (II.2.2) of Definition II.2.1] to them, we have

$$\begin{aligned} \mathbf{U}(\mathbf{T}(X)+\mathbf{T}(Y)) &= \mathbf{U}(\mathbf{T}(X))+\mathbf{U}(\mathbf{T}(Y)) \\ & \qquad\quad\downarrow \qquad\qquad\quad \downarrow \\ &= \mathbf{U} \circ \mathbf{T}(X)+\mathbf{U} \circ \mathbf{T}(Y) && \text{Definition II.1.5 of } \mathbf{U} \circ \mathbf{T}. \end{aligned} \qquad \text{(II.4.21)}$$

Setting (II.4.20) equal to (II.4.21) reveals that for any pair of vectors $X, Y \in U$,

$$\mathbf{U} \circ \mathbf{T}(X+Y) = \mathbf{U} \circ \mathbf{T}(X) + U \circ \mathbf{T}(Y)$$

which is all that we require to establish condition (II.2.2) of Definition II.2.1. The function $\mathbf{U} \circ \mathbf{T}$ is a *linear* transformation, and the proof is done.

EXERCISES

In Exercises II.4.11 through II.4.18, we are given

$$\mathbf{T}_1\begin{pmatrix} x_1 \\ x_2 \end{pmatrix} = \begin{pmatrix} -x_2+x_1 \\ 3x_1 \end{pmatrix} \qquad \text{for all } \begin{pmatrix} x_1 \\ x_2 \end{pmatrix} \in V_2$$

$$\mathbf{T}_2\begin{pmatrix} x_1 \\ x_2 \end{pmatrix} = \begin{pmatrix} x_1-x_2 \\ x_1+x_2 \\ 2x_1 \end{pmatrix} \qquad \text{for all } \begin{pmatrix} x_1 \\ x_2 \end{pmatrix} \in V_2$$

$$\mathbf{T}_3\begin{pmatrix} x_1 \\ x_2 \\ x_3 \end{pmatrix} = \begin{pmatrix} x_1-x_2+x_3 \\ x_2-x_3 \end{pmatrix} \qquad \text{for all } \begin{pmatrix} x_1 \\ x_2 \\ x_3 \end{pmatrix} \in V_3.$$

II.4.11 Describe the operator $\mathbf{T}_1 \circ \mathbf{T}_1$.

II.4.12 Describe the operator $\mathbf{T}_1 \circ \mathbf{T}_3$.

II.4.13 Describe the operator $(\mathbf{T}_3) \circ (\mathbf{T}_2 \circ \mathbf{T}_1)$; note that the parentheses indicate that the operator $\mathbf{T}_3$ follows the operator $(\mathbf{T}_2 \circ \mathbf{T}_1)$.

II.4.14 Describe the operator $(\mathbf{T}_3 \circ \mathbf{T}_2) \circ (\mathbf{T}_1)$; note that the parentheses indicate that the operator $(\mathbf{T}_3 \circ \mathbf{T}_2)$ follows the operator $\mathbf{T}_1$ (see Exercise II.4.13).

II.4.15 Describe the operator $\mathbf{T}_2 \circ \mathbf{T}_3$.

II.4.16 Describe the operator $\mathbf{T}_3 \circ \mathbf{T}_2$ (compare with Exercise II.4.15).

II.4.17 Describe the operator $(\mathbf{T}_2) \circ (\mathbf{T}_3 \circ \mathbf{T}_2)$.

II.4.18 Describe the operator $(\mathbf{T}_2 \circ \mathbf{T}_3) \circ (\mathbf{T}_2)$ (compare with Exercise II.4.17).

II.4.19 If all has gone well, you will have noticed that the apparently different operators of Exercises II.4.13 and II.4.14 agree as do the operators of Exercises II.4.17 and II.4.18. The purpose of this exercises is to prove that we were not just lucky. In fact, let the linear operators $\mathbf{T}_1$, $\mathbf{T}_2$, and $\mathbf{T}_3$ belong to the vector-spaces $\mathscr{L}(U,V)$, $\mathscr{L}(V,W)$, and $\mathscr{L}(W,Z)$, respectively, where U, V, W, and Z are vector-spaces. Prove that

(a) $$\mathbf{T}_3 \circ (\mathbf{T}_2 \circ \mathbf{T}_1) = (\mathbf{T}_3 \circ \mathbf{T}_2) \circ \mathbf{T}_1.$$

This property (a) is called the *associative property* of the multiplication (composition of operators) operation $\circ$. That is, the operation $\circ$ acts only between *pairs* of operators (see Definition II.1.5 and Exercise II.1.31) so that parentheses are necessary to "associate" the appropriate pairs of operators in the above expression. On the left-hand side, $\mathbf{T}_2$ is associated (combined) with $\mathbf{T}_1$ and then this operator $\mathbf{T}_2 \circ \mathbf{T}_1$ is associated with $\mathbf{T}_3$. On the right-hand side, however, $\mathbf{T}_1$ is associated with the operator $\mathbf{T}_3 \circ \mathbf{T}_2$. The property that is to be proven is that no matter how one associates pairs of operators in a triple product, the end result is always the same. Is linearity necessary for (a)?

In what follows, we shall consider the vector-space $\mathscr{L}(V,W)$ for the very special case where $V = W$. Why on earth would we want to do that? One good reason is that we do not have to worry as to whether or not we can follow 1 linear operator of this set with another. This is *always* possible. In other words, if $\mathbf{T}_1$ and $\mathbf{T}_2$ belong to $\mathscr{L}(V,W)$, then not only do $\mathbf{T}_1 \circ \mathbf{T}_2$ and $\mathbf{T}_2 \circ \mathbf{T}_1$ *exist*, but each of these composite operators is, again, an element of $\mathscr{L}(V,V)$. A more formal way of saying this is that the vector-space $\mathscr{L}(V,V)$ is *closed under the operation of composition (multiplication)* $\circ$. Recall that in establishing the fact that $\mathscr{L}(V,V)$ was a vector-space, we showed that (among other things) $\mathscr{L}(V,V)$ was closed under the operations of addition $\dot{+}$ (Theorem II.4.2) and scalar multiplication (Theorem II.4.4).

We will show that $\mathscr{L}(V,V)$ has structural properties in addition to the ordinary vector-space properties. The operations of $\dot{+}$ (addition) and $\circ$ (composition), each of which sends a pair of linear transformations to a third linear transformations, are, in fact, related! This is the content of the next theorem.

Theorem II.4.8 Given the vector-space $\mathscr{L}(V,V)$ of all linear transformations sending vector-space V to itself, then for all $\mathbf{T}, \mathbf{U}, \mathbf{V} \in \mathscr{L}(V,V)$, the following relations obtain:

$$\mathbf{T} \circ (\mathbf{U} \dot{+} \mathbf{V}) = \mathbf{T} \circ \mathbf{U} \dot{+} \mathbf{T} \circ \mathbf{V}. \qquad \text{(II.4.22a)}$$

$$(\mathbf{T} \dot{+} \mathbf{U}) \circ \mathbf{V} = \mathbf{T} \circ \mathbf{V} \dot{+} \mathbf{U} \circ \mathbf{V}. \qquad \text{(II.4.22b)}$$

Equation (II.4.22a) says that $\mathbf{T}$ "distributes itself through the parentheses" from the left and (II.4.22b) says that $\mathbf{V}$ "distributes itself through the parentheses" from the right.

Proof: We shall prove only (II.4.22a), leaving (II.4.22b) as an exercise. Thus, we are obliged to show that for all $X \in V$, and all $\mathbf{T}, \mathbf{U}, \mathbf{V} \in \mathscr{L}(V, V)$,

$$\mathbf{T} \circ (\mathbf{U} \dotplus \mathbf{V})(X) = (\mathbf{T} \circ \mathbf{U} \dotplus \mathbf{T} \circ \mathbf{V})(X). \tag{II.4.23}$$

Let us begin with the left-hand side of (II.4.24) and in a step-by-step fashion (using only legal methods), transform it to the right-hand side of (II.4.24). Now,

$$\begin{aligned}
\mathbf{T} \circ (\mathbf{U} \dotplus \mathbf{V})(X) &= \mathbf{T}((\mathbf{U} \dotplus \mathbf{V})(X)) && \text{[Definition II.1.5 applied to } \mathbf{T} \circ (\mathbf{U} \dotplus \mathbf{V})] && \text{(II.4.24)}\\
&= \mathbf{T}(\mathbf{U}(X) + \mathbf{V}(X)) && \text{(Definition II.4.1 applied to } \mathbf{U} \dotplus \mathbf{V})\\
&= \mathbf{T}(\mathbf{U}(X)) + \mathbf{T}(\mathbf{V}(X)) && \text{[Equation (II.2.2) of Definition II.2.1 obtains for } \mathbf{T} \text{ which is linear]}\\
&= \mathbf{T} \circ \mathbf{U}(X) + \mathbf{T} \circ \mathbf{V}(X) && \text{(Definition II.1.5)}\\
&= (\mathbf{T} \circ \mathbf{U} \dotplus \mathbf{T} \circ \mathbf{V})(X) && \text{(Definition II.4.1).} && \text{(II.4.25)}
\end{aligned}$$

We have done as we have promised. Expressions (II.4.24) and (II.4.25) are equal; hence, (II.4.23) obtains, which is to say, (II.4.22a) obtains. Except for establishing (II.4.22b), the student's share of the proof (see Exercise II.4.26), the proof is done.

You have probably noticed by now that we have spoken of linear transformations from 2 points of view. On one hand, we considered the results of applying a linear transformation to elements of a vector-space (compare Propositions II.2.2, II.3.1, II.3.4, II.3.5, Theorem II.3.6, and so on); on the other hand, we also *ignored* the underlying vector-space and considered the effects of combining a linear operator with *other* linear operators (compare Theorems II.4.2, II.4.4, II.4.6, II.4.7, and II.4.8). Roughly, then, we can (and do) study linear transformations by their relationships (a) to the underlying vectors or (b) to other linear operators. In defining a certain operator, the identity operator $\mathbf{1} \in \mathscr{L}(V, V)$, we shall follow viewpoint (b). It is an exercise for the student (Exercise II.4.27) to establish the fact that there exists a viewpoint (a) definition for $\mathbf{1}$, and that this definition is *equivalent* to Definition II.4.9.

Definition II.4.9 Consider the vector-space of linear operators $\mathscr{L}(V,V)$. If we can find a linear operator, call it **1**, with the (feeble multiplication) property that

$$\mathbf{1} \circ \mathbf{T} = \mathbf{T} \circ \mathbf{1} = \mathbf{T} \quad \text{for every} \quad \mathbf{T} \in \mathscr{L}(V,V), \tag{II.4.26}$$

Then **1** is called the **identity operator**, **identity transformation** or **identity** of $\mathscr{L}(V,V)$.

REMARK. We notice that the preceding definition releases 2 (seemingly awkward) points, which we now discuss.

First, we state "*If* we can find" an operator **1** that does such-and-such; but can it be that *no* operator in $\mathscr{L}(V,V)$ obliges itself to act as (II.4.26) says it should? Even if existence *is* assured [as it shall be in Exercise II.4.27(b)], then we are faced with the second awkward point; the possibility of oversupply, or nonuniqueness. If there *is* such an identity, then must there be only 1 such? The student will validly be able to make the claim of *uniqueness* . . . once he has completed Exercise II.4.27(c).

In advancing to this point, we have unearthed and studied certain properties of n-tuples (vectors), linear transformations acting *on* these n-tuples, and collections $\mathscr{L}(V,W)$ of linear transformations. We retrace our steps so as to introduce certain mathematical terminology for the various structures that we have *already* encountered.

Definition II.4.10

GROUP. Any collection of objects upon which a pairwise addition is defined in accordance with axioms A.1, A.3, A.4, and A.5 of Definition I.5.2 is called a **group**. If A.2 obtains, as well, then the group is called a **commutative group (abelian group)**. The integers form a (commutative) group under the usual addition.

VECTOR-SPACE. Any collection of objects for which a pairwise addition is defined in accordance with axioms A.1 through A.5 of Definition I.5.2 (that is, any commutative group) which accepts the further property of scalar *multiplication* as per axioms B.1 through B.5 of Definition I.5.2, is called a **vector-space**. For example, the set of n-tuples of real numbers forms a vector-space with the addition and scalar multiplication defined in Definitions I.1.2 and I.1.3. Also, $\mathscr{L}(V,W)$ is a real vector-space according to Theorem II.4.6.

RING. A set R of objects which is a commutative group under some pairwise addition $+$, and accepts a pairwise multiplication operation $\cdot$ which is associative and distributive, is called a **ring**. That is, the following axioms obtain:

A.1 through A.5 of Definition I.5.2 (the commutative group axioms)

C.1 For all x and y of the set R, which we are calling a ring, the product $x \cdot y$ is also an element of the ring R (closure axiom for $\cdot$)

C.2 For all (not necessarily distinct) x, y, and z of the ring R,

$$x \cdot (y \cdot z) = (x \cdot y) \cdot z \qquad \text{(associativity axiom for } \cdot).$$

C.3 For all x, y, and z of the ring R,

$$x \cdot (y + z) = (x \cdot y) + (x \cdot z)$$

and

$$(y + z) \cdot x = (y \cdot x) + (z \cdot x) \qquad \text{(distributivity).}$$

If, moreover, a (feeble) element $\mathbf{1}$ exists in R such that

C.4 For all x of the ring R,

$$\mathbf{1} \cdot x = x \cdot \mathbf{1} = x \qquad \text{(existence of a multiplicative identity } \mathbf{1})$$

then we call the set R a **ring with identity**.

As an example, the set of all integers, with the usual $+$ and $\times$ operations, is a ring with identity **1**, while the set of all *even* integers is a ring (without identity). The set $\mathscr{L}(V, V)$ is a ring (of linear transformations) with identity **1**.

[*Remark.* We have not mentioned the notion of scalar multiplication in connection with a ring. Loosely speaking, a commutative group which accepts a scalar multiplication is a vector-space; analagously (and just as loosely), any ring which accepts a scalar multiplication is an **algebra**. Equivalently, we might say that an algebra is a vector-space which accepts a multiplication. To be precise, we define *algebra*.]

ALGEBRA. An **algebra** is any collection of objects on which an addition and a multiplication are defined on pairs of objects, and on which scalar multiplication is defined in accordance with the following:

Axioms A.1 through A.5 of Definition I.5.2

Axioms B.1 through B.5 of Definition I.5.2

Axioms C.1 through C.3 in above definition of "ring."

The set of all real numbers with the usual addition and multiplication is an algebra, as is the set $\mathscr{L}(V,V)$.

We remark that C.3 is not an axiom concerning *only* addition or *only* multiplication; rather, it is this axiom that *relates* the 2 operations.

Figure 2.14 summarizes these definitions.

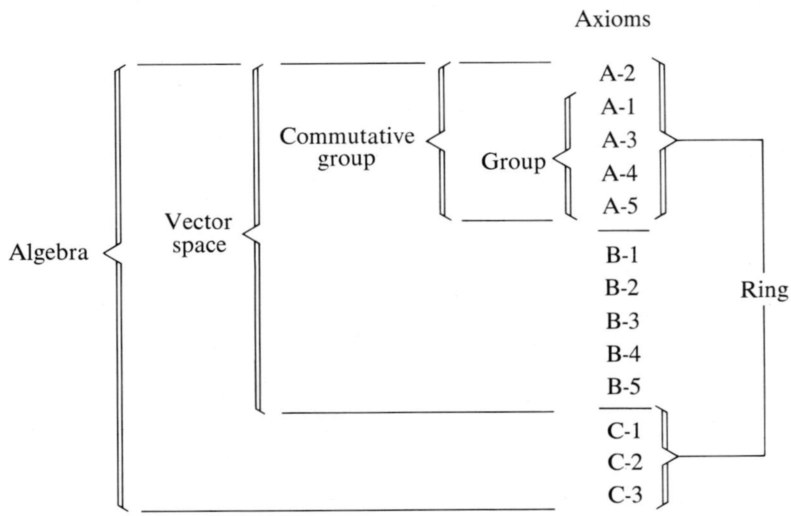

Figure 2.14

EXERCISES

For Exercises II.4.20 through II.4.25, let **T**, **U**, **V** $\in \mathscr{L}(V_2,V_2)$ be defined as follows:

$$\mathbf{T}\begin{pmatrix} x_1 \\ x_2 \end{pmatrix} = \begin{pmatrix} 0 \\ x_1 \end{pmatrix}, \mathbf{U}\begin{pmatrix} x_1 \\ x_2 \end{pmatrix} = \begin{pmatrix} x_1 + x_2 \\ x_1 - x_2 \end{pmatrix}, \mathbf{V}\begin{pmatrix} x_1 \\ x_2 \end{pmatrix} = \begin{pmatrix} -2x_1 \\ x_1 - x_2 \end{pmatrix} \quad \text{for all } \begin{pmatrix} x_1 \\ x_2 \end{pmatrix} \in V_2.$$

II.4.20 Find $\mathbf{T} \circ (\mathbf{U} \circ \mathbf{V})\begin{pmatrix} x_1 \\ x_2 \end{pmatrix}$ for all $\begin{pmatrix} x_1 \\ x_2 \end{pmatrix} \in V_2$.

II.4.21 Find $\mathbf{U} \circ (\mathbf{T} \circ \mathbf{V})\begin{pmatrix} x_1 \\ x_2 \end{pmatrix}$ for all $\begin{pmatrix} x_1 \\ x_2 \end{pmatrix} \in V_2$.

II.4.22 Find $(\mathbf{T} \circ \mathbf{U}) \circ \mathbf{V}\begin{pmatrix} x_1 \\ x_2 \end{pmatrix}$ for all $\begin{pmatrix} x_1 \\ x_2 \end{pmatrix} \in V_2$.

II.4.23 Find $\mathbf{T} \circ (\mathbf{U} + \mathbf{V})\begin{pmatrix} x_1 \\ x_2 \end{pmatrix}$

and

$$\mathbf{T} \circ \mathbf{U}\begin{pmatrix} x_1 \\ x_2 \end{pmatrix} + \mathbf{T} \circ \mathbf{V}\begin{pmatrix} x_1 \\ x_2 \end{pmatrix} \quad \text{for all } \begin{pmatrix} x_1 \\ x_2 \end{pmatrix} \in V_2.$$

II.4.24 Find $(\mathbf{U} \dotplus \mathbf{V}) \circ \mathbf{T}\begin{pmatrix} x_1 \\ x_2 \end{pmatrix}$

and

$$\mathbf{U} \circ \mathbf{T}\begin{pmatrix} x_1 \\ x_2 \end{pmatrix} \dotplus \mathbf{V} \circ \mathbf{T}\begin{pmatrix} x_1 \\ x_2 \end{pmatrix} \quad \text{for all } \begin{pmatrix} x_1 \\ x_2 \end{pmatrix} \in V_2.$$

II.4.25 Find $(\mathbf{U} \dotplus \mathbf{T}) \circ \mathbf{V}\begin{pmatrix} y \\ z \end{pmatrix}$

and

$\mathbf{U} \circ \mathbf{V}\begin{pmatrix} y \\ z \end{pmatrix} \dotplus \mathbf{T} \circ \mathbf{V}\begin{pmatrix} y \\ z \end{pmatrix}$ for all $\begin{pmatrix} y \\ z \end{pmatrix} \in V_2$. Compute $\mathbf{V} \circ (\mathbf{U} \dotplus \mathbf{T})\begin{pmatrix} y \\ z \end{pmatrix}$ and $\mathbf{V} \circ \mathbf{U}\begin{pmatrix} y \\ z \end{pmatrix} \dotplus \mathbf{V} \circ \mathbf{T}\begin{pmatrix} y \\ z \end{pmatrix}$ for all $\begin{pmatrix} y \\ z \end{pmatrix} \in V_2$.

II.4.26 (See Theorem II.4.8.) Show that for all linear operators $\mathbf{T}, \mathbf{U}, \mathbf{V} \in \mathscr{L}(V,V)$, Equation (II.4.22b) obtains; that is, $(\mathbf{T} \dotplus \mathbf{U}) \circ \mathbf{V} = \mathbf{T} \circ \mathbf{V} \dotplus \mathbf{U} \circ \mathbf{V}$.

II.4.27 (See Definition II.4.9.) Let $\mathbf{A}$ be a function from vector-space V to itself which "leaves every vector intact," that is, for every vector $X \in V$, $\mathbf{A}(X) = X$.

(a) Show that $\mathbf{A}$ is a *linear* transformation sending V to itself (see Definition II.2.1). That is, prove that $\mathbf{A} \in \mathscr{L}(V,V)$.

(b) Prove that $\mathbf{A}$ satisfies the condition (II.4.26) of an identity operator of $\mathscr{L}(V,V)$ (thus, the *existence* of at least 1 identity for $\mathscr{L}(V,V)$ is assured).

(c) Let $\mathbf{B}_1$ and $\mathbf{B}_2$ be 2 (apparently different) linear operators in $\mathscr{L}(V,V)$ which are identities for $\mathscr{L}(V,V)$, that is, for all $\mathbf{T} \in \mathscr{L}(V,V)$, $\mathbf{B}_1 \circ \mathbf{T} = \mathbf{T} \circ \mathbf{B}_1 = \mathbf{T}$ and $\mathbf{B}_2 \circ \mathbf{T} = \mathbf{T} \circ \mathbf{B}_2 = \mathbf{T}$. Prove that $\mathbf{B}_1 = \mathbf{B}_2$ (thus, the existence of *at most 1* identity is assured).

We note that part (b) shows us that *at least* 1 identity exists, while part (c) has it that *at most 1* identity exists for $\mathscr{L}(V,V)$. These 2 remarks blend to the single statement that a *unique* identity *exists* for $\mathscr{L}(V,V)$.

II.4.28 (a) Do the integers form a group under the usual addition?

(b) Do the integers form a vector-space under the usual addition?

(c) Do the integers form a ring under the usual addition and multiplication?

(d) Do the integers form a ring under the usual addition, and under the (unusual) multiplication $*$ defined between pairs of integers by

$$a * b = 0$$

for all integers a, b? If not, precisely which ring properties of Definition II.4.10 do *not* hold?

II.4.29 (a) Prove that $\mathscr{L}(V_n, V_n)$ is an algebra over the scalars.

(b) Describe the (action of the) linear functions $\mathbf{0}$, and $\mathbf{1} \in \mathscr{L}(V_n, V_n)$, the additive (feeble) identity and the multiplicative (feeble) identity of $\mathscr{L}(V_n, V_n)$, respectively.

II.4.30 Is $\mathscr{L}(V_n, V_n)$ an algebra over the scalars under the addition $\hat{\dotplus}$ defined by

$$\mathbf{T} \mathbin{\hat{\dotplus}} \mathbf{V} = \mathbf{0} \quad \text{for all } \mathbf{T}, \mathbf{V} \in \mathscr{L}(V_n, V_n),$$

and under the usual multiplication (that is, composition) of transformations? If not, precisely which algebra properties of Definition II.4.10 *fail* in this example?

5. *Matrices of Linear Transformations (the 2 × 2 case)*

One of our observations about the set of linear transformations $\mathscr{L}(V,W)$ is that $\mathscr{L}(V,W)$ is an (abstract) vector-space (Theorem II.4.6) in accordance with Definition I.5.2. That is, each linear transformation from V to W is, in its own right, a *vector*. We recall that at the outset, we defined an n-vector to be an ordered n-tuple of real numbers (Definition I.1.1); an ordered n-tuple, such as $\begin{pmatrix} 3 \\ -1 \end{pmatrix}$, may be thought of as a concrete "picture" representation of a vector in n-dimensional space V_n. Of course, $\begin{pmatrix} 3 \\ -1 \end{pmatrix}$ represents a vector in V_2. It would appear as though a "picture" representation of our newly discovered vectors [linear transformations in $\mathscr{L}(V,W)$] would be desirable. It is the purpose of this section to present such a number-picture, consisting of a certain arrangement of numbers, for vectors which happen to be linear transformations. Thus, we will have a concrete picture form for linear transformations as well as for n-vectors. For a first pass at developing this construction, we shall consider the problem of building a number-picture (which we shall call a *matrix*) for those linear transformations in $\mathscr{L}(V_2,V_2)$, sending 2-space V_2 to itself.

Toward this end, we let $\mathscr{X} = \{X_1,X_2\}$ and $\mathscr{Y} = \{Y_1,Y_2\}$ be 2 (not necessarily distinct) ordered bases for V_2, and let $\mathbf{T}$ be a linear transformation sending V_2 to V_2.

For convenience, Figure 2.15 illustrates the domain and range of $\mathbf{T}$ as being disjoint sets. We recall that *any* vector of V_2 can be represented in one and only one way as a linear combination of basis vectors (Theorem I.6.8). In particular, we are justified in using this happy property to

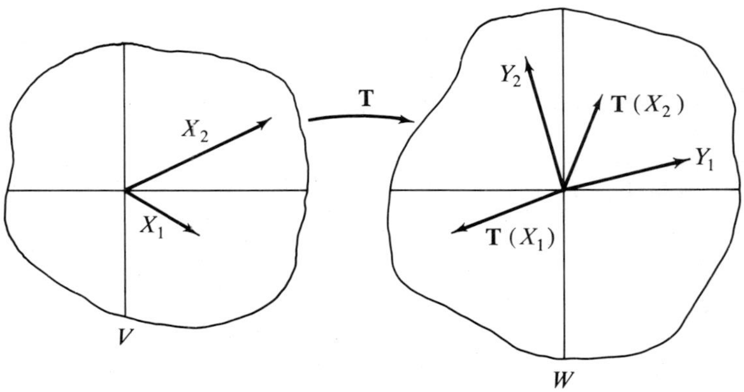

Figure 2.15

express the vectors $\mathbf{T}(X_1)$ and $\mathbf{T}(X_2)$ in terms of the basis vectors Y_1 and Y_2. And so it is, that there are *unique* scalars a_{11}, a_{21}, a_{21}, a_{22} with the property that

$$\begin{aligned} \mathbf{T}(X_1) &= \overbrace{a_{11} \cdot Y_1 + a_{21}}^{\text{column 1}} \cdot Y_2 \\ \mathbf{T}(X_2) &= \overbrace{a_{12} \cdot Y_1 + a_{22}}^{\text{column 1}} \cdot Y_2 \end{aligned} \tag{II.5.1}$$

Now look at that! Having chosen a pair of ordered bases (1 basis for the domain V, another for the impact set W), and having chosen a linear transformation $\mathbf{T}$, we set down the equations of (II.5.1) which offer us 4 *unique* scalars. These scalars are shaded in (II.5.1). Let us seize these unique scalars which grew from our choice of ordered bases $\mathscr{X}$ and $\mathscr{Y}$, and linear transformation $\mathbf{T}$, and isolate them. Thus, we shall have our matrix (number-picture) for $\mathbf{T}$. Toward this end, we peel the scalars a_{11} and a_{21} from the $\mathscr{Y}$ representation for $\mathbf{T}(X_1)$ [see (II.5.1)]; let this pair of scalars constitute our first matrix *column* (see Figure 2.16). Similarly, by lifting the scalars a_{12} and a_{22} from the $\mathscr{Y}$ representation of $\mathbf{T}(X_2)$ [see (II.5.1)], we obtain our second matrix *column* (see Figure 2.16). We graphically present this matrix-building procedure in Figure 2.16.

Definition II.5.1 The rectangular arrangement of scalars (right-hand side of Figure 2.16) is the matrix of $\mathbf{T}$ which we symbolize as

$$mx\mathbf{T}_{\mathscr{X},\mathscr{Y}} = \begin{pmatrix} a_{11} & a_{12} \\ a_{21} & a_{22} \end{pmatrix}, \tag{II.5.2}$$

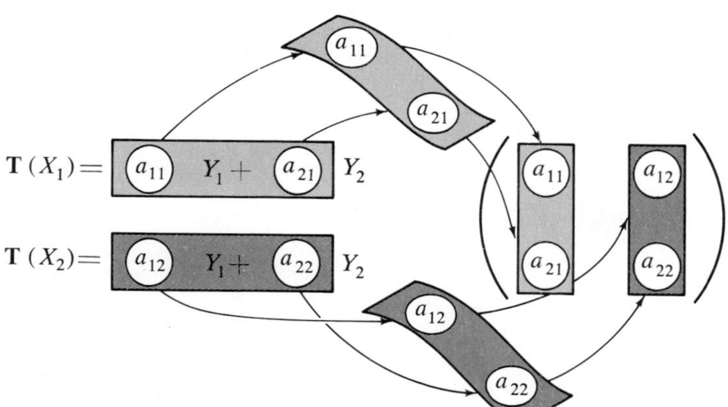

Figure 2.16

where the symbol $mx\mathbf{T}_{\mathscr{X},\mathscr{Y}}$ is read, the **matrix of T relative to the ordered bases $\mathscr{X}$ and $\mathscr{Y}$**. We must be mindful of the fact that this matrix depends upon the *pair of bases* $\mathscr{X}$ and $\mathscr{Y}$, and their *orderings* as well as the transformation **T**. Thus, if we change either of these bases, or the *ordering* of a basis, we should not be surprised if we wind up with a different matrix. The following examples will illustrate this fact.

Example Let $\mathbf{T} \in \mathscr{L}(V_2, V_2)$ be defined by the equation

$$\mathbf{T}\begin{pmatrix} x_1 \\ x_2 \end{pmatrix} = \begin{pmatrix} x_1 + x_2 \\ 0 \end{pmatrix}, \quad \text{for all } X = \begin{pmatrix} x_1 \\ x_2 \end{pmatrix} \in V_2.$$

Let us choose the following pair of bases for V_2.

$$\mathscr{X} = \left\{ \underset{\substack{\uparrow \\ X_1}}{\begin{pmatrix} 1 \\ 1 \end{pmatrix}}, \underset{\substack{\uparrow \\ X_2}}{\begin{pmatrix} 1 \\ 0 \end{pmatrix}} \right\}, \qquad \mathscr{Y} = \left\{ \underset{\substack{\uparrow \\ Y_1}}{\begin{pmatrix} 3 \\ 0 \end{pmatrix}}, \underset{\substack{\uparrow \\ Y_2}}{\begin{pmatrix} 1 \\ 2 \end{pmatrix}} \right\}.$$

The student should verify the fact that the sets $\mathscr{X}$ and $\mathscr{Y}$ are, indeed, bases for V_2. Now the scheme outlined in Figure 2.16 tells us to consider the images under **T** of the basis vectors X_1 and X_2, expressed in terms of the basis vectors Y_1 and Y_2. That is, for which unique scalars a_{11}, a_{12}, a_{21}, and a_{22} is it true that

(a) $$\mathbf{T}(X_1) = \mathbf{T}\begin{pmatrix} 1 \\ 1 \end{pmatrix} = \begin{pmatrix} 2 \\ 0 \end{pmatrix} = a_{11}Y_1 + a_{21}Y_2,$$

and

(b) $$\mathbf{T}(X_2) = \mathbf{T}\begin{pmatrix} 1 \\ 0 \end{pmatrix} = \begin{pmatrix} 1 \\ 0 \end{pmatrix} = a_{12}Y_1 + a_{22}Y_2? \qquad \text{(See Figure 2.16.)}$$

Let us attack (a) first. Since $Y_1 = \begin{pmatrix} 3 \\ 0 \end{pmatrix}$ and $Y_2 = \begin{pmatrix} 1 \\ 2 \end{pmatrix}$, we are asking for scalars a_{11} and a_{21} such that

$$\begin{pmatrix} 2 \\ 0 \end{pmatrix} = a_{11}\begin{pmatrix} 3 \\ 0 \end{pmatrix} + a_{21}\begin{pmatrix} 1 \\ 2 \end{pmatrix},$$

which, by Definition I.1.2 and I.1.3, gives us

$$\begin{pmatrix} 2 \\ 0 \end{pmatrix} = \begin{pmatrix} 3 \cdot a_{11} + 1 \cdot a_{21} \\ 0 \cdot a_{11} + 2 \cdot a_{21} \end{pmatrix}.$$

Now 2 vectors are equal if and only if they agree in each entry (Definition I.1.4) so that

$$2 = 3a_{11} + a_{21}$$
$$0 = 0 \quad + 2a_{21}.$$

The second equation tells us that $2 \cdot a_{21}$ equals zero, so that a_{21} must equal zero. But if $a_{21} = 0$, then the first equation $2 = 3 \cdot a_{11} + \not{a}^{0}_{21}$ obtains only when $a_{11} = \frac{2}{3}$. Thus, we have $a_{11} = \frac{2}{3}$ and $a_{21} = 0$.

Let us consider Equation (b) so as to discover the true identities of the scalars a_{12} and a_{22}. Now a_{12} and a_{22} have the property that

$$\mathbf{T}(X_2) = \mathbf{T}\begin{pmatrix}1\\0\end{pmatrix} = \begin{pmatrix}1\\0\end{pmatrix} = a_{12}\cdot Y_1 + a_{22}\cdot Y_2.$$

Equivalently, since $Y_1 = \begin{pmatrix}3\\0\end{pmatrix}$ and $Y_2 = \begin{pmatrix}1\\2\end{pmatrix}$,

$$\begin{pmatrix}1\\0\end{pmatrix} = a_{12}\begin{pmatrix}3\\0\end{pmatrix} + a_{22}\begin{pmatrix}1\\2\end{pmatrix},$$

or

$$\begin{pmatrix}1\\0\end{pmatrix} = \begin{pmatrix}3\cdot a_{12} + 1\cdot a_{22}\\0\cdot a_{12} + 2\cdot a_{22}\end{pmatrix}.$$

Upon matching entries of this last equation, we obtain the equations

$$\begin{aligned}1 &= 3\cdot a_{12} + a_{22}\\0 &= 0 \qquad + 2a_{22}.\end{aligned}$$

The last of these equations reveals to us that $a_{22} = 0$. When this value is substituted into the equation previous to the last, we obtain $1 = 3\cdot a_{12} + 0$, which says that $a_{12} = \frac{1}{3}$. We have found, in accordance with the scheme of (II.5.1) and Figure 2.16, that

column 1

$$\mathbf{T}(X_1) = \boxed{2/3}\cdot Y_1 + \boxed{0}\cdot Y_2,$$

and

column 2

$$\mathbf{T}(X_2) = \boxed{1/3}\cdot Y_1 + \boxed{0}\cdot Y_2.$$

If we extract these scalars (shaded above) according to the procedure outlined in Figure 2.16, we obtain our matrix (number-picture) of the linear transformation **T** relative to the given ordered bases $\mathscr{X}$ and $\mathscr{Y}$, namely,

▶ $$mx(\mathbf{T})_{\mathscr{X},\mathscr{Y}} = \begin{pmatrix}\frac{2}{3} & \frac{1}{3}\\0 & 0\end{pmatrix},$$

where

$$\mathscr{X} = \left\{\begin{pmatrix}1\\1\end{pmatrix}, \begin{pmatrix}1\\0\end{pmatrix}\right\}, \quad \mathscr{Y} = \left\{\begin{pmatrix}3\\0\end{pmatrix}, \begin{pmatrix}1\\2\end{pmatrix}\right\}$$

and $\mathbf{T}\begin{pmatrix}x_1\\x_2\end{pmatrix} = \begin{pmatrix}x_1 + x_2\\0\end{pmatrix}$ for all $X = \begin{pmatrix}x_1\\x_2\end{pmatrix} \in V_2$.

Example In this example, we shall take the same linear transformation as in the previous example, changing only the bases. To make life easy, let us consider the standard basis for V_2, that is, set

$$\mathscr{X} = \left\{ \underset{\substack{\uparrow \\ X_1}}{\begin{pmatrix}1\\0\end{pmatrix}}, \underset{\substack{\uparrow \\ X_2}}{\begin{pmatrix}0\\1\end{pmatrix}} \right\} \quad \text{and} \quad \mathscr{Y} = \left\{ \underset{\substack{\uparrow \\ Y_1}}{\begin{pmatrix}1\\0\end{pmatrix}}, \underset{\substack{\uparrow \\ Y_2}}{\begin{pmatrix}0\\1\end{pmatrix}} \right\},$$

where (as before) $\mathbf{T}\begin{pmatrix}x_1\\x_2\end{pmatrix} = \begin{pmatrix}x_1 + x_2\\0\end{pmatrix}$ for all $X = \begin{pmatrix}x_1\\x_2\end{pmatrix} \in V_2$. Thus $\mathbf{T}(X_1) = \mathbf{T}\begin{pmatrix}1\\0\end{pmatrix} = \begin{pmatrix}1\\0\end{pmatrix}$ and $\mathbf{T}(X_2) = \mathbf{T}\begin{pmatrix}0\\1\end{pmatrix} = \begin{pmatrix}1\\0\end{pmatrix}$. That is,

$$\mathbf{T}(X_1) = 1 \cdot \begin{pmatrix}1\\0\end{pmatrix} + 0 \cdot \begin{pmatrix}0\\1\end{pmatrix} = \underbrace{\textcircled{1} \cdot Y_1 + \textcircled{0}}_{\text{col 1}} \cdot Y_2$$

and

$$\mathbf{T}(X_2) = 1 \cdot \begin{pmatrix}1\\0\end{pmatrix} + 0 \cdot \begin{pmatrix}0\\1\end{pmatrix} = \underbrace{\textcircled{1} \cdot Y_1 + \textcircled{0}}_{\text{col 2}} \cdot Y_2$$

so that

▶ $$mx(\mathbf{T})_{\mathscr{X}\mathscr{Y}} = \begin{pmatrix}1 & 1\\0 & 0\end{pmatrix}$$

where

$$\mathscr{X} = \left\{ \begin{pmatrix}1\\0\end{pmatrix}, \begin{pmatrix}0\\1\end{pmatrix} \right\}, = \left\{ \begin{pmatrix}1\\0\end{pmatrix}, \begin{pmatrix}0\\1\end{pmatrix} \right\} \quad \text{and } \mathbf{T}\begin{pmatrix}x_1\\x_2\end{pmatrix} = \begin{pmatrix}x_1 + x_2\\0\end{pmatrix}$$

for all $\begin{pmatrix}x_1\\x_2\end{pmatrix} \in V_2$.

Example Again, we shall choose $\mathbf{T}$ as before and change only the *ordering* of the second basis $\mathscr{Y}$ in the preceding example, that is,

$$\mathbf{T}\begin{pmatrix}x_1\\x_2\end{pmatrix} = \begin{pmatrix}x_1 + x_2\\0\end{pmatrix} \quad \text{and} \quad \mathscr{X} = \underset{\substack{\uparrow \\ X_1}}{\begin{pmatrix}1\\0\end{pmatrix}}, \underset{\substack{\uparrow \\ X_2}}{\begin{pmatrix}0\\1\end{pmatrix}}, \mathscr{Y} = \underset{\substack{\uparrow \\ Y_1}}{\begin{pmatrix}0\\1\end{pmatrix}}, \underset{\substack{\uparrow \\ Y_2}}{\begin{pmatrix}1\\0\end{pmatrix}}.$$

Now,

$$\mathbf{T}(X_1) = \mathbf{T}\begin{pmatrix}1\\0\end{pmatrix} = \begin{pmatrix}1\\0\end{pmatrix} = 0 \cdot \begin{pmatrix}0\\1\end{pmatrix} + 1 \cdot \begin{pmatrix}1\\0\end{pmatrix} = \underbrace{\textcircled{0} \cdot Y_1 + \textcircled{1}}_{\text{col 1}} \cdot Y_2,$$

and

col 2

$$\mathbf{T}(X_2) = \mathbf{T}\begin{pmatrix}0\\1\end{pmatrix} = \begin{pmatrix}1\\0\end{pmatrix} = 0 \cdot \begin{pmatrix}0\\1\end{pmatrix} + 1 \cdot \begin{pmatrix}1\\0\end{pmatrix} = \boxed{0} \cdot Y_1 + \boxed{1} \cdot Y_2.$$

Consulting Figure 2.16 and Definition II.5.1, we have

$$mx(\mathbf{T})_{\mathscr{X},\mathscr{Y}} = \begin{pmatrix}0 & 0\\1 & 1\end{pmatrix},$$

where $\mathscr{X} = \left\{\begin{pmatrix}1\\0\end{pmatrix},\begin{pmatrix}0\\1\end{pmatrix}\right\}$, $\mathscr{Y} = \left\{\begin{pmatrix}0\\1\end{pmatrix},\begin{pmatrix}1\\0\end{pmatrix}\right\}$ and $\mathbf{T}\begin{pmatrix}x_1\\x_2\end{pmatrix} = \begin{pmatrix}x_1 + x_2\\0\end{pmatrix}$ for all $X = \begin{pmatrix}x_1\\x_2\end{pmatrix} \in V_2$.

The previous examples considered the same linear transformation **T** throughout. We obtained a different matrix (number-picture) for **T** each time we changed bases $\mathscr{X}$ and $\mathscr{Y}$ or the ordering of a basis.

EXERCISES

In the following exercises, find the matrices (plural of matrix) of the given linear transformations in $\mathscr{L}(V_2,V_2)$, where the ordered bases $\mathscr{X}$ and $\mathscr{Y}$ are given.

II.5.1 Take $\mathbf{T} \in \mathscr{L}(V_2,V_2)$ equal to the identity transformation $\mathbf{T}(X) = X$ for all $X \in V_2$. Put $\mathscr{X} = \left\{\begin{pmatrix}1\\0\end{pmatrix},\begin{pmatrix}0\\1\end{pmatrix}\right\}$ and

$$\mathscr{Y} = \left\{\begin{pmatrix}1\\0\end{pmatrix},\begin{pmatrix}0\\1\end{pmatrix}\right\}.$$

II.5.2 As before, $\mathbf{T}(X) = X$ for all $X \in V_2$. Let

$$\mathscr{X} = \left\{\begin{pmatrix}1\\-1\end{pmatrix},\begin{pmatrix}0\\1\end{pmatrix}\right\} \quad \text{and} \quad \mathscr{Y} = \left\{\begin{pmatrix}1\\-1\end{pmatrix},\begin{pmatrix}0\\1\end{pmatrix}\right\}.$$

II.5.3 $\mathbf{T}(X) = X$ for all $X \in V_2$. Let

$$\mathscr{X} = \left\{\begin{pmatrix}\frac{1}{2}\\-1\end{pmatrix},\begin{pmatrix}3\\0\end{pmatrix}\right\} \quad \text{and} \quad \mathscr{Y} = \left\{\begin{pmatrix}\frac{1}{2}\\-1\end{pmatrix},\begin{pmatrix}3\\0\end{pmatrix}\right\}.$$

II.5.4 $\mathbf{T}(X) = X$ for all $X \in V_2$. Let

$$\mathscr{X} = \left\{\begin{pmatrix}-1\\-2\end{pmatrix},\begin{pmatrix}6\\1\end{pmatrix}\right\} \quad \text{and} \quad = \left\{\begin{pmatrix}-1\\-2\end{pmatrix},\begin{pmatrix}6\\1\end{pmatrix}\right\}.$$

II.5.5 After doing Exercises II.5.1 through II.5.4, you might have noticed a persistent lack of change to the matrix of the identity transformation $\mathbf{T}(X) = X$ for all $X \in V_2$. Indeed, show that whenever the ordered (domain) basis $\mathscr{X}$ equals the ordered range basis $\mathscr{Y}$ (as is the case in the previous exercises), then the resulting matrix for the identity transformation is always the same. More specifically, if $\mathscr{X} = \{X_1,X_2\}$ and $\mathscr{Y} = \{X_1,X_2\}$, then $mx(\mathbf{T})_{\mathscr{X},\mathscr{Y}} = \begin{pmatrix}1 & 0\\0 & 1\end{pmatrix}$.

REMARK. It happens that scalar multiples of the identity transformation $\mathbf{T}(X) = \alpha \cdot X$ are the *only* operators in $\mathscr{L}(V_2, V_2)$ whose matrices are always the same for *all* ordered bases $\mathscr{X}$ and $\mathscr{Y}$ where $\mathscr{X} = \mathscr{Y}$.

II.5.6 Again, we consider the identity operator $\mathbf{T}(X) = X$ for all $X \in V_2$, only the ordered bases $\mathscr{X}$ and $\mathscr{Y}$ will be different from each other.

$$\text{Put } \mathscr{X} = \left\{\begin{pmatrix}3\\5\end{pmatrix}, \begin{pmatrix}-1\\-1\end{pmatrix}\right\} \quad \text{and} \quad \mathscr{Y} = \left\{\begin{pmatrix}1\\0\end{pmatrix}, \begin{pmatrix}0\\1\end{pmatrix}\right\}.$$

II.5.7 $\mathbf{T}(X) = X$ for all $X \in V_2$ and

$$\mathscr{X} = \left\{\begin{pmatrix}1\\0\end{pmatrix}, \begin{pmatrix}0\\1\end{pmatrix}\right\}, \mathscr{Y} = \left\{\begin{pmatrix}0\\1\end{pmatrix}, \begin{pmatrix}1\\0\end{pmatrix}\right\}$$

(the bases $\mathscr{X}$ and $\mathscr{Y}$ are *not* "the same," inasmuch as the *orderings* differ).

II.5.8 $\mathbf{T}(X) = X$ for all $X \in V_2$ where

$$\mathscr{X} = \left\{\begin{pmatrix}0\\1\end{pmatrix}, \begin{pmatrix}1\\0\end{pmatrix}\right\} \quad \text{and} \quad \mathscr{Y} = \left\{\begin{pmatrix}1\\0\end{pmatrix}, \begin{pmatrix}0\\1\end{pmatrix}\right\}.$$

II.5.9 Let $\mathbf{T}$ be the operator that multiplies every vector by the scalar p, that is, $\mathbf{T}(X) = p \cdot X$ for all X (if $p = 1$, then $\mathbf{T}$ is the identity operator). Show that if the basis $\mathscr{X}$ equals the basis $\mathscr{Y}$ (their elements and ordering of the elements agree), then

$$mx(\mathbf{T})_{\mathscr{X},\mathscr{Y}} = \begin{pmatrix}p & 0\\0 & p\end{pmatrix}.$$

II.5.10 Let $\mathbf{T}\begin{pmatrix}x_1\\x_2\end{pmatrix} = \begin{pmatrix}x_1\\-x_2\end{pmatrix}$ for all $\begin{pmatrix}x_1\\x_2\end{pmatrix} = X \in V_2$.

$$\text{Put } \mathscr{X} = \left\{\begin{pmatrix}1\\0\end{pmatrix}, \begin{pmatrix}0\\1\end{pmatrix}\right\} \quad \text{and} \quad \mathscr{Y} = \left\{\begin{pmatrix}1\\0\end{pmatrix}, \begin{pmatrix}0\\1\end{pmatrix}\right\}.$$

II.5.11 Let $\mathbf{T} \in \mathscr{L}(V_2, V_2)$ be as given in Exercise II.5.10. Let

$$\mathscr{X} = \left\{\begin{pmatrix}-2\\1\end{pmatrix}, \begin{pmatrix}3\\0\end{pmatrix}\right\} \quad \text{and} \quad \mathscr{Y} = \left\{\begin{pmatrix}1\\0\end{pmatrix}, \begin{pmatrix}0\\1\end{pmatrix}\right\}.$$

II.5.12 $\mathbf{T} \in \mathscr{L}(V_2, V_2)$ is given in Exercise II.5.10. Let

$$\mathscr{X} = \left\{\begin{pmatrix}-2\\1\end{pmatrix}, \begin{pmatrix}3\\0\end{pmatrix}\right\} \quad \text{and} \quad \mathscr{Y} = \left\{\begin{pmatrix}0\\1\end{pmatrix}, \begin{pmatrix}1\\0\end{pmatrix}\right\}.$$

II.5.13 Let $\mathbf{T}\begin{pmatrix}x_1\\x_2\end{pmatrix} = \begin{pmatrix}x_2\\x_1 + x_2\end{pmatrix}$ for all $\begin{pmatrix}x_1\\x_2\end{pmatrix} = X \in V_2$. Let

$$\mathscr{X} = \left\{\begin{pmatrix}1\\0\end{pmatrix}, \begin{pmatrix}1\\1\end{pmatrix}\right\}, \mathscr{Y} = \left\{\begin{pmatrix}1\\0\end{pmatrix}, \begin{pmatrix}0\\1\end{pmatrix}\right\}.$$

II.5.14 With $\mathbf{T} \in \mathscr{L}(V_2, V_2)$ as in Exercise II.5.13, let

$$\mathscr{X} = \left\{\begin{pmatrix}1\\0\end{pmatrix}, \begin{pmatrix}0\\1\end{pmatrix}\right\}, \mathscr{Y} = \left\{\begin{pmatrix}1\\0\end{pmatrix}, \begin{pmatrix}0\\1\end{pmatrix}\right\}.$$

II.5.15 With $\mathbf{T} \in \mathscr{L}(V_2, V_2)$ as in Exercise II.5.13, let

$$\mathscr{X} = \left\{\begin{pmatrix}0\\1\end{pmatrix}, \begin{pmatrix}1\\0\end{pmatrix}\right\}, \mathscr{Y} = \left\{\begin{pmatrix}1\\0\end{pmatrix}, \begin{pmatrix}0\\1\end{pmatrix}\right\}.$$

At this point, where do we stand? Well, we have shown that every time we choose a linear operator $\mathbf{T} \in \mathscr{L}(V_2, V_2)$, and a pair of ordered bases for V_2, we can construct a *unique* matrix for $\mathbf{T}$ as per Definition II.5.1. It so happened that in our case, the matrices were always of the square form $\begin{pmatrix}a_{11} & a_{12}\\ a_{21} & a_{22}\end{pmatrix}$; such matrices are called 2×2 (read, "two by two") matrices. But what are we really saying? For a pair of ordered bases (which shall be *fixed* throughout), we are saying that to every linear transformation $\mathbf{T} \in \mathscr{L}(V_2, V_2)$, we may assign one and *only* one 2×2 matrix. Doesn't this sentence smack of the very notion of *function* (Definition II.1.1)? Yes it does, for if we regard $\mathbf{T}$ as an "input," then our "output" elements are simply 2×2 matrices! (See Figure 2.17.) In goes a linear operator, and

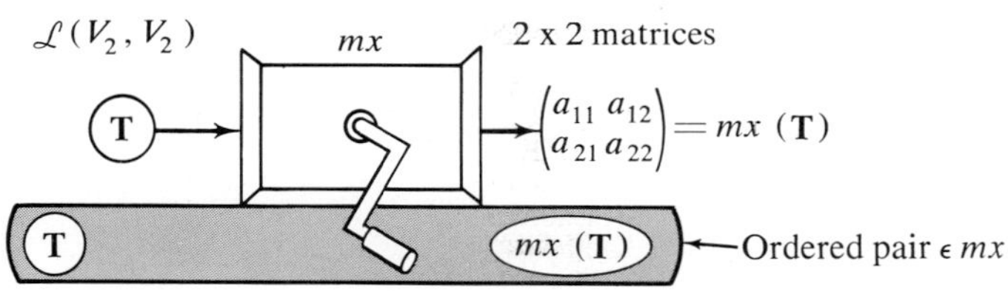

Figure 2.17

out comes one and only one 2×2 matrix, and since there is *only* 1 output for each input, the set of ordered pairs $(\mathbf{T}, mx(\mathbf{T}))$ constitutes a bonafide function (see Definition II.1.1), which we denote by the symbol "mx." Moreover, this function, which sends $\mathscr{L}(V_2, V_2)$ to the set of all 2×2 matrices, is 1-1 (see Definition II.1.6); that is, no 2 distinct linear transformations can give rise to the same 2×2 matrix (we establish this property in Theorem II.5.2). We are now encountering a situation where we *naturally* bump into a *function*, the situation where matrices are associated with linear transformations.

QUESTION. Is the 1-1 function mx *onto*; that is does *every* 2×2 matrix $\begin{pmatrix}a_{11} & a_{12}\\ a_{21} & a_{22}\end{pmatrix}$ come from some linear transformation in $\mathscr{L}(V_2, V_2)$? The answer, as we shall see, is yes.

Theorem II.5.2 The function mx which sends $\mathscr{L}(V_2, V_2)$ to the set of 2×2 matrices is 1-1 and onto.

Proof:

THE ONE-ONE PROPERTY. We have already established the 1-1 property of the function mx, but there is no harm done if we re-establish the argument more formally. Now suppose our ordered bases are $\mathscr{X} = \{X_1, X_2\}$ and $\mathscr{Y} = \{Y_1, Y_2\}$ and that 2 (distinct?) linear transformations, **T** and **U**, give rise to the *same* 2×2 matrix. Thus,

$$\begin{aligned} \mathbf{T}(X_1) &= a_{11}Y_1 + a_{21}Y_2 \qquad & \mathbf{U}(X_1) &= a_{11}Y_1 + a_{21}Y_2 \\ \mathbf{T}(X_2) &= a_{12}Y_1 + a_{22}Y_2 \qquad & \mathbf{U}(X_2) &= a_{12}Y_1 + a_{22}Y_2. \end{aligned}$$

But this means that the actions of both **U** and **T** *agree* on the basis vectors X_1 and X_2, that is, $\mathbf{T}(X_1) = \mathbf{U}(X_1)$ and $\mathbf{T}(X_2) = \mathbf{U}(X_2)$. Now **T** and **U** are not just *any* functions, but are *linear* functions, so that for *any* X, which equals $aX_1 + bX_2$ for some (unique) scalars a and b (see Theorem I.6.8), we have

$$\begin{aligned} \mathbf{T}(X) &= \mathbf{T}(aX_1 + bX_2) & & \\ &= a\mathbf{T}(X_1) + b\mathbf{T}(X_2) & & \text{since } \mathbf{T} \text{ is linear (see Proposition II.2.2)} \\ & \quad\ \downarrow \qquad\quad\ \downarrow & & \\ &= a\mathbf{U}(X_1) + b\mathbf{U}(X_2) & & \text{since } \mathbf{T}(X_1) = \mathbf{U}(X_1),\ \mathbf{T}(X_2) = \mathbf{U}(X_2) \\ &= \mathbf{U}(aX_1 + bX_2) & & \text{since } \mathbf{U} \text{ is linear (see Proposition II.2.2)} \\ &= \mathbf{U}(X). & & \end{aligned}$$

So it is, that for *all* X (not only for basis elements), $\mathbf{T}(X) = \mathbf{U}(X)$, that is, the 2 linear transformations **T** and **U**, which give rise to the same 2×2 matrix, are in fact, one and the same transformation.

THE ONTO PROPERTY. We must show that any and every 2×2 matrix can be obtained from a certain linear transformation in $\mathscr{L}(V_2, V_2)$. Toward this end, let $\begin{pmatrix} a_{11} & a_{12} \\ a_{21} & a_{22} \end{pmatrix}$ be an arbitrary, unfettered, unrestricted matrix of your own choosing. If this *were* a matrix of a certain linear transformation, **T**, say, then the first column $\begin{pmatrix} a_{11} \\ a_{21} \end{pmatrix}$ of the matrix would reveal the coefficients of $\mathbf{T}(X_1)$ expressed in terms of the ordered basis $\{Y_1, Y_2\}$, and the second column $\begin{pmatrix} a_{12} \\ a_{22} \end{pmatrix}$ of the matrix would reveal the coefficients of $\mathbf{T}(X_2)$ relative to the ordered basis $\{Y_1, Y_2\}$. In other words, the matrix $\begin{pmatrix} a_{11} & a_{12} \\ a_{21} & a_{22} \end{pmatrix}$ (if it is to come from a linear transformation at all), must come from a linear transformation **T** with the properties:

$$\begin{aligned}\mathbf{T}(X_1) &= a_{11}Y_1 + a_{21}Y_2 \\ \mathbf{T}(X_2) &= a_{21}Y_1 + a_{22}Y_2.\end{aligned} \tag{II.5.3}$$

QUESTION. *Is* there such a linear transformation? Notice that we are not through, because our function **T** has so far been defined for 2 (input) elements of V_2, namely, for X_1 and X_2. But we can *extend* **T** so that it will be defined for *all* vectors $X \in V_2$, and not just for the 2 basis vectors X_1 and X_2. That is, for an *arbitrary* X of V_2, we are obliged to describe the (output) vector $\mathbf{T}(X)$. Very easy. Since X can be written in terms of the basis elements X_1 and X_2 (that is, $X = b_1X_1 + b_2X_2$ for some unique scalars b_1 and b_2; Theorem I.6.8), we extend our function **T** to *all* X as follows:

For *any* X (which can be set equal to $b_1X_1 + b_2X_2$ for certain scalars b_1 and b_2), we define $\mathbf{T}_0(X)$ by

$$\mathbf{T}_0(X) = b_1\mathbf{T}(X_1) + b_2\mathbf{T}(X_2). \tag{II.5.4}$$

Notice, that the scalars b_1 and b_2 come from the $\mathscr{X}$-basis representation for X, while the 2 vectors $\mathbf{T}(X_1)$ and $\mathbf{T}(X_2)$ are the already known quantities $a_{11}Y_1 + a_{21}Y_2$ and $a_{21}Y_1 + a_{22}Y_2$ [see (II.5.3)], respectively. Thus, $\mathbf{T}_0(X)$ is a well-defined, unambiguous vector of V_2, for *each* and *every* $X \in V_2$. We conclude that $\mathbf{T}_0$ is a function (linear?) from *all* of V_2 to V_2.

There are 2 small gaps remaining which are left as Exercises II.5.16 and II.5.17.

GAP 1. Does the function T_0, defined for *all* X in (II.5.4), really *extend* the function **T** of (II.5.3), which is defined only for the two basis vectors X_1 and X_2? That is, is it true that $\mathbf{T}_0(X_1) = \mathbf{T}(X_1)$ and that $\mathbf{T}_0(X_2) = \mathbf{T}(X_2)$? Notice, that the left-hand sides of these equations are defined in (II.5.4), while the right-hand sides are defined (via the matrix) in (II.5.3). The answer is yes.

GAP 2. Is this extended function $\mathbf{T}_0$ (which agrees with **T** on the vectors X_1 and X_2) a *linear* function? Thus, $\mathbf{T}_0$ must be *tested* against Equations (II.2.1) and (II.2.2) of Definition II.2.1. The answer is that $\mathbf{T}_0$ *is* a genuine, *linear* function from V_2 to V_2, that is, $\mathbf{T}_0 \in \mathscr{L}(V_2, V_2)$.

We conclude (after filling the gaps) that for any and every 2×2 matrix $\begin{pmatrix} a_{11} & a_{12} \\ a_{21} & a_{22} \end{pmatrix}$, there *is* a linear transformation $\mathbf{T}_0 \in \mathscr{L}(V_2, V_2)$ such that $mx(\mathbf{T}_0)_{\mathscr{X},\mathscr{Y}} = \begin{pmatrix} a_{11} & a_{12} \\ a_{21} & a_{22} \end{pmatrix}$. We have seen that $\mathbf{T}_0$ must be just that linear transformation with the property that, at least for the vectors X_1 and X_2,

$$\mathbf{T}_0(X) = a_{11}Y_1 + a_{21}Y_2 \quad \text{and} \quad \mathbf{T}(X_2) = a_{21}Y_1 + a_{22}Y_2.$$

Example Let our fixed ordered bases be the standard basis in each case, that is, let $\mathscr{X} = \left\{\begin{pmatrix}1\\0\end{pmatrix}, \begin{pmatrix}0\\1\end{pmatrix}\right\}$ and $\mathscr{Y} = \left\{\begin{pmatrix}1\\0\end{pmatrix}, \begin{pmatrix}0\\1\end{pmatrix}\right\}$. Consider the 2×2 matrix $\begin{pmatrix}1 & 3\\-4 & 2\end{pmatrix}$. Describe the unique linear transformation of $\mathscr{L}(V_2, V_2)$ which gives rise to this matrix. Equivalently, we must describe $\mathbf{T}(X)$, the "output" element, for any and every $X \in V_2$. (See Figure 2.18.)

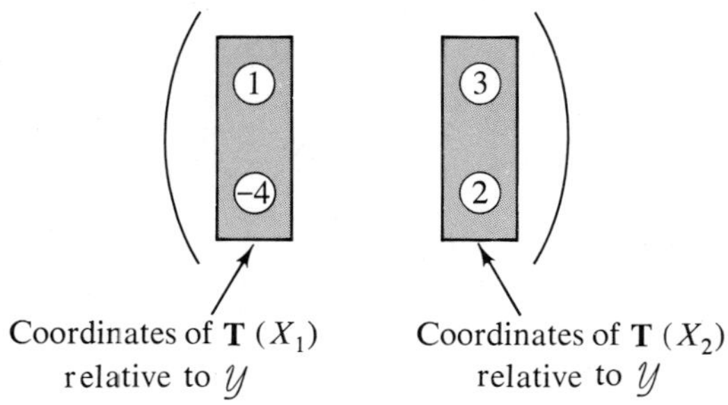

Figure 2.18

Our $\mathbf{T}$ must have the property that

$$\mathbf{T}(X_1) = (1)\cdot Y_1 + (-4)Y_2$$

and (II.5.5)

$$\mathbf{T}(X_2) = (3)\cdot Y_1 + (2)Y_2.$$

Now $X_1 = Y_1 = \begin{pmatrix}1\\0\end{pmatrix}$ and $X_2 = Y_2 = \begin{pmatrix}0\\1\end{pmatrix}$, so that the previous pair of equations becomes

$$\mathbf{T}\begin{pmatrix}1\\0\end{pmatrix} = 1\cdot\begin{pmatrix}1\\0\end{pmatrix} + (-4)\cdot\begin{pmatrix}0\\1\end{pmatrix} = \begin{pmatrix}1\\-4\end{pmatrix},$$

and (II.5.5a)

$$\mathbf{T}\begin{pmatrix}0\\1\end{pmatrix} = 3\cdot\begin{pmatrix}1\\0\end{pmatrix} + 2\cdot\begin{pmatrix}0\\1\end{pmatrix} = \begin{pmatrix}3\\2\end{pmatrix}.$$

Now what *linear* function has the property that it sends $\begin{pmatrix}1\\0\end{pmatrix}$ to $\begin{pmatrix}1\\-4\end{pmatrix}$ and sends $\begin{pmatrix}0\\1\end{pmatrix}$ to $\begin{pmatrix}3\\2\end{pmatrix}$ as per (II.5.5a) above? If we but mimic our proof of Theorem II.5.2, we shall have the answer, for, if $X = \begin{pmatrix}x_1\\x_2\end{pmatrix}$ is arbitrary, then

$$\mathbf{T}_0\begin{pmatrix}x_1\\x_2\end{pmatrix} = \mathbf{T}_0\left(x_1\cdot\begin{pmatrix}1\\0\end{pmatrix} + x_2\cdot\begin{pmatrix}0\\1\end{pmatrix}\right), \text{ since } \begin{pmatrix}x_1\\x_2\end{pmatrix} = x_1\cdot\begin{pmatrix}1\\0\end{pmatrix} + x_2\cdot\begin{pmatrix}0\\1\end{pmatrix}. \qquad \text{(II.5.6)}$$

If $\mathbf{T}_0$ is *linear*, then the right-hand side (Proposition II.2.2) would become

$$= x_1 \cdot \mathbf{T}_0\begin{pmatrix}1\\0\end{pmatrix} + x_2 \cdot \mathbf{T}_0\begin{pmatrix}0\\1\end{pmatrix}$$

$$\downarrow \qquad\qquad \downarrow$$

$$= x_1 \cdot \begin{pmatrix}1\\-4\end{pmatrix} + x_2 \begin{pmatrix}3\\2\end{pmatrix} \qquad \text{from (II.5.5a)}$$

$$= \begin{pmatrix}x_1\\-4x_1\end{pmatrix} + \begin{pmatrix}3x_2\\2x_2\end{pmatrix} \qquad \text{from Definition I.1.3}$$

$$= \begin{pmatrix}x_1 + 3x_2\\-4x_1 + 2x_2\end{pmatrix} \qquad \text{from Definition I.1.2.} \tag{II.5.7}$$

Thus, in setting (II.5.6) equal to II.5.7), we have the fact that for *any* $X = \begin{pmatrix}x_1\\x_2\end{pmatrix} \in V_2$,

$$\mathbf{T}_0\begin{pmatrix}x_1\\x_2\end{pmatrix} = \begin{pmatrix}x_1 + 3x_2\\-4x_1 + 2x_2\end{pmatrix}. \tag{II.5.8}$$

It is an easy matter to see that this function $\mathbf{T}_0$ in (II.5.8) is a true extension of the $\mathbf{T}$ in (II.5.5) by setting $X = \begin{pmatrix}1\\0\end{pmatrix}$ or $\begin{pmatrix}0\\1\end{pmatrix}$. (See Gap 1 of our proof in Theorem II.5.2 and Exercise II.5.16.) Moreover, $\mathbf{T}_0$, as given by (II.5.8) is *linear* (see Gap 2 1f our proof of Theorem II.5.2 and Exercise II.5.17). Since our extended function $\mathbf{T}_0$ in (II.5.8) satisfies the conditions of (II.5.5), we have [from inspecting (II.5.5) and peeling away coefficients] in accordance with Definition II.5.1, that

$$mx(\mathbf{T}_0)_{\mathscr{X},\mathscr{Y}} = \begin{pmatrix}1 & 3\\-4 & 2\end{pmatrix}.$$

Example Let $\mathscr{X} = \left\{\begin{pmatrix}2\\0\end{pmatrix}, \begin{pmatrix}1\\1\end{pmatrix}\right\}$ and $\mathscr{Y} = \left\{\begin{pmatrix}0\\1\end{pmatrix}, \begin{pmatrix}1\\0\end{pmatrix}\right\}$. For which $\mathbf{T} \in \mathscr{L}(V_2, V_2)$ is it true that $mx(\mathbf{T})_{\mathscr{X},\mathscr{Y}} = \begin{pmatrix}1 & 2\\-1 & 3\end{pmatrix}$? Again, we must describe $\mathbf{T}(X)$. The first column $\begin{pmatrix}1\\-1\end{pmatrix}$ of this matrix yields the coordinates of $\mathbf{T}(X_1)$ relative to the basis $\mathscr{Y}$; the second column yields the coordinates of $\mathbf{T}(X_2)$ relative to $\mathscr{Y}$ (see Definition II.5.1). Thus,

$$\mathbf{T}(X_1) = (1) \cdot Y_1 + (-1) \cdot Y_2$$

and (II.5.9a)

$$\mathbf{T}(X_2) = (2) \cdot Y_1 + (3) \cdot Y_2.$$

In this case, it is given that $X_1 = \begin{pmatrix}2\\0\end{pmatrix}$, $X_2 = \begin{pmatrix}1\\1\end{pmatrix}$, $Y_1 = \begin{pmatrix}0\\1\end{pmatrix}$ and $Y_2 = \begin{pmatrix}1\\0\end{pmatrix}$, so that (II.5.9a) becomes

$$\mathbf{T}\begin{pmatrix}2\\0\end{pmatrix} = (1) \cdot \begin{pmatrix}0\\1\end{pmatrix} + (-1) \cdot \begin{pmatrix}1\\0\end{pmatrix} \quad \text{which equals} \quad \begin{pmatrix}-1\\1\end{pmatrix},$$

and (II.5.9b)

$$\mathbf{T}\begin{pmatrix}1\\1\end{pmatrix} = (2) \cdot \begin{pmatrix}0\\1\end{pmatrix} + (3) \cdot \begin{pmatrix}1\\0\end{pmatrix} \quad \text{which equals} \quad \begin{pmatrix}3\\2\end{pmatrix}.$$

Now we must extend **T**, so that $\mathbf{T}\begin{pmatrix} x_1 \\ x_2 \end{pmatrix}$ is defined for *all* $\begin{pmatrix} x_1 \\ x_2 \end{pmatrix}$, and not just for the 2 basis vectors $\begin{pmatrix} 2 \\ 0 \end{pmatrix} = X_1$ and $\begin{pmatrix} 1 \\ 1 \end{pmatrix} = X_2$. But for any $X = \begin{pmatrix} x_1 \\ x_2 \end{pmatrix} \in V_2$, we observe that $\begin{pmatrix} x_1 \\ x_2 \end{pmatrix} = b_1X_1 + b_2X_2$ for certain scalars b_1 and b_2, which follows, since $\{X_1, X_2\}$ is a basis. More specifically,

$$\begin{pmatrix} x_1 \\ x_2 \end{pmatrix} = \underset{\substack{\uparrow \\ b_1}}{\left(\frac{x_1 - x_2}{2}\right)} \cdot \underset{\substack{\uparrow \\ X_1}}{\begin{pmatrix} 2 \\ 0 \end{pmatrix}} + \underset{\substack{\uparrow \\ b_2}}{(x_2)} \cdot \underset{\substack{\uparrow \\ X_2}}{\begin{pmatrix} 1 \\ 1 \end{pmatrix}} \qquad \text{(II.5.10)}$$

so that $b_1 = (x_1 - x_2)/2$ and $b_2 = x_2$. We extend **T** by defining, for *any* $X = \begin{pmatrix} x_1 \\ x_2 \end{pmatrix}$,

$$\begin{aligned}
\mathbf{T}_0(X) &= \mathbf{T}_0\begin{pmatrix} x_1 \\ x_2 \end{pmatrix} \\
&= \mathbf{T}_0\left(\frac{(x_1 - x_2)}{2} \cdot \begin{pmatrix} 2 \\ 0 \end{pmatrix} + (x_2) \cdot \begin{pmatrix} 1 \\ 1 \end{pmatrix}\right) && \text{from (II.5.10)} \\
&= \frac{(x_1 - x_2)}{2} \cdot \mathbf{T}_0\begin{pmatrix} 2 \\ 0 \end{pmatrix} + (x_2) \cdot \mathbf{T}_0\begin{pmatrix} 1 \\ 1 \end{pmatrix} && \text{a necessary condition if } \mathbf{T}_0 \text{ is to be linear} \qquad \text{(II.5.11)} \\
&\qquad\qquad\qquad\quad \downarrow \qquad\qquad\qquad\qquad \downarrow \\
&= \frac{(x_1 - x_2)}{2} \cdot \begin{pmatrix} -1 \\ 1 \end{pmatrix} + (x_2) \cdot \begin{pmatrix} 3 \\ 2 \end{pmatrix} && \text{from (II.5.9b)} \\
&= \begin{pmatrix} -\frac{1}{2}x_1 + \frac{7}{2}x_2 \\ \frac{1}{2}x_1 + \frac{3}{2}x_2 \end{pmatrix}, && \text{from Definitions I.1.2 and I.1.3.} \qquad \text{(II.5.12)}
\end{aligned}$$

Equating (II.5.11) to (II.5.12) tells us that for all $X = \begin{pmatrix} x_1 \\ x_2 \end{pmatrix}$,

$$\mathbf{T}_0(X) = \mathbf{T}_0\begin{pmatrix} x_1 \\ x_1 \end{pmatrix} = \begin{pmatrix} -\frac{1}{2}x_1 + \frac{7}{2}x_2 \\ \frac{1}{2}x_1 + \frac{3}{2}x_2 \end{pmatrix}. \qquad \text{(II.5.13)}$$

Again, $\mathbf{T}_0$ as given in (II.5.13), satisfies the equations of (II.5.9b) (the student should check this), which derive from the matrix $\begin{pmatrix} 1 & 2 \\ -1 & 3 \end{pmatrix}$ and from Definition II.5.1. Moreover, $\mathbf{T}_0$ is easily seen to be *linear* (Definition II.2.1). Thus, for the $\mathbf{T}_0$ described in (II.5.13), which is a member of $\mathscr{L}(V_2, V_2)$, we have that $mx(\mathbf{T}_0)_{\mathscr{X},\mathscr{Y}} = \begin{pmatrix} 1 & 2 \\ -1 & 3 \end{pmatrix}$ (see Definition II.5.1).

EXERCISES

II.5.16 (See Gap 1, proof of Theorem II.5.2.) Let $\{X_1, X_2\}$ be a basis for V_2. Suppose T is a function sending $\{X_1, X_2\}$ into V_2, that is, **T** is defined *only* on the 2 (input) vectors X_1 and X_2, and the 2, not necessarily distinct image (output) elements are $\mathbf{T}(X_1)$ and $\mathbf{T}(X_2)$, respectively. [More precisely, $\mathbf{T}$ is the set of ordered pairs $\{X_1,\mathbf{T}(X_1)), (X_2,\mathbf{T}(X_2))\}$; see Definition II.1.1.]

We construct a function $\mathbf{T}_0$, whose domain is *all* of V_2, as follows. Now for any $X \in V_2$, X has a unique representation $X = b_1X_1 + b_2X_2$, for certain scalars b_1 and b_2 (Theorem I.6.8). With this in mind, we define $\mathbf{T}_0$ by setting

$$\mathbf{T}_0(X) = b_1\mathbf{T}(X_1) + b_2\mathbf{T}(X_2), \quad \text{for all } X \in V_2.$$

Prove that $\mathbf{T}_0$ agrees with $\mathbf{T}$ for the basis vectors X_1 and X_2, that is, $\mathbf{T}_0(X_1) = \mathbf{T}(X_1)$ and $\mathbf{T}_0(X_2) = \mathbf{T}(X_2)$, so that $\mathbf{T}_0$ "extends" $\mathbf{T}$.

II.5.17 Given $\mathbf{T}_0$ as defined in Exercise II.5.16, show that $\mathbf{T}_0$ is a *linear* function, that is, show that $\mathbf{T}_0 \in \mathscr{L}(V_2,V_2)$.

In Exercises II.5.18 through II.5.23, a function $\mathbf{T}$ is defined only on a pair of basis vectors X_1 and X_2 for V_2. In each case, find the unique *linear* transformation, $\mathbf{T}_0$ say, which agrees with $\mathbf{T}$ on X_1 and X_2, but is defined on *all* of V_2.

II.5.18 $\mathbf{T}\begin{pmatrix}1\\0\end{pmatrix} = \begin{pmatrix}2\\3\end{pmatrix}$ and $\mathbf{T}\begin{pmatrix}0\\1\end{pmatrix} = \begin{pmatrix}1\\0\end{pmatrix}$.

II.5.19 $\mathbf{T}\begin{pmatrix}1\\1\end{pmatrix} = \begin{pmatrix}0\\0\end{pmatrix}$ and $\mathbf{T}\begin{pmatrix}1\\-1\end{pmatrix} = \begin{pmatrix}0\\0\end{pmatrix}$.

II.5.20 $\mathbf{T}\begin{pmatrix}1\\0\end{pmatrix} = \begin{pmatrix}1\\0\end{pmatrix}$ and $\mathbf{T}\begin{pmatrix}0\\1\end{pmatrix} = \begin{pmatrix}1\\0\end{pmatrix}$.

II.5.21 $\mathbf{T}\begin{pmatrix}1\\0\end{pmatrix} = \begin{pmatrix}1\\0\end{pmatrix}$ and $\mathbf{T}\begin{pmatrix}1\\0\end{pmatrix} = \begin{pmatrix}1\\1\end{pmatrix}$.

II.5.22 $\mathbf{T}\begin{pmatrix}1\\0\end{pmatrix} = \begin{pmatrix}0\\1\end{pmatrix}$ and $\mathbf{T}\begin{pmatrix}0\\1\end{pmatrix} = \begin{pmatrix}1\\0\end{pmatrix}$.

II.5.23 $\mathbf{T}\begin{pmatrix}1\\0\end{pmatrix} = \begin{pmatrix}a\\0\end{pmatrix}$ and $\mathbf{T}\begin{pmatrix}0\\1\end{pmatrix} = \begin{pmatrix}0\\a\end{pmatrix}$.

In the remaining exercises, a 2×2 matrix is given along with a pair of ordered bases $\mathscr{X}$ and $\mathscr{Y}$ of V_2. In each exercise, find the unique linear operator $\mathbf{T} \in \mathscr{L}(V_2,V_2)$ which gives rise to the given matrix, relative to the given bases $\mathscr{X}$ and $\mathscr{Y}$; that is, find the $\mathbf{T} \in \mathscr{L}(V_2,V_2)$ such that $mx(\mathbf{T})_{\mathscr{X},\mathscr{Y}}$ is the given matrix (see Definition II.5.1 and the 2 examples following the proof of Theorem II.5.2).

II.5.24 The matrix is $\begin{pmatrix}1 & 0\\0 & 0\end{pmatrix}$, $\mathscr{X} = \left\{\begin{pmatrix}1\\0\end{pmatrix},\begin{pmatrix}0\\1\end{pmatrix}\right\}$ and $\mathscr{Y} = \left\{\begin{pmatrix}1\\0\end{pmatrix},\begin{pmatrix}0\\1\end{pmatrix}\right\}$.

II.5.25 The matrix is $\begin{pmatrix}1 & 1\\0 & 0\end{pmatrix}$, $\mathscr{X} = \left\{\begin{pmatrix}1\\0\end{pmatrix},\begin{pmatrix}0\\1\end{pmatrix}\right\}$ and $\mathscr{Y} = \left\{\begin{pmatrix}1\\0\end{pmatrix},\begin{pmatrix}0\\1\end{pmatrix}\right\}$.

II.5.26 The matrix is $\begin{pmatrix}1 & 2\\3 & 0\end{pmatrix}$, $\mathscr{X} = \left\{\begin{pmatrix}1\\0\end{pmatrix},\begin{pmatrix}0\\1\end{pmatrix}\right\}$ and $\mathscr{Y} = \left\{\begin{pmatrix}1\\0\end{pmatrix},\begin{pmatrix}1\\1\end{pmatrix}\right\}$.

II.5.27 The matrix is $\begin{pmatrix}1 & 2\\3 & 4\end{pmatrix}$, $\mathscr{X} = \left\{\begin{pmatrix}1\\0\end{pmatrix},\begin{pmatrix}1\\1\end{pmatrix}\right\}$ and $\mathscr{Y} = \left\{\begin{pmatrix}1\\0\end{pmatrix},\begin{pmatrix}0\\1\end{pmatrix}\right\}$.

II.5.28 The matrix is $\begin{pmatrix}1 & 1\\1 & 1\end{pmatrix}$, $\mathscr{X} = \left\{\begin{pmatrix}1\\0\end{pmatrix},\begin{pmatrix}0\\1\end{pmatrix}\right\}$, $\mathscr{Y} = \left\{\begin{pmatrix}1\\0\end{pmatrix},\begin{pmatrix}0\\2\end{pmatrix}\right\}$.

II.5.29 The matrix is $\begin{pmatrix} -2 & 5 \\ 0 & 3 \end{pmatrix}$, $\mathscr{X} = \left\{\begin{pmatrix} 1 \\ 1 \end{pmatrix}, \begin{pmatrix} -2 \\ 0 \end{pmatrix}\right\}$, $\mathscr{Y} = \left\{\begin{pmatrix} 1 \\ 0 \end{pmatrix}, \begin{pmatrix} 0 \\ 1 \end{pmatrix}\right\}$.

II.5.30 The matrix is $\begin{pmatrix} 1 & 0 \\ 0 & 1 \end{pmatrix}$, $\mathscr{X} = \left\{\begin{pmatrix} 1 \\ 0 \end{pmatrix}, \begin{pmatrix} 1 \\ 2 \end{pmatrix}\right\}$, $\mathscr{Y} = \left\{\begin{pmatrix} 1 \\ 0 \end{pmatrix}, \begin{pmatrix} 1 \\ 2 \end{pmatrix}\right\}$.

II.5.31 The matrix is $\begin{pmatrix} 1 & 0 \\ 0 & 1 \end{pmatrix}$, $\mathscr{X} = \begin{Bmatrix} x_1 & x_3 \\ x_2 & x_1 \end{Bmatrix}$, $\mathscr{Y} = \left\{\begin{pmatrix} x_1 \\ x_2 \end{pmatrix}, \begin{pmatrix} x_3 \\ x_4 \end{pmatrix}\right\}$.

II.5.32 The matrix is $\begin{pmatrix} 1 & 0 \\ 0 & 1 \end{pmatrix}$, $\mathscr{X} = \left\{\begin{pmatrix} 1 \\ 0 \end{pmatrix}, \begin{pmatrix} 1 \\ 2 \end{pmatrix}\right\}$, $\mathscr{Y} = \left\{\begin{pmatrix} 1 \\ 1 \end{pmatrix}, \begin{pmatrix} 0 \\ -1 \end{pmatrix}\right\}$.

II.5.33 *Addition on Pairs of* 2×2 *Matrices.* We impose, or force an addition $+$ between pairs of matrices as follows:

Given $A = \begin{pmatrix} a_{11} & a_{12} \\ a_{21} & a_{22} \end{pmatrix}$, $B = \begin{pmatrix} b_{11} & b_{12} \\ b_{21} & b_{22} \end{pmatrix}$. Let $\mathscr{X}, \mathscr{Y}$ be a pair of fixed bases for V_2 and let $\mathbf{T}$ and $\mathbf{U} \in \mathscr{L}(V_2, V_2)$ be those unique linear operators for which

(a) $$A = mx\mathbf{T}_{\mathscr{X}\mathscr{Y}}, \quad \text{and} \quad B = mx\mathbf{U}_{\mathscr{X}\mathscr{Y}}.$$

The *sum* of A and B, denoted $A + B$, is *defined* to be the 2×2 matrix $mx(\mathbf{T} \dotplus \mathbf{U})_{\mathscr{X}\mathscr{Y}}$, that is,

(b) $$A + B = mx(\mathbf{T} \dotplus \mathbf{U})_{\mathscr{X}\mathscr{Y}}.$$

(Substituting (a) into (b), yields the defining equation

(c) $$mx(\mathbf{T})_{\mathscr{X}\mathscr{Y}} + mx(\mathbf{U})_{\mathscr{X}\mathscr{Y}} = mx(\mathbf{T} \dotplus \mathbf{U})_{\mathscr{X}\mathscr{Y}}.$$

Show that regardless of bases $\mathscr{X}$ and $\mathscr{Y}$ chosen, (b) implies that

(d) $$\begin{pmatrix} a_{11} & a_{12} \\ a_{21} & a_{22} \end{pmatrix} + \begin{pmatrix} b_{11} & b_{12} \\ b_{21} & b_{22} \end{pmatrix} = \begin{pmatrix} a_{11} + b_{11} & a_{12} + b_{12} \\ a_{21} + b_{21} & a_{22} + b_{22} \end{pmatrix}.$$

II.5.34 [See (d) of Exercise II.5.33.] If $A = \begin{pmatrix} 1 & 3 \\ -2 & 3 \end{pmatrix}$ and $B = \begin{pmatrix} 0 & -2 \\ 2 & 0 \end{pmatrix}$, find $A + B$ and $B + A$.

II.5.35 If $A = \begin{pmatrix} -\frac{2}{3} & \frac{1}{2} \\ 1 & 0 \end{pmatrix}$ and $B = \begin{pmatrix} 9 & \frac{2}{3} \\ \frac{4}{9} & 0 \end{pmatrix}$, find

$$A + B \quad \text{and} \quad B + A.$$

II.5.36 If $A = \begin{pmatrix} -3 & -4 \\ -\frac{1}{2} & -2 \end{pmatrix}$ and $B = \begin{pmatrix} 3 & -4 \\ -\frac{1}{2} & -2 \end{pmatrix}$, find

$$A + B \quad \text{and} \quad B + A.$$

In Exercises II.5.37 through II.5.40, matrices A, B, and C are given. In each case, find $A + B$, $B + C$, $A + C$, $(A + B) + C$ and $A + (B + C)$.

II.5.37 $A = \begin{pmatrix} 3 & 0 \\ -1 & 2 \end{pmatrix}$, $B = \begin{pmatrix} -1 & 3 \\ 4 & 0 \end{pmatrix}$, $C = \begin{pmatrix} -2 & -3 \\ 3 & -2 \end{pmatrix}$.

II.5.38 $A = \begin{pmatrix} 1 & \frac{1}{2} \\ 2 & 3 \end{pmatrix}$, $B = \begin{pmatrix} -1 & \frac{1}{2} \\ -2 & -3 \end{pmatrix}$, $C = \begin{pmatrix} 0 & 5 \\ 5 & 0 \end{pmatrix}$.

II.5.39 $A = \begin{pmatrix} -1 & -1 \\ -1 & -1 \end{pmatrix}$, $B = \begin{pmatrix} 2 & 2 \\ 2 & 2 \end{pmatrix}$, $C = \begin{pmatrix} 3 & 3 \\ 3 & 3 \end{pmatrix}$.

II.5.40 $A = \begin{pmatrix} 0 & 2 \\ -1 & 3 \end{pmatrix}$, $B = \begin{pmatrix} \frac{1}{2} & \frac{1}{2} \\ \frac{1}{2} & \frac{1}{2} \end{pmatrix}$, $C = \begin{pmatrix} \frac{3}{2} & \frac{5}{3} \\ 1 & 3 \end{pmatrix}$.

II.5.41 We know that $\mathscr{L}(V_2, V_2)$ is a vector-space, and hence possesses an additive identity $\mathbf{0}$, that is, $\mathbf{0} + \mathbf{T} = \mathbf{T}$ for all $\mathbf{T} \in \mathscr{L}(V_2, V_2)$; see Theorem II.4.6 and Exercises II.4.3 and II.4.10. Show by using Definition II.5.1, that no matter which ordered bases $\mathscr{X}$ and $\mathscr{Y}$ we choose for V_2,

$$mx(\mathbf{0})_{\mathscr{X}\mathscr{Y}} = \begin{pmatrix} 0 & 0 \\ 0 & 0 \end{pmatrix}.$$

II.5.42 *"Concrete" approach to verifying A.3 of Definition I.5.2.* By use of the concrete representation for $mx(\mathbf{0})_{\mathscr{X}\mathscr{Y}}$ in Exercise II.5.41 and by use of Exercise II.5.33(d), show that $mx(\mathbf{0})_{\mathscr{X}\mathscr{Y}}$ is our (necessarily unique) additive identity matrix. Once this is done, condition A.3 of Definition I.5.2 will have been confirmed.

II.5.43 *"Abstract" approach to verifying A.3 of Definition I.10.* Using only the algebraic property of $\mathbf{0}$ in $\mathscr{L}(V_2, V_2)$ [that is, $\mathbf{0} + \mathbf{T} = \mathbf{T}$ for every $\mathbf{T}$ in $\mathscr{L}(V_2, V_2)$], and using the "abstract" definition of matrix addition (Exercise II.5.33), show that relative to any ordered bases $\mathscr{X}$ and $\mathscr{Y}$, $mx(\mathbf{0})_{\mathscr{X}\mathscr{Y}}$ is an additive identity for the set of all real 2×2 matrices.

II.5.44 *Concrete form for $mx(\mathbf{T}')$, given $mx(\mathbf{T})$.* The vector-space $\mathscr{L}(V_2, V_2)$ has the property that each of its elements $\mathbf{T}$, we may assign a negative (that is, additive inverse) $\mathbf{T}'$ in $\mathscr{L}(V_2, V_2)$. (That is, for each $\mathbf{T}$ in $\mathscr{L}(V_2, V_2)$, there exists a $\mathbf{T}'$ in $\mathscr{L}(V_2, V_2)$ such that $\mathbf{T} + \mathbf{T}' = \mathbf{0}$; see Theorem II.4.6 and Exercises II.4.4 and II.4.10. Suppose that $\mathbf{T}$ is a linear transformation in $\mathscr{L}(V_2, V_2)$, and that relative to the ordered bases $\mathscr{X}$ and $\mathscr{Y}$,

$$mx(\mathbf{T}) = \begin{pmatrix} a_{11} & a_{12} \\ a_{21} & a_{22} \end{pmatrix}.$$

Show that relative to this pair of ordered bases $\mathscr{X}$ and $\mathscr{Y}$

$$mx(\mathbf{T}') = \begin{pmatrix} -a_{11} & -a_{12} \\ -a_{21} & -a_{22} \end{pmatrix}.$$

II.5.45 *"Concrete" approach to verifying A.4 of Definition I.5.2.* By use of the concrete representation for $mx(\mathbf{T}')$, in Exercise II.5.44, and by use of Exercise II.5.33(d), show that $mx(\mathbf{T}')$ is our (necessarily unique) additive inverse for $mx(\mathbf{T})$.

II.5.46 *"Abstract" approach to verifying A.4 of Definition I.5.2.* Let $\mathbf{T} \in \mathscr{L}(V_2, V_2)$ be arbitrary but fixed throughout. Using only the algebraic property of $\mathbf{T}'$ in $\mathscr{L}(V_2, V_2)$ [that is, $\mathbf{T} + \mathbf{T}' = \mathbf{0}$ in $\mathscr{L}(V_2, V_2)$], and using Exercise II.5.33 which defines addition of matrices, show that relative to any fixed pair of ordered bases, $mx(\mathbf{T}')$ is the additive inverse for $mx(\mathbf{T})$ relative to $+$ defined on 2×2 matrices.

II.5.47 For each of the matrices of Exercises II.5.37 through II.5.46, find the additive inverse; see A.4 of Definition I.5.2.

II.5.48 *"Concrete" approach to verifying A.5 of Definition I.5.2.* Using the "concrete" representation [Exercise II.5.33(d)] for addition between *pairs* (addition acts only on pairs) of matrices, verify A.5 of Definition I.5.2 for

$$A = \begin{pmatrix} a_{11} & a_{12} \\ a_{21} & a_{22} \end{pmatrix}, \quad B = \begin{pmatrix} b_{11} & b_{12} \\ b_{21} & b_{22} \end{pmatrix}, \quad C = \begin{pmatrix} c_{11} & c_{12} \\ c_{21} & c_{22} \end{pmatrix}.$$

II.5.49 *"Abstract" approach to verifying A.5 of Definition I.5.2.* $\mathscr{L}(V_2, V_2)$ is a real vector-space (Theorem II.4.6) and hence the axiom of distributivity, A.5 of Definition I.5.2, holds (see Exercises II.4.5 and II.4.10). Using this fact, and using the definition for pairwise matrix addition [Exercise II.5.33(b)], verify A.5 of Definition I.5.2; that is, show that the distributive property of $\mathscr{L}(V_2, V_2)$ is forced onto the set of all 2×2 matrices through the function "*mx*."

II.5.50 *Scalar multiplication on 2×2 matrices.* We induce a scalar multiplication on 2×2 matrices as follows:

Let $A = \begin{pmatrix} a_{11} & a_{12} \\ a_{21} & a_{22} \end{pmatrix}$, and let α be any scalar. (Which 2×2 matrix shall represent $\alpha \cdot A$?) Let $\mathscr{X}$, $\mathscr{Y}$ be a fixed pair of ordered bases for V_2 and let $\mathbf{T} \in \mathscr{L}(V_2, V_2)$ be that unique linear operator such that

(a) $$mx(\mathbf{T})_{\mathscr{X}\mathscr{Y}} = A.$$

Then $\alpha \cdot A$ is defined by

(b) $$\alpha \cdot A = mx(\alpha\mathbf{T})_{\mathscr{X}\mathscr{Y}}$$

where $\alpha\mathbf{T}$ is the linear operator defined by Definition II.4.3. (Combining (a) and (b), yields

(c) $$\alpha \cdot mx(\mathbf{T})_{\mathscr{X}\mathscr{Y}} = mx(\alpha\mathbf{T})_{\mathscr{X}\mathscr{Y}})$$

Show that regardless of the bases $\mathscr{X}$, $\mathscr{Y}$ chosen (b) implies

(d) $$\alpha \cdot \begin{pmatrix} a_{11} & a_{12} \\ a_{21} & a_{22} \end{pmatrix} = \begin{pmatrix} \alpha a_{11} & \alpha a_{12} \\ \alpha a_{21} & \alpha a_{22} \end{pmatrix}.$$

II.5.51 *"Concrete" verification of B.2 of Definition I.5.2.* Using the concrete representation for scalar multiplication [Exercise II.5.50(d)], show that for all scalars α and β, and for all matrices $A = \begin{pmatrix} a_{11} & a_{12} \\ a_{21} & a_{22} \end{pmatrix}$,

$$(\alpha\beta) \cdot A = \alpha \cdot (\beta \cdot A).$$

Note the apparent difference between the 2 sides of this equation; the *left*-hand side has the matrix A multiplied by the single scalar $\alpha\beta$, while the *right*-hand side has the matrix $\beta \cdot A$ multiplied by the scalar α.

II.5.52 *"Abstract" approach to verifying B.2 of Definition I.5.2.* Axiom B.2 of Definition I.5.2 holds in $\mathscr{L}(V_2, V_2)$ since it is a vector-space; see Exercises II.4.7 and II.4.10. Using the "abstract" definition of scalar multiplication on real 2×2 matrices [Exercise II.5.50(b)], show that B.2 of Definition I.5.2 holds.

II.5.53 *"Concrete" approach to verifying B.3 of Definition I.5.2.* With the help of Exercise II.5.50(d), verify B.3 of Definition I.5.2.

II.5.54 *"Abstract" approach to verifying B.3 of Definition I.5.2.* Note that since $\mathscr{L}(V_2, V_2)$ is a real vector-space, B.3 of Definition I.5.2 obtains, that is, $\mathbf{1} \circ \mathbf{T} = \mathbf{T}$ for all $\mathbf{T} \in \mathscr{L}(V_2, V_2)$ (see Exercise II.4.10). Use this property along with Exercise II.5.50(b) to show that B.3 of Definition I.5.2 obtains for the set of all real 2×2 matrices.

II.5.55 *"Concrete" approach to verifying B.4 of Definition I.5.2.* Using the "concrete" representations for matrix addition and scalar multiplication [Exercise II.5.33(d), Exercise II.5.50(d)], verify B.4 of Definition I.5.2 for the set of 2×2 matrices.

II.5.56 *"Abstract" approach to verifying B.4 of Definition I.5.2.* Recall that B.5 of Definition I.5.2 obtains for the vector-space $\mathscr{L}(V_2, V_2)$. With the aid of the "abstract" definitions of matrix addition and scalar multiplication [Exercise II.5.33(b), Exercise II.5.50(b)], show that B.4 of Definition I.5.2 obtains for the set of 2×2 matrices.

II.5.57 Given matrices A and B along with scalars α and β compute the matrix $\alpha \cdot A + \beta \cdot B$.

(a) $A = \begin{pmatrix} 3 & 0 \\ 0 & 1 \end{pmatrix}, \quad B = \begin{pmatrix} -3 & 0 \\ 0 & -1 \end{pmatrix}, \alpha = 2, \beta = 2$

(b) $A = \begin{pmatrix} -1 & 2 \\ -3 & 0 \end{pmatrix}, B = \begin{pmatrix} \frac{1}{2} & \frac{1}{2} \\ 2 & 2 \end{pmatrix}, \alpha = \frac{1}{2}, \beta = -1$

(c) $A = \begin{pmatrix} 3 & 3 \\ 3 & 3 \end{pmatrix}, \quad B = \begin{pmatrix} 0 & 0 \\ 0 & 0 \end{pmatrix}, \alpha = 0, \beta = 2$

(d) $A = \begin{pmatrix} 0 & -2 \\ 2 & 3 \end{pmatrix}, B = \begin{pmatrix} 5 & 7 \\ -2 & 3 \end{pmatrix}, \alpha = \frac{1}{3}, \beta = \frac{2}{3}$

(e) $A = \begin{pmatrix} 9 & -7 \\ \frac{1}{4} & 3 \end{pmatrix}, B = \begin{pmatrix} \frac{1}{4} & 3 \\ 9 & -7 \end{pmatrix}, \alpha = 1, \beta = 2$

(f) $A = \begin{pmatrix} 1 & 3 \\ 5 & 7 \end{pmatrix}, \quad B = \begin{pmatrix} 1 & 3 \\ 5 & 7 \end{pmatrix}, \alpha = \frac{1}{4}, \beta = \frac{3}{4}.$

II.5.58 *"Concrete" verification of B.5 of Definition I.5.2.* Use the concrete representations for matrix addition and scalar multiplication [Exercises II.5.33(d), and II.5.50(d)] to show that

$$\alpha \cdot (A + B) = \alpha \cdot A + \alpha \cdot B$$

for all pairs of 2×2 matrices A, B, and for all scalars α.

II.5.59 *"Abstract" verification of B.5 Definition I.5.2.* We know that $\mathscr{L}(V_2, V_2)$ is a vector-space (Theorem II.4.6), so that in particular, the equation

$$\alpha \cdot (\mathbf{T} \dotplus \mathbf{U}) = \alpha \cdot \mathbf{T} \dotplus \alpha \cdot \mathbf{U}$$

is true for all $\mathbf{T}$, $\mathbf{U} \in \mathscr{L}(V_2, V_2)$ and for all scalars α. Use the "abstract" definition of matrix addition and scalar multiplication [Exercises II.5.53(b) and II.5.50(b)] to show that correspondingly,

$$\alpha \cdot (A + B) = \alpha \cdot A + \alpha \cdot B$$

for all pairs of 2×2 matrices A, B, and for all scalars α.

II.5.60 *A multiplication between pairs of 2×2 matrices.* We induce a product defined on pairs of 2×2 matrices as follows:

Given $A = \begin{pmatrix} a_{11} & a_{12} \\ a_{21} & a_{22} \end{pmatrix}$, $B = \begin{pmatrix} b_{11} & b_{12} \\ b_{21} & b_{22} \end{pmatrix}$, and $\mathscr{X}$ a fixed ordered basis of V_2. Let $\mathbf{T}$ and $\mathbf{U}$ be the unique linear operators of $\mathscr{L}(V_2, V_2)$ such that

(a) $$A = mx(\mathbf{T})_{\mathscr{X}\mathscr{X}}, \quad \text{and} \quad B = mx(\mathbf{U})_{\mathscr{X}\mathscr{X}}.$$

The *product* of A with B, denoted $A \cdot B$, is defined to be the 2×2 matrix $mx(\mathbf{T} \circ \mathbf{U})_{\mathscr{X}\mathscr{X}}$, that is,

(b) $$A \cdot B = mx(\mathbf{T} \circ \mathbf{U})_{\mathscr{X}\mathscr{X}},$$

where $\mathbf{T} \circ \mathbf{U}$ is the composition of the linear operators $\mathbf{T}$ and $\mathbf{U}$ (see Definition II.1.5 and Theorem II.4.7).

Substituting (a) into (b) yields

(c) $$mx(\mathbf{T})_{\mathscr{X}\mathscr{X}} \cdot mx(\mathbf{U})_{\mathscr{X}\mathscr{X}} = mx(\mathbf{T} \circ \mathbf{U})_{\mathscr{X}\mathscr{X}}.$$

Show that regardless of the basis $\mathscr{X}$ chosen, matrix multiplication is realized concretely according to the equation

(d) $$\begin{pmatrix} a_{11} & a_{12} \\ a_{21} & a_{22} \end{pmatrix} \cdot \begin{pmatrix} b_{11} & b_{12} \\ b_{21} & b_{22} \end{pmatrix} = \begin{pmatrix} a_{11}b_{11} + a_{12}b_{21} & a_{11}b_{12} + a_{12}b_{22} \\ a_{21}b_{11} + a_{22}b_{21} & a_{21}b_{12} + a_{22}b_{22} \end{pmatrix}.$$

II.5.61 Given the 2×2 matrices A and B, find the matrix products $A \cdot B$ and $B \cdot A$ [see Exercise II.5.60(d)].

(a) $A = \begin{pmatrix} 2 & 3 \\ -1 & 2 \end{pmatrix}$, $B = \begin{pmatrix} -1 & \frac{3}{2} \\ -1 & \frac{3}{2} \end{pmatrix}$

(b) $A = \begin{pmatrix} 0 & 1 \\ 2 & 3 \end{pmatrix}$, $B = \begin{pmatrix} 0 & 3 \\ 2 & 1 \end{pmatrix}$

(c) $A = \begin{pmatrix} 3 & 0 \\ 0 & 3 \end{pmatrix}$, $B = \begin{pmatrix} 17 & 2 \\ -1 & -2 \end{pmatrix}$

(d) $A = \begin{pmatrix} \alpha & 0 \\ 0 & \alpha \end{pmatrix}$, $B = \begin{pmatrix} 5 & 7 \\ 9 & 2 \end{pmatrix}$, α is a fixed scalar.

(e) $A = \begin{pmatrix} \alpha & 0 \\ 0 & \alpha \end{pmatrix}$, $B = \begin{pmatrix} b_{11} & b_{12} \\ b_{21} & b_{22} \end{pmatrix}$, α, b_{ij}'s are scalars.

(f) $A = \begin{pmatrix} 1 & 0 \\ 0 & 0 \end{pmatrix}$, $B = \begin{pmatrix} -2 & \frac{1}{2} \\ \frac{3}{2} & 0 \end{pmatrix}$

(g) $A = \begin{pmatrix} 0 & 0 \\ 0 & 0 \end{pmatrix}$, $B = \begin{pmatrix} b_{11} & b_{12} \\ b_{21} & b_{22} \end{pmatrix}$, b_{ij}'s are scalars.

(h) $A = \begin{pmatrix} 1 & -1 \\ -1 & 1 \end{pmatrix}$, $B = \begin{pmatrix} 3 & 4 \\ 3 & 4 \end{pmatrix}$

(i) $A = \begin{pmatrix} -2 & -2 \\ -3 & -1 \end{pmatrix}$, $B = \begin{pmatrix} -1 & -2 \\ 0 & -1 \end{pmatrix}$

(j) $A = \begin{pmatrix} 7 & 5 \\ -1 & 0 \end{pmatrix}$, $B = \begin{pmatrix} 7 & 5 \\ -1 & 0 \end{pmatrix}$.

II.5.62 The solution of Exercise II.5.61(d) should have revealed that

$$\begin{pmatrix} \alpha & 0 \\ 0 & \alpha \end{pmatrix} \cdot B = B \cdot \begin{pmatrix} \alpha & 0 \\ 0 & \alpha \end{pmatrix}.$$

Where B is an arbitrary 2×2 matrix, that is, the so-called *diagonal scalar* matrix $\begin{pmatrix} \alpha & 0 \\ 0 & \alpha \end{pmatrix}$ commutes with any and every 2×2 matrix. In this exercise, show that the diagonal scalar matrices are the *only* matrices that commute with every matrix, that is, if the 2×2 matrix A_0 has the property that $A_0 \cdot B = B \cdot A_0$ for any and every 2×2 matrix B, then, necessarily, $A_0 = \begin{pmatrix} \alpha & 0 \\ 0 & \alpha \end{pmatrix}$, where α is a scalar.

II.5.63 [See Exercise II.5.60(b)]. Consider the following linear transformations **T** and **U** in $\mathscr{L}(V_2, V_2)$ defined by

$$\mathbf{T}\begin{pmatrix} x_1 \\ x_2 \end{pmatrix} = \begin{pmatrix} 3x_1 + x_2 \\ x_1 - x_2 \end{pmatrix}, \quad \mathbf{U}\begin{pmatrix} x_1 \\ x_2 \end{pmatrix} = \begin{pmatrix} x_1 + x_2 \\ x_1 + 2x_2 \end{pmatrix},$$

for all $\begin{pmatrix} x_1 \\ x_2 \end{pmatrix} \in V_2$. Relative to the standard ordered basis $= \left\{\begin{pmatrix} 1 \\ 0 \end{pmatrix}, \begin{pmatrix} 0 \\ 1 \end{pmatrix}\right\}$,

(a) find $mx(\mathbf{T})$, $mx(\mathbf{U})$, $mx(\mathbf{T} \circ \mathbf{U})$ and $mx(\mathbf{U} \circ \mathbf{T})$.

(b) Compute [see Exercise II.5.60(d)] $mx(\mathbf{T}) \cdot mx(\mathbf{U})$ and $mx(\mathbf{U}) \cdot mx(\mathbf{T})$, and with the results of part (a), show that in this special case, $mx(\mathbf{T}) \cdot mx(\mathbf{U}) = mx(\mathbf{T} \circ \mathbf{U})$, and $mx(\mathbf{U} \circ \mathbf{T}) = mx(\mathbf{U}) \cdot mx(\mathbf{T})$.

This result should confirm the validity of Exercise II.5.60(d), at least for the special operators **T** and **U** defined in this exercise.

II.5.64 [See Exercises II.5.33(d) and II.5.60(d).] Given the 2×2 matrices A, B, and C, compute the matrix product $A \cdot (B + C)$. Compare this with the matrix sum $(A \cdot B) + (A \cdot C)$. Similarly, compute $(B + C) \cdot A$, the product of the matrices $(B + C)$ and A, and compare this result with $(B \cdot A) + (C \cdot A)$, the sum of the matrices $B \cdot A$ and $C \cdot A$.

(a) $A = \begin{pmatrix} 1 & 1 \\ 3 & 2 \end{pmatrix}, \quad B = \begin{pmatrix} 0 & -1 \\ 2 & 3 \end{pmatrix}, \quad C = \begin{pmatrix} 1 & 1 \\ 3 & -1 \end{pmatrix}.$

(b) $A = \begin{pmatrix} 1 & 2 \\ 3 & 6 \end{pmatrix}, \quad B = \begin{pmatrix} -1 & -4 \\ 1 & 3 \end{pmatrix}, C = \begin{pmatrix} -1 & 0 \\ 0 & -1 \end{pmatrix}.$

(c) $A = \begin{pmatrix} -1 & 3 \\ -2 & 1 \end{pmatrix}, B = \begin{pmatrix} 5 & 7 \\ 0 & 3 \end{pmatrix}, \quad C = \begin{pmatrix} -\frac{1}{2} & \frac{3}{4} \\ 1 & -2 \end{pmatrix}.$

(d) $A = \begin{pmatrix} -1 & 0 \\ 2 & 3 \end{pmatrix}, B = \begin{pmatrix} 3 & -\frac{1}{2} \\ 5 & 0 \end{pmatrix}, \quad C = \begin{pmatrix} 1 & 7 \\ -2 & -2 \end{pmatrix},$

(e) $A = \begin{pmatrix} 1 & 1 \\ 1 & 1 \end{pmatrix}, \quad B = \begin{pmatrix} 3 & 3 \\ 3 & 3 \end{pmatrix}, \quad C = \begin{pmatrix} 4 & 4 \\ 4 & 4 \end{pmatrix}.$

II.5.65 *"Concrete" approach to verify two-sided distributivity for matrices.* Consider arbitrary matrices A, B, and C in "concrete" form, that is,

$$A = \begin{pmatrix} a_{11} & a_{12} \\ a_{21} & a_{22} \end{pmatrix}, B = \begin{pmatrix} b_{11} & b_{12} \\ b_{21} & b_{12} \end{pmatrix}, C = \begin{pmatrix} c_{11} & c_{12} \\ c_{21} & c_{22} \end{pmatrix}.$$

Use Exercises II.5.33(d) and II.5.60(d) to prove that

$$A \cdot (B + C) = (A \cdot B) + (A \cdot C),$$

and that

$$(B + C) \cdot A = B \cdot A + C \cdot A.$$

II.5.66 "*Abstract*" *approach to verifying 2-sided distributivity for matrices.* Use the fact that C.2 of Definition II.4.10 holds for the algebra $\mathscr{L}(V_2, V_2)$, along with the Definitions of Exercises II.5.33(b) and II.5.60(b), to show that for arbitrary 2×2 matrices A, B, and C,

$$A \cdot (B + C) = (A \cdot B) + (A \cdot C)$$

and

$$(B + C) \cdot A = (B \cdot A) + (C \cdot A).$$

II.5.67 Using the specific 2×2 matrices, A, B, and C in each of the 5 parts of Exercise II.5.64, verify the associative law for matrix multiplication [see Exercise II.5.60(d)]. That is, show that in these special cases, the matrix $A \cdot (B \cdot C)$ is equal to the matrix $(A \cdot B) \cdot C$.

II.5.68 [See Exercises II.5.67 and II.5.60(d)]. Represent matrices A, B, C in arbitrary but concrete form, that is, set

$$A = \begin{pmatrix} a_{11} & a_{12} \\ a_{21} & a_{22} \end{pmatrix}, \quad B = \begin{pmatrix} b_{11} & b_{12} \\ b_{21} & b_{22} \end{pmatrix}, \quad \text{and} \quad C = \begin{pmatrix} c_{11} & c_{12} \\ c_{21} & c_{22} \end{pmatrix}.$$

Show that

$$A \cdot (B \cdot C) = (A \cdot B) \cdot C.$$

6. *Matrices of Linear Transformations (the $m \times n$ case)*

The purpose of this section is to extend the results of the previous section to the most general linear transformations $\mathscr{L}(V_n, V_m)$ (see Definition II.4.5), that is, we no longer require that $m = n = 2$. As you will see, the *ideas* involved will *not* be new; we will be treading over familiar ground. However, new *notation* will have to be introduced to assist us in the problem of extending our ideas. Toward this end, we introduce the sigma notation.

We will have to consider sums of numbers, vectors, or matrices, and our problem will be to devise a shorthand for these sums. A typical problem might be:

What is a convenient shorthand notation for representing the following sum of real numbers?

$$\tfrac{1}{1} + \tfrac{1}{2} + \tfrac{1}{3} + \tfrac{1}{4} + \tfrac{1}{5} + \tfrac{1}{6} + \tfrac{1}{7} + \tfrac{1}{8} + \tfrac{1}{9} + \tfrac{1}{10}. \qquad \text{(II.6.1a)}$$

As the terms of (II.6.1a) are ordered, we see that the *first* term is $\frac{1}{1}$, the *second* terms is $\frac{1}{2}$, and so on. But what does "and so on" mean in this example? In this case, we can discern an overall *pattern* for the terms in (II.6.1a), for if k represents any integer from 1 through 10, we may describe the kth term in terms of k itself; in fact, the kth term is just the number $1/k$. Hence, (II.6.1a) may be rewritten as

The sum of all numbers of the form $1/k$, where the integer k varies from 1 through 10. (II.6.1b)

Now watch the shorthand for (II.6.1b) [which is another way of stating (II.6.1a)] develop. We shall simply replace the italicized phrase "*The sum of all numbers of the form*" by the symbol "$\sum$," which is the uppercase Greek letter "sigma." In other words, with the convention that

$$\textit{The sum of all numbers of the form} = \sum, \tag{II.6.1c}$$

display (II.6.1b) becomes

$$\sum \frac{1}{k}, \text{ where the integer } k \text{ varies from 1 through 10.} \tag{II.6.1d}$$

Now to abbreviate (II.6.1d) even more, we shall do away with the cumbersome phrase "where the integer k varies from 1 through 10," and attach this information to the symbol "$\sum$" as follows:

$$\sum_{k=1}^{10} \frac{1}{k}, \tag{II.6.1e}$$

which is to be read, "the sum of all numbers of the form $1/k$, where the (indexing) integer k varies from 1 through 10," that is, exactly as (II.6.1b) reads. Hence, (II.6.1e) gives us a more compact *notation* for (II.6.1a); note that the difference between (II.6.1a) and (II.6.1e) is *only in the notation* of the statement and not in the statement *itself.*

There are 2 remarks we wish to make regarding (II.6.1e) before we proceed to a *general* definition for the sigma notation.

REMARK. The indexing symbol k used in (II.6.1e) is called a **dummy variable**; one reason for this is that k may be replaced by any *other* convenient symbol, and the essential shorthand message of (II.6.1e) remains intact. Thus the following notations

$$\sum_{k=1}^{10} \frac{1}{k}, \qquad \sum_{n=1}^{10} \frac{1}{n}, \qquad \sum_{p=1}^{10} \frac{1}{p}$$

may all be used to express one and the same idea, namely, the sum

$$1 + \tfrac{1}{2} + \tfrac{1}{3} + \tfrac{1}{4} + \tfrac{1}{5} + \tfrac{1}{6} + \tfrac{1}{7} + \tfrac{1}{8} + \tfrac{1}{9} + \tfrac{1}{10}.$$

REMARK. In our specific example, the number $1/k$ represents the kth term in our sum; for a *general* sum, we might indicate the kth term by a symbol involving the "term-number" k; $a_{(k)}$, for example. In other words, if $a_{(k)}$ represents the kth number in a sum, then the symbol $\sum_{k=1}^{5} a_{(k)}$ can be read "the sum of all numbers of the form $a_{(k)}$, where the integer k varies from 1 through 5." Thus $\sum_{k=1}^{5} a_{(k)}$ stands for the expression $a_{(1)} + a_{(2)} + a_{(3)} + a_{(4)} + a_{(5)}$.

There are times when the index (or dummy variable) will not start with the number 1. In the most general case, we will not even demand that only numbers shall be added. After all, some measure of hard work has gone into defining "sums" of n-vectors, linear operators, and 2×2 matrices, so the sigma notation shall find its use in the additions of these objects as well.

We consider some examples and conclude with a formal definition.

Example Let us choose the symbol u_j to represent the number $\dfrac{j+1}{j+2}$, which is to be the jth term of our sum. We then denote the 4-term sum, where j varies from 0 through 3, by

$$\sum_{j=0}^{3} u_j \text{ or, equivalently, } \sum_{j=0}^{3} \frac{j+1}{j+2}.$$

These symbols stand for the sum of all numbers of the form $u_j = \dfrac{j+1}{j+2}$, as the integer j varies from 0 through 3, that is,

$$\sum_{j=0}^{3} u_j = u_0 + u_1 + u_2 + u_3.$$

Now our first sentence of this example *defines* u_j to be the number $\dfrac{j+1}{j+2}$ for each j.

Hence $u_0 = \dfrac{0+1}{0+2} = \dfrac{1}{2}$, $u_1 = \dfrac{1+1}{1+2} = \dfrac{2}{3}$, $u_2 = \dfrac{2+1}{2+2} = \dfrac{3}{4}$ and $u_3 = \dfrac{3+1}{3+2} = \dfrac{4}{5}$.
Finally,

$$\sum_{j=0}^{3} u_j = \sum_{j=0}^{3} \frac{j+1}{j+2} = \frac{1}{2} + \frac{2}{3} + \frac{3}{4} + \frac{4}{5} \left(\text{which equals } \frac{163}{60}\right).$$

REMARK. Once again, we point out that the indexing symbol (dummy variable) j could be replaced by any other symbol, so that the following expressions would each denote the sum given in this example.

$$\sum_{p=0}^{3} u_p, \qquad \sum_{k=0}^{3} \frac{k+1}{k+2}, \qquad \sum_{s=0}^{3} \frac{s+1}{s+2}, \qquad \sum_{t=0}^{3} u_t.$$

Example In what follows, let us consider the sigma notation for the sum of the four 2×2 matrices

$$\underset{\substack{\text{first term} \\ i=1}}{\underset{\uparrow}{\begin{pmatrix} 1 & 0 \\ 1 & 2 \end{pmatrix}}} + \underset{\substack{\text{second term} \\ i=2}}{\underset{\uparrow}{\begin{pmatrix} 2 & 0 \\ 1 & 3 \end{pmatrix}}} + \underset{\substack{\text{third term} \\ i=3}}{\underset{\uparrow}{\begin{pmatrix} 3 & 0 \\ 1 & 4 \end{pmatrix}}} + \underset{\substack{\text{fourth term} \\ i=4}}{\underset{\uparrow}{\begin{pmatrix} 4 & 0 \\ 1 & 5 \end{pmatrix}}}. \tag{II.6.2a}$$

In this case, it is easy to express each matrix in terms of its "term-number," call it i. That is, we observe that if we *define*

$$A_i = \begin{pmatrix} i & 0 \\ 1 & i+1 \end{pmatrix} \quad \text{for } i = 1, 2, 3, \text{ or } 4,$$

then A_1 is the *first* term of (II.6.2a), A_2 is the *second* term of (II.6.2a) and, in fact, A_i is the ith term of (II.6.2a) for *each* and *every* integer i we choose from 1 through 4. Thus, (II.6.2a) can be collapsed into either of the more compact expressions

$$\sum_{i=1}^{4} A_i, \quad \text{or} \quad \sum_{i=1}^{4} \begin{pmatrix} i & 0 \\ 1 & i+1 \end{pmatrix}. \tag{II.6.2b}$$

This matrix sum, incidentally, yields the matrix $\begin{pmatrix} 10 & 0 \\ 4 & 14 \end{pmatrix}$.

REMARK. We have chosen, in our examples, rather scrawny sums consisting of no more than 10 terms. The convenience of the sigma notation of (II.6.2b) for example, would become boldy apparent if (II.6.2a) were given as a 1000-term sum. Fortunately, we have the symbol $\sum_{i=1}^{1000} \begin{pmatrix} i & 0 \\ 1 & i+1 \end{pmatrix}$ that spares us the time and trouble of writing out a complete 1000-term matrix sum.

The following definition delineates the use of the sigma notation.

Definition II.6.1 Let $a_1, a_2, \ldots, a_i, \ldots, a_n$ denote n objects upon which a pairwise, commutative and associative addition $+$ (compare A.2 and A.5 of Definition I.5.5) is defined. Then we use the symbol

$$\sum_{i=1}^{n} a_i = a_1 + a_2 + \cdots + a_i + \cdots + a_n$$

to denote the sum of these n objects; each object a_i is called a **term** of the sum.

EXERCISES

Exercises II.6.1 through II.6.5, write all the terms of the indicated sum explicitly and perform the addition.

II.6.1 $\sum_{s=1}^{6} s.$

II.6.2 $\sum_{k=3}^{10} \frac{k}{k-2}.$

II.6.3 $\sum_{r=0}^{3} v_r$ where $v_r = \begin{pmatrix} 2r \\ r+1 \end{pmatrix}$, a real 2-vector.

II.6.4 $\sum_{r=0}^{4} \frac{ni}{i+1}$ where n is some fixed integer. How does this quantity compare with $n \cdot \sum_{i=1}^{4} \frac{i}{i+1}$?

II.6.5 $\sum_{j=0}^{4} \begin{pmatrix} j & 2j \\ 0 & 3 \end{pmatrix}.$

II.6.6 $\sum_{j=0}^{5} \alpha_i \cdot X_i$, where for each integer i, α_i is the scalar $3i$, and X_i is the real 3-vector

$$\begin{pmatrix} i+1 \\ i-1 \\ 0 \end{pmatrix}.$$

II.6.7 Suppose that $X_1, X_2, \ldots, X_i, \ldots, X_n$ are n real vectors (see Definition I.5.2), of the same vector space: for example, each X_i might be a number, k-vector or a matrix. If c is any scalar (number) which is fixed throughout, prove that

$$\sum_{i=1}^{n} c \cdot X_i = c \cdot \sum_{i=1}^{n} X_i.$$

Use Definition II.6.1 and Definition I.5.2 to support your proof. For a special case of this result, see Exercise II.6.4, where "n" assumes the role of the a fixed scalar.

In the remaining exercises, write the sums in compact form by means of the sigma notation.

II.6.8 $1 + 2 + 3 + 4 + 5.$

II.6.9 $3 + 4 + 5 + 6 + 7 + 8.$

II.6.10 $1 + 2 + 3 + 4 + \cdots + 516 + 517.$

II.6.11 $2 + 4 + 6 + 8 + 10 + 12 + 14.$

II.6.12 $\frac{1}{2} + \frac{1}{4} + \frac{1}{6} + \frac{1}{8} + \frac{1}{10}.$

II.6.13 $\begin{pmatrix} 1 \\ 0 \end{pmatrix} + \begin{pmatrix} 2^2 \\ 0 \end{pmatrix} + \begin{pmatrix} 3^2 \\ 0 \end{pmatrix} + \begin{pmatrix} 4^2 \\ 0 \end{pmatrix} + \begin{pmatrix} 5^2 \\ 0 \end{pmatrix}.$

II.6.14

$$\frac{1}{3+a} + \frac{1}{4+a} + \frac{1}{5+a} + \frac{1}{6+a} + \frac{1}{7+a} + \frac{1}{8+a},$$

where a is some fixed, positive integer.

II.6.15

$$\frac{1}{3+a}+\frac{1}{5+a}+\frac{1}{7+a}+\frac{1}{9+a}+\cdots+\frac{1}{2n+1+a},$$

where a and n are fixed positive integers; assume n is greater than 4.

The development of the remainder of this section will parallel that of Section 5 of this chapter. Our objective will be to obtain a number-picture (matrix) for a linear operator **T**. $\mathbf{T} \in \mathscr{L}(V_n, V_m)$; we understand that the symbols V_n and V_m represent the vector-spaces of n-tuples and m-tuples, respectively.

Now let **T** denote an arbitrary but fixed linear operator in $\mathscr{L}(V_n, V_m)$. Now let $\mathscr{X} = \{X_1, X_2, \ldots, X_n\}$ and $\mathscr{Y} = \{Y_1, Y_2, \ldots, Y_m\}$ be ordered *bases* for V_n and V_m, respectively (see Theorem I.6.12). We note that the following n elements

$$\mathbf{T}(X_1), \mathbf{T}(X_2), \ldots, \mathbf{T}(X_n)$$

belong to V_m, which contains the range of **T**, that is, V_m is an impact set for **T**. Since each $\mathbf{T}(X_i)$ $i = 1, 2, \ldots, n$ is in V_m, $\mathbf{T}(X_i)$ can be expressed in 1 and *only* 1 way as a linear combination of the V_m basis elements $Y_1, Y_2, \ldots, Y_m$ (see Theorem I.6.8). Thus, we may write the following equations:

$$\begin{aligned}
\mathbf{T}(X_1) &= a_{11} \cdot Y_1 + a_{21} \cdot Y_2 + \cdots + a_{m1} \cdot Y_m \\
\mathbf{T}(X_2) &= a_{12} \cdot Y_1 + a_{22} \cdot Y_2 + \cdots + a_{m2} \cdot Y_m \\
&\ \ \vdots \\
\mathbf{T}(X_i) &= \boxed{a_{1i}} \cdot Y_1 + \boxed{a_{2i}} \cdot Y_2 + \cdots + \boxed{a_{mi}} \cdot Y_m \qquad \text{(}i\text{th column of } mx\mathbf{T}\text{)} \\
&\ \ \vdots \\
\mathbf{T}(X_n) &= a_{1n} \cdot Y_1 + a_{2n} \cdot Y_2 + \cdots + a_{mn} \cdot Y_m.
\end{aligned} \qquad \text{(II.6.3a)}$$

We may, if we choose, use the sigma notation to denote the statements of (II.6.3a) in more compact form. In fact, we can say exactly the same thing by writing

$$\mathbf{T}(X_i) = \sum_{j=1}^{m} a_{ji} \cdot Y_j \qquad \text{for all } i = 1, 2, \ldots, n. \qquad \text{(II.6.3b)}$$

Equation (II.6.3a) or (II.6.3b) is a generalization of Equation (II.5.1). The same conclusions obtain, namely, that each of the scalars a_{ij} which appears in (II.6.3a and b) is *uniquely determined* due to Theorem I.6.8.

Also, we should sense by now that these (unique) scalars a_{ij} in (II.6.3a and b) somehow tell us everything we need to know about the linear operator **T**, once we understand which ordered bases $\mathscr{X}$ and $\mathscr{Y}$ we are working with. We shall extract these scalars [as we did in (II.5.1)] by taking the scalars in the ith row of (II.6.3a) for each $i = 1, 2, \ldots, n$, and arrange them in order as the ith *column* of a rectangular array of scalars; our final display will be called the $m \times n$ (*read* "*m by n*") *matrix of* **T** *relative to the ordered bases* $\mathscr{X}$ *and* $\mathscr{Y}$, and is denoted by the symbol $mx(\mathbf{T})_{\mathscr{X}\mathscr{Y}}$. That is, we have developed Definition II.6.2.

Definition II.6.2 Let $\mathscr{X} = \{X_1, X_2, \ldots, X_n\}$ and $\mathscr{Y} = \{Y_1, Y_2, \ldots, Y_m\}$ be ordered bases of V_n and V_m, respectively. Let **T** by any linear operator sending V_n and V_m. If for each X_j, we have $\mathbf{T}(X_j) = a_{1j} \cdot Y_1 + a_{2j} \cdot Y_2 + \cdots + a_{mj} \cdot Y_m$, then the following rectangular array of the scalars a_{ij}

$$\begin{array}{c} \\ \\ \\ i\text{th row} \rightarrow \\ \\ \\ \end{array}\begin{pmatrix} a_{11} & a_{12} & \cdots & a_{1j} & \cdots & a_{1n} \\ a_{21} & a_{22} & \cdots & a_{2j} & \cdots & a_{2n} \\ \vdots & \vdots & \ddots & & & \vdots \\ a_{i1} & a_{i2} & \cdots & a_{ij} & \cdots & a_{in} \\ \vdots & \vdots & & \vdots & & \vdots \\ a_{m1} & a_{m2} & \cdots & a_{mj} & \cdots & a_{mn} \end{pmatrix} = mx(\mathbf{T})_{\mathscr{X}\mathscr{Y}} \qquad \text{(II.6.4)}$$

(with the jth column marked "jth column ↓")

is called the $\boldsymbol{m \times n}$ **matrix of T** relative to the ordered bases $\mathscr{X}$ and $\mathscr{Y}$. The following examples serve to illustrate specific constructions of matrices.

Example We chose a linear operator $\mathbf{T} : V_2 \rightarrow V_4$ from the real vector-space $\mathscr{L}(V_2, V_4)$ defined by

$$\blacktriangleright \qquad \mathbf{T}\begin{pmatrix} x_1 \\ x_2 \end{pmatrix} = \begin{pmatrix} x_1 \\ 3x_1 + x_2 \\ 0 \\ x_1 + 2x_1 \end{pmatrix} \quad \text{for all } X = \begin{pmatrix} x_1 \\ x_2 \end{pmatrix} \text{ in } V_2.$$

Let us choose the standard ordered bases (Definition I.6.11) for V_2 and V_4, that is, set

$$\mathscr{X} = \left\{\begin{pmatrix} 1 \\ 0 \end{pmatrix}, \begin{pmatrix} 0 \\ 1 \end{pmatrix}\right\} = \{X_1, X_2\} \quad \text{and} \quad \mathscr{Y} = \left\{\begin{pmatrix} 1 \\ 0 \\ 0 \\ 0 \end{pmatrix}, \begin{pmatrix} 0 \\ 1 \\ 0 \\ 0 \end{pmatrix}, \begin{pmatrix} 0 \\ 0 \\ 1 \\ 0 \end{pmatrix}, \begin{pmatrix} 0 \\ 0 \\ 0 \\ 1 \end{pmatrix}\right\} = \{Y_1, Y_2, Y_3, Y_4\}.$$

It is easily verified that the function $\mathbf{T}$ is *linear* (as per Definition II.2.1). In order to construct the 2×4 matrix for $\mathbf{T}$ relative to these ordered bases, we need only write down $\mathbf{T}\begin{pmatrix}1\\0\end{pmatrix}$ and $\mathbf{T}\begin{pmatrix}0\\1\end{pmatrix}$ in terms of the $\mathscr{Y}$ basis vectors (see II.6.3a). From the very definition of $\mathbf{T}$ above, we have

$$\begin{aligned}\mathbf{T}(X_1) &= \mathbf{T}\begin{pmatrix}1\\0\end{pmatrix} = \begin{pmatrix}1\\3\\0\\1\end{pmatrix}\\ \mathbf{T}(X_2) &= \mathbf{T}\begin{pmatrix}0\\1\end{pmatrix} = \begin{pmatrix}0\\1\\0\\2\end{pmatrix}\end{aligned} \tag{II.6.5}$$

Now we must expand $\mathbf{T}(X_1)$ and $\mathbf{T}(X_2)$ in terms of the $\mathscr{Y}$ basis vectors

$$Y_1 = \begin{pmatrix}1\\0\\0\\0\end{pmatrix},\ Y_2 = \begin{pmatrix}0\\1\\0\\0\end{pmatrix},\ Y_3 = \begin{pmatrix}0\\0\\1\\0\end{pmatrix} \text{ and } Y_4 = \begin{pmatrix}0\\0\\0\\1\end{pmatrix}.$$

Thus, (II.6.5) becomes

$$\begin{aligned}\mathbf{T}(X_1) &= \begin{pmatrix}1\\3\\0\\1\end{pmatrix} = \overbrace{1 \cdot Y_1 + 3 \cdot Y_2 + 0 \cdot Y_3 + 1 \cdot Y_4}^{\text{first column}}\\ \mathbf{T}(X_2) &= \begin{pmatrix}0\\1\\0\\2\end{pmatrix} = \overbrace{0 \cdot Y_1 + 1 \cdot Y_2 + 0 \cdot Y_3 + 2 \cdot Y_4}^{\text{second column}}\end{aligned} \tag{II.6.6}$$

which clearly delineates the first and second columns of $mx(\mathbf{T})_{\mathscr{X}\mathscr{Y}}$. Finally,

$$mx(\mathbf{T})_{\mathscr{X}\mathscr{Y}} = \begin{pmatrix}1 & 0\\3 & 1\\0 & 0\\1 & 2\end{pmatrix}. \tag{II.6.7}$$

Example In what follows, we choose the same linear operator $\mathbf{T}: V_2 \to V_4$ as in the previous example. We shall also retain the ordered basis $\mathscr{Y} = \{Y_1, Y_2, Y_3, Y_4\}$ for V_4 but this time we select $\mathscr{X} = \left\{\begin{pmatrix}1\\2\end{pmatrix}, \begin{pmatrix}1\\-1\end{pmatrix}\right\} = \{X_1, X_2\}$ for our ordered basis of

V_2. With (II.6.3a) as our guide, we write

$$\mathbf{T}(X_1) = \mathbf{T}\begin{pmatrix}1\\2\end{pmatrix} = \begin{pmatrix}1\\5\\0\\5\end{pmatrix}$$
$$\mathbf{T}(X_2) = \mathbf{T}\begin{pmatrix}1\\-1\end{pmatrix} = \begin{pmatrix}1\\2\\0\\-1\end{pmatrix} \tag{II.6.8}$$

which is just a consequence of the definition of **T**. Thus,

$$\mathbf{T}(X_1) = \overbrace{1 \cdot Y_1 + 5 \cdot Y_2 + 0 \cdot Y_3 + 5 \cdot Y_4}^{\text{column 1}}$$
$$\mathbf{T}(X_2) = \overbrace{1 \cdot Y_1 + 2 \cdot Y_2 + 0 \cdot Y_3 + (-1) \cdot Y_4}^{\text{column 2}} \tag{II.6.9}$$

so that from Definition II.6.2 we have

$$mx(\mathbf{T})_{\mathscr{X}\mathscr{Y}} = \begin{pmatrix}1 & 1\\5 & 2\\0 & 0\\5 & -1\end{pmatrix}$$

Example With $\mathbf{T}: V_2 \to V_4$ as above, set

$$\mathscr{X} = \left\{\begin{pmatrix}0\\1\end{pmatrix}, \begin{pmatrix}-2\\0\end{pmatrix}\right\} = \{X_1, X_2\}$$

and

$$\mathscr{Y} = \left\{\begin{pmatrix}1\\0\\0\\1\end{pmatrix}, \begin{pmatrix}0\\1\\0\\1\end{pmatrix}, \begin{pmatrix}0\\0\\2\\0\end{pmatrix}, \begin{pmatrix}0\\0\\0\\1\end{pmatrix}\right\} = \{Y_1, Y_2, Y_3, Y_4\}.$$

The reader should check that the ordered sets $\mathscr{X}$ and $\mathscr{Y}$ are *bases* for V_2 and V_4, respectively (see Definition I.6.7 and Theorem I.6.8). We see from the definition of **T** that

$$\mathbf{T}(X_1) = \mathbf{T}\begin{pmatrix}0\\1\end{pmatrix} = \begin{pmatrix}0\\1\\0\\2\end{pmatrix}$$
$$\mathbf{T}(X_2) = \mathbf{T}\begin{pmatrix}-2\\0\end{pmatrix} = \begin{pmatrix}-2\\-6\\0\\-2\end{pmatrix}, \tag{II.6.10}$$

which is to say

$$\mathbf{T}(X_1) = \underbrace{0 \cdot Y_1 + 1 \cdot Y_2 + 0 \cdot Y_3 + 1 \cdot Y_4}_{\text{column 1}}$$

$$\mathbf{T}(X_2) = \underbrace{(-2) \cdot Y_1 + (-6) \cdot Y_2 + 0 \cdot Y_3 + 6 \cdot Y_4}_{\text{column 2}}. \qquad \text{(II.6.11)}$$

[Why are these the *only* scalars that may be used in (II.6.11)?]

The reader should now replace the Y_i's in (II.6.11) with the specific 4-vectors that they represent; subsequent computation should then show that (II.6.11) and (II.6.10) agree.

According to our definition, the 4×2 matrix for $\mathbf{T}$ is given by

$$mx(\mathbf{T})_{\mathscr{X}\mathscr{Y}} = \begin{pmatrix} 0 & -2 \\ 1 & -6 \\ 0 & 0 \\ 1 & 6 \end{pmatrix}$$

relative to the ordered bases $\mathscr{X}$ and $\mathscr{Y}$ given above.

In all these examples, we dealt with the same linear transformation $\mathbf{T}$ in $\mathscr{L}(V_2, V_4)$. Only the bases were varied and, consequently, the resulting matrices varied from example to example.

EXERCISES

In the following exercises, find the matrices of the given linear transformation in $\mathscr{L}(V_n, V_m)$, where the ordered bases $\mathscr{X}$ and $\mathscr{Y}$ are given.

II.6.16 Let $\mathbf{T} \in \mathscr{L}(V_3, V_3)$ be the identity transformation, that is, $\mathbf{T}(X) = X$ for all $X \in V_3$. Let

$$\mathscr{X} = \left\{\begin{pmatrix}1\\0\\0\end{pmatrix}, \begin{pmatrix}0\\1\\0\end{pmatrix}, \begin{pmatrix}0\\0\\1\end{pmatrix}\right\} \quad \text{and} \quad \mathscr{Y} = \left\{\begin{pmatrix}1\\0\\0\end{pmatrix}, \begin{pmatrix}0\\1\\0\end{pmatrix}, \begin{pmatrix}0\\0\\1\end{pmatrix}\right\}.$$

II.6.17 Exercise II.6.16. Let $\mathbf{T}(X) = X$ for all $X \in V_3$. Let the 2 ordered bases for V_3 be the same, that is, $\mathscr{X} = \mathscr{Y} = \{X_1, X_2, X_3\}$. Show that

$$mx(\mathbf{T})_{\mathscr{X}\mathscr{Y}} = \begin{pmatrix} 1 & 0 & 0 \\ 0 & 1 & 0 \\ 0 & 0 & 1 \end{pmatrix}.$$

In other words, $mx(\mathbf{T})_{\mathscr{X}\mathscr{Y}}$ is a 3×3 matrix with 1's on the diagonal and zeros elsewhere.

II.6.18 (See Exercise II.6.2). Let $\mathbf{T} \in \mathscr{L}(V_n, V_n)$ be the identity operator, that is, $\mathbf{T}(X) = X$ for all $X \in V_n$. Show that whenever the ordered bases $\mathscr{X}$ and $\mathscr{Y}$ are the same, the matrix for $\mathbf{T}$, $mx(\mathbf{T})_{\mathscr{X}\mathscr{Y}}$ is an $n \times n$ (square) matrix with 1's down the diagonal and zeros elsewhere.

II.6.19 Give an example of a pair of ordered bases of V_3 for which the matrix of the identity transformation is not in the diagonal form described in Exercises II.6.18.

REMARK. (See Exercises II.5.1 through II.5.5.) In general, scalar multiples of the identity transformation in $\mathscr{L}(V_n, V_n)$ (operators of the form $\mathbf{T}(X) = \alpha \cdot X$ for all X in V_n) are the *only* operators in $\mathscr{L}(V_n, V_n)$ whose matrices are always the same for all ordered bases $\mathscr{X}$ and $\mathscr{Y}$, provided $\mathscr{X} = \mathscr{Y}$. In fact $mx(\mathbf{T})_{\mathscr{X}\mathscr{Y}}$ has α's along the diagonal and zeros elsewhere.

II.6.20 (See Remark above.) Let $\mathbf{T}(X) = \alpha \cdot X$ for all $X \in V_n$, where α is a fixed scalar. Show that for any (and every) ordered basis $\mathscr{X} = \{X_1, X_2, \ldots, X_n\}$ of V_n, $mx(\mathbf{T})_{\mathscr{X}\mathscr{Y}}$ is the $n \times n$ diagonal matrix with α's down the main diagonal and zeros elsewhere (see Exercise II.5.9).

II.6.21 We define $\mathbf{T} : V_2 \to V_3$ as follows:

$$\mathbf{T}\begin{pmatrix} x_1 \\ x_2 \end{pmatrix} = \begin{pmatrix} x_1 + 2x_2 \\ -x_1 \\ 3x_2 \end{pmatrix} \quad \text{for all} \quad X = \begin{pmatrix} x_1 \\ x_2 \end{pmatrix} \quad \text{in } V_2.$$

Find $mx(\mathbf{T})_{\mathscr{X}\mathscr{Y}}$ where

$$\mathscr{X} = \left\{\begin{pmatrix} 1 \\ 1 \end{pmatrix}, \begin{pmatrix} 0 \\ 1 \end{pmatrix}\right\}, \mathscr{Y} = \left\{\begin{pmatrix} 1 \\ 0 \\ 0 \end{pmatrix}, \begin{pmatrix} 0 \\ 1 \\ 0 \end{pmatrix}, \begin{pmatrix} 0 \\ 0 \\ 1 \end{pmatrix}\right\}.$$

II.6.22 With $\mathbf{T}$ defined in Exercise II.6.21, find $mx(\mathbf{T})_{\mathscr{X}\mathscr{Y}}$ where

$$\mathscr{X} = \left\{\begin{pmatrix} 2 \\ 0 \end{pmatrix}, \begin{pmatrix} 0 \\ 1 \end{pmatrix}\right\} \quad \text{and} \quad \mathscr{Y} = \left\{\begin{pmatrix} 1 \\ 0 \\ 2 \end{pmatrix}, \begin{pmatrix} 0 \\ 3 \\ 2 \end{pmatrix}, \begin{pmatrix} 0 \\ 0 \\ 1 \end{pmatrix}\right\}.$$

II.6.23 Define $\mathbf{T} : V_3 \to V_1$ be setting

$$\mathbf{T}\begin{pmatrix} x_1 \\ x_2 \\ x_3 \end{pmatrix} = (4x_1 + 2x_2 - x_3), \quad \text{for all} \begin{pmatrix} x_1 \\ x_2 \\ x_3 \end{pmatrix} \text{ in } V_3.$$

Verify that $\mathbf{T}$ is linear (Definition II.2.1). Find

$$mx(\mathbf{T})_{\mathscr{X}\mathscr{Y}} \quad \text{if} \quad = \mathscr{X}\left\{\begin{pmatrix} 0 \\ 1 \\ 0 \end{pmatrix}, \begin{pmatrix} 1 \\ 0 \\ 1 \end{pmatrix}, \begin{pmatrix} 0 \\ 0 \\ 1 \end{pmatrix}\right\} \quad \text{and} \quad \mathscr{Y} = \{(2)\}.$$

II.6.24 $\mathbf{T} \in \mathscr{L}(V_3, V_1)$ is given in Exercise II.6.23. Let

$$\mathscr{X} = \left\{\begin{pmatrix} 1 \\ 0 \\ 0 \end{pmatrix}, \begin{pmatrix} 0 \\ 0 \\ 1 \end{pmatrix}, \begin{pmatrix} 0 \\ 1 \\ 0 \end{pmatrix}\right\} \quad \text{and} \quad \mathscr{Y} = \{(1)\}.$$

Find $mx(\mathbf{T})_{\mathscr{X}\mathscr{Y}}$.

II.6.25 $\mathbf{T} \in \mathscr{L}(V_3, V_1)$ is given in Exercise II.6.23. Let

$$\mathscr{X} = \left\{\begin{pmatrix} 1 \\ 0 \\ 0 \end{pmatrix}, \begin{pmatrix} 0 \\ 1 \\ 0 \end{pmatrix}, \begin{pmatrix} 0 \\ 0 \\ 1 \end{pmatrix}\right\}, \mathscr{Y} = \{(1)\}.$$

Find $mx(\mathbf{T})_{\mathscr{X}\mathscr{Y}}$.

II.6.26 $\mathbf{T} \in \mathscr{L}(V_1, V_2)$ is defined by putting $\mathbf{T}(x) = \begin{pmatrix} 2x \\ 5x \end{pmatrix}$ for every 1-vector $(x) \in V_1$.

Let $\mathscr{X} = \{(1)\}$ and $\mathscr{Y} = \left\{\begin{pmatrix} 1 \\ 0 \end{pmatrix}, \begin{pmatrix} 0 \\ 1 \end{pmatrix}\right\}$. Find $mx(\mathbf{T})_{\mathscr{X}\mathscr{Y}}$.

II.6.27 With $T : V_1 \to V_2$ given in Exercise II.6.26, let $\mathscr{X} = \{(2)\}$ and $\mathscr{Y} = \left\{\begin{pmatrix} 1 \\ 1 \end{pmatrix}, \begin{pmatrix} 0 \\ -1 \end{pmatrix}\right\}$. Find $mx(\mathbf{T})_{\mathscr{X}\mathscr{Y}}$.

II.6.28 Suppose we are given $\mathbf{T} \in \mathscr{L}(V_n, V_n)$ and ordered bases $\mathscr{X} = \{X_1, X_2, \ldots, X_n\}$ and $\mathscr{Y} = \{Y_1, Y_2, \ldots, Y_m\}$ for V_n and V_m, respectively.

(a) In considering Display (II.6.4) in Definition II.6.2, how can we be sure that *at least 1* such set of mn scalars a_{ij} exists which makes the representation of $\mathbf{T}(X_i)$ in terms of Y_j's possible?

(b) In considering Display (II.6.4) in Definition II.6.2, how can we be sure that *at most 1* such set of mn scalars a_{ij} exists?

Note: Exercise II.6.28(a) asks you to verify *existence* of an $m \times n$ matrix for the operator $\mathbf{T}$. Exercise II.6.28(b) asks you to establish *uniqueness* of an $m \times n$ matrix for the operator $\mathbf{T}$.

Figure 2.19 depicts the fact that "mx" can be considered as a *function* which sends linear operators [in $\mathscr{L}(V_2, V_2)$] to square 2×2 matrices. However, Definition II.6.2 (which generalizes Definition II.5.1) lends itself to a like interpretation. Suppose that $\mathscr{X} = \{X_1, X_2, \ldots, X_n\}$ and $\mathscr{Y} = \{Y_1, Y_2, \ldots, Y_m\}$ are ordered bases for V_n and V_m, respectively. $\mathscr{X}$ and $\mathscr{Y}$ will be assumed to be fixed throughout. Now let $\mathbf{T}$ be an arbitrary linear operator in $\mathscr{L}(V_n, V_m)$. We note that Definition II.6.2 associates an $m \times n$ matrix with each and every such $\mathbf{T} \in \mathscr{L}(V_n, V_m)$. As Exercise II.6.28 attempts to point out, to every $\mathbf{T} \in \mathscr{L}(V_n, V_m)$ there can be associated (via Definition II.6.2) *1 and only 1* $m \times n$ matrix of scalars a_{ij}. In our definition of function (Definition II.1.1), 1 and only 1 "output" element was associated with each of the "input" elements. If we regard the linear operator $\mathbf{T}$ as such as "input" element and its unique $m \times n$ matrix $mx(\mathbf{T})_{\mathscr{X}\mathscr{Y}}$ as the corresponding "output" element, then we may, indeed, regard "mx" as a function which sends the set $\mathscr{L}(V_n, V_m)$ to (onto?) the set of $m \times n$ matrices. Figure 2.19 below informally illustrates the function $\mathbf{T} \to mx(\mathbf{T})$.

We now repair to Section 5 (which dealt with 2×2 matrices) to see whether the questions raised there can be posed in this section (here we discuss the more general $m \times n$ matrix case). For example, is the **1-1** function

$$mx : \mathscr{L}(V_n, V_m) \to m \times n \text{ matrices,}$$

equivalently, $$mx : \mathbf{T} \longrightarrow mx(\mathbf{T})$$

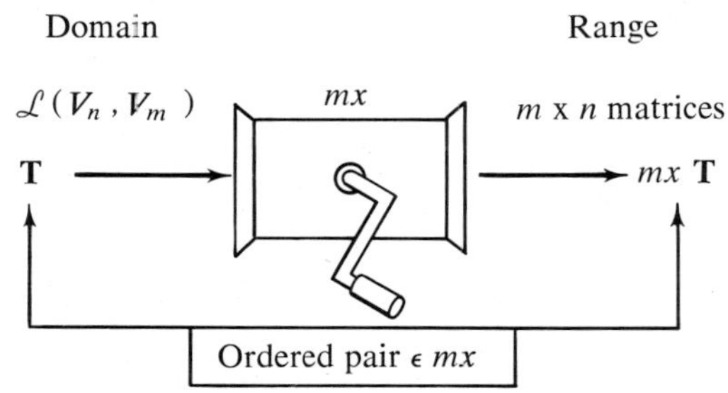

Figure 2.19

an *onto* function. This question has been answered for the 2×2 case (Theorem II.5.2).

We may also ask whether we are able to induce an algebraic structure on the set of all $m \times n$ matrices as we did for the 2×2 matrices (Exercises II.5.33, II.5.50, and II.5.60).

In answering this question, we shall meet no new ideas; rather, bookkeeping and notation will be our greatest problems. For instance, we ought, in good conscience, to establish a compact symbol to represent the phrase "the set of all $m \times n$ matrices whose entries are scalars." Toward this end, we set down Notation II.6.3.

Notation II.6.3 The symbol $\mathcal{M}_{m,n}$ will represent the set of all $m \times n$ matrices whose entries are scalars. We are in a position to state the first theorem of the section.

Theorem II.6.4 The function

$$mx : \mathcal{L}(V_n, V_m) \to \mathcal{M}_{m,n}$$

which is defined by (II.6.4) of Definition II.6.2 is 1-1 and onto.

Proof: (see also, proof of Theorem II.5.2).

THE ONE-ONE PROPERTY. Throughout this proof, we will suppose that $\mathcal{X} = \{X_1, X_2, \ldots, X_n\}$ and $\mathcal{Y} = \{Y_1, Y_2, \ldots, Y_n\}$ are our fixed ordered bases for V_n and V_m, respectively. Let **T** and **U** be 2 (distinct?) linear transformations in $\mathcal{L}(V_n, V_m)$ which give rise to the same $m \times n$ matrix. That is,

$$\mathbf{T}(X_i) = a_{1i}Y_1 + a_{2i}Y_2 + \cdots + a_{mi}Y_m = \sum_{k=1}^{m} a_{ki}Y_k,$$

and

$$\mathbf{U}(X_i) = a_{1i}Y_1 + a_{2i}Y_2 + \cdots + a_{mi}Y_m = \sum_{k=1}^{m} a_{ki}Y_k$$

for each $i = 1, 2, \ldots, m$. These equations follow from Definition II.6.2. Thus, if $\mathbf{T}$ and $\mathbf{U}$ give rise to the same matrix, then $\mathbf{T}$ and $\mathbf{U}$ agree on each of the basis vectors X_i, that is, $\mathbf{T}(X_i) = \mathbf{U}(X_i)$, for all i. In order to show that $\mathbf{T} = \mathbf{U}$, we must show that $\mathbf{T}(X) = \mathbf{U}(X)$ for *all* X in V_n, and not only for basis vectors $X = X_i$ (see Exercise II.3.1). Toward this end, we arbitrarily select $X \in V_n$. By virtue of Theorem I.6.8,

$$X = b_1X_1 + b_2X_2 + \cdots + b_nX_n \tag{II.6.12}$$

for some unique sets of scalars $\{b_1, b_2, \ldots, b_n\}$. Now $\mathbf{T}$ and $\mathbf{U}$ are *linear* functions, after all, so that we have

$$\begin{aligned}
\mathbf{T}(X) &= \mathbf{T}(b_1X_1 + b_2X_2 + \cdots + b_nX_n) && \text{from (II.6.12)}\\
&= b_1\mathbf{T}(X_1) + b_2\mathbf{T}(X_2) + \cdots + b_n\mathbf{T}(X_n) && \text{since } \mathbf{T} \text{ is linear (see Proposition II.2.2)}\\
&\qquad\downarrow \qquad\qquad\quad \downarrow \qquad\qquad\qquad\quad \downarrow\\
&= b_1\mathbf{U}(X_1) + b_2\mathbf{U}(X_2) + \cdots + b_n\mathbf{U}(X_n) && \text{since } \mathbf{T}(X_i) = \mathbf{U}(X_i) \text{ for all } i \quad \text{(II.6.13)}\\
&= \mathbf{U}(b_1X_1 + b_2X_2 + \cdots + b_nX_n) && \text{since } \mathbf{U} \text{ is linear see (Proposition II.2.2)}\\
&= \mathbf{U}(X) && \text{from (II.6.12).}
\end{aligned}$$

We see from (II.6.13) that for *all* X, $\mathbf{T}(X)$ is the same as $\mathbf{U}(X)$, whenever $\mathbf{T}$ and $\mathbf{U}$ give rise to the same $m \times n$ matrices. In other words, if $mx(\mathbf{T})_{\mathscr{X}\mathscr{Y}} = mx(\mathbf{U})_{\mathscr{X}\mathscr{Y}}$, then, necessarily, $\mathbf{T} = \mathbf{U}$. That is, mx is a 1-1 function (see Definition II.1.4) from $\mathscr{L}(V_n, V_m)$ to $\mathscr{M}_{m,n}$.

THE ONTO PROPERTY. We set down the following arbitrary (but fixed from here on) $m \times n$ matrix.

$$\begin{pmatrix}
a_{11} & a_{12} & \cdots & a_{1j} & \cdots & a_{1n}\\
a_{21} & a_{22} & \cdots & a_{2j} & \cdots & a_{2n}\\
\vdots & \vdots & \ddots & \vdots & & \vdots\\
a_{i1} & a_{i2} & \cdots & a_{ij} & \cdots & a_{in}\\
\vdots & \vdots & & \vdots & & \vdots\\
a_{m1} & a_{m2} & \cdots & a_{mj} & \cdots & a_{mn}
\end{pmatrix} = A. \tag{II.6.14}$$

$$\begin{matrix} \uparrow & \uparrow & \uparrow & \uparrow \\ \text{column } 1 & \text{column } 2 & \text{column } j & \text{column } n \end{matrix}$$

To show that the impact set $\mathscr{M}_{m,n}$ is, in fact, the range of the function mx, we must show that there is at least one $\mathbf{T}$ in $\mathscr{L}(V_n, V_m)$ such that $mx(\mathbf{T})_{\mathscr{X}\mathscr{Y}}$ is exactly the $m \times n$ matrix of (II.6.14).

Suppose, optimistically, that such a $\mathbf{T}$ exists. Then from (II.6.4) of Definition II.6.2, we see that the *first* column of (II.6.14) would reveal the coefficients of $\mathbf{T}(X_1)$, the *second* column would reveal the coefficients of $\mathbf{T}(X_2)$, and, in general, the jth column would reveal the coefficients of $\mathbf{T}(X_j)$. That is, if such a $\mathbf{T}$ exists, then it must have the property that

$$\mathbf{T}(X_j) = a_{1j}Y_1 + a_{2j}Y_2 + \cdots + a_{mj}Y_m \quad \text{for each } j = 1, 2, \ldots, n. \tag{II.6.15}$$

Now, we have in (II.6.15) a legitimate function; but it is defined *only* for the n (basis) vectors $X_1, X_2, \ldots, X_n$. What we need is a *linear* function $\mathbf{T}_0$ defined on *all* vectors which still obeys (II.6.15). But this extension is easy since the set $\{X_1, X_2, \ldots, X_n\}$ is a *basis*. In fact, since *any* X in V_n can be written $X = b_1X_1 + b_2X_2 + \cdots + b_nX_n$, for some (unique) scalars $b_1, b_2, \ldots, b_n$ (Theorem I.6.8). We define a linear transformation $\mathbf{T}_0$ in $\mathscr{L}(V_n, V_m)$ by

$$\mathbf{T}_0(X) = b_1\mathbf{T}(X_1) + b_2\mathbf{T}(X_2) + \cdots + b_n\mathbf{T}(X_n) \tag{II.6.16}$$

for all X in V_n.

Thus, we do, indeed, have a function $\mathbf{T}_0$ which is defined for *all* vectors $X \in V_n$.

We draw attention to the same questions which remained at the end of the proof of Theorem II.5.2; namely, does $\mathbf{T}_0$ *extend* $\mathbf{T}$ in the sense that $\mathbf{T}_0(X_i) = \mathbf{T}(X_i)$, and is $\mathbf{T}_0$ a bonafide *linear* function? These questions are resolved via Exercises II.6.29 and II.6.30.

Example Let

$$\mathscr{X} = \left\{ \begin{pmatrix}1\\0\\0\\1\end{pmatrix}, \begin{pmatrix}0\\1\\0\\2\end{pmatrix}, \begin{pmatrix}0\\0\\1\\0\end{pmatrix}, \begin{pmatrix}0\\0\\0\\1\end{pmatrix} \right\} = \{X_1, X_2, X_3, X_4\}$$

be our ordered basis for V_4 and let $\mathscr{Y} = \left\{ \begin{pmatrix}1\\1\end{pmatrix}, \begin{pmatrix}0\\1\end{pmatrix} \right\} = \{Y_1, Y_2\}$ be our ordered basis for V_2. Relative to these fixed bases, then, we wish to describe the unique linear operator $\mathbf{T} \in \mathscr{L}(V_4, V_2)$ with the property that

$$mx(\mathbf{T})_{\mathscr{X}\mathscr{Y}} = \begin{pmatrix} 1 & 0 & -1 & 4 \\ 2 & 0 & 3 & 5 \end{pmatrix}. \tag{II.6.17}$$

$$\begin{matrix} \uparrow & \uparrow & \uparrow & \uparrow \\ \text{column 1} & \text{column 2} & \text{column 3} & \text{column 4} \end{matrix}$$

According to Definition II.6.2, the jth column of a matrix represents an ordered listing of the $\mathscr{Y}$ coordinates of $\mathbf{T}(X_j)$, that is,

$$
\begin{aligned}
\mathbf{T}_0(X_1) &= 1 \cdot Y_1 + 2 \cdot Y_2 = \begin{pmatrix} 1 \\ 2 \end{pmatrix} \\
\mathbf{T}_0(X_2) &= 0 \cdot Y_1 + 0 \cdot Y_2 = \begin{pmatrix} 0 \\ 0 \end{pmatrix} \\
\mathbf{T}_0(X_3) &= (-1) \cdot Y_1 + 3 \cdot Y_2 = \begin{pmatrix} -1 \\ 2 \end{pmatrix} \\
\mathbf{T}_0(X_4) &= 4 \cdot Y_1 + 5 \cdot Y_2 = \begin{pmatrix} 4 \\ 9 \end{pmatrix}.
\end{aligned}
\tag{II.6.18}
$$

So now we know $\mathbf{T}_0(X)$ at least for the 4 cases $X = X_1, X_2, X_3, X_4$. But it is easily verified that for *any* $X \in V_4$,

$$
X = \begin{pmatrix} x_1 \\ x_2 \\ x_3 \\ x_4 \end{pmatrix} = x_1 \cdot X_1 + x_2 \cdot X_2 + x_3 \cdot X_3 + (x_4 - 2x_2 - x_1) \cdot X_4. \tag{II.6.19}
$$

Of course, the set $\{X_1, X_2, X_3, X_4\}$ forms a basis for V_4 so that the scalar coefficients on the right-hand side of (II.6.19) are uniquely determined.

At this point, we define our fully extended linear function $\mathbf{T}_0 \in \mathscr{L}(V_4, V_2)$ by

$$
\begin{aligned}
\mathbf{T}_0(X) &= \mathbf{T}_0(x_1 \cdot X_1 + x_2 \cdot X_2 + x_3 \cdot X_3 + (x_4 - 2x_2 - x_1) \cdot X_4) \quad \text{from (II.6.19)} \\
&= x_1\mathbf{T}_0(X_1) + x_2\mathbf{T}_0(X_2) + x_3\mathbf{T}_0(X_3) + (x_4 - 2x_2 - x_1)\mathbf{T}_0(X_4) \\
&\qquad \downarrow \quad \text{since } \mathbf{T}_0 \text{ is linear} \quad \downarrow \\
&= x_1\begin{pmatrix} 1 \\ 2 \end{pmatrix} + x_2\begin{pmatrix} 0 \\ 0 \end{pmatrix} + x_3\begin{pmatrix} -1 \\ 2 \end{pmatrix} + (x_4 - 2x_2 - x_1)\begin{pmatrix} 4 \\ 9 \end{pmatrix} \quad \text{from (II.6.19)} \\
&= \begin{pmatrix} -3x_1 - 8x_2 - x_3 + 4x_4 \\ -7x_1 - 18x_2 + 2x_3 + 9x_4 \end{pmatrix} \quad \text{from Definitions I.1.2 and I.1.3.}
\end{aligned}
\tag{II.6.20}
$$

That is, (II.6.20) would have us believe that the linear operator

$$
\mathbf{T}_0\begin{pmatrix} x_1 \\ x_2 \\ x_3 \\ x_4 \end{pmatrix} = \begin{pmatrix} -3x_1 - 8x_2 - x_3 + 4x_4 \\ -7x_1 - 18x_2 + 2x_3 + 9x_4 \end{pmatrix}
$$

is that unique linear operator in $\mathscr{L}(V_4, V_2)$ whose matrix, relative to the ordered bases $\mathscr{X}$ and $\mathscr{Y}$, is given in (II.6.17)... and this is, indeed, the case. Also, see the examples following Theorem II.5.2.

EXERCISES

II.6.29 (See Exercise II.5.16.) Let $\mathscr{X} = \{X_1, X_2, \ldots, X_n\}$ be an (ordered) basis for V_n and suppose $\mathbf{T}$ is a function from V_n to V_M defined *only* on the vectors $X_1, X_2, \ldots, X_n$, that is, $\mathbf{T}$ is the *finite* set of ordered pairs $\{(X_1, \mathbf{T}(X_1)), (X_2, \mathbf{T}(X_2)), \ldots, (X_n, \mathbf{T}(X_n))\}$ where each $\mathbf{T}(X_i) \in V_m$ (see Definition II.1.1). Now for *any* $X \in V_n$, we have n unique scalars $b_1, b_2, \ldots, b_n$ such that $X = b_1 \cdot X_1 + b_2 \cdot X_2 + \cdots + b_n \cdot X_n$. (Why?) If $\mathbf{T}_0$ is defined on *all* of V_n by setting

(a) $$\mathbf{T}_0(X) = b_1 \cdot \mathbf{T}(X_1) + b_2 \cdot \mathbf{T}(X_2) + \cdots + b_n \cdot \mathbf{T}(X_n).$$

Then show that $\mathbf{T}_0$ extends $\mathbf{T}$ in the sense that $\mathbf{T}_0(X_i) = \mathbf{T}(X_i)$ for all $i = 1, 2, \ldots, n$.

II.6.30 With $\mathbf{T}_0$ defined in Exercise II.6.29(a), show that $\mathbf{T}_0$ is a *linear* function from V_n to V_m, that is, $\mathbf{T}_0 \in \mathscr{L}(V_n, V_m)$.

In Exercises II.6.31 through II.6.36, a function $\mathbf{T}$ is defined only on the vectors $X_1, X_2, \ldots, X_n$ which form a basis for V_m. In each case, find the unique linear transformation $\mathbf{T}_0$ which is defined on all of V_n, yet agrees with $\mathbf{T}$ on X_i, for all $i = 1, 2, \ldots, n$ (see Exercises II.5.18 through II.5.23).

II.6.31 $V_n = V_3$ and $\mathbf{T}\begin{pmatrix}1\\0\\0\end{pmatrix} = \begin{pmatrix}1\\3\\1\end{pmatrix}$, $\mathbf{T}\begin{pmatrix}0\\1\\2\end{pmatrix} = \begin{pmatrix}2\\1\\0\end{pmatrix}$, $\mathbf{T}\begin{pmatrix}0\\0\\1\end{pmatrix} = \begin{pmatrix}2\\1\\0\end{pmatrix}$.

II.6.32 $V_n = V_4$ and $\mathbf{T}\begin{pmatrix}1\\0\\0\\1\end{pmatrix} = \begin{pmatrix}2\\1\end{pmatrix}$, $\mathbf{T}\begin{pmatrix}0\\1\\0\\1\end{pmatrix} = \begin{pmatrix}1\\0\end{pmatrix}$,

$$\mathbf{T}\begin{pmatrix}0\\0\\1\\1\end{pmatrix} = \begin{pmatrix}-1\\0\end{pmatrix}, \quad \mathbf{T}\begin{pmatrix}0\\0\\0\\1\end{pmatrix} = \begin{pmatrix}0\\0\end{pmatrix}.$$

II.6.33 $V_n = V_3$ and $\mathbf{T}\begin{pmatrix}0\\1\\0\end{pmatrix} = (2)$, $\mathbf{T}\begin{pmatrix}1\\0\\1\end{pmatrix} = (1)$, $\mathbf{T}\begin{pmatrix}1\\0\\0\end{pmatrix} = (3)$.

II.6.34 n is arbitrary and $\{X_1, X_2, \ldots, X_n\}$ is a basis for V_n. $\mathbf{T}(X_1) = O$, $\mathbf{T}(X_2) = X_1, \ldots, \mathbf{T}(X_i) = X_{i-1}, \ldots, \mathbf{T}(X_n) = X_{n-1}$.

II.6.35 $\{X_1, X_2, \ldots, X_n\}$ is a basis for V_n. Y_0 is a fixed vector in V_m. $\mathbf{T}(X_i) = Y_0$ for all $i = 1, 2, \ldots, n$.

II.6.36 $\{X_1, X_2, \ldots, X_n\}$ is a basis for V_n. For each $i = 1, 2, \ldots, n$, we have $\mathbf{T}(X_i) = i \cdot X_i$.

II.6.37 In each of the Exercises II.6.31 through II.6.36, linear transformations $\mathbf{T}_0$ have been defined in $\mathscr{L}(V_n, V_m)$ for various n's and m's. In each case, choose the standard (ordered) bases for V_n and V_m (Definition I.6.11), and find $mx(\mathbf{T}_0)$ relative to this pair of bases (see Definition II.6.2).

In the remaining exercises, an $m \times n$ matrix is given along with a pair $\mathscr{X}$ and $\mathscr{Y}$ of ordered bases for V_n and V_m, respectively. In each exercise, find that (unique) $\mathbf{T} \in \mathscr{L}(V_n, V_m)$ which gives rise to the given matrix, relative to the ordered bases $\mathscr{X}$ and $\mathscr{Y}$ (see Exercises II.5.24 through II.5.32).

II.6.38
$$\mathscr{X} = \left\{ \begin{pmatrix} 1 \\ 0 \\ 0 \end{pmatrix}, \begin{pmatrix} 0 \\ 1 \\ -1 \end{pmatrix}, \begin{pmatrix} 0 \\ 0 \\ 1 \end{pmatrix} \right\}, \mathscr{Y} = \{(2)\}$$

and the matrix is (1,2,3).

II.6.39
$$\mathscr{X} = \left\{ \begin{pmatrix} 1 \\ 0 \\ 0 \end{pmatrix}, \begin{pmatrix} 0 \\ 1 \\ 0 \end{pmatrix}, \begin{pmatrix} 0 \\ 0 \\ 1 \end{pmatrix} \right\}, \mathscr{Y} = \left\{ \begin{pmatrix} 1 \\ 0 \end{pmatrix}, \begin{pmatrix} 0 \\ 1 \end{pmatrix} \right\}$$

and the matrix is

$$\begin{pmatrix} 3 & 0 & -1 \\ 1 & 5 & 0 \end{pmatrix}.$$

II.6.40
$$\mathscr{X} = \left\{ \begin{pmatrix} 0 \\ 1 \\ 0 \end{pmatrix}, \begin{pmatrix} 1 \\ 0 \\ 0 \end{pmatrix}, \begin{pmatrix} 0 \\ 0 \\ 1 \end{pmatrix} \right\}, \mathscr{Y} = \left\{ \begin{pmatrix} 1 \\ 0 \end{pmatrix}, \begin{pmatrix} 0 \\ 1 \end{pmatrix} \right\}$$

and the matrix is

$$\begin{pmatrix} 3 & 0 & -1 \\ 1 & 5 & 0 \end{pmatrix}.$$

II.6.41
$$\mathscr{X} = \{(1)\}, \mathscr{Y} = \left\{ \begin{pmatrix} 2 \\ 1 \\ 0 \end{pmatrix}, \begin{pmatrix} 1 \\ 1 \\ 0 \end{pmatrix}, \begin{pmatrix} 0 \\ 0 \\ 1 \end{pmatrix} \right\}$$

and the matrix is

$$\begin{pmatrix} 4 \\ -1 \\ 2 \end{pmatrix}.$$

II.6.42
$$\mathscr{X} = \left\{ \begin{pmatrix} 1 \\ 0 \\ 0 \end{pmatrix}, \begin{pmatrix} 0 \\ 1 \\ 0 \end{pmatrix}, \begin{pmatrix} 1 \\ 0 \\ 2 \end{pmatrix} \right\}, \mathscr{Y} = \left\{ \begin{pmatrix} 1 \\ 0 \\ 0 \end{pmatrix}, \begin{pmatrix} 0 \\ 1 \\ 0 \end{pmatrix}, \begin{pmatrix} 0 \\ 0 \\ 1 \end{pmatrix} \right\}$$

and the matrix is

$$\begin{pmatrix} 1 & 0 & 0 \\ 0 & 1 & 0 \\ 0 & 0 & 1 \end{pmatrix}.$$

II.6.43 $\mathscr{X}$ and $\mathscr{Y}$ are the standard bases for V_4 and V_5, respectively. The matrix is

$$\begin{pmatrix} 3 & 0 & 1 & 1 \\ 0 & 1 & 0 & 0 \\ 7 & 0 & 2 & 2 \\ 0 & 0 & 0 & 0 \\ 0 & 0 & 0 & 0 \end{pmatrix}.$$

II.6.44 $\mathscr{X} = \mathscr{Y}$ is the standard basis for V_3 and the matrix is

$$\begin{pmatrix} 0 & 0 & 1 \\ 0 & 1 & 0 \\ 1 & 0 & 0 \end{pmatrix}.$$

II.6.45 $\mathscr{X} = \mathscr{Y}$ is the standard basis for V_4 and the matrix is

$$\begin{pmatrix} 1 & 0 & 0 & 0 \\ 0 & 1 & 0 & 0 \\ 0 & 0 & 0 & 0 \\ 0 & 0 & 0 & 0 \end{pmatrix}.$$

The induced algebraic structure on $\mathscr{M}_{m,n}$, the set of all $m \times n$ matrices. (See Figure 2.20.)

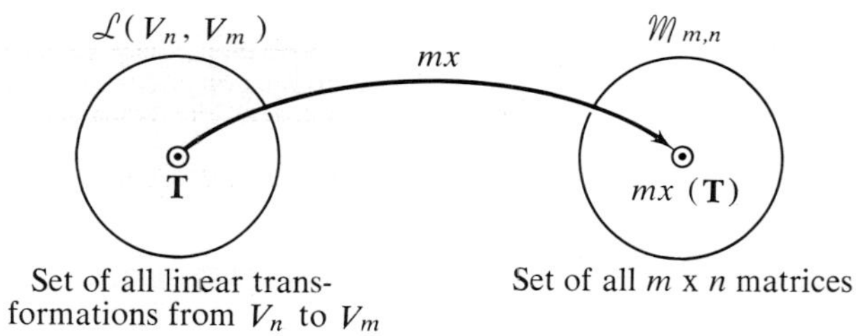

Figure 2.20

Relative to fixed bases $\mathscr{X}$ and $\mathscr{Y}$ for V_n and V_m, respectively, Theorem II.6.4 says that to each $\mathbf{T} \in \mathscr{L}(V_n, V_m)$, we may assign 1 and only 1 $m \times n$ matrix $mx(\mathbf{T})$ in $\mathscr{M}_{m,n}$; moreover, for *each* and *every* $m \times n$ matrix A in $\mathscr{M}_{m,n}$, there is a unique $\mathbf{T}$ in $\mathscr{L}(V_n, V_m)$ such that $A = mx(\mathbf{T})$, that is, $\mathscr{M}_{m,n} = \text{rg}\,(mx)$. We shall take advantage of this 1-1 and onto property of the function mx to force a closed addition operation $+$, and a scalar multiplication $\cdot$ on the set $\mathscr{M}_{m,n}$. As we have seen in Section 5, we may do this by taking the addition and scalar multiplication that we *already have* on $\mathscr{L}(V_n, V_m)$, and reflect it through the function mx onto the heretofore *un*structured set $\mathscr{M}_{m,n}$ (see Exercise II.5.33).

The addition on $\mathscr{M}_{m,n}$. Given any 2 (not necessarily distinct) $m \times n$ matrices A and B in $\mathscr{M}_{m,n}$, we are to specify which matrix (also in $\mathscr{M}_{m,n}$)

is to stand for $A + B$. We know from Theorem II.6.4 that relative to the fixed bases $\mathscr{X}$ and $\mathscr{Y}$, there exists unique **T** and **U** in $\mathscr{L}(V_n, V_m)$ such that

$$mx(\mathbf{T}) = A \quad \text{and} \quad mx(\mathbf{U}) = B.$$

We are to decide, then, which $m \times n$ matrix is to stand for $mx(\mathbf{T}) + mx(\mathbf{U})$; and we choose the (third) matrix $mx(\mathbf{T} \dotplus \mathbf{U})$. Accordingly, then, we come to Definition II.6.5.

Definition II.6.5 Let A and B be $m \times n$ matrices. For arbitrary but fixed ordered bases $\mathscr{X}$ and $\mathscr{Y}$ for V_n and V_m respectively, let **T** and **U** be those unique linear operators in $\mathscr{L}(V_n, V_m)$ such that $mx(\mathbf{T}) = A$ and $mx(\mathbf{U}) = B$. Then $A + B$, the **matrix sum of A and B**, is the $m \times n$ matrix defined by

$$A + B \quad (\text{that is, } mx(\mathbf{T}) + mx(\mathbf{U})) = mx(\mathbf{T} \dotplus \mathbf{U}). \qquad \text{(II.6.21)}$$

We shall take Definition II.6.5 and see how we may interpret this addition in terms of *specific* $m \times n$ matrices. Prior to this, however, the following notation will prove useful.

Notation II.6.6 Suppose A is an $m \times n$ matrix whose ijth entry is some number; let us call it a_{ij}, that is, a_{ij} appears in the ith row and jth column of A. Then we denote A by

$$A = (a_{ij})_{ij}.$$

REMARK. The symbols of Notation II.6.6 may be related by the following schematic commentary:

$$A = (a_{ij}) \;\; {}_{ij}\,.$$

$\nearrow$ the scalar $\quad$ $\nwarrow$ its corresponding entry

Example If $i = 1, 2, 3$ and $j = 1, 2, 3, 4$, then

$$(i + j^2)_{ij} = \begin{pmatrix} 1+1^2 & 1+1^2 & 1+3^2 & 1+4^2 \\ 2+1^2 & 2+2^2 & 2+3^2 & 2+4^2 \\ 3+1^2 & 3+2^2 & 3+3^2 & 3+4^2 \end{pmatrix}$$

$$= \begin{pmatrix} 2 & 5 & 10 & 17 \\ 3 & 6 & 11 & 18 \\ 4 & 7 & 12 & 19 \end{pmatrix}.$$

In this case, our ijth entry is the scalar $a_{ij} = i + j^2$.

Example Suppose we are given $m \cdot n$ scalars a_{ij} where $i = 1, 2, \ldots, m$ and $j = 1, 2, \ldots, n$. Suppose we construct the following $n \times m$ matrix:

$$A = \begin{pmatrix} a_{11} & a_{21} & \cdots & a_{j1} & \cdots & a_{m1} \\ a_{12} & a_{22} & \cdots & a_{j2} & \cdots & a_{m2} \\ \vdots & \vdots & \ddots & \vdots & & \vdots \\ a_{1i} & a_{2i} & \cdots & a_{ji} & \cdots & a_{mi} \\ \vdots & \vdots & & \vdots & \ddots & \vdots \\ a_{1n} & a_{2n} & \cdots & a_{jn} & \cdots & a_{mn} \end{pmatrix} \leftarrow i\text{th row.}$$

$$\uparrow$$
$$j\text{th column}$$

We ask ourselves: Which scalar appears in the ijth entry (in the ith row and jth column) of A? According to the way A is defined, exactly the scalar a_{ji} (and not a_{ij}) appears in the ijth entry. We denote this matrix by $A = (a_{ji})_{ij}$, that is,

$$A = (a_{ji})\ {}_{ij}\,.$$

$\nearrow$ the scalar $\quad$ $\nwarrow$ which appears in this entry

In our next theorem, which describes the mechanics of performing the addition on any pair of $m \times n$ matrices, we should encounter no surprises. We have from Exercise II.5.33(d) that if $m = n = 2$, then addition is effected by adding "entrywise." The same statement may be made for arbitrary m and n. We state the precise meaning of this phrase with Theorem II.6.7.

Theorem II.6.7 Let $A = (a_{ij})_{ij}$ and $B = (b_{ij})_{ij}$ be $m \times n$ matrices, that is, $i = 1, 2, \ldots, m$ and $j = 1, 2, \ldots, n$. Then the $m \times n$ matrix $A + B$ is exactly the matrix

$$A \quad + \quad B \quad = (a_{ij} + b_{ij})_{ij}, \quad \text{or equivalently,}$$
$$\downarrow \qquad \downarrow$$
$$(a_{ij})_{ij} + (b_{ij})_{ij} = (a_{ij} + b_{ij})_{ij}.$$

Example If $A = \begin{pmatrix} 1 & 3 & -2 \\ 7 & 0 & 1 \end{pmatrix}$ and $B = \begin{pmatrix} -4 & -3 & 1 \\ -2 & 6 & 9 \end{pmatrix}$, then by adding corresponding entries of A and B, we obtain the (third) matrix

$$A + B = \begin{pmatrix} 1 - (-4) & 3 + (-3) & -2 + 1 \\ 7 - (-2) & 0 + \quad 6 & 1 + 9 \end{pmatrix} = \begin{pmatrix} -3 & 0 & -1 \\ 5 & 6 & 10 \end{pmatrix}.$$

Proof of Theorem II.6.7: Let us choose arbitrary (but fixed throughout) ordered bases $\mathscr{X} = \{X_1, X_2, \ldots, X_n\}$ and $\mathscr{Y} = \{Y_1, Y_2, \ldots, Y_m\}$ for V_n and V_m, respectively. By virtue of Theorem II.6.4, we have unique linear transformations $\mathbf{T}$ and $\mathbf{U}$ in $\mathscr{L}(V_n, V_m)$ such that relative to bases $\mathscr{X}$ and $\mathscr{Y}$,

$$mx(\mathbf{T}) = (a_{ij})_{ij} \quad \text{and}$$
$$mx(\mathbf{U}) = (b_{ij})_{ij}.$$

But Definition II.6.2 tells us that this is equivalent to saying

$$\text{(a)}\quad \mathbf{T}(X_j) = \sum_{i=1}^{m} a_{ij} Y_i$$

and (II.6.22)

$$\text{(b)}\quad \mathbf{U}(X_j) = \sum_{i=1}^{m} b_{ij} Y_i \qquad \text{[see Equations (II.6.3a) and (II.6.3b)];}$$

Equations (II.6.22a) and (II.6.22b) are true for each $j = 1, 2, \ldots, n$.

Now, Definition II.6.5 forces us to construct $mx(\mathbf{T} \dotplus \mathbf{U})$ (which is our desired $m \times n$ matrix $A + B$). Since we have no choice, we had best get to this piece of business. Recall that for all $X \in V_n$, $(\mathbf{T} \dotplus \mathbf{U})X = \mathbf{T}(X) + \mathbf{U}(X)$ from Definition II.4.1. Now $mx(\mathbf{T} \dotplus \mathbf{U})$ will be ours . . . once we know $(\mathbf{T} \dotplus \mathbf{U})X_j$ for each and every $j = 1, 2, \ldots, n$ (see Definition II.6.2). Accordingly, then, for any and every $j = 1, 2, \ldots, n$,

$$\begin{aligned}
(\mathbf{T} \dotplus \mathbf{U})X_j &= \mathbf{T}(X_j) + \mathbf{U}(X_j) && \text{Definition II.4.1} \\
&= \left(\sum_{i=1}^{m} a_{ij} Y_i\right) + \left(\sum_{i=1}^{m} b_{ij} Y_i\right) && \text{from (II.6.22a) and (II.6.22b)} \\
& && \text{(II.6.23)} \\
&= \sum_{i=1}^{m} \underbrace{(a_{ij} + b_{ij})}_{ij\text{th entry of } mx(\mathbf{T} \dotplus \mathbf{U})} \cdot Y_i && \text{from Definition II.6.1.}
\end{aligned}$$

But do you realize what we've done?! In setting down (II.6.25), we have *all* the information we need in order to construct $mx(\mathbf{T} \dotplus \mathbf{U})$; for, if we look at Definition II.6.2 [replacing $\mathbf{T}$ with $(\mathbf{T} \dotplus \mathbf{U})$], then we conclude that the ijth entry of $mx(\mathbf{T} \dotplus \mathbf{U})$ appears in the expansion for $(\mathbf{T} \dotplus \mathbf{U})X_j$ and is just that scalar which is attached to the vector Y_i! In this case, we see in (II.6.23) that this scalar is $(a_{ij} + b_{ij})$. That is,

$$mx(\mathbf{T} \dotplus \mathbf{U}) = (a_{ij} + b_{ij})_{ij}.$$

If we couple this remark with the fact that

$$mx(\mathbf{T} \dotplus \mathbf{U}) = A + B \text{ [or } (a_{ij})_{ij} + (b_{ij})_{ij}] \quad \text{Definition II.6.5,}$$

then we have

$$A + B \quad \text{(that is, } (a_{ij})_{ij} + (b_{ij})_{ij}) = (a_{ij} + b_{ij})_{ij},$$

which concludes the proof of our theorem.

We have, to this point, the fact that the set $\mathscr{M}_{m,n}$ is *closed* under the matrix addition $+$; this can be seen directly from Definition II.6.5 or from Theorem II.6.7 which we have just proved. That is, axiom A.1 of Definition I.5.2 obtains.

In showing that $\mathscr{M}_{m,n}$ is a commutative group under the operation $+$ given by Definition II.6.5, we must verify axioms A.2 through A.5 of Definition I.5.2 (see also Definition II.4.10). Thus, it remains to show that for matrices in $\mathscr{M}_{m,n}$:

(a) $A + B = B + A$ for all matrices A and B (commutivity axiom A.2)

(b) there exists a (feeble) matrix, call it O, such that $A + O = A$ for all matrices A (existence of additive identity axiom A.3)

(c) for any matrix A, there is a matrix, call it $(-A)$, such that $A + (-A) = O$ (existence of additive inverse axiom A.4)

(d) $(A + B) + C = A + (B + C)$ for all matrices A, B, and C (associativity axiom A.5). (II.6.24)

That the conditions of (II.6.24) do obtain, is the content of Theorem II.6.8.

Theorem II.6.8 The set $\mathscr{M}_{m,n}$ of all $m \times n$ matrices satisfies the 5 addition axioms (A.1 through A.5 of Definition I.5.2) of a vector-space with addition $+$ of matrices defined in Definition II.6.5; that is, $\mathscr{M}_{m,n}$ is a commutative group under addition $+$.

Proof: Verification of closure, axiom A.1 follows instantly from Definition II.6.5 itself. There are 2 routes that may be taken in verifying the axioms A.2 through A.5, namely, the "abstract" route which employs Definition II.6.5 or the "concrete" route which uses the matrix addition which is *explicitly* given to us in Theorem II.6.7. We shall prove that (II.6.24b) (axiom A.3) obtains for $\mathscr{M}_{m,n}$ using both avenues of approach; the remaining properties are straightforwardly established and are left as Exercises II.6.60 through II.6.65.

First (abstract) proof of (b) (using Definition II.6.5): To show the existence of a (feeble) $m \times n$ matrix $O_{m,n}$, say, such that $A + O_{m,n} = A$ for all $m \times n$ matrices A, we *define* $O_{m,n}$ by

$$O_{m,n} = mx(\mathbf{O}), \qquad \text{(II.6.25a)}$$

where $\mathbf{O}$ is the zero *linear operator* in $\mathscr{L}(V_n, V_m)$. Theorem II.4.6 tells us that $\mathscr{L}(V_n, V_m)$ is a vector-space so that, necessarily, $\mathscr{L}(V_n, V_m)$ obeys axiom A.3 itself and contains its own unique "feeble" operator $\mathbf{O}$ where $\mathbf{T} \dotplus \mathbf{O} = \mathbf{T}$ for each and every $\mathbf{T}$ in $\mathscr{L}(V_n, V_m)$. Thus, we need not worry as to whether $\mathbf{O}$ in $\mathscr{L}(V_n, V_m)$ exists. . . it does, and hence, so does $mx(\mathbf{O})$, which is the $m \times n$ matrix denoted by $O_{m,n}$. Does $O_{m,n}$ [that is, $mx(\mathbf{O})$] have the appropriate (feeble) property that we hope it does? Let us see.

Now for any $m \times n$ matrix $A \in \mathscr{M}_{m,n}$, there is 1 and only 1 $\mathbf{T} \in \mathscr{L}(V_n, V_m)$ such that

$$A = mx(\mathbf{T}) \tag{II.6.25b}$$

relative to some fixed ordered bases $\mathscr{X}$ and $\mathscr{Y}$ of V_n and V_m, respectively. Thus, for any $A \in \mathscr{M}_{mn}$, we have

$$\begin{aligned}
A + O_{m,n} &= mx(\mathbf{T}) + mx(\mathbf{O}) && \text{from (II.6.25a) and (II.6.25b)} \\
&= mx(\mathbf{T} \dotplus \mathbf{O}) && \text{Definition II.6.5} \\
&= mx(\mathbf{T}) && \text{Property of } \mathbf{O} \text{ in } \mathscr{L}(V_n, V_m)\text{, that is, } \mathbf{T} \dotplus \mathbf{O} = \mathbf{T} \text{ for all } \mathbf{T} \\
&= A && \text{Definition of } \mathbf{T}.
\end{aligned} \tag{II.6.26}$$

That does it! Equation (II.6.26) is a confirmation of the fact that (II.6.24b) (axiom A.3) obtains for the set $\mathscr{M}_{m,n}$.

Second (concrete) proof of (b) (using Theorem II.6.7): Were we to compute the entries of the $m \times n$ matrix $O_{m,n}$ defined by (II.6.25a), we would see that $O_{m,n}$ would accept only zero in each of its entries, that is,

$$O_{m,n} = (0)_{ij}, \quad \text{the } m \times n \text{ matrix whose every entry is zero.}$$

Regardless of how we stumbled upon $(0)_{ij}$ as our suspected (feeble) additive identity, we are obliged only to verify that our suspicions are correct. Hence, we write for arbitrary A,

$$A = (\mathrm{a}_{ij})_{ij} \in \mathscr{M}_{m,n}, \tag{II.6.27}$$

the following sequence of equalities:

$$\begin{aligned}
A + (0)_{ij} &= (a_{ij})_{ij} + (0)_{ij} && \text{Equation (II.6.27)} \\
&= (a_{ij} + 0)_{ij} && \text{Theorem II.6.7} \\
&= (a_{ij})_{ij} && \text{property of the scalar 0, that is, } a_{ij} + 0 = a_{ij} \\
&= A && \text{Equation (II.6.27).}
\end{aligned} \tag{II.6.28}$$

Equation (II.6.28) is our so-called concrete proof of property (II.6.24b) (axiom A.3).

The scalar multiplication on $\mathscr{M}_{m,n}$

Which $m \times n$ matrix shall we assign to the symbol $\alpha \cdot A$, where α is a scalar and A is an $m \times n$ matrix in $\mathscr{M}_{m,n}$? As in Exercise II.5.50(b), which dealt with the case $m = n = 2$, we employ the 1-1 onto function mx so as to take advantage of the scalar multiplication that *already* exists on $\mathscr{L}(V_n, V_m)$. Specifically, we set down Definition II.6.9.

Definition II.6.9 Let α be a scalar and let $A \in \mathscr{M}_{m,n}$. Let $\mathscr{X}$ and $\mathscr{Y}$ be arbitrary but fixed ordered bases for V_n and V_m, respectively. The symbol $\mathbf{T}$ will denote that unique linear operator in $\mathscr{L}(V_n, V_m)$ such that $mx(\mathbf{T}) = A$ relative to the bases $\mathscr{X}$ and $\mathscr{Y}$. Then the $m \times n$ matrix $\boldsymbol{\alpha} \cdot \boldsymbol{A}$, the **scalar product of $\boldsymbol{\alpha}$ and $\boldsymbol{A}$**, is defined by

$$\alpha \cdot A \qquad = mx(\alpha \mathbf{T}),$$

that is,

$$\alpha \cdot mx(\mathbf{T}) = mx(\alpha \mathbf{T}).$$

REMARK. Since $\mathscr{L}(V_n, V_m)$ is a vector-space (Theorem II.4.6), the symbol $\alpha\mathbf{T}$ represents a bonafide linear operator in $\mathscr{L}(V_n, V_m)$. Thus, the second equation in Definition II.6.9 is meaningful.

Secondly, our definition apparently depends upon the choice of bases $\mathscr{X}$ and $\mathscr{Y}$. This dependence, however, is illusory insofar as computing the specific matrix $\alpha \cdot A$ is concerned. In our next theorem, we shall see that the choice of bases does not affect or alter the final result. This theorem follows.

Theorem II.6.10 Let $A = (a_{ij})_{ij}$ and let α be any scalar. Then $\alpha \cdot A$ is the matrix $(\alpha a_{ij})_{ij}$, that is,

$$\alpha \cdot (a_{ij})_{ij} = (\alpha a_{ij})_{ij}.$$

The theorem says, in effect, that in order to multiply a matrix A by a scalar α (that is, to find $\alpha \cdot A$), simply multiply each and every entry of A by that scalar; the resulting $m \times n$ matrix is the matrix which has been preordained by Definition II.6.9, as the "answer." Thus, Theorem II.6.10 would imply the statements of the following example:

Example

$$2 \cdot \begin{pmatrix} 1 & 0 & 5 & 6 \\ 0 & 1 & -2 & 4 \\ 3 & -1 & 7 & 6 \end{pmatrix} \quad \text{should equal} \quad \begin{pmatrix} 2\cdot 1 & 2\cdot 0 & 2\cdot 5 & 2\cdot 6 \\ 2\cdot 0 & 2\cdot 1 & 2\cdot(-2) & 2\cdot 4 \\ 2\cdot 3 & 2\cdot(-1) & 2\cdot 7 & 2\cdot 6 \end{pmatrix}$$

$$= \begin{pmatrix} 2 & 0 & 10 & 12 \\ 0 & 2 & -4 & 8 \\ 6 & -2 & 14 & 12 \end{pmatrix}.$$

Proof of Theorem II.6.10: [See Exercise II.5.50(d).] Let $\mathscr{X} = \{X_1, X_2, \ldots, X_n\}$ and $\mathscr{Y} = \{Y_1, Y_2, \ldots, Y_m\}$ be arbitrary but fixed ordered bases for V_n and V_m, respectively. Let $A = (a_{ij})_{ij}$ be a typical matrix in $\mathscr{M}_{m,n}$, and let α be an arbitrary scalar.

We denote by **T** that unique linear transformation in $\mathscr{L}(V_n, V_m)$ with the property that $mx(\mathbf{T}) = (a_{ij})_{ij}$ relative to the ordered bases $\mathscr{X}$ and $\mathscr{Y}$. This means, of course, that $\mathbf{T}(X_j) = \sum_{i=1}^{m} a_{ij} \cdot Y_i$, for all $j = 1, 2, \ldots, n$ (see Definition II.6.2). Do not forget that our objective is to compute $mx(\alpha\mathbf{T})$; and in order to do this, all we need to know is which scalars appear when we write $\alpha\mathbf{T}(X_j)$ in terms of the Y_i's (see Definition II.6.2). Accordingly then, for each and every $j = 1, 2, \ldots, n$, we have

$$\begin{aligned}
\alpha\mathbf{T}(X_j) &= \alpha \cdot (\mathbf{T}(X_j)) && \text{from Definition II.4.3 for } \alpha\mathbf{T} \\
&= \alpha \cdot \left(\sum_{i=1}^{m} a_{ij} \cdot Y_i\right) && \text{Definition II.6.2 for } mx(\mathbf{T}) \\
&= \sum_{i=1}^{m} \alpha \cdot (a_{ij} \cdot Y_i) && \text{Definition I.5.2, axiom B.5. See Exercise II.6.7.} \\
&= \sum_{i=1}^{m} (\alpha a_{ij}) \cdot Y_i && \text{Definition I.5.2, axiom B.2.}
\end{aligned} \tag{II.6.29}$$

Now if we read Equation (II.6.29) in conjunction with Definition II.6.2, we see that the ijth entry of $mx(\alpha\mathbf{T})$ is the scalar attached to the vector Y_i, that is, the ijth entry of $mx(\alpha\mathbf{T})$ is none other than the scalar αa_{ij}. In other words, $mx(\alpha\mathbf{T}) = (\alpha a_{ij})_{ij}$. Since we *defined* $mx(\alpha\mathbf{T})$ to represent $\alpha \cdot mx(\mathbf{T})$, we conclude that $\alpha \cdot mx(\mathbf{T})$ [that is, $\alpha \cdot (a_{ij})_{ij}$] equals $(\alpha a_{ij})_{ij}$, which completes our proof.

The natural question to raise at this point is whether our newly defined scalar multiplication on $\mathscr{M}_{m,n}$ is good enough so as to allow $\mathscr{M}_{m,n}$ to qualify as a bonafide vector-space. That is, we are obliged to thumb back to Definition I.5.2, and test $\mathscr{M}_{m,n}$, with its addition and scalar multiplication, against the axioms A.1 through A.5 and B.1 through B.5. We have already indicated that A.1 through A.5 do hold for $\mathscr{M}_{m,n}$ (Theorem II.6.8), and the fact that $\mathscr{M}_{m,n}$ is closed under scalar multiplication (axiom B.1) can be read out of Definition II.6.9 or the "concrete" representation theorem, Theorem II.6.10. The only conditions remaining which cry out for verification, then, are the following:

$$
\begin{array}{llll}
\text{(a)} & (\alpha\beta)\cdot A = \alpha\cdot(\beta\cdot A) & \text{for all scalars } \alpha \text{ and } \beta, \text{ and all matrices } A \text{ (axiom B.2)} & \\
\text{(b)} & 1\cdot A = A & \text{for all matrices } A \text{ (axiom B.3)} & \text{(II.6.30)} \\
\text{(c)} & (\alpha+\beta)\cdot A = \alpha\cdot A + \beta\cdot A & \text{for all scalars } \alpha \text{ and } \beta, \text{ and all matrices } A \text{ (axiom B.4)} & \\
\text{(d)} & \alpha\cdot(A+B) = \alpha\cdot A + \alpha\cdot B & \text{for all scalars } \alpha, \text{ and all matrices } A \text{ and } B \text{ (axiom B.5).} &
\end{array}
$$

That the 4 properties of (II.6.30) do, indeed, hold true for $\mathscr{M}_{m,n}$ is the content of Theorem II.6.11.

Theorem II.6.11 The set $\mathscr{M}_{m,n}$ of all real $m \times n$ matrices with the operations of addition and scalar multiplication (delineated via Definitions II.6.5 and II.6.9) forms a vector-space (in accordance with Definition I.5.2).

Proof: We shall establish only one of the properties of our list, and we shall do it by 2 different means; we choose to attack property (a) of (II.6.30).

First proof of (a) (using Definition II.6.9): Let A be an arbitrary $m \times n$ matrix in $\mathscr{M}_{m,n}$, and let α and β be any scalars. If $\mathscr{X}$ and $\mathscr{Y}$ are 2 fixed ordered bases for V_n and V_m, respectively, then let $\mathbf{T}$ denote that unique linear transformation in $\mathscr{L}(V_n, V_m)$ such that

$$mx(\mathbf{T}) = A \qquad \text{relative to } \mathscr{X} \text{ and } \mathscr{Y}. \tag{II.6.31}$$

In case we have lost sight of our objective, it is simply this: to show $(\alpha\beta)\cdot A = \alpha\cdot(\beta A)$. Now

$$
\begin{array}{rll}
(\alpha\beta)\cdot A &= (\alpha\beta)\cdot mx(\mathbf{T}) & \text{from (II.6.31)} \\
&= mx((\alpha\beta)\mathbf{T}) & \text{Definition II.6.9 (substitute } \alpha\beta \text{ for } \alpha) \\
&= mx(\alpha(\beta\mathbf{T})) & \text{Theorem II.4.6, since } \mathscr{L}(V_n, V_m) \text{ is a vector space, axiom B.2 holds} \qquad \text{(II.6.32)} \\
&= \alpha\cdot mx(\beta\mathbf{T}) & \text{Definition II.6.9 (substitute } \beta\mathbf{T} \text{ for } \mathbf{T}) \\
&= \alpha\cdot(\beta\cdot mx\mathbf{T}) & \text{Definition II.6.9 again (substitute } \beta \text{ for } \alpha) \\
&= \alpha\cdot(\beta\cdot A) & \text{from (II.6.31).}
\end{array}
$$

That completes our "abstract" proof of the fact that axiom B.2 of Definition I.5.2 holds in $\mathcal{M}_{m,n}$.

Second proof of (a) (using Theorem II.6.10): Let $A = (a_{ij})_{ij}$ be a typical $m \times n$ matrix in $\mathcal{M}_{m,n}$. If α and β are arbitrary scalars, then

$$\begin{aligned}
(\alpha\beta) \cdot A &= (\alpha\beta) \cdot ((a_{ij})_{ij}) && \text{since } A = (a_{ij})_{ij} \\
&= (\alpha\beta \cdot a_{ij})_{ij} && \text{Theorem II.6.10 (substitute } \alpha\beta \text{ for } \alpha) \\
&= (\alpha \cdot (\beta \cdot a_{ij}))_{ij} && \text{since the scalars form a vector-space, axiom B.2 obtains there, that is, } \alpha\beta \cdot a_{ij} = \alpha \cdot \beta a_{ij} \qquad \text{(II.6.33)} \\
&= \alpha \cdot ((\beta a_{ij})_{ij}) && \text{Theorem II.6.10 (substitute } \beta a_{ij} \text{ for } a_{ij}) \\
&= \alpha \cdot (\beta \cdot (a_{ij})_{ij}) && \text{Theorem II.6.10 (substitute } \beta \text{ for } \alpha) \\
&= \alpha \cdot (\beta \cdot A) && \text{since } A = (a_{ij})_{ij}.
\end{aligned}$$

So we conclude the "concrete" proof of axiom B.2 relative to $\mathcal{M}_{m,n}$. Properties (II.6.30) (b), (c), (d) are to be verified by the student in Exercises II.6.60 through II.6.65.

EXERCISES

II.6.46 Suppose we have $2m$ elements $u_1, u_2, \ldots, u_m, v_1, v_2, \ldots, v_m$ belonging to some set satisfying **A**.1 through A.5, Definition I.5.2, so that there is an addition $+$ defined between pairs of elements.

(a) Prove that

$$\left(\sum_{i=1}^{m} u_i\right) + \left(\sum_{j=1}^{m} v_j\right) = \sum_{k=1}^{m} (u_k + v_k);$$

of course, Definition II.6.1 must enter into any supporting arguments for (a).

(b) Show how this result is used in II.6.23, which is used in proving Theorem II.6.7.

In Exercises II.6.47 through II.6.49, find $A + B$ and $B + A$, where possible.

II.6.47 $A = \begin{pmatrix} 2 & 0 & 3 & 4 \\ -1 & 0 & 5 & 6 \end{pmatrix}$ $\qquad B = \begin{pmatrix} -2 & 1 & -2 & 1 \\ 1 & 3 & \frac{1}{2} & 3 \end{pmatrix}$

II.6.48 $A = \begin{pmatrix} 9 & 0 & 2 & 1 \\ 21 & 0 & 1 & 2 \\ 4 & 1 & 0 & 1 \end{pmatrix}$ $\qquad B = \begin{pmatrix} -8 & 1 & -1 & 0 \\ -20 & 1 & 0 & -1 \\ -3 & 0 & 1 & .0 \end{pmatrix}$

II.6.49 $A = \begin{pmatrix} 2 & 1 & 5 \\ 0 & 0 & 1 \end{pmatrix}$ $\qquad B = \begin{pmatrix} 0 & 2 \\ 0 & 1 \\ 1 & 5 \end{pmatrix}.$

In Exercises II.6.50 through II.6.53, matrices A, B, and C are given. In each case, find $A + B$, $B + C$, $A + C$, $(A + B) + C$, and $A + (B + C)$.

II.6.50 $A = (3), \qquad B = (2), \qquad C = (4).$

II.6.51 $$A = \begin{pmatrix} 1 & 3 & 1 \\ 2 & 3 & 1 \\ -1 & 2 & 0 \\ 0 & 1 & 5 \end{pmatrix}, \quad B = \begin{pmatrix} -1 & -3 & -1 \\ -2 & -3 & -1 \\ 1 & -2 & 0 \\ 0 & -1 & -5 \end{pmatrix}, \quad C = \begin{pmatrix} 1 & 0 & 0 \\ 0 & 1 & 0 \\ 0 & 0 & 1 \\ 0 & 0 & 0 \end{pmatrix}.$$

II.6.52 $$A = \begin{pmatrix} 3 & 7 & 19 & 6 \\ 5 & 1 & 3 & 2 \end{pmatrix}, \quad B = \begin{pmatrix} 1 & -2 & -3 & 2 \\ 0 & 1 & 0 & 1 \end{pmatrix}, \quad C = \begin{pmatrix} 7 & 5 & 0 & 1 \\ 2 & 4 & -1 & -1 \end{pmatrix}.$$

II.6.53 $$A = \begin{pmatrix} 2 & \frac{1}{2} & -\frac{1}{2} \\ 0 & 0 & \frac{1}{2} \\ \frac{1}{2} & \frac{1}{2} & 1 \end{pmatrix}, \quad B = \begin{pmatrix} 1 & 0 & 3 \\ 0 & 1 & 2 \\ 5 & 5 & 3 \end{pmatrix}, \quad C = \begin{pmatrix} 9 & 8 & 1 \\ 2 & 5 & 1 \\ -1 & -1 & -1 \end{pmatrix}.$$

In Exercises II.6.54 through II.6.59, linear operators **T** and **U** in $\mathscr{L}(V_n, V_m)$ are given. In each case,

(a) characterize the linear operator $\mathbf{T} + \mathbf{U}$, that is, find (describe) $(\mathbf{T} + \mathbf{U})(X)$ for each and every $X \in V_n$

(b) using the standard ordered bases for V_n and V_m (see Definition I.6.11), find $mx(\mathbf{T})$, $mx(\mathbf{U})$, and $mx(\mathbf{T} + \mathbf{U})$ relative to these bases; use Definition II.6.5 and *not* Theorem II.6.7 for finding $mx(\mathbf{T} + \mathbf{U})$

(c) find $mx(\mathbf{T} + \mathbf{U})$ by use of Theorem II.6.7 and compare this result with the corresponding result in (b).

II.6.54 $$\mathbf{T}\begin{pmatrix} x_1 \\ x_2 \\ x_3 \end{pmatrix} = \begin{pmatrix} x_1 \\ x_2 \\ 0 \end{pmatrix}, \quad \mathbf{U}\begin{pmatrix} x_1 \\ x_2 \\ x_3 \end{pmatrix} = \begin{pmatrix} 0 \\ 0 \\ x_3 \end{pmatrix} \quad \text{for all } \begin{pmatrix} x_1 \\ x_2 \\ x_3 \end{pmatrix} \in V_3.$$

II.6.55 $$\mathbf{T}\begin{pmatrix} x_1 \\ x_2 \\ x_3 \end{pmatrix} = \begin{pmatrix} x_1 - x_2 \\ 2x_1 \end{pmatrix}, \quad \mathbf{U}\begin{pmatrix} x_1 \\ x_2 \\ x_3 \end{pmatrix} = \begin{pmatrix} x_2 - x_1 \\ -x_1 \end{pmatrix} \quad \text{for all } \begin{pmatrix} x_1 \\ x_2 \\ x_3 \end{pmatrix} \in V_3.$$

II.6.56 $$\mathbf{T}(x) = \begin{pmatrix} x \\ 2x \\ 3x \end{pmatrix}, \quad \mathbf{U}(x) = \begin{pmatrix} 2x \\ -x \\ -x \end{pmatrix} \quad \text{for all } (x) \in V_1.$$

II.6.57 $$\mathbf{T}\begin{pmatrix} x_1 \\ x_2 \end{pmatrix} = (2x_1 - x_2), \quad \mathbf{U}\begin{pmatrix} x_1 \\ x_2 \end{pmatrix} = (x_1 + 2x_2) \quad \text{for all } \begin{pmatrix} x_1 \\ x_2 \end{pmatrix} \in V_2.$$

II.6.58 $$\mathbf{T}\begin{pmatrix} x_1 \\ x_2 \\ \vdots \\ x_n \end{pmatrix} = \begin{pmatrix} 0 \\ x_1 \\ x_2 \\ \vdots \\ x_{n-1} \end{pmatrix}, \quad \mathbf{U}\begin{pmatrix} x_1 \\ x_2 \\ \vdots \\ x_n \end{pmatrix} = \begin{pmatrix} x_2 \\ x_3 \\ \vdots \\ x_n \\ 0 \end{pmatrix} \quad \text{for all } \begin{pmatrix} x_1 \\ x_2 \\ \vdots \\ x_n \end{pmatrix} \in V_n.$$

II.6.59 $\mathbf{T}(X) = 2 \cdot X, \qquad \mathbf{U}(X) = 3 \cdot X \qquad$ for all $X \in V_n$.

II.6.60 *"Concrete" verification of* (II.6.24)(a) *for* $\mathscr{M}_{m,n}$. Using Theorem II.6.7, along with certain properties of the real numbers, show that $A + B =$

$B + A$ for all matrices $A, B \in \mathscr{M}_{m,n}$. Let $(a_{ij})_{ij}$ and $(b_{ij})_{ij}$ denote A and B, respectively.

II.6.61 *"Abstract" verification of* (II.6.24)(a) *for* $\mathscr{M}_{m,n}$. Using Definition II.6.5 and certain properties of $\mathscr{L}(V_n, V_m)$, show that for all matrices $A, B \in \mathscr{M}_{m,n}$, we have $A + B = B + A$. With this proof, don't even *bother* with specific matrix entries a_{ij} as you did in Exercise II.6.46, that is, do not use Theorem II.6.7.

II.6.62 *"Concrete" verification of* (II.6.24)(a) *for* $\mathscr{M}_{m,n}$. Using Theorem II.6.7 and certain properties of the real numbers, show that for any matrix $A = (a_{ij})_{ij} \in \mathscr{M}_{m,n}$, there is, indeed, (at least one) matrix $A' = (?)_{ij}$ with the property that $A + A' = O_{m,n}$, the zero matrix in $\mathscr{M}_{m,n}$ (see proof of Theorem II.6.8 for information regarding $O_{m,n}$).

II.6.63 *"Abstract" verification of* (II.6.24)(c) *for* $\mathscr{M}_{m,n}$. Using Definition II.6.5 and certain properties of $\mathscr{L}(V_n, V_n)$, show that for any matrix $A \in \mathscr{M}_{m,n}$, there exists a matrix A' in $\mathscr{M}_{m,n}$ with the property that $A + A' = O_{m,n}$. Do not use Theorem II.6.8 in this exercise; the result is to be established *without* reference to the matrix entries a_{ij}.

II.6.64 *"Concrete" verification of* (II.6.24)(d) *for* $\mathscr{M}_{m,n}$. Using Theorem II.6.7 and certain properties of the real numbers, show that for arbitrary matrices $A = (a_{ij})_{ij}$, $B = (b_{ij})_{ij}$ and $C = (c_{ij})_{ij}$ in $\mathscr{M}_{m,n}$, we have

(a) $(A + B) + C = A + (B + C)$.

Recall that $+$ is defined only between *pairs* of matrices; hence, some pairwise bracketing with parentheses is necessary in (a) so that a sum of 3 matrices becomes meaningful.

II.6.65 *"Abstract" verification of* (II.6.24)(d) *for* $\mathscr{M}_{m,n}$. Using Definition II.6.5 and certain properties of $\mathscr{L}(V_n, V_m)$, show that for all $A, B, C \quad \mathscr{M}_{m,n}$, we have

$$(A + B) + C = A + (B + C).$$

Do not use Theorem II.6.7 in your proof.

In Exercises II.6.66 through II.6.69, linear operators $\mathbf{T}$ and $\mathbf{U}$ in $\mathscr{L}(V_n, V_m)$ are given along with scalars α and β. For each exercise,

(a) characterize the linear operators $(\alpha + \beta)\mathbf{T}$, $\alpha\mathbf{T}$, $\beta\mathbf{T}$, $\alpha(\mathbf{T} + \mathbf{U})$ and $\alpha\mathbf{U}$ (see Definitions II.4.1 and II.4.3), that is, describe the effect (output) that each of these 5 operators produces with an *arbitrary* X (input) in V_n

(b) relative to the standard bases for V_n and V_m, find the matrix of each of 5 linear operators given in (a)

(c) using the 5 matrices from (b), and using Theorems II.6.7 and II.6.10, verify that in these cases (at least),

(1) $mx((\alpha + \beta)\mathbf{T}) = mx(\alpha\mathbf{T}) + mx(\beta\mathbf{T})$

and

(2) $mx(\alpha(\mathbf{T} + \mathbf{U})) = mx(\alpha\mathbf{T}) + mx(\alpha\mathbf{U})$.

II.6.66 $$\mathbf{T}\begin{pmatrix} x_1 \\ x_2 \end{pmatrix} = \begin{pmatrix} 3x_1 + x_2 \\ x_1 - x_2 \\ x_1 + x_2 \end{pmatrix}, \mathbf{U}\begin{pmatrix} x_1 \\ x_2 \end{pmatrix} = \begin{pmatrix} 0 \\ 2x_1 \\ 2x_2 \end{pmatrix} \quad \text{for all } \begin{pmatrix} x_1 \\ x_2 \end{pmatrix} \in V_2;$$

$$\alpha = 3, \beta = -2.$$

II.6.67 $$\mathbf{T}\begin{pmatrix}x_1\\x_2\\x_3\end{pmatrix}=\begin{pmatrix}2x_1+x_2-x_3\\x_1+x_2+x_3\\x_1+x_2+x_3\end{pmatrix},\ \mathbf{U}\begin{pmatrix}x_1\\x_2\\x_3\end{pmatrix}=\begin{pmatrix}x_1-x_2\\x_2-x_3\\x_3-x_1\end{pmatrix}$$

for all

$$\begin{pmatrix}x_1\\x_2\\x_3\end{pmatrix}\in V_3;\ \alpha=1,\beta=4.$$

II.6.68 $\mathbf{T}(X)=6X,\ \mathbf{U}(X)=-4X$ for all $X\in V_n$ $\quad\alpha=-1,\beta=2.$

II.6.69 $\mathbf{T}\begin{pmatrix}x_1\\x_2\end{pmatrix}=\begin{pmatrix}x_1-x_2\\x_2\end{pmatrix},\ \mathbf{U}\begin{pmatrix}x_1\\x_2\end{pmatrix}=\begin{pmatrix}x_2\\x_1\end{pmatrix}$ $\quad\alpha=1,\beta=-1.$

II.6.70 "*Concrete*" *verification of* (II.6.30)(b) *for* $\mathscr{M}_{m,n}$. Using Theorem II.6.7 and certain properties of the real numbers, show that for any arbitrary matrix

$$A=(a_{ij})_{ij}\in\mathscr{M}_{m,n},\quad\text{that } 1\cdot A=A.$$

II.6.71 "*Abstract*" *verification of* (II.6.30)(b) *for* $\mathscr{M}_{m,n}$. Using Definition II.6.5 and the fact that axiom B.3 (Definition I.5.2) obtains for the vector-space $\mathscr{L}(V_m, V_n)$, show that for any matrix $A\in\mathscr{M}_{m,n}$ $1\cdot A=A$.

II.6.72 "*Concrete*" *verification of* (II.6.30)(c) *for* $\mathscr{M}_{m,n}$. Using Theorem II.6.7, and certain properties of the real numbers, show that for any matrix $A=(a_{ij})_{ij}\in\mathscr{M}_{m,n}$, and for any pair of scalars α,β, that

$$(\alpha+\beta)\cdot A=\alpha\cdot A+\beta\cdot A.$$

II.6.73 "*Abstract*" *verification of* (II.6.30)(c) *for* $\mathscr{M}_{m,n}$. Using Definition 11.6.5 and and axiom B.4 which holds for $\mathscr{L}(V_n,V_m)$ show that $(\alpha+\beta)\cdot A=\alpha\cdot A+\beta\cdot A$ for all scalars α and β, and all matrices $A\in\mathscr{M}_{m,n}$.

II.6.74 "*Concrete*" *verification of* (II.6.30)(d) *for* $\mathscr{M}_{m,n}$. Use Theorem II.6.7 and axiom B.5, which holds for the vector-space of scalars, to show that $\alpha\cdot(A+B)=\alpha\cdot A+\alpha\cdot B$ for any scalar α and any pair of matrices $A=(a_{ij})_{ij}$ and $B=(b_{ij})_{ij}$ in $\mathscr{M}_{m,n}$.

II.6.75 "*Abstract*" *verification of* (II.6.30)(d) *for* $\mathscr{M}_{m,n}$. Use Definition II.6.5 and axiom B.5, which obtains for the vector-space $\mathscr{L}(V_n,V_m)$, to show that $\alpha\cdot(A+B)=\alpha\cdot A+\alpha\cdot B$ for any and all scalars α and all pairs of matrices A,B $\mathscr{M}_{m,n}$.

II.6.76 Given scalars α and β, and $m\times n$ matrices A and B, find

(a) $\alpha\cdot A$	(e) $(\alpha+\beta)\cdot A$
(b) $\beta\cdot B$	(f) $(\alpha+\beta)\cdot B$
(c) $\alpha\cdot B$	(g) $\alpha\cdot(A+B)$
(d) $\beta\cdot A$	(h) $\beta\cdot(A+B)$

(1) $\alpha=1,\beta=-1,\ A=\begin{pmatrix}3&1&-1&0\\4&2&3&0\end{pmatrix},\quad B=\begin{pmatrix}-1&2&0&1\\0&1&2&4\end{pmatrix}.$

(2) $\alpha=-1,\beta=2,\ A=\begin{pmatrix}1&4&-1\\0&2&0\\0&0&3\end{pmatrix},\quad B=\begin{pmatrix}1&-2&1\\1&3&1\\1&3&1\end{pmatrix}.$

(3) $\alpha = 2, \beta = -2, A = \begin{pmatrix} -1 & \frac{1}{2} & \frac{1}{3} & 1 & 0 \\ 2 & 3 & 0 & \frac{1}{2} & 1 \\ -1 & 0 & 0 & 1 & \frac{1}{2} \end{pmatrix}, B = \begin{pmatrix} 1 & -\frac{1}{2} & -\frac{2}{3} & 1 & 1 \\ \frac{1}{2} & 0 & 1 & 0 & 1 \\ 2 & 1 & 3 & 1 & -1 \end{pmatrix}.$

(4) $\alpha = 1, \beta = 1, A = (1 \quad 3 \quad 4 \quad 2), B = (1 \quad 3 \quad 2 \quad 4).$

(5) You will have noticed that part (a) + part (c) = part (g) and that part (a) + part (d) = part (e) in the previous 4 examples. Is this always true? (Why?) Or is this simply an accident of this particular problem? (Why?)

Matrix Multiplication In Section 5, a multiplication was forced on the set of 2×2 matrices $\mathscr{M}_{2,2}$ only because pairs of operators in $\mathscr{L}(V_2, V_2)$ could, themselves, be "multiplied", or composed (see Exercise II.5.60). Let us consider the more general $m \times n$ matrices and see whether a composition of operators is meaningful for the corresponding vector-space $\mathscr{L}(V_n, V_m)$. Curses! In the general case, given arbitrary $\mathbf{T}$ and $\mathbf{U}$ in $\mathscr{L}(V_n, V_m)$, the composition $\mathbf{T} \circ \mathbf{U}$ [defined by $\mathbf{T} \circ \mathbf{U}(X) = \mathbf{T}(\mathbf{U}(X))$] has *no* meaning unless the range of the operator $\mathbf{U}$ lies in the domain of the operator $\mathbf{T}$. To be consistent, we must accept the fact that no meaningful multiplication between pairs of $m \times n$ matrices can be defined if $m \neq n$.

But if the idea of composition of linear transformations is so important in forcing a matrix multiplication, why not consider vector-spaces of linear operators where we *have* a bonafide composition (see the example following Figure 2.13 and Theorem II.4.7)? As we have seen in the 2×2 matrix case, (Exercise II.5.60), we may define our multiplication on matrices so that it is simply a reflection of multiplication (that is, composition) of linear transformation. To be more specific, we set down Definition II.6.12. (See Figure 2.21.)

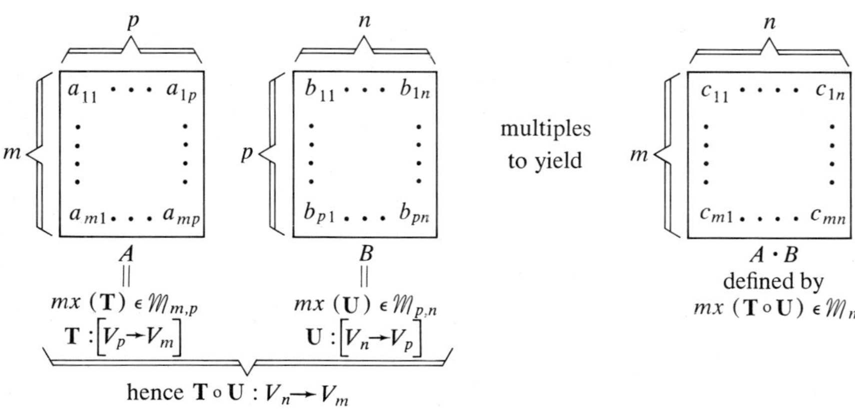

Figure 2.21

Definition II.6.12 Let $A \in \mathcal{M}_{m,p}$ be an $m \times p$ matrix and let $B \in \mathcal{M}_{p,n}$ be a $p \times n$ matrix. We define the **matrix product, A · B**, to be an $m \times n$ matrix as follows (roughly speaking, we "convert" matrices A and B to appropriate linear transformations **T** and **U**, and then "reconvert" the well-known function product $\mathbf{T} \circ \mathbf{U}$ back to a matrix which is to stand for $A \cdot B$): Let

$$\mathscr{X} = \{X_1, X_2, \ldots, X_n\}, \ \mathscr{Z} = \{Z_1, Z_2, \ldots, Z_p\} \text{ and } \mathscr{Y} = \{Y_1, Y_2, \ldots, Y_m\}$$

be arbitrary but fixed bases for vector-spaces V_n, V_p and V_m, respectively. Now let $\mathbf{T} \in \mathscr{L}(V_p, V_m)$ and $\mathbf{U} \in \mathscr{L}(V_n, V_p)$ be those unique linear transformations such that

$$mx(\mathbf{T})_{\mathscr{Z},\mathscr{Y}} = A \quad \text{and} \quad mx(\mathbf{U})_{\mathscr{X},\mathscr{Z}} = B.$$

Noting that $\mathbf{T} \circ \mathbf{U}$, the product (composition) of the linear functions **T** and **U**, is meaningful, we *define* $A \cdot B$ to be the $m \times n$ matrix $mx(\mathbf{T} \circ \mathbf{U})_{\mathscr{X}\mathscr{Y}}$. That is,

$$A \cdot B \quad (\text{that is, } mx(\mathbf{T})_{\mathscr{Z}\mathscr{Y}} \cdot mx(\mathbf{U})_{\mathscr{X}\mathscr{Z}}) = mx(\mathbf{T} \circ \mathbf{U})_{\mathscr{X}\mathscr{Y}} \qquad \text{(II.6.34)}$$

(see Exercise II.5.60b).

Figure 2.21 relates the objects which play a role in Definition II.6.12 and may be considered as a pictorial summary of the definition. It would appear, at first glance, that this definition inflicts upon us a great woe. Observe that under this definition, we cannot multiply two $m \times n$ *matrices* together unless $m = n$ (we have an $m \times p$ matrix, when multiplied on the right by a $p \times n$ matrix, yields a certain $m \times n$ matrix). But this is a tolerable situation once we realize that this is merely a reflection of the fact that we can't compose (multiply) 2 *linear transformations* in $\mathscr{L}(V_n, V_m)$ unless $m = n$.

Now if someone were to give us a particular $m \times p$ matrix A and a particular $p \times n$ matrix B, would Definition II.6.12 enable us to compute the *specific* $m \times n$ matrix $A \cdot B$? It certainly will, after we establish Theorem II.6.13.

Theorem II.6.13 (See Exercise II.5.60.) Let $A = (a_{ik})_{ik} \in \mathcal{M}_{m,p}$ $i = 1, 2, \ldots, m$; $k = 1, 2, \ldots, p$, and let $B = (b_{lj})_{lj} \in \mathcal{M}_{p,n}$ $l = 1, 2, \ldots, p$; $u = 1, 2, \ldots, n$. Then $A \cdot B = (\sum_{k=1}^{p} a_{ik}b_{kj})_{ij}$. That is, the ijth entry of the $m \times n$ matrix $A \cdot B$ is the scalar $\sum_{k=1}^{p} a_{ik}b_{kj}$, which we denote by c_{ij}. That is

$$(a_{ik})_{ik} \cdot (b_{lj})_{lj} = \left(\sum_{k=1}^{p} a_{ik}b_{kj} \right)_{ij}.$$

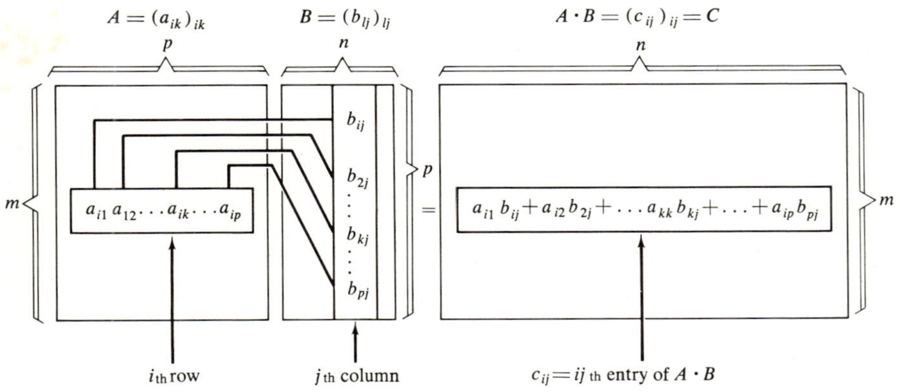

Figure 2.22

Figure 2.22 describes a geometric "row-column" procedure for multiplying a pair of matrices together. Figure 2.22 illustrates the following procedure: In order to obtain c_{ij}, the ijth entry of the product $A \cdot B$, you take the ith row of A, the jth column of B, multiply their corresponding entries (connected by lines), and simply add all these products together.

But behold! The computational result of Figure 2.22 and Theorem II.6.13 has already been encountered in Definition I.2.3! And this *same* mechanical means of multiplying matrices arises here as a *matrix* reflection of the composition of *linear transformations*. Thus, we have another good reason for living with this albeit oddball row-column ritual for multiplying matrices.

We now present some examples, and then the proof of Theorem II.6.13.

Example

$$\begin{pmatrix} 1 & 0 & 2 \\ 4 & -2 & 3 \end{pmatrix} \cdot \begin{pmatrix} 6 & -1 \\ 1 & 5 \\ 2 & 3 \end{pmatrix} = \begin{pmatrix} 1 \cdot 6 + 0 \cdot 1 + 2 \cdot 2 & 1 \cdot (-1) + 0 \cdot 5 + 2 \cdot 3 \\ 4 \cdot 6 + (-2) \cdot 1 + 3 \cdot 2 & 4 \cdot (-1) + (-2) \cdot 5 + 3 \cdot 3 \end{pmatrix}$$

$$= \begin{pmatrix} 10 & 5 \\ 28 & -5 \end{pmatrix}.$$

As expected, a 2×3 matrix, when multiplied on the right by a 3×2 matrix, yields a 2×2 matrix.

Example As it turns out, we may multiply the matrices of our previous example in reverse order. Reversing order is not *always* possible, since the number of columns of the left-hand factor must *always* equal the number of rows of the right-hand factor.

$$\begin{pmatrix} 6 & -1 \\ 1 & 5 \\ 2 & 3 \end{pmatrix} \cdot \begin{pmatrix} 1 & 0 & 2 \\ 4 & -2 & 3 \end{pmatrix} = \begin{pmatrix} 6\cdot 1 + (-1)\cdot 4 & 6\cdot 0 + (-1)\cdot(-2) & 6\cdot 2 + (-1)\cdot 3 \\ 1\cdot 1 + 5\cdot 4 & 1\cdot 0 + 5\cdot(-2) & 1\cdot 2 + 5\cdot 3 \\ 2\cdot 1 + 3\cdot 4 & 2\cdot 0 + 3\cdot(-2) & 2\cdot 2 + 3\cdot 3 \end{pmatrix}$$

$$= \begin{pmatrix} 2 & 2 & 9 \\ 21 & -10 & 17 \\ 14 & -6 & 13 \end{pmatrix}.$$

Example

$$\begin{matrix} A & & B & & A\cdot B = C \end{matrix}$$

$$\begin{pmatrix} -1 & 8 & 2 & 3 \\ 4 & 5 & 7 & 6 \end{pmatrix} \cdot \begin{pmatrix} 2 \\ 3 \\ 0 \\ 1 \end{pmatrix} = \begin{pmatrix} (-1)\cdot 2 + 8\cdot 3 + 2\cdot 0 + 3\cdot 1 \\ 4\cdot 2 + 5\cdot 3 + 7\cdot 0 + 6\cdot 1 \end{pmatrix} = \begin{pmatrix} 25 \\ 29 \end{pmatrix}.$$

Let us now prove that the technique of matrix multiplication which was used in these examples, is the unavoidable concrete consequence of Definition II.6.12.

Proof of Theorem II.6.13: Let $A = (a_{ik})_{ik} \in \mathcal{M}_{m,p}$ and $B = (b_{lj})_{lj} \in \mathcal{M}_{p,n}$ be typical $m \times p$ and $p \times n$ real matrices, respectively. As per Definition II.6.12, we select arbitrary but fixed ordered bases $\mathscr{X} = \{X_1, X_2, \ldots, X_n\}$, $\mathscr{Z} = \{Z_1, Z_2, \ldots, Z_p\}$ and $\mathscr{Y} = \{Y_1, Y_2, \ldots, Y_m\}$ for the vector-spaces V_n, V_p, and V_m, respectively. This means that there exists unique $\mathbf{T} \in \mathscr{L}(V_p, V_m)$ and unique $\mathbf{U} \in \mathscr{L}(V_n, V_p)$ (see Theorem II.6.4) such that

(a) $$mx(\mathbf{T})_{\mathscr{Z}\mathscr{Y}} = A = (a_{ik})_{ik}$$

and

(b) $$mx(\mathbf{U})_{\mathscr{X}\mathscr{Z}} = B = (b_{lj})_{lj}.$$

Now everybody knows what (a) and (b) really mean; according to Definition II.6.2 [see Equations (II.6.3) (a) and (b) and (II.6.4)], this is just another way of saying

(a′) $$\mathbf{T}(Z_k) = \sum_{i=1}^{m} a_{ik} \cdot Y_i \quad \text{for each } k = 1, 2, \ldots, p$$

(b′) $$\mathbf{U}(X_j) = \sum_{l=1}^{p} b_{lj} \cdot Z_l \quad \text{for each } j = 1, 2, \ldots, n.$$

The stage is set. The big game we are after is $mx(\mathbf{T} \circ \mathbf{U})_{\mathscr{X}\mathscr{Y}}$ (see Definition II.6.12), which actually defines $A \cdot B$. Thus, all we need for each $j = 1$,

$2, \ldots, n$, is to know what $\mathbf{T} \circ \mathbf{U}(X_j)$ looks like as a vector in V_m. Accordingly,

$$\begin{aligned}
\mathbf{T} \circ \mathbf{U}(X_j) &= \mathbf{T}(\mathbf{U}(X_j)) && \text{(Definition II.1.5)} \\
&= \mathbf{T}\left(\sum_{k=1}^{p} b_{kj} \cdot Z_k\right) && \text{from (b}'\text{); here we use the dummy variable } k \text{ instead of } l \text{ which appears in (b}'\text{)} \\
&= \sum_{k=1}^{p} \mathbf{T}(b_{kj} \cdot Z_k) && \text{since } \mathbf{T} \text{ is linear} \\
&= \sum_{k=1}^{p} b_{kj} \cdot \mathbf{T}(Z_k) && \text{since } \mathbf{T} \text{ is linear } \mathbf{T}(b_{kj} \cdot Z_k) = b_{kj} \cdot \mathbf{T}(Z_k); \text{ see Definition II.2.1, Equation (II.2.1)} \\
&= \sum_{k=1}^{p} b_{kj} \cdot \left(\sum_{i=1}^{m} a_{ik} \cdot Y_i\right) && \text{from (a}'\text{)} \qquad \text{(II.6.35)} \\
&= \sum_{k=1}^{p} \left(\sum_{i=1}^{m} b_{kj}(a_{ik} \cdot Y_i)\right) && \text{Distributivity, axiom B.5, Definition I.5.2} \\
&= \sum_{k=1}^{p} \left(\sum_{i=1}^{m} (b_{kj}a_{ik}) \cdot Y_i\right) && \text{axiom B.2, Definition I.5.2} \\
&= \sum_{i=1}^{m} \left[\left(\sum_{k=1}^{p} a_{ik}b_{kj}\right) \cdot Y_i\right] && \text{since } b_{jk}a_{ik} = a_{ik}b_{kj} \text{ and order of summation can be reversed.}
\end{aligned}$$

Now according to Definition II.6.2, a_{ij}, the ijth entry of the matrix $C = mx(\mathbf{U} \circ \mathbf{T})$ (which defines $A \cdot B$), is *exactly* that coefficient in the expression for $\mathbf{T} \circ \mathbf{U}(X_j)$ which is "attached" to the basis vector Y_i. The last expression in (II.6.35) tells us that this coefficient is the very scalar $\sum_{k=1}^{p} a_{ik}b_{kj} = a_{i1}b_{1j} + a_{i2}b_{2j} + \cdots + a_{ip}b_{pj}$. With this observation, we conclude the proof of Theorem II.6.13.

Theorem II.6.14 The set $\mathcal{M}_{n,n}$ of all $n \times n$ matrices whose entries are real numbers, with the operations of addition, scalar multiplication and matrix multiplication defined above, forms an *algebra*.

Proof: In consulting Definition II.4.10 for the list of properties which characterize algebras, we see that almost all of these properties have already been established for $\mathcal{M}_{n,n}$. Theorem II.6.11 tells us that $\mathcal{M}_{n,n}$ is a vector-space, so that *all* of the axioms, A.1 through A.5 and B.1 through

B.5 of Definition II.4.10 (see, also Definition I.5.2), hold. Thus, only the axioms C.1 (closure of multiplication), C.2 (associativity of multiplication) and C.3 (distributive law) need any further attention.

Now axiom C.1, the closure axiom, can be verified by even a casual glance at either Definition II.6.12 or Theorem II.6.13 which shows that an $n \times n$ matrix can be multiplied by an $n \times n$ matrix; moreover, the end result is, again, an $n \times n$ matrix. So we may agree that $\mathscr{M}_{n,n}$ is *closed* under multiplication, that is, axiom C.1 does indeed, hold for $\mathscr{M}_{n,n}$.

Axiom C.2, the associativity axiom, is left as Exercise II.6.81. A "concrete" proof of the associativity axiom C.2, dealing with the specific matrix entries a_{ij}, b_{ij}, and c_{ij}, is left for the student also as Exercise II.6.81(a). In Exercise II.6.81(b), the "abstract" proof of C.2 becomes a consequence of a slightly more general result.

Axiom C.3, the distributivity axiom, is also left for the student to verify; see Exercises II.6.83 and II.6.84.

Concluding Remarks: At this point, we have forced a structure upon $\mathscr{M}_{m,n}$, that is, an addition, a scalar multiplication and, in case $m = n$, a matrix multiplication. Since this structure was transmitted to $\mathscr{M}_{m,n}$ from $\mathscr{L}(V_n, V_m)$ through the function mx (see Definitions II.6.5, II.6.9, and II.6.12), we might say that $\mathscr{M}_{m,n}$ is a kind of mathematical twin of $\mathscr{L}(V_n, V_m)$. A more precise way of saying this is that $\mathscr{L}(V_n, V_m)$ are isomorphic as *vector-spaces*; in case $m = n$ (so that a closed multiplication exists in both $\mathscr{L}(V_n, V_m)$ and in $\mathscr{M}_{m,n}$), we have that $\mathscr{L}(V_n, V_n)$ and $\mathscr{M}_{n,n}$ are even isomorphic as *algebras* (Definitions II.6.15 follows). The following Figure 2.23 summarizes the property of structure-preserving maps (functions).

REMARK. We hasten to add that this is not the first time we have taken 2 apparently different sets of objects, and devised a 1-1 onto linear transformation sending 1 set to the other. In fact, Exercises I.1.14, I.1.15, I.1.16, and I.1.17, we defined the function "col," which sent certain polynomials with real coefficients onto the set of n-vectors. The exercises following this definition, consisted in verifying the very fact that the function "col" was *linear*, as per Definition II.2.1 (which was to come later).

We must be mindful, however, that the linear function mx did more than preserve the operations of addition and scalar multiplication; for the special case where $m = n$. Definition II.6.12 is a formal way of saying that the function mx preserves multiplication, or preserves the multiplicative structure of $\mathscr{L}(V_n, V_n)$.

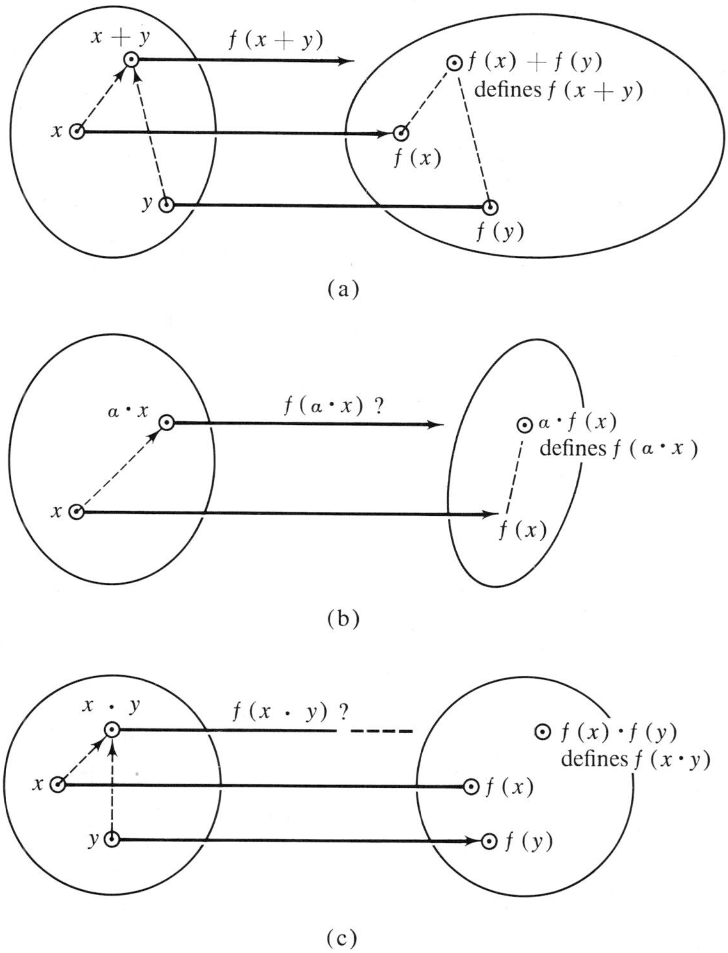

Figure 2.23 ***(a) f preserves addition if $f(x+y) = f(x) + f(y)$, all x, y. (b) f preserves scalar multiplication. (c) f preserves pairwise multiplication.***

These 2 real-life examples of the 1-1 onto linear functions mx and "col" give us good reason to think well of such linear functions. They not only preserve the algebraic structure (that is, a *vector-space* has a *vector-space* as its image; an *algebra* has an *algebra* as its image, and so on), but they do it in a 1-1 fashion. Intuitively, we may think of these structure-preserving, 1-1 functions as the most distortion-free functions which send (or transform) 1 structured space to another. Such functions are called *isomorphisms*; we refer back to Definition II.4.10 to define this term more precisely in the following Definition II.6.15.

Definition II.6.15 Let $\mathscr{R}$ and $\mathscr{S}$ be groups, that is, for any x and any y in the group there exists an addition $+$ so that the element $x + y$ is in the group also (see Figure 2.14 of Definition II.4.10). Suppose $f : \mathscr{R} \to \mathscr{S}$ is a function sending $\mathscr{R}$ to $\mathscr{S}$ which preserves this group structure (that is, the additive structure). In other words

(a) $$f(x + y) = f(x) + f(y) \quad \text{for all } x, y \in \mathscr{R}.$$

Then f is called a **group homomorphism** from $\mathscr{R}$ to $\mathscr{S}$. If, moreover, f is 1-1 and onto, then f is called a **group isomorphism** from $\mathscr{R}$ onto $\mathscr{S}$.

Suppose $\mathscr{R}$ and $\mathscr{S}$ each has the structure of a vector-space, that is, on each set $\mathscr{R}$ and $\mathscr{S}$, there is a commutative addition $+$ defined between pairs of elements and a scalar multiplication $\cdot$ defined between scalars and elements (vectors) of $\mathscr{R}$ and $\mathscr{S}$. Suppose $f : \mathscr{R} \to \mathscr{S}$ is a function sending $\mathscr{R}$ to $\mathscr{S}$ which preserves the vector-space structure, that is,

(b) $$f(x + y) = f(x) + f(y) \quad \text{for all } x,y \in \mathscr{R},$$

and

$$f(\alpha \cdot x) = \alpha \cdot f(x) \quad \text{for all } x \in \mathscr{R}, \text{ all scalars } \alpha.$$

Then f is said to be a **vector-space homomorphism** from $\mathscr{R}$ to $\mathscr{S}$. If, moreover, f is 1-1 and onto, then f is called a **vector-space isomorphism** from $\mathscr{R}$ onto $\mathscr{S}$.

Suppose $\mathscr{R}$ and $\mathscr{S}$ each has the structure of a ring, that is, on each set $\mathscr{R}$ and $\mathscr{S}$, there is a commutative addition $+$ defined between pairs of elements, and a multiplication $\circ$ operation defined between pairs of elements such that axioms A.1 through A.5 and C.1 through C.3, Definition II.4.10 are satisfied. Suppose $f : \mathscr{R} \to \mathscr{S}$ is a function sending $\mathscr{R}$ to $\mathscr{S}$, which preserves this ring structure, that is,

(c) $$f(x + y) = f(x) + f(y) \quad \text{for all } x, y \in \mathscr{R},$$

and

$$f(x \circ y) = f(x) \circ f(y) \quad \text{for all } y, y \in \mathscr{R}.$$

Then f is said to be a **ring homomorphism** from $\mathscr{R}$ to $\mathscr{S}$. If, moreover, f is 1-1 and onto, then f is called a **ring isomorphism** from $\mathscr{R}$ onto $\mathscr{S}$.

Suppose $\mathscr{R}$ and $\mathscr{S}$ each has the structure of an algebra (see Definition II.4.10). Thus, we have on each set $\mathscr{R}$ and $\mathscr{S}$, a certain commutative, pairwise addition $+$, a certain pairwise multiplication $\circ$, and a certain scalar multiplication $\cdot$. Let $f : \mathscr{R} \to \mathscr{S}$ be a function sending $\mathscr{R}$ to $\mathscr{S}$ where f preserves the algebra structure, that is,

(d) $$f(x + y) = f(x) + f(y) \quad \text{for all } x, y \in \mathscr{R},$$
$$f(x \circ y) = f(x) \circ f(y) \quad \text{for all } x, y \in \mathscr{R},$$

and

$$f(\alpha \cdot x) = \alpha \cdot f(x) \qquad \text{for all } x \in \mathscr{R}, \text{ all scalars } \alpha.$$

Then f is called an **algebra homomorphism** from $\mathscr{R}$ to $\mathscr{S}$. If, moreover, f is 1-1 and onto, then f is said to be an **algebra isomorphism** from $\mathscr{R}$ onto $\mathscr{S}$.

In any of these 4 cases (whether $\mathscr{R}$ and $\mathscr{S}$ are a pair of groups, a pair of vector-spaces, a pair of rings, or a pair of algebras), if at least 1 structure-preserving function f exists as per condition (a), (b), (c), or (d), then the sets $\mathscr{R}$ and $\mathscr{S}$ are said to be **homomorphic to each other,** or simply, **homomorphic.** Sometimes the *type* of homomorphism is stated; for example, if a function f were to exist satisfying condition (c), and $\mathscr{R}$ and $\mathscr{S}$ are rings, then we might say that $\mathscr{R}$ and $\mathscr{S}$ are **ring homomorphic** to each other.

If, moreover, f exists as a 1-1 and onto function, then the sets $\mathscr{R}$ and $\mathscr{S}$ are said to be **isomorphic to each other,** or simply, **isomorphic.**

We note that in the above list of structure-preserving functions, an old friend reappears; what we have previously defined to be a *linear transformation* or a *linear function* from one vector-space to another (see Definition II.2.1), is defined here as a *vector-space homorphism* [see (b) above].

Consider the following loose threads:

(a) $mx : \mathscr{L}(V_n, V_m) \to \mathscr{M}_{m,n}$ is a 1-1 and onto function (Theorem II.6.4)

(b) $mx(\mathbf{T} + \mathbf{U}) = mx(\mathbf{T}) + mx(\mathbf{U})$ (Definition II.6.5)

(c) $mx(\alpha\mathbf{T}) = \alpha \cdot mx(\mathbf{T})$ (Definition II.6.9)

(d) if $m = n$ (whence $\mathbf{T} \circ \mathbf{U} \in \mathscr{L}(V_n, V_m)$ whenever $\mathbf{T}, \mathbf{U} \in \mathscr{L}(V_n, V_m)$) then $mx(\mathbf{T} \circ \mathbf{U}) = mx(\mathbf{T}) \cdot mx(\mathbf{U})$.

Conditions (a), (b), and (c) are equivalent to the statement that $\mathscr{L}(V_n, V_m)$ and $\mathscr{M}_{m,n}$ are vector-space isomorphic to each other [see (b) of Definition II.6.15]. If $m = n$, so that both vector-spaces $\mathscr{L}(V_n, V_m)$ and $\mathscr{M}_{m,n}$ become *algebras*, then conditions (a), (b), (c), and (d) obtain, that is, $\mathscr{L}(V_n, V_n)$ and $\mathscr{M}_{n,n}$ are *algebra* isomorphic to each other. Thus, we have argued our way through a proof of Theorem II.6.16.

Theorem II.6.16 (Fundamental Isomorphism Theorem) The real vector-spaces $\mathscr{L}(V_n, V_m)$ and $\mathscr{M}_{m,n}$ are isomorphic to each other as vector-spaces [(as per Definition II.6.15 (b)].

If $m = n$, then the real algebras $\mathscr{L}(V_n, V_n)$ and $\mathscr{M}_{n,n}$ are isomorphic to each other as real algebras. In each case, the function $mx : \mathscr{L}(V_n, V_m) \to \mathscr{M}_{m,n}$ (see Definition II.6.2) is such an isomorphism between these 2 spaces.

A loose interpretation of this theorem is the following: The set $\mathscr{L}(V_n, V_m)$ of all possible linear transformations sending V_n to V_m, is a "carbon copy" of the set $\mathcal{M}_{m,n}$ of all possible scalar-entried $m \times n$ matrices. One set is, structurally, exactly like the other. In fact, as you have probably observed by now (and will observe in the future), our so-called "abstract" proofs have always been set in the space $\mathcal{L}(V_n, V_m)$ of linear operators, while our "concrete" proofs have always been set in the space $\mathcal{M}_{m,n}$ of all $m \times n$ matrices. This terminology is due to the author's own prejudice toward accepting matrices as objects which are (somehow) more real and touchable than linear functions are.

REMARK. Believe it or not, we have *already* come across matrix multiplication (as described in Theorem II.6.13) in another context!

If you refer back to Theorem II.2.3, Equation (II.2.4), you will see that this structure theorem, which completely characterizes a typical linear transformation in $\mathcal{L}(V_n, V_m)$, lends itself to a matrix interpretation. If we were to regard the n-vector X as just another $n \times 1$ matrix, then Theorem II.2.3 says exactly this:

▶ For any linear transformation $\mathbf{T} \in \mathcal{L}(V_n, V_m)$, there is a unique $m \times n$ scalar matrix $A = (a_{ij})_{ij}$ such that

(a) $$\mathbf{T}(X) = A \cdot X \quad \text{for all } n\text{-vectors } X \in V_n.$$

Notice that the left-hand side of (a) is read "$\mathbf{T}$ of X," or "$\mathbf{T}$ evaluated at X," while the right-hand side of (a) is read "the $m \times n$ matrix A multiplied by the $n \times 1$ *matrix* X."

As it turns out, the $m \times n$ matrix A will be the same as $mx(\mathbf{T})_{\mathscr{X}\mathscr{Y}}$ (see Definition II.6.2) if the ordered bases $\mathscr{X}$ and $\mathscr{Y}$ are the standard bases for V_n and V_m, respectively. That is, Theorem II.2.3 may be roughly interpreted as follows:

▶ Any and every linear transformation on V_n to V_m may be affected by merely multiplying each n-vector ($n \times 1$ matrix) $X \in V_n$ by a suitable fixed $m \times n$ matrix A. Moreover, (and this observation does not appear explicitly in Theorem II.2.3), this $m \times n$ matrix A is exactly $mx(\mathbf{T})$ relative to the *standard* ordered bases for V_n and V_m.

Exercises II.6.87, II.6.88 and II.6.89 establish the fact that a left- (or right-) hand multiplication operator *always* proves to be a *linear* operator even when we are not multiplying column ($n \times 1$) matrices. This fact is little more than a restatement of the distributive law for matrix multiplication (see Exercises II.6.83 and II.6.84).

EXERCISES

II.6.77 With the given matrices, compute the desired products.

$$A = \begin{pmatrix} 2 & 0 \\ 1 & 4 \end{pmatrix}, \quad B = \begin{pmatrix} 3 & 0 & 1 \\ -2 & 4 & 1 \end{pmatrix}, \quad C = \begin{pmatrix} 3 \\ 2 \\ -1 \end{pmatrix}.$$

Find (1) $M = A \cdot B$, (2) $N = B \cdot C$, (3) $M \cdot C$, (4) $A \cdot N$, (5) $A^2 = A \cdot A$, (6) $A^2 \cdot B$, (7) $A \cdot M$.
Which property, if any, of matrix multiplication assures you in advance, that (3) and (4) will be equal, and that (6) and (7) will be equal?

II.6.78 Given matrices

$$A = \begin{pmatrix} 1 & -2 & 1 \\ 0 & 3 & 4 \end{pmatrix}, \quad B = \begin{pmatrix} 5 & -2 \\ 0 & 1 \\ 1 & 1 \end{pmatrix}, \quad C = \begin{pmatrix} 6 & 1 \\ 1 & 2 \end{pmatrix}.$$

Find (1) $M = A \cdot B$, (2) $N = B \cdot C$, (3) $M \cdot C$, (4) $A \cdot N$, (5) $P = C \cdot A$, (6) $C \cdot M$, (7) $P \cdot B$, (8) C^2, (9) $C^2 \cdot A$, (10) $B \cdot A$, (11) $(B \cdot A) \cdot B$.

II.6.79 (See Exercise II.5.62.) Let A and B be square matrices in $\mathscr{M}_{n,n}$, where A is diagonal scalar and B is arbitrary. To say that A is diagonal scalar is to say

$$A = \begin{pmatrix} \alpha & 0 & \cdots & 0 \\ 0 & \alpha & & \cdot \\ \cdot & & \ddots & \cdot \\ \cdot & & & \cdot \\ 0 & \cdot & \cdot & \alpha \end{pmatrix}$$

is a square matrix whose every entry is zero except for the diagonal entries, which are each equal to some scalar α; $A = (a_{ij})_{ij}$, where $a_{ij} = 0$ whenever $i \neq j$ and $a_{ii} = \alpha$ for each i.

(a) Show that $A \cdot B = B \cdot A$ for any and every $B \in \mathscr{M}_{n,n}$.

(b) Show that the diagonal scalar matrices are the *only* matrices in $\mathscr{M}_{n,n}$ which commute with each and every (other) matrix $B \in \mathscr{M}_{n,n}$, that is $X \cdot B = B \cdot X$ for all $B \in \mathscr{M}_{n,n} \rightarrow X$ is diagonal scalar.

II.6.80 Given $\mathbf{T} \in \mathscr{L}(V_2, V_3)$, $\mathbf{U} \in \mathscr{L}(V_3, V_3)$ and $\mathbf{V} \in \mathscr{L}(V_2, V_2)$ as follows:

$$\mathbf{T}\begin{pmatrix} x_1 \\ x_2 \end{pmatrix} = \begin{pmatrix} x \\ x_3 \\ x_1 + x_2 \end{pmatrix} \quad \text{for all } X = \begin{pmatrix} x_1 \\ x_2 \end{pmatrix} \in V_2$$

$$\mathbf{U}\begin{pmatrix} x_1 \\ x_2 \\ x_3 \end{pmatrix} = \begin{pmatrix} x_1 + x_3 \\ x_2 + x_3 \\ x_1 + x_2 \end{pmatrix} \quad \text{for all } X = \begin{pmatrix} x_1 \\ x_2 \\ x_3 \end{pmatrix} \in V_3$$

$$\mathbf{V}\begin{pmatrix} x_1 \\ x_2 \end{pmatrix} = \begin{pmatrix} x_2 \\ -x_1 \end{pmatrix} \quad \text{for all } X = \begin{pmatrix} x_1 \\ x_2 \end{pmatrix} \in V_2.$$

(a) Relative to ordered bases

$$\mathscr{X} = \left\{\begin{pmatrix}1\\0\end{pmatrix},\begin{pmatrix}1\\1\end{pmatrix}\right\} \quad\text{and}\quad \mathscr{Y} = \begin{pmatrix}1\\0\\0\end{pmatrix},\begin{pmatrix}0\\1\\0\end{pmatrix},\begin{pmatrix}0\\1\\1\end{pmatrix} \quad \text{for } V_2 \text{ and } V_3,$$

respectively, compute $A = mx(\mathbf{T})_{\mathscr{X}\mathscr{Y}}$, $B = mx(\mathbf{U})_{\mathscr{X}\mathscr{X}}$ and $C = mx(\mathbf{V})_{\mathscr{X}\mathscr{X}}$ (see Definition II.6.2).

(b) Describe the operators $\mathbf{U} \circ \mathbf{T} \in \mathscr{L}(V_2,V_3)$ and $\mathbf{T} \circ \mathbf{V} \in \mathscr{L}(V_2,V_3)$ that is, describe

$$\mathbf{U} \circ \mathbf{T}\begin{pmatrix}x_1\\x_2\end{pmatrix} \quad\text{and}\quad \mathbf{T} \circ \mathbf{V}\begin{pmatrix}x_1\\x_2\end{pmatrix}.$$

(c) Using Definition II.6.2, find $D = mx(\mathbf{U} \circ \mathbf{T})_{\mathscr{X}\mathscr{Y}}$ and $E = mx(\mathbf{T} \circ \mathbf{V})_{\mathscr{X}\mathscr{Y}}$.

(d) Using Theorem II.6.13, compute $B \cdot A$ and $A \cdot C$. In this case, $B \cdot A$ [that is, $mx(\mathbf{U}) \cdot mx(\mathbf{T})$] should equal D [that is, $mx(\mathbf{U} \circ \mathbf{T})$] and $A \cdot C$ [that is, $mx(\mathbf{T}) \cdot mx(\mathbf{V})$] should equal E [that is, $mx(\mathbf{T} \circ \mathbf{V})$].

(e) Is it true that $B \cdot (A \cdot C) = (B \cdot A) \cdot C$?

II.6.81 *Matrix multiplication is associative* (see proof of Theorem II.6.14). We are to show that if A,B, and C are any real matrices, then $(A \cdot B) \cdot C = A \cdot (B \cdot C)$ whenever the matrix dimensions allow the products to exist. Specifically, we require that $A \in \mathscr{M}_{m,p}$, $B \in \mathscr{M}_{p,q}$, and $C \in \mathscr{M}_{q,n}$.

(a) "*Concrete*" *proof of associativity.* Let $A = (a_{ij})_{ij}$, $B = (b_{kl})_{kl}$, and $C = (c_{st})_{st}$. Use Theorem II.6.13 (more than once) and the associativity of the scalars to show that $(A \cdot B) \cdot C = A \cdot (B \cdot C)$.

(b) "*Abstract*" *proof of associativity.* We must first check that if $\mathbf{V} \in \mathscr{L}(V_n,V_q)$, $\mathbf{U} \in \mathscr{L}(V_q,V_p)$, and $\mathbf{T} \in \mathscr{L}(V_p,V_m)$, then function multiplication (composition) is associative, that is, use Definition II.1.5 (more than once) to prove that $(\mathbf{T} \circ \mathbf{U}) \circ \mathbf{V}$, which belongs to $\mathscr{L}(V_n,V_m)$, is the same linear operator as $\mathbf{T} \circ (\mathbf{U} \circ \mathbf{V})$, also a member of $\mathscr{L}(V_n,V_m)$ (see Exercise II.4.19).

Use Definitions II.6.2 and II.6.12, along with the result of the previous paragraph, to show that if (relative to appropriate bases) $A = mx(\mathbf{T})$, $B = mx(\mathbf{U})$ and $C = mx(\mathbf{V})$, then unavoidably,

$$(A \cdot B) \cdot C = A \cdot (B \cdot C).$$

II.6.82 (See Exercise II.5.64.) Given matrices A, B, and C, compute the desired products and sums.

(a) $$A = \begin{pmatrix}3 & -1 & 2\\1 & 4 & 0\end{pmatrix},\ B = \begin{pmatrix}-1\\-1\\1\end{pmatrix},\ C = \begin{pmatrix}2\\1\\2\end{pmatrix}.$$

Find (1) $M = B + C$; (2) $A \cdot M = A \cdot (B + C)$; (3) $A \cdot B$, (4) $A \cdot C$; (5) Is it true that (2) equals (3) plus (4) that is, that in this case, the product $A \cdot (B + C)$ equals the sum $(A \cdot B) + (A \cdot C)$?

(b) $$A = \begin{pmatrix}1 & 2\\0 & 1\\2 & 1\end{pmatrix},\ B = \begin{pmatrix}5 & 7\\1 & 1\\2 & 2\end{pmatrix},\ C = \begin{pmatrix}3 & 1\\2 & 4\end{pmatrix}.$$

Find (1) $M = A + B$; (2) $M \cdot C = (A + B) \cdot C$; (3) $A \cdot C$. (4) Is the product $(A + B) \cdot C$ equal to the sum $(A \cdot C) + (B \cdot C)$?

II.6.83 (See Exercise II.6.65.) "*Concrete*" *proof of distributivity for matrices.* Let $A \in \mathscr{M}_{m,p}$ and $B,C \in \mathscr{M}_{p,q}$, that is, $A = (a_{ij})_{ij}$ $i = 1, 2, \ldots, m, j = 1, 2, \ldots, p$, and $B = (b_{kl})_{kl}, C = (c_{kl})_{kl}$ where $k = 1, 2, \ldots, p, l = 1, 2, \ldots, q$. Use Theorems II.6.7 and II.6.13 along with distributivity for the scalars to show that $A \cdot (B + C) = (A \cdot B) + (A \cdot C)$.

II.6.84 (See Exercise II.5.66.) "*Abstract*" *proof of distributivity for matrices.*

(a) We must first prove a distributivity property for linear transformations. Hence, if $\mathbf{U},\mathbf{V} \in \mathscr{L}(V_q, V_p)$ and $\mathbf{T} \in \mathscr{L}(V_p, V_m)$, then prove that $\mathbf{T} \circ (\mathbf{U} + \mathbf{V}) = (\mathbf{T} \circ \mathbf{U}) + (\mathbf{T} \circ \mathbf{V})$, is an element of $\mathscr{L}(V_q, V_m)$. That is, show that for every $X \in V_q$, $\mathbf{T} \circ (\mathbf{U} + V)(X) = ((\mathbf{T} \circ \mathbf{U}) + (\mathbf{T} \circ \mathbf{V}))X$ using Definitions II.1.5 and II.4.1 (see Theorem II.4.8 and Exercise II.4.26).

(b) Using appropriate bases (as per Definition II.6.2), let $A = mx(\mathbf{T})$, $B = mx(\mathbf{U})$, and $C = mx(\mathbf{V})$, where $A \in \mathscr{M}_{m,p}$ and $B,C \in \mathscr{M}_{p,q}$. From the truth of (a) and from Definitions II.6.5 and II.6.12 prove that the following $m \times q$ matrices are equal:

$$A \cdot (B + C) \quad \text{and} \quad (A \cdot B) + (A \cdot C).$$

Note. Of course, the techniques used in Exercises II.6.83 and II.6.84 can be used to establish the validity of a right-hand distributivity for matrices. Also, the proof of Theorem II.6.14, which rests on these exercises, considers only the matrices which are square, that is, that all vector-spaces in our "abstract" proofs have the same dimension.

II.6.85 We shall consider the transpose function, denoted **tps**, which sends $\mathscr{M}_{2,2}$ to $\mathscr{M}_{2,2}$ as follows:

For any matrix $\begin{pmatrix} a_{11} & a_{12} \\ a_{21} & a_{22} \end{pmatrix} \in \mathscr{M}_{2,2}$,

$$\mathbf{tps}\begin{pmatrix} a_{11} & a_{12} \\ a_{22} & a_{21} \end{pmatrix} = \begin{pmatrix} a_{11} & a_{21} \\ a_{12} & a_{22} \end{pmatrix}.$$

(a) Show that **tps** is a 1-1 and onto function.

(b) Show that for all matrices $A,B \in \mathscr{M}_{2,2}$ and for any scalar α,

(i) $\mathbf{tps}(A + B) = \mathbf{tps}(A) + \mathbf{tps}(B)$

and

(ii) $\mathbf{tps}(\alpha \cdot A) = \alpha \cdot \mathbf{tps}(A)$.

Statements (a) and (b) together are equivalent to saying that the function $\mathbf{tps} : \mathscr{M}_{2,2} \to \mathscr{M}_{2,2}$ is a vector space isomorphism from $\mathscr{M}_{2,2}$ onto itself [see Definition II.5.15(b)].

(c) Show that for any and every pair of 2×2 matrices A and B, $\mathbf{tps}(A \cdot B) = \mathbf{tps}(B) \cdot \mathbf{tps}(A)$. With this fact in mind, we see that *all* conditions of Definition II.6.15(d) are *not* fulfilled, so that **tps** is not an *algebra* isomorphism of $\mathscr{M}_{2,2}$ with itself. This is *not* to say that $\mathscr{M}_{2,2}$ is not algebra isomorphic with itself; rather, that **tps** is not such an algebra isomorphism [the identity function $\mathbf{I}(A) = A$ for all A, *is* such an algebra isomorphism, however].

II.6.86 We define the tranpose function $\mathbf{tps}: \mathcal{M}_{n,n} \to \mathcal{M}_{n,n}$ for general square matrices as follows:

For all $A = (a_{ij})_{ij} \in \mathcal{M}_{n,n}$, $\mathbf{tps}((a_{ij})_{ij}) = (a_{ji})_{ij}$. Geometrically, the transpose function reflects (or revolves 180 degrees) each entry about the main diagonal so that columns "become" rows and rows "become" columns. Show that properties (a), (b), and (c) listed in Exercise II.6.85, still obtain in this general setting.

II.6.87 (a) Given $A_0 = \begin{pmatrix} 2 & 1 & 3 \\ 1 & 0 & 1 \end{pmatrix}$. Let

$$X = \begin{pmatrix} x_{11} & x_{12} \\ x_{21} & x_{22} \\ x_{31} & x_{32} \end{pmatrix} \quad \text{and} \quad Y = \begin{pmatrix} y_{11} & y_{12} \\ y_{21} & y_{22} \\ y_{31} & y_{32} \end{pmatrix}$$

be typical elements of $\mathcal{M}_{3,2}$. Compute (1) $A_0 \cdot X$; (2) $A_0 \cdot Y$; (3) $(X + Y)$; (4) $A_0 \cdot (X + Y)$. (5) Is it true that part (1) plus part (2) equals part (4), that is, that the matrix sum $A_0 \cdot X + A_0 \cdot Y$ equals the matrix product $A_0 \cdot (X + Y)$? (6) Is $\alpha \cdot (A \cdot X)$ always equal to $A \cdot (\alpha \cdot X)$ for all scalars α?

(b) Given

$$A_0 = \begin{pmatrix} 2 & 4 \\ -1 & 5 \\ 0 & 6 \end{pmatrix}.$$

Let $X = \begin{pmatrix} x_1 \\ x_2 \end{pmatrix}$ and $Y = \begin{pmatrix} y_1 \\ y_2 \end{pmatrix}$ be typical elements of $\mathcal{M}_{2,1}$. Compute (1) $A_0 \cdot X$; (2) $A_0 \cdot Y$; (3) $(X + Y)$; (4) $A_0 \cdot (X + Y)$. (5) Is it true that the matrix sum $A_0 \cdot X + A_0 Y$ equals the matrix product $A_0 \cdot (X + Y)$? (6) Is $\alpha \cdot (A \cdot X)$ always equal to $A \cdot (\alpha \cdot X)$ for all scalars α?

II.6.88 (See Exercise II.6.84.) We define a (left-hand multiplication) function $\mathbf{L}_{A_0}: \mathcal{M}_{q,n} \to \mathcal{M}_{m,n}$ as follows: For a fixed $m \times q$ matrix $A_0 \in \mathcal{M}_{m,q}$, we put

$$\mathbf{L}_{A_0}(X) = A_0 \cdot X \quad \text{for each and every } q \times n \text{ matrix } X \in \mathcal{M}_{q,n}.$$

(Note that $\mathbf{L}_{A_0}$ does, indeed, transform $q \times n$ matrices X to $m \times n$ matrices $A_0 \cdot X$.) Prove that $\mathbf{L}_{A_0}$ is a vector-space homomorphism (that is, a linear function) from vector space $\mathcal{M}_{q,n}$ to vector-space $\mathcal{M}_{m,n}$. That is,

$$\mathbf{L}_{A_0}(X + Y) = \mathbf{L}_{A_0}(X) + \mathbf{L}_{A_0}(Y) \quad \text{for all } X, Y \in \mathcal{M}_{q,n},$$

and $\mathbf{L}_{A_0}(\alpha \cdot X) = \alpha \cdot \mathbf{L}_{A_0}(X)$ for all $X \in \mathcal{M}_{q,n}$ and all scalars α [see Definitions II.2.1 and II.5.12(b)].

II.6.89 (See Exercise II.6.84.) We define a (right-hand multiplication) function $\mathbf{R}_{A_0}: \mathcal{M}_{m,q} \to \mathcal{M}_{m,n}$ as follows: For a fixed $q \times n$ matrix $A_0 \in \mathcal{M}_{q,n}$, we put $\mathbf{R}_{A_0}(X) = X \cdot A_0$ for each and every $m \times q$ matrix $X \in \mathcal{M}_{m,q}$. Prove that $\mathbf{R}_{A_0}$ is a vector-space homomorphism of vector-space $\mathcal{M}_{m,q}$ to vector-space $\mathcal{M}_{m,n}$.

II.6.90 Show by specific example, that even if $m = q = n$ in Exercises II.6.88 and II.6.89 (so that $\mathcal{M}_{m,q}, \mathcal{M}_{q,n}$, and $\mathcal{M}_{m,n}$ are all *algebras*), the linear operators $\mathbf{L}_{A_0}$ and $\mathbf{R}_{A_0}$ are not algebra homomorphisms, that is, it is not *necessarily* true that $\mathbf{L}_{A_0}(X \cdot Y) = \mathbf{L}_{A_0}(X) \cdot \mathbf{L}_{A_0}(Y)$ and $\mathbf{R}_{A_0}(X \cdot Y) = \mathbf{R}_{A_0}(X) \cdot \mathbf{R}_{A_0}(Y)$ for all square matrices $X, Y \in \mathcal{M}_{m,n}$.

II.6.91 Let the symbol Z denote the set of all natural integers, that is

$$Z = \{0, \pm 1, \pm 2, \ldots, \pm n, \ldots\}.$$

Let Z be a ring where the operations of addition and (pairwise) multiplication are defined in the usual, sensible way. In what follows, functions $f : Z \to Z$, from Z to itself are defined for you. In each case, decide whether the function f preserves addition $(f(x+y) = f(x) + f(y))$, and whether it preserves multiplication $(f(x \cdot y) = f(x) \cdot f(y))$. Recall that if addition is preserved, then f is a group homomorphism; if both addition and multiplication are preserved, then f is a ring homomorphism (see Definition II.5.12).

(a) f squares every integer, that is, $f(x) = x \cdot x$ for all integers $x \in Z$.

(b) f multiplies every integer by 3, that is, $f(x) = 3x$ for all integers $x \in Z$.

(c) $f(x) = 0$ for all integers $x \in Z$.

(d) $f(x) = 1$ for all integers $x \in Z$.

(e) $f(x) = n_0 x$ for all $x \in Z$. Here, n_0 is some fixed integer [see part (b)].

(f) $f(x) = n_0 x + m_0$ for all $x \in Z$. Here n_0 and m_0 are fixed integers.

(g) $f(x) = x$ for all integers $x \in Z$.

II.6.92 Let $\mathbf{f}$ be the function sending 2×2 matrices to 2×2 matrices, which multiplies "input" matrices on the left by some fixed matrix A_0, that is, $\mathbf{f}(B) = A_0 \cdot B$ for all 2×2 matrices B.

(a) Show that $\mathbf{f}$ preserves matrix addition [$\mathbf{f}(B + C) = \mathbf{f}(B) + \mathbf{f}(C)$ for all matrices B and C].

(b) By choosing a specific matrix for A_0 show that $\mathbf{f}$ does not necessarily preserve multiplication, that is, it is not always the case that

$$\mathbf{f}(B \cdot C) = \mathbf{f}(B) \cdot \mathbf{f}(C).$$

(c) Show that $\mathbf{f}$ preserves scalar multiplication $\mathbf{f}(\alpha \cdot B) = \alpha \cdot \mathbf{f}(B)$ for all scalars α, and all matrices B.

To establish conditions (a) and (c) above, is to establish the fact that $\mathbf{f} : B \to A_0 \cdot B$ is a vector-space homomorphism. Thus, even though the 2×2 matrices form an *algebra*, only the vector-space structure (addition, scalar multiplication) is necessarily preserved by the given function $\mathbf{f}$.

(d) Prove or disprove that statements (a), (b), and (c) above remain valid if the right-hand multiplication function $\mathbf{g}(B) = B \cdot A_0$ for all matrices B, replaces the function $\mathbf{f}$.

Chapter II Summary

Chapter II opens with a review of the notion of a function. In Section 2, we study very special functions which send 1 vector-space to another. These are exactly the functions which preserve the precious 2 operations on the vector-space, that is, the functions preserve the addition and the scalar multiplication of the vector-space (Definition II.2.1). Section 3 continues the analysis of these structure-preserving functions on vector-spaces, called linear functions (or linear transformations), by introducing the notions of rank and nullity of a linear transformation (Definition II.3.8). The important relation between these ideas is given by the statement that the rank plus the nullity of a linear transformation is always equal to the dimension of its domain (Theorem II.3.6). A simple consequence of this fact is that linear transformations must either maintain or reduce dimension (Corollary II.3.7). Section 3 concludes by exhibiting conditions which are equivalent to invertibility of a linear transformation (Theorem II.3.9, Corollary II.3.10). It is also shown that if $\mathbf{T}^{-1}$, the inverse of a linear transformation $\mathbf{T}$ exists as a function, then $\mathbf{T}^{-1}$ is automatically a *linear transformation*, too. In Section 4, the algebraic vector-space properties of linear transformations are developed. Just as we could add vectors, so we have a closed, associative, commutative addition between pairs of linear transformations (Definition II.4.1). A means for multiplying linear maps with scalars is given (Definition II.4.3), as well as a multiplication between pairs of certain linear transformations (see Theorem II.4.7). After all this structure is set down on the set of *all* linear transformations on a vector-space, we conclude that these operations of addition, scalar multiplication, and composition are good enough and blend together well enough, to yield a new vector-space (or algebra, if composition is permitted) (Theorem II.4.6, Exercise II.4.30). We next devise a function mx which converts abstract linear transformations into concrete, rectangular matrices (Definition II.6.2). Through this 1-1, onto function mx, we force or induce structure on the set of all fixed-dimension rectangular matrices, namely, an addition (Definition II.6.5, Theorem II.6.7), a scalar multiplication (Definition II.6.9, Theorem II.6.10), and where applicable, a pairwise multiplication (Definition II.6.12, Theorem II.6.13). Thus, the set of all rectangular (respectively, square) matrices becomes a vector-space (respectively, an algebra) which is a concrete isomorphism of the set of all linear transformations which send one vector-space to another (respectively, to itself) (Theorem II.6.16).

CHAPTER III

Systems of Linear Equations

1. *Row- and Column-echelon Matrices*

In Chapter I, Section 3, we encountered a system of m equations in n unknowns of the form

$$\begin{array}{cccc} a_{11}x_1 + a_{12}x_2 + \cdots + a_{1n}x_n = y_1 \\ a_{21}x_1 + a_{22}x_2 + \cdots + a_{2n}x_n = y_2 \\ \vdots \qquad \vdots \qquad \quad \vdots \quad \vdots \\ a_{m1}x_1 + a_{m2}x_2 + \cdots + a_{mn}x_n = y_m. \end{array} \tag{III.1.1}$$

Display (III.1.1) can be abbreviated considerably if written as a matrix equation. In fact, to say that (III.1.1) holds, is to say that the 2 concrete m-vectors (III.1.2a) and (III.1.2b),

$$\underset{\text{(a)}}{\begin{pmatrix} a_{11}x_1 + a_{12}x_2 + \cdots + a_{1n}x_n \\ a_{21}x_1 + a_{22}x_2 + \cdots + a_{2n}x_n \\ \vdots \\ a_{m1}x_1 + a_{m2}x_2 + \cdots + a_{mn}x_n \end{pmatrix}}, \quad \underset{\text{(b)}}{\begin{pmatrix} y_1 \\ y_2 \\ \vdots \\ y_m \end{pmatrix}} \tag{III.1.2}$$

are *equal* (see Definition I.1.4).

Now the m-vector (III.1.2a) is observed to be the matrix product of the $m \times n$ matrix $A = (a_{ij})_{ij}$, $i = 1, 2, \ldots, m$, $j = 1, 2, \ldots, n$, with the $n \times 1$ column vector X whose kth entry is x_k. That is, the system (III.1.1) of m separate equations is equivalent to the single matrix equation

$$\underbrace{\begin{pmatrix} a_{11} & a_{12} & \cdots & a_{1n} \\ a_{21} & a_{22} & \cdots & a_{2n} \\ \vdots & \vdots & & \vdots \\ a_{m1} & a_{m2} & \cdots & a_{mn} \end{pmatrix}}_{\substack{A \\ \uparrow \\ m \times n}} \cdot \underbrace{\begin{pmatrix} x_1 \\ x_2 \\ \vdots \\ x_n \end{pmatrix}}_{\substack{X \\ \uparrow \\ n \times 1}} = \underbrace{\begin{pmatrix} y_1 \\ y_2 \\ \vdots \\ y_m \end{pmatrix}}_{\substack{Y \\ \uparrow \\ m \times 1}}. \tag{III.1.3}$$

When studying Gaussian Elimination previously in Chapter I, we loosely defined a *solution* to the system (III.1.1) as an ordered n-tuple of scalars $x_1, x_2, \ldots, x_n$, that gave us the "right answers." It will profit us now to think of such an ordered set of n scalars as a concrete *n-vector* as per Definition I.1.1. This affords us the vocabulary with which to define a *solution* to a system of equations, which we now do.

Definition III.1.1 Let $A = (a_{ij})_{ij}$ $i = 1, 2, \ldots, m$, $j = 1, 2, \ldots, n$, be an $m \times n$ matrix and let Y_0 be a specific fixed concrete m-vector ($m \times 1$ matrix). A **system of linear equations** will be a matrix equation

$$\underbrace{A}_{m \times n} \cdot \underbrace{X}_{n \times 1} = \underbrace{Y_0}_{m \times 1} \tag{III.1.4}$$

where

$$X = \begin{pmatrix} x_1 \\ x_2 \\ \vdots \\ x_n \end{pmatrix}$$

is an unspecified column n-vector. A particular n-vector X_0 for which it is true that

$$A \cdot X_0 = Y_0$$

is called a **solution (solution vector) for the system** (III.1.4). The $m \times n$ matrix $A = (a_{ij})_{ij}$ is called the **coefficient matrix** of the system $A \cdot X = Y_0$, while the $m \times (n + 1)$ matrix $A : Y$, defined as

$$\begin{pmatrix} a_{11} & a_{12} & \cdots & a_{1n} & y_1 \\ a_{21} & a_{22} & \cdots & a_{2n} & y_2 \\ \vdots & \vdots & & \vdots & \vdots \\ a_{m1} & a_{m2} & \cdots & a_{mn} & y_m \end{pmatrix} = A : Y$$

is called the **augmented coefficient matrix** of the system.

Our objective is to transform the 3 legitimate steps used in **Gaussian Elimination** to the language of matrices. Recall that we were permitted to use 3 operations on systems of equations which altered (and simplified) the *appearance* but *not* the solution vector. These were the operations of

(a) switching equations

▶(b) multiplying an equation through by a nonzero scalar

(c) adding a multiple of 1 equation to another.

Now what do these operations mean if we view the system of equations

$$\begin{aligned} a_{11}x_1 + a_{12}x_2 + \cdots + a_{1n}x_n &= y_1 \\ a_{21}x_1 + a_{22}x_2 + \cdots + a_{2n}x_n &= y_2 \\ \vdots \qquad \vdots \qquad\qquad \vdots \quad &\quad \vdots \\ a_{m1}x_1 + a_{m2}x_2 + \cdots + a_{mn}x_n &= y_m \end{aligned}$$

in terms of its augmented coefficient matrix (see Definition III.1.1)

$$\begin{pmatrix} a_{11} & a_{12} & \cdots & a_{1n} & y_1 \\ a_{21} & a_{22} & \cdots & a_{2n} & y_2 \\ \vdots & \vdots & & \vdots & \vdots \\ a_{m1} & a_{m2} & \cdots & a_{mn} & y_m \end{pmatrix} = A : Y? \tag{III.1.5}$$

The "Gaussian" operations now read [relative to (III.1.5)]

(a)′ interchange 2 rows of the augmented matrix $A : Y$

▶(b)′ multiply a row of $A : Y$ through by a nonzero scalar

(c)′ add a scalar multiple of 1 row to another row of $A : Y$.

Hence (a)′, (b)′, and (c)′ represent (a), (b), and (c) in a *matrix* setting. We are ready for some notation describing the operations for arbitrary matrices.

Definition III.1.2 Let $A = (a_{ij})_{ij}$ be an $m \times n$ matrix. We use the symbols

$$A_i \quad \text{and} \quad A^j$$

to denote the ith-column-vector and the jth-row-vector of A, respectively. Accordingly, the matrix A will be denoted by either of the symbols

$$A = [A_1 A_2 \cdots A_n] \quad \text{or} \quad A = \begin{bmatrix} A^1 \\ A^2 \\ \vdots \\ A^m \end{bmatrix}.$$

The **elementary column operations** C_{ij}, $C_i(a)$, and $C_j{}^i(a)$, along with the corresponding **elementary row operations** R_{ij}, $R_i(a)$, and $R_j{}^i(a)$ are functions sending $\mathscr{M}_{m,n}$ to itself defined by:

(a) $C_{ij}[A_1A_2\cdots A_i\cdots A_j\cdots A_n]=[A_1A_2\cdots \underbrace{A_j}_{i\text{th place}}\cdots \underbrace{A_i}_{j\text{th place}}\cdots A_n]$

(C_{ij} interchanges positions of column-vectors A_i and A_j)

(b) $C_i(a)[A_1A_2\cdots A_i\cdots A_n]=[A_1A_2\cdots aA_i\cdots A_n]$

[$C_i(a)$ multiplies column-vector A_i by the scalar "a"]

(c) $C_j{}^i(a)[A_1A_2\cdots A_i\cdots A_j\cdots A_n]=[A_1A_2\cdots \underbrace{A_i+aA_j}_{i\text{th place}}\cdots A_j\cdots A_n]$

[$C_j{}^i(a)$ adds column-vector aA_j to column-vector A_i in the ith column].

Similarly, R_{ij} interchanges row-vector A^i with row-vector A^j; $R_i(a)$ multiplies row-vector A^i by the scalar "a"; $R_i{}^j(a)$ adds the row-vector aA^i to the row-vector A^j in the jth row; that is, only the jth row is altered.

Example Let

$$A=\begin{pmatrix}1&0&1&-1\\2&1&6&2\\3&4&2&7\end{pmatrix}.$$

Then

$$C_{24}(A)=C_{24}\begin{pmatrix}1&0&1&-1\\2&1&6&2\\3&4&2&7\end{pmatrix}=\begin{pmatrix}1&-1&1&0\\2&2&6&1\\3&7&2&4\end{pmatrix}$$

$$R_{23}(A)=R_{23}\begin{pmatrix}1&0&1&-1\\2&1&6&2\\3&4&2&7\end{pmatrix}=\begin{pmatrix}1&0&1&-1\\3&4&2&7\\2&1&6&2\end{pmatrix}$$

$$C_3(-2)(A)=C_3(-2)\begin{pmatrix}1&0&1&-1\\2&1&6&2\\3&4&2&7\end{pmatrix}=\begin{pmatrix}1&0&-2&-1\\2&1&-12&2\\3&4&-4&4\end{pmatrix}$$

$$R_1(6)(A)=R_1(6)\begin{pmatrix}1&0&1&-1\\2&1&6&2\\3&4&2&7\end{pmatrix}=\begin{pmatrix}6&0&6&-6\\2&1&6&2\\3&4&2&7\end{pmatrix}$$

$$C_1{}^2(2)(A)=\begin{pmatrix}1&2\cdot1+0&1&-1\\2&2\cdot2+1&6&2\\3&2\cdot3+4&2&7\end{pmatrix}=\begin{pmatrix}1&2&1&-1\\2&5&6&2\\3&10&2&7\end{pmatrix}$$

$$R_2{}^1(-\tfrac{1}{2})(A)=\begin{pmatrix}1+(-\frac{1}{2})\cdot2&0+(-\frac{1}{2})\cdot1&1+(-\frac{1}{2})\cdot6&-1+(-\frac{1}{2})\cdot2\\2&1&6&2\\3&4&2&7\end{pmatrix}$$

$$=\begin{pmatrix}0&-\frac{1}{2}&-2&-2\\2&1&6&2\\3&4&2&7\end{pmatrix}.$$

It is not pertinent at the moment, but the student may easily check that each of these elementary column and row operations, or functions, sending the vector-space $\mathscr{M}_{m,n}$ to itself, are true *linear* operators, that is, functions which preserve the vector-space operations of addition and scalar multiplication (see Definition II.2.1).

EXERCISES

III.1.1 Put

$$A = \begin{pmatrix} 3 & 0 & 1 \\ 2 & 1 & 3 \\ 1 & -1 & -2 \end{pmatrix}.$$

Find: (a) $C_{13}(A), C_{31}(A), R_{13}(A), R_{31}(A)$.
(b) $C_{13}(C_{31}(A))$ and $R_{13}(R_{13}(A))$.
(c) $C_3{}^1(-3)(A), C_3{}^2(-2)(A)$.
(d) $R_3{}^1(-3)(A), R_2{}^3(-\frac{1}{2})(A)$.
(e) $R_2(-3)(A), C_3(\frac{1}{2})(A)$.

III.1.2

$$A = \begin{pmatrix} 1 & 0 & 4 & 2 & 6 \\ 1 & 8 & 0 & -2 & 0 \\ 1 & 4 & 2 & 0 & 3 \end{pmatrix}.$$

Find: (a) $C_2{}^3(-1)(A), \quad C_4{}^5(6)(A)$.
(b) $R_2{}^3(-2)(A), \quad R_3{}^2(-2)(A)$.
(c) $C_1{}^3(C_2(-1)(A)), \quad R_3{}^1(-1)_1 \circ R_3{}^1(3)(A)$.
(d) $C_4(4)(A), \quad C_4(\frac{1}{4}) \circ C_4(4)(A), \quad C_4(-4) \circ C_4(4)(A)$.
(e) $C_{23} \circ R_1(2) \circ C_1(\frac{1}{2})(A)$.

III.1.3 Given

$$A = \begin{pmatrix} a_{11} & a_{12} & a_{13} & a_{14} \\ a_{21} & a_{22} & a_{23} & a_{24} \\ a_{31} & a_{32} & a_{33} & a_{34} \end{pmatrix},$$

a typical 3×4 matrix.

Find: (a) $C_{24}(A), \quad C_{42}(A), \quad C_{24} \circ C_{24}(A)$.
(b) $R_{12}(A), \quad R_{12} \circ R_1(3)(A), \quad R_1(3) \circ R_{12}(A)$.
(c) $C_4{}^1(c)(A), \quad C_4{}^1(-c) \circ C_4{}^1(c)(A)$.
(d) $R_2{}^3\left(-\dfrac{a_{31}}{a_{21}}\right) \circ R_2{}^1\left(-\dfrac{a_{11}}{a_{21}}\right)(A), R_3(2) \circ R_{12} \circ C_1{}^2(2)(A)$.

III.1.4 The purpose of this exercise is to establish the fact that the elementary row and column operations are *reversible*. That is to say, for every elementary row or column operation which alters the form of a matrix, there is a corresponding elementary row or column operation which "undoes" the work of the original operation; that is, after applying a row or column operation to a matrix A which produces an "altered" matrix A', we may then apply a corresponding row or column operation to A' which will return us to our original matrix A. More formally, to every elementary row (column) operation (call it D), we will see that a corresponding row (column) operation D' exists such that $D' \circ D(A) = A$ for *all* matrices A. Compare this equation with Definition

II.1.8; the elementary operation D' qualifies as a bonafide *inverse* for the elementary operator (that is, function) D. For this reason, D' will be denoted as D^{-1}, the inverse function to D.

(a) Show that $C_{ij}^{-1} = C_{ij}$ and $R_{ij}^{-1} = R_{ij}$, that is, the inverse of the elementary row or column operator which interchanges row (respectively, column) i with row (respectively, column) ij is its *own* inverse. You are to show that $C_{ij} \circ C_{ij}(A) = A$ and $R_{ij} \circ R_{ij}(A) = A$ for all $m \times n$ matrices A, that is, $C_{ij} \circ C_{ij} = R_{ij} \circ R_{ij} = 1$, the identity operator.

(b) If $a \neq 0$, then show that $C_i(a)^{-1} = C_i(a^{-1})$ and $R_i(a)^{-1} = R_i(a^{-1})$. That is, prove that $R_i(a^{-1}) \circ R_i(a)(A) = A$ and $C_i(a^{-1}) \circ C_i(a)(A) = A$ for all $m \times n$ matrices A.

(c) Show that $C_i^{\,j}(a)^{-1} = C_i^{\,j}(-a)$ and $R_i^{\,j}(a)^{-1} = R_i^{\,j}(-a)$. That is, prove that $C_i^{\,j}(-a) \circ C_i^{\,j}(a)(A) = A$ and $R_i^{\,j}(-a) \circ R_i^{\,j}(a)(A) = A$ for all $m \times n$ matrices A.

Thus, in establishing (a), (b), and (c), we have shown that every elementary row or column operation has an inverse which is, again, an elementary row or column operation, respectively.

III.1.5 Regard the elementary row and column operations as *functions* $\mathcal{M}_{m,n} \to \mathcal{M}_{m,n}$ sending the $m \times n$ matrices $\mathcal{M}_{m,n}$ to itself. Prove that the impact set $\mathcal{M}_{m,n}$ for each of these elementary operations is, in fact, the *range* of each of these elementary operations. That is to say, for arbitrary but fixed elementary operators C_{ij}, $C_i(a)$, $C_i^{\,j}(a)$, R_{ij}, $R_i(a)$, $R_i^{\,j}(a)$, the function is *onto* (Definition II.1.4).

III.1.6 If matrix A has at least 3 rows, then show that $R_{32} \circ R_{12} = R_{12} \circ R_{13}$. From this, deduce that $R_{12} \circ R_{32} \circ R_{12} = R_{13}$.

Let us recall now, the transparent form

$$\begin{array}{rl}
x_1 + 0 + 0 + \cdots + 0 & + c_{11}x_{n+1} + \cdots + c_{1k}x_{n+k} = b_1 \\
x_2 + 0 + \cdots + 0 & + c_{21}x_{n+1} + \cdots + c_{2k}x_{n+k} = b_2 \\
x_3 + \cdots + 0 & + c_{31}x_{n+1} + \cdots + c_{3k}x_{n+k} = b_3 \\
\ddots \quad \vdots & \qquad\qquad\qquad\qquad\qquad\qquad \vdots \\
x_n & + c_{n1}x_{n+1} + \cdots + c_{nk}x_{n+k} = b_n
\end{array} \tag{III.1.6}$$

which resulted from the process of Gaussian Elimination [compare (I.3.16)]. The augmented coefficient matrix for (III.1.6), the "simplified system" is (Definition III.1.1)

$$\begin{pmatrix}
1 & 0 & \cdots & 0 & c_{11} & \cdots & c_{1k} & b_1 \\
0 & 1 & \cdots & 0 & c_{21} & \cdots & c_{2k} & b_2 \\
\vdots & & \ddots & \vdots & \vdots & & \vdots & \vdots \\
0 & & \cdots & 1 & c_{n1} & \cdots & c_{nk} & b_n
\end{pmatrix}. \tag{III.1.7}$$

Note that the diagonal of ones could always be achieved by renumbering (reordering) the unknowns $x_1, x_2, \cdots, x_{n+k}$. We propose to formally describe "simplified" matrices [such as (III.1.7)] with the following definition of *row-echelon* and *column-echelon* matrices.

Definition III.1.3 Let A be an $m \times n$ matrix whose rows are (n-vectors) $A^1, A^2, \ldots, A^m$, and whose columns are (m-vectors) $A_1, A_2, \ldots, A_n$. A is said to be a **row-echelon matrix with row-echelon rank r** (respectively, **column-echelon matrix with column-echelon rank r**) if and only if all of the following conditions obtain:

(a) The first r rows $A^1, A^2, \ldots, A^r$ (columns $A_1, A_2, \ldots, A_r$) are nonzero if any nonzero rows (columns) exist at all, while the remaining rows (columns) are the zero rows (columns).

(b) The first entry (coordinate) of each of the nonzero rows (columns) is the number 1.

(c) The leading entries (that is, the earliest 1's that appear) in the r nonzero rows (columns) appear successively later as we pass to later rows (columns). That is to say, if the leading entry 1 appears in the k_ith entry of row A^i (of column A_i), then all entries, up to and including the k_ith entry of every *later* row $A^{i+1}, A^{i+2}, \ldots, A^m$ (*later* column $A_{i+1}, A_{i+2}, \ldots, A_n$) are necessarily zero. Zeros appear below *and* above (to right *and* the left of) all leading entries.

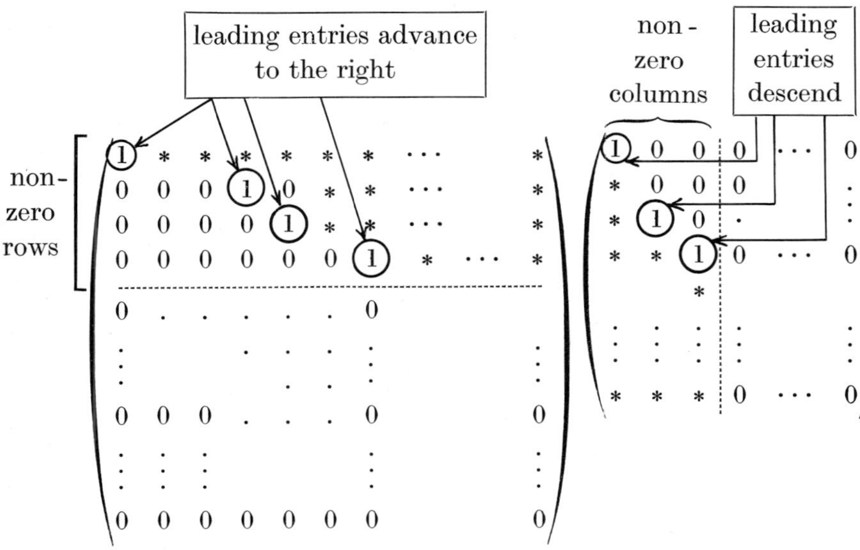

Row-echelon Matrix (zeros everywhere below main diagonal)

Column-echelon Matrix (zeros everywhere above main diagonal).

(III.1.8)

Experience with simultaneous systems of linear equations would seem to indicate that every $m \times n$ matrix can be reduced (via elementary row operations; see Definition III.1.2) to a row-echelon matrix. It seems reasonable to suspect that a column-echelon form can always be generated by a succession of elementary column operations only. So as to state the validity of this conjecture formally, we set down Theorem III.1.4.

Theorem III.1.4 Every $m \times n$ matrix can be reduced to a row (column) echelon matrix by means of the 3 elementary row (column) operations.

Informal Proof: We will indicate how an arbitrary $m \times n$ matrix may be reduced to a row-echelon matrix only; a similar argument would obtain in showing that reduction to a column-echelon matrix is always possible.

We understand, now, that we have only the 3 elementary-row operations at our disposal. Let us consider the $m \times n$ matrix $A = (a_{ij})_{ij}$ whose m rows are $A^1 = (a_{11}a_{12}\ldots a_{1n})$, $A^2 = (a_{21}a_{22}\ldots a_{2n}), \ldots, A^m = (a_{m1}a_{m2}\ldots a_{mn})$. Now we scan the nonzero rows (if any) and fix our attention on the one which has the earliest nonzero entry. That is, we look at all the *first* entries of the rows; are they all zero or is there a nonzero first entry? If they are all zero, we pass to all *second* entries of the rows $A^1, A^2, \ldots, A^n$. We halt this search only when we find a row with a nonzero entry which occurs no later than any other row's nonzero entry. By the elementary operation of interchanging positions of this chosen row with the first row (if necessary) we now have a matrix whose *first* row is the row with the earliest nonzero entry, call it c. We normalize the entry c by applying operation $R_1(c^{-1})$ to this matrix; we now have a matrix whose first row is the row with an earliest nonzero entry; moreover this entry is the number 1. Now if there are any other rows *below* this row possessing an entry which is "as early" as row one's "early" entry, we then proceed to annihilate that competing early entry. For example, if row one is of the form $A^1 = (0\ 0\ 1\ a_{14}\ \ldots\ a_{1n})$ and row p say, is of the form $A^p = (0\ 0\ a_{p3}\ a_{p4}\ \ldots\ a_{pn})$ where $a_{p3} \neq 0$ (so that A^p has "as early" a nonzero entry, namely, a_{p3}, as does A^1), then by adding $-a_{p3}A^1$ to the row A^p [that is, applying $R_1{}^p(-a_{p3})$ to the matrix] we remove any trace of so early a nonzero entry for A^p. To this point, we have a matrix of the form

"Earliest" nonzero entry of any row

$$\left(\begin{array}{ccc:c:cccc}
0 & \cdots & 0 & 1 & * & * & \cdots & * \\
\hdashline
0 & \cdots & 0 & 0 & * & * & \cdots & * \\
0 & \cdots & 0 & 0 & * & * & \cdots & * \\
\vdots & & \vdots & \vdots & \vdots & \vdots & X & \vdots \\
0 & \cdots & 0 & 0 & * & * & \cdots & *
\end{array}\right) \qquad \text{(III.1.9)}$$

Possible zeros

The following remark will certainly justify the use of the adjective "informal" in describing this proof. We simply say that if we repeat this procedure; that is, (1) select a row with an "earliest" nonzero entry, (2) switch its position to the top line, (3) annihilate *all* other "equally early" nonzero entries of other rows above *and* below the new leading entry in submatrix X of (III.1.9), then we obtain a matrix (if possible) with 2 rows whose leading entries are 1, and the leading entry 1 of the second row appears *later* than the leading entry 1 of the first row. We continue in this manner [see (1), (2), and (3) above] until we have only zero rows left, or until we consume all the rows of the matrix. We have a matrix with exactly the 3 properties (a), (b), and (c) required by Definition III.1.3.

A similar argument reveals that the 3 elementary *column* operations can reduce any matrix to *column*-echelon form. This ends our very informal proof.

EXERCISES

In Exercises III.1.7 through III.1.15, certain matrices are given. Using elementary row operations only, reduce each matrix to a row-echelon matrix; similarly, by means of elementary column operations only, reduce the matrices to a column-echelon matrix. Find both the row-echelon rank and the column-echelon rank for each echelon matrix (see Definition III.1.3) derived from A.

III.1.7 Let A be the 5×6 matrix of Example 3.1 (reduce A only to row-echelon form and find its row-echelon rank).

III.1.8 $$A = \begin{pmatrix} 1 & 0 & 0 & 0 & 7 & 6 & -2 \\ 0 & 0 & 1 & 0 & 3 & 1 & 6 \\ 0 & 0 & 0 & 1 & 1 & -1 & 3 \\ 0 & 0 & 0 & 0 & 0 & 0 & 0 \end{pmatrix}.$$

III.1.9 $$B = \begin{pmatrix} 3 & 1 & 0 & 0 \\ 0 & 5 & 1 & 2 \\ 1 & 2 & 0 & 1 \\ 6 & 1 & 1 & 1 \end{pmatrix}.$$

III.1.10 $C = B \cdot A$ where A and B are the matrices of Exercises III.1.8 and III.1.9, respectively.

III.1.11 B^2 and B^3 (that is, $B \cdot B$ and $B \cdot B \cdot B$), where matrix B is given in Exercise III.1.9.

III.1.12 $$D = \begin{pmatrix} 1 & 2 & 3 & 4 \\ 3 & 4 & 6 & 7 \end{pmatrix}.$$

III.1.13 $D \cdot B$ where D is given in Exercise III.1.12 and B appears in Exercise III.1.9.

III.1.14 $D \cdot A$ where D and A are described in Exercises III.1.12 and III.1.8, respectively.

III.1.15 $D \cdot C$ where D and C appear in Exercises III.1.12 and III.1.10, respectively.

III.1.16 In light of the previous exercises, what conjecture can you make as to how *row* rank of an echelon matrix is related to the *column* of the matrix?

III.1.17 Let ech(A) denote a (row or column) echelon matrix derived from the matrix A. From the results of Exercises III.1.10 through III.1.15, can you conjecture how the row-echelon and /or column-echelon ranks of the following echelon matrices compare:

$$\text{ech}(A), \qquad \text{ech}(B), \qquad \text{ech}(A \cdot B)?$$

The following exercises, Exercises III.1.18 through III.1.23, extend the notion of row and column rank from echelon matrices to arbitrary matrices.

Definition(*). Let A be an $m \times n$ matrix with columns $A_1, A_2, \ldots, A_n$ and rows $A^1, A^2, \ldots, A^m$. By the **column space of A** (respectively **row space of A**), we will mean sp$[A_1, A_2, \ldots, A_n]$ (respectively sp$[A^1, A^2, \ldots, A^m]$), the concrete vector-space spanned by the n columns (respectively, m rows) of A (see Definition I.7.3).

The **column rank** of A and the **row rank** of A are defined to be the dimensions of the vector-spaces sp$[A_1, A_2, \ldots, A_n]$ and sp$[A^1, A^2, \ldots, A^m]$, respectively (we shall soon see that echelon rank- = rank).

III.1.18 Suppose that A is a column-echelon (respectively rows echelon) matrix. Prove that the row-echelon rank (respectively column-echelon rank) of Definition III.1.3 for echelon matrices is the same as the row rank (respectively column rank) of Definition (*) above.

III.1.19 Let A be an arbitrary $m \times n$ matrix with columns $A_1, A_2, \ldots, A_n$ and rows $A^1, A^2, \ldots, A^m$. Show that the column space sp$[A_1, A_2, \ldots, A_n]$ (respectively row space sp$[A^1, A^2, \ldots, A^m]$) is unaltered by any of the elementary column (respectively elementary row) operations. That is, show that for the $m \times n$ matrices,

$$A, \qquad C_{ij}(A), \qquad C_i(a)(A), \qquad C_i^j(a)(A), \; a \neq 0,$$

the column spaces are all the same, and for the $m \times n$ matrices

$$A, \qquad R_{ij}(A), \qquad R_i(a)(A), \qquad R_i^j(a)(A), \; a \neq 0,$$

the row spaces are all the same.

III.1.20 (Corollary to Exercise III.1.19.) Let A be an $m \times n$ matrix. Then the column ranks of the following matrices are equal:

$$A, \qquad C_{ij}(A), \qquad C_i(a)(A), \qquad C_i^j(a)(A).$$

Similarly, the row ranks of the following matrices are equal:

$$A, \qquad R_{ij}(A), \qquad R_i(a)(A), \qquad R_i^j(a)(A).$$

Prove the truth of this corollary to Exercise III.1.19.

III.1.21 (Corollary to Exercises III.1.19, III.1.20, and Theorem III.1.4.) Given an $m \times n$ matrix A. Then the column space (respectively row space) of A is the same as the column space (respectively row space) of any column-echelon (respectively row-echelon) matrix obtained from A. Prove this.

2. *Systems of Equations Revisited*

Homogeneous Systems of Equations In Section 1, we combined the mechanics of Gaussian Elimination (elementary operations) with the notion of the dimension of a vector-space (row rank and column rank). We shall now consider the rôle of vector-space dimension relative to systems of equations $A \cdot X = Y_0$. For example, we shall see that the so-called *degrees of freedom* of Chapter I, Section 3, is really the *dimension* of the vector-space of solutions $N = \{X : A \cdot X = O\}$.

Definition III.2.1 The system of linear equations

$$\begin{aligned} a_{11}x_1 + a_{12}x_2 + \cdots + a_{1n}x_n &= y_1 \\ a_{21}x_1 + a_{22}x_2 + \cdots + a_{2n}x_n &= y_2 \\ \vdots \qquad \vdots \qquad\qquad \vdots \quad &\quad \vdots \\ a_{m1}x_1 + a_{m2}x_2 + \cdots + a_{mn}x_n &= y_m, \end{aligned}$$

equivalently denoted $A \cdot X = Y_0$, is called **homogeneous** if and only if $y_1 = y_2 = \cdots = y_m = 0$, that is, if and only if $Y_0 = O$, the zero m-vector in V_m.

We shall now exhibit some qualities of the set of all solutions to a homogeneous system, that is, the set of all $X \in V_n$ such that $A \cdot X = Y_0$. In fact, we shall soon see:

▶ The set of all solution vectors X (where $A \cdot X = Y_0$) is not just a *set*; it is a *vector-space* (a subspace of V_n if A is $m \times n$). Moreover, the dimension of this subspace of solutions is exactly $n -$ (row rank of A), where the row rank of A is the dimension of the subspace of row vectors spanned by the m rows of the $m \times n$ matrix A. (III.2.1)

From (III.2.1), we shall be able to prove the following remarkable fact about matrices (Theorem III.2.5).

▶ For any and every $m \times n$ matrix A, the row rank always equals column rank. (III.2.2)

This last property (III.2.2) was heretofore stated only for the very special echelon matrices (Exercise III.1.18).

The following theorem has appeared as Exercise III.1.19.

Theorem III.2.2 Let A be the $m \times n$ matrix with row vectors A^1, $A^2, \ldots, A^m$ and column vectors $A_1, A_2, \ldots, A_n$. The vector-spaces sp $[A^1,A^2,\ldots,A^m]$ and sp $[A_1,A_2,\ldots,A_n]$ (see Definition I.7.3), will be called the **row space** and the **column space** of A, respectively. We claim that the row spaces of the following 4 matrices are the same:

$$A, \quad R_{ij}(A), \quad R_i(a)(A), \quad R_i{}^j(a)(A), \quad (a \neq 0).$$

Similarly, elementary column operations do not alter the column space of A.

Proof: Let us show that the column space is not affected by the elementary column operations C_j, $C_i(a)$ and $C_i{}^j(a)$, $(a \neq 0)$.

(1) C_{ij} *leaves column space invariant.* This is really easy, since if $A = [A_1A_2 \cdots A_i \cdots A_j \cdots A_n]$, then $C_{ij}(A) = [A_1A_2 \cdots A_j \cdots A_i \cdots A_n]$ (see Definition III.1.2). Thus, the columns of both matrices are identical, only their ordering has been changed. In any case, identical columns span identical column spaces.

(2) $C_i(a)$ *leaves column space invariant.* In this case,

$$A = [A_1A_2 \cdots A_i \cdots A_n]$$

while

$$C_i(a)(A) = [A_1A_2 \cdots aA_i \cdots A_n].$$

We first show sp $[A_1,A_2,\ldots,A_i,\ldots,T_n] \subset$ sp $[A_1,A_2,\ldots,aA_i,\ldots,A_n]$, that is, column space of $A \subset$ column space of $C_i(a)(A)$. We will then show the opposite inclusion to establish equality. Certainly we can agree that every column $A_1, A_2,\ldots, A_i,\ldots, A_n$ is in sp $[A_1,A_2,\ldots,aA_i,\ldots,A_n]$ since each A_i, $i = 1, 2,\ldots, n$ can be written as a linear combination of the column vectors $A_1, A_2,\ldots,aA_i,\ldots, A_n$. Surely

$$A_1 = 1 \cdot A_1 + 0 \cdot A_2 + \cdots + 0 \cdot (aA_i) + \cdots + 0 \cdot A_n,$$

and even

$$\underbrace{A_i = 0 \cdot \overset{\uparrow}{A_1} + 0 \cdot \overset{\uparrow}{A_2} + \cdots + \frac{1}{a}\overset{\uparrow}{(aA_i)} + \cdots + 0 \cdot \overset{\uparrow}{A_n}.}_{\text{generators of column space of } C_i(a)(A)}$$

Now since $\{A_1,A_2,\ldots,A_i,\ldots,A_n\} \subset$ sp $[A_1,A_2,\ldots,aA_i,\ldots,A_n]$ we are assured that

$$\text{sp}\,[A_1,A_2,\ldots,A_i,\ldots,A_n] \subset \text{sp}\,[A_1,A_2,\ldots,aA_i,\ldots,A_n]. \quad \text{(III.2.3)}$$

That is, column space of $A \subset$ column space of $Ci(a)(A)$.

Conversely, each $A_1, A_2, \ldots, aA_i, \ldots, A_n$ (generators of sp $[A_1,A_2,\ldots,aA_i,\ldots,A_n]$) is in sp $[A_1,A_2,\ldots,A_i,\ldots,A_n]$. Surely

$$A_1 = 1 \cdot A_1 + 0 \cdot A_2 + \cdots + 0 \cdot A_i + \cdots + 0 \cdot A_n$$

and

$$\underbrace{aA_i = 0 \cdot \overset{}{A_1} + 0 \cdot A_2 + \cdots + a \cdot A_i + \cdots + 0 \cdot A_n.}_{\text{generators of column space of } A}$$

Since $\{A_1,A_2,\ldots,aA_i,\ldots,A_n\} \subset \text{sp}\,[A_1,A_2,\ldots,A_i,\ldots,A_n]$ we have that

$$\text{sp}\,[A_1,A_2,\ldots,aA_i,\ldots,A_n] \subset \text{sp}\,[A_1,A_2,\ldots,A_i,\ldots,A_n]. \qquad \text{(III.2.4)}$$

With (III.2.3) and (III.2.4), we have that the column spaces of A and $C_i(a)(A)$ are equal.

(3) *$C_i{}^j(a)$ leaves column space invariant.* Here

$$A = [A_1 \cdots A_i \cdots A_j \cdots A_n]$$

and

$$C_i{}^j(a)(A) = [A_1 \cdots A_i \cdots \underset{\uparrow\ j\text{th column of } C_i{}^j(a)(A).}{(aA_i + A_j)} \cdots A_n].$$

Now the columns of both matrices above are exactly the same—except for the jth column. But the jth column of A can be expressed as a linear combination of columns of $C_i{}^j(a)(A)$ as follows:

$$\underbrace{A_j = 0 \cdot A_1 + \cdots + (-a)A_i + \cdots + (1) \cdot (aA_i + A_j) + \cdots + 0 \cdot A_n.}_{\text{generators of column space of } C_i{}^j(a)(A)}$$

Thus,

$$\{A_1,A_2,\ldots,A_i,\ldots,A_j,\ldots,A_n\} \subset \text{sp}\,[A_1,\ldots,A_i,\ldots,aA_i + A_j,\ldots,A_n],$$

so that

$$\text{column space of } A \subset \text{column space of } C_i{}^j(a)(A). \qquad \text{(III.2.5)}$$

Similarly, the jth column of $C_i{}^j(a)(A)$ can be expressed in terms of the columns of A by writing

$$\underbrace{aA_i + A_j = 0 \cdot A_1 + \cdots + a \cdot A_i + \cdots + 1 \cdot A_j + \cdots + 0 \cdot A_n.}_{\text{generators of column space of } A}$$

Thus, $\{A_1,\ldots,A_i,\ldots,aA_i + A_j,\ldots,A_n\} \subset \text{sp}[A_1,A_2,\ldots,A_i,\ldots,A_j,\ldots,A_n]$, so that

$$\text{column space of } C_j{}^i(a)(A) \subset \text{column space of } A. \qquad \text{(III.2.6)}$$

With (III.2.5) and (III.2.6) together, we have the fact that the column spaces of A and $C_i{}^j(a)(A)$ agree.

A similar argument obtains relative to elementary row operations. The proof is done.

REMARK. We come now to a corollary to Theorem III.2.2 which gives us an easy and practical means for finding either the row rank or the column rank of an arbitrary matrix A. In fact, if we but reduce A by means of elementary row operations to row-echelon form, then the row rank of A becomes as obvious as an elephant in a bathtub; it is simply going to be the number of nonzero rows which appear in the row-echelon matrix obtained from A. A similar truism holds for the column rank of the matrix A. Let us develop this now.

Corollary III.2.3 The row rank of a matrix A is the same as the row rank of any row-echelon matrix A' obtained from A via elementary row operations. The row rank of a row-echelon matrix, in turn, is just the number of nonzero rows which appear (that is, the echelon rank). A similar statement relative to column rank also holds true.

Proof: Since the elementary row operations leave the row space of A invariant (Theorem III.2.2), certainly the dimension of the row space is unaltered. Consequently, the row rank of A is the same as the row rank of any row-echelon matrix A' obtained from A (see Theorem III.1.4). Thanks to the peculiar form of row-echelon matrices, the row rank of A' always equals the number of nonzero rows which appear (this is so because the rows are linearly independent and are therefore a *basis* for its row space, which is the same as the row space of A; see Exercise III.2.1). That is, row rank of A = row rank of echelon A' = number of nonzero rows of A'. This ends the proof once we observe that a similar argument applies to the column rank of A.

REMARK. Note that in the proof above, we have established the following: The echelon rank of an *echelon* matrix in Definition III.1.3 is the same as the rank of an *arbitrary* matrix defined prior to Exercise III.1.18 and in Theorem III.2.2.

EXERCISES

III.2.1 Let A be an $m \times n$ matrix which reduces to the row-echelon matrix A' (after applying elementary row operations only). Suppose the row rank of A' is r, that is, the first r rows of A' are nonzero while the remaining rows of

A' are zero. Prove that the nonzero rows of the echelon matrix A' form a *basis* for the row space of A'. (That is, prove that the first r rows of A' *span* the row space and are *linearly independent*.)

III.2.2 Let A be an arbitrary $m \times n$ matrix, which gives rise to the row-echelon matrix A' whose row rank is r. Prove that the r nonzero rows of A' form a basis for the row space of the *original* matrix A. (*Hint:* Use Exercise III.2.1 and Theorem III.2.2.)

In the following exercises, use Exercise III.2.2 to construct a basis for sp $[X_1,X_2,\ldots,X_k]$, where $\{X_1,X_2,\ldots,X_k\}$ is an arbitrary set of concrete n-vectors. That is, construct a matrix whose rows (or columns) are the vectors. $X_1, X_2,\ldots, X_k$. Then reduce this matrix to row (or column) echelon form. Behold the basis in the echelon form! More specifically, in Exercises III.2.3 through III.2.8, you are given a set of vectors. In each case

(a) find the dimension of the vector-space spanned by these vectors, and

(b) from these given vectors, select a basis for the vector space.

III.2.3 $$\begin{pmatrix}2\\0\\0\\1\\2\\1\end{pmatrix},\begin{pmatrix}4\\0\\0\\2\\4\\2\end{pmatrix},\begin{pmatrix}0\\1\\1\\1\\0\\0\end{pmatrix},\begin{pmatrix}0\\1\\0\\1\\0\\0\end{pmatrix},\begin{pmatrix}0\\0\\1\\0\\0\\0\end{pmatrix}.$$

III.2.4 $$\begin{pmatrix}1\\1\end{pmatrix},\begin{pmatrix}1\\-1\end{pmatrix},\begin{pmatrix}3\\0\end{pmatrix},\begin{pmatrix}4\\0\end{pmatrix},\begin{pmatrix}2\\2\end{pmatrix}.$$

III.2.5 $$\begin{pmatrix}2\\0\\1\\0\end{pmatrix},\begin{pmatrix}-2\\0\\-1\\0\end{pmatrix}.$$

III.2.6 $$\begin{pmatrix}1\\0\\0\\0\\0\end{pmatrix},\begin{pmatrix}1\\1\\0\\0\\0\end{pmatrix},\begin{pmatrix}1\\1\\1\\0\\0\end{pmatrix},\begin{pmatrix}0\\0\\1\\0\\0\end{pmatrix}.$$

III.2.7 $$\begin{pmatrix}3\\a\\0\\1\end{pmatrix},\begin{pmatrix}3\\0\\0\\1\end{pmatrix},\begin{pmatrix}3\\0\\0\\0\end{pmatrix}.$$

Does the value of (a) affect dimension?

III.2.8 $$\begin{pmatrix}x_1\\1\\0\\0\end{pmatrix},\begin{pmatrix}0\\x_2\\0\\0\end{pmatrix},\begin{pmatrix}0\\0\\1\\0\end{pmatrix},\begin{pmatrix}0\\1\\0\\0\end{pmatrix}.$$

Do the values of x_1 and x_2 affect dimension?

In Exercises III.2.9 through III.2.12, you are given certain row- and column-echelon matrices. For each row-echelon matrix, reduce it to column-echelon form; for each column echelon matrix, reduce it to row-echelon form. Thus, for every echelon matrix given, compare the row rank and the column rank.

III.2.9
$$\begin{pmatrix} 1 & 0 & 0 & 0 & 0 & 5 \\ 0 & 0 & 1 & 0 & 0 & 5 \\ 0 & 0 & 0 & 0 & 1 & 2 \\ 0 & 0 & 0 & 0 & 0 & 0 \end{pmatrix}.$$

III.2.10
$$\begin{pmatrix} 1 & 0 & 0 & 0 \\ 0 & 1 & 0 & 0 \\ 0 & 0 & 1 & 0 \\ 2 & 1 & 1 & 0 \\ 1 & 3 & 0 & 0 \end{pmatrix}.$$

III.2.11
$$\begin{pmatrix} 1 & 0 & 0 \\ 0 & 1 & 0 \\ 0 & 0 & 0 \end{pmatrix}.$$

III.2.12
$$\begin{pmatrix} 1 & 5 & 1 & 3 & 2 \\ 0 & 0 & 0 & 0 & 0 \\ 0 & 0 & 0 & 0 & 0 \end{pmatrix}.$$

Reduce to row-echelon form.

III.2.13
$$\begin{pmatrix} 1 & 0 & 0 & 0 & 0 & 0 \\ 7 & 1 & 0 & 0 & 0 & 0 \\ 0 & 5 & 0 & 0 & 0 & 0 \\ 3 & 7 & 1 & 0 & 0 & 0 \\ 1 & 3 & 2 & 1 & 0 & 0 \\ 2 & 3 & 2 & 1 & 1 & 0 \end{pmatrix}.$$

Reduce to column-echelon form.

III.2.14
$$\begin{pmatrix} 1 & \mathrm{a} & \mathrm{b} & c \\ 0 & 1 & \mathrm{d} & e \\ 0 & 0 & 1 & f \\ 0 & 0 & 0 & 1 \end{pmatrix}.$$

III.2.15 (See Exercises III.2.9 through III.2.12.) Prove that for row- or column-echelon matrices, the row rank always equals the column rank.

III.2.16 Herewith is presented a fallacious argument which "proves" that for *any* matrix, row rank equals column rank. Find the fallacy.

(a) Given a matrix A, we can always reduce it to a row-echelon matrix A' by means of elementary row operations (Theorem III.1.4).

(b) The row ranks of A and of the echelon matrix A' are the same (Exercise III.1.20).

(c) But the row rank of A' equals the column rank of A'.

(d) But the column rank of A' is the same as the column rank of the original matrix A; hence, row rank of A equals column rank of A, for any A.

Let us return to homogeneous systems of linear equations $A \cdot X = O$ (see Definition III.2.1). We are now ready to establish the vector-space property of the solution set of (column) vectors and its remarkable consequence (III.2.2)—the universal equality of row and column rank. The proof of (III.2.1) rests in Theorem III.2.4 which follows.

Theorem III.2.4 Let $A \cdot X = O_m$ represent a homogeneous system of linear equations, where $A = (a_{ij})_{ij}$ is the $m \times n$ coefficient matrix for the system, O_m is the zero column vector of V_m and $X = \begin{pmatrix} x_1 \\ x_2 \\ \cdot \\ \cdot \\ \cdot \\ x_n \end{pmatrix}$ is an $n \times 1$ matrix (or column vector in V_n).

Suppose the row rank of A is r_{row}. We conclude from these hypotheses that the set of all X for which $A \cdot X = O_m$ forms an $(n - r_{row})$-dimensional subspace of V_n.

Before going into a proof, let us go through an example which illustrates the theorem. The techniques used in our *special* case shall set a pattern for our proof in the *general* case.

Example (three equations, five unknowns).
Consider the homogeneous system

$$\begin{aligned}
&\text{(a)} \quad -1 \cdot x_1 + 2 \cdot x_2 - 1 \cdot x_3 + 1 \cdot x_4 + 2 \cdot x_5 = 0 \\
&\text{(b)} \quad 2 \cdot x_1 - 2 \cdot x_2 + 1 \cdot x_3 + 1 \cdot x_4 + 0 \cdot x_5 = 0 \\
&\text{(c)} \quad 3 \cdot x_1 - 4 \cdot x_2 + 2 \cdot x_3 + 0 \cdot x_4 - 2 \cdot x_5 = 0.
\end{aligned} \tag{III.2.7}$$

Our 3×5 coefficient matrix for the system is

$$A = \begin{pmatrix} -1 & 2 & -1 & 1 & 2 \\ 2 & -2 & 1 & 1 & 0 \\ 3 & -4 & 2 & 0 & -2 \end{pmatrix}.$$

Recall that the elementary row operations performed on our original *system* of linear equations does not alter the solution(s) to that system; that is, in (III.2.7) we may interchange any 2 equations, multiply both sides of any equation by a scalar, or add any multiple of 1 equation to another. The "new" system will enjoy the *same* solutions as does the "original" system of (III.2.7).

The central idea is this: We reduce our system (III.2.7) to row-echelon form. In so doing, we alter the *form* but not the *solutions* that we ultimately seek. There are 2 distinct advantages to this procedure:

(1) We may easily discern the row rank r_{row} of the coefficient matrix A (Corollary III.2.3), and
(2) We almost always have a simpler form with our "new" system when we have a row-echelon form.

So much for the small talk. Let us row-echelonize our system (III.2.7).

It is easy to check that the sequence of row operations $R_2{}^1(1)$, $R_1{}^2(-2)$, $R_1{}^3(-3)$, $R_2{}^3(-2)$, and $R_2(-\frac{1}{2})$, in that order, reduce our coefficient matrix A to the row-echelon matrix

$$A' = \begin{pmatrix} 1 & 0 & 0 & 2 & 2 \\ 0 & 1 & -\frac{1}{2} & \frac{3}{2} & 2 \\ 0 & 0 & 0 & 0 & 0 \end{pmatrix}, \text{ whose row (and column) rank is 2.} \qquad \text{(III.2.8)}$$

The "new" coefficient matrix A' corresponds to the homogeneous system

$$\begin{aligned} &\text{(a)} \qquad x_1 + 0 \cdot x_2 + 0 \cdot x_3 + 2 \cdot x_4 + 2 \cdot x_5 = 0 \\ &\text{(b)} \qquad x_2 + -\tfrac{1}{2} \cdot x_3 + \tfrac{3}{2} \cdot x_4 + 2 \cdot x_5 = 0. \end{aligned} \qquad \text{(III.2.9)}$$

From the appearance of our "new" reduced system (III.2.9) [which still has the same solutions as the original system (III.2.7)], we may select *arbitrary values* for the unknowns x_3, x_4, and x_5; once this is done, the value of x_1 is fixed [see Equation (III.2.9)(a)], and so, too, is the value of x_2 automatically fixed [see Equation (III.2.9)(b)]. Intuitively, at least, we sense that there are exactly 3 "degrees of freedom" in designating our solution vector

$$X = \begin{pmatrix} x_1 \\ x_2 \\ x_3 \\ x_4 \\ x_5 \end{pmatrix},$$

namely, the 3 arbitrary choices we make on x_3, x_4, and x_5 (which thereafter *fix* the values of x_1 and x_2). Intuitively, again, we sense that these 3 degrees of freedom give rise to a subspace of solutions in V_5 of (III.2.9) [and hence of (III.2.7)] whose dimension is 3.

That these semimystical observations which agree with the statement of Theorem III.2.4 are indeed correct, is the content of the following remarks.

Regardless of the values chosen for x_3, x_4, and x_5, Equations (III.2.9) (a) and (b) force us to conclude that

$$x_1 = -2 \cdot x_4 - 2 \cdot x_5$$

and (III.2.10)

$$x_2 = \tfrac{1}{2} \cdot x_3 - \tfrac{3}{2} \cdot x_4 - 2 \cdot x_5.$$

Now (III.2.10) tells us that if we are to present a bonafide solution to systems (III.2.7) and (III.2.9), we must do more than write

$$X = \begin{pmatrix} x_1 \\ x_2 \\ x_3 \\ x_4 \\ x_5 \end{pmatrix},$$

for, in fact, x_1 and x_2 depend upon the values chosen for x_3, x_4, x_5.

An honest statement would be that the vector X is solution to the homogeneous system (III.2.9) [hence, also to (III.2.7)] if and only if

$$X = \begin{pmatrix} -2 \cdot x_4 - 2 \cdot x_5 \leftarrow x_1 \\ \frac{1}{2} \cdot x_3 - \frac{3}{2} \cdot x_4 - 2 \cdot x_5 \leftarrow x_2 \\ x_3 \\ x_4 \\ x_5 \end{pmatrix} \begin{matrix} \Big\} \text{ from (III.2.10)} \\ \\ \\ \end{matrix} \qquad \text{(III.2.11)}$$

where the scalars x_3, x_4, and x_5 are chosen arbitrarily and *without restriction.*

There is a pleasant property of our general solution vector X in (III.2.11). It is easily verified that each of the variable scalars x_3, x_4, and x_5 can be "factored out" in the following way relative to our typical solution vector X:

$$X = \begin{pmatrix} -2 \cdot x_4 - 2 \cdot x_5 \\ \frac{1}{2} \cdot x_3 - \frac{3}{2} \cdot x_4 - 2 \cdot x_5 \\ x_3 \\ x_4 \\ x_5 \end{pmatrix} = x_3 \cdot \underbrace{\begin{pmatrix} 0 \\ \frac{1}{2} \\ 1 \\ 0 \\ 0 \end{pmatrix}}_{X_1} + x_4 \cdot \underbrace{\begin{pmatrix} -2 \\ -\frac{3}{2} \\ 0 \\ 1 \\ 0 \end{pmatrix}}_{X_2} + x_5 \cdot \underbrace{\begin{pmatrix} -2 \\ -2 \\ 0 \\ 0 \\ 1 \end{pmatrix}}_{X_3}. \qquad \text{(III.2.12)}$$

Observe: Equation (III.2.12) tells us that the 3 (constant) vectors X_1, X_2, and X_3 *span* the subspace of all possible solutions to our system (III.2.7) and (III.2.9). Recall that x_3, x_4, and x_5 are each arbitrary, so that our solution space consists of *all* possible linear combinations of the vectors X_1, X_2, and X_3.

On the other hand, one can easily see (Theorem I.6.6) that the set of vectors $\{X_1,X_2,X_3\}$ is *linearly independent*. In fact, it is the nature of just the last 3 entries that makes this so, for *any* 3 vectors of the form

$$\begin{pmatrix} * \\ * \\ 1 \\ 0 \\ 0 \end{pmatrix}, \quad \begin{pmatrix} * \\ * \\ 0 \\ 1 \\ 0 \end{pmatrix}, \quad \begin{pmatrix} * \\ * \\ 0 \\ 0 \\ 1 \end{pmatrix}$$

are linearly independent, regardless of the first and second (asterisks) entries. Having established linear independence and the spanning property for the vectors X_1, X_2, X_3 we see that $\{X_1,X_2,X_3\}$ is a *basis* for the subspace of solutions to (III.2.7).

Our problem is completely solved. We have shown that a vector X is a solution to the homogeneous sytem $A \cdot X = O$ of (III.2.7) if and only if $X \in \text{sp}\,[X_1,X_2,X_3]$ where $\{X_1,X_2,X_3\}$ is a basis for the 3-dimensional space of solutions $\text{sp}\,[X_1,X_2,X_3]$, and the vectors X_1, X_2, and X_3 are defined in (III.2.12).

Indeed, we see from the row-echelon matrix A' for A, that row rank $A = 2$. Our solution space $\text{sp}\,[X_1,X_2,X_3]$ has dimension 3; we shall say, then, that the system has *nullity* 3. In this case, $2 + 3 = 5$, the number of "unknowns" in our homogeneous system. This result is in agreement with (but does not *prove*) our Theorem III.2.4, that matrix rank + nullity = the number of unknowns (see Theorem II.3.6). But the number of unknowns is precisely the dimension of V_5, the domain of the linear operator $\mathbf{T} : X \to A \cdot X$, where $mx(\mathbf{T}) = A$ relative to the standard bases. In terms of linear operator notation, we shall soon see that Theorem III.2.4 is exactly Theorem II.3.6—only the terminology differs.

Proof of Theorem III.2.4: We are given the $m \times n$ coefficient matrix A, whose row rank is r_{row}. We proceed to reduce A to row-echelon form (Theorem III.1.4) by means of elementary row operations. Conceivably, our echelon matrix will assume the form

$$\begin{pmatrix} 0 & 1 & * & * & * & * & * & \cdots & * \\ 0 & 0 & 0 & 1 & * & * & * & \cdots & * \\ 0 & 0 & 0 & 0 & 0 & 1 & * & \cdots & * \\ 0 & 0 & 0 & . & . & . & . & & 0 \\ \vdots & \vdots & \vdots & & & & & & \vdots \\ 0 & 0 & 0 & . & . & . & . & & 0 \end{pmatrix}$$

where the leading 1's of the r_{row} nonzero rows do not all fall on the main diagonal. This is 1 condition we must not tolerate. To rectify this, we appeal to the one elementary *column* operation that leaves the solution space to the homogeneous system intact, namely, C_{ij}, the interchanging of 2 columns. Thus, we may renumber the unknowns $x_1, x_2, \ldots, x_n$ (if necessary) and interchange columns until our row-echelon matrix assumes the special form

$$\underbrace{\begin{pmatrix} 1 & * & * & * & \cdots & * \\ 0 & 1 & * & * & \cdots & * \\ \vdots & & \ddots & & & \vdots \\ 0 & \cdots & 1 & * & \cdots & * \\ \hline 0 & \cdots & \cdots & \cdots & \cdots & 0 \\ \vdots & & & & & \vdots \\ 0 & \cdots & \cdots & \cdots & \cdots & 0 \end{pmatrix}}_{n} \begin{matrix} \left.\vphantom{\begin{matrix}1\\1\\1\\1\end{matrix}}\right\} r_{row} \\ \left.\vphantom{\begin{matrix}1\\1\\1\end{matrix}}\right\} m - r_{row} \end{matrix}$$

where the leading 1's of the first r_{row} nonzero rows fall on the main diagonal. Since the matrix has this form, we may clean it up further through elementary row operations so that there are zeros in each entry above the first r_{row} 1's on the main diagonal. Our final echelon form, then, for our $m \times n$ coefficient matrix A with row rank r_{row} is

$$\underbrace{\begin{pmatrix} 1 & 0 & \cdots & 0 & b_{1r+1} & \cdots & b_{1n} \\ 0 & 1 & \cdots & 0 & b_{2r+1} & \cdots & b_{2n} \\ \vdots & \vdots & \ddots & \vdots & & & \\ 0 & 0 & \cdots & 1 & b_{rr+1} & \cdots & b_{rn} \\ 0 & \cdots & \cdots & \cdots & \cdots & \cdots & 0 \\ \vdots & & & & & & \vdots \\ 0 & \cdots & \cdots & \cdots & \cdots & \cdots & 0 \end{pmatrix}}_{n} \begin{matrix} \left.\vphantom{\begin{matrix}1\\1\\1\\1\end{matrix}}\right\} r = r_{row} \\ \left.\vphantom{\begin{matrix}1\\1\\1\end{matrix}}\right\} m - r \end{matrix} \tag{III.2.13}$$

For convenience, we shall often use r in place of r_{row} to denote the row rank.

Now this coefficient matrix (III.2.13) corresponds to the homogeneous system

$$\begin{aligned} x_1 \qquad\qquad\qquad &+ b_{1r+1}x_{r+1} + \cdots + b_{1n}x_n = 0 \\ x_2 \qquad\qquad &+ b_{2r+1}x_{r+1} + \cdots + b_{2n}x_n = 0 \\ \ddots \qquad & \\ x_r &+ b_{rr+1}x_{r+1} + \cdots + b_{rn}x_n = 0 \end{aligned}$$

[compare (III.2.9)] which has the same solution space as our original system $A \cdot X = O$. We see from (III.2.13) that values for the $n-r$ scalars $x_{r+1}, x_{r+2}, \ldots, x_n$ can be chosen at will, without restriction; once that is done, values for $x_1, x_2, \ldots, x_r$ become fixed since they depend upon the choice of $x_{r+1}, x_{r+2}, \ldots, x_n$. In writing our most general solution vector

$$X = \begin{pmatrix} x_1 \\ \vdots \\ x_r \\ x_{r+1} \\ \vdots \\ x_n \end{pmatrix}$$

and using (III.2.13) to solve for each of the scalars $x_1, x_2, \ldots, x_r$ in terms of the scalars $x_{r+1}, x_{r-2}, \ldots, x_n$, we have a typical solution vector X in the form

$$X = \begin{matrix} x_1 \rightarrow \\ x_2 \rightarrow \\ \\ x_r \rightarrow \\ \\ \\ \\ \\ \end{matrix} \begin{pmatrix} -b_{1r+1}x_{r+1} - b_{1r+2}x_{r+2} - \cdots - b_{1n}x_n \\ -b_{2r+1}x_{r+1} - b_{2r+2}x_{r+2} - \cdots - b_{2n}x_n \\ \vdots \\ -b_{rr+1}x_{r+1} - b_{rr+2}x_{r+2} - \cdots - b_{rn}x_n \\ x_{r+1} \\ x_{r+2} \\ \vdots \\ x_n \end{pmatrix} \begin{matrix} \left.\begin{matrix} \\ \\ \\ \\ \end{matrix}\right\} r \text{ row} \\ \\ \\ \\ \\ \end{matrix}.$$

If we "factor out" the $n-r$ variables $x_{r+1}, x_{r+2}, \ldots, x_n$ [see (III.2.12)] we see that the above equation decomposes into the following sum of $n-r$ vectors:

$$X = x_{r+1} \cdot \underbrace{\begin{pmatrix} -b_{1r+1} \\ -b_{2r+1} \\ -b_{rr+1} \\ \hdashline 1 \\ 0 \\ \vdots \\ 0 \end{pmatrix}}_{X_1} + x_{r+2} \cdot \underbrace{\begin{pmatrix} -b_{1r+2} \\ -b_{2r+2} \\ -b_{rr+2} \\ \hdashline 0 \\ 1 \\ 0 \\ \vdots \\ 0 \end{pmatrix}}_{X_2} + \cdots + x_n \cdot \underbrace{\begin{pmatrix} -b_{1n} \\ -b_{2n} \\ -b_{nn} \\ \hdashline 0 \\ 0 \\ \vdots \\ 0 \\ 1 \end{pmatrix}}_{X_{n-r}}. \tag{III.2.14}$$

Thus, (III.2.14) defines $n-r$ vectors $X_1, X_2, \ldots, X_{n-r}$ which *span* the solution space for the system $A \cdot X = O$. Moreover, these vectors are linearly independent since *any* set of $n-r$ vectors of the form

$$\begin{matrix} r \text{ entries} \\ \\ n-r \text{ entries} \end{matrix} \quad \begin{pmatrix} * \\ \vdots \\ * \\ \hline 1 \\ 0 \\ \vdots \\ 0 \end{pmatrix}, \begin{pmatrix} * \\ \vdots \\ * \\ \hline 0 \\ 1 \\ \vdots \\ 0 \end{pmatrix}, \ldots, \begin{pmatrix} * \\ \vdots \\ * \\ \hline 0 \\ \vdots \\ 0 \\ 1 \end{pmatrix}$$

proves to be LI regardless of the first r entries.

Thus we have proved that if $A \cdot X = O$ represents a homogeneous system of simultaneous linear equations, and if the row rank of the coefficient matrix A is r_{row}, then the subspace of solutions (the null space for) the system is spanned by certain LI n-vectors $X_1, X_2, \ldots, X_{n-r}$, that is, the dimension of the solution space is n (the number of variables) minus r_{row} (the row rank of A).

Let us prove now that row rank and column rank always agree, regardless of the matrix A.

Theorem III.2.5 For any $m \times n$ matrix $A = (a_{ij})_{ij}$, column rank always equals row rank.

Proof: Starting with the arbitrary but fixed matrix A, we construct the homogeneous system $A \cdot X = O$ of equations for which A is the coefficient matrix (we "imbed" A into the situation described in Theorem III.2.4). Hence, we extract from Theorem III.2.4 the following:

(a) $\quad$ row rank of $A = n - \dim(N_A)$,

where N_A is the subspace (of V_n) of all solution vectors $X \in V_n$ such that $A \cdot X = O_m$.

On the other hand, consider the function $\mathbf{T}$ sending (column) vectors of V_n to (column) vectors of V_m defined by

$$\mathbf{T}(X) = A \cdot X \quad \text{for all } X \in V_n.$$

Now $\mathbf{T}$ is not just a *function*, but is a *linear* function, that is, $\mathbf{T} \in \mathscr{L}(V_n, V_m)$. (In fact, A then becomes $A = mx\mathbf{T}_{\mathscr{E}\mathscr{E}'}$, the matrix of $\mathbf{T}$ induced by the standard ordered bases $\mathscr{E}$ and $\mathscr{E}'$ of V_n and V_m, respectively—see Theorem

II.2.3.) Now if the $m \times n$ matrix $A = [A_1 A_2 \cdots A_n]$ with column vectors $A_1, A_2, \ldots, A_n \in V_m$, then for every $n \times 1$ matrix

$$X = \begin{pmatrix} x_1 \\ x_2 \\ \vdots \\ x_n \end{pmatrix} \in V_n,$$

the $m \times 1$ matrix $A \cdot X$ assumes the following revealing form:

$$\underbrace{\begin{pmatrix} a_{11} & a_{12} & \cdots & a_{1n} \\ a_{21} & a_{22} & \cdots & a_{2n} \\ \vdots & \vdots & & \vdots \\ a_{m1} & a_{m2} & \cdots & a_{mn} \end{pmatrix}}_{A_1 \; A_2 \; \cdots \; A_n} \cdot \underbrace{\begin{pmatrix} x_1 \\ x_2 \\ \vdots \\ x_n \end{pmatrix}}_{X} = \underbrace{x_1 \underbrace{\begin{pmatrix} a_{11} \\ a_{21} \\ \vdots \\ a_{m1} \end{pmatrix}}_{A_1} + x_2 \underbrace{\begin{pmatrix} a_{12} \\ a_{22} \\ \vdots \\ a_{m2} \end{pmatrix}}_{A_2} + \cdots + x_n \underbrace{\begin{pmatrix} a_{1n} \\ a_{n2} \\ \vdots \\ a_{mn} \end{pmatrix}}_{A_n}}_{\text{linear combination of the columns } A_1, A_2, \ldots, A_n}. \qquad \text{(III.2.15a)}$$

Behold! From nothing more than the definition of matrix multiplication (compare Definition I.2.3, Theorem II.6.13), we have (III.2.15a) which says that

▶ The $m \times 1$ matrix $A \cdot X$ is always a linear combination of the columns (not the rows!) of the matrix A (that is, $A \cdot X$ is always a vector in $\text{sp}\,[A_1, A_2, \ldots, A_n]$). (III.2.15b)

To say the same thing another way,

The range of $\mathbf{T}$, $\text{rg}\mathbf{T}$, is just $\text{sp}\,[A_1, A_2, \ldots, A_n]$, the column space of A, where $\mathbf{T}(X) = A \cdot X$. Thus,

$$\begin{aligned} \dim(\text{rg}\,\mathbf{T}) &= \dim \text{sp}\,[A_1, A_2, \ldots, A_n] \\ &= \textit{column} \text{ rank of } A. \end{aligned} \qquad \text{(III.2.16)}$$

As usual, dim $(\cdot)$ stands for "the dimension of $(\cdot)$."

Now, Theorem II.3.6 would have us believe

$$\dim(\text{dom}\,\mathbf{T}) = \dim(\text{rg}\,\mathbf{T}) + \dim(\ker \mathbf{T}). \qquad \text{(III.2.17)}$$

But dom $\mathbf{T} = V_n$; hence

(a) $$\dim(\text{dom } \mathbf{T}) = n.$$

From (III.2.16),

(b) $$\dim(\text{rg } \mathbf{T}) = \text{column rank of } A.$$

Finally, ker $\mathbf{T} =$ all $X \in V_n$ for which $\mathbf{T}(X)\ (= A \cdot X) = O$, is exactly the solution space of vectors for the homogeneous system $A \cdot X = O$. Our previous theorem, Theorem III.2.4, says

(c) $$\dim(\ker \mathbf{T}) = n - (\text{row rank of } A).$$

Substituting (a), (b), and (c) into (III.2.17) yields

$$\begin{aligned} \overset{(a)}{\dim(\text{dom } \mathbf{T})} &= \overset{(b)}{\dim(\text{rg } \mathbf{T})} + \overset{(c)}{\dim(\ker \mathbf{T})} \\ n &= (\text{column rank of } A) + (n - (\text{row rank of } A)). \end{aligned} \tag{III.2.18}$$

Thus, column rank of $A =$ row rank of A for all A. The theorem is done.

REMARK. With Theorem III.2.5 in hand, we no longer need to distinguish between row rank and column rank of a matrix. Thus, the single word "rank" will suffice. Accordingly, we present Definition III.2.6.

Definition III.2.6 The **rank** of a matrix A will mean the row (or column) rank of that matrix.

EXERCISES

III.2.17 Prove directly that the set of all n-vectors X satisfying the matrix equation $A \cdot X = O_m$ (A is some fixed $m \times n$ matrix, O_m is the $m \times 1$ matrix whose every entry is zero), forms a vector-space in accordance with Definition I.5.2.

For the homogeneous systems of linear equations in Exercises III.2.18 through III.2.33, find a basis for the vector-space of solutions $\{X : A \cdot X = O_m\}$.

III.2.18 $3 \cdot x_1 = 0.$

III.2.19 $1 \cdot x_1 + 0 \cdot x_2 = 0$

$0 \cdot x_1 + 1 \cdot x_2 = 0.$

III.2.20 $1 \cdot x_1 + 0 \cdot x_2 + 3 \cdot x_3 = 0$

$0 \cdot x_1 + 1 \cdot x_2 - 1 \cdot x_3 = 0.$

III.2.21 $1 \cdot x_1 + 0 \cdot x_2 + 2 \cdot x_3 - 2 \cdot x_4 = 0$

$0 \cdot x_1 + 1 \cdot x_2 + 0 \cdot x_3 - 1 \cdot x_4 = 0.$

III.2.22
$$\begin{aligned} 1\cdot x_1+0\cdot x_2+0\cdot x_3-x_4&=0\\ 0\cdot x_1+1\cdot x_2+0\cdot x_3+x_4&=0\\ 0\cdot x_1+0\cdot x_2+1\cdot x_3+2\cdot x_4&=0. \end{aligned}$$

III.2.23 $1\cdot x_1-2\cdot x_2+x_3-x_4=0.$

III.2.24 $2\cdot x_1-3\cdot x_2+1\cdot x_3-1\cdot x_4-1\cdot x_5=0.$

III.2.25
$$\begin{aligned} 2\cdot x_1-4\cdot x_2+2\cdot x_3-6\cdot x_4+2\cdot x_5&=0\\ 1\cdot x_1+2\cdot x_2+3\cdot x_3-4\cdot x_4+1\cdot x_5&=0. \end{aligned}$$

III.2.26
$$\begin{aligned} 2\cdot x_2+3\cdot x_3-4\cdot x_4&=0\\ 4\cdot x_2+6\cdot x_3+8\cdot x_4&=0. \end{aligned}$$

III.2.27
$$\begin{aligned} 1\cdot x_1+2\cdot x_2-3\cdot x_3-x_4&=0\\ 0\cdot x_1-1\cdot x_2+1\cdot x_3+1\cdot x_4&=0\\ 1\cdot x_1+0\cdot x_2-1\cdot x_3+1\cdot x_4&=0. \end{aligned}$$

III.2.28
$$\begin{aligned} 2\cdot x_1+3\cdot x_2-4\cdot x_3&=0\\ 4\cdot x_1+6\cdot x_2-8\cdot x_3&=0\\ 6\cdot x_1+9\cdot x_2-12\cdot x_3&=0. \end{aligned}$$

III.2.29
$$\begin{aligned} 2\cdot x_1+3\cdot x_2+x_3+x_4+x_5+x_6&=0\\ x_1-x_2+x_3-x_4+x_5+x_6&=0\\ 3x_1+2x_2+2x_3+0\cdot x_4+2x_5+2\cdot x_6&=0\\ x_1+5x_2+3x_3+x_4+3x_5+3x_6&=0\\ x_1+6x_2+2x_3+x_4+2x_5+2x_6&=0. \end{aligned}$$

III.2.30
$$\begin{aligned} x_1+x_2+0\cdot x_3+0\cdot x_4&=0\\ 0\cdot x_1+x_2+x_3+0\cdot x_4&=0\\ 0\cdot x_1+0\cdot x_2+x_3+x_4&=0. \end{aligned}$$

III.2.31
$$\begin{aligned} x_1-x_2&=0\\ x_2+x_2&=0\\ x_1+2x_2&=0\\ 2x_1+x_2&=0. \end{aligned}$$

III.2.32
$$\begin{aligned} 2x_2+3\cdot x_2-x_3&=0\\ x_1-2\cdot x_2+x_3&=0\\ x_1+x_2+x_3&=0\\ x_1-x_2-x_3&=0\\ 2\cdot x_1+2\cdot x_2-x_3&=0. \end{aligned}$$

III.2.33
$$\begin{aligned} 2x_1+3x_2-x_3&=0\\ x_1-2x_2+x_3&=0. \end{aligned}$$

III.2.34 Suppose we have a homogeneous system of linear equations $A\cdot X=O$ where there are more unknowns than there are equations, that is, if A is an $m \times n$ matrix, then $m < n$. Prove that there are always nonzero solution vectors X for such a system. (*Hint:* See Theorems III.2.4 and III.2.5.)

III.2.35 (See Exercise III.2.34.) If we have a homogeneous system of linear equations $A \cdot X = O$ where there are fewer unknowns than there are equations, it is necessarily true that (at least) 1 nonzero solution vector X exists for such a system? If so, prove it; if not, then prove your case by giving an example of a homogeneous system $A \cdot X = O$, where more equations than unknowns appear, and yet $X = O$. is the only *possible* solution.

Nonhomogeneous Systems of Equations We have studied the homogeneous system $A \cdot X = O$ and found that there is an entire vector subspace full of solutions. Shall we now be confronted with an equally vast set of solutions for the *general* system $A \cdot X = Y$, where Y is not necessarily the zero vector? A truly interesting property will emerge (Theorem III.2.7 below) and it is this: For a given $m \times n$ system of equations $A \cdot X = Y$, suppose that we know what all the solutions are to the corresponding homogeneous system $A \cdot X = O$. Suppose, too, that we are lucky enough to find but one X_0 as a solution to the general system, that is, $A \cdot X_0 = Y$. The conclusion is that we would then have *all* possible solutions in hand, because any *other* solution to the general system turns out to be just $X_0 + Z$, where X_0 is our first "lucky" solution to the general system $A \cdot X = Y$, and Z is some solution to the homogeneous system $A \cdot X = O$. We prove this now.

Theorem III.2.7 Let $A \cdot X = Y$ represent a system of m equations and n unknowns. Let X_0 be one solution to this system, that is, $A \cdot X_0 = Y$. If X_1 is *any* other solution to this system, then $X_1 = X_0 + Z$ where Z is a solution to the corresponding homogeneous system of equations, that is, $A \cdot Z = O$.

Proof: Let X_0 and X_1 be 2 solutions to the general system of equations. Using our matrix multiplication notation (see Definition III.1.1), this means that

(a) $$A \cdot X_0 = Y$$

and

(b) $$A \cdot X_1 = Y.$$

If we subtract Equation (b) from Equation (a), we obtain

(c) $$A \cdot X_1 - A \cdot X_0 = O.$$

Since matrix multiplication is distributive (see Exercises II.6.83 and II.6.84), (c) becomes

(d) $$A \cdot (X_1 - X_0) = O.$$

That is, the vector $X_1 - X_0$, which we denote by Z, is a solution to the corresponding *homogeneous* system $A \cdot X = O$. Finally,

$$X_1 - X_0 = Z,$$

or

$$X_1 = X_0 + Z$$

which proves Theorem III.2.7.

Theorem III.2.7 tells us that, in a certain sense, once we have found *1* solution to a general system of equations, we have found them *all*, assuming that the solutions to the corresponding homogeneous system are all known. The theorem does not tell us how to *find* the general solution. However, the procedure for doing so is not very different from the row-echelon method used in the example for the homogeneous case, and we illustrate it now.

Example

(a)
$$\begin{aligned} x_1 - x_2 \quad + x_3 \quad + x_4 &= 2 \\ x_1 + x_2 \quad + 2x_3 \quad + x_4 &= 3 \\ 2 \cdot x_1 + 0 \cdot x_2 + 3 \cdot x_3 + 2 \cdot x_4 &= 5. \end{aligned}$$

In this case, we extract the 3×5 augmented matrix of this system

$$\left(\begin{array}{cccc:c} \multicolumn{4}{c}{\overbrace{\hphantom{1 \quad -1 \quad 1 \quad 1}}^{3 \times 4 \text{ coefficient matrix } A.}} \\ 1 & -1 & 1 & 1 & 2 \\ 1 & 1 & 2 & 1 & 3 \\ 2 & 0 & 3 & 2 & 5 \end{array}\right)$$

and by means of the 5 elementary row operations $R_1{}^2(-1)$, $R_1{}^3(-2)$, $R_2{}^3(-1)$, $R_2(\frac{1}{2})$ and $R_1{}^2(1)$, we reduce this matrix to the row-echelon form

(b)
$$\left(\begin{array}{cccc:c} 1 & 0 & \frac{3}{2} & 1 & \frac{5}{2} \\ 0 & 1 & \frac{1}{2} & 0 & \frac{1}{2} \\ 0 & 0 & 0 & 0 & 0 \end{array}\right).$$

This row-echelon matrix corresponds to the system (c) [which shares the same solutions as our original system (a)];

(c)
$$\begin{aligned} x_1 \qquad + \tfrac{3}{2}x_3 + x_4 &= \tfrac{5}{2} \\ x_2 + \tfrac{1}{2}x_3 \qquad &= \tfrac{1}{2}. \end{aligned}$$

Now let us characterize the vector-space of solutions to the corresponding *homogeneous* system as we did before. Thus, we consider the system

$$\begin{aligned} x_1 \qquad + \tfrac{3}{2}\, x_3 + x_4 &= 0 \\ x_2 + \tfrac{1}{2}\, x_3 \qquad &= 0 \end{aligned}$$

which tells us that values for x_3 and x_4 may be chosen arbitrarily; each choice of scalars for x_3 and x_4 fix values for x_1 and x_2. In fact, the above equations tell us that, indeed

$$x_1 = -\tfrac{3}{2}\, x_3 - x_4$$

and

$$x_2 = -\tfrac{1}{2}\, x_3.$$

Thus, a 4-vector

$$X = \begin{pmatrix} x_1 \\ x_2 \\ x_3 \\ x_4 \end{pmatrix}$$

is a solution to the *homogeneous* system $A \cdot X = O$ if and only if

$$X = \begin{pmatrix} -\tfrac{3}{2}\, x_3 - x_4 \\ -\tfrac{1}{2}\, x_3 \\ x_3 \\ x_4 \end{pmatrix} \quad \text{were } x_3 \text{ and } x_4 \text{ are arbitrary.}$$

In other words, X is a solution to the corresponding *homogeneous* system if and only if

$$X = x_3 \cdot \underbrace{\begin{pmatrix} -\tfrac{3}{2} \\ -\tfrac{1}{2} \\ 1 \\ 0 \end{pmatrix}}_{X_1} + x_4 \cdot \underbrace{\begin{pmatrix} -1 \\ 0 \\ 0 \\ 1 \end{pmatrix}}_{X_2}$$

which finally defines a basis $\{X_1, X_2\}$ (there are *many* bases) for the vector-space of solutions to the homogeneous system.

But what have we done? We have "thrown away" the column

$$Y = \begin{pmatrix} 2 \\ 3 \\ 5 \end{pmatrix}$$

in our original system (a) $A \cdot X = Y$, and replaced it with the zero column.

But Theorem III.2.7 tells us that we have not wasted our time, for all we need to do now is find *1* solution vector X_0 for our general system (a), and we may generate *all* solutions from this one by adding all possible solutions of the *homogeneous* system to it.

So let us find *1* solution to (a), that is, 1 solution to the (equivalent) system (c). This is easy to find in (c) for all we need to do is set $x_3 = x_4 = 0$;

this tells us that $x_1 = \frac{5}{2}$ and $x_2 = \frac{1}{2}$. If you don't believe this, simply take this vector

$$X_0 = \begin{pmatrix} \frac{5}{2} \\ \frac{1}{2} \\ 0 \\ 0 \end{pmatrix}$$

and *try* it out in (a) [or in (c)]. You will see that, indeed,

$$\underbrace{\begin{pmatrix} 1 & -1 & 1 & 1 \\ 1 & 1 & 2 & 1 \\ 2 & 0 & 3 & 2 \end{pmatrix}}_{A} \cdot \underbrace{\begin{pmatrix} \frac{5}{2} \\ \frac{1}{2} \\ 0 \\ 0 \end{pmatrix}}_{X_0} = \underbrace{\begin{pmatrix} 2 \\ 3 \\ 5 \end{pmatrix}}_{Y}$$

which is in agreement with (a).

Thus, we have found a *particular* solution X_0 for the system (a) [by reducing (a) to the simpler equivalent system (c)] and we have found a basis $\{X_1, X_2\}$ for the vector-space of solutions to the corresponding homogeneous system. Theorem III.2.7 tells us that *any* solution to the system (a) of linear equations is of the form

$$X = \underbrace{\begin{pmatrix} -\frac{5}{2} \\ \frac{1}{2} \\ 0 \\ 0 \end{pmatrix}}_{X_0} + a \cdot \underbrace{\begin{pmatrix} -\frac{3}{2} \\ -\frac{1}{2} \\ 1 \\ 0 \end{pmatrix}}_{X_1} + b \cdot \underbrace{\begin{pmatrix} -1 \\ 0 \\ 0 \\ 1 \end{pmatrix}}_{X_2}$$

where a and b are arbitrary scalars. This can be checked by substituting the X above into (a).

REMARK. If the basis vectors X_1 and X_2 above are unappealing to you, you may alter them without altering their span. For example, multiplying X_1 by -2 would eliminate the negative fraction entries and we could just as easily cite the vectors

$$\begin{pmatrix} 3 \\ 1 \\ -2 \\ 0 \end{pmatrix} \quad \begin{pmatrix} -1 \\ 1 \\ 0 \\ 1 \end{pmatrix}$$

as a basis for the vector-space of solutions to the homogeneous system.

REMARK. If you observe the 3×5 echelon matrix (b) for the original system of the previous example, you will see that if the last column were eliminated, then you would have a 3×4 row-echelon matrix for the corresponding *homogeneous* system. Note that in both cases, only 2 nonzero rows appear, that is, the rank of the 3×4 coefficient matrix A and the rank of the 3×5 augmented matrix agree, and this is precisely the condition which is necessary and sufficient to guarantee the existence of solutions for the system (Exercise III.2.37).

This last remark would suggest that certain systems $A \cdot X = Y$ have *no* solution. Can this be? If $Y \neq O$, then this certainly can occur as evidenced in the following example.

Example

$$x_1 + x_2 = 2$$
$$x_1 + x_2 = 3$$

$$\underbrace{\begin{pmatrix} 1 & 1 \\ 1 & 1 \end{pmatrix}}_{A} \cdot \underbrace{\begin{pmatrix} x_1 \\ x_2 \end{pmatrix}}_{X} = \underbrace{\begin{pmatrix} 2 \\ 3 \end{pmatrix}}_{Y}.$$

Although solutions are always obtainable for the corresponding *homogeneous* system $A \cdot X = O$ (since $X = O$ *always* works), *no* solution is obtainable for the *general* system given above. This is clear intuitively, at least in this example, since the scalar $x_1 + x_2$ cannot equal both 2 *and* 3 as our system demands. But to look at this more formally, let us reduce the 2×3 augmented matrix $A : Y$ to row-echelon form, bearing in mind that the first 2 columns comprise the row-echelon form for the 2×2 coefficient matrix A. This is important; since at least one solution vector X exists for the system $A \cdot X = Y$ if and only if the ranks of the two matrices A and $[A : Y]$ agree (Exercise III.2.37)!

$$A : Y = \left(\begin{array}{cc|c} 1 & 1 & 2 \\ 1 & 1 & 3 \end{array}\right)$$
$$\downarrow {}_1R_1(-1)$$
$$\underbrace{\left(\begin{array}{cc|c} 1 & 1 & 2 \\ 0 & 0 & 1 \end{array}\right)}_{\text{row-echelon matrix for } A}.$$

Clearly the rank of A is 1, while the rank of the augmented matrix $[A : Y]$ is 2; *no* solution for the system $A \cdot X = Y$ can then exist.

So we see that rank has everything to do with existence or nonexistence of solutions to systems of linear equations. Moreover, the row-echelon reduction techniques, when applied to certain coefficient matrices, give us a means for actually finding solutions (assuming they exist). We contend that our original problem (see beginning of this chapter) has been sufficiently analyzed so that the reader may grapple with systems of simultaneous linear equations to the point of solution.

We present a theorem which describes circumstances under which a system $A \cdot X = Y$ has at *most* 1 (that is, unique) solution vector X.

As we have mentioned, conditions for which $A \cdot X = Y$ has *at least* 1 solution can be found in Exercise III.2.37.

Proposition III.2.8 Let $A \cdot X = Y$ represent a system of m linear equations in n unknowns (that is, A is an $m \times n$ matrix). Assume that a solution X_0 exists for this system. Then X_0 is the *only* solution for the system if and only if rank $(A) = n$, the number of unknowns for the system.

Moreover, if rank (A) equals n, then of necessity, $m \geq n$, so that there cannot be fewer equations than unknowns if a *unique* solution for the system $A \cdot X = Y$ does exist.

Proof: We note that we are not going to prove the *existence* of a solution X_0 for the system, for existence is one of our hypotheses. The last example shows us that not every system has a solution. We concern ourselves only with *uniqueness*.

Now Theorem III.2.7 tells us that any 2 solutions X_0 and X_1 say, must differ by a vector Z, which is a solution to the *homogeneous* system $A \cdot X = O$. That is to say, if X_0 and X_1 are *any* 2 solutions, then $X_0 - X_1 = Z$ where $A \cdot Z = O$. But then any 2 solutions X_0 and X_1 will always be equal if and only if $X_1 - X_0 = O$, or if and only if $Z = O$ is the *only* solution to the homogeneous system $A \cdot X = O$. Now Theorem III.2.4 tells us when just this happy state of affairs occurs. We see that O is the *only* solution for the homogeneous system if and only if n, the number of unknowns equals r, the rank of A. Thus, we have shown

▶ X_0 is a unique solution for $A \cdot X = Y \Leftrightarrow O$ is the only solution for $A \cdot X = O$ $\Leftrightarrow$ rank $(A) = n$,

which proves the first part of Proposition III.2.8.

In proving the "moreover" phrase of Proposition III.2.8 just notice that we are now considering an $m \times n$ matrix A whose rank is n. Row rank equals column rank (Corollary III.2.5), so that we must agree that the row rank equals n, that is, the dimension of the subspace sp $[{}^1A, {}^1A, \ldots, {}^mA]$ spanned by the m rows of A, equals n. Thus, *at least* n row vectors A^i are needed to achieve such a dimension. That is, $m \geq n$ and Proposition III.2.8 is proved.

This final example illustrates an application of the theory of simultaneous linear equations to linear operators (see Exercises III.2.49 through III.2.53).

Example Consider the basis

$$\left\{ \begin{pmatrix} 2 \\ 1 \\ 0 \end{pmatrix}, \begin{pmatrix} 0 \\ 1 \\ 1 \end{pmatrix}, \begin{pmatrix} 3 \\ 1 \\ 0 \end{pmatrix} \right\}$$

for V_3. Suppose we are given a linear operator $\mathbf{T} \in \mathscr{L}(V_3, V_2)$ whose values on those basis vectors are defined by the equations

$$\mathbf{T}\begin{pmatrix}2\\1\\0\end{pmatrix} = \begin{pmatrix}0\\1\end{pmatrix}, \quad \mathbf{T}\begin{pmatrix}0\\1\\1\end{pmatrix} = \begin{pmatrix}2\\2\end{pmatrix}, \quad \mathbf{T}\begin{pmatrix}3\\1\\0\end{pmatrix} = \begin{pmatrix}4\\1\end{pmatrix}. \tag{III.2.19}$$

(a) Given the typical vector

$$Y = \begin{pmatrix}y_1\\y_2\\y_3\end{pmatrix}$$

we shall first attempt to find the unique scalars α_1, α_2, and α_3 such that

$$\alpha_1 \cdot \underbrace{\begin{pmatrix}2\\1\\0\end{pmatrix}}_{X_1} + \alpha_2 \cdot \underbrace{\begin{pmatrix}0\\1\\1\end{pmatrix}}_{X_2} + \alpha_3 \cdot \underbrace{\begin{pmatrix}3\\1\\0\end{pmatrix}}_{X_3} = \begin{pmatrix}y_1\\y_2\\y_3\end{pmatrix} = Y. \tag{III.2.20}$$

The statement (III.2.20) can be recast into the following form:

$$\begin{aligned} 2 \cdot \alpha_1 + 0 \cdot \alpha_2 + 3 \cdot \alpha_3 &= y_1 \\ 1 \cdot \alpha_1 + 1 \cdot \alpha_2 + 1 \cdot \alpha_3 &= y_2 \\ 0 \cdot \alpha_1 + 1 \cdot \alpha_2 + 0 \cdot \alpha_3 &= y_3 \end{aligned} \tag{III.2.21}$$

which is simply a system of 3 equations in the 3 unknowns α_1, α_2, and α_3.

Let us proceed, then, to solve for the scalars, α_1, α_2, and α_3 by reducing the augmented matrix

$$\left(\begin{array}{ccc|c} 2 & 0 & 3 & y_1 \\ 1 & 1 & 1 & y_2 \\ 0 & 1 & 0 & y_3 \end{array}\right)$$

to the row-echelon matrix [try R_{23}, R_{31}, $R_1{}^3(-2)$, $R_2{}^3(2)$, $R_2{}^1(-1)$, and $R_3{}^1(-1)$]

$$\left(\begin{array}{ccc|c} 1 & 0 & 0 & -y_1 + 3y_2 - 3y_3 \\ 0 & 1 & 0 & y_3 \\ 0 & 0 & 1 & y_1 - 2y_2 + 2y_3 \end{array}\right)$$

which corresponds to the equivalent system (that is, a system with the same set of solutions)

$$\begin{aligned} \alpha_1 \qquad\qquad &= -y_1 + 3y_2 - 3y_3 \\ \alpha_2 \qquad &= \quad y_3 \\ \alpha_3 &= \quad y_1 - 2y_2 + 2y_3. \end{aligned} \tag{III.2.22}$$

The equations of (III.2.22) give us the unique values for α_1, α_2, and α_3 (in terms of y_1, y_2, and y_3) and so we have our solution for (III.2.20), and hence, for (III.2.21).

(b) Using linearity of **T**, find $\mathbf{T}(Y)$ for a typical vector $Y \in V_3$.
We use part (a). Since for any and every $Y \in V_3$ we may write

$$Y = \begin{pmatrix} y_1 \\ y_2 \\ y_3 \end{pmatrix} \quad \underbrace{(-y + 3y_2 - 3y_3)}_{\alpha_1} \cdot \underbrace{\begin{pmatrix} 2 \\ 1 \\ 0 \end{pmatrix}}_{X_1} + \underbrace{y_3}_{\alpha_2} \cdot \underbrace{\begin{pmatrix} 0 \\ 1 \\ 1 \end{pmatrix}}_{X_2} + \underbrace{(y_1 - 2y_2 + 2y_3)}_{\alpha_3} \cdot \underbrace{\begin{pmatrix} 3 \\ 1 \\ 0 \end{pmatrix}}_{X_3}$$

from which it follows that

$$\mathbf{T}(Y) = (-y_1 + 3y_2 - 3y_3)\mathbf{T}(X_1) + y_3(\mathbf{T}(X_2) + (y_1 - 2y_2 + 2y_3)\mathbf{T}(X_3). \tag{III.2.23}$$

But we know the vectors $\mathbf{T}(X_1)$, $\mathbf{T}(X_2)$, and $\mathbf{T}(X_3)$ from our defining Equations (III.2.19) for **T**. Substituting these values into (III.2.23) yields

$$\mathbf{T}\begin{pmatrix} y_1 \\ y_2 \\ y_3 \end{pmatrix} = (-y_1 + 3y_2 - 3y_3) \cdot \begin{pmatrix} 0 \\ 1 \end{pmatrix} + y_3 \cdot \begin{pmatrix} 2 \\ 2 \end{pmatrix} + (y_1 - 2y_2 + 2y_3) \cdot \begin{pmatrix} 4 \\ 1 \end{pmatrix}$$

which simplifies to

$$\mathbf{T}\begin{pmatrix} y_1 \\ y_2 \\ y_3 \end{pmatrix} = \begin{pmatrix} 4y_1 - 8y_2 + 10y_3 \\ y_2 + y_3 \end{pmatrix} \tag{III.2.24}$$

for all

$$Y = \begin{pmatrix} y_1 \\ y_2 \\ y_3 \end{pmatrix} \in V_3.$$

Fortunately, we may easily check whether our computations leading to (III.2.24) are error-free. We simply compute $\mathbf{T}(Y)$ according to (III.2.24) when Y assumes the values of X_1, X_2, and X_3, respectively. The results must either be in perfect agreement with (III.2.19) or else an unfortunate error has infested our computations.

(c) Find $mx(\mathbf{T})_{\mathscr{Z}\mathscr{Y}}$, the matrix of **T** relative to the ordered bases $\mathscr{Z} = \{Z_1, Z_2, Z_3\}$ and the standard ordered basis $\mathscr{Y} = \left\{\begin{pmatrix} 1 \\ 0 \end{pmatrix}, \begin{pmatrix} 0 \\ 1 \end{pmatrix}\right\}$, where

$$Z_1 = \begin{pmatrix} 1 \\ 0 \\ 0 \end{pmatrix}, \; Z_2 = \begin{pmatrix} 1 \\ 1 \\ 0 \end{pmatrix}, \text{ and } Z_3 = \begin{pmatrix} 1 \\ 1 \\ 1 \end{pmatrix}.$$

This part depends upon part (b). In order to find $mx(\mathbf{T})_{\mathscr{Z}\mathscr{Y}}$, all we need is to know $\mathbf{T}(Z_1)$, $\mathbf{T}(Z_2)$, and $\mathbf{T}(Z_3)$ for columns 1, 2, and 3, respectively, of

$mx\mathbf{T}$ (Definition II.6.2). Now (III.2.24) tells us that

$$\mathbf{T}(Z_1) = \mathbf{T}\begin{pmatrix}1\\0\\0\end{pmatrix} = \begin{pmatrix}4\\0\end{pmatrix} = 4 \cdot E_1 + 0 \cdot E_2$$

$$\mathbf{T}(Z_2) = \mathbf{T}\begin{pmatrix}1\\1\\0\end{pmatrix} = \begin{pmatrix}-4\\1\end{pmatrix} = -4 \cdot E_1 + 1 \cdot E_2 \qquad \text{(III.2.25)}$$

$$\mathbf{T}(Z_3) = \mathbf{T}\begin{pmatrix}1\\1\\1\end{pmatrix} = \begin{pmatrix}6\\2\end{pmatrix} = 6 \cdot E_1 + 2 \cdot E_2.$$

Using (III.2.25) and Definition II.6.2, we see that

$$mx(\mathbf{T})_{\mathscr{X}\mathscr{Y}} = \begin{pmatrix}4 & -4 & 6\\0 & 1 & 2\end{pmatrix}.$$

EXERCISES

III.2.36 Prove that the set of all solution vectors, provided that such vectors exist for the system $A \cdot X = Y$, where $Y \neq O$, is *not* a vector-space.

III.2.37 Where Proposition III.2.8 speaks of *uniqueness* of solutions to the system $A \cdot X = Y$, we now discuss *existence*. Prove that a necessary and sufficient condition for (at least) 1 solution vector X to exist for the system $A \cdot X = Y$, is that rank A = rank $[A : Y]$, where $[A : Y]$ the augmented matrix of the system (see Definition III.1.1). [*Hint:* Reduce the augmented matrix (via elementary row operations only) to row-echelon form. Pose the same question in this "simplified" form.]

In Exercises III.2.38 through III.2.47 find the solutions (if any exist) for the given system of simultaneous linear equations.

III.2.38 $2x_1 + 3x_2 - x_3 = 6.$

How can you tell at a glance that the solution exists (see Exercise III.2.37) and that it is not unique (see Proposition III.2.8)?

III.2.39
$$\begin{aligned} 2x_1 + 3x_2 - x_3 &= 6\\ x_1 + x_2 + x_3 &= 2\\ x_1 + 4x_2 - 2x_3 &= 3. \end{aligned}$$

III.2.40
$$\begin{aligned} 2x_1 + 3x_2 - x_3 &= 0\\ x_1 - x_2 + x_3 &= 0\\ x_1 + 4x_2 - 2x_3 &= 0. \end{aligned}$$

III.2.41
$$\begin{aligned} x_1 + x_2 &= 0\\ x_1 - x_2 &= 1\\ 2x_1 &= 1\\ 3x_1 - x_2 &= 2. \end{aligned}$$

III.2.42
$$\begin{aligned} x_1 + x_2 &= 0 \\ x_1 - x_2 &= 1 \\ 2x_1 \qquad &= 0 \\ 3x_1 - x_2 &= 2. \end{aligned}$$

III.2.43
$$\begin{aligned} 3x_1 - 2x_2 + x_3 + x_4 &= 0 \\ x_1 - x_2 + x_3 + x_4 &= 0. \end{aligned}$$

Give a 1-sentence (that is, brief) argument to show, without computation, that the solution for the above system exists, but is not unique.

III.2.44
$$\begin{aligned} 3x_1 - 2x_2 + x_3 + x_4 &= 0 \\ x_1 - x_2 + x_3 + x_4 &= 0 \\ x_2 + x_2 + 0 \cdot x_3 - x_4 &= 1. \end{aligned}$$

III.2.45
$$\begin{aligned} 5x + 6y - z + w + 2u &= 3 \\ 4x + y + 0 \cdot z + 0 \cdot w - u &= 2. \end{aligned}$$

III.2.46 Is it generally true that if a nonhomogeneous system has more unknowns than it has equations, then at least 1 solution always exists? Consider the system

$$\begin{aligned} x + y + z + w &= 1 \\ x - y + 0 + 0 &= 0 \\ 3x + y + 2z + 2w &= 1. \end{aligned}$$

(Also see Exercise III.2.44.)

III.2.47
$$\begin{aligned} x_1 - 2x_2 + x_3 \qquad - x_5 &= 2 \\ x_1 + x_2 \qquad + x_4 \qquad &= 1 \\ x_2 - x_3 + x_4 \qquad &= 0 \\ 2x_1 + x_3 \qquad + x_4 + x_5 &= 3. \end{aligned}$$

III.2.48 Given the vectors

$$X = \begin{pmatrix} a_{11} \\ a_{21} \\ \vdots \\ a_{m1} \end{pmatrix}, X_2 = \begin{pmatrix} a_{12} \\ a_{22} \\ \vdots \\ a_{m2} \end{pmatrix}, \ldots, X_i = \begin{pmatrix} a_{1i} \\ a_{2i} \\ \vdots \\ a_{mi} \end{pmatrix}, \ldots, X_n = \begin{pmatrix} a_{1n} \\ a_{2n} \\ \vdots \\ a_{mn} \end{pmatrix}.$$

Prove that the m-vectors $X_1, X_2, \ldots, X_n$ are LI if and only if the $m \times n$ matrix whose ith column is X_i, has rank n.

As a corollary to this result, prove that any $m+1$ separate vectors $X_1, X_2, \ldots, X_{m+1}$ in V_m are necessarily LD.

In the following exercises, a basis $\mathscr{X} = \{X_1, X_2, \ldots, X_n\}$ is given for the vector-space V_n. Only the vectors $\mathbf{T}(X_1), \mathbf{T}(X_2), \ldots, \mathbf{T}(X_n)$ are given for a *linear* transformation $\mathbf{T} \in \mathscr{L}(V_n, V_m)$. Thus, the values of $\mathbf{T}(X)$ can be found for *any* $X \in V_n$ since $\mathbf{T}$ is linear (see Exercise II.3.1). You are asked to do the following:

(a) Given an arbitrary vector

$$Y = \begin{pmatrix} y_1 \\ y_2 \\ \cdot \\ \cdot \\ \cdot \\ y_n \end{pmatrix}$$

in V_n, find the (unique) scalars $\alpha_1, \alpha_2, \ldots, \alpha_n$ (which shall each depend upon the scalars $y_1, y_2, \ldots, y_n$) such that, relative to the given bases vectors $X_1, X_2, \ldots, X_n$

$$\alpha_1 \cdot X_1 + \alpha_2 \cdot X_2 + \cdots + \alpha_n X_n = Y.$$

(b) Using linearity of $\mathbf{T}$, find $\mathbf{T}(Y)$ for the typical vector Y in V_n.
(c) Relative to the second given ordered basis $\mathscr{Z} = \{Z_1, Z_2, \ldots, Z_n\}$ for V_n, and the standard ordered basis $\mathscr{Y} = \{E_1, E_2, \ldots, E_m\}$ for V_m (which is an impact set for $\mathbf{T}$), find $mx(\mathbf{T})_{\mathscr{Z}\mathscr{Y}}$.

III.2.49 $\mathscr{X} = \left\{\begin{pmatrix}1\\0\end{pmatrix}, \begin{pmatrix}1\\1\end{pmatrix}\right\}$ and $\mathbf{T}\begin{pmatrix}1\\0\end{pmatrix} = \begin{pmatrix}2\\2\end{pmatrix}, \quad \mathbf{T}\begin{pmatrix}1\\1\end{pmatrix} = \begin{pmatrix}0\\3\end{pmatrix}.$

For part (c), take $\mathscr{Z} = \left\{\begin{pmatrix}2\\1\end{pmatrix}, \begin{pmatrix}0\\1\end{pmatrix}\right\}.$

III.2.50 $\mathscr{X} = \left\{\begin{pmatrix}1\\0\end{pmatrix}, \begin{pmatrix}1\\1\end{pmatrix}\right\}$ and $\mathbf{T}\begin{pmatrix}1\\0\end{pmatrix} = \begin{pmatrix}1\\2\\0\\1\end{pmatrix}, \quad \mathbf{T}\begin{pmatrix}1\\1\end{pmatrix} = \begin{pmatrix}1\\0\\2\\1\end{pmatrix}.$

For part (c), take $\mathscr{Z} = \left\{\begin{pmatrix}0\\1\end{pmatrix}, \begin{pmatrix}1\\2\end{pmatrix}\right\}.$

III.2.51 $\mathscr{X} = \left\{\begin{pmatrix}0\\1\\0\end{pmatrix}, \begin{pmatrix}1\\1\\0\end{pmatrix}, \begin{pmatrix}2\\1\\1\end{pmatrix}\right\}$ and

$$\mathbf{T}\begin{pmatrix}0\\1\\0\end{pmatrix} = \begin{pmatrix}1\\1\end{pmatrix}, \quad \mathbf{T}\begin{pmatrix}1\\1\\0\end{pmatrix} = \begin{pmatrix}1\\1\end{pmatrix}, \quad \mathbf{T}\begin{pmatrix}2\\1\\1\end{pmatrix} = \begin{pmatrix}0\\0\end{pmatrix}.$$

For part (c), take

$$\mathscr{Z} = \left\{\begin{pmatrix}0\\0\\1\end{pmatrix}, \begin{pmatrix}0\\2\\2\end{pmatrix}, \begin{pmatrix}3\\0\\0\end{pmatrix}\right\}.$$

III.2.52 $\mathscr{X} = \left\{\begin{pmatrix}1\\0\\0\\0\end{pmatrix}, \begin{pmatrix}0\\2\\0\\1\end{pmatrix}, \begin{pmatrix}0\\0\\1\\0\end{pmatrix}, \begin{pmatrix}0\\0\\0\\3\end{pmatrix}\right\}$

and

$$\mathbf{T}\begin{pmatrix}1\\0\\0\\0\end{pmatrix}=\begin{pmatrix}1\\2\\1\end{pmatrix},\quad \mathbf{T}\begin{pmatrix}0\\2\\0\\1\end{pmatrix}=\begin{pmatrix}-1\\0\\0\end{pmatrix},\quad \mathbf{T}\begin{pmatrix}0\\0\\1\\0\end{pmatrix}=\begin{pmatrix}0\\0\\1\end{pmatrix}$$

and $\mathbf{T}\begin{pmatrix}0\\0\\0\\3\end{pmatrix}=\begin{pmatrix}1\\0\\0\end{pmatrix}$.

For part (c), take

$$\mathscr{X}=\left\{\begin{pmatrix}0\\0\\1\\1\end{pmatrix},\begin{pmatrix}0\\1\\0\\1\end{pmatrix},\begin{pmatrix}1\\0\\0\\1\end{pmatrix},\begin{pmatrix}0\\3\\1\\2\end{pmatrix}\right\}.$$

3. Elementary Matrices and Invertibility

We come upon a very simple, but important, observation, namely, that in order to effect any elementary row operation on an $m \times n$ matrix A, all we need to do is multiply A on the left by a certain unique $m \times m$ matrix which does not even depend on A. Similarly, any elementary column operation on A is effected by a right-hand multiplication of A by a certain unique $n \times n$ matrix which is independent of A. What we are saying then, is that elementary row (respectively, column) operations are in reality certain left hand (respectively, right hand) multiplication operators; this fact, in conjunction with Exercises II.6.88 and II.6.89, establishes that the elementary operations (which send $m \times n$ matrices to $m \times n$ matrices) are all *linear* operators from $\mathscr{M}_{m,n}$ to $\mathscr{M}_{m,n}$.

Let us suppose, for example, that there *is* such a magical $n \times n$ matrix C with the following properties: For any $m \times n$ (m is unrestricted) matrix A, $A \cdot C$ looks just like A except that the first 2 columns are interchanged (for this reason, let us suppose $n \geq 2$). In other words, we presume the existence of an $n \times n$ matrix C, such that for any and all $m \times n$ matrices A, we have

$$A \cdot C = C_{12}(A). \tag{III.3.1}$$

Truly a remarkable feat is (III.3.1); if only we could find this $n \times n$ matrix C. Now *if* such a C does exist, then certainly it must impose its column-switching effects on *all* $m \times n$ (do not forget m is unrestricted) matrices, including I_n, the $n \times n$ identity matrix. Thus, by choosing $A = I_n$ in (III.3.1), we obtain

$$C = C_{12}(I_n) \tag{III.3.2}$$

which nicely isolates the $n \times n$ matrix C for us. From the definition of C_{12}, we see that the right-hand side of (III.3.2) is simply the identity matrix after its first 2 columns are switched. Thus, we see that

$$C = \begin{pmatrix} 0 & 1 & 0 & 0 & \cdots & 0 \\ 1 & 0 & 0 & 0 & \cdots & 0 \\ 0 & 0 & 1 & 0 & \cdots & 0 \\ 0 & 0 & 0 & 1 & \cdots & 0 \\ \vdots & \vdots & \vdots & \vdots & \ddots & \vdots \\ \vdots & \vdots & \vdots & \vdots & & 0 \\ 0 & 0 & 0 & 0 & \cdots & 1 \end{pmatrix}. \tag{III.3.3}$$

If you were to check the performance of C in (III.3.3) for the special case $n = 3$, say, you would see that, indeed,

$$\underbrace{\begin{pmatrix} a_{11} & a_{12} & a_{13} \\ a_{21} & a_{22} & a_{23} \\ \vdots & \vdots & \vdots \\ a_{k1} & a_{k2} & a_{k3} \\ \vdots & \vdots & \vdots \\ a_{m1} & a_{m2} & a_{m3} \end{pmatrix}}_{A} \cdot \underbrace{\begin{pmatrix} 0 & 1 & 0 \\ 1 & 0 & 0 \\ 0 & 0 & 1 \end{pmatrix}}_{C} = \underbrace{\begin{pmatrix} a_{12} & a_{11} & a_{13} \\ a_{22} & a_{21} & a_{23} \\ \vdots & \vdots & \vdots \\ a_{k2} & a_{k1} & a_{k3} \\ \vdots & \vdots & \vdots \\ a_{m2} & a_{m1} & a_{m3} \end{pmatrix}}_{A\text{, with first 2 columns interchanged.}}. \tag{III.3.4}$$

What we have seen, then, is that *if* such a column-switching matrix C exists at all, then it must look like (III.3.3); (III.3.4) only *verifies* its promised performance relative to arbitrary $m \times 3$ matrices, at least. So as not to get embroiled in endless detail for each of the elementary row and column operations, we shall allow this foregoing discussion to represent a heuristic "proof" by example of the following theorem:

Theorem III.3.1 For each of the elementary row and column operations R_{ij}, $R_i(a)$ $(a \neq 0)$, $R_j{}^i(a)$, C_{ij}, $C_i(a)$ $(a \neq 0)$, $C_j{}^i(a)$ which acts on an $m \times n$ matrix A, there exists a square matrix, independent of A, such that

(a) $R_{ij} \cdot A = R_{ij}(A)$ (d) $A \cdot C_{ij} = C_{ij}(A)$
(b) $R_i(a) \cdot A = R_i(a)(A)$ (e) $A \cdot C_i(a) = C_i(a)(A)$
(c) $R_j{}^i(a) \cdot A = R_{ji}(a)(A)$ (f) $A \cdot C_{ji}(a) = C_{ji}(a)(A)$. (III.3.5)

Note that the left-hand side of each of the 6 equations in (III.3.5) represents the product of 2 *matrices*; by abuse of notation, we are using the same symbol for the square matrix as for the elementary row or column operation that it effects.

Moreover, each of the square matrices may be obtained by effecting the corresponding elementary row (respectively, column) operation on the identity matrix I_m (respectively, I_n). The row rank and column rank of these matrices is the maximum possible, that is, column rank $=$ row rank $= m$ for the $m \times m$ matrices R_{ij}, $R_j(a)$ $(a \neq 0)$, $R_j{}^i(a)$ and column rank $=$ row rank $= n$ for the $n \times n$ matrices C_{ij}, $C_i(a)$ $(a \neq 0)$, $C_{ji}(a)$.

Proof: We shall only list the matrices and leave the following verifications for the student:

(a) For any $m \times n$ matrix A, (III.3.5) obtains.

(b) The row rank $=$ column rank $=$ the maximum possible for each of these square matrices.

Accordingly, by applying the elementary operations to the appropriate identity matrix we see that

(a)

$$R_{ij} = \begin{array}{r} \\ \\ \text{—}i\rightarrow \\ \\ \text{—}j\rightarrow \\ \\ \\ \end{array} \begin{pmatrix} 1 & & & & \\ & \ddots & & & \\ & & 0 \cdots 1 & & \\ & & \vdots \ \ddots \ \vdots & & \\ & & 1 \cdots 0 & & \\ & & & \ddots & \\ & & & & 1 \end{pmatrix}$$

all main diagonal entries are 1 except for the *ii*th and *jj*th entries; all off-diagonal entries are zero except for the *ij*th and *ji*th entries.

R_{ij} is $m \times m$. Column Rank $=$ Row Rank $= m$.

(b)

$$a \neq 0 \quad R_i(a) = \begin{array}{r} \\ \\ \\ \text{—}i\rightarrow \\ \\ \\ \\ \end{array} \begin{pmatrix} 1 & & & & & & \\ & \ddots & & & & \bigcirc & \\ & & 1 & & & & \\ & & & a & & & \\ & & & & 1 & & \\ & \bigcirc & & & & \ddots & \\ & & & & & & 1 \end{pmatrix}$$

all off-diagonal entries are zero.

$R_i(a)$ is $m \times m$. Column Rank $=$ Row Rank $= m$. (III.3.6)

(c)

$$R_j{}^i(a) = \begin{array}{r} \\ \\ \text{—}i\rightarrow \\ \\ \text{—}j\rightarrow \\ \\ \\ \end{array}\begin{pmatrix} 1 & & & & & & \\ & \ddots & & & & & \\ & & 1 & \cdots & a & & \\ & & \vdots & \ddots & \vdots & & \\ & & & & 1 & & \\ & & & & & \ddots & \\ & & & & & & 1 \end{pmatrix}$$

all main diagonal entries are 1; all off-diagonal entries are zero except for ijth entry.

$R_j{}^i(a)$ is $m \times m$. Column Rank $=$ Row Rank $= m$.

(d)

$$C_{ij} = \begin{array}{c} \begin{array}{ccccccccccc} & & & \overset{i}{\downarrow} & & & & \overset{j}{\downarrow} & & & \end{array} \\ \begin{pmatrix} 1 & & & & & & & & & & \\ & \ddots & & & & & & & & & \\ & & 1 & & & & & & & & \\ & & & 0 & & \cdots & & 1 & & & \\ & & & & 1 & & & & & & \\ & & & \vdots & & \ddots & & \vdots & & & \\ & & & & & & 1 & & & & \\ & & & 1 & & \cdots & & 0 & & & \\ & & & & & & & & 1 & & \\ & & & & & & & & & \ddots & \\ & & & & & & & & & & 1 \end{pmatrix} \end{array}$$

all main diagonal entries are 1 except for the iith and jjth entries; all off-diagonal entries are zero except for the ijth and ijth entries.

C_{ij} is $n \times n$. Column Rank $=$ Row Rank $= n$.

(e)

$$a \neq 0,\ C_i(a) = \begin{pmatrix} 1 & & & & & & & & \\ & \ddots & & & & & & & \\ & & 1 & & & & & & \\ & & & \ddots & & & & & \\ & & & & \overset{\overset{i}{\downarrow}}{a} & & & & \\ & & & & & \ddots & & & \\ & & & & & & 1 & & \\ & & & & & & & \ddots & \\ & & & & & & & & 1 \end{pmatrix}$$

all off-diagonal entries are zero.

$C_i(a)$ is $n \times n$. Column Rank $=$ Row Rank $= n$.

(f)

$$C_j{}^i(a) = \begin{pmatrix} 1 & & \overset{\overset{i}{\downarrow}}{} & & \overset{\overset{j}{\downarrow}}{} & & \\ & \ddots & & & & & \\ & & 1 & & & & \\ & & \vdots & \ddots & & & \\ & & a & \cdots & 1 & & \\ & & & & & \ddots & \\ & & & & & & 1 \end{pmatrix}$$

all main diagonal entries are 1; all off-diagonal entries are zero except for the jith entry.

$C_j{}^i(a)$ is $n \times n$. Column Rank $=$ Row Rank $= n$.

The matrices of (III.3.6) will prove to be more than a curiosity; for this reason, we highlight their existence through Definition III.3.2.

Definition III.3.2 The square matrices described in (III.3.6) are called **elementary matrices**.

The following example illustrates how we may actually find a square matrix R such that if A is an $m \times n$ matrix and A' is a row-echelon reduction of A, then $R \cdot A = A$ (see Theorems III.1.4 and III.3.1).

Example For A, let us choose the 5×6 matrix

$$A = \begin{pmatrix} 2 & 2 & 0 & 4 & 0 & 4 \\ 0 & 2 & 4 & 4 & 0 & 2 \\ 1 & 1 & 6 & 1 & 0 & 0 \\ 0 & 0 & 0 & 0 & 0 & 0 \\ -1 & 1 & -2 & 3 & 0 & 2 \end{pmatrix}.$$

We apply the following elementary row operations to A: $R_1(\frac{1}{2})$, $R_1{}^3(-1)$, $R_1{}^5(1)$, $R_2(\frac{1}{2})$, $R_2{}^5(-2)$, $R_3{}^5(1)$ and finally $R_3(\frac{1}{6})$; we thus obtain the row-echelon matrix

$$A' = \begin{pmatrix} 1 & 1 & 0 & 2 & 0 & 2 \\ 0 & 1 & 2 & 2 & 0 & 1 \\ 0 & 0 & 1 & -\frac{1}{6} & 0 & -\frac{1}{3} \\ 0 & 0 & 0 & 0 & 0 & 0 \\ 0 & 0 & 0 & 0 & 0 & 0 \end{pmatrix}.$$

We are to find a 5×5 matrix R such that $R \cdot A = A'$. Now Theorem III.3.1 tells us that each of the 7 required row operations above may be effected by a left-hand multiplication by some 5×5 elementary matrix R_i, that is,

(a) $$R_7 \cdots R_2 \cdot R_1 \cdot A = A'$$

where each R_i denotes 1 of the particular 7 elementary row operations used on A to produce A'. We see from (a) that we are looking for the matrix

$$\text{(b)} \qquad R = R_7 \cdots R_2 \cdot R_1$$

which is the desired left-hand multiplier on A. We could, of course, write down each of the 7 matrices of (b) and multiply them together, but there is a much more merciful way. Recall (proof of Theorem III.3.1) that in order to find our first elementary matrix R_1 (which corresponds to $R_1(\frac{1}{2})$ in our example), we simply perform that very row operation on I_5, the 5×5 identity matrix. There is no future in finding the matrix indicing $R_1{}^3(-1)$, our second elementary row operation; what is really needed is the *product*, $R_1{}^3(-1) \cdot R_1(\frac{1}{2})$. But this can easily be obtained by effecting the operation $R_1{}^3(\frac{1}{2})$ on our *first* matrix $R_1(\frac{1}{2})$. Similarly (by avoiding the pitfall of finding the *individual* matrices for the row operations), we compute the *product* $R_1{}^5(1) \cdot R_1{}^3(-1) \cdot R_1(\frac{1}{2})$ by simply applying the row operation $R_1{}^5(1)$ to the matrix (which we now have from the previous step) $R_1{}^3(-1) \cdot R_1(\frac{1}{2})$. We continue to successively "build up" our product until our last (seventh, in our case) elementary row operation is consumed. Thus, we have the following rule.

▶ ***Rule:*** To find the matrix R (respectively C) such that $R \cdot A = A'$ (respectively, $A \cdot C = A'$), a row (respectively column) echelon matrix for A, apply the same elementary row (respectively column) operations in sequence to the identity matrix as are applied to the original matrix A.

We illustrate this rule by writing 2 columns of matrices. In the left-hand column we apply our sequence of row operations to A; in the right-hand column we apply the same operations to I_5. The final result in the left-hand column will be A', the row-echelon matrix for A; the final result in the right-hand column will be R in (b).

$$A = \begin{pmatrix} 2 & 2 & 0 & 4 & 0 & 4 \\ 0 & 2 & 4 & 4 & 0 & 2 \\ 1 & 1 & 6 & 1 & 0 & 0 \\ 0 & 0 & 0 & 0 & 0 & 0 \\ -1 & 1 & -2 & 3 & 0 & 2 \end{pmatrix} \qquad I_5 = \begin{pmatrix} 1 & 0 & 0 & 0 & 0 \\ 0 & 1 & 0 & 0 & 0 \\ 0 & 0 & 1 & 0 & 0 \\ 0 & 0 & 0 & 1 & 0 \\ 0 & 0 & 0 & 0 & 1 \end{pmatrix}$$

$$\downarrow \boxed{R_1(\tfrac{1}{2})} \downarrow$$

$$\begin{pmatrix} 1 & 1 & 0 & 2 & 0 & 2 \\ 0 & 2 & 4 & 4 & 0 & 2 \\ 1 & 1 & 6 & 1 & 0 & 0 \\ 0 & 0 & 0 & 0 & 0 & 0 \\ -1 & 1 & -2 & 3 & 0 & 2 \end{pmatrix} \qquad \begin{pmatrix} \frac{1}{2} & 0 & 0 & 0 & 0 \\ 0 & 1 & 0 & 0 & 0 \\ 0 & 0 & 1 & 0 & 0 \\ 0 & 0 & 0 & 1 & 0 \\ 0 & 0 & 0 & 0 & 1 \end{pmatrix}$$

$R_1{}^3(-1)$ and $R_1{}^5(1)$

$$\begin{pmatrix} 1 & 1 & 0 & 2 & 0 & 2 \\ 0 & 2 & 4 & 4 & 0 & 2 \\ 0 & 0 & 6 & -1 & 0 & -2 \\ 0 & 0 & 0 & 0 & 0 & 0 \\ 0 & 2 & -2 & 5 & 0 & 4 \end{pmatrix} \qquad \begin{pmatrix} \frac{1}{2} & 0 & 0 & 0 & 0 \\ 0 & 1 & 0 & 0 & 0 \\ -\frac{1}{2} & 0 & 1 & 0 & 0 \\ 0 & 0 & 0 & 1 & 0 \\ \frac{1}{2} & 0 & 0 & 0 & 1 \end{pmatrix}$$

$R_2(\frac{1}{2})$

$$\begin{pmatrix} 1 & 1 & 0 & 2 & 0 & 2 \\ 0 & 1 & 2 & 2 & 0 & 1 \\ 0 & 0 & 6 & -1 & 0 & -2 \\ 0 & 0 & 0 & 0 & 0 & 0 \\ 0 & 2 & -2 & 5 & 0 & 4 \end{pmatrix} \qquad \begin{pmatrix} \frac{1}{2} & 0 & 0 & 0 & 0 \\ 0 & \frac{1}{2} & 0 & 0 & 0 \\ -\frac{1}{2} & 0 & 1 & 0 & 0 \\ 0 & 0 & 0 & 1 & 0 \\ \frac{1}{2} & 0 & 0 & 0 & 1 \end{pmatrix}$$

$R_2{}^5(-2)$ and $R_3{}^5(1)$

$$\begin{pmatrix} 1 & 1 & 0 & 2 & 0 & 2 \\ 0 & 1 & 2 & 2 & 0 & 1 \\ 0 & 0 & 6 & -1 & 0 & -2 \\ 0 & 0 & 0 & 0 & 0 & 0 \\ 0 & 0 & 0 & 0 & 0 & 0 \end{pmatrix} \qquad \begin{pmatrix} \frac{1}{2} & 0 & 0 & 0 & 0 \\ 0 & \frac{1}{2} & 0 & 0 & 0 \\ -\frac{1}{2} & 0 & 1 & 0 & 0 \\ 0 & 0 & 0 & 1 & 0 \\ 0 & -1 & 1 & 0 & 1 \end{pmatrix}$$

$R_3(\frac{1}{6})$

$$A' = \begin{pmatrix} 1 & 1 & 0 & 2 & 0 & 2 \\ 0 & 1 & 2 & 2 & 0 & 1 \\ 0 & 0 & 1 & -\frac{1}{6} & 0 & -\frac{1}{3} \\ 0 & 0 & 0 & 0 & 0 & 0 \\ 0 & 0 & 0 & 0 & 0 & 0 \end{pmatrix} \qquad R = \begin{pmatrix} \frac{1}{2} & 0 & 0 & 0 & 0 \\ 0 & \frac{1}{2} & 0 & 0 & 0 \\ -\frac{1}{12} & 0 & \frac{1}{6} & 0 & 0 \\ 0 & 0 & 0 & 1 & 0 \\ 0 & -1 & 1 & 0 & 1 \end{pmatrix}.$$

The student may verify by matrix multiplication that the R we have constructed, without resorting to matrix multiplication of any kind, is the R for which (a) and (b) obtain.

EXERCISES

In Exercises III.3.1 through III.3.5, rectangular matrices are given and certain matrix products are indicated. Compute the ranks of these matrices along with the ranks of the products.

III.3.1 $A = \begin{pmatrix} 1 & 0 & 0 \\ 0 & 0 & 0 \end{pmatrix} \quad B = \begin{pmatrix} 0 & 0 & 0 \\ 0 & 1 & 0 \\ 0 & 0 & 0 \end{pmatrix}$

$A \cdot B, \quad A \cdot B \cdot B, \quad B^2, \quad B^3, \quad B^n, n = 1, 2, 3, \ldots$

III.3.2 $A = \begin{pmatrix} 1 & 0 & 0 \\ 0 & 0 & 0 \end{pmatrix} \quad B = \begin{pmatrix} 1 & 0 \\ 0 & 0 \\ 0 & 0 \end{pmatrix}$

$A \cdot B, \quad A \cdot B \cdot A, \quad A \cdot B \cdot A \cdot B, \quad B \cdot A, \quad B \cdot A \cdot B.$

III.3.3 $A = \begin{pmatrix} 1 & 2 \\ 0 & 1 \end{pmatrix} \quad B = \begin{pmatrix} 1 & 5 & 6 \\ 1 & 5 & 6 \end{pmatrix}$

$A \cdot B, \quad A^2 \cdot B, \quad A^3 \cdot B.$

III.3.4 $A = \begin{pmatrix} 0 & 1 & 0 & 5 \\ 0 & 0 & 1 & 2 \\ 0 & 0 & 0 & 1 \\ 0 & 0 & 0 & 0 \end{pmatrix}, \quad A^2, \quad A^3, \quad A^4, \quad A^5, \quad A^n.$

III.3.5 $A = \begin{pmatrix} 0 & 0 & 0 & 0 \\ 1 & 0 & 0 & 0 \\ a & 1 & 0 & 0 \\ b & c & 1 & 0 \end{pmatrix}, \quad A, \quad A^2, \quad A^3, \quad A^4, \quad A^5, \quad A^n.$

In Exercises III.3.6 through III.3.11, an $m \times n$ matrix A is given along with a matrix B which is obtained from A by applying some elementary row or column operation. In each case, find the elementary matrix X such that

$X \cdot A = B$ for an elementary row operation or

$A \cdot X = B$ for an elementary column operation.

III.3.6 $A = \begin{pmatrix} 1 & 5 & 7 & -9 & 2 \\ 8 & 0 & 1 & 3 & 0 \\ 5 & 6 & 1 & 4 & 0 \end{pmatrix} \quad B = \begin{pmatrix} 1 & 5 & 12 & 9 & 2 \\ 8 & 0 & 1 & 3 & 0 \\ 5 & 6 & 7 & 4 & 0 \end{pmatrix}.$

III.3.7 $A = \begin{pmatrix} 1 & 0 & 0 \\ 2 & 1 & 0 \\ 3 & 2 & 1 \\ 4 & 0 & 0 \end{pmatrix} \quad B = \begin{pmatrix} 1 & 0 & 2 \\ 2 & 1 & 4 \\ 3 & 2 & 7 \\ 4 & 0 & 8 \end{pmatrix}.$

III.3.8 $A = \begin{pmatrix} 1 & 1 & 1 & 1 \\ 0 & 0 & 1 & 1 \\ 0 & 0 & 0 & 1 \end{pmatrix} \quad B = \begin{pmatrix} 1 & 1 & 0 & 0 \\ 0 & 0 & 1 & 1 \\ 0 & 0 & 0 & 1 \end{pmatrix}.$

III.3.9 $A = \begin{pmatrix} 0 & 2 & 4 & 10 \\ 0 & 1 & 2 & 1 \\ 1 & 0 & 3 & 2 \\ 1 & 3 & 0 & 2 \end{pmatrix} \qquad B = \begin{pmatrix} 0 & 2 & 4 & 10 \\ 0 & 1 & 2 & 1 \\ 1 & 1 & 5 & 7 \\ 1 & 3 & 0 & 2 \end{pmatrix}.$

III.3.10 $A = \begin{pmatrix} 1 & 0 & 0 & 0 \\ 0 & 5 & 0 & 0 \\ 0 & 0 & 1 & 0 \\ 0 & 0 & 0 & 1 \end{pmatrix} \qquad B = \begin{pmatrix} 0 & 0 & 1 & 0 \\ 0 & 5 & 0 & 0 \\ 1 & 0 & 0 & 0 \\ 0 & 0 & 0 & 1 \end{pmatrix}.$

III.3.11 $A = \begin{pmatrix} 0 & 2 & 4 & 10 \\ 0 & 11 & 2 & 1 \\ 1 & 0 & 4 & 2 \\ 1 & 3 & 0 & 2 \end{pmatrix} \qquad B = \begin{pmatrix} 0 & 2 & 2 & 10 \\ 0 & 1 & 1 & 1 \\ 1 & 0 & 2 & 2 \\ 1 & 3 & 0 & 2 \end{pmatrix}.$

In the following exercises, a rectangular matrix A is given. In each case reduce A first to row-echelon from A' and find the square matrix R such that $R \cdot A = A'$. Then reduce A to column-echelon from A'' and find the square matrix C such that $A \cdot C = A''$.

III.3.12 $A = \begin{pmatrix} 0 & 1 & 0 & 2 & 0 \\ 0 & 0 & 1 & 0 & 0 \\ 0 & 0 & 0 & 0 & 1 \end{pmatrix}.$

III.3.13 $A = \begin{pmatrix} 1 & 0 & 0 & 0 & 0 \\ 0 & 1 & 0 & 0 & 0 \\ 0 & 0 & 0 & 2 & 0 \\ 0 & 0 & 1 & 0 & 1 \end{pmatrix}.$

III.3.14 $A = (1 \quad 2 \quad 1 \quad 3).$

III.3.15 $A = \begin{pmatrix} 1 & a \\ 0 & 1 \end{pmatrix}.$

III.3.16 $A = \begin{pmatrix} 3 & 0 & 6 \\ 2 & 0 & 0 \\ 1 & 0 & 0 \end{pmatrix}.$

III.3.17 $A = \begin{pmatrix} 2 & 3 \\ 2 & 3 \\ 1 & 0 \\ 0 & 1 \end{pmatrix}.$

III.3.18 $A = \begin{pmatrix} 0 & 0 & 3 & 1 \\ 2 & 1 & 1 & 1 \\ 0 & 1 & 1 & 1 \\ 1 & 0 & 0 & 2 \\ 2 & 0 & 1 & 0 \\ 3 & 0 & 0 & 0 \end{pmatrix}.$

III.3.19 Prove that elementary-column (respectively, row) operations [which leave column (respectively, row) rank fixed, by Corollary III.2.3] always leave row (respectively, column) rank fixed. (*Hint:* See Theorem III.2.5.)

III.3.20 Note that Definition III.3.2 in no way distinguished between elementary matrices inducing row operations and those inducing column operations (except to indicate that left multiplication resulted in a row operation and right multiplication yielded a column operation). Show that, indeed, there is "essentially" no difference, that is, show that

(a) $R_{ij} = C_{ij}$

(b) $R_i(a) = C_i(a)$

(c) $R_i{}^j(a) = C_j{}^i(a)$,

where the symbols in (a), (b), and (c) represent the $n \times n$ elementary matrices.

Applications of Elementary Operations to Invertibility of Matrices

The elementary row and column operations, which proved useful for computing row and column ranks of $m \times n$ matrices, will soon provide us with a means of answering the following question relative to square $n \times n$ matrices:

As usual, let

$$I_n = \begin{pmatrix} 1 & 0 & \cdots & 0 \\ 0 & 1 & \ddots & \vdots \\ \vdots & & \ddots\, 1\, \ddots & \vdots \\ \vdots & & & \cdot\, 0 \\ 0 & \cdots\cdots 0 & & 1 \end{pmatrix}$$

be the $n \times n$ (multiplicative) identity matrix, that is, $I_n = (\delta_{ij})_{ij}$ where $\delta_{ij} = 0$ if $i \neq j$ and $\delta_{ij} = 1$ if $i = j$. It is easy to see that I_n has the "feeble" multiplicative property that for any and every $n \times n$ matrix A, $I_n \cdot A = A \cdot I_n = A$ (see Theorem II.6.13). We would wish to know, then, which $n \times n$ matrices A (if any) possess an $n \times n$ *multiplicative* inverse, which we shall denote by A^{-1}; the $n \times n$ matrix A^{-1} would be defined by the property that when it is multiplied by the original matrix A (on the left *or* on the right), we are brought back to the identity matrix I_n. Thus, we come to Definition III.3.3.

Definition III.3.3 For any $n \times n$ matrix A, the multiplicative inverse, A^{-1}, is exactly that $n \times n$ matrix with the property that

$$A^{-1} \cdot A = A \cdot A^{-1} = I_n.$$

Matrices for which such inverses exist are called **invertible,** or **nonsingular.** All other matrices are called **noninvertible,** or **singular.**

Not only would we wish to know which $n \times n$ matrices A possess an inverse, but also whether there is a reasonable means of actually computing these inverse matrices A^{-1}.

REMARKS. It is not a bad idea to observe the parallel properties of the additive identity (usually denoted by 0; see axiom A.3 of Definition I.5.2) and the multiplicative identity (matrix) I_n described above. Each identity is "feeble" in the sense that relative to its own operation, that is, addition or multiplication, it leaves every *other* element intact and unaltered. In each case, the identity leads to the definition of an *inverse* for certain elements of the system; these are the elements that "return us" to the identity element. Thus, a' is an *additive* inverse for a if $a' + a = 0$, the additive identity, and a'' is a *multiplicative* inverse for a if $a'' \cdot a = 1$, the multiplicative identity. Of course, a' and a'' are usually denoted as $-a$ and a^{-1}, respectively. As we have seen before in our discussion of additive identities and inverses, these identities and inverses, if they exist at all, are *necessarily unique*. Thus, we speak of *the* additive or multiplicative identity and not of *an* additive or multiplicative identity. Similarly, we speak of *the* inverse of an element and not of *an* inverse of an element. Amazingly, this uniqueness of identity and inverse is not a characteristic, then, of $n \times n$ matrices per se. It is a consequence *only* of the fact that the operation defined between pairs of elements is *closed*, and *associative*. See axioms A.1 and A.5 of Definition I.5.2 (see also Exercises III.3.26 and III.3.27).

Example Let $A = \begin{pmatrix} 1 & 2 \\ 0 & 3 \end{pmatrix}$. We claim that the matrix $\begin{pmatrix} 1 & -\frac{2}{3} \\ 0 & \frac{1}{3} \end{pmatrix}$ is A^{-1}, *the* multiplicative inverse of A. One may quickly check that

$$\begin{pmatrix} 1 & -\frac{2}{3} \\ 0 & \frac{1}{3} \end{pmatrix} \cdot \begin{pmatrix} 1 & 2 \\ 0 & 3 \end{pmatrix} = \begin{pmatrix} 1 & 2 \\ 0 & 3 \end{pmatrix} \cdot \begin{pmatrix} 1 & -\frac{2}{3} \\ 1 & \frac{1}{3} \end{pmatrix} = \begin{pmatrix} 1 & 0 \\ 0 & 1 \end{pmatrix} = I_2.$$

Thus, $\begin{pmatrix} 1 & -\frac{2}{3} \\ 0 & \frac{1}{3} \end{pmatrix}$ meets the "test" of Definition III.3.3.

Example Let $B = \begin{pmatrix} 1 & 0 \\ 0 & 0 \end{pmatrix}$. We claim that B has *no* multiplicative inverse, for if we put $B' = \begin{pmatrix} b_{11} & b_{12} \\ b_{21} & b_{22} \end{pmatrix}$ and try to force the equation $B \cdot B' = \begin{pmatrix} 1 & 0 \\ 0 & 1 \end{pmatrix}$, then we are asking that

$$\underbrace{\begin{pmatrix} 1 & 0 \\ 0 & 0 \end{pmatrix} \cdot \begin{pmatrix} b_{11} & b_{12} \\ b_{21} & b_{22} \end{pmatrix}} = \underbrace{\begin{pmatrix} 1 & 0 \\ 0 & 1 \end{pmatrix}}, \text{ or that}$$

$$\begin{pmatrix} b_{11} & b_{12} \\ 0 & 0 \end{pmatrix} = \begin{pmatrix} 1 & 0 \\ 0 & 1 \end{pmatrix}.$$

Clearly this is not possible. No matter how cleverly we choose the scalars b_{11}, b_{12}, b_{21}, and b_{22}, the lower-right-hand corner entries of the last 2 matrices will *never* agree. Some matrices have multiplicative inverses, then, and some do not. All matrices A, however, possess *additive* inverses, usually denoted $-A$.

EXERCISES

III.3.21 If $A = \begin{pmatrix} a_{11} & a_{12} \\ 0 & 0 \end{pmatrix}$ and $B = \begin{pmatrix} 0 & a_{12} \\ 0 & a_{22} \end{pmatrix}$, then show that A and B do *not* have multiplicative inverses, that is, A^{-1} and B^{-1} do not exist as per Definition III.3.3.

III.3.22 If $A = \begin{pmatrix} a_{11} & k \cdot a_{11} \\ a_{21} & k \cdot a_{21} \end{pmatrix}$ and $B = \begin{pmatrix} a_{11} & a_{12} \\ ka_{11} & ka_{12} \end{pmatrix}$, then prove that multiplicative inverses for A and B do not exist.

III.3.23 Exercise III.3.21 is a special case of the following results which you are asked to prove:

If any column or any row of an $n \times n$ matrix A, consists entirely of zero entries, then A does not possess a multiplicative inverse. *Hint:* If the ith row (column) of A is the zero row (column), then show that for any typical $n \times n$ matrix $B = (b_{ij})_{ij}$, the iith entry, that is, the ith main diagonal entry, of $A \cdot B$ (of $B \cdot A$) is always zero. Thus, for no B is it possible that $A \cdot B = I_n$ since the iith entry of I_n is always 1.

III.3.24 Exercise III.3.22 is a special case of the following result which you are asked to prove: If any column (respectively, any row) of an $n \times n$ matrix A is a scalar multiple of any other column (respectively row) of A, then A does not possess a multiplicative inverse. *Hint:* If the jth row (column) of A is k times the ith row (column) of A then show that for any $n \times n$ matrix B, the jth row (column) of the product $A \cdot B$ (of $B \cdot A$) is *also* just k times the ith row (column) of $A \cdot B$ (of $B \cdot A$). Thus $A \cdot B$ ($B \cdot A$) could not possibly equal I_n, since no row or column of I_n is a scalar multiple of any other row or column of I_n.

III.3.25 Verify that if $k = ad - bc \neq 0$, then

$$\begin{pmatrix} a & b \\ c & d \end{pmatrix}^{-1} = \begin{pmatrix} d/k & -b/k \\ -c/k & a/k \end{pmatrix}.$$

III.3.26 (Uniqueness of identity.) Let $\mathscr{S}$ be a set of elements endowed with a closed, associative operation $*$, which we call "gleep," defined between all pairs of elements of $\mathscr{S}$. That is,

(a) For all $a, b \in \mathscr{S}$, $a * b$ (read "a gleep b") belongs to $\mathscr{S}$.

(b) For all $a, b, c \in \mathscr{S}$, $(a * b) * c$ is the same elements as $a * (b * c)$.

Suppose we hypothesize the existence of *at least* one 2-sided gleep identity which we denote by e, that is, e has the "gleep-feeble" property that $e * a = a * e = a$ for any and every $a \in \mathscr{S}$. Prove that $\mathscr{S}$ contains *at most* one such 2-sided gleep identity. [*Hint:* Assume the existence of 2 (distinct for the moment) gleep identities, e_1 and e_2. Describe $e_1 * e_2$ in 2 ways.]

III.3.27 (Uniqueness of inverses.) Suppose some elements of $\mathscr{S}$ (see Exercise III.3.26) possess *at least one* 2-sided gleep inverse which returns us to the (unique) gleep identity. That is, an element of $\mathscr{S}$, a', is called a 2-sided gleep inverse for a if $a' * a = a * a' = e$. Prove that if an element a possesses at *most* one 2-sided gleep inverse. *Hint*: Assume the existence of 2 (distinct for the moment) gleep inverses a' and a''. Compare the elements $(a' * a) * a''$ and $a' * (a * a'')$; use associativity of $*$ [see Exercise III.3.26(b)].

Note: Define a *left gleep inverse* a' for a to be any element such that $a' * a = e$, the gleep identity; define a *right gleep inverse* a'' for a to be any element such that $a * a'' = e$. Then Exercise III.3.27 really asks you to show that if left and right inverses for a exists, then any *left* (gleep) inverse for a is equal to any and every *right* inverse for a, and vice versa. Consequently, there can only be 1 such (left and right) 2-sided gleep inverse. In fact, suppose x is an element with 2 left inverses, a_1 and a_2, and at least 1 right inverse, b.

Thus,

$$
\begin{array}{ll}
\text{(a)} & a_1 * x = e \\
\text{(b)} & a_2 * y = e \\
\text{(c)} & \quad x * b = e.
\end{array}
$$

But any right inverse b is equal to *any* left inverse (a_1 and a_2). Thus $b = a_1$ and $b = a_2$, which implies that $a_1 = a_2$. Inverses are unique if left and right inverses both exist.

We shall now state a theorem which will eventually produce all the information needed to find 2-sided inverses for certain square matrices.

Theorem III.3.4 Given an $n \times n$ (square) matrix A. Then A can be reduced to the identity matrix I_n by means of elementary row operations *only* (also by means of elementary column operations *only*), if and only if the rank of the $n \times n$ matrix A is equal to n.

Proof: ($\Rightarrow$) Let us prove the "only if" part of the theorem. Thus, we assume that by applying *only* elementary column operations to A and also by applying *only* elementary row operations to A, the matrix I_n can, in both cases, be obtained; from this assumption we are to show that necessarily, the rank of A is n.

But this is easy.

Everybody knows that elementary row operations and column operations do not alter either row rank or column rank (Corollary III.2.3 and Theorem III.2.5). This means that under our assumption, the rank of our original matrix A is the same as the rank of I_n, which in turn, is n.

($\Leftarrow$) Let us prove the "if" part of the theorem. Thus, after showing (above) that

$$\text{reducibility of } A \text{ to } I_n \Rightarrow \text{rank } A = n,$$

we are *now* obliged to show that

$$\text{rank } A = n \Rightarrow \text{reducibility of } A \text{ to } I_n.$$

Let us, then, start with a matrix whose rank is n. Let us reduce it to either row-echelon or column-echelon form by means of elementary row operation or elementary column operations, respectively (Theorem

III.1.4). We must, of necessity, arrive at a square $n \times n$ echelon matrix which has precisely n nonzero rows or n nonzero columns, respectively, since the number of nonzero rows (respectively columns) in any row (respectively column) echelon matrix must match the rank of the original matrix A. Now any $n \times n$ echelon matrix with all n rows (respectively columns) nonzero (so that it has rank n), has got to be exactly I_n, the $n \times n$ identity matrix! Thus, rank $A = n \Rightarrow$ reducibility of A to I_n. The theorem is proved.

To this point, what do we have? Our theorem above tells us that if an $n \times n$ matrix A has rank n, then this is exactly equivalent to saying that (through elementary row or column operations only), I_n will be the echelon matrix for A. The stage is set, then, for our next important theorem.

Theorem III.3.5 Let A be an $n \times n$ matrix. Then the following statements are equivalent to each other:

(a) A has rank n
(b) A unique (2-sided) $n \times n$ inverse A^{-1} exists for A, that is, A is nonsingular (or invertible)
(c) every 1-sided inverse B ($B \cdot A = I_n$ or $A \cdot B = I_n$) is 2-sided ($B \cdot A = I_n$ *and* $A \cdot B = I_n$).

Proof: Our argument essentially rests on the "observation" mentioned in the proof of Theorem III.2.5. That is, as a consequence of Theorem III.3.1, a succession of elementary row (column) operations on A corresponds to a succession of left-hand (right-hand) multiplications on A by corresponding elementary matrices. That is, for any row-echelon matrix A' and column-echelon matrix A'' for A, there is an $n \times n$ matrix R and an $n \times n$ matrix C such that

$$R \cdot A = A' \tag{1}$$

and

$$A \cdot C = A'', \tag{2}$$

where R and C are products of elementary matrices.

So much for review.

Now let us use our hypothesis (a) of Theorem III.3.5. Thus, we assume at this point that A has rank n. Theorem III.3.4 assures us that this condition is *equivalent* to the condition that I_n is both a row-echelon *and* a column-echelon matrix for A. Very good. Let us now replace A' in (1) and A'' in (2) with I_n to obtain the statement:

For every $n \times n$ matrix A whose rank is n, there exist $n \times n$ matrices R and C such that

$$R \cdot A = I_n \tag{1$'$}$$

and

$$A \cdot C = I_n. \tag{2$'$}$$

Statements (1)′ and (2)′ allow us to conclude that A has rank n if and only if both a left inverse R and a right inverse C exist for A. Now, through no virtue of matrices or of matrix multiplication, we shall conclude that the left inverse R must be equal to the right inverse. The truth of this statement follows first from the fact that a left and right inverse do exist for A [see (a)′ and (b)′], and secondly, from the fact that our matrix multiplication is associative (see note at end of Exercise III.3.27 and Exercise II.6.81). For those skeptics who still refuse to believe that necessarily $R = C$, just watch this:

$$\begin{aligned} R &= R \cdot I_n && \text{property of } I_n \\ &= R \cdot (A \cdot C) && \text{from (2)}' \\ &= (R \cdot A) \cdot C && \text{associativity of } \cdot \\ &= I_n \cdot C && \text{from (1)}' \\ &= C && \text{property of } I_n. \end{aligned} \tag{III.3.7}$$

Thus, every left and right inverse is a 2-sided inverse as per Definition III.3.3. This is a proof of condition (c).

The truly pretty part of proving that every left inverse R is the same as every right inverse C, is that the reasoning in (III.3.7) does *not* lean on any messy matrix multiplication computations; only the structure of *associativity* of that multiplication is needed.

As to showing that only 1 left (hence right, hence 2-sided) inverse A^{-1} exists, we actually argue according to (III.3.7) all over again. For suppose A_1^{-1} and A_2^{-2} are two (distinct) 2-sided inverses. Then

$$\begin{aligned} A_1^{-1} &= A_1^{-1} \cdot I_n = A_1^{-1}(A \cdot A_2^{-1}) = (A_1^{-1} \cdot A) \cdot A_2^{-1} \\ &= I_n \cdot A_2^{-1} = A_2^{-1}. \end{aligned} \tag{III.3.8}$$

Since we have established existence [see (1)′, (2)′, and (III.3.7)] and uniqueness [see (III.3.8)] of A^{-1}, our proof is complete.

REMARK. For an example of an element with a left inverse and no right inverse, see Exercise III.3.35.

The following example illustrates what is perhaps the easiest means of computing matrix inverses. The technique to be used, however, is already familiar to you through an example. In finding a matrix A^{-1} such that $A^{-1} \cdot A = I_n$, we are simply looking for that product of elementary (row) matrices (since A^{-1} appears on the left of A) which reduces A to the canonical (row) echelon matrix I_n. Since we are simply repeating the problem that the example illustrates (for the case where our echelon matrix is I_n), we employ the techniques used there.

Example Find the (2-sided) inverse for the nonsingular matrix

$$A = \begin{pmatrix} 1 & 2 & 0 & 4 \\ 0 & 3 & 1 & 0 \\ 0 & 1 & 0 & 4 \\ 1 & 1 & 1 & 4 \end{pmatrix}.$$

We proceed to reduce A to a column-echelon matrix (as an alternative, we could reduce A to a row-echelon matrix). If we eventually obtain I_4, the identity matrix, then we know that this is completely equivalent to the fact that A has a (2-sided) inverse A^{-1} (see Remark above). Thus, by applying the same sequence of elementary column operations to A and to I_4 separately, we obtain from A, the echelon matrix I_4, while form I_4, we obtain A^{-1}.

$$A = \begin{pmatrix} 1 & 2 & 0 & 4 \\ 0 & 3 & 1 & 0 \\ 0 & 1 & 0 & 4 \\ 1 & 1 & 1 & 4 \end{pmatrix} \qquad I_4 = \begin{pmatrix} 1 & 0 & 0 & 0 \\ 0 & 1 & 0 & 0 \\ 0 & 0 & 1 & 0 \\ 0 & 0 & 0 & 1 \end{pmatrix}$$

clear first row of A $\quad$ $C_1{}^2(-2), C_1{}^4(-4)$

$$\begin{pmatrix} 1 & 0 & 0 & 0 \\ 0 & 3 & 1 & 0 \\ 0 & 1 & 0 & 4 \\ 1 & -1 & 1 & 0 \end{pmatrix} \qquad \begin{pmatrix} 1 & -2 & 0 & -4 \\ 0 & 1 & 0 & 0 \\ 0 & 0 & 1 & 0 \\ 0 & 0 & 0 & 1 \end{pmatrix}$$

clear second row $\quad$ C_{23}; then $C_2{}^3(-3)$

$$\begin{pmatrix} 1 & 0 & 0 & 0 \\ 0 & 1 & 0 & 0 \\ 0 & 0 & 1 & 4 \\ 1 & 1 & -4 & 0 \end{pmatrix} \qquad \begin{pmatrix} 1 & 0 & -2 & -4 \\ 0 & 0 & 1 & 0 \\ 0 & 1 & -3 & 0 \\ 0 & 0 & 0 & 1 \end{pmatrix}$$

clear third row $\quad \boxed{C_3{}^4(-4)}$

$$\begin{pmatrix} 1 & 0 & 0 & 0 \\ 0 & 1 & 0 & 0 \\ 0 & 0 & 1 & 0 \\ 1 & 1 & -4 & 16 \end{pmatrix} \qquad \begin{pmatrix} 1 & 0 & -2 & 4 \\ 0 & 0 & 1 & -4 \\ 0 & 1 & -3 & 12 \\ 0 & 0 & 0 & 1 \end{pmatrix}$$

clear last row $\quad \boxed{C_4(\tfrac{1}{16});\ \text{then}\ C_4{}^1(-1), C_4{}^2(-1), C_4{}^3(4)}$

$$I_4 = \begin{pmatrix} 1 & 0 & 0 & 0 \\ 0 & 1 & 0 & 0 \\ 0 & 0 & 1 & 0 \\ 0 & 0 & 0 & 1 \end{pmatrix} \qquad A^{-1} = \begin{pmatrix} \frac{3}{4} & -\frac{1}{4} & -1 & \frac{1}{4} \\ \frac{1}{4} & \frac{1}{4} & 0 & -\frac{1}{4} \\ -\frac{3}{4} & \frac{1}{4} & 0 & \frac{3}{4} \\ -\frac{1}{16} & -\frac{1}{16} & \frac{1}{4} & \frac{1}{16} \end{pmatrix}.$$

We may use actual matrix multiplication to verify that $A^{-1} \cdot A = A \cdot A^{-1} = I_4$, where A^{-1} is the last matrix of our second column.

We are on the threshold of discovering a very interesting property of nonsingular matrices, namely,

Theorem III.3.6 To say A is invertible is exactly equivalent to saying A can be written as the product of elementary $n \times n$ matrices.

Proof: Let us recall the equation of lines (1)′ and (2)′ from the proof of Theorem III.3.5. Thus, for A nonsingular,

(1)′
$$\underbrace{R_k \cdot \cdots \cdot R_2 \cdot R_1}_{R} \cdot A = I_n$$

and

(2)′
$$A \cdot \underbrace{C_1 \cdot C_2 \cdots C_j}_{C} = I_n$$

where each of the R_i's and each of the C_i's is an elementary matrix.

We now use the fact that *any* elementary matrix has an inverse which is, again, an *elementary matrix* (see Exercise III.1.4). Thus, the $n \times n$ matrices $R_k^{-1}, \ldots, R_2^{-2}, R_1^{-1}$ exist and are elementary matrices.

Let us multiply both sides of Equation (1)′ on the left by R_k^{-1}. This yields

$$\overbrace{(R_k^{-1} \quad R_k)}^{I_n} \cdot R_{k-1} \cdots R_2 \cdot R_1 \cdot A = R_k^{-1} \cdot I_n,$$

or

$$R_{k-1} \cdots R_2 \cdot R_1 \cdot A = R_k^{-1}.$$

We simply continue this left-hand multiplication by the (elementary) matrices $R_{k-1}^{-1}, \ldots, R_2^{-1}, R_1^{-1}$ in that order to obtain

$$A = R_1^{-1} \cdot R_2^{-1} \cdots R_{k-1}^{-1} \cdot R_k^{-1}, \qquad \text{(III.3.9)}$$

which displays A as a product of the elementary matrices $R_1^{-1}, R_2^{-1}, \ldots, R_k^{-1}$.

A similar argument using Equation (2)′ and right-hand multiplications by C_i^{-1} matrices produces the same conclusion. Our proof is finished.

Editorially, we note that Theorem III.3.6 tells us that the elementary matrices, far from being freaks of matrices, are the very stuff from which *all* nonsingular matrices are constructed. They represent all the invertible matrices there are in the sense that elementary matrices and products of elementary matrices account for *every* invertible matrix known to man (or to anybody else, for that matter).

EXERCISES

III.3.28 By means of (further) elementary row operations only, reduce the following square row-echelon matrices to the square identity matrix:

(a) $\begin{pmatrix} 1 & 5 & 3 & 0 & 2 \\ 0 & 1 & -1 & 0 & 3 \\ 0 & 0 & 1 & 5 & 1 \\ 0 & 0 & 0 & 1 & 0 \\ 0 & 0 & 0 & 0 & 1 \end{pmatrix}$ (c) $\begin{pmatrix} 1 & \text{a} & \text{b} & \text{c} \\ 0 & 1 & \text{d} & \text{e} \\ 0 & 0 & 1 & \text{f} \\ 0 & 0 & 0 & 1 \end{pmatrix}$

(b) $\begin{pmatrix} 1 & -1 & 0 \\ 0 & 1 & 8 \\ 0 & 0 & 1 \end{pmatrix}$ (d) $\begin{pmatrix} 1 & 1 & 1 & 1 & 1 & 1 \\ 0 & 1 & 1 & 1 & 1 & 1 \\ 0 & 0 & 1 & 1 & 1 & 1 \\ 0 & 0 & 0 & 1 & 1 & 1 \\ 0 & 0 & 0 & 0 & 1 & 1 \\ 0 & 0 & 0 & 0 & 0 & 1 \end{pmatrix}$

III.3.29 By means of (further) elementary column operations *only*, reduce the following square column-echelon matrices to the square identity matrix:

(a) $\begin{pmatrix} 1 & 0 & 0 \\ 2 & 1 & 0 \\ 3 & 4 & 1 \end{pmatrix}$ (c) $\begin{pmatrix} 1 & 0 & 0 & 0 \\ \text{a} & 1 & 0 & 0 \\ \text{b} & \text{c} & 1 & 0 \\ \text{d} & \text{e} & \text{f} & 1 \end{pmatrix}$

(b) $\begin{pmatrix} 1 & 0 & 0 & 0 & 0 & 0 \\ 2 & 1 & 0 & 0 & 0 & 0 \\ 3 & 2 & 1 & 0 & 0 & 0 \\ 4 & 3 & 2 & 1 & 0 & 0 \\ 5 & 4 & 3 & 2 & 1 & 0 \\ 6 & 5 & 4 & 3 & 2 & 1 \end{pmatrix}$ (d) $\begin{pmatrix} 1 & 0 & 0 & 0 & 0 \\ -1 & 1 & 0 & 0 & 0 \\ 3 & 2 & 1 & 0 & 0 \\ 7 & 6 & -1 & 1 & 0 \\ 3 & 2 & 8 & 5 & 1 \end{pmatrix}$

III.3.30 If possible, find the 2-sided inverses for the following matrices:

(a) $\begin{pmatrix} \frac{1}{2} & 0 \\ \frac{1}{2} & 2 \end{pmatrix}$

(b) $\begin{pmatrix} 1 & 0 & 0 & 0 \\ 0 & 0 & 1 & 0 \\ 0 & 1 & 0 & 0 \\ 0 & 0 & 0 & 1 \end{pmatrix}$

(c) $\begin{pmatrix} 0 & \frac{1}{2} & 0 & 0 \\ 2 & 0 & 0 & 0 \\ 0 & 0 & 0 & 4 \\ 0 & 0 & 1 & 0 \end{pmatrix}$

(d) $\begin{pmatrix} 0 & 0 & 1 \\ 1 & 2 & 1 \\ 1 & 2 & 1 \end{pmatrix}$

(e) $\begin{pmatrix} 3 & 1 & 2 \\ 2 & 1 & 3 \\ 1 & 2 & 3 \end{pmatrix}$

(f) $\begin{pmatrix} 1 & b & 0 \\ a & 1 & c \\ 0 & 0 & 1 \end{pmatrix}$

(g) $\begin{pmatrix} 2 & 1 & 0 & 0 \\ 1 & 5 & 1 & 0 \\ 3 & 0 & 0 & 2 \\ 4 & 0 & 0 & 2 \end{pmatrix}$

(h) $\begin{pmatrix} 1 & 1 & 1 & 1 & 1 \\ 0 & 1 & 0 & 1 & 0 \\ 1 & 0 & 0 & 1 & 0 \\ 0 & 1 & 0 & 1 & 1 \\ 0 & 1 & 0 & 0 & 0 \end{pmatrix}$

(i) $\begin{pmatrix} 2 & 0 & 2 & 0 \\ 1 & 0 & 2 & 0 \\ 3 & 1 & 2 & 0 \\ 4 & 1 & 0 & 1 \end{pmatrix}$

(j) $\begin{pmatrix} 6 & 0 & 6 & 1 \\ 0 & 6 & 1 & 1 \\ 2 & 1 & 0 & 0 \\ 1 & 0 & 0 & 0 \end{pmatrix}$

III.3.31 Let $A = (1)_{ij}$, $i, j = 1, 2, \ldots, n$. That is, A is the $n \times n$ matrix whose every entry is 1. Give a brief argument (1 sentence should do) to show that A^{-1} does not exist.

III.3.32 Compute the inverses for the nonsingular echelon matrices of Exercises III.3.28 and III.3.29.

III.3.33 (a) Find the inverse of each of the elementary matrices (see Definition III.3.2).

(b) Show that the inverse of an elementary matrix is always an elementary matrix.

III.3.34 Suppose the $n \times n$ matrices A and B each have rank n (that is, are nonsingular by Theorem III.3.5). Prove that the $n \times n$ product $A \cdot B$ is also nonsingular, so that the existence of an inverse $(A \cdot B)^{-1}$ for $A \cdot B$ does indeed exist. In order to be more specfic, prove that this inverse is the matrix $B^{-1} \cdot A^{-1}$, that is, prove that for $n \times n$ matrices A and B,

(a) A^{-1} and B^{-1} exist $\Rightarrow (A \cdot B)^{-1}$ exists and equals $B^{-1} \cdot A^{-1}$.
(*Hint:* Simply test the suspected inverse $B^{-1} \cdot A^{-1}$ as per Definition III.3.3.)
Now prove a converse, namely,

(b) The inverse of the product $A \cdot B$ exists $\Rightarrow$ the inverse of each of the factors exists.

Caution: Do not *assume* what you are to prove in (b). The hypothesis here does *not* say that the inverse of the product $A \cdot B$ equals $B^{-1} \cdot A^{-1}$; we are told only that the inverse of $A \cdot B$ exists and we are not told what it is.

III.3.35 An element A_0 with a left inverse and no right inverse. Let S be the infinite set $S = \{x_1, x_2, \ldots, x_i, \ldots\}$. Let $\mathscr{F}(S)$ be the set of all functions $A : S \to S$ sending S to itself. Let us define a "product" between functions, as usual, by $A \cdot B(x_i) = A(B(x_i))$ for all $x_i \in S$. Prove

(a) For all $A, B \in \mathscr{F}(S)$, $A \cdot B \in \mathscr{F}(S)$ (closure of $\mathscr{F}(S)$ under the product).

(b) For all $A, B, C \in \mathscr{F}(S)$, $A \cdot (B \cdot C) = (A \cdot B) \cdot C$ (associativity of the product).

(c) If $I \in \mathscr{F}(S)$ is defined by $I(x_i) = x_i$, then I is a (feeble) identity in $\mathscr{F}(S)$, that is, $I \cdot A = A \cdot I$ for all $A \in \mathscr{F}(S)$.

Properties (a), (b), and (c) are properties of functions, regardless of the "number of elements" in S, for example, linear functions on the set $S = V_n$.

Define $A_0 \in \mathscr{F}(S)$ by $A_0(x_i) = x_{i+1}$ for each $i = 1, 2, \ldots$.

(d) Show that A_0 has many left inverses. In fact, if for each $k = 1, 2, \ldots,$ we define $A_k \in \mathscr{F}(S)$ by

$$A_k(x_i) = x_{i-1}, \text{ for each } i = 2, 3, \ldots, \text{ and}$$
$$A_k(x_1) = x_k,$$

then A_k is a left inverse for A_0 (that is, $A_k \cdot A_0 = I$).

(e) Prove that A_0 is 1-1 but not onto, and A_k is onto but not 1-1.

(f) Prove that no right inverse for A_0 can exist in $\mathscr{F}(S)$, that is, no $B \in \mathscr{F}(S)$ exists in $\mathscr{F}(S)$ with the property that $A_0 \cdot B = I$.

Chapter III Summary

We incorporate the notion of vector-space dimension in our analysis of system of equations $A \cdot X = Y$. Thus, the vector-space of all X, where $A \cdot X = 0$ for the $m \times n$ matrix A must have dimension $n -$ rank A (Theorem III.2.4). From this, the solution is unique (hence, $X = O$ is the only possible solution) for $A \cdot X = O$, if and only if rank $A = n$. Uniqueness and existence, for the general system $A \cdot X = Y$ is discussed in Proposition III.2.8 and Exercise III.2.37. Again the notion of rank is central in this discussion. After applications of system of equations to linear transformations, we find that the elementary matrices (Definition III.3.2), which allowed us to achieve the echelon forms for systems of equations, actually characterize the invertible matrices. In fact, A^{-1} exists if and only if A is a product of these elementary matrices (Theorem III.3.6). A distinguishing feature of matrices is that 1-sided inverses are always 2-sided (Theorem III.3.5). That this is not true for all mathematical systems is revealed through Exercise III.3.35.

CHAPTER IV

Determinants: A Classical Approach

1. Permutations

We are about to encounter a function, denoted by "det," which sends *square* matrices to scalars. As we shall see, a study of this 1 function will lead us to a more complete understanding of matrices and their corresponding linear transformations. This function is called the *determinant function.*

Before we are able to isolate some properties of the determinant function via a classical approach, we must first understand at least some of the structure of a certain class of functions called *permutations*. Our first section, then, devotes itself to a study of the permutation functions, which we define now.

Definition IV.1.1 Any 1-1 onto function σ which sends the finite set of integers $\{1,2,\ldots,n\}$ onto itself, is called a **permutation on *n* letters**. The symbol S_n denotes the set of all possible permutations on n letters.

REMARK. It is to be observed that since $\{1,2,\ldots,n\}$ has only a finite number of elements, any function sending $\{1,2,\ldots,n\}$ *onto* $\{1,2,\ldots,n\}$ is automatically 1-1. Also, any 1-1 function from $\{1,2,\ldots,n\}$ to itself must be *onto*. Thus, Definition IV.1.1 could be equivalently stated by defining a permutation to be any 1-1 (hence onto) function of $\{1,2,\ldots,n\}$ to itself (see Exercise IV.1.1).

Example Consider the 5-element set $\{1,2,3,4,5\}$. We define 2 permutation functions σ and τ (on 5 letters) by the equations

$$\begin{aligned}
&\text{(a)} \quad \sigma(1) = 3, \quad \sigma(2) = 1, \quad \sigma(3) = 5, \quad \sigma(4) = 4, \quad \sigma(5) = 2\\
&\text{(b)} \quad \tau(1) = 5, \quad \tau(2) = 3, \quad \tau(3) = 4, \quad \tau(4) = 1, \quad \tau(5) = 2.
\end{aligned} \qquad \text{(IV.1.1)}$$

Notice that the domains (input sets) and the ranges (output sets) for each of the functions σ and τ are all equal to the 5-element set $\{1,2,3,4,5\}$, that is, dom $(\sigma) =$ rg $(\sigma) =$ dom $(\tau) =$ rg $(\tau) = \{1,2,3,4,5\}$. Just for the fun of it, we might wish to compose these 2 functions σ and τ; this is always possible since the range of 1 is a subset of (in fact equal to) the domain of the other (see Definition II.1.5). Recalling that the composition function $\sigma \circ \tau$ is defined by $\sigma \circ \tau(i) = \sigma(\tau(i))$ for each and every $i \in \{1,2,3,4,5\}$, we see that

$$\begin{aligned} \sigma \circ \tau(1) &= \sigma(\tau(1)) && \text{Definition II.1.5} \\ &= \sigma(5) && \text{since } \tau(1) = 5 \text{ from (IV.1.1b).} \\ &= 2 && \text{since } \sigma(5) = 2 \text{ from (IV.1.1a).} \end{aligned}$$

Continuing in this manner, we arrive at the following:

$$\sigma \circ \tau(1) = 2, \qquad \sigma \circ \tau(2) = 5, \qquad \sigma \circ \tau(3) = 4, \qquad \sigma \circ \tau(4) = 3, \qquad \sigma \circ \tau(5) = 1.$$

It is to be observed that our "new" function $\sigma \circ \tau$ is not just a function; it is even a *permutation*. That is, it is a 1-1 function (no 2 "input" elements are sent to the same "output" element by $\sigma \circ \tau$) sending $\{1,2,3,4,5\}$ *onto* itself.

Now each permutation σ in S_n is a function (that is, a certain set of ordered pairs; see Definition II.1.1) whose domain is the *finite* set $\{1,2,\ldots,n\}$. Thus, σ is, by definition, the (finite) set of ordered pairs

$$\sigma = \{(1,\sigma(1)),(2,\sigma(2)),\ldots,(n,\sigma(n))\}.$$

A notation for the permutation σ has developed which actually *lists* the ordered pairs of σ. This somewhat universally accepted notation happens to list each ordered pair $(k,\sigma(k))$ in a *vertical* manner, as $\genfrac{}{}{0pt}{}{k}{\sigma(k)}$; in any case, we are dealing with a pair of integers which is ordered. That is, we understand which element is *first* (an "input" element) and which element is *second* (an "output" element). We display this notation now.

Notation IV.1.2 Let σ be the permutation of S_n such that for each $k \in \{1,2,3,\ldots,n\}$, $\sigma : k \to \sigma(k)$, that is, σ sends k to $\sigma(k)$, or $(k,\sigma(k)) \in \sigma$. Then we write

input element k

$$\sigma = \begin{pmatrix} 1 & 2 & 3 & \cdots & k & \cdots & n \\ \sigma(1) & \sigma(2) & \sigma(3) & \cdots & \sigma(k) & \cdots & \sigma(n) \end{pmatrix}$$

output element $\sigma(k)$. ordered pair.

The evaluation of σ at the integer k may be written $\sigma(k)$, σk, or σ_k.

Example For σ, τ, and $\sigma \circ \tau \in S_5$ defined in the previous Example, we may write

$$\sigma = \begin{pmatrix} 1 & 2 & 3 & 4 & 5 \\ 3 & 1 & 5 & 4 & 2 \end{pmatrix}$$

$$\tau = \begin{pmatrix} 1 & 2 & 3 & 4 & 5 \\ 5 & 3 & 4 & 1 & 2 \end{pmatrix}$$

$$\sigma \circ \tau = \begin{pmatrix} 1 & 2 & 3 & 4 & 5 \\ 2 & 5 & 4 & 3 & 1 \end{pmatrix}.$$

In computing $\tau \circ \sigma$ (we already know $\sigma \circ \tau$), we would obtain

$$\tau \circ \sigma(1) = 4, \qquad \tau \circ \sigma(2) = 5, \qquad \tau \circ \sigma(3) = 2, \qquad \tau \circ \sigma(4) = 1, \qquad \tau \circ \sigma(5) = 3,$$

which we denote

$$\tau \circ \sigma = \begin{pmatrix} 1 & 2 & 3 & 4 & 5 \\ 4 & 5 & 2 & 1 & 3 \end{pmatrix}.$$

Incidentally, we see that in this example, $\sigma \circ \tau \neq \tau \circ \sigma$.

One easily sees that the function e which leaves every integer intact, is the (necessarily unique) identity (feeble) element of S_n. That is to say, if $e(k) = k$ for each $k \in \{1,2,\ldots,n\}$, then $e \circ \sigma = \sigma \circ e$ for any and every permutation σ.

As an incidental remark, we observe that in our case, $(\sigma \circ \tau)^{-1}$ happens to equal $\tau^{-1} \circ \sigma^{-1}$. But if all the inverses σ^{-1}, τ^{-1} and $(\sigma \circ \tau)^{-1}$ exist, then we *always* have that $(\sigma \circ \tau)^{-1} = \tau^{-1} \circ \sigma^{-1}$. This can be proven algebraically just from the definition of inverse and the (feeble) property of the identity (see Exercise III.3.34), or it can be verified directly in treating σ^{-1}, τ^{-1} and $(\sigma \circ \tau)^{-1}$ as functions in particular cases (see Exercise IV.1.5).

EXERCISES

IV.1.1 Let σ be a function with domain and impact set equal to the finite set of integers $\{1,2,\ldots,n\}$.

(a) Suppose σ is 1-1 [that is, $\sigma(i) = \sigma(j)$ *only* for $i = j$]. Prove that σ is necessarily *onto* (that is, for each and every $k \in \{1,2,\ldots,n\}$, there is a $j \in \{1,2,\ldots,n\}$ such that $\sigma(j) = k$).

(b) Suppose σ is *onto*. Prove that σ is necessarily 1-1.

Hence, if the domain of a function σ is finite, the 1-1 property is *equivalent* to the onto property for σ.

IV.1.2 Given $\sigma = \begin{pmatrix} 1 & 2 & 3 & 4 \\ 1 & 3 & 2 & 4 \end{pmatrix}$, $\tau = \begin{pmatrix} 1 & 2 & 3 & 4 \\ 4 & 1 & 2 & 3 \end{pmatrix}$. Find

$\sigma \circ \sigma$, σ^{-1}, $\tau \circ \tau$, τ^{-1}, $\sigma \circ \tau$, $\tau \circ \sigma$, $(\tau \circ \sigma)^{-1}$, $(\sigma \circ \tau)^{-1}$.

IV.1.3 Given $\sigma = \begin{pmatrix} 1 & 2 & 3 & 4 & 5 & 6 \\ 1 & 4 & 6 & 3 & 2 & 5 \end{pmatrix}$, $\tau = \begin{pmatrix} 1 & 2 & 3 & 4 & 5 & 6 \\ 6 & 5 & 3 & 4 & 1 & 2 \end{pmatrix}$. Find

$\sigma \circ \sigma$, σ^{-1}, $\tau \circ \tau$, τ^{-1}, $\sigma \circ \tau$, $\tau \circ \sigma$, $(\sigma \circ \tau)^{-1}$, $(\tau \circ \sigma)^{-1}$.

IV.1.4 Given $\sigma = \begin{pmatrix} 1 & 2 & \cdots & k & \cdots & n \\ n & n-1 & \cdots & n-k+1 & \cdots & 1 \end{pmatrix} \in S_n$. Prove that σ^2 (that is, $\sigma \circ \sigma) = e$, the identity permutation.

IV.1.5 Given $\sigma = \begin{pmatrix} 1 & 2 & 3 \\ 3 & 2 & 1 \end{pmatrix}$, $\tau = \begin{pmatrix} 1 & 2 & 3 \\ 1 & 3 & 2 \end{pmatrix}$, $\mu = \begin{pmatrix} 1 & 2 & 3 \\ 3 & 1 & 2 \end{pmatrix}$. Find

$$\sigma \circ \tau, \quad (\sigma \circ \tau) \circ \mu, \quad \tau \circ \mu, \quad \sigma \circ (\tau \circ \mu).$$

Note: Compare $(\sigma \circ \tau) \circ \mu$ with $\sigma \circ (\tau \circ \mu)$.

IV.1.6 List all the permutations possible on 2 letters, that is, list all the functions in S_2.

IV.1.7 List all the permutations in S_3.

IV.1.8 How many permutations are there in S_4?
Hint: For a permutation σ, $\sigma(1)$ can take on any of the 4 values 1, 2, 3, or 4. So far, then, there are 4 types of permutations, namely, those σ where $\sigma(1) = 1$, $\sigma(1) = 2$, $\sigma(1) = 3$ or $\sigma(1) = 4$. Now once a value for $\sigma(1)$ is *fixed*, there are only $4 - 1 = 3$ values possible for $\sigma(2)$. Thus, we have accounted for $4 \cdot 3 = 12$ distinct permuations since there are 3 permutations that can branch out from each of the 4 permutation types previously described.

$\sigma(2) = 2$	$\sigma(2) = 1$	$\sigma(2) = 1$	$\sigma(2) = 1$
or	or	or	or
$\sigma(1) = 1 \to \sigma(2) = 3$	$\sigma(1) = 2 \to \sigma(2) = 3$	$\sigma(1) = 3 \to \sigma(2) = 2$	$\sigma(1) = 4 \to \sigma(=) = 2$
or	or	or	or
$\sigma(2) = 4$	$\sigma(2) = 4$	$\sigma(2) = 4$	$\sigma(2) = 3$

To *each* of these 12 permutations, how many values are left for $\sigma(3)$ to assume? After $\sigma(1)$, $\sigma(2)$, and $\sigma(3)$ are chosen, how many ways may $\sigma(4)$ be selected?

IV.1.9 How many permutations are there in S_n for n an arbitrary integer (see Exercise IV.1.8)?

Our task now is to classify the permutations of S_n, that is, we shall invent 2 classes such that each permutation shall fall into exactly 1 of these classes. In fact, we shall soon be able to take an arbitrary permutation in hand and label it as either *even* or *odd*. Before we define this notion of evenness and oddness (the parity) for a permutation, we must somehow measure how badly a permutation σ scrambles the *order* of the domain set $\{1,2,\ldots,n\}$ (note that heretofore, no ordering on the domain or range of σ was ever mentioned or required; it is introduced here for the first time). The next definition formalizes this notion of "scrambling" and describes conditions under which a permutation is to be called "even" or "odd." Examples then follow which illustrate the notions of the definition.

Definition IV.1.3 Let $\sigma = \begin{pmatrix} 1 & 2 & \cdots & i & \cdots & j & \cdots & n \\ \sigma(1) & \sigma(2) & \cdots & \sigma(i) & \cdots & \sigma(j) & \cdots & \sigma(n) \end{pmatrix}$ denote a typical permutation of S_n.

We will say that σ induces an **inversion** (*relative to a pair of integers* (i,j)) each time we have the condition that $i < j$ and yet $\sigma(i) > \sigma(j)$.

If the total number of inversions induced by σ is even, then σ is called an **even permutation.**

If the total number of inversions induced by σ is odd, then σ is called an **odd permutation.**

When we refer to the **parity of a permutation** σ we refer to its classification as an even or odd permutation.

REMARK. Intuitively, then, we consider an inversion to be an instance where σ does *not* respect the relative order of a given pair of integers. Thus, for integers i and j, where $i < j$ only 2 possibilities can occur, namely, either $\sigma(i) < \sigma(j)$ or else $\sigma(i) > \sigma(j)$. (Why?) If $\sigma(i) < \sigma(j)$ we say that σ respects or preserves the relative order of i and j since $i < j$. In the opposite case, where $\sigma(i) > \sigma(j)$, σ does *not* respect the order of i and j since $i < j$ and yet $\sigma(i) > \sigma(j)$; thus, the ordered pair $(\sigma(i), \sigma(j))$ would represent an inversion induced by σ.

Example Let $= \begin{pmatrix} 1 & 2 & 3 & 4 \\ 3 & 1 & 4 & 2 \end{pmatrix}$. We extract from this notation, the *ordered* range of σ, $\{3, 1, 4, 2\}$. Let us count the number of inversions, that is:

> The number of times that i is *less* than j, and yet the corresponding image elements $\sigma(i)$ and $\sigma(j)$, are such that $\sigma(i)$ is *greater* than $\sigma(j)$.

To say that i *is less than* j, however, is only to say that physically, $\sigma(i)$ appears *to the left of* $\sigma(j)$ in the ordered range of σ. Hence, to count the numbers of inversions (that is, the number of times that σ reverses order of a pair of "input" elements) is to count the number of pairs of integers (p,q) in $\{3,1,4,2\}$, the ordered range of σ, where p is to the left of q and yet p is *greater* than q (p and q are images of some pair of integers, i and j, respectively, where i is less than j; in fact, p is in the ith place and q is in the jth place of the ordered range of σ).

We now list *all* pairs of integers which we extract from the ordered range of σ, $\{3,1,4,2\}$, being careful to maintain the relative ordering:

(3,1)*	(1,4)	(4,2)*
(3,4)	(2,1)	
(3,2)*		

The asterisks denote the 3 inversions induced by σ since these pairs are in reverse order relative to their entry numbers in the ordered range of σ. σ is an odd permutation; that is, 3 inversions account for an odd permutation.

Example Let $\sigma = \begin{pmatrix} 1 & 2 & 3 & 4 & 5 & 6 & 7 & 8 \\ 5 & 3 & 1 & 8 & 6 & 4 & 2 & 7 \end{pmatrix}$. The ordered range of σ is $\{5,3,1,8,6,4,2,7\}$. We see that 5 is to the left of 4, 3, 2, 1; this accounts for the 4 inversions (5,4), (5,3), (5,2), and (5,1). Passing to the second integer 3, we see that 3 precedes 2 and 1; this accounts for 2 more inversions. Passing to the third integer 1, we see that 1 precedes only numbers which are *greater* than itself. Hence 1 accounts for *no* inversions. We continue to compare each integer with entries to the *right*, and select those smaller numbers to the right to account for our inversions. The complete list of inversions for σ follows:

$$(5,3),\ (5,1),\ (5,4),\ (5,2),\ (3,1),\ (3,2),\ (8,6),\ (8,4),\ (8,2),\ (8,7),\ (6,4),\ (6,2),\ (4,2).$$

Hence, there are 13 inversions (or inverted subpairs) out of a totality of 28 pairs, incidentally. Thirteen is odd, and so we agree to say that σ is an *odd permutation*, or that the *parity* of σ is *odd*.

At this point, we choose to concentrate on what seems to be a very special type of permutation, namely, those permutations which leave fixed every element except for two. Such permutations are called *transpositions*. We will show that these permutations are odd. Also, the importance of these "peculiar" permutations is evident once it is realized that, in a sense, these are *all* the permutations there are! More specifically, any and every permutation can be obtained by a succession (composition) of these "pairwise switching," or transposition permutations.

Definition IV.1.4 A **transposition** is any permutation $\sigma \in S_n$ with the property that for exactly 1 pair of distinct integers, $i,j \in \{1,2,\ldots,n\}$, which we shall call *nonfixed points of* σ, $\sigma(i) = j$ and $\sigma(j) = i$. All other integers remain fixed under the action of σ, that is, for all (other) integers k, $k \neq i$, $k \neq j$, we require that $\sigma(k) = k$.

Example

$$\sigma = \begin{pmatrix} 1 & 2 & 3 & 4 & 5 \\ 1 & 4 & 3 & 2 & 5 \end{pmatrix}$$

and

$$\tau = \begin{pmatrix} 1 & 2 & 3 & 4 & 5 \\ 4 & 2 & 3 & 1 & 5 \end{pmatrix}$$

represent 2 transpositions in S_5. σ transposes or interchanges 2 and 4; τ transposes or interchanges 1 and 4, which are the nonfixed points for σ.

Just for fun, we observe that

$$\sigma^{-1} = \begin{pmatrix} 1 & 2 & 3 & 4 & 5 \\ 1 & 4 & 3 & 2 & 5 \end{pmatrix} \text{ and } \tau^{-1} = \begin{pmatrix} 1 & 2 & 3 & 4 & 5 \\ 4 & 2 & 3 & 1 & 5 \end{pmatrix}.$$

That is, $\sigma = \sigma^{-1}$ and $\tau = \tau^{-1}$. This is generally true for transpositions (Exercise IV.1.15).

Let σ_0 be a typical but fixed permutation which induces k_0 inversions in its ordered range. Our next theorem tells us that if σ_0 is followed by any transposition permutation, then the number of inversions is changed by an odd number. In other words, transpositions *reverse* the parity of any permutation.

Theorem IV.1.5 Let σ_0 be an arbitrary permutation in S_n, and suppose that σ_0 is even (respectively, odd). Then for any transposition $\tau \in S_n$, the permutation $\tau \circ \sigma_0$ is odd (respectively, even).

Proof: We first consider the special case where p and q, the 2 nonfixed points for the transposition τ, are physically adjacent to each other in $\{\sigma_0(1),\ldots,\sigma_0(n)\}$, the ordered range of σ_0. Such a transposition shall be referred to as an "adjacent" transposition, and is illustrated in (IV.1.2) below.

$$\begin{array}{lll}
 & \{\ 1,\ldots\ldots\ldots\ldots\ldots\ldots\ldots\ldots,n\ \} & \\
 & \sigma_0 \downarrow \quad \downarrow \quad \downarrow\downarrow \quad \downarrow \quad \downarrow & \\
\text{(a)} & \{\sigma_0(1),\ldots,i,\ldots,q,p,\ldots,j,\ldots,\sigma_0(n)\} \text{ ordered range of } \sigma_0 & \text{(IV.1.2)} \\
 & \tau \downarrow \quad \downarrow \quad \downarrow\downarrow \quad \downarrow \quad \downarrow & \\
\text{(b)} & \{\sigma_0(1),\ldots,i,\ldots,q,p,\ldots,j,\ldots,\sigma_0(n)\} \text{ ordered range of } \tau\circ\sigma_0. &
\end{array}$$

Let us tabulate the *changes* in the number of inversions as we pass from the ordered range of σ_0 [(IV.1.2), line (a)] to the ordered range of $\tau \circ \sigma_0$ (IV.1.2, line (b)). We observe that any of the fixed integers i which precedes both p and q in line (a) of (IV.1.2), *still* precedes p and q in line (b) of (IV.1.2). Hence, the number of inversions among all such pairs (i,p) and (i,q) is the same both before and after the action of τ.

Similarly, any of the fixed integers j, which is to the right of both p and q in line (a) of (IV.1.2), is *still* to the right of both p and q in line (b) of (IV.1.2). The number of inversions, or reversals of order among all pairs of the form (p,j) and (q,j) remains unchanged by the action of τ.

So far, we have compared the number of inversions relative to pairs of integers associating a *fixed* point of τ with one of the *nonfixed* points (that is, with p or q) of τ. We have yet to see how the inversions caused by a pair of integers change under the action of τ when *both* integers are fixed, or when *both* integers are *nonfixed* under τ.

Obviously, for *fixed* integers i and j [that is, $\tau(i) = i$ and $\tau(j) = j$], the *same relative ordering* (i,j) or (j,i) occurs between i and j in *both* lines (a) and (b) of (IV.1.2). τ does not alter the number of inversions induced by the association of a pair of *fixed* points.

The *only* case remaining concerns the inversion change (if any) for the pair of nonfixed points of τ; p and q. Regardless of whether p is less than q, we have p physically *preceding* q in line (a) of (IV.1.2), while p is to the *right* of q in line (b) of (IV.1.2), after the action of τ. Therefore, there is a *change* of 1 inversion in this case. If $p < q$, then we go from (p,q) in (a), a noninversion, to (q,p) in (b), an inversion. If $p > q$, we *start* with an inversion (p,q) in (a) of (IV.1.2) and after the action of τ, we finish with

(q,p) in (b) of (IV.1.2), which does *not* represent an inversion. Thus, we have shown, after considering all associations of integer pairs and their changes in inversion under τ, that any "adjacent" transposition *changes the number of inversions by 1*, that is, we have 1 more or 1 less inversion in the ordered range of $\tau \circ \sigma_0$, than we do in the ordered range of σ_0.

We are ready to tackle the general transposition τ that *does not* require that the nonfixed points p and q are adjacent in the ordered range of σ_0. Happily, such a general transposition can always be effected by applying a succession of our "adjacent" permutations (which interchange only physically adjacent integers). We will show, in fact, that *any* transposition permutation can be obtained by applying an *odd* number of "adjacent" transpositions; the first part of our proof tells us that "adjacent" transpositions change the number of inversions by plus or minus 1. Hence, an *odd* number of adjacent transpositions will have to account for a total of an *odd* number of changes in inversions.

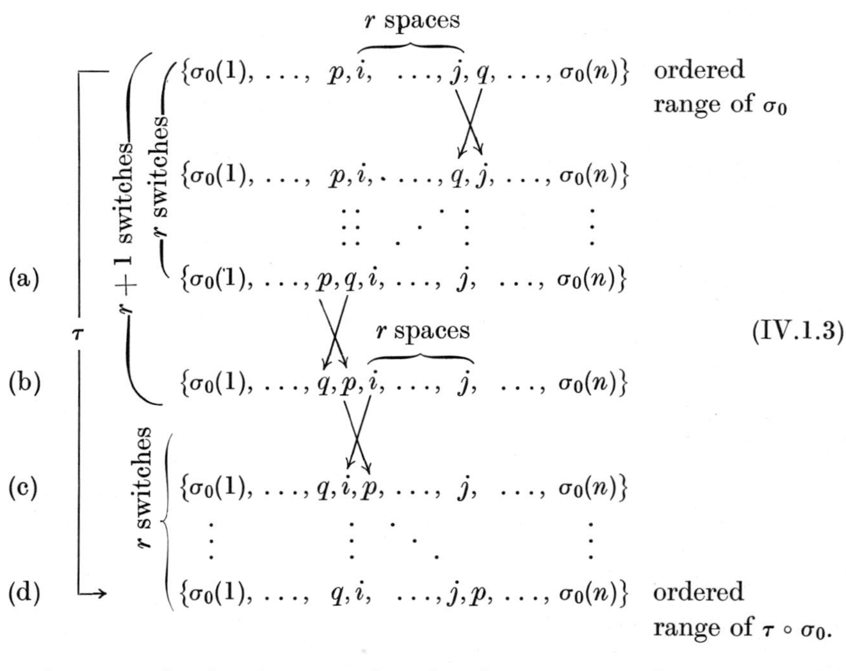

(IV.1.3)

Let τ have nonfixed points p and q, that is, $\tau(p) = q$ and $\tau(q) = p$ and let there be exactly r integers physically *between* p and q in the ordered range of σ_0 [see (IV.1.3)]. Our first "adjacent" transposition interchanges q with j, the integer to its immediate left. Our second transposition interchanges q with the integer now to its immediate left. After r such adjacent transpositions, our range looks like line (a) of (IV.1.3) where p and q are adjacent. Interchange p with q and we are on line (b) of (IV.1.3). To this point, we have consumed $r+1$ "adjacent" transpositions. We now interchange the integer p with integers to the immediate right; doing this r times takes us from line (b) to line (d) of (IV.1.3). Altogether, then, we are able to effect τ by applying $2r+1$ "adjacent" transpositions, and $2r+1$ is always an *odd* integer. How many changes of *inversions* has occurred, then? Each "adjacent" transposition accounts for a change of 1 inversion. Hence, the *total* number of inversion changes must be

$$\underbrace{(\pm 1)+(\pm 1)+(\pm)+\cdots+(\pm 1)}_{2r+1 \text{ terms}} = \text{an } odd \text{ number.} \qquad \text{(IV.1.4)}$$

But any sum of terms which are either 1 or -1, occurring an *odd* number of times, yields an *odd* number. Consequently, *any* transposition permutation changes the number of inversions by an odd number.

Corollary IV.1.6 Transpositions are always odd permutations.

Proof: Take σ_0 in Theorem IV.1.5 to be e, the identity permutation which happens to be an even permutation.

Example Let $\sigma = \begin{pmatrix} 1 & 2 & 3 & 4 & 5 & 6 & 7 \\ 4 & 2 & 1 & 7 & 6 & 5 & 3 \end{pmatrix}$ and let $\tau = \begin{pmatrix} 1 & 2 & 3 & 4 & 5 & 6 & 7 \\ 1 & 6 & 3 & 4 & 5 & 2 & 7 \end{pmatrix}$ represent the transposition whose nonfixed points are 2 and 6. Considering the ordered range of σ, we see that out of all the 21 possible ordered pairs therein, only the 10 ordered pairs

$$(4,2),\ (4,1),\ (4,3),\ (2,1),\ (7,6),\ (7,5),\ (7,3),\ (6,5),\ (6,3),\ (5,3),$$

account for inversions. Ten is an *even* number so we call σ an *even* permutation.

If we follow σ with the transposition τ, we obtain $\tau \circ \sigma = \begin{pmatrix} 1 & 2 & 3 & 4 & 5 & 6 & 7 \\ 4 & 6 & 1 & 7 & 2 & 5 & 3 \end{pmatrix}$. Out of 21 possible ordered pairs from the ordered range of $\tau \circ \sigma$, only the 11 ordered pairs

$$(4,1),\ (4,2),\ (4,3),\ (6,1),\ (6,2),\ (6,5),\ (6,3),\ (7,2),\ (7,5),\ (7,3),\ (5,3),$$

account for inversions. In any case, $\tau \circ \sigma$ is an *odd* permutation inasmuch as 11 is an *odd* integer. Indeed, $\tau \circ \sigma$ has the opposite parity of σ in this example.

EXERCISES

IV.1.10 Compute the parity of the given permutations.

(a) $\sigma = \begin{pmatrix} 1 & 2 & 3 & 4 & 5 & 6 & 7 & 8 \\ 1 & 3 & 6 & 7 & 2 & 4 & 8 & 5 \end{pmatrix}$

(b) $\tau = \begin{pmatrix} 1 & 2 & 3 & 4 & 5 & 6 & 7 & 8 \\ 1 & 2 & 4 & 3 & 5 & 7 & 6 & 8 \end{pmatrix}$

(c) $\sigma \circ \tau$
(d) $\tau \circ \sigma$
(e) σ^{-1}
(f) τ^{-1}

IV.1.11 Compute the parity of the given permutations.

(a) $\sigma = \begin{pmatrix} 1 & 2 & 3 & 4 & 5 \\ 2 & 3 & 5 & 1 & 4 \end{pmatrix}$

(b) $\tau = \begin{pmatrix} 1 & 2 & 3 & 4 & 5 \\ 5 & 3 & 1 & 2 & 1 \end{pmatrix}$

(c) $\sigma \circ \tau$
(d) $\tau \circ \sigma$
(e) $\tau \circ \tau \circ \sigma$
(f) $\sigma \circ \tau \circ \sigma$
(g) σ^{-1}
(h) τ^{-1}

IV.1.12 (a) List the permutations in S_2 and compute the parities.
(b) Is the parity of each permutation σ equal to the parity of the inverse permutation σ^{-1}?

IV.1.13 (a) List the 6 permutations in S_3 and compute the parities.
(b) Is the parity of each permutation σ in S_3, equal to the parity of the inverse permutation σ^{-1}?

IV.1.14 Given the transposition $\tau = \begin{pmatrix} 1 & 2 & 3 & 4 & 5 & 6 & 7 & 8 & 9 & 10 \\ 1 & 2 & 8 & 4 & 5 & 6 & 7 & 3 & 9 & 10 \end{pmatrix}$ in S_{10}. List the 9 "adjacent" transpositions (see Proof of Theorem IV.1.5) $\tau_1, \tau_2, \ldots, \tau_9$ in S_{10} such that $\tau = \tau_9 \circ \cdots \circ \tau_2 \circ \tau_1$.

IV.1.15 Show that for any transposition permutation τ, is its own inverse, that is, $\tau^{-1} = \tau$,

IV.1.16 Given the permutation σ and the transposition τ, find the number of inversions induced by σ and the number of inversions induced by $\tau \circ \sigma$ (see Theorem IV.1.5).

(a) $\sigma = \begin{pmatrix} 1 & 2 & 3 & 4 & 5 & 6 \\ 6 & 5 & 4 & 3 & 2 & 1 \end{pmatrix}, \qquad \tau = \begin{pmatrix} 1 & 2 & 3 & 4 & 5 & 6 \\ 1 & 5 & 3 & 4 & 2 & 6 \end{pmatrix}.$

(b) $\sigma = \begin{pmatrix} 1 & 2 & 3 & 4 & 5 & 6 \\ 5 & 1 & 3 & 2 & 6 & 4 \end{pmatrix}, \qquad \tau = \begin{pmatrix} 1 & 2 & 3 & 4 & 5 & 6 \\ 1 & 2 & 3 & 4 & 6 & 5 \end{pmatrix}.$

(c) $\sigma = \begin{pmatrix} 1 & 2 & 3 & 4 \\ 1 & 3 & 2 & 4 \end{pmatrix}, \qquad \tau = \begin{pmatrix} 1 & 2 & 3 & 4 \\ 1 & 2 & 4 & 3 \end{pmatrix}.$

IV.1.17 Transpositions are not the only permutations which are their own inverses (see Exercise IV.1.15).

(a) Show that if $\sigma = \begin{pmatrix} 1 & 2 & 3 & 4 & 5 \\ 5 & 4 & 3 & 2 & 1 \end{pmatrix}$, then $\sigma = \sigma^{-1}$.

(b) Show that if $\sigma = \begin{pmatrix} 1 & 2 & \cdots & k & \cdots & n \\ n & n-1 & \cdots & n-k+1 & \cdots & 1 \end{pmatrix}$, then $\sigma = \sigma^{-1}$.

V.1.18 Show that for the permutation

$$\sigma = \begin{pmatrix} 1 & 2 & \cdots & k & \cdots & n-1 & n \\ n & n-1 & \cdots & n-k+1 & \cdots & 2 & 1 \end{pmatrix},$$

$$\begin{aligned} \text{the parity of } \sigma &= \text{the parity of the integer } \frac{n}{2} \text{ if } n \text{ is even,} \\ &= \text{the parity of the integer } \frac{n-1}{2} \text{ if } n \text{ is odd.} \end{aligned}$$

IV.1.19 Given $\sigma \in S_n$, write σ as a composition of transpositions, that is, write $\sigma = \tau_k \circ \cdots \circ \tau_2 \circ \tau_1$ where each τ_i is a transposition. Compare the parity of the number of transpositions with the parity of σ.

(a) $\sigma = \begin{pmatrix} 1 & 2 & 3 & 4 \\ 3 & 1 & 2 & 4 \end{pmatrix}$

(b) $\sigma = \begin{pmatrix} 1 & 2 & 3 & 4 \\ 4 & 3 & 2 & 1 \end{pmatrix}$

(c) $\sigma = \begin{pmatrix} 1 & 2 & 3 & 4 & 5 & 6 \\ 1 & 4 & 3 & 6 & 5 & 2 \end{pmatrix}$

IV.1.20 (Generalization of Exercise IV.1.19.)

(a) Prove by induction on n that any permutation $\sigma \in S_n$ can always be written as a product (composition) of transposition permutations in S_n, that is, for each $\sigma \in S_n$,

$$\sigma = \tau_k \circ \tau_{k-1} \circ \cdots \circ \tau_2 \circ \tau_1, \tag{1}$$

where each τ_i is a transposition. (*Hint*: See Appendix A.) Let $P(n)$ be the statement "Each $\sigma \in S_n$ can be written as a product of certain transpositions in S_n." Surely $P(1)$ is true. It is easy to check that $P(2)$ is true. Now assume the truth of $P(n)$ for some arbitrary (but fixed) n. You are to show that the truth of $P(n)$ will force the truth of $P(n+1)$. In fact, let $\sigma \in S_{n+1}$ be a typical element of S_{n+1} with ordered range

$$\operatorname{rg}(\sigma) = \{\sigma(1),\sigma(2),\ldots,\underset{\substack{\uparrow \\ i\text{th } place}}{n+1},\ldots,\sigma(n),\sigma(n+1)\}, \tag{2}$$

where the integer $n+1$ appears in the ith place, that is, $\sigma(i) = n+1$. Let us restore $n+1$ to its rightful place by following σ with the transposition τ_0 which has $n+1$ and $\sigma(n+1)$ as its nonfixed points. From (2) we see that

$$\operatorname{rg}(\tau_0 \circ \sigma) = \{\sigma(1),\sigma(2),\ldots,\sigma(n+1),\ldots,\sigma(n),\underset{\substack{\uparrow \\ n+1\text{st } place}}{n+1}\}. \tag{3}$$

From (3) we see that $\tau_0 \circ \sigma$ leaves $n+1$ fixed and so may be thought of as a permutation only on the first n letters $\{1,2,\ldots,n\}$; that is, $\tau_0 \circ \sigma \in S_n$ and lends itself to form (1) by our assumption that $P(n)$ is true. Prove now that $P(n+1)$ is true.

(b) From part (a), any $\sigma \in S_n$ is of the form (1). Prove that the parity (evenness or oddness) of k, the number of transpositions, is always the same as the parity of the number of inversions in the ordered range of σ. In other words, if $\sigma = \tau_k \circ \tau_{k+1} \circ \cdots \circ \tau_2 \circ \tau_1$, then the parity of k equals the parity of σ. (*Hint*: Use Theorem IV.1.5.)

2. *Definition of the Determinant*

Before we formally define the function det : $\mathcal{M}_{n,n} \to$ scalars, we should be prepared for a minor surprise. Our definition will consist of a sum of certain terms $a(\sigma)$, say, each term being dependent upon a *permutation* σ in S_n. There will be 1 term in our sum assigned to each and every permutation σ in S_n. Hence, we shall encounter a summation $\sum_{\sigma \in S_n} a(\sigma)$, which employs the symbol $\sum_{\sigma \in S_n}$. Thus, instead of letting our indexing symbol (or dummy variable) be an *integer*, our indexing symbol (or dummy variable) is a *permutation* in S_n. The symbol $\sum_{\sigma \in S_n} a(\sigma)$ may be read, "the sum of all numbers of the form $a(\sigma)$, as σ varies over all the possible permutations in the set S_n."

Of course, once we know what this peculiar scalar $a(\sigma)$ is, which depends on the permutation σ, our definition is complete. Being forewarned about the summation over permutations, as opposed to integers, we state the definition of the function det which sends $\mathcal{M}_{n,n}$, the algebra of all $n \times n$ matrices, to the set of scalars.

Definition IV.2.1 Let $A = (a_{ij})_{ij}$ be an $n \times n$ matrix in $\mathcal{M}_{n,n}$. The function det [or det $(\cdot)$], the **determinant** function, is denoted

$$\det((a_{ij})_{ij}) = \det \begin{pmatrix} a_{11} & a_{12} & \cdots & a_{ij} & \cdots & a_{1n} \\ a_{21} & a_{22} & \cdots & a_{2j} & \cdots & a_{2n} \\ \vdots & \vdots & & \vdots & & \vdots \\ a_{i1} & a_{i2} & \cdots & a_{ij} & \cdots & a_{in} \\ \vdots & \vdots & & \vdots & & \vdots \\ a_{n1} & a_{n2} & \cdots & a_{nj} & \cdots & a_{nn} \end{pmatrix}$$

and is defined to be equal to the sum

$$\sum_{\sigma \in S_n} \operatorname{sgn} \begin{pmatrix} 1 & 2 & \cdots & i & \cdots & n \\ \sigma(1) & \sigma(2) & \cdots & \sigma(i) & \cdots & \sigma(n) \end{pmatrix} \cdot \overbrace{a_{1\sigma(1)} a_{2\sigma(2)} \cdots a_{1\sigma(i)} \cdots a_{n\sigma(n)}}^{\text{entry from row } i, \text{ column } \sigma(i)} \tag{IV.2.1}$$

(same ordering: $1, 2, \ldots, i, \ldots, n$ in the top row of the permutation correspond to the row indices of $a_{1\sigma(1)}, a_{2\sigma(2)}, \ldots, a_{1\sigma(i)}, \ldots, a_{n\sigma(n)}$)

where $\operatorname{sgn}\begin{pmatrix}1 & 2 \cdots i \cdots n\\ \sigma(1) & \sigma(2)\cdots\sigma(i)\cdots\sigma(n)\end{pmatrix}$, which may be denoted sgn (σ) and read **the sign of $\boldsymbol{\sigma}$**, is defined by

$$\operatorname{sgn}\begin{pmatrix}1 & 2 \cdots i \cdots n\\ \sigma(1) & \sigma(2)\cdots\sigma(i)\cdots\sigma(n)\end{pmatrix} = +1 \text{ if } \sigma \text{ is even,}$$

$$\operatorname{sgn}\begin{pmatrix}1 & 2 \cdots i \cdots n\\ \sigma(1) & \sigma(2)\cdots\sigma(i)\ldots\sigma(n)\end{pmatrix} = -1 \text{ if } \sigma \text{ is odd.}$$

Our next 2 examples apply Definition IV.2.1 to find det (A) explicitly for the case of a typical 2×2 matrix, and for a typical 3×3 matrix.

Example Let $A = \begin{pmatrix}a & b\\ c & d\end{pmatrix}$ represent a typical 2×2 matrix. We are to find the value for $\det\begin{pmatrix}a & b\\ c & d\end{pmatrix}$. Now (IV.2.1) of Definition IV.2.1 tells us that we are summing over all the permutations in S_2. But S_2 consists of just the 2 permutations

$$e = \begin{pmatrix}1 & 2\\ 1 & 2\end{pmatrix} \quad\text{and}\quad \sigma = \begin{pmatrix}1 & 2\\ 2 & 1\end{pmatrix}.$$

Now $\begin{pmatrix}1 & 2\\ 1 & 2\end{pmatrix}$ is even and $\begin{pmatrix}1 & 2\\ 2 & 1\end{pmatrix}$ is odd, so that $\operatorname{sgn}\begin{pmatrix}1 & 2\\ 1 & 2\end{pmatrix} = +1$ and $\operatorname{sgn}\begin{pmatrix}1 & 2\\ 2 & 1\end{pmatrix} = -1$. Thus, the sum (IV.1.2) reads

$$\det(A) = \sum_{\sigma\in S_2}\operatorname{sgn}\begin{pmatrix}1 & 2\\ \sigma(1) & \sigma(2)\end{pmatrix} a_{1\sigma(1)}\cdot a_{2\sigma(2)}$$

$$= \underbrace{\operatorname{sgn}\begin{pmatrix}1 & 2\\ 1 & 2\end{pmatrix}}_{+1} a_{11}\cdot a_{22} + \underbrace{\operatorname{sgn}\begin{pmatrix}1 & 2\\ 2 & 1\end{pmatrix}}_{-1} a_{12}\cdot a_{21} \tag{IV.2.2}$$

$$= \underbrace{a_{11}}_{\substack{\text{row 1}\\ \text{col 1}}} \cdot \underbrace{a_{22}}_{\substack{\text{row 2}\\ \text{col 1}}} - \underbrace{a_{12}}_{\substack{\text{row 1}\\ \text{col 2}}} \cdot \underbrace{a_{21}}_{\substack{\text{row 2}\\ \text{col 1}}}.$$

The 4 sets of subscripts (1,1), (2,2), (1,2), and (2,1), which appear above, refer *only* to the entries of the *matrix A* from which the scalars a_{11}, a_{22}, a_{12}, and a_{21} are extracted. Since $A = \begin{pmatrix}a & b\\ c & d\end{pmatrix}$, we have

$$a_{11} = a,\ a_{22} = d,\ a_{12} = b, \quad\text{and}\quad a_{21} = c.$$

Substituting these values into the last equality of (IV.2.2) yields

$$\det\begin{pmatrix}a & b\\ c & d\end{pmatrix} = a\cdot d - b\cdot c$$

for any 2×2 matrix $\begin{pmatrix}a & b\\ c & d\end{pmatrix}$.

Example Let

$$A = \begin{pmatrix}a_{11} & a_{12} & a_{13}\\ a_{21} & a_{22} & a_{23}\\ a_{31} & a_{32} & a_{33}\end{pmatrix}$$

represent a typical 3×3 matrix in $\mathscr{M}_{3,3}$. According to Definition IV.2.1, we will eventually have to worry about each permutation σ in S_3, so as to determine its parity, and hence, the value for $\operatorname{sgn}(\sigma)$. Let us do this first by listing the 6 permutations which comprise S_3.

$$\begin{array}{cc}
\sigma & \sigma \\
\left[\left.\begin{array}{c}\begin{pmatrix}1&2&3\\1&2&3\end{pmatrix}\\[6pt]\begin{pmatrix}1&2&3\\3&1&2\end{pmatrix}\\[6pt]\begin{pmatrix}1&2&3\\2&3&1\end{pmatrix}\end{array}\right\}\text{even}\right. &
\left[\left.\begin{array}{c}\begin{pmatrix}1&2&3\\2&1&3\end{pmatrix}\\[6pt]\begin{pmatrix}1&2&3\\3&2&1\end{pmatrix}\\[6pt]\begin{pmatrix}1&2&3\\1&3&2\end{pmatrix}\end{array}\right\}\text{odd.}\right. \\
\operatorname{sgn}(\sigma) = +1 & \operatorname{sgn}(\sigma) = -1
\end{array}$$

The left-hand column lists the 3 *even* permutations (which includes the identity permutation); $\operatorname{sgn}(\sigma) = +1$ for these permutations. The right-hand column lists the 3 *odd* permutations of S_3. Each one of these, incidentally, happens to be a transposition (see Corollary IV.I.6); $\operatorname{sgn}(\sigma) = -1$ for each σ in this category.

Let us write the 6 terms of the form $a_{1\sigma(1)}a_{2\sigma(2)}a_{3\sigma(3)}$, in accordance with (IV.2.1) of Definition IV.2.1. We remember to keep the *first* subscripts (the row indices) in natural order $\{1,2,3\}$, and the second subscripts (the column indices) in exactly the same order as the ordered range of the permutation σ, $\{\sigma(1), \sigma(2), \sigma(3)\}$. We attach a plus 1 or a minus 1 to each $a_{1\sigma(1)}a_{2\sigma(2)}a_{3\sigma(3)}$ according as whether σ is even or odd. Thus,

$$\begin{aligned}
\det\begin{pmatrix}a_{11}&a_{12}&a_{13}\\a_{21}&a_{22}&a_{23}\\a_{31}&a_{32}&a_{33}\end{pmatrix} &= \overset{+1}{\operatorname{sgn}\begin{pmatrix}1&2&3\\1&2&3\end{pmatrix}}a_{11}\cdot a_{22}\cdot a_{33}\\
&\quad + \overset{+1}{\operatorname{sgn}\begin{pmatrix}1&2&3\\3&1&2\end{pmatrix}}a_{13}\cdot a_{21}\cdot a_{32}\\
&\quad + \overset{+1}{\operatorname{sgn}\begin{pmatrix}1&2&3\\2&3&1\end{pmatrix}}a_{12}\cdot a_{23}\cdot a_{31}\\
&\quad + \overset{-1}{\operatorname{sgn}\begin{pmatrix}1&2&3\\2&1&3\end{pmatrix}}a_{12}\cdot a_{21}\cdot a_{33}\\
&\quad + \overset{-1}{\operatorname{sgn}\begin{pmatrix}1&2&3\\3&2&1\end{pmatrix}}a_{13}\cdot a_{22}\cdot a_{31}\\
&\quad + \overset{-1}{\operatorname{sgn}\begin{pmatrix}1&2&3\\1&3&2\end{pmatrix}}a_{11}\cdot a_{23}\cdot a_{32}\\
&= a_{11}\cdot a_{22}\cdot a_{33} + a_{13}\cdot a_{21}\cdot a_{32} + a_{12}\cdot a_{23}\cdot a_{31}\\
&\quad - a_{12}\cdot a_{21}\cdot a_{33} - a_{13}\cdot a_{22}\cdot a_{31} - a_{11}\cdot a_{23}\cdot a_{32}.
\end{aligned} \tag{IV.2.3}$$

Now the determinant of a 4×4 matrix can be similarly computed. As there are 24 permutations in S_4, there will be 24 terms of the form $a_{1\sigma(1)}\, a_{2\sigma(2)}\, a_{3\sigma(3)}\, a_{4\sigma(4)}$ to contend with in that case: A very good problem for someone stranded on a desert island with nothing else to do.

For the actual computations of determinants of "large" square matrices, some clever tricks will have to be employed. The tricks that we shall use will entail an altering of the original matrix in question in such a way that the determinant of the altered matrix is easier to compute, and relates in a "natural" way to the determinant of the original matrix.

Before we proceed, however, let us offer some remarks on the definition of the determinant function and the notation used.

REMARKS. As we have pointed out, the typical term in the sum of (IV.2.1), $\operatorname{sgn}(\sigma)\; a_{1\sigma(1)}\; a_{2\sigma(2)} \ldots a_{i\sigma i)} \ldots a_{n\sigma(n)}$, depends upon a typical permutation $\sigma = \begin{pmatrix} 1 & 2 \;\ldots\; i \;\ldots\; n \\ \sigma(1) & \sigma(2) \ldots \sigma(i) \ldots \sigma(n) \end{pmatrix}$ in S_n.

(1) The subscript pairs that appear, $1\sigma(1)$, $2\sigma(2), \ldots, i\sigma(i), \ldots, n\sigma(n)$, refer *only* to the entry number of the row and column from which each scalar is extracted.

(2) Note that Definition IV.2.1 requires, as a matter of form, that the *first* subscripts (that is, the row indices for the scalars) are always in the *natural* order $\{1,2,\ldots,i,\ldots,n\}$. However, the *second* subscripts (that is, the column indices for the scalars) are the first n integers, but their order is *scrambled*. In fact, this scrambled set of second subscripts $\{\sigma(1),\sigma(2),\ldots, \sigma(i),\ldots,\sigma(n)\}$ is just the *ordered range of the permutation*

$$\sigma = \begin{pmatrix} 1 & 2 \;\ldots\; i \;\ldots\; n \\ \sigma(1) & \sigma(2) \ldots \sigma(i) \ldots \sigma(n) \end{pmatrix}.$$

(3) $\operatorname{sgn}(\sigma)$ has its value (plus or minus 1) wholly dependent, then, upon the order of the set of *second* subscripts, once the first subscripts (row indices) appear in the *natural* order. In fact, $\operatorname{sgn}(\sigma) = +1$ if there are an even number of inverted pairs in the ordered set $\{\sigma(1),\sigma(2),\ldots,\sigma(i),\ldots, \sigma(n)\}$ of second subscripts, and $\operatorname{sgn}(\sigma) = -1$ when an odd number of inverted pairs appears in the ordered set of second entries.

(4) In considering a typical term $\operatorname{sgn}(\sigma) \cdot a_{1\sigma(1)} \cdot a_{2\sigma(2)} \cdots a_{i\sigma(i)} \cdots a_{n\sigma(n)}$, we see that the first factor $a_{1\sigma(1)}$ is an element of the first row, the second factor $a_{2\sigma(2)}$ is an element of the second row, and in general, the ith factor, is an element of the ith row. Thus we have n factors, 1 from each 1 of the n possible rows. Similarly, the second subscripts $\sigma(1)$, $\sigma(2)$,

$\ldots, \sigma(i), \ldots, \sigma(n)$ are just the first n integers in some order, which indicates that the n factors are also extracted from *each* of the n *columns*. Thus, $a_{1\sigma(1)} \cdot a_{2\sigma(2)} \cdots a_{i\sigma(i)} \cdots a_{n\sigma(n)}$ represents a product of n scalars, 1 chosen from *each* row and *each* column of the $n \times n$ matrix A.

There is a simple graphical device which allows us to compute the determinant of a 2×2 or a 3×3 matrix. The previous two examples gave us the determinant of such matrices explicitly in terms of the entries of the matrix. Therefore, we shall cite (IV.2.2) and (IV.2.3) to justify the validity of the graphical method which follows (Figure 4.1).

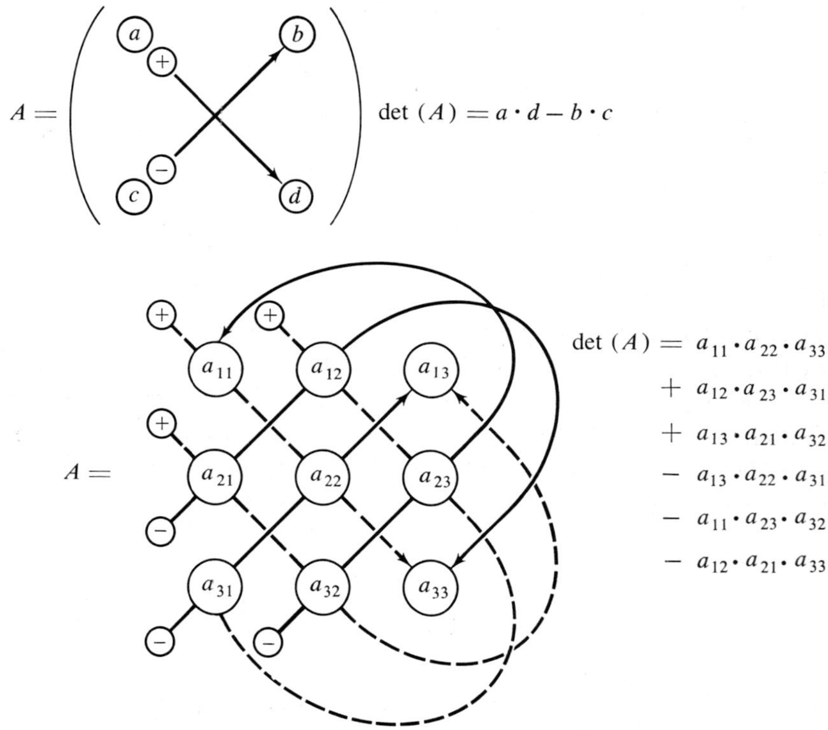

Figure 4.1

The matrices A are written in such a way that any straight line $\nearrow$ touching at least 2 entries of A from lower left to upper right, captures those entries which multiply together, with a *minus* sign attached. Similarly, any straight line $\searrow$ passing through at least 2 entries from upper left to lower right captures those entries which, when multiplied together, account for a $+1$ factor in the sum for $\det(A)$. This procedure does *not* extend to $n \times n$ matrices where n is greater than 3.

EXERCISES

IV.2.1 Given the 2×2 matrix $A=\begin{pmatrix} a & b \\ c & d \end{pmatrix}$. If A^t, the transpose of A, is defined to be $A^t=\begin{pmatrix} a & c \\ b & d \end{pmatrix}$, then show that $\det(A)=\det(A^t)$.

IV.2.2 Given the 3×3 matrix $A = (a_{ij})_{ij}$, $i,j = 1,2,3$. If A^t, the transpose of A, is defined by $A^t = (a_{ji})_{ij}$, $i,j = 1,2,3$, then prove that $\det(A) = \det(A^t)$.

IV.2.3 Find the value of the determinant function on each of the 2×2 elementary matrices (see Definition III.3.2), R_{12}, $R_i(a)$ where $i = 1$ or 2, $R_2{}^1(a)$, C_{12}, $C_i(a)$ where $i = 1$ or 2, $C_1{}^2(a)$.

IV.2.4 (See Exercise IV.2.3). Find the value of the determinant function on each of the 3×3 elementary matrices $R_{1,3}$, $R_{2,3}$, $R_1(-6)$, $R_1{}^3(7)$, $C_{2,3}$, $C_2(a)$, $C_2{}^1(a)$, $C_2{}^3(a)$, $C_3{}^1(a)$.

IV.2.5 Find

$$\det\begin{pmatrix} 1 & 4 & 6 \\ 0 & 2 & 2 \\ 0 & 0 & 3 \end{pmatrix}.$$

IV.2.6 Find

$$\det\begin{pmatrix} a & p & q \\ 0 & b & r \\ 0 & 0 & c \end{pmatrix}.$$

IV.2.7 In generalizing Exercises IV.2.5 and IV.2.6, show that for *any* triangular square matrix

$$A = \begin{pmatrix} a_{11} & * & \cdots & * \\ 0 & a_{22} & \cdots & * \\ \vdots & & \ddots & \vdots \\ 0 & \cdots & 0 & a_{nn} \end{pmatrix},$$

whose each entry below the main diagonal is zero, $\det(A) = a_{11}a_{22}\ldots a_{nn}$, the product of the main diagonal elements. Hence, as an easy special case, $\det(I_n) = 1$. See (4) of Remarks section on page 297.

IV.2.8 Find the value of the determinant on the following matrices:

(a) $\begin{pmatrix} 1 & 1 \\ 2 & 2 \end{pmatrix}$, (b) $\begin{pmatrix} 1 & 2 & 3 \\ 4 & -1 & 2 \\ 5 & 1 & 5 \end{pmatrix}$, (c) $\begin{pmatrix} 3 & 0 & 1 \\ 0 & 1 & 3 \\ 2 & 1 & 1 \end{pmatrix}$,

(d) $\begin{pmatrix} 1 & 7 & 6 \\ 2 & 0 & 0 \\ 1 & -7 & -6 \end{pmatrix}$, (e) $\begin{pmatrix} 0 & 1 & 2 \\ 3 & 4 & 1 \\ 1 & 2 & 1 \end{pmatrix}$.

IV.2.9 Given the square matrices A and B, compute (a) $\det(A)$, (b) $\det(B)$, (c) $\det(A \cdot B)$, (d) $\det(B \cdot A)$, and (e) $\det(A) \cdot \det(B)$.

(a) $A = \begin{pmatrix} 1 & 0 & 0 \\ 0 & 0 & 0 \\ 0 & 0 & 0 \end{pmatrix}$, $B = \begin{pmatrix} 0 & 1 & 0 \\ 0 & 0 & 0 \\ 0 & 0 & 0 \end{pmatrix}$.

(b) $A = \begin{pmatrix} 1 & 0 \\ 0 & 0 \end{pmatrix}, \quad B = \begin{pmatrix} a & b \\ c & d \end{pmatrix}.$

(c) $A = \begin{pmatrix} 0 & 1 \\ 1 & 0 \end{pmatrix}, \quad B = \begin{pmatrix} a & b \\ c & d \end{pmatrix}.$

(d) $A = \begin{pmatrix} 0 & a \\ b & 0 \end{pmatrix}, \quad B = \begin{pmatrix} c & 3 \\ 0 & d \end{pmatrix}.$

(e) $A = \begin{pmatrix} 4 & 2 & 1 \\ 0 & 1 & 1 \\ 0 & 0 & 3 \end{pmatrix}, B = \begin{pmatrix} 3 & 0 & 2 \\ 0 & 6 & 1 \\ 0 & 0 & 1 \end{pmatrix}.$

IV.2.10 Prove that for 3×3 matrices

$$\det \begin{pmatrix} p_1 + q_1 & a_{12} & a_{13} \\ p_2 + q_2 & a_{22} & a_{23} \\ p_3 + q_3 & a_{32} & a_{33} \end{pmatrix} = \det \begin{pmatrix} p_1 & a_{12} & a_{13} \\ p_2 & a_{22} & a_{23} \\ p_3 & a_{32} & a_{33} \end{pmatrix} + \det \begin{pmatrix} q_1 & a_{12} & a_{13} \\ q_2 & a_{22} & a_{23} \\ q_3 & a_{32} & a_{33} \end{pmatrix}.$$

IV.2.11 Prove that for 3×3 matrices

$$\det \begin{pmatrix} a_{11} & a_{12} & a_{13} \\ p_1 + q_1 & p_2 + q_2 & p_3 + q_3 \\ a_{31} & a_{32} & a_{33} \end{pmatrix} = \det \begin{pmatrix} a_{11} & a_{12} & a_{13} \\ p_1 & p_2 & p_3 \\ a_{31} & a_{32} & a_{33} \end{pmatrix} + \det \begin{pmatrix} a_{11} & a_{12} & a_{13} \\ q_1 & q_2 & q_3 \\ a_{31} & a_{32} & a_{33} \end{pmatrix}.$$

IV.2.12 Find

$$\det \begin{pmatrix} 2+3 & 1+6 & 0+1 \\ 5 & 1 & 0 \\ 2 & 1\ 1 & 1 \end{pmatrix}$$

and verify, by computation, that this value is equal to

$$\det \begin{pmatrix} 2 & 1 & 0 \\ 5 & 1 & 0 \\ 2 & 1 & 1 \end{pmatrix} + \det \begin{pmatrix} 3 & 6 & 1 \\ 5 & 1 & 0 \\ 2 & 1 & 1 \end{pmatrix}.$$

In what follows, our objective is to determine how the value of the determinant is affected when we alter our matrix in a certain way; these "alterations" will eventually allow us to reduce an apparently unwieldy $n \times n$ matrix to an $n \times n$ matrix whose determinant is easier to compute.

What exactly *are* these alterations? Briefly stated, we shall alter a matrix by applying at least one of the 3 elementary row operations, R_{ij}, $R_i(a)$ where $a \neq 0$, or $R_j{}^i(a)$ (Definition III.1.2). Evidently, we shall ask how the determinant of a matrix A changes after the matrix is multiplied on the left by any of the elementary matrixes (Theorem III.3.1 and Definition III.3.2). The theorems which speak to this point follow.

Theorem IV.2.2 Let $A = (a_{ij})_{ij}$ be any $n \times n$ matrix. If A' is the matrix obtained from A by interchanging the pth and qth rows, (that is, $A' = R_{pq} \cdot A$), then

$$\det(A') = -\det(A).$$

Equivalently, (IV.2.4)

$$\det(R_{pq} \cdot A) = -\det(A).$$

Proof: From the hypothesis, we have

$$A = \begin{pmatrix} a_{11} & a_{12} & \dots & a_{1n} \\ \vdots & \vdots & & \vdots \\ a_{p1} & a_{p2} & \dots & a_{pn} \\ \vdots & \vdots & & \vdots \\ a_{q1} & a_{q2} & \dots & a_{qn} \\ \vdots & \vdots & & \vdots \\ a_{n1} & a_{n2} & \dots & a_{nn} \end{pmatrix} \begin{matrix} \\ \\ \leftarrow p\text{th row of } A \\ \\ \leftarrow q\text{th row of } A \\ \\ \\ \end{matrix}$$

$$A' = \begin{pmatrix} a_{11} & a_{12} & \dots & a_{1n} \\ \vdots & \vdots & & \vdots \\ a_{q1} & a_{q2} & \dots & a_{qn} \\ \vdots & \vdots & & \vdots \\ a_{p1} & a_{p2} & \dots & a_{pn} \\ \vdots & \vdots & & \vdots \\ a_{n1} & a_{n2} & \dots & a_{nn} \end{pmatrix} \begin{matrix} \\ \\ \leftarrow p\text{th row of } A' \\ \\ \leftarrow q\text{th row of } A'. \\ \\ \\ \end{matrix}$$

From the definition of the determinant function, we must agree that

$$\det(A') = \sum_{\sigma \in S_n} \operatorname{sgn} \begin{pmatrix} 1 & 2 \dots & i & \dots & n \\ \sigma(1) & \sigma(2)\dots & \sigma(i) & \dots & \sigma(n) \end{pmatrix} \times a_{1\sigma(1)} \cdots \underbrace{a_{q\sigma(p)}}_{\substack{\text{element} \\ \text{in row } p, \\ \text{col } \sigma(p) \\ \text{of } A'}} \cdots \underbrace{a_{p\sigma(q)}}_{\substack{\text{element} \\ \text{in row } q, \\ \text{col } \sigma(q) \\ \text{of } A'}} \cdots a_{n\sigma(n)}.$$

So as to force the above expression into form (IV.2.1), we must have the first subscripts in *natural* order. As the subscripts now stand the first subscripts are in natural order, except that the integers p and q $(p < q)$ are in inverse order. Therefore, we interchange the positions of the scalars $a_{p\sigma(q)}$ and $a_{q\sigma(p)}$ to obtain

$$\det(A') = \sum_{\sigma \in S_n} \operatorname{sgn} \begin{pmatrix} 1 & 2 \dots & p & \dots & q & \dots & n \\ \sigma(1) & (2)\dots & \sigma(p) & \dots & \sigma(q) & \dots & \sigma(n) \end{pmatrix} \times a_{1\sigma(1)} \cdots a_{p\sigma(q)} \cdots a_{q\sigma(p)} \cdots a_{n\sigma(n)}$$

so that the set of first subscripts are in the natural order.

How is the permutation $\sigma = \begin{pmatrix} 1 & \dots & p & \dots & q & \dots & n \\ \sigma(1) & \dots & \sigma(p) & \dots & \sigma(q) & \dots & \sigma(n) \end{pmatrix}$ (which appears above) related to the permutation

$$\tau = \begin{pmatrix} 1 & \dots & q & \dots & p & \dots & n \\ \sigma(1) & \dots & \sigma(q) & \dots & \sigma(p) & \dots & \sigma(n) \end{pmatrix}?$$

τ is simply the permutation σ *followed* by the transposition permutation $\tau_{\sigma(p),\sigma(q)}$ whose nonfixed points are the integers $\sigma(p)$ and $\sigma(q)$. That is, $\tau = \tau_{\sigma(p),\sigma(q)} \circ \sigma$, and so the parities of these 2 permutations, σ and τ, must be different (Theorem IV.1.5). This says that sgn $(\sigma) = -$ sgn (τ) from the very definition of sgn (·) (see Definition IV.1.2). We now replace sgn (σ) with its equivalent, $-$sgn (τ) in our last equation to obtain

$$\det(A') = -\sum_{\sigma \in S_n} \operatorname{sgn} \overbrace{\begin{pmatrix} 1 & 2 & \dots & p & \dots & q & \dots & n \\ \sigma(1) & \sigma(2) & \dots & \sigma(q) & \dots & \sigma(p) & \dots & \sigma(n) \end{pmatrix}}^{\tau = \tau_{\sigma(p),\sigma(q)} \circ \sigma}$$

same ordering

$$\times\, a_{1\sigma(1)} a_{2\sigma(2)} \cdots a_{p\sigma(q)} \cdots a_{q\sigma(p)} \cdots a_{n\sigma(n)},$$

which finally gives us the form of (IV.2.1) for $-\det(A)$, that is, this is the form [see (IV.2.1)] that demands that the *first* subscripts in the product $a_{1\sigma(1)} \cdots a_{p\sigma(q)} \cdots a_{q\sigma(p)} \cdots a_{n\sigma(n)}$ appear in natural order (we're O.K. here) and that *whatever* the order of the second subscripts, this set should coincide with the ordered range of the permutation which appears after the symbol "sgn" (and we're O.K. here). Thus, our last equation shows that det (A') equals $-$det (A) where A' is obtained from A by interchanging rows p and q (that is, $A' = R_{p,q} \cdot A$). That ends the proof of Theorem IV.2.2.

If you are willing to believe that the only number which is equal to its negative is the number zero, then you will find this next corollary to be a painless consequence of Theorem IV.2.2.

Corollary IV.2.3 Let A be an $n \times n$ matrix such that at least 2 rows are the same. Then det $(A) = 0$.

Proof: If A' is the matrix obtained from A by interchanging 2 *identical* rows, then $A' = A$. But Theorem IV.2.2 tells us that $\det(A') = -\det(A)$. Putting these 2 facts together, we have

$$-\det(A) = \det(A') = \det(A),$$

from Theorem IV.2.2 since $A = A'$. Thus, $\det(A)$ equals its own negative, $-\det(A)$, and hence, equals zero, which was to be proved.

Our next theorem tells us that in multiplying each entry of a row by a fixed scalar a, we effectively multiply the determinant by a. To be more precise, we present Theorem IV.2.4.

Theorem IV.2.4 Let $A = (a_{ij})_{ij}$ be any $n \times n$ matrix. If A' is the matrix obtained from A by multiplying each entry of the pth row by the scalar a [that is, $A' = R_p(a) \cdot A$], then

$$\det(A') = a \cdot \det(A).$$

Equivalently,

$$\det(R_p(a) \cdot A) = a \cdot \det(A). \tag{IV.2.5}$$

Proof: Note that the pth row of the original matrix A consists of the scalars $(a_{p1}, a_{p2}, \ldots, a_{pj}, \ldots, a_{pn})$. The pth row of A' consists of the scalars $(a \cdot a_{p1}, a \cdot a_{p2}, \ldots, a \cdot a_{pj} \ldots, a \cdot a_{pn})$. Hence, from Definition IV.2.1, we have

$$\det(A') = \sum_{\sigma \in S_n} \operatorname{sgn}\begin{pmatrix} 1 & 2 \ldots & p & \ldots & n \\ \sigma(1) & \sigma(2) \ldots & \sigma(p) & \ldots & \sigma(n) \end{pmatrix} \times a_{1\sigma(1)} \cdot a_{2\sigma(2)} \cdots \overbrace{(a \cdot a_{p\sigma(p)})}^{\text{entry from row } p\text{, column } \sigma(p) \text{ of } A'} \cdots a_{n\sigma(n)}$$

$$= a \cdot \left[\sum_{\sigma \in S_n} \operatorname{sgn}\begin{pmatrix} 1 & 2 \ldots & p & \ldots & n \\ \sigma(1) & \sigma(2) \ldots & \sigma(p) & \ldots & \sigma(n) \end{pmatrix} \times a_{1\sigma(1)} a_{2\sigma(2)} \cdots a_{p\sigma(p)} \cdots a_{n\sigma(n)} \right]$$

(since the scalar a factors out from each term after the symbol. See Exercise II.6.7).

$$= a \cdot \det((a_{ij})_{ij}) = a \cdot \det(A).$$

This completes our proof.

The next theorem is one which is almost hard to believe. It tells us that the most *elaborate* of our elementary row operations, that of adding an arbitrary multiple of one row to any other row, in no way affects the determinant. Let's see this now.

Theorem IV.2.5 Let $A = (a_{ij})_{ij}$ be any $n \times n$ matrix. If A' is the matrix obtained from A by adding a times the pth row of A to the qth row of A, where $p \neq q$, [that is, $A' = R_p{}^q(a) \cdot A$], then

$$\det(A') = \det(A).$$

Equivalently,

$$\det(R_p^q(a) \cdot A) = \det(A). \tag{IV.2.6}$$

Proof: Before we compute $\det(A')$, let us note that except for the qth row of A', the rows of A and A' are identical. The qth row of A' consists of the scalars $(a \cdot a_{p1} + a_{q1}, a \cdot a_{p2} + a_{q2}, \ldots, a \cdot a_{pn} + a_{qn})$. Thus, we may write

$$\det(A')$$

$$= \sum_{\sigma \in S_n} \operatorname{sgn}\begin{pmatrix} 1 & \ldots & p & \ldots & q & \ldots & n \\ \sigma(1) & \ldots & \sigma(p) & \ldots & \sigma(q) & \ldots & \sigma(n) \end{pmatrix} \times a_{1\sigma(1)} \cdots a_{p\sigma(p)} \cdots \overbrace{(a \cdot a_{p\sigma(q)} + a_{q\sigma(q)})}^{\text{entry from row } q,\ \text{column } \sigma(q) \text{ of } A'} \cdots a_{n\sigma(n)}$$

$$= (\mathrm{a}) \sum_{\sigma \in S_n} \operatorname{sgn}\begin{pmatrix} 1 & \ldots & p & \ldots & q & \ldots & n \\ \sigma(1) & \ldots & \sigma(p) & \ldots & \sigma(q) & \ldots & \sigma(n) \end{pmatrix} \times a_{1\sigma(1)} \cdots a_{p\sigma(p)} \cdots (a \cdot a_{p\sigma(q)}) \cdots a_{n\sigma(n)}$$

$$\underbrace{+ (\mathrm{b}) \sum_{\sigma \in S_n} \operatorname{sgn}\begin{pmatrix} 1 & \ldots & p & \ldots & q & \ldots & n \\ \sigma(1) & \ldots & \sigma(p) & \ldots & \sigma(q) & \ldots & \sigma(n) \end{pmatrix} a_{1\sigma(1)} \cdots a_{p\sigma(p)} \cdots a_{q\sigma(q)} \cdots a_{n\sigma(n)}}_{\det((a_{ij})_{ij}) = \det(A).}$$

$$= (\mathrm{a})\ a \cdot \sum_{\sigma \in S_n} \operatorname{sgn}\begin{pmatrix} 1 & \ldots & p & \ldots & q & \ldots & n \\ \sigma(1) & \ldots & \sigma(p) & \ldots & \sigma(q) & \ldots & \sigma(n) \end{pmatrix} a_{1\sigma(1)} \cdots \overbrace{a_{p\sigma(p)}}^{\text{entry from row } p,\ \text{column } \sigma(p)} \cdots \overbrace{a_{p\sigma(q)}}^{\text{entry from row } q,\ \text{column } \sigma(q)} \cdots a_{n\sigma(n)}.$$

We see that the sum (b) is a direct copy of (IV.2.1) which defines $\det((a_{ij})_{ij}) = \det(A)$. Our problem is to look at (a) in such a way that we can *interpret* this sum as a bonafide *determinant* of some $n \times n$ matrix! But which matrix? Let us but construct a matrix, call it B, which agrees with our original matrix A *except* that the pth row and the qth row are *identical*; moreover, these 2 rows must agree with the pth row $(a_{p1}\, a_{p2} \ldots a_{pj} \ldots a_{pn})$ of matrix A. Then (a) is exactly $\det(B)$, the determinant of this matrix B.

But the matrix B has 2 identical rows, and so from Corollary IV.2.3, $\det(B) = 0$, that is, the summation (a) equals zero. We have no choice now. We *must* conclude that

$$\det(A') = \det(A)$$

where A' is obtained from A by adding any multiple of 1 row to any other row.

We offer an example which employs the results of our previous theorems in such a way that we reduce an $n \times n$ matrix to one whose determinant is easier to compute. After this reduction is effected, and the determinant computed, the determinant of our *original* matrix will be known. Let us follow this development now.

Example Let A be the 5×5 matrix

$$A = \begin{pmatrix} 1 & 0 & 1 & 2 & 1 \\ 2 & 1 & 0 & 3 & 1 \\ 0 & 1 & 4 & 1 & 0 \\ 1 & 0 & 2 & 1 & 1 \\ 0 & 2 & 1 & 3 & 1 \end{pmatrix}.$$

One "simple" form that A may assume is its row-echleon-like form; in fact, we strive toward a *triangular* form. This form may always be achieved through elementary row operation, that is, through left-hand multiplications by elementary matrices (Theorems III.1.4 and III.3.1). Our objective is to generate zeros everywhere below (or else everywhere above) the main diagonal.

Observe that this can be accomplished, in part, by annihilating all nonzero entries below the diagonal in the first column by applying the elementary row operations. $R_1{}^2(-2)$ and $R_1{}^4(-1)$ to A. These particular operations do not alter the value of the determinant (Theorem IV.2.5). Our next step is to clear the below-diagonal entries from the second column, and so on, until triangular form flows from our original matrix A. The determinant of a triangular matrix is computed by simply multiplying the n main-diagonal entries together (Exercise IV.2.7). We shall comment further upon the validity of this procedure when we actually come to use it in this example.

$$\det \begin{pmatrix} 1 & 0 & 1 & 2 & 1 \\ 2 & 1 & 0 & 3 & 1 \\ 0 & 1 & 4 & 1 & 0 \\ 1 & 0 & 2 & 1 & 1 \\ 0 & 2 & 1 & 3 & 1 \end{pmatrix} \xrightarrow[\text{(Theorem IV.2.5)}]{R_1{}^2(-2),\ R_1{}^4(-1)} = \det \begin{pmatrix} 1 & 0 & 1 & 2 & 1 \\ 0 & 1 & -2 & -1 & -1 \\ 0 & 1 & 4 & 1 & 0 \\ 0 & 0 & 1 & -1 & 0 \\ 0 & 2 & 1 & 3 & 1 \end{pmatrix}$$

$$\xrightarrow[\text{(Theorem IV.2.5)}]{R_2{}^3(-1)\ R_2{}^5(-2)} = \det \begin{pmatrix} 1 & 0 & 1 & 2 & 1 \\ 0 & 1 & -2 & -1 & -1 \\ 0 & 0 & 6 & 2 & 1 \\ 0 & 0 & 1 & -1 & 0 \\ 0 & 0 & 5 & 5 & 3 \end{pmatrix}$$

$$\xrightarrow[\text{(Theorem IV.2.2)}]{R_{34}} = (-1)\det \begin{pmatrix} 1 & 0 & 1 & 2 & 1 \\ 0 & 1 & -2 & -1 & -1 \\ 0 & 0 & 1 & -1 & 0 \\ 0 & 0 & 6 & 2 & 1 \\ 0 & 0 & 5 & 5 & 3 \end{pmatrix}$$

$$\xrightarrow[\text{(Theorem IV.2.5)}]{R_3{}^4(-6),\ R_3{}^5(-5)} = (-1)\cdot\det\begin{pmatrix} 1 & 0 & 1 & 2 & 1 \\ 0 & 1 & -2 & -1 & -1 \\ 0 & 0 & 1 & -1 & 0 \\ 0 & 0 & 0 & 8 & 1 \\ 0 & 0 & 0 & 10 & 3 \end{pmatrix}$$

$$\xrightarrow[R_4{}^5(-10/8)]{\text{(Theorem IV.2.5)}} = (-1)\cdot\det\begin{pmatrix} \textcircled{1} & 0 & 1 & 2 & 1 \\ 0 & \textcircled{1} & -2 & -1 & -1 \\ 0 & 0 & \textcircled{1} & -1 & 0 \\ 0 & 0 & 0 & \textcircled{8} & 1 \\ 0 & 0 & 0 & 0 & \textcircled{\frac{14}{8}} \end{pmatrix}$$

What, then, is the determinant of such a triangular 5×5 matrix? As we have observed before [see Remark (4) on page 297], the typical term, $\operatorname{sgn}(\sigma)\, a_{1\sigma(1)}\, a_{2\sigma(2)} \cdots a_{n\sigma(n)}$, in (IV.2.1) chooses for its factors, the n scalars, $a_{1\sigma(1)}, a_{2\sigma(2)}, \ldots, a_{n\sigma(n)}$, 1 from *each* row and *each* column. If any 1 of these factors $a_{i\sigma(i)}$ equals zero, then, of course, the product, $\operatorname{sgn}(\sigma) = a_{1\sigma(1)} \ldots a_{i\sigma(i)} \ldots a_{n\sigma(n)}$, equals zero.

For our triangular matrix, let us select n entries, 1 from each row and each column so that we may multiply them together, and develop the sum of *all* such products to obtain the value of the determinant.

From the first column, which entry shall we choose? *Only* the topmost entry, a_{11}, the entry in row 1, has any chance of being nonzero. Hence, any typical term of (IV.2.1) must be of the form $\operatorname{sgn}\begin{pmatrix} 1 & \sigma(2) \ldots \sigma(n) \\ 1 & 2 \ldots n \end{pmatrix}$ $a_{11}\, a_{2\sigma(2)} \cdots a_{n(n)}$.

From the second column, which entry can we choose? Only the 2 topmost entries, a_{12} and a_{22} (the entries in the first 2 rows), can be nonzero. However, our previous choice of a_{11} already accounts for a first-row selection; only the second row may then be used for selection in this second column, that is, only a_{22} is available to use. From this point on, we have that any term in (IV.2.1) is of the form

$$\operatorname{sgn}(\sigma)\, a_{11} \cdot a_{22} \cdot a_{3\sigma(3)} \cdots a_{n\sigma(n)}.$$

We proceed in this way so that in selecting a nonzero entry from the jth column, only the j topmost entries $a_{ij}, a_{2j}, \cdots, a_{jj}$ (the entries in the first j rows) can be nonzero. However, selections from rows $1, 2, \ldots, j-1$ will *already* have been made since $a_{11}, a_{22}, \cdots, a_{j-1,j-1}$ will have been chosen. This leaves the entry from the jth row, a_{jj}, as our only (possibly) nonzero candidate in column j.

The conclusion from all this, is that our defining sum (IV.2.1) turns out to be not a sum at all, in the sense that precisely *one* term,

$$\operatorname{sgn}\begin{pmatrix} 1 & 2 & \dots & j & \dots & n \\ 1 & 2 & \dots & j & \dots & n \end{pmatrix} a_{11}a_{22}\dots a_{jj}\dots a_{nn},$$

remains as the candidate for being the nonzero term. Thus, if A is triangular, then $\det(A) = a_{11}\,a_{22}\dots a_{nn}$, the product of the main diagonal entries. Note that $\sigma = \begin{pmatrix} 1 & 2 & \dots & n \\ 1 & 2 & \dots & n \end{pmatrix} = e$, the identity permutation so that $\operatorname{sgn}(\sigma) = +1$ in this case.

For A given above, we now have

$$\begin{aligned} \det(A) &= (-1)\cdot\det\begin{pmatrix} 1 & 0 & 1 & 2 & 1 \\ 0 & 1 & -2 & -1 & -1 \\ 0 & 0 & 1 & -1 & 0 \\ 0 & 0 & 0 & 8 & 1 \\ 0 & 0 & 0 & 0 & \frac{14}{8} \end{pmatrix} \\ &= (-1)\cdot(1\cdot 1\cdot 1\cdot 8\cdot \tfrac{14}{8}) \\ &= -14. \end{aligned}$$

Example Find the determinant for A, where

$$A = \begin{pmatrix} 0 & 1 & 0 & 1 & 0 & 1 \\ 0 & 2 & 1 & 0 & 1 & 2 \\ 1 & 0 & 2 & 1 & 1 & 0 \\ 1 & 2 & 3 & 1 & 2 & 2 \\ 1 & 4 & 4 & 1 & 3 & 4 \\ 0 & 3 & 1 & 1 & 1 & 3 \end{pmatrix}.$$

If we but effect the elementary row operation $R_3{}^4(-1)$, which clears the 1 from the fourth row in column 1, we see that row 4 would then be (0,2,1,0,1,2), which exactly agrees with row 2. The row operation $R_3(-1)$ in no way alters the determinant (Theorem III.2.5) so that

$$\begin{aligned} \det(A) &= \det(R_3{}^4(-1)(A)), \text{ which, in turn,} \\ &= 0 \quad \text{since 2 rows agree (Corollary IV.2.3).} \end{aligned}$$

Had this phenomenon of the 2 agreeing rows not been noticed, the continuing attempt to bring A to triangular form would have resulted in the last 3 rows each being equal to (0,0,0,0,0,0), the zero row. This follows because rank A happens to equal 3. Again, since at least 1 row, the jth row, equals zero [or else because 2 (zero) rows agree; Corollary IV.2.3], for all a,

$$\begin{aligned} a\cdot\det(A) &= \det(R(a)A) \qquad \text{Theorem IV.2.4} \\ &= \det(A). \quad \text{since row } j \text{ equals } (0,0,0,0,0,0). \end{aligned}$$

But $a\cdot\det(A) = \det$ for *all* a if and only if $\det(A) = 0$. In fact, set $a = 0$.

No matter how A is attacked, its determinant is zero.

EXERCISES

IV.2.13 Use (IV.2.1) of Definition IV.2.1 to prove that

$$\det\begin{pmatrix} (p_1+q_1) & (p_2+q_2) & \cdots & (p_j+q_j) & \cdots & (p_n+q_n) \\ a_{21} & a_{2j} & \cdots & a_{23} & \cdots & a_{2n} \\ \vdots & \vdots & & \vdots & & \vdots \\ a_{i1} & a_{i2} & \cdots & a_{ij} & \cdots & a_{in} \\ \vdots & & & & & \vdots \\ a_{n1} & a_{n2} & \cdots & a_{nj} & \cdots & a_{nn} \end{pmatrix}$$

$$= \det\begin{pmatrix} p_1 & p_2 & \cdots & p_j & \cdots & p_n \\ a_{21} & a_{22} & \cdots & a_{2j} & \cdots & a_{2n} \\ \vdots & \vdots & \vdots & \vdots & \vdots & \vdots \\ a_{i1} & a_{i2} & \cdots & a_{ij} & \cdots & a_{in} \\ \vdots & & & & & \\ a_{n1} & a_{n2} & \cdots & a_{nj} & \cdots & a_{nn} \end{pmatrix} + \det\begin{pmatrix} q_1 & q_2 & \cdots & q_j & \cdots & q_n \\ a_{21} & a_{22} & \cdots & a_{2j} & \cdots & a_{2n} \\ \vdots & \vdots & & \vdots & & \vdots \\ a_{i1} & a_{i2} & \cdots & a_{ij} & \cdots & a_{in} \\ \vdots & \vdots & & \vdots & & \vdots \\ a_{n1} & a_{n2} & \cdots & a_{nj} & \cdots & a_{nn} \end{pmatrix}.$$

IV.2.14 With the aid of Theorem IV.2.2 show that a similar result, corresponding to Exercise IV.2.13, obtains for the ith row of an $n \times n$ matrix, that is, show that

$$\det\begin{pmatrix} * & * & \cdots & * \\ \vdots & \vdots & & \vdots \\ (p_1+q_1) & (p_2+q_2) & \cdots & (p_n+q_n) \\ \vdots & \vdots & & \vdots \\ * & * & \cdots & * \end{pmatrix}$$

$$= \det\begin{pmatrix} * & * & \cdots & * \\ \vdots & \vdots & & \vdots \\ p_1 & p_2 & \cdots & p_n \\ \vdots & \vdots & & \vdots \\ * & * & \cdots & * \end{pmatrix} + \det\begin{pmatrix} * & * & \cdots & * \\ \vdots & \vdots & & \vdots \\ q_1 & q^2 & \cdots & q_n \\ \vdots & \vdots & & \vdots \\ * & * & \cdots & * \end{pmatrix},$$

where the respective rows of the 3 matrices above are identical to each other *except* for the ith rows, which are explicitly denoted.

IV.2.15 Prove that if λA is the $n \times n$ matrix obtained from A by multiplying each and every entry by a scalar λ [that is, $A = (a_{ij})_{ij} \Rightarrow \lambda A = (\lambda a_{ij})_{ij}$], then $\det(\lambda A) = \lambda^n \det A$.

IV.2.16 Find the values of the determinants for the following matrices:

(a) $\begin{pmatrix} 0 & 0 & 0 & 0 & 0 & 1 \\ 1 & 0 & 0 & 0 & 0 & 0 \\ 0 & 1 & 0 & 0 & 0 & 0 \\ 0 & 0 & 1 & 0 & 0 & 0 \\ 0 & 0 & 0 & 1 & 0 & 0 \\ 0 & 0 & 0 & 0 & 1 & 0 \end{pmatrix}$

(b) $\begin{pmatrix} p & 0 & 0 & 0 \\ 0 & a & b & 0 \\ 0 & c & d & 0 \\ 0 & 0 & 0 & q \end{pmatrix}$

(c) $\begin{pmatrix} 2 & 0 & 0 & 0 & 1 \\ 0 & 6 & 1 & 1 & 0 \\ 0 & 0 & 1 & 0 & 0 \\ 0 & 1 & 2 & 2 & 0 \\ 1 & 0 & 0 & 0 & 2 \end{pmatrix}$

(d) $\begin{pmatrix} 1 & 0 & 2 & 0 & 1 \\ 2 & 2 & 1 & 1 & 1 \\ 1 & 1 & 0 & 0 & 0 \\ 3 & 1 & 0 & 0 & 0 \\ 0 & 0 & 1 & 2 & 3 \end{pmatrix}$

Our previous theorems and the corollary (Theorems IV.2.2, IV.2.4, and IV.2.5, and Corollary IV.2.3) tells us how elementary *row* operations affect the determinant of a typical square matrix A. However, there is another, almost spectacular way of viewing these *same* theorems which allows us to deduce a product theorem for pairs of square matrices, when at least 1 of the factors happens to be an elementary matrix. That is, we shall prove that for 2 square $n \times n$ matrices F and A, $\det(F \cdot A) = \det(F) \cdot \det(A)$ whenever F is an elementary matrix (Corollary IV.2.7).

As long as we are going to make a statement involving the determinants of elementary matrices in particular, it behooves us to know the precise value of $\det(F)$ for any and all elementary matrices F. Do not let this prospect alarm you. We do *not* have to deal with the cumbersome sum (IV.2.1) which defines the determinant and compute the result for each class of elementary matrices. In fact, our work has *already* been done in our most recent theorems and corollaries. To see how this question has been answered, we set down Corollary IV.2.6.

Corollary IV.2.6 Suppose R_{pq}, $R_p(a)$ $(a \neq 0)$, and $R_p(a)$ are typical elementary matrices (Definition III.3.2). Then

$$\begin{aligned} &\text{(a)}\quad \det(R_{pq}) = \det(R_{pq}^t) = -1 \\ &\text{(b)}\quad \det(R_p(a)) = \det(R_p(a)^t) = a \quad (a \neq 0) \\ &\text{(c)}\quad \det(R_{p^q}(a)) = \det(R_{p^q}(a)^t) = 1 \quad \text{for all } a. \end{aligned} \tag{IV.2.7}$$

As usual, A^t denotes the transpose of the matrix A, that is, if $A = (a_{ij})_{ij}$, then $A^t = (a_{ji})_{ij}$.

Proof: The proof of this corollary must surely rank among the 10 easiest proofs of all time.

We note that $\det(I_n) = 1$, where I_n is the $n \times n$ identity matrix. If we set $A = I_n$ in (IV.2.4) of Theorem IV.2.2, we obtain

$$\det(R_{pq}) = -\det(I_n) = -1.$$

Also, $R^t_{pq} = R_{pq}$ (compare Exercises IV.2.2, IV.2.3, IV.2.4, and IV.2.22)

$$\det(R_q) = \det(R^t_{pq}) = -1$$

which proves (IV.2.7)(a).

To prove (IV.2.7)(b), replace A by I_n in (IV.2.5) to obtain

$$\det(R_p(a)) = a \det(I_n) = a.$$

This equation, coupled with the fact that $R_p(a) = R_p(a)^t$ (Exercise IV.2.22) gives us (IV.2.7)(b).

Similarly, in (IV.2.6) we replace A by I_n to obtain

$$\det(R_p{}^q(a)) = \det(I_n) = 1;$$

also note that $R_p{}^q(a)^t = R_q{}^p(a)$ (Exercise IV.2.22), an elementary matrix of the *same type*, so that

$$\det(R_p{}^q(a)^t) = \det(R_q{}^p(a)) = 1.$$

This proves (IV.2.7)(c), and hence, the corollary.

We have seen all the results, which we shall now blend into an important product theorem for determinants when the left-hand factor is an elementary matrix. To wit, Corollary IV.2.7.

Corollary IV.2.7 Let A be any $n \times n$ matrix. Then

$$\det(F \cdot A) = \det(F)\det(B)$$

for all $n \times n$ elementary matrices F.

Proof: We consider the case $F = R_{pq}$.

$$\begin{aligned} \det(F \cdot A) &= \det(R_{pq} \cdot A) && \text{Definition of } F \\ &= \underbrace{(-1)}_{\downarrow} \det(A) && \text{Theorem IV.2.2} \\ &= \det(R_{pq}) \cdot \det(A) && \text{Corollary IV.2.6, Display (IV.2.7)(a)} \\ &= \det(F) \cdot \det(A) && \text{Definition of } F. \end{aligned}$$

The remaining cases $F = R_p(a)$, $a \neq 0$, and $F = R_p{}^q(a)$, are left as Exercise IV.2.25; the proof is similar to that above.

The following theorem tells us that every theorem concerning rows of A applies equally well to columns. In fact, we shall see that for any $n \times n$ matrix A, det (A^t), the determinant of the transpose of A, always equals det (A). Thus, rows of A^t become columns of A; we then apply Theorems IV.2.2, IV.2.4, and IV.2.5 to (rows of) A^t to obtain identical results for *columns* of A.

Theorem IV.2.8 Let $A = (a_{ij})_{ij}$ be an $n \times n$ matrix, and let $A^t = (a_{ji})_{ij}$ be its transpose. Then

$$\det(A) = \det(A^t).$$

Proof: We shall use the following facts:

(a) If F is an elementary matrix, so is its transpose F^t (Exercise IV.2.22). Moreover, det $(F) =$ det (F^t) (Corollary IV.2.6).

(b) The transpose operation reverses the order of a product, that is, $(AB)^t = B^tA^t$ for all $n \times n$ matrices A and B (Exercise IV.2.23).

CASE 1. *A is invertible.* In this case, A is simply a product of elementary matrices $F_1, F_2, \ldots, F_k$ (Theorem III.3.6). Thus,

$\det A = \det(F_1 \cdot F_2 \cdots F_k)$	F^t is elementary
$= \det(F_1)\det(F_2 \cdots F_k)$	Corollary IV.2.7
$\vdots$	$\vdots$
$= \det(F_1)\det(F_2)\cdots\det(F_k)$	Corollary IV.2.7
$= \det(F_k)\cdots\det(F_2)\det(F_1)$	Scalars det (F_i) commute
$= \det(F_k{}^t)\cdots\det(F_2{}^t)\det(F_1{}^t)$	See (a) above
$= \det(F_k^t F_{k-1}^t)\det(F_{k-2}^t)\cdots \det(F_2{}^t)\det(F_1{}^t)$	Corollary IV.2.7
$\vdots$	$\vdots$
$= \det(F_k^t F_{k-1}^t \cdots F_2{}^t \cdot F_1{}^t)$	Corollary IV.2.7
$= \det((F_1 \cdot F_2 \cdots F_k)^t)$	See (b) above
$= \det(A^t)$	Since $A = F_1 \cdot F_2 \ldots F_k$.

CASE 2. *A is not invertible* (rank $A < n$). Since A is not invertible, that is, A has rank $<n$ (Theorem III.3.5), we can reduce A to an $n \times n$ column echelon form A_{ech} by means of successive right-hand multiplications by elementary matrices $F_1, F_2, \ldots, F_p$ (Theorems III.1.4 and III.3.1). Now bear in mind that A_{ech} is of *triangular* form with zeros everywhere above the main diagonal. Equivalently, $(A_{\text{ech}})^t$, the transpose of A_{ech}, is a *row*-echelon $n \times n$ matrix. Since rank $A < n$, it follows that the last row

of $(A_{\text{ech}})^t$ consists entirely of zeros; hence at least 1 zero lies in the main diagonal of the triangular matrix $(A_{\text{ech}})^t$, which implies

$$\det((A_{\text{ech}})^t) = 0. \tag{IV.2.8}$$

This equality also follows from Exercise IV.2.7 or from Theorem IV.2.4 where $a = 0$ and $p = n$. Now for certain elementary $n \times n$ matrices $F_1, F_2, \ldots, F_p$,

$$A_{\text{ech}} = A \cdot F_1 \cdot F_2 \ldots F_p \text{ (Theorems III.1.4 and III.3.1)}$$

so that

$$(A_{\text{ech}})^t = F_p{}^t \ldots F_2{}^t \cdot F_1{}^t \cdot A. \quad \text{See (b) above.} \tag{IV.2.9}$$

Taking determinants of both sides of (IV.2.9), and using Corollary IV.2.7 p times, we obtain

$$\begin{aligned} 0 &= \det((A_{\text{ech}})^t) \quad \text{from (IV.2.8)} \\ &= \det(F_p{}^t \ldots F_2{}^t \cdot F_1{}^t \cdot A) \qquad \text{(IV.2.10)} \\ &= \det(F_p{}^t) \ldots \det(F_2{}^t) \det(F_1{}^t) \det(A) \quad \text{Corollary IV.2.7.} \end{aligned}$$

↑ ↑ ↑

nonzero factors

Now the determinants of elementary matrices, $\det(F_i{}^t)$, are never zero (compare Corollary IV.2.6). Thus, if rank $A < n$, we must conclude from (IV.2.10) that $\det(A) = 0$.

Fortunately for us, the row rank and column rank of A agree (Theorem III.2.5). Thus, rank $(A) =$ rank (A^t) since the row rank of A is exactly the column rank of A^t. But if we agree that rank $A^t < n$, then we must similarly conclude from (IV.2.10) that $\det(A^t) = 0$ as well. Hence, $\det(A) = \det(A^t)$ for *all* $n \times n$ matrices A, and the theorem is done.

Before the coals get cold, let us extract another result from the proof of Theorem IV.2.8.

Theorem IV.2.9 Let A be an $n \times n$ matrix. Then $\det(A) \neq 0$ if and only if A^{-1} exists (if and only if rank $(A) = n$).

Proof: Let A_{ech} be a column-echelon form of A so that in rewriting (IV.2.9), we have once again

$$(A_{\text{ech}})^t = F_p{}^t \ldots F_2{}^t \cdot F_1{}^t \cdot A$$

for certain elementary matrices $F_p{}^t, \ldots, F_2{}^t, F_1{}^t$. By applying Corollary IV.2.7 p times, we obtain

$$\det((A_{\text{ech}})^t) = \det(F_p{}^t) \ldots \det(F_2{}^t) \cdot \det(F_1{}^t) \cdot \det(A) \tag{IV.2.11}$$

↑ ↑ ↑

nonzero scalars

where det $(F_i{}^t) \neq 0$ since each $F_i{}^t$ is an elementary matrix (Corollary IV.2.6). From (IV.2.11), then, det $(A) \neq 0$ if and only if det $((A_{\text{ech}})^t) \neq 0$. But the only echelon matrix (echelon matrices are always triangular) which has nonzero determinant, is the identity matrix I_n; I_n has rank n. That is, det $(A) \neq 0$ if and only if rank $A = n$ (that is, if and only if A^{-1} exists from Theorem III.3.5). This proves the desired result.

As a formality, we restate Theorems IV.2.2, IV.2.4, IV.2.5, and Corollary IV.2.3 without the restriction of limiting our elementary operations only to rows.

Theorem IV.2.10 Let A be an $n \times n$ matrix $(a_{ij})_{ij}$. Then for elementary row operations $R_{pq}(\cdot)$, $R_p(a)(\cdot)$ $(a \neq 0)$, $R_p{}^q(a)(\cdot)$ and elementary column operations $C_{pq}(\cdot)$, $C_p(a)(\cdot)$ $(a \neq 0)$, $C_p{}^q(a)(\cdot)$, regarded as functions sending matrices to matrices,

(a) det $(R_{pq}(A))=$ det $(C_{pq}(A)) = -$ det (A)
(switching rows or columns negates the determinant);

(b) det $(R_p(a)(A)) =$ det $(C_p(a)(A)) = a$ det (A)
(multiplying a row or a column by a multiplies the determinant by a);

(c) det $(R_p{}^q(a)(A)) =$ det $(C_p{}^q(a)(A)) =$ det (A)
(adding a multiple of 1 row (or column) to another row (or column) does not affect the determinant);

(d) If 2 rows or 2 columns of A are the same then det $(A) = 0$.

Proof: Statements (a), (b), and (c) follow from Theorems IV.2.2, IV.2.4, and IV.2.5 along with Theorem IV.2.8. We also use the fact that in all cases, an elementary column operation followed by the transpose yields the same matrix as the transpose of a matrix followed by the corresponding elementary row operation. For example,

$$C_{pq}(A)^t = R_{pq}(A^t). \quad \text{Thus,}$$

[w] det $(C_{pq}(A))$	$=$ det $(R_{pq}(A^t)^t)$	transpose both sides of equation above; take the determinants of both sides
	$=$ det $(R_{pq}(A^t))$	from Theorem IV.2.8
	$= -$det (A^t)	replace A with A^t in Theorem IV.2.2
	$= -$det (A)	Theorem IV.2.8
[v]	$=$ det $(R_{pq}(A))$	Theorem IV.2.2.

Equality of [**w**] and [**v**], above, establishes statement (a). Statements (b), (c), and (d) also use something like (IV.2.12) with previous theorems. These are left as Exercise IV.2.26.

EXERCISES

IV.2.17 Prove that scalars "factor out" of rows of matrices in the sense that

$$\det\begin{pmatrix} a_{11} & a_{12} & \cdots & a_{1n} \\ \vdots & \vdots & & \vdots \\ \lambda a_{i1} & \lambda a_{i2} & \cdots & \lambda a_{in} \\ \vdots & \vdots & & \vdots \\ a_{n1} & a_{n2} & \cdots & a_{nn} \end{pmatrix} = \lambda \det\begin{pmatrix} a_{11} & a_{12} & \cdots & a_{1n} \\ \vdots & \vdots & & \vdots \\ a_{i1} & a_{i2} & \cdots & a_{in} \\ \vdots & \vdots & & \vdots \\ a_{n1} & a_{n2} & \cdots & a_{nn} \end{pmatrix}.$$

$$\det\begin{pmatrix} a_{11} & \cdots & \lambda a_{ij} & \cdots & a_{1n} \\ \vdots & & \vdots & & \vdots \\ a_{21} & \cdots & \lambda a_{2j} & \cdots & a_{2n} \\ \vdots & & \vdots & & \vdots \\ a_{11} & \cdots & \lambda a_{nj} & \cdots & a_{nn} \end{pmatrix} = \lambda \det\begin{pmatrix} a_{11} & \cdots & a_{ij} & \cdots & a_{1n} \\ \vdots & & \vdots & & \vdots \\ a_{21} & \cdots & a_{2j} & \cdots & a_{2n} \\ \vdots & & \vdots & & \vdots \\ a_{11} & \cdots & a_{nj} & \cdots & a_{nn} \end{pmatrix}.$$

IV.2.18 Prove that if any row or column of an $n \times n$ matrix consists entirely of zero entries, then $\det(A) = 0$.

IV.2.19 Prove the slightly more general form of Corollary IV.2.3: If 1 row (respectively column), of an $n \times n$ matrix is just a scalar multiple, λ, of some other row (respectively, column) of that matrix, then $\det(A) = 0$ (put $\lambda = 1$ and the statement above is the statement of Corollary IV.2.3).

IV.2.20 Compute the determinants of the given square matrices (see Exercise IV.2.14).

(a) $\begin{pmatrix} p_1 + q_1 & c \\ p_2 + q_2 & d \end{pmatrix}$, $\begin{pmatrix} p_1 & c \\ p_2 & d \end{pmatrix}$, $\begin{pmatrix} q_1 & c \\ q_2 & d \end{pmatrix}$.

(b) $\begin{pmatrix} 1 & 4 & 7 \\ 3 & 1 & \\ 2+3 & 7+(-7) & 6+(-6) \end{pmatrix}$, $\begin{pmatrix} 1 & 4 & 7 \\ 3 & 1 & 2 \\ 2 & 7 & 6 \end{pmatrix}$, $\begin{pmatrix} 1 & 4 & 7 \\ 3 & 1 & 2 \\ 3 & -7 & -6 \end{pmatrix}$.

(c) $\begin{pmatrix} 2 & 1 & 1+1 \\ 1 & 1 & 1+2 \\ 1 & 1 & 1+1 \end{pmatrix}$, $\begin{pmatrix} 2 & 1 & 1 \\ 1 & 1 & 1 \\ 1 & 1 & 1 \end{pmatrix}$, $\begin{pmatrix} 2 & 1 & 1 \\ 1 & 1 & 2 \\ 1 & 1 & 1 \end{pmatrix}$.

IV.2.21 Compute the determinants of the given matrices (see Theorems IV.2.2, IV.2.4, and IV.2.5).

(a) $A = \begin{pmatrix} a & b \\ c & d \end{pmatrix}$, $\quad R_2{}^1(5)(A)$, $\quad C_1{}^2(2)(A)$.

(b) $A = \begin{pmatrix} 0 & 1 & 1 \\ 1 & 1 & 0 \\ 1 & 0 & 1 \end{pmatrix}$, $\quad R_{13}(A)$, $\quad C_3{}^1(3)(A)$.

IV.2.22 Let R_{pq}, $R_p(a)$ $(a \neq 0)$, $R_p{}^q(a)$, C_{pq}, $C_p(a)$ $(a \neq 0)$, and $C_p{}^q(a)$ be $n \times n$ elementary matrices (Definition III.3.2). If A^t denotes the transpose of any matrix A, then prove

(a) $R^t_{pq} = R_{pq} = C_{pq} = C^t_{pq}$

(b) $R_p(a)^t = R_p(a) = C_p(a) = C_p(a)^t$

(c) $R_p{}^q(a)^t = R_q{}^p(a) = C_p{}^q(a) = C_q{}^p(a)^t$.

IV.2.23 Prove that for all square matrices A and B, $(A \cdot B)^t = B^t \cdot A^t$, where A. denotes the transpose of A.

IV.2.24 Compute the determinants of the following matrices by using elementary row and/or column operations, to reduce them to triangular form.

(a) $\begin{pmatrix} 1 & 2 \\ 3 & 4 \end{pmatrix}$ (b) $\begin{pmatrix} 1 & 2 & 3 \\ 4 & 5 & 6 \\ 7 & 8 & 9 \end{pmatrix}$

(c) $\begin{pmatrix} 1 & 0 & 0 & 0 \\ 2 & 0 & 8 & 0 \\ 1 & 1 & 1 & 1 \\ 1 & 0 & 1 & 0 \end{pmatrix}$ (d) $\begin{pmatrix} a & 5 & 1 & 2 \\ 0 & b & 3 & 4 \\ 0 & 0 & c & 0 \\ 0 & 0 & 7 & d \end{pmatrix}$

(e) $\begin{pmatrix} 1 & 3 & 5 & 1 \\ 2 & 1 & 0 & 1 \\ -1 & 0 & 1 & 2 \\ 0 & 0 & 2 & -1 \end{pmatrix}$ (f) $\begin{pmatrix} 1 & 2 & -1 & 1 \\ 2 & 4 & -2 & 2 \\ 3 & 6 & -3 & 3 \\ 1 & 1 & 1 & 1 \end{pmatrix}$

IV.2.25 (See proof of Corollary IV.2.7). Prove that for any $n \times n$ matrix A,

$$\det(F \cdot A) = \det(F)\det(A),$$

where

(a) F is the elementary matrix $R_p(a)$, $a \neq 0$.

(b) F is the elementary matrix $R_p{}^q(a)$.

IV.2.26 Prove statements (b), (c), and (d) of Theorem IV.2.10.

3. *Expansion by Minors*

So far, which means are available for computing determinants for $n \times n$ matrices? For 2×2 or 3×3 matrices we have at least a graphical device which tells us what the typical product in the sum (IV.2.1) looks like (Figure 4.1). For $n \times n$ matrices in general, we may avoid the mental cruelties which would come from computing (IV.2.1) directly; we may reduce A to triangular form and then multiply diagonal elements to obtain the determinant.

The motivation for developing an expansion by minors (whatever that is) lies in the fact that we are able to reduce the determinant of an $n \times n$ matrix to a sum involving n smaller $(n-1) \times (n-1)$ matrices and their determinants. In fact, if we play our cards well, we may always reduce the determinant of an $n \times n$ matrix to the problem of finding the determinant of only 1 (smaller) $(n-1) \times (n-1)$ matrix.

By this time, your curiosity must surely be provoked, so we go directly to describing this method of expansion (of determinants) by minors.

We pose the following notation, which we shall use later.

Notation IV.3.1 If A is an $n \times n$ matrix, then the symbol $\boldsymbol{A(i,j)}$ will denote that $(n-1) \times (n-1)$ matrix which is obtained from A by striking out the ith row and jth column. We call the scalar, $\det(A(i,j))$, a **minor** of A (of order $n-1$).

The following theorem simply rewrites the defining sum (IV.2.1) of det in Definition IV.2.1 in such a way that minors of A (of order $n-1$) appear.

Theorem IV.3.2 (Expansion of $\det(A)$ by the first row.) If $A = (a_{ij})_{ij}$ is an $n \times n$ matrix, then

$$\det(A)$$

$$= a_{11}\det(A(1,1)) - a_{12}\det(A(1,2)) + \cdots + (-1)^{k+1}\, a_{1k}\det(A(1,k)) + \cdots$$

$$+ (-1)^{n+1} a_{1n}\det(A(1,n)) = \sum_{k=1}^{n} (-1)^{k+1} a_{1k}\det(A(1,k)). \qquad \text{(IV.3.1)}$$

Proof: Consider the defining sum (IV.2.1) for $\det(A)$;

$$\sum_{\sigma\in S_n} \operatorname{sgn}\begin{pmatrix}1 & 2 & & i & & n\\ \sigma(1) & \sigma(2) & \ldots & \sigma(i) & \ldots & \sigma(n)\end{pmatrix} \underbrace{a_{1\sigma(1)}}_{\text{first factor}} a_{2\sigma(2)}\cdots a_{i\sigma(i)}\cdots a_{n\sigma(n)}.$$

There are only n possible first factors $a_{1\sigma(1)}$ which may appear in any typical product, namely, a_{11}, or a_{12} or,..., or a_{ij} or,..., *or* a_{1n}. Another way of saying the same thing is that there are only n possible values for $\sigma(1)$, the second subscript of the first factor, namely, $\sigma(1) = 1$, or $\sigma(1) = 2$, or $\cdots$ or $\sigma(1) = j$, or $\cdots$ or $\sigma(1) = n$. Our sum will then decompose into n separate sums, each sum containing 1 of the possible first factors $a_{1\sigma(1)}$. Accordingly, we write (IV.2.1) as

$$\det(A) = \sum_{\substack{\sigma\in S_n\\ \sigma 1=(1)}} \operatorname{sgn}\begin{pmatrix}1 & 2 & \ldots & n\\ 1 & \sigma(2) & \ldots & \sigma(n)\end{pmatrix} a_{11}a_{2\sigma(2)}\cdots a_{n\sigma(n)}$$

$$+ \sum_{\substack{\sigma\in S_n\\ \sigma(1)=2}} \operatorname{sgn}\begin{pmatrix}1 & 2 & \ldots & n\\ 2 & \sigma(2) & \ldots & \sigma(n)\end{pmatrix} a_{12}a_{2\sigma(2)}\cdots a_{n\sigma(n)}$$

$$\vdots$$

$$\sum_{\substack{\sigma \in S_n \\ \sigma(1)=k}} \operatorname{sgn}\begin{pmatrix} 1 & 2 & \dots & n \\ k & \sigma(2) & \dots & \sigma(n) \end{pmatrix} a_{1k}a_{2\sigma(2)}\cdots a_{n\sigma(n)}$$

$$\vdots$$

$$+ \sum_{\substack{\sigma \in S_n \\ \sigma(1)=n}} \operatorname{sgn}\begin{pmatrix} 1 & 2 & \dots & n \\ n & \sigma(2) & \dots & \sigma(n) \end{pmatrix} a_{1n}a_{2\sigma(2)}\cdots a_{n\sigma(n)}.$$

Since the factors $a_{11}, a_{12}, \dots, a_{1k}, \dots, a_{1n}$ are *fixed* scalars in each of the n sums above, we factor out each a_{1k} (from the kth sum; $k = 1, 2, \dots, n$) to obtain

$$\begin{aligned}
\det(A) = {} & a_{11} \underbrace{\sum_{\substack{\sigma \in S_n \\ \sigma(1)=1}} \operatorname{sgn}\begin{pmatrix} 1 & 2 & \dots & n \\ 1 & \sigma(2) & \dots & \sigma(n) \end{pmatrix} a_{2\sigma(2)}a_{3\sigma(3)}\cdots a_{n\sigma(n)}}_{A_{11}} \\
& + a_{12} \underbrace{\sum_{\substack{\sigma \in S_n \\ \sigma(1)=2}} \operatorname{sgn}\begin{pmatrix} 1 & 2 & \dots & n \\ 2 & \sigma(2) & \dots & \sigma(n) \end{pmatrix} a_{2\sigma(2)}a_{3\sigma(3)}\cdots a_{n\sigma(n)}}_{A_{12}} \\
& \vdots \\
& + a_{1k} \underbrace{\sum_{\substack{\sigma \in S_n \\ \sigma(1)=k}} \operatorname{sgn}\begin{pmatrix} 1 & 2 & \dots & n \\ k & \sigma(2) & \dots & \sigma(n) \end{pmatrix} a_{2\sigma(2)}a_{3\sigma(3)}\cdots a_{n\sigma(n)}}_{A_{1k}} \\
& \vdots \\
& + a_{1n} \underbrace{\sum_{\substack{\sigma \in S_n \\ \sigma(1)=n}} \operatorname{sgn}\begin{pmatrix} 1 & 2 & \dots & n \\ n & \sigma(2) & \dots & \sigma(n) \end{pmatrix} a_{2\sigma(2)}a_{3\sigma(3)}\cdots a_{n\sigma(n)}}_{A_{1n}}.
\end{aligned} \tag{IV.3.2}$$

From (IV.3.2) above, you will see that we have used an ad hoc notation A_{1k}, $k = 1, 2, \dots, n$, to symbolize each of the n sums, that is, we have put

$$A_{1k} = \sum_{\sigma \in S_n} \operatorname{sgn}\begin{pmatrix} 1 & 2 & \dots & k & \dots & n \\ k & \sigma(2) & \dots & \sigma(k) & \dots & \sigma(n) \end{pmatrix} a_{2\sigma(2)}a_{3\sigma(3)}\cdots a_{k\sigma(k)}\cdots a_{n\sigma(n)}. \tag{IV.3.3}$$

We must now look at (IV.3.3) in such a clever way that each of the sums A_{1k} in (IV.3.2) represents an actual *determinant* for some $(n-1) \times (n-1)$ matrix—but for *which* $(n-1) \times (n-1)$ matrix?

We claim that for each k, $k = 1, 2, \dots, n$, $A_{1k} = (-1)^{k+1} \det(A(1,k))$, where $A(1,k)$ is the $(n-1) \times (n-1)$ matrix obtained from A by deleting the first row and kth column:

How can this statement be justified?

In looking at a typical term, $\operatorname{sgn}\begin{pmatrix}1 & 2 \dots & n \\ k & \sigma(2)\dots & \sigma(n)\end{pmatrix} a_{2\sigma(2)}\, a_{3\sigma(3)} \cdots a_{n\sigma(n)}$, of the sum defining A_{1k} in (IV.3.2), we see that we are off to a very good start—since $\sigma(j) \neq k$ for $j = 2, 3, \dots, n$, the subscripts of the factors $a_{2\sigma(2)}, a_{3\sigma(3)}, \dots, a_{n\sigma(n)}$ reveal that these are scalars which come 1 from each row of A *except* for the first row, and 1 from each *column* of A except for the kth column. That is, each term of A_{1k} involves a product of $n - 1$ entries of $A(1,k)$, 1 chosen from each row and each column.

Now the ordered set of second subscripts

(a) $$\{\sigma(2),\sigma(3),\dots,\sigma(n)\}$$

and its number of *inverted* pairs, must be related to the inversions of the set

(b) $$\{k,\sigma(2),\sigma(3),\dots,\sigma(n)\},$$

the ordered range of $\sigma = \begin{pmatrix}1 & 2 & 3 \dots & n \\ k & \sigma(2) & \sigma(3)\dots & \sigma(n)\end{pmatrix}$. And this is easy.

The only *difference* in the number of inverted pairs in each of the sets (a) and (b), occurs where the integer k precedes the $k - 1$ *smaller* integers $1, 2, \dots, k - 1$ in (b). That is to say, in (b), we have the $k - 1$ inversions $(k,1), (k,2), \dots, (k,k - 1)$ which *do not* appear in (a) [k is not even *among* the integers $\sigma(2), \sigma(3), \dots, \sigma(n)$, since $\sigma(1) = k$]. Otherwise, *any* inversion which appears in (a) *also* appears in (b) since we are then comparing pairs of integers, *neither* of which equals k.

Consider the following functions, the first of which we have already seen:

(c) $\operatorname{sgn}\begin{pmatrix}1 & 2 & 3 \dots & n \\ k & \sigma(2) & \sigma(3)\dots & \sigma(n)\end{pmatrix} = +1$ if the number of inversions in $\{k,\sigma(2),\sigma(3),\dots,\sigma(n)\}$ is *even*

$= -1$ if the number of inversions in $\{k,\sigma(2),\sigma(3),\dots,\sigma(n)\}$ is *odd*.

(d) $\operatorname{sgn}\begin{pmatrix}2 & 3 \dots & n \\ \sigma(2) & \sigma(3)\dots & \sigma(n)\end{pmatrix} = +1$ if the number of inversions in $\{\sigma(2),\sigma(3),\dots,\sigma(n)\}$ is *even*

$= -1$ if the number of inversions in $\{\sigma(2),\sigma(3),\dots,\sigma(n)\}$ is *odd*.

Now if (b) were to account for an *odd* number of inversions *more* than (a), we would have (c) equal to the *negative* of (d) (one would account for an even number of inversions, the other would account for an odd number of inversions). If (b) were to account for an *even* number of inversions in *excess* to those of (a) then (c) and (d) would be *equal* (since *both* would account for an even number of inversions or *both* would account for an odd number of inversions). In any event, we have shown that (b) always accounts for $k-1$ *more* inversions [namely, $(k,1),(k,2),\ldots,(k,k-1)$] than (a). Hence,

$$\operatorname{sgn}\begin{pmatrix}1 & 2 & 3 & \ldots & n\\ k & \sigma(2) & \sigma(3) & \ldots & \sigma(n)\end{pmatrix} = (-1)^{k-1}\operatorname{sgn}\begin{pmatrix}2 & 3 & \ldots & n\\ \sigma(2) & \sigma(3) & \ldots & \sigma(n)\end{pmatrix}. \quad \text{(IV.3.4)}$$

Equation (IV.3.4) can be easily verified for a specific case or two. If $k=3$, then $3-1=2$ is the difference in the number of inversions between (a) and (b); 2 is an *even* number so that (c) and (d) are *equal*. The factor $(-1)^{3+1}=+1$ in (IV.3.4) agrees with this. If $k=8$, for example, then the number of inversions in (b) would be $8-1=7$ *more* than the number of inversions in (a); 7 is an *odd* number, so that (c) and (d) should be *negatives* of each other. The factor $(-1)^{8+1}=(-1)$ in (IV.3.14) agrees with this observation.

We substitute (IV.3.4) into (IV.3.3) to obtain

$$A_{1k} = (-1)^{k+1}\underbrace{\sum_{\substack{\sigma\in S_n\\ \sigma(1)=k}}\operatorname{sgn}\begin{pmatrix}2 & 3 & \ldots & n\\ \sigma(2) & \sigma(3) & \ldots & \sigma(n)\end{pmatrix}a_{2\sigma(2)}a_{3\sigma(3)}\cdots a_{n\sigma(n)}}_{\det(A(1,k))}. \quad \text{(IV.3.5)}$$

Compare the summation in (IV.3.5) with the defining sum (IV.2.1) in Definition IV.2.1, we see that $A_{1,k}$ coincides with $(-1)^{k+1}$ times the determinant of the $(n-1)\times(n-1)$ matrix $A(1,k)$; this fact is indicated in (IV.3.5) above. Setting $A_{1,k}$ equal to $(-1)^{k+1}\det(A(1,k))$ in (IV.3.2), for each k, then, allows us to rewrite (IV.3.2) as

$$\begin{aligned}\det(A) &= a_{11}\det(A(1,1))\\ &= -a_{12}\det(A(1,2))\\ &\quad\vdots\\ &\quad+(-1)^{k+1}a_{1k}\det(A(1,k)) \qquad \text{(IV.3.6)}\\ &\quad\vdots\\ &\quad+(-1)^{n+1}a_{1n}\det(A(1,n)) = \sum_{k-1}^{n}(-1)^{k+1}a_{1k}\det(a(1,k)),\end{aligned}$$

which was to be (and is now) proven.

Example Expand $\det(A)$ in terms of the first row for the 4×4 matrix

$$A = \begin{pmatrix}2 & 5 & 3 & 4\\ 1 & 0 & 1 & 1\\ 2 & 0 & 1 & 1\\ 3 & 3 & 1 & 1\end{pmatrix}.$$

We shall have use for the four 3×3 matrices which are obtained from A by deleting the first row and jth column, $j = 1, 2, 3, 4$. These are the matrices

$$A(1,1) = \begin{pmatrix} 0 & 1 & 1 \\ 0 & 1 & 1 \\ 3 & 1 & 1 \end{pmatrix} \qquad A(1,3) = \begin{pmatrix} 1 & 0 & 1 \\ 2 & 0 & 1 \\ 3 & 3 & 1 \end{pmatrix}$$

$$A(1,2) = \begin{pmatrix} 1 & 1 & 1 \\ 2 & 1 & 1 \\ 3 & 1 & 1 \end{pmatrix} \qquad A(1,4) = \begin{pmatrix} 1 & 0 & 1 \\ 2 & 0 & 1 \\ 3 & 3 & 1 \end{pmatrix}.$$

The determinants of these 3×3 matrices will be needed. But $\det(A(1,1)) = \det(A(1,2)) = 0$, since the matrices $A(1,1)$ and $A(1,2)$ each has a matching pair of columns [Theorem IV.2.10(d)]. From Figure 4.1, for example, we compute $\det(A(1,3)) = \det(A(1,4)) = +3$.

Respecting the alternating plus and minus sign pattern given us in Theorem III.3.2, we write

$$\begin{aligned} \det(A) &= +a_{11}\det(A(1,1)) - a_{12}\det(A(1,2)) + a_{13}\det(A(1,3)) - a_{14}\det(A(1,4)) \\ &= +2\cdot 0 - 5\cdot 0 + 3\cdot 3 - 4\cdot 3 \\ &= -3. \end{aligned}$$

The above method is a bit simpler than that of reducing A to triangular form.

Theorm IV.3.2 extends quite easily to an expansion theorem involving the general lth row, along with the $(n-1) \times (n-1)$ "deleted" matrices $A(l,k)$.

Theorem IV.3.3 (Expansion of $\det(A)$ by the lth row). If $A = (a_{ij})_{ij}$ is an $n \times n$ matrix, then

$$\begin{aligned} \det(A) &= (-1)^{l+1}a_{l1}\det(A(l,1)) + (-1)^{l+2}a_{l2}\det(A(l,2)) + \cdots \\ &\quad + (-1)^{l+k}a_{lk}\det(A(l,k)) + \cdots + (-1)^{l+n}\det(A(l,n)) \qquad \text{(IV.3.7)} \\ &= \sum_{k=1}^{n}(-1)^{l+k}a_{lk}\det(A(l,k)). \end{aligned}$$

Proof: We simply reduce A (if $l \neq 1$) to the case of Theorem IV.3.2 as follows

$$\underset{A}{\begin{pmatrix} a_{11} & a_{12} & \cdots & a_{1n} \\ \vdots & \vdots & & \vdots \\ a_{l1} & a_{l1} & \cdots & a_{ln} \\ \vdots & \vdots & & \vdots \\ a_{n1} & a_{n2} & \cdots & a_{nn} \end{pmatrix}} \xrightarrow{R_{1,2}\cdots R_{l-2,l-1}R_{l-1,l}} \underset{A'}{\begin{pmatrix} a_{l1} & a_{l2} & \cdots & a_{ln} \\ \vdots & \vdots & & \vdots \\ a_{11} & a_{12} & \cdots & a_{1n} \\ \vdots & \vdots & & \vdots \\ a_{n1} & a_{n2} & \cdots & a_{nn} \end{pmatrix}}.$$

That is, by successive interchanges of the lth row $(a_{l1},a_{l2},\ldots,a_{ln})$ with the row immediately above, we eventually create a matrix, A', whose *first* row is $(a_{l1},a_{l2},\ldots,a_{ln})$, and whose remaining rows are in the same *relative* order *to each other* as in the original matrix A. How many successive adjacent interchanges (applications of elementary operations $R_{i-1,i}$) are required to bring row l to the top row? Exactly $l-1$ such interchanges are required. Since each interchange of rows alters the value of the determinant by a factor of (-1) (Theorem IV.2.2), we have

$$\det(A) = (-1)^{l-1}\det(A'). \tag{IV.3.8}$$

In expanding A' by the first row, Theorem IV.3.2 tells us that

$$\det(A') = \sum_{k=1}^{n}(-1)^{k+1}\overbrace{a_{l1}}^{\text{entry from row 1, column } k \text{ of } A'}\det(A'(1,k)) \tag{IV.3.9}$$

where $A'(1,k)$ is the $(n-1)\times(n-1)$ matrix obtain by deleting the first row and kth column of A'. Aha! But this is exactly the same $(n-1)\times(n-1)$ matrix that results by deleting the lth row and kth column of our *original* matrix A, that is,

$$A'(1,k) = A(l,k).$$

Effecting this substitution into (IV.3.9) yields

$$\det(A') = \sum_{k=1}^{n}(-1)^{k+1}a_{lk}\det(A(l,k)). \tag{IV.3.10}$$

But (IV.3.8) tells us that to obtain $\det(A)$ from $\det(A')$, we simply multiply $\det(A')$ by $(-1)^{l-1}$. Applying this factor to (IV.3.10) gives us

$$\det(A)\ [\text{that is, } (-1)^{l-1}\det(A')] = \sum_{k=1}^{n}(-1)^{k+l}a_{kl}\det(A(k,l)) \tag{IV.3.11}$$

which completes the proof.

For the sake of completeness, we wish not to omit expansions of determinants by columns. Theorem IV.2.8 tells us that our work is done. We need only state Theorem IV.3.4.

Theorem IV.3.4 [Expansion of $\det(A)$ by the kth column.] If $A = (a_{ij})_{ij}$ is an $n\times n$ matrix, then

$$\begin{aligned}\det(A) &= (-1)^{1+k}a_{1k}\det(A(1,k)) + (-1)^{2+k}a_{2k}\det(A(2,k)) + \ldots \\ &\quad + (-1)^{l+k}a_{lk}\det(A(l,k)) + \cdots + (-1)^{n+k}a_{nk}\det(A(n,k)) \\ &= \sum_{l=1}^{n}(-1)^{l+k}a_{lk}\det(A(l,k)).\end{aligned} \tag{IV.3.12}$$

Proof: Let us expand the determinant of $A^t = (a_{ji})_{ij}$ by its kth row, a procedure well defined in Theorem IV.3.3. Thus,

$$\det(A) = \det(A^t) \qquad \text{Theorem IV.2.8}$$

$$= \sum_{l=1}^{n} (-1)^{l+k} \underbrace{a_{l,k}}_{\substack{\text{entry from (the fixed)}\\ \text{row } k\text{, column } l \text{ of } A^t}} \det(\tilde{A}(k,l))$$

where $\tilde{A}(k,l)$ is the $(n-1) \times (n-1)$ matrix obtained by striking out the kth column and lth row of A and then taking the transpose; that is, $\tilde{A}(k,l) = (A(l,k))^t$.

Effecting this substitution into the last summation, leads us to the fact that

$$\det(A) = \sum_{l=1}^{n} (-1)^{l+k} a_{l,k} \det((A(l,k))^t)$$

$$= \sum_{l=1}^{n} (-1)^{l+k} a_{l,k} \det(A(l,k)) \quad \text{Theorem IV.2.8}$$

which is precisely the statement of our theorem. The proof is done.

Figure 4.2(a) illustrates an $n \times n$ matrix with its kth row and lth column blocked in; the residue entries comprise the $(n-1) \times (n-1)$ matrix $A(k,l)$.

In regard to the factor $(-1)^{k+l}$ which appears in (IV.3.7) and (IV.3.12), we see that this effects an alternating, or a checkerboard, pattern on the n^2 entries of any $n \times n$ matrix as illustrated in Figure 4.2(b). Thus, when computing the appropriate sign, plus or minus, which must be "attached"

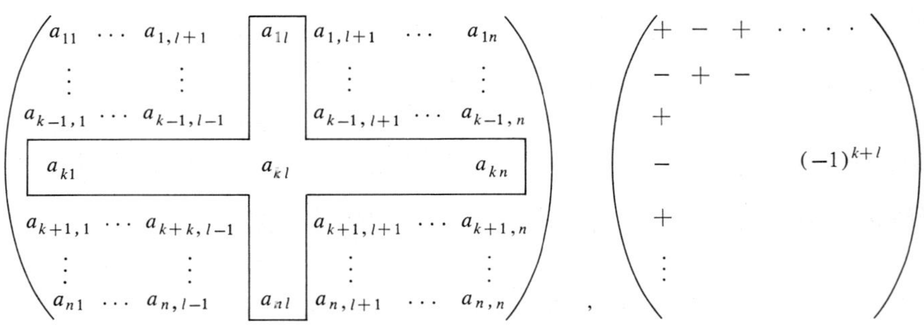

(a) (b)

Figure 4.2

to the form $a_{kl}\det(A(k,l))$ in either (IV.3.7) or (IV.3.12), the graphical checkerboard pattern provides the answer; that is, employ the sign, $+1$ or -1, that resides in the entry that the scalar a_{kl} resides in.

Example Compute the determinant of the matrix A, where

$$A = \begin{pmatrix} 1^{+} & 0^{-} & 1^{-} & 2^{-} & 0 \\ 0 & 2 & 1 & 0^{+} & 0 \\ 0 & 0 & 1 & 0^{-} & 2 \\ 3 & 4 & 1 & 0^{+} & 1 \\ 0 & 2 & 1 & 0^{-} & 1 \end{pmatrix}$$

and each entry is a nonnegative integer (note the alternating plus-minus pattern). Expansion by the fourth column will allow us 4 *zero* entries and only 1 nonzero entry. So much the better for computation, for then we apply Theorem IV.3.2 where $k = 4$, to obtain

$$\begin{aligned} \det(A) &= -\overset{2}{a_{14}} \cdot \det(A(1,4)) + \overset{0}{a_{24}} \cdot \det(A(2,4)) \\ &\quad - \overset{0}{a_{34}} \cdot \det(A(3,4)) + \overset{0}{a_{44}} \cdot \det(A(4,4)) \\ &\quad - \overset{0}{a_{54}} \cdot \det(A(5,4)) \qquad \text{(IV.3.13)} \\ &= -2\det \underbrace{\begin{pmatrix} 0 & 2 & 1 & 0 \\ 0 & 0 & 1 & 2 \\ 3 & 4 & 1 & 1 \\ 0 & 2 & 1 & 1 \end{pmatrix}}_{A(1,4) = A'}. \end{aligned}$$

We now hunt for a row or column with many zeros; rows 1 and 2 of the above 4×4 matrix A_1 each has 2, but column 1 has 3 zeros! We expand the 4×4 matrix A_1 of (IV.3.13) by its first column, to obtain

$$\begin{aligned} &-2\det\begin{pmatrix} 0^{+} & 2 & 1 & 0 \\ 0^{-} & 0 & 1 & 2 \\ 3^{+} & 4 & 1 & 1 \\ 0^{-} & 2 & 1 & 1 \end{pmatrix} \\ &= -2[\overset{0}{a_{11}} \det(A'(1,1)) - \overset{0}{a_{21}} \cdot \det(A'(2,1)) + \overset{3}{a_{31}} \det(A'(3,1)) - \overset{0}{a_{41}} \cdot \det(A'(4,1))] \qquad \text{(IV.3.14)} \\ &= -2 \cdot 3 \det \underbrace{\begin{pmatrix} 2 & 1 & 0 \\ 0 & 1 & 2 \\ 2 & 1 & 1 \end{pmatrix}}_{A'(3,1) = A''}. \end{aligned}$$

Now the determinant of the 3×3 matrix A'' in (IV.3.14) can be computed in many ways. We should feel quite relieved that we have been able to unburden ourselves of a cumbersome, tedious, determinant involving the 5×5 matrix A, in favor of a determinant of a relatively harmless 3×3 matrix A''. Nonetheless, to illustrate the methods of Theorems IV.3.3 and IV.3.4, we shall expand A'' by the first row. Why the first row? After we effect the elementary column operation $C_2{}^1(-2)$, we shall enjoy 2 zeros in the first row, and the determinant will be unchanged in doing so (Theorem IV.2.10(d). From (IV.3.14), we advance to

$$-2 \cdot 3 \cdot \det \underbrace{\begin{pmatrix} 2 & 1 & 0 \\ 0 & 1 & 2 \\ 2 & 1 & 1 \end{pmatrix}}_{A''} = -2 \cdot 3 \cdot \det \underbrace{\begin{pmatrix} 0^+ & 1^- & 0^+ \\ 2^- & 1 & 2 \\ 0^+ & 1 & 1 \end{pmatrix}}_{A''' = A'' \cdot C_2{}^1(-2)} \quad \text{Theorem IV.2.10(d)} \tag{IV.3.15}$$

$$\begin{aligned} &= -2 \cdot 3[\overset{0}{a_{11}} \cdot \det (A'''(1,1)) - \overset{1}{a_{12}} \cdot \det (A'''(1,2)) + \overset{0}{a_{13}} \cdot \det (A'''(1,3))] \\ &= -2 \cdot 3 \cdot 1 \cdot \det \underbrace{\begin{pmatrix} -2 & 2 \\ 0 & 1 \end{pmatrix}}_{A'''(1,2)} \\ &= -2 \cdot 3 \cdot 1 \cdot (-2) \\ &= 12. \end{aligned}$$

Starting from (IV.3.13) and reading through (IV.3.15), we see that $\det (A) = 12$.

EXERCISES

IV.3.1 For the following matrices, find the n^2 minors $\det (A(i,j))$, $i, j = 1, 2, \ldots, n$.

(a) $\begin{pmatrix} 1 & 0 & 3 \\ 1 & 2 & 1 \\ -1 & 4 & 3 \end{pmatrix}$ (b) $\begin{pmatrix} 1 & 0 & 0 & 0 \\ 0 & 1 & 0 & 0 \\ 0 & 0 & 2 & 0 \\ 0 & 0 & 0 & 3 \end{pmatrix}$

(c) $\begin{pmatrix} 1 & 0 & 0 & 2 \\ 0 & 2 & 1 & 0 \\ 0 & 0 & 3 & 0 \\ 0 & 0 & 0 & 0 \end{pmatrix}$ (d) $\begin{pmatrix} x & 0 & 0 & 1 \\ 0 & y & 0 & 1 \\ u & v & 0 & 1 \\ w & x & 0 & 1 \end{pmatrix}$

IV.3.2 Expand the typical 3×3 matrix $A = (a_{ij})_{ij}$ (a) by the first row. (b) By the second column. Show, by actual computation, that the results are equal. That is, if

$$A = \begin{pmatrix} a_{11} & a_{12} & a_{13} \\ a_{21} & a_{22} & a_{13} \\ a_{31} & a_{32} & a_{33} \end{pmatrix},$$

then for

(1) $\det_1(A) = a_{11} \det \begin{pmatrix} a_{22} & a_{23} \\ a_{32} & a_{33} \end{pmatrix} - a_{12} \det \begin{pmatrix} a_{21} & a_{23} \\ a_{31} & a_{33} \end{pmatrix} + a_{13} \det \begin{pmatrix} a_{21} & a_{22} \\ a_{31} & a_{32} \end{pmatrix}$

(2) $\det_2(A) = -a_{12} \det \begin{pmatrix} a_{21} & a_{23} \\ a_{31} & a_{33} \end{pmatrix} + a_{22} \det \begin{pmatrix} a_{11} & a_{13} \\ a_{31} & a_{33} \end{pmatrix} - a_{32} \det \begin{pmatrix} a_{11} & a_{13} \\ a_{21} & a_{23} \end{pmatrix},$

we have $\det_1(A) = \det_2(A) = \det(A)$.

IV.3.3 Expand the 4×4 triangular matrix

$$A = \begin{pmatrix} a_{11} & a_{12} & a_{13} & a_{14} \\ 0 & a_{22} & a_{23} & a_{24} \\ 0 & 0 & a_{33} & a_{34} \\ 0 & 0 & 0 & a_{44} \end{pmatrix}$$

by the first column, to obtain the result that $\det(A) = a_{11}a_{22}a_{33}a_{44}$, the product of the main diagonal entries.

IV.3.4 Use the result of Exercise IV.3.3 to prove that any 5×5 triangular matrix has its determinant equal to the product of its 5 main diagonal entries.

IV.3.5 Compute the determinants of the following matrices:

(a) $\begin{pmatrix} 0 & 6 & 1 & 0 & 1 \\ 1 & 1 & 2 & 2 & 1 \\ 0 & 2 & 3 & 1 & 1 \\ 1 & 0 & 2 & 1 & 0 \\ 0 & 0 & 4 & 2 & 1 \end{pmatrix}$ (b) $\begin{pmatrix} 0 & 1 & 0 & 1 & 0 & 1 & 0 \\ 1 & 0 & 1 & 0 & 1 & 0 & 1 \\ 1 & 1 & 1 & 0 & 0 & 0 & 0 \\ 0 & 0 & 1 & 0 & 0 & 0 & 0 \\ 2 & 1 & 3 & 1 & 0 & 0 & 0 \\ 0 & 1 & 1 & 0 & 0 & 1 & 1 \\ 0 & 0 & 0 & 0 & 0 & 0 & 1 \end{pmatrix}$

(c) $\begin{pmatrix} 2 & 3 & 4 & 5 & 6 \\ 3 & 4 & 5 & 6 & 7 \\ 4 & 5 & 6 & 7 & 8 \\ 5 & 6 & 7 & 8 & 9 \\ 1 & 2 & 3 & 4 & 5 \end{pmatrix}$ (d) $\begin{pmatrix} 1 & 3 & 0 & 3 & 2 \\ 2 & 1 & 5 & 7 & 8 \\ 0 & 0 & 1 & 0 & 0 \\ 0 & 0 & 0 & 1 & 1 \\ 0 & 0 & 2 & 0 & 1 \end{pmatrix}$

(e) $\left(\begin{array}{cc:ccc} 1 & 3 & a & b & c \\ 2 & 1 & d & e & f \\ \hdashline 0 & 0 & 1 & 0 & 0 \\ 0 & 0 & 0 & 1 & 1 \\ 0 & 0 & 2 & 0 & 1 \end{array}\right)$

IV.3.6 (A generalization of Exercises IV.3.2) and IV.3.4.) Show that if

$$D = \left(\begin{array}{cc:ccc} a_{11} & a_{12} & * & * & * \\ a_{21} & a_{22} & * & * & * \\ \hdashline 0 & 0 & b_{11} & b_{12} & b_{13} \\ 0 & 0 & b_{21} & b_{22} & b_{23} \\ 0 & 0 & b_{31} & b_{32} & b_{33} \end{array}\right) = \begin{pmatrix} A & C \\ 0 & B \end{pmatrix}$$

where

$$A = \begin{pmatrix} a_{11} & a_{12} \\ a_{21} & a_{22} \end{pmatrix} \qquad B = \begin{pmatrix} b_{11} & b_{12} & b_{13} \\ b_{21} & b_{22} & b_{23} \\ b_{31} & b_{32} & b_{33} \end{pmatrix}$$

and C is an *arbitrary* 2×3 matrix, then $\det(D) = \det(A) \cdot \det(B)$ [see Exercises IV.3.5(d) and IV.3.5(e) for examples of such matrices].

IV.3.7 Let us regard the determinant function $\det(\cdot)$, which is defined for any $n \times n$ matrix A, as a function on the n *columns* A_i, $i = 1, 2, \ldots, n$, of $A = [A_1, A_2, \ldots, A_n]$. That is, $\det(\cdot)$ can be considered as a function of n "variables," each "variable" being a column A_i, of the matrix A. The object of this exercise lies in proving that the function $A \to \det(A)$, that is, $[A_1,A_2,\ldots,A_n] \to \det[A_1,A_2,\ldots,A_n]$, is *linear* in *each* argument (see Definition II.2.1 for a function linear in 1 argument). To state this precisely, you are asked to prove that if $A_1, A_2,\ldots, A_j,\ldots, A_n, B$ represent columns (n-vectors), then

(1) For all scalars λ,

$$\det(A_1,A_2,\ldots,\lambda A_j,\ldots,A_n) = \lambda \det(A_1,A_2,\ldots,A_j,\ldots,A_n)$$

[see Theorem IV.2.10(b) and Exercise IV.2.17].

(2) $$\det(A_1,A_2,\ldots,A_j + B,\ldots,A_n) = \det(A_1,A_2,\ldots,A_j,\ldots,A_n) + \det(A_1,A_2,\ldots,B,\ldots,A_n).$$

(see Exercises IV.2.13, IV.2.14).

Hint: For (2), set

$$A_j = \begin{pmatrix} a_{ij} \\ a_{2j} \\ \vdots \\ a_{ij} \\ \vdots \\ a_{nj} \end{pmatrix}$$

for each $j = 1, 2,\ldots, n$, and put

$$B = \begin{pmatrix} b_1 \\ b_2 \\ \vdots \\ b_n \end{pmatrix}.$$

Then expand $\det(A_1,A_2,\ldots,A_j + B,\ldots,A_n)$ by its jth column

$$\begin{pmatrix} a_{1j} + b_1 \\ a_{2j} + b_2 \\ \vdots \\ a_{nj} + b_n \end{pmatrix}.$$

IV.3.8 Given the $n \times n$ matrix $A = (i + j - 1)_{ij}$, that is,

$$A = \begin{pmatrix} 1 & 2 & 3 & & n \\ 2 & 3 & 4 & \cdots & n+1 \\ 3 & 4 & 5 & \cdots & n+2 \\ \vdots & \vdots & \vdots & & \vdots \\ n & n+1 & n+2 & \cdots & 2n-1 \end{pmatrix}.$$

Show that $\det(A) = 0$ unless $n = 1$ or 2.

IV.3.9 (a) Show that if the $n \times n$ matrix A is skew-symmetric (that is, $A^t = -A$), and if n is odd, then necessarily, $\det(A) = 0$.

(b) Give an example of an $n \times n$ skew-symmetric matrix A such that $\det(A) \neq 0$ [from (a), n is necessarily *even*?].

(c) Give an example of an $n \times n$ skew-symmetric matrix A, where n is *even*, and yet, $\det(A) = 0$.

(d) Does (c) conflict with (a)? Why or why not?

IV.3.10 Show that for the Vandermonde matrix

$$A = \begin{pmatrix} 1 & 1 & 1 \\ a & b & c \\ a^2 & b^2 & c^2 \end{pmatrix}, \qquad \det(A) = (c-a)(c-b)(b-a).$$

4. *The Product Theorem and Applications of det $(\cdot)$*

We seek to extend the half-hearted product theorem (namely, Corollary IV.2.7) to arbitrary square matrices. Without delay, we present Theorem IV.4.1.

Theorem IV.4.1 Let A and B be any pair of $n \times n$ matrices. Then $\det(A \cdot B) = \det(B \cdot A) = \det(A)\det(B)$.

Proof: There are only 2 cases to consider: (1) when at least one of the matrices A, B has no inverse (that is, rank A or rank $B < n$: see Theorem III.3.5), or (2) when none of the matrices A, B is singular (that is, both A and B are invertible).

CASE 1. If either A or B has no inverse (see Definition III.3.3), then we claim that $A \cdot B$ cannot have an inverse either. Let us bear in mind the fact that (Theorem III.3.5)

Every 1-sided matrix inverse for A is always a 2-sided inverse. (IV.4.1)

Now suppose that A has *no* inverse. This means:

For no $n \times n$ matrix M on the face of this (or any other) earth is either of the following equations true: (IV.4.2)

$$\text{(a)}\quad M \cdot A = I_n \qquad \text{(b)}\quad A \cdot M = I_n.$$

Note that (IV.4.2) uses (IV.4.1) in that nonexistence of a 1-sided inverse implies nonexistence of a 2-sided inverse.

Let us now foolishly suppose that $A \cdot B$ *has* an $n \times n$ inverse, call it N. We shall see that this induces nonsense, for in assuming

$$(A \cdot B) \cdot N = I_n, \qquad \text{definition of inverse}$$

we are automatically assuming

$$A \cdot \underbrace{(B \cdot N)}_{M} = I_n. \quad \text{Associativity of matrix multiplication.} \tag{IV.4.3}$$

But (IV.4.3) tells us that the matrix $B \cdot N$ is a right (hence, 2-sided) inverse of A. This contradicts our basic assumption (IV.4.2), that A has *no* inverse. Hence, we must discard our second assumption, namely, $A \cdot B$ has an inverse N. A similar argument shows: B has no inverse $\Rightarrow$ $A \cdot B$ has no inverse (Exercise IV.4.10).

Now we have established the following:

$$\begin{array}{ccccc} \underbrace{A \text{ has no inverse,}} & \text{or} & \underbrace{B \text{ has no inverse}} & \Rightarrow & \underbrace{A \cdot B \text{ has no inverse}} \\ \downarrow \text{Theorem IV.2.9} & & \downarrow \text{Theorem IV.2.9} & & \downarrow \text{Theorem IV.2.9} \\ \det(A) = 0 & \text{or} & \det(B) = 0 & \Rightarrow & \det(A \cdot B) = 0. \end{array} \tag{IV.4.4}$$

Finally, from (IV.4.4) we deduce the fact that if A has no inverse, or if B has no inverse, then

$$\begin{array}{ccc} \det(A \cdot B) & = & \det(A) \cdot \det(B) \\ \uparrow & & \nwarrow \quad \nearrow \\ \text{zero} & & \text{at least 1 factor} \\ & & \text{equals zero.} \end{array} \tag{IV.4.5}$$

CASE 2. Both A and B have inverses, A^{-1} and B^{-1}, respectively. Then

$$A = R_1 \cdot R_2 \cdots R_k$$

and

$$B = S_1 \cdot S_2 \cdots S_l$$

for certain elementary matrices $R_1, R_2, \ldots, R_k, S_1, S_2, \ldots, S_l$ (Theorem III.3.6). Using our product theorem for elementary matrices (Corollary IV.2.7), we have

$$\begin{aligned} \det(A \cdot B) &= \det(R_1 \cdot R_2 \cdots R_k \cdot S_1 \cdot S_2 \cdots S_l) \\ &= \det(R_1) \det(R_2 \cdots R_k \cdot S_1 \cdot S_2 \cdots S_l) \\ &\;\;\vdots \qquad\qquad\qquad \vdots \\ &= [\det(R_1) \cdots \det(R_k)][\det(S_1) \cdots \det(S_l)] \qquad \text{(IV.4.6)} \\ &\;\;\vdots \\ &= \underbrace{\det(R_1 \cdot R_2 \cdots R_k)}_{A} \; \underbrace{\det(S_1 \cdot S_2 \cdots S_k)}_{B} \\ &= \det(A) \cdot \det(B). \end{aligned}$$

The proof is done, since (IV.4.5) and (IV.4.6) hold true.

One very easy result which flows from the product theorem concerns the determinant of an invertible matrix A (which is nonzero, by Theorem IV.2.9), and that of its inverse A^{-1}. In fact, the determinant of the inverse, $\det(A^{-1})$, will be shown to be $[\det(A)]^{-1}$, the inverse of the determinant.

Corollary IV.4.2 Let A be a nonsingular $n \times n$ matrix, that is, A^{-1} exists. Then

$$\det(A^{-1}) = [\det(A)]^{-1}.$$

Proof: Note that the superscript $^{-1}$ above is used in 2 ways. On the left-hand side of the equation, it denotes the multiplicative inverse of the $n \times n$ *matrix*, A; on the right-hand side, it denotes the multiplicative inverse of the *scalar*, $\det(A)$.

Now $\quad A^{-1} \cdot A = I_n \quad$ (Definition III.3.3),

so that $\quad \det(A^{-1} \cdot A) = \det(I_n) = 1. \quad$ (IV.4.7)

That is, $\quad$ Theorem IV.4.1

$$\det(A^{-1}) \cdot \det(A) = 1$$

Now whenever you have two scalars multiplying together to yield 1 [as in (IV.4.7)], one factor is called the inverse of the other. That is, $\det(A^{-1})$ is the inverse of (the scalar) $\det(A)$, or, to say the same thing symbolically,

$$\det(A^{-1}) = [\det(A)]^{-1}.$$

That proves our corollary.

Now that we know that a nonzero determinant of a matrix is equivalent to the invertibility of that matrix (Theorem IV.2.9), we are able to use our expansion theorems to develop *another* construction for inverses of $n \times n$ matrices.

We describe this method, but first we define for each $n \times n$ matrix A, another $n \times n$ matrix, $\operatorname{adj}(A)$, called the **adjoint of A**. If A^{-1} exists, then we shall see that this easily defined matrix, $\operatorname{adj}(A)$, is "almost" the matrix A^{-1} that we seek.

A loose definition for $\operatorname{adj}(A)$ goes like this: Starting with your original matrix $A = (a_{ij})_{ij}$, follow these 3 steps:

▶ (1) Build an $n \times n$ matrix by placing the scalar $\det(A(i,j))$ in the i,jth entry, that is, in row i and column j. Recall that $A(i,j)$ is the $(n-1) \times (n-1)$ matrix obtained from A by striking out the ith row and jth column of the $n \times n$ matrix A.

▶ (2) To each of these entries attach the appropriate factor $+1$ or -1 according to the checkerboard pattern, starting with $+1$ in the upper left-hand corner. That is, attach the factor $(-1)^{i+j}$ to the scalar $\det(A(i,j))$ which already sits in the ijth entry.

▶ (3) Take the transpose of whatever comes from (1) and (2); you have now constructed adj (A). That is, $\text{adj}(A) = ((-1)^{i+j} \det(A(j,i)))_{ij}$.

Example Let us choose for our matrix, the 3×3 matrix

$$A = \begin{pmatrix} 1 & 0 & 2 \\ 3 & 1 & 4 \\ 7 & 7 & 7 \end{pmatrix}.$$

Our task is to construct adj (A). To perform step (1) above, we must place the 9 scalars $\det(A(i,j))$. in the corresponding row and column, that is, in row i and column j.

$$\text{Step (1)} \quad \begin{pmatrix} \det\begin{pmatrix} 1 & 4 \\ 7 & 7 \end{pmatrix} & \det\begin{pmatrix} 3 & 4 \\ 7 & 7 \end{pmatrix} & \det\begin{pmatrix} 3 & 1 \\ 7 & 7 \end{pmatrix} \\ \det\begin{pmatrix} 0 & 2 \\ 7 & 7 \end{pmatrix} & \det\begin{pmatrix} 1 & 2 \\ 7 & 7 \end{pmatrix} & \det\begin{pmatrix} 1 & 0 \\ 7 & 7 \end{pmatrix} \\ \det\begin{pmatrix} 0 & 2 \\ 1 & 4 \end{pmatrix} & \det\begin{pmatrix} 1 & 2 \\ 3 & 4 \end{pmatrix} & \det\begin{pmatrix} 1 & 0 \\ 3 & 1 \end{pmatrix} \end{pmatrix}$$

$$= \begin{pmatrix} -21 & -7 & 14 \\ -14 & -7 & 7 \\ -2 & -2 & 1 \end{pmatrix}.$$

To perform Step (2), we affix the factors $+1$ or -1 in a checkboard pattern, starting with $+1$ in the upper left-hand corner. Thus, we obtain

$$\text{Step (2)} \qquad \begin{pmatrix} -21 & 7 & 14 \\ 14 & -7 & -7 \\ -2 & 2 & 1 \end{pmatrix}.$$

Step (3) requires only that we take the transpose, which brings us to adj (A).

$$\text{Step (3)} \qquad \begin{pmatrix} -21 & 14 & -2 \\ 7 & -7 & 2 \\ 14 & -7 & 1 \end{pmatrix}.$$

So what?

We here made mention of the fact that adj (A) is "almost" the inverse for the matrix A, if A^{-1} exists, that is, if $\det(A) \neq 0$. [Were we to compute $\det(A)$, we would find that $\det(A) = 7$, so A^{-1} *does* exist; see Theorem IV.2.9.] Let us compute the product $A \cdot \text{adj}(A)$ and see how close we get to obtaining I_3.

$$A \cdot \text{adj}(A) = \begin{pmatrix} 1 & 0 & 2 \\ 3 & 1 & 4 \\ 7 & 7 & 7 \end{pmatrix} \cdot \begin{pmatrix} -21 & 14 & -2 \\ 7 & -7 & 2 \\ 14 & -7 & 1 \end{pmatrix}$$

$$= \begin{pmatrix} 7 & 0 & 0 \\ 0 & 7 & 0 \\ 0 & 0 & 7 \end{pmatrix}$$

$$= 7 \cdot \begin{pmatrix} 1 & 0 & 0 \\ 0 & 1 & 0 \\ 0 & 0 & 1 \end{pmatrix}.$$

That number 7—we've seen it before as det (A). We have in this special case that $A \cdot \operatorname{adj}(A) = \det(A) I_3$. As we shall show, this relationship *always* obtains. In fact $A \cdot \operatorname{adj}(A) = \operatorname{adj}(A) \cdot A = \det(A) I_n$ for *all* $n \times n$ matrices A.

Since $\det(A) = 7 \neq 0$, we may divide by det (A) to obtain

$$A \cdot \frac{\operatorname{adj}(A)}{\det(A)} = I_3,$$

which is to say

$$\frac{\operatorname{adj}(A)}{\det(A)} = A^{-1},$$

the inverse of A.

A *formal* definition for adj (A) now seems in order.

Definition IV.4.3 Given the $n \times n$ matrix A. Then **adj (*A*)**, the **adjoint of *A***, is the $n \times n$ matrix defined by

$$\operatorname{adj}(A) = ((-1)^{k+j} \det(A(j,k)))_{k,j}.$$

We will now prove that the product of A with its adjoint, adj (A), yields the diagonal scalar matrix $\det(A) I_n$.

Theorem IV.4.4 For any $n \times n$ matrix A,

$$A \cdot \operatorname{adj}(A) = \operatorname{adj}(A) \cdot A = \det(A) I_n.$$

Proof: The proof involves little more than a routine manipulation of matrix products along with an understanding of the expansion of determinants by minors.

Now we have $A = (a_{ik})_{ik}$ and $\operatorname{adj}(A) = ((-1)^{k+j} \det(A(j,k)))_{kj}$ from Definition IV.4.3. Thus (Theorem II.6.13),

$$A \cdot \operatorname{adj}(A) = \left(\sum_{k=1}^{n} a_{ik} (-1)^{k+j} \det(A(j,k)) \right)_{ij}, \quad i,j = 1, 2, \ldots, n. \qquad \text{(IV.4.8)}$$

Notice that the summation (over the dummy index k) in (IV.4.8) represents the ijth entry of our product $A \cdot \operatorname{adj}(A)$. We shall try to interpret this sum as a *determinant* of some $n \times n$ matrix expanded by a row or column.

Accordingly, let us see what we can make of the main diagonal (that is, the jjth) entries of $A \cdot \operatorname{adj}(A)$ as given in (IV.4.8). That is, put $i = j$ in (IV.4.8) to obtain the fact that the jjth diagonal entry of $A \cdot \operatorname{adj}(A)$ is precisely the sum

$$\sum_{k=1}^{n} a_{jk}(-1)^{k+j} \det(A(j,k)) = \det(A). \quad \text{Theorem IV.3.3} \tag{IV.4.9}$$

[If in (IV.3.7) of Theorem IV.3.3, we put $l = j$, then we see that (IV.4.9) represents $\det(A)$ expanded by the jth row.]

What about the off-diagonal elements of $A \cdot \operatorname{adj}(A)$? For $i \neq j$, we see from (IV.4.8) that the ijth entry of $A \cdot \operatorname{adj}(A)$ is the sum

$$\sum_{k=1}^{n} \underbrace{a_{ik}}_{\text{entry from row } j\text{, column } k \text{ of } A'}(-1)^{k+j} \det(A(j,k)) = \det(A'). \tag{IV.4.10}$$

But if we construct the $n \times n$ matrix A' so that each row except for the jth row agrees with A, and the jth row equals the ith row of A, that is, A' is represented by

$$A' = \begin{pmatrix} a_{11} & a_{12} & \cdots & a_{1k} & \cdots & a_{1n} \\ \vdots & \vdots & & \vdots & & \vdots \\ a_{i1} & a_{i2} & \cdots & a_{ik} & \cdots & a_{in} \\ \vdots & \vdots & & \vdots & & \vdots \\ a_{i1} & a_{i2} & \cdots & a_{ik} & \cdots & a_{in} \\ \vdots & \vdots & & \vdots & & \vdots \\ a_{n1} & a_{n2} & \cdots & a_{nk} & \cdots & a_{nn} \end{pmatrix} \begin{matrix} \\ \\ \leftarrow \text{row } i \\ \\ \leftarrow \text{row } j \\ \\ \\ \end{matrix} \tag{IV.4.11}$$

then we see that (IV.4.10) can be interpreted as the determinant of A' (not of A) expanded about its jth row (see IV.4.11) in accordance with (IV.3.7) of Theorem IV.3.3. But A' has 2 identical rows, namely, row i and row j; hence, $\det(A') = 0$ (Corollary IV.2.3).

We must conclude, then, that the diagonal entries of $A \cdot \operatorname{adj}(A)$ equal $\det(A)$, [see (IV.4.9)] and the off-diagonal entries are all zero [see (IV.4.10)]. Thus, we have proved that

$$A \cdot \operatorname{adj}(A) = \det(A) I_n.$$

To prove that $\operatorname{adj}(A) \cdot A$ also equals $\det(A) I_n$, we may proceed in a similar manner by inspecting sums which will then be interpreted as determinants of certain matrices expanded by columns instead of rows.

We employ Theorem IV.4.4, which relates A to its adjoint, adj (A), under multiplication and Theorem IV.2.9 to arrive at the following corollary. This corollary gives the explicit representation for an inverse of a matrix in terms of the adjoint.

Corollary IV.4.5 If A is nonsingular, then $A^{-1} = (\det(A))^{-1} \cdot \operatorname{adj}(A)$.

Proof: A is nonsingular, so that we are just saying that $\det(A) \neq 0$ (Theorem IV.2.9). That is, the inverse $(\det(A))^{-1}$ of the scalar $\det(A)$ exists.

For any $n \times n$ matrix A (regardless of its determinant), Theorem IV.4.4 allows us to write

$$\operatorname{adj}(A) \cdot A = \det(A) \cdot I_n.$$

Multiply both sides by $(\det(A))^{-1}$, and the result is

$$\underbrace{[(\det(A))^{-1} \cdot \operatorname{adj}(A)]}_{\text{left inverse of } A} \cdot A = I_n.$$

The left-hand factor is, by Definition III.3.3, the left inverse of A. We are through with our proof of this corollary once we notice that for matrices, at least, left inverse are also right inverses (Theorem III.3.5); hence the left-hand factor $\det(A))^{-1} \cdot \operatorname{adj}(A)$ is actually A^{-1}, the 2-sided inverse of A.

REMARK. Is this method easier for computing inverses than the technique of Chapter III, which used elementary operations to reduce A to the identity? Since, for an $n \times n$ matrix A, there are n^2 determinants to compute in constructing adj (A), each determinant involving an $(n - 1) \times (n - 1)$ submatrix of A, we can say that generally this is *not* a practical method. For $n \leq 4$, it seems fairly convenient, however.

EXERCISES

IV.4.1 Prove that if an $n \times n$ matrix A is nonsingular, then no finite product of A can be singular, that is, for *no* positive integer p, is it true that $A^p = A \cdot A \cdot \cdots \cdot A$ is singular.

IV.4.2 Show by direct computation that for the matrices A and B given below, $\det(A \cdot B) = \det(B \cdot A) = \det(A)\det(B)$.

(a) $A = \begin{pmatrix} 1 & 2 \\ 3 & 4 \end{pmatrix}$, $\quad B = \begin{pmatrix} 1 & -1 \\ 1 & 1 \end{pmatrix}$.

(b) $A = \begin{pmatrix} 1 & 0 & 0 \\ 0 & 0 & 7 \\ 0 & 1 & 0 \end{pmatrix}$, $\quad B = \begin{pmatrix} 0 & 1 & 0 \\ 3 & 0 & 0 \\ 0 & 0 & 1 \end{pmatrix}$.

(c) $A = \begin{pmatrix} 1 & 0 & 0 \\ 0 & a & b \\ 0 & c & d \end{pmatrix}$, $B = \begin{pmatrix} p & q & 0 \\ r & s & 0 \\ 0 & 0 & 1 \end{pmatrix}$.

(d) $A = \begin{pmatrix} a & d & e \\ 0 & b & f \\ 0 & 0 & c \end{pmatrix}$, $B = \begin{pmatrix} p & 0 & 0 \\ s & q & 0 \\ u & t & r \end{pmatrix}$.

IV.4.3 Given only that det $(A \cdot B) = \det(A) \det(B)$ (Theorem IV.4.1) for all square matrices A and B, show that necessarily, det $(A \cdot B) = \det(B \cdot A)$.

IV.4.4 What is wrong with the following "proof" of the fact that all pairs of matrices commute with each other under matrix multiplication, that is, that $A \cdot B = B \cdot A$ for all square matrices A and B:

"From the properties of determinants, one may show that for *any* pair of square matrices, A and B, det $(A \cdot B)$ equals det $(B \cdot A)$. But this happens if $A \cdot B = B \cdot A$. Therefore, for *any* pair of square matrices, $A \cdot B = B \cdot A$."

IV.4.5 (1) For the 2×2 matrix $A = \begin{pmatrix} a & b \\ c & d \end{pmatrix}$, compute the matrix adj (A).

(2) Verify Theorem IV.4.4 in this case by direct computation, that is, show that $A \cdot \text{adj}(A) = \text{adj}(A) \cdot A = \begin{pmatrix} \det(A) & 0 \\ 0 & \det(A) \end{pmatrix}$.

(3) If det $(A) \neq 0$, (if A^{-1} exists), then use the adjoint matrix, adj(A), to show that

$$A^{-1} = \begin{pmatrix} d/\det(A) & -b/\det(A) \\ -c/\det(A) & a/\det(A) \end{pmatrix}$$

(see Corollary IV.4.5).

IV.4.6 Prove that for the $n \times n$ matrix adj (A), det (adj $(A)) = \det(A)^{n-1}$.

IV.4.7 Prove that if the $n \times n$ matrix P is invertible, then for all $n \times n$ matrices A, det $(P^{-1}AP) = \det(A)$.

IV.4.8 (1) For the following matrices A, use both the adjoint matrix adj (A) and the elementary operations of Chapter III, to compute A^{-1}, the inverse matrix.

(2) Compute det (A) and det (A^{-1}) *directly* in each case. Compare your results with the statement of Corollary IV.4.2.

(a) $A = \begin{pmatrix} 1 & 5 & 2 \\ 0 & 2 & 0 \\ 0 & 0 & 1 \end{pmatrix}$

(b) $A = \begin{pmatrix} 1 & 1 & 0 \\ 2 & 2 & 1 \\ 0 & 1 & 1 \end{pmatrix}$

(c) $A = \begin{pmatrix} 3 & 0 & 1 & 0 \\ 0 & 2 & 2 & 2 \\ 0 & 0 & 5 & 1 \\ 0 & 0 & 0 & 7 \end{pmatrix}$

(d) $A = \begin{pmatrix} 2 & 0 & 0 & 1 \\ 3 & 1 & 0 & 2 \\ 1 & 0 & 1 & 1 \\ 0 & 0 & 0 & 1 \end{pmatrix}$

IV.4.9 Let X and Y be the typical 3-vectors

$$\begin{pmatrix} x_1 \\ x_2 \\ x_3 \end{pmatrix} \text{ and } \begin{pmatrix} y_1 \\ y_2 \\ y_3 \end{pmatrix},$$

respectively. We define the 3-vector $X \wedge Y$ by

$$X \wedge Y = \begin{pmatrix} \det \begin{pmatrix} x_1 & y_1 \\ x_2 & y_2 \end{pmatrix} \\ \det \begin{pmatrix} x_1 & y_1 \\ x_3 & y_3 \end{pmatrix} \\ \det \begin{pmatrix} x_2 & y_2 \\ x_3 & y_3 \end{pmatrix} \end{pmatrix}.$$

Prove:

(a) $X \wedge Y = -(Y \wedge X)$.

(b) $(\lambda X) \wedge Y = X \wedge (\lambda Y) = \lambda(X \wedge Y)$ for all scalars λ.

(c) $(X + Y) \wedge Z = (X \wedge Z) + (Y \wedge Z)$ for all vectors X, Y, and Z.

(d) $X \wedge Y = \begin{pmatrix} 0 \\ 0 \\ 0 \end{pmatrix}$ if and only if X is a scalar multiple of Y, that is, if and only if $X = \lambda Y$ for some scalar λ.

IV.4.10 Prove that if the $n \times n$ matrix B has no (2-sided) inverse, then neither can $A \cdot B$, where A is any $n \times n$ matrix.

One application of Theorem IV.4.4 [which states that $A \cdot \operatorname{adj}(A) = \operatorname{adj}(A) \cdot A = \det(A)I_n$ for all A] is Cramer's Rule. This is a device that allows us to solve for the $n \times 1$ matrix X in the matrix equation $A \cdot X = Y$, where $A = (a_{ij})_{ij}$ is any $n \times n$ matrix whose determinant is nonzero. As you may recall, $A \cdot X = Y$ may also be interpreted as a system of a simultaneous linear equations in n unknowns. As it turns out, Cramer's Rule will be *less* efficient than the method of solution we have already encountered (see the examples, following Theorem III.2.7) which employed elementary row operations. Also, Cramer's Rule is severely restricted to the very special situation where the coefficient matrix A is *square*, that is, there are an equal number of equations and unknowns.

Now, Proposition III.2.8 tells us when the system $A \cdot X = Y$ has a unique solution vector X. For n equations and n unknowns, a *unique* solution occurs if and *only* if the rank of A is maximal, that is, equal to n. That is, a unique solution X exists if and only if A is invertible (Theorem III.3.5), which is equivalent to saying that $\det(A) \neq 0$ (Theorem IV.2.9).

From these observations, we are justified in approaching the problem of finding a unique X for $A \cdot X = Y$ as follows:

With the matrix equation $A \cdot X = Y$ in hand, compute $\det(A)$. If $\det(A) \neq 0$, then one and only one $n \times 1$ matrix X can satisfy the matrix equation $A \cdot X = Y$. . . no *matter* what Y is. If, on the other hand, $\det(A) = 0$, then we may or may not have a solution. If a solution X_0, say, *does* exist, then it will not be unique, for any nonzero null vector N $(A \cdot N = O)$ can be added to X_0 giving $X_0 + N$ as *another* solution (Theorem III.2.7). Thus, for $\det(A) = 0$, we shy away from the whole affair; methods previously devised handle this situation easily and do not require an *equal* number of equations and unknowns.

For the sake of illustration, we present examples of systems $A \cdot X = Y$ where $\det(A) = 0$. In one case *no* solution exists, and in the other, infinitely many solutions exist. As we have said, if $\det(A) = 0$, these are the *only* 2 cases possible; we shall *never* have a *unique* solution X for $A \cdot X = Y$, regardless of the value of the vector Y.

Example CASE 1. (no solution exists)

$$\underbrace{\begin{pmatrix} 1 & 0 \\ 0 & 0 \end{pmatrix}}_{A} \cdot \underbrace{\begin{pmatrix} x_1 \\ x_2 \end{pmatrix}}_{X} = \underbrace{\begin{pmatrix} 0 \\ 1 \end{pmatrix}}_{Y}.$$

Surely $\det(A) = 0$, and when the left-hand side of the equation above is multiplied out, we obtain

$$\begin{pmatrix} x_1 \\ 0 \end{pmatrix} = \begin{pmatrix} 0 \\ 1 \end{pmatrix}.$$

No matter *which* vector $X = \begin{pmatrix} x_1 \\ x_2 \end{pmatrix}$ we choose, the truth of this last equation is not to be realized.

CASE 2. (infinitely many solutions exist)

$$\underset{A}{\begin{pmatrix} 1 & 0 \\ 0 & 0 \end{pmatrix}} \cdot \underset{X}{\begin{pmatrix} x_1 \\ x_2 \end{pmatrix}} = \underset{Y}{\begin{pmatrix} 1 \\ 0 \end{pmatrix}}.$$

Here $\det(A) = 0$. When the left-hand side of this equation is multiplied out, we obtain

$$\begin{pmatrix} x_1 \\ 0 \end{pmatrix} = \begin{pmatrix} 1 \\ 0 \end{pmatrix},$$

the truth of which places 1 and *only* 1 constraint on any solution vector $X = \begin{pmatrix} x_1 \\ x_2 \end{pmatrix}$; namely, that $x_1 = 1$. The value assumed by x_2, the second coordinate, can be arbitrary. Thus, the solution(s) to $A \cdot X = Y$ in this case is the set of *all* X of the form $X = \begin{pmatrix} 1 \\ x \end{pmatrix}$. Surely there is more than 1, *unique* solution, in this case.

We now describe Cramer's Rule, present an example of its application, and then prove its validity.

CRAMER'S RULE

Given the n equations and n unknowns $A \cdot X = Y$, that is,

$$\begin{array}{ccccccccc}
a_{11}a_1 & + a_{12}x_2 & + \cdots + & a_{1k}x_k & + \cdots + & a_{1n}x_n & = & y_1 \\
a_{21}x_1 & + a_{22}x_2 & + \cdots + & a_{2k}x_k & + \cdots + & a_{2n}x_n & = & y_2 \\
\vdots & \vdots & & \vdots & & \vdots & & \vdots \\
a_{n1}x_1 & + a_{n2}x_2 & + \cdots + & a_{nk}x_k & + \cdots + & a_{nn}x_n & = & y_n,
\end{array}$$

where $\det(A) \neq 0$. Then for each $k = 1, 2, \ldots, n$,

$$x_k = \frac{\det\begin{pmatrix}
a_{11} & a_{12} & \cdots & \overset{k\text{th column}}{\overset{\downarrow}{y_1}} & \cdots & a_{1n} \\
a_{21} & a_{22} & \cdots & y_2 & \cdots & a_{2n} \\
\vdots & \vdots & & \vdots & & \\
a_{n1} & a_{n2} & \cdots & y_n & \cdots & a_{nn}
\end{pmatrix}}{\det(A)}. \tag{IV.4.12}$$

Notice that the $n \times n$ matrix which appears in the numerator of (IV.4.12) is obtained from the coefficient matrix $A = (a_{ij})_{ij}$, by replacing the kth column with the $n \times 1$ column Y.

Example Solve for (the unique vector)

$$X = \begin{pmatrix} x_1 \\ x_2 \\ x_3 \end{pmatrix}$$

in

$$\overbrace{\begin{pmatrix} 3 & 0 & 1 \\ 1 & 2 & 3 \\ 4 & 0 & 1 \end{pmatrix}}^{A} \cdot \overbrace{\begin{pmatrix} x_1 \\ x_2 \\ x_3 \end{pmatrix}}^{X} = \overbrace{\begin{pmatrix} 6 \\ 14 \\ 7 \end{pmatrix}}^{Y}.$$

Equivalently, find the scalars x_1, x_2, x_3 such that

$$\begin{aligned} 3x_1 + x_3 &= 6 \\ x_1 + 2x_2 + 3x_3 &= 14 \\ 4x_1 + 3x_3 &= 7. \end{aligned}$$

Now det $(A) = -2$. From (IV.4.12),

$$x_1 = \frac{\overset{-2}{\det\begin{pmatrix} 6 & 0 & 1 \\ 14 & 2 & 3 \\ 7 & 0 & 1 \end{pmatrix}}}{-2},\quad x_2 = \frac{\overset{-4}{\det\begin{pmatrix} 3 & 6 & 1 \\ 1 & 14 & 3 \\ 4 & 7 & 1 \end{pmatrix}}}{-2},\quad x_3 = \frac{\overset{-6}{\det\begin{pmatrix} 3 & 0 & 6 \\ 1 & 2 & 14 \\ 4 & 0 & 7 \end{pmatrix}}}{-2}.$$

Thus, we obtain $x_1 = 1$, $x_2 = 2$ and $x_3 = 3$ as the unique solution set for our system of 3 equations in 3 unknowns.

Theroem IV.4.6 (Cramer's Rule) Suppose $A \cdot X = Y$ represents a system of n equations in n unknowns $x_1, x_2, \ldots, x_k, \ldots, x_n$. Suppose $\det(A) \neq 0$. Then the unique solutions for the system are given according to Equation (IV.4.12).

Proof: Since det $(A) \neq 0$, A^{-1} exists (Theorem IV.2.9). Multiply both sides of the system equation $A \cdot X = Y$ on the left by A^{-1} to isolate X, thereby obtaining

$$X = A^{-1} \cdot Y. \tag{IV.4.13}$$

(In fact, "testing" this value of $A^{-1} \cdot Y$ for X in the original equation $A \cdot X = Y$, reveals that, indeed,

$$A \cdot (\overset{X}{A^{-1} \cdot Y}) = (A \cdot A^{-1}) \cdot Y = I \cdot Y = Y.$$

That is, $A^{-1} \cdot Y$ "works," so this is *the* X we wish to find.) But $A^{-1} = (1/\det(A))\,\text{adj}\,(A)$, where

$$\text{adj}\,(A) = ((-1)^{i+j} \det\,(A(j,i)))_{ij} \quad \text{(Corollary IV.4.5).}$$

Thus, (IV.4.13) becomes

$$X = \frac{\overbrace{\text{adj}\,(A)}^{n \times n} \cdot \overbrace{Y}^{n+1}}{\det\,(A)}. \tag{IV.4.14}$$

In seeking x_k, we seek the kth entry of the $n \times 1$ matrix

$$[1/\det\,(A)] \cdot \text{adj}\,(A) \cdot Y$$

in (IV.4.14). But this is obtained by multiplying the kth row of the $n \times n$ matrix adj (A) by the column vector Y, and dividing the result by det (A).

$$\frac{1}{\text{t}\,(A)}\underbrace{\begin{pmatrix} * & * & & * & & * \\ \vdots & \vdots & & \vdots & & \vdots \\ (-1)^{k+1}\det(A(1,k)) & (-1)^{k+2}\det(A(2,k)) & \cdots & (-1)^{k+i}\det(A(i,k)) & \cdots & (-1)^{k+n}\det(A(n,k)) \\ \vdots & \vdots & & \vdots & & \vdots \\ * & * & & * & & * \end{pmatrix}}_{\text{adj }(A)}\underbrace{\begin{pmatrix} y_1 \\ y_2 \\ \vdots \\ y_i \\ \vdots \\ y_n \end{pmatrix}}_{Y}$$

That is,

$$x_k = \frac{\sum_{i=1}^{n}(-1)^{k+1}\, y_i \det(A(i,k))}{\det(A)},$$

which is exactly the expansion of a certain $n \times n$ matrix by its kth column (see Theorem IV.3.4; replace the dummy index l there by i to achieve the form above). But what is this "certain" matrix? The scalars y_i determine the kth column; therefore, the numerator of the last equation equals the determinant (expanded by the kth column) which agrees with A except for the kth column, which agrees with the column vector Y. The theorem is proved, and (IV.4.12) is validated.

EXERCISES

IV.4.11 Solve the following vector equations $A \cdot X = Y$, for X (solve the system of equations for $x_1, x_2, \ldots, x_n$) by 3 methods:

(1) Cramer's Rule.

(2) By computing A^{-1} and applying it to the vector Y to obtain $X = A^{-1} \cdot Y$.

(3) By using elementary row operations, which reduce the system to echelon form (see examples following Theorem III.2.7).

(a)
$$\begin{aligned} 3x_1 + 2x_2 &= 1 \\ 4x_1 + 5x_2 &= 1. \end{aligned}$$

(b)
$$\begin{aligned} 2x_1 \qquad\quad + x_3 &= 3 \\ 3x_1 + 2x_2 - x_3 &= 2 \\ x_2 + 2x_3 &= 2. \end{aligned}$$

(c)
$$\begin{aligned} x_1 + x_2 + x_3 &= 2 \\ 3x_1 - x_2 + x_3 &= -2 \\ x_1 + x_2 - x_3 &= 6. \end{aligned}$$

(d)
$$\begin{aligned} x_1 + 2x_2 \qquad\quad + x_4 &= 4 \\ 2x_2 + x_3 \qquad &= 2 \\ x_1 - x_2 + x_3 + x_4 &= 1 \\ x_2 + 2x_3 - x_4 &= 1. \end{aligned}$$

Chapter IV Summary

Section 1 is devoted to 1-1 functions sending a finite set to itself, that is, to permutations. The inverse and parity (evenness, oddness) of permutations is studied especially for its applications in the classical development of determinants. After developing properties of determinants (compare Theorems IV.2.2, IV.2.4, IV.2.5), the product theorem is proven under the constraint that at least 1 matrix factor is elementary (Corollary IV.2.7). Hence, Section 3 introduces a refinement and elaboration on determinant manipulation (expansion by minors). In Section 2, the remarkable fact that invertibility of an $n \times n$ matrix is equivalent to the nonzeroness of its determinant, is presented (Theorem IV.2.9). This leads us to the general, unconstrained product theorem (Theorem IV.4.1). Properties of the determinant and matrix inverses are developed (Corollary IV.4.2), and Cramer's rule is an example of determinants applied to systems of n equations in n unknowns (Theorem IV.4.6).

CHAPTER V

Determinants: A Nonclassical Approach

1. Definition of det $(\cdot)$

In our "classical" approach to determinants, a theory of permutations was required. In the nonclassical approach which follows, the much less elaborate idea of a *multilinear* functional is employed to define the determinant function.

What is a multilinear functional?

We are familiar with the idea of a *linear function* f [which may also be symbolized $f(\cdot)$] sending 1 vector-space to another (Definition II.2.1). A linear function f is a function of only 1 variable. A multilinear functional will differ from a garden-variety linear function (transformation) in the following 2 respects:

(1) While a linear function acts on single vector X, a multilinear functional acts on an *ordered k-tuple* of vectors $(X_1,X_2,\ldots,X_k)$ (hence, the term "multi"). That is, a multilinear functional is a function of k variables instead of just 1 variable. Moreover, the multilinear functional is linear in *each* argument; this means that when any $k-1$ vectors of the ordered (input) k-tuple are fixed, then the multilinear functional behaves like a familiar linear function relative to the 1 remaining argument (hence, the term "linear").

(2) Unlike a linear function whose range lies in some vector-space, a multilinear functional *must* have the *scalars* for an impact set (hence, the term "function*al*").

We now present a definition of a multilinear functional.

Definition V.1.1 A function f which sends ordered k-tuples of vectors $(X_1,X_2,\ldots,X_k)$ to the scalars is called a **multlinear functional,** or a

***k*-linear functional** for some fixed integer $k \geq 1$, if and only if

(a) $f(X_1,\ldots,\alpha X_i,\ldots,X_k) = \alpha f(X_1,\ldots,X_i,\ldots,X_k)$ for each $i = 1, 2, \ldots, k$, and all ordered k-tuples $(X_1,X_2,\ldots,X_i,\ldots,X_k)$, and all scalars α.

(b) $$f(X_1,\ldots,X_i + Y_i,\ldots,X_k)$$
$$= f(X_1,\ldots,X_i,\ldots,X_k) + f(X_1,\ldots,Y_i,\ldots,X_k),$$

for each $i = 1, 2,\ldots, k$, for all ordered k-tuples $(X_1,\ldots, X_i,\ldots,X_k)$ and $(X_1,\ldots,Y_i,\ldots,X_k)$ (see Definition II.2.1).

Example We define a 3-linear functional (that is, a function on ordered triples of vectors) as follows:

For any and all ordered triples of 2-vectors

$$(X,Y,Z) = \left(\begin{pmatrix} x_1 \\ x_2 \end{pmatrix}, \begin{pmatrix} y_1 \\ y_2 \end{pmatrix}, \begin{pmatrix} z_1 \\ z_2 \end{pmatrix}\right)$$

put

$$f\left(\begin{pmatrix} x_1 \\ x_2 \end{pmatrix}, \begin{pmatrix} y_1 \\ y_2 \end{pmatrix}, \begin{pmatrix} z_1 \\ z_2 \end{pmatrix}\right) = 3x_1 \cdot y_2 \cdot z_2 + 2x_2 \cdot y_1 \cdot z_1.$$

To see whether f is a multilinear functional, we need only test it against Definition V.1.1.

Certainly f sends 3-tuples to scalars, so that the domain and impact set for f are in accordance with the definition.

We should test that scalars "factor out" of each of the 3 arguments, as per Definition V.1.1(a). We will content ourselves to show only that scalars factor out of the first argument, leaving the verification for the other 2 arguments to the reader. What are we to show? We shall need to verify Definition V.1.1(a); that is, for any ordered triple (X,Y,Z), and any scalar α, $f(\alpha X,Y,Z) = \alpha f(X,Y,Z)$. But

$$\begin{aligned}
f(\alpha X,Y,Z) &= f\left(\begin{pmatrix} \alpha x_1 \\ \alpha x_2 \end{pmatrix}, \begin{pmatrix} y_1 \\ y_2 \end{pmatrix}, \begin{pmatrix} z_1 \\ z_2 \end{pmatrix}\right) \\
&= 3(\alpha x_1) \cdot y_2 \cdot z_2 + 2(\alpha x_2) \cdot y_1 \cdot z_1 \quad \text{Definition of } f \\
&= \alpha(3x_1 \cdot y_2 \cdot z_2 + 2x_2 \cdot y_1 \cdot z_1) \\
&= \alpha f\left(\begin{pmatrix} x_1 \\ x_2 \end{pmatrix}, \begin{pmatrix} y_1 \\ y_2 \end{pmatrix}, \begin{pmatrix} z_1 \\ z_2 \end{pmatrix}\right) \quad \text{Definition of } f \\
&= \alpha f(X,Y,Z),
\end{aligned}$$

which says that V.1.1(a) obtains . . . at least for the first argument of the given function f.

Similar computations yield property (a) of Definition V.1.1 relative to the second and third arguments.

We should verify Definition V.1.1(b) for the given f; we content ourselves to attend only to the second argument. That is, we must show that

$$f(X,Y + W,Z) = f(X,Y,Z) + f(X,W,Z)$$

for all vectors $X = \begin{pmatrix} x_1 \\ x_2 \end{pmatrix}$, $Y = \begin{pmatrix} y_1 \\ y_2 \end{pmatrix}$, $Z = \begin{pmatrix} z_1 \\ z_2 \end{pmatrix}$ and $W = \begin{pmatrix} w_1 \\ w_2 \end{pmatrix}$. Now

$$
\begin{aligned}
f(X, Y + W, Z) &= f\left(\begin{pmatrix} x_1 \\ x_2 \end{pmatrix}, \begin{pmatrix} y_1 + w_1 \\ y_2 + w_2 \end{pmatrix}, \begin{pmatrix} z_1 \\ z_2 \end{pmatrix}\right) \\
&= \underbrace{3x_1 \cdot (y_2 + w_2) \cdot z_2}_{\downarrow} + \underbrace{2x_2 \cdot (y_1 + w_1) \cdot z_1}_{\searrow} \quad \text{Definition of } f \\
&= (3x_1 \cdot y_2 \cdot z_2 + 3x_1 \cdot w_2 \cdot z_2) + (2x_2 \cdot y_1 \cdot z_1 + 2x_2 \cdot w_1 \cdot z_1) \\
&\quad \downarrow \qquad\qquad\qquad\qquad\qquad\qquad \downarrow \\
&= \underbrace{3x_1 \cdot y_2 \cdot z_2 + 2x_2 \cdot y_1 \cdot z_1}_{\downarrow} + \underbrace{3x_1 \cdot w_2 \cdot z_2 + 2x_2 \cdot w_1 \cdot z_1}_{\downarrow} \\
&= f\left(\begin{pmatrix} x_1 \\ x_2 \end{pmatrix}, \begin{pmatrix} y_1 \\ y_2 \end{pmatrix}, \begin{pmatrix} z_1 \\ z_2 \end{pmatrix}\right) + f\left(\begin{pmatrix} x_1 \\ x_2 \end{pmatrix}, \begin{pmatrix} w_1 \\ w_2 \end{pmatrix}, \begin{pmatrix} z_1 \\ z_2 \end{pmatrix}\right) \\
&= f(X, Y, Z) + f(X, W, Z),
\end{aligned}
$$

which establishes Definition V.1.1(b) for f relative to the second argument, at least.

The validity of property (b) of Definition V.1.1 relative to the first and third arguments, is established in a similar manner.

EXERCISES

Test the following functions against properties (a) and (b) of Definition V.1.1. If both properties obtain, then the function qualifies as a bonafide multilinear functional. Do the values of $f(X_1, \ldots, X_n)$ change when pairs of the variables are interchanged?

V.1.1 (a) $f(x) = x$, for all scalars x.

(b) $f(x,y) = x$, for all scalars x and y.

(c) $f\left(\begin{pmatrix} x_1 \\ x_2 \end{pmatrix}, \begin{pmatrix} y_1 \\ y_2 \end{pmatrix}\right) = x_1 x_2 y_1$, for all pairs of 2-vectors $\begin{pmatrix} x_1 \\ x_2 \end{pmatrix}, \begin{pmatrix} y_1 \\ y_2 \end{pmatrix}$.

(d) $f\left(\begin{pmatrix} x_1 \\ x_2 \end{pmatrix}, \begin{pmatrix} y_1 \\ y_2 \end{pmatrix}\right) = 2 \cdot \begin{pmatrix} x_1 \\ x_2 \end{pmatrix} + \begin{pmatrix} y_1 \\ y_2 \end{pmatrix}$, for all pairs of 2-vectors $\begin{pmatrix} x_1 \\ x_2 \end{pmatrix}, \begin{pmatrix} y_1 \\ y_2 \end{pmatrix}$.

(e) $f\left(\begin{pmatrix} x_1 \\ x_2 \end{pmatrix}, \begin{pmatrix} y_1 \\ y_2 \end{pmatrix}\right) = \det \begin{pmatrix} x_1 & y_1 \\ x_2 & y_2 \end{pmatrix} = x_1 y_2 - x_2 y_1$.

(f) $f\left(\begin{pmatrix} x_1 \\ x_2 \end{pmatrix}, \begin{pmatrix} y_1 \\ y_2 \end{pmatrix}\right) = x_1 y_2 + x_2 y_1$.

(g) $f\left(\begin{pmatrix} x_1 \\ x_2 \\ x_3 \end{pmatrix}, \begin{pmatrix} y_1 \\ y_2 \\ y_3 \end{pmatrix}\right) = x_1 y_2 - x_2 y_1 + x_3 y_3$.

(h) $f\left(\begin{pmatrix} x_1 \\ x_2 \\ x_3 \end{pmatrix}, \begin{pmatrix} y_1 \\ y_2 \\ y_3 \end{pmatrix}\right) = x_1^2 y_1$.

(i) $f(X_1, X_2, \ldots, X_k) = 0$ for all $X_1, X_2, \ldots, X_k \in V_n$.

(j) $f\left(\begin{pmatrix} x_1 \\ x_2 \end{pmatrix}, \begin{pmatrix} y_1 \\ y_2 \end{pmatrix}, \begin{pmatrix} z_1 \\ z_2 \end{pmatrix}, \begin{pmatrix} w_1 \\ w_2 \end{pmatrix}\right) = x_1 y_2 + z_2 w_1$.

V.1.2 If

$$X_1 = \begin{pmatrix} a_{11} \\ a_{21} \\ \vdots \\ a_{n1} \end{pmatrix}, X_2 = \begin{pmatrix} a_{12} \\ a_{22} \\ \vdots \\ a_{n2} \end{pmatrix}, \ldots, X_j = \begin{pmatrix} a_{ij} \\ a_{2j} \\ \vdots \\ a_{nj} \end{pmatrix}, \ldots, X_n = \begin{pmatrix} a_{1n} \\ a_{2n} \\ \vdots \\ a_{nn} \end{pmatrix}$$

then $f(X_1,X_2,\ldots,X_j,\ldots,X_n)$ is defined by

$$f(X_1,X_2,\ldots,X_j,\ldots X_n,) = \det\,(X_1,X_2,\ldots,X_j,\ldots,X_n)$$

$$= \det\,((a_{ij})_{ij}).$$

Prove that det, when considered as a function of the n ordered *column* vectors of an $n \times n$ matrix, is indeed a multilinear functional. (See Exercise V.1.1(e), IV.2.13, and IV.2.14.)

We should be convinced by now that the classical function det, considered as a function of the ordered columns of an $n \times n$ matrix, is a multilinear functional. The verification of this fact depends wholly upon our classical definition of determinant (Definition IV.2.1) and is lubricated by such computational aids as the expansion of minors (see Theorems IV.3.3, IV.3.4, and Exercises IV.2.13, IV.2.14, and IV.3.6).

What we propose, in our nonclassical approach to determinants, is to *start* by defining a determinant function as a certain multilinear functional acting on ordered n-tuples $(X_1,X_2,\ldots,X_n)$, where each X_i is an n-vector (or a column of some $n \times n$ matrix).

What are the advantages of such an approach?

One advantage lies in its inherent "cleanliness." Many properties of determinants follow without recourse to a clumsy sum over permutations. In fact, we shall be able to prove the product theorem for determinants (Theorem IV.4.1) which says that $\det\,(A \cdot B) = \det\,(A) \cdot \det\,(B)$ for all square matrices A and B, without the help of elementary matrices. Moreover, we will not have to consider cases, where 1 factor is nonsingular or not; rather, we shall present a brief proof which accounts for all cases simultaneously (Theorem V.2.1).

Are there other results which can be more easily obtained via the nonclassical definition-results we have not yet encountered?

Yes. We can easily *generalize* or strengthen the following result: To compute the determinant of an $n \times n$ triangular matrix [zeros appear everywhere above (respectively below) the main diagonal], simply multiply the n main diagonal entries together (see the Example following Theorem

IV.2.5 and Exercise IV.2.7). We shall show that if a matrix is block-triangular, that is, of the form

$$P = \begin{pmatrix} A & * & \cdots & * \\ 0 & B & & \cdot \\ \vdots & & \ddots & \cdot \\ \vdots & & & * \\ 0 & 0 & \cdots & K \end{pmatrix},$$

where the symbols $A, B, \ldots, K$ represent square matrices of various sizes, and where zeros appear everywhere below these blocks, then $\det(P) = \det(A) \cdot \det(B) \cdots \det(K)$, the product of the determinants of the matrices on the diagonal. If *each* of the diagonal matrices $A, B, \ldots, K$ is a 1×1 matrix, then we obtain our original result concerning triangular matrices (Theorem V.2.3).

We offer our nonclassical definition of the determinant function now.

Definition V.1.2 The **determinant** function, **det**, is the function acting on ordered n-tuples $(X_1, X_2, \ldots, X_i, \ldots, X_n)$ of n-vectors X_i such that

(1) *det is multilinear*. That is,

$$\det(X_1, \ldots, \alpha X_i, \ldots, X_n) = \alpha \det(X_1, \ldots, X_i, \ldots, X_n),$$

and

$$\det(X_1, \ldots, X_i + Y_i, \ldots, X_n) = \det(X_1, \ldots, X_i, \ldots, X_n) + \det(X_1, \ldots, Y_i, \ldots, X_n),$$

for all $i = 1, 2, \ldots, n$.

(2) *det is antisymmetric*. That is, when any pair of vector arguments is interchanged, the value of det changes by a factor of -1. In other words,

$$\det(X_1, \ldots, \underset{\substack{\uparrow \\ i\text{th place}}}{X_i}, \ldots, \underset{\substack{\uparrow \\ j\text{th place}}}{X_j}, \ldots, X_n) = -\det(X_1, \ldots, \underset{\substack{\uparrow \\ i\text{th place}}}{X_j}, \ldots, \underset{\substack{\uparrow \\ j\text{th place}}}{X_i}, \ldots, X_n),$$

for all i, j.

(3) *det* $(E_1, E_2, \ldots, E_i, \ldots, E_n) = 1$, where $\{E_1, E_2, \ldots, E_i, \ldots, E_n\}$ is the standard ordered basis for V_n, that is, E_i is the n-vector whose ith entry is 1, and all other entries are zero.

The scalar $\det(X_1, X_2, \ldots, X_i, \ldots, X_n)$ is also denoted $\det(A)$, where A is the $n \times n$ matrix whose ith column is the n-vector X_i, $i = 1, 2, \ldots, n$. Property (3) would then read $\det(I_n) = 1$.

QUESTION. We notice that the definition above describes *the* determinant function according to properties (1), (2), and (3). How do we know that there are not *many* multilinear, antisymmetric functionals that send I_n to 1? How do we know that our classical determinant function (Definition IV.2.1) is the *only* such function? Or is it?

It is not at all obvious that the multilinear definition above will eventually lead us to 1 and *only* 1 qualified functional, and that this functional will be our classical friend, det, of Definition IV.2.1. However, for the cases $n = 2$ and $n = 3$, this fact is easily obtained, as is shown in the following examples.

Example Using Definition V.1.2 above, show that $\det\left(\begin{pmatrix}a\\c\end{pmatrix}, \begin{pmatrix}b\\d\end{pmatrix}\right)$ $\left[\text{alternate notation, } \det\left(\begin{pmatrix}a & b\\c & d\end{pmatrix}\right)\right] = ad - bc$, which agrees with det defined in Definition IV.2.1.

Now the *only* vectors for which det is *specifically* evaluated in Definition V.1.2, are the vectors E_1, E_2 $\left[\text{that is, } \begin{pmatrix}1\\0\end{pmatrix}, \begin{pmatrix}0\\1\end{pmatrix}\right]$. It behooves us, then, to force our general vectors $\begin{pmatrix}a\\c\end{pmatrix}, \begin{pmatrix}b\\d\end{pmatrix}$ into a form involving E_1 and E_2. Thus, we write

$$\begin{pmatrix}a\\c\end{pmatrix} = a\begin{pmatrix}1\\0\end{pmatrix} + c\begin{pmatrix}0\\1\end{pmatrix} = aE_1 + cE_3,$$

$$\begin{pmatrix}b\\d\end{pmatrix} = b\begin{pmatrix}1\\0\end{pmatrix} + d\begin{pmatrix}0\\1\end{pmatrix} = bE_1 + dE_2.$$

This leads us to

$$\begin{aligned}
\det\begin{pmatrix}a & b\\c & d\end{pmatrix} &= \det\left(\begin{pmatrix}a\\c\end{pmatrix}, \begin{pmatrix}b\\d\end{pmatrix}\right)\\
&= \det\,(aE_1 + cE_2, bE_1 + dE_2)\\
&= a\det\,(E_1, bE_1 + dE_2) + c\det\,(E_2, bE_1 + dE_2)
\end{aligned}$$

(linearity of det in first argument; the second argument fixed)

$$\begin{aligned}
&= a[b\det\,(E_1,E_1) + d\det\,(E_1,E_2)]\\
&\quad + c[b\det\,(E_2,E_1) + d\det\,(E_2,E_2)]
\end{aligned} \tag{V.1.1}$$

(linearity of det in second argument; the first argument fixed)

1 [Definition V.1.2 (3)]

$$= ab\det\,(E_1,E_1) + ad\det\,(E_1,E_2) + cb\det\,(E_2,E_1) + cd\det\,(E_2,E_2).$$

How can this 4-term sum be simplified?

So far, we have not used the property of antisymmetry [Definition V.1.2(2)], or the fact that $\det\,(E_1,E_2) = 1$ [Definition V.1.2(3)]. These 2 properties can be used to determine det (E_i,E_j), $i, j = 1, 2$ above. In fact, by antisymmetry, interchanging (the only) 2 arguments, introduces a factor of -1, that is,

(a) $\det(E_1,E_1) = -\det(E_1,E_1)$ (hence, equals 0)

(b) $\det(E_2,E_2) = -\det(E_2,E_2)$ (hence, equals 0)

(c) $\det(E_2,E_1) = -\det(E_1,E_2)$ (hence, equals -1).

Now det (E_1,E_2) and det (E_2,E_2) are each equal to their own negatives, according to (a) and (b) above; the only scalar equal to its own negative is zero. Thus, then any 2 arguments of an antisymmetric functional agree, the value of the functional is zero (Proposition V.1.3).

As for (c), we have that det $(E_2,E_1) = -1$, the negative of det (E_1,E_2) which equals 1 by Definition V.1.2(3).

These facts allow us to continue (V.1.1) as follows:

$$\det\begin{pmatrix} a & c \\ b & d\end{pmatrix} = ab \overset{0}{\det(E_1,E_1)} + ad \overset{+1}{\det(E_1,E_2)} + cb \overset{-1}{\det(E_2,E_1)} + cd \overset{0}{\det(E_2,E_2)}$$

$$= ad - bc$$

which was to be proved.

Before we show that Definition V.1.2 describes the classical determinant relative to 3×3 matrices, we isolate a property of antisymmetric functionals that we used toward the end of the last example, namely, that if 2 arguments (variables) agree, then the value of the functional is zero (compare with Corollary IV.2.3).

Proposition V.1.3 Suppose f is an antisymmetric function sending ordered k-tuples $(X_1,X_2,\ldots,X_k)$ of vectors to the scalars. We conclude that if (at least) 2 of the vectors $X_1, X_2,\ldots, X_k$ are identical, then $f(X_1,X_2,\ldots,X_k) = 0$.

Proof: Suppose that for some i and j, $X_i = X_j$. Then $f(X_1,\ldots,X_i,\ldots,X_j,\ldots,X_k) = f(X_1,\ldots,X_j,\ldots,X_i,\ldots,X_k)$ since the ordered k-tuple of vectors is identical in both sides of this equation. On the other hand,

$$f(X_1,\ldots,X_i,\ldots,X_j,\ldots,X_k) = -f(X_1,\ldots,X_j,\ldots,X_i,\ldots,X_k)$$

since the switching of X_i with X_j introduces a factor of (-1) (antisymmetry property).

Thus, $f(X_1,\ldots,X_i,\ldots,X_j,\ldots,X_n)$ equals its own negative, and hence, equals zero whenever 2 vector arguments are identical. The proof is done.

REMARK. We have just shown that if some 2 vectors of the k-tuple $(X_1,X_2,\ldots,X_k)$ happen to be the same, then

$$f \text{ antisymmetric} \Rightarrow f(X_1,X_2,\ldots,X_k) = 0.$$

Is the converse true? That is, if some 2 vectors of the k-tuple $(X_1,X_i,\ldots,X_k)$ happen to be the same, then

$$f(X_1,X_2,\ldots,X_k)=0 \quad \Rightarrow \quad f \text{ is antisymmetric?}$$

The answer is yes, *if* f happens to be multilinear at the same time. Note that f was *not* presumed to be multilinear in Proposition V.1.3. Exercise V.1.4, then asks you to prove this converse proposition, which establishes the fact that for multilinear functionals the properties of antisymmetry and that of sending nondistinct k-tuples of vectors (2 of which are the same) to zero, are *equivalent*. One property is as strong as the other, or 1 property implies the other.

Example Find

$$\det \begin{pmatrix} a_{11} & a_{12} & a_{13} \\ a_{21} & a_{22} & a_{23} \\ a_{31} & a_{32} & a_{33} \end{pmatrix}$$

for det defined in Definition V.1.2.

We write

$$\begin{pmatrix} a_{11} \\ a_{21} \\ a_{31} \end{pmatrix} = a_{11}E_1 + a_{21}E_2 + a_{31}E_3$$

$$\begin{pmatrix} a_{12} \\ a_{22} \\ a_{32} \end{pmatrix} = a_{12}E_1 + a_{22}E_2 + a_{32}E_3$$

$$\begin{pmatrix} a_{13} \\ a_{23} \\ a_{33} \end{pmatrix} = a_{13}E_1 + a_{23}E_2 + a_{33}E_3,$$

which allows us to write

$$\det \left(\begin{pmatrix} a_{11} \\ a_{21} \\ a_{31} \end{pmatrix}, \begin{pmatrix} a_{12} \\ a_{22} \\ a_{32} \end{pmatrix}, \begin{pmatrix} a_{13} \\ a_{23} \\ a_{33} \end{pmatrix} \right) = \det\,((a_{ij})_{ij}) \qquad \text{(V.1.2)}$$

$$= \det\,(a_{11}E_1 + a_{21}E_2 + a_{31}E_3,\ a_{12}E_1 + a_{22}E_2 + a_{32}E_3,\ a_{13}E_1 + a_{23}E_2 + a_{33}E_3).$$

Now using the fact that det is linear in each argument will produce a sum of 27 terms of the form

$$a_{1i_1},\ a_{2i_2}a_{3i_3} \cdot \det\,(E_{i_1},\ E_{2_2},\ E_{i_3}),\ i_1,\ i_2,\ i_3 = 1,\ 2,\ 3.$$

Happily, most of these terms are equal to zero. In fact, as Proposition V.1.3 would have us believe, whenever any 2 vectors of the ordered triple $(E_{i_1},E_{i_2},E_{i_3})$ agree, $\det\,(E_{i_1},E_{i_2},E_{i_3}) = 0$. The only *non*zero terms, then, can only come from the ordered triples $(E_{i_1},E_{i_2},E_{i_3})$ where the subscripts are all *different*. Thus, the nonzero terms which are permitted to survive in (V.1.2) are the 6 terms

$$\begin{aligned}\det((a_{ij})_{ij}) &= a_{11}a_{22}a_{33}\underbrace{\det(E_1,E_2,E_3)}_{=1\ [\text{Definition V.1.2(3)}]} + a_{12}a_{23}a_{31}\det(E_2,E_3,E_1)\\ &\quad + a_{13}a_{21}a_{32}\det(E_3,E_1,E_2) + a_{11}a_{23}a_{32}\det(E_1,E_3,E_2)\\ &\quad + a_{13}a_{22}a_{31}\det(E_3,E_2,E_1) + a_{12}a_{21}a_{33}\det(E_2,E_1,E_3).\end{aligned} \tag{V.1.3}$$

Now we may alter 5 of the terms $a_{1i_1}a_{2i_2}a_{3i_3}\det(E_{i_1},E_{i_2},E_{i_3})$, by a judicious succession of pairwise interchanges of the arguments. Using Definition V.1.2(2) and (3), allows us to write

$$\begin{aligned}\det(E_2,E_3,E_1) &= -\det(E_3,E_2,E_1) = \det(E_1,E_2,E_3) = 1\\ \det(E_3,E_1,E_2) &= -\det(E_3,E_2,E_1) = \det(E_1,E_2,E_3) = 1\\ \det(E_1,E_3,E_2) &= -\det(E_1,E_2,E_3) = -1\\ \det(E_3,E_2,E_1) &= -\det(E_1,E_2,E_3) = -1\\ \det(E_2,E_1,E_3) &= -\det(E_1,E_2,E_3) = -1.\end{aligned}$$

By substituting these values into V.1.3, we obtain

$$\det\begin{pmatrix} a_{11} & a_{12} & a_{13}\\ a_{21} & a_{22} & a_{23}\\ a_{31} & a_{32} & a_{33}\end{pmatrix} = a_{11}a_{22}a_{33} + a_{12}a_{23}a_{31} + a_{13}a_{21}a_{32} - a_{11}a_{23}a_{32} - a_{13}a_{22}a_{31} - a_{12}a_{21}a_{33};$$

which agrees *exactly* with the value of $\det((a_{ij})_{ij})$ according to our classical definition Definition IV.2.1 (see also Figure 4.1).

The previous examples showed us that for $n = 2$ and $n = 3$, 1 and *only* 1 function exists which conforms to the 3 requirements of Definition V.1.2. We propose to show uniqueness of det for all n. Before we do this, we shall need to know the value of $\det(E_{i_1},E_{i_2},\ldots,E_{i_n})$, where E_{i_k} is the i_kth standard basis vector and the subscripts i_k (hence the basis vectors E_{i_k}) are *distinct* from each other. That is, the ordered set of subscripts $\{i_1,i_2,\ldots,i_k,\ldots,i_n\}$ is just the set of integers $\{1,2,\ldots,n\}$, in some scrambled order. To settle this question, we present the slightly more general Proposition V.1.4.

Proposition V.1.4 Let g be an antisymmetric (not necessarily multilinear) function sending ordered n-tuples of vectors to the scalars. If $\{E_1,E_2,\ldots,E_k,\ldots,E_n\}$ is any set of n *distinct* vectors, then

$$g(E_{i_1},E_{i_2},\ldots,E_{i_k},\ldots,E_{i_n}) = \operatorname{sgn}\begin{pmatrix}1 & 2\cdots k\cdots n\\ i_1 & i_2\cdots i_k\cdots i_n\end{pmatrix} g(E_1,E_2,\ldots,E_k,\ldots,E_n),$$

where $\{i_1,i_2,\ldots,i_k,\ldots,i_n\}$ is the set of integers $\{1,2,\ldots,n\}$ is some order. Equivalently stated, $g(E_{i_1},E_{i_2},\ldots,E_{i_k},\ldots,E_{i_n})$ is just $+1$ or -1. times

$g(E_1,E_2,\ldots,E_k,\ldots,E_n)$, according to whether the ordered set of subscripts $\{i_1,i_2,\ldots,i_k,\ldots,i_n\}$ accounts for an even number or an odd number of inversions (see Definition IV.1.3).

Proof: The basic idea is to "unscramble" $g(E_{i_1},E_{i_2},\ldots,E_{i_k},\ldots,E_{i_n})$ by a succession of pairwise interchanges on the ordered arguments (variables). The final objective would then be to obtain $g(E_1,E_2,\ldots,E_k,\ldots,E_n)$.

Now each time such an interchange, or pairwise switching, of variables occurs, the sign of the value of g changes [Definition V.1.2(2)]. On the other hand, a pairwise switching of variables may be thought of as just a pairwise switching on the ordered set of *subscripts* $\{i_1,i_2,\ldots,i_k,\ldots,i_n\}$, that is, a transposition. The question of how many pairwise switches of variables are needed to unscramble the ordered set of vectors $(E_{i_1},E_{i_2},\ldots,E_{i_k},\ldots,E_{i_n})$, reduces, then, to the question of how many *transpositions* are needed to unscramble the ordered, or permuted set of subscripts $\{i_1,i_2,\ldots,i_k,\ldots,i_n\}$. This question has arisen in Exercises IV.1.19 and IV.1.20, where we are told that any permutation

$$\left[\text{such as } i = \begin{pmatrix} 1 & 2\cdots & k\cdots & n \\ i_1 & i_2\cdots & i_k\cdots & i_n \end{pmatrix}\right]$$

can *always* be realized as a succession of transpositions, and that the parity of the number of transpositions required equals the parity of the permutation i, that is, the parity of the number of inversions in $\{i_1,i_2,\ldots,i_k,\ldots,i_n\}$, the ordered range of the permutation i.

What does this mean?

We have observed that the parity of inversions of the ordered set of subscripts $\{i_1,i_2,\ldots,i_k,\ldots,i_n\}$ exactly equals the parity of the number of pairwise switches required to take the ordered set $(E_1,E_2,\ldots,E_k,\ldots,E_n)$ to the scrambled (permuted) set $(E_{i_1},E_{i_2},\ldots,E_{i_k},\ldots,E_{i_n})$. By reversing the order of switchings, the parity of the number of inversions of $\{i_1,i_2,\ldots,i_k,\ldots,i_n\}$ equals the parity of the number of switches required to carry us from the scrambled set $(E_{i_1},E_{i_2},\ldots,E_{i_k},\ldots,E_{i_n})$ back to the naturally ordered set $(E_1,E_2,\ldots,E_k,\ldots,E_n)$ [see Exercise IV.1.20(b)].

We have observed that each such pairwise switching of variables introduces a factor of minus 1 (a sign change). There are as many -1 factors produced as there are pairwise switches. Thus, for p transpositions sending $(E_{i_1},E_{i_2},\ldots,E_{i_k},\ldots,E_{i_n})$ to $(E_1,E_2,\ldots,E_k,\ldots,E_n)$,

$$g(E_{i_1},E_{i_2},\ldots,E_{i_k},\ldots,E_{i_n}) = (-1)^p g(E_1,E_2,\ldots,E_k,\ldots,E_n). \quad \text{(V.1.4)}$$

Now p, the number of transpositions, is odd (respectively, even) if and only if the number of inversions in the ordered set $\{i_1,i_2,\ldots,i_k,\ldots,i_n\}$ is odd (respectively even) [Exercise IV.1.20(b)]. Thus

$$(-1)^p = \operatorname{sgn}\begin{pmatrix} 1 & 2 \cdots k \cdots n \\ i_1 & i_2 \cdots i_k \cdots i_n \end{pmatrix}.$$

Substituting this quantity into (V.1.4), gives us

$$g(E_{i_1},E_{i_2},\ldots,E_{i_k},\ldots,E_{i_n}) = \operatorname{sgn}\begin{pmatrix} 1 & 2 \cdots k \cdots i \\ i_1 & i_2 \cdots i_k \cdots i_n \end{pmatrix} g(E_1,E_2,\ldots,E_k,\ldots,E_n)$$

which completes our proof.

Our next result will prove to be of fundamental importance. Briefly, it says that *any* multilinear, antisymmetric functional g, acting on ordered n-tuples of n-vectors $(X_1,X_2,\ldots,X_k,\ldots,X_n)$ [that is, only properties (1) and (2) of Definition V.1.2 obtain on g; it is not necessary that $g(I_n) = 1$], is "essentially" the determinant function. In fact, such a multilinear, antisymmetric functional is always just the classical determinant function of Definition IV.2.1, multiplied by some scalar; the scalar, incidentally, is the number $g(I_n)$. This important theorem is given now.

Theorem V.1.5 Let g be a functional sending ordered n-tuples of n-vectors $(X_1,X_2,\ldots,X_k,\ldots,X_n)$ to scalars. Suppose g is (1) multilinear, and (2) antisymmetric. Then necessarily,

$$g(X_1,X_2,\ldots,X_k,\ldots,X_n) = \underbrace{g(E_1,E_2,\ldots,E_k,\ldots,E_n)}_{\text{equals the constant } g(I_n)} \det_0 (X_1,X_2,\ldots,X_k,\ldots,X_n),$$

for all ordered n-tuples of n-vectors $(X_1,X_2,\ldots,X_k,\ldots,X_n)$, where $\det_0$ has the properties:

(1) $\det_0$ is a multilinear functional
(2) $\det_0$ is antisymmetric
(3) $\det_0(E_1,E_2,\ldots,E_k,\ldots,E_n) = 1$.

Moreover, $\det_0$ is unique and does not depend on g. That is, any and every multilinear, antisymmetric functional is just a scalar multiple of the single, fixed functional $\det_0$. As it turns out, this "generating functional" is just the classical determinant function det introduced in Definition IV.2.1.

If A is the $n \times n$ matrix whose kth column is the n-vector X_k, then the statement of the theorem can be written

$$g(A) = g(I_n) \det_0 (A) \quad \text{for all } n \times n \text{ matrices } A.$$

Proof: We resort to writing each X_k in terms of the standard ordered basis $\{E_1,E_2,\ldots,E_k,\ldots,E_n\}$. Now if

$$X_1 = \begin{pmatrix} a_{11} \\ a_{12} \\ \vdots \\ a_{1i_1} \\ \vdots \\ a_{1n} \end{pmatrix}, X_2 = \begin{pmatrix} a_{21} \\ a_{22} \\ \vdots \\ a_{2i_2} \\ \vdots \\ a_{2n} \end{pmatrix}, \ldots, X_k = \begin{pmatrix} a_{k1} \\ a_{k2} \\ \vdots \\ a_{ki_k} \\ \vdots \\ a_{kn} \end{pmatrix}, \ldots, X_n = \begin{pmatrix} a_{n1} \\ a_{n2} \\ \vdots \\ a_{ni_n} \\ \vdots \\ a_{nn} \end{pmatrix},$$

then this is equivalent to saying that

$$X_k = \sum_{i_k=1}^{n} a_{ki_k} E_{i_k}, \quad \text{for each } X_k,\ k = 1, 2, \ldots, n.$$

Note that the dummy variable for the sum above has been denoted as i_k instead of i; this is to underscore the fact that i_k belongs to the sum representing the kth vector, X_k. Also, observe that the scalars a_{ki_k} which appear are necessarily *unique* since $\{E_1,E_2,\ldots,E_n\}$ is a *basis* for V_n (Theorem I.6.8).

Now that notation is established we write

$$\begin{aligned}
g(&X_1,X_2,\ldots,X_k,\ldots,X_n) \\
&= g\left(\sum_{i_1=1}^{n} a_{1i_1}E_{i_1}, \sum_{i_2=1}^{n} a_{2i_2}E_{i_2},\ldots, \sum_{i_k=1}^{n} a_{ki_k}E_{i_k},\ldots, \sum_{i_n=1}^{n} a_{ni_n}E_{i_n}\right) \\
&= \sum_{i_1=1}^{n} a_{1i_1} g\left(E_{i_1}, \sum_{i_2=1}^{n} a_{2i_2}E_{i_2},\ldots, \sum_{i_k=1}^{n} a_{ki_k}E_{i_k},\ldots, \sum_{i_n=1}^{n} a_{ni_n}E_{i_n}\right) \qquad \text{(V.1.5)} \\
&\qquad \text{(linearity of } g \text{ in her first argument)} \\
&= \sum_{i_1=1}^{n}\sum_{i_2=1}^{n} a_{1i_1}a_{2i_2} g\left(E_{i_1},E_{i_2}, \sum_{i_3=1}^{n} a_{3i_3}E_{i_3},\ldots, \sum_{i_k=1}^{n} a_{ki_k}E_{i_k},\ldots, \sum_{i_n=1}^{n} a_{ni_n}E_{i_n}\right) \\
&\qquad \text{(linearity of } g \text{ in her second argument)} \\
&= \sum_{i_1=1}^{n}\sum_{i_2=1}^{n}\cdots\sum_{i_k=1}^{n}\cdots\sum_{i_n=1}^{n} a_{1i_1}a_{2i_2}\cdots a_{ki_k}\cdots a_{ni_n} g(E_{i_1},E_{i_2},\ldots, E_{i_k},\ldots, E_{i_n}). \\
&\qquad \text{(linearity of } g \text{ in her } n \text{ arguments)}
\end{aligned}$$

Now that is quite a sum! Fortunately, most of the terms are zero. In fact if any 2 of the standard vectors of the n-tuple $(E_{i_1},E_{i_2},\ldots,E_{i_k},\ldots,E_{i_n})$ are the *same*, then $g(E_{i_1},E_{i_2},\ldots,E_{i_k},\ldots,E_{i_n}) = 0$ (Proposition V.I.3). All the nonzero terms *must* occur when the vectors $E_{i_1},E_{i_2},\ldots,E_{i_k},\ldots,E_{i_n}$ are *distinct*, which is to say, when the subscripts $i_1, i_2,\ldots, i_k,\ldots, i_n$ are distinct.

Now $g(E_{i_1},E_{i_2},\ldots,E_{i_k},\ldots,E_{i_n})$ is just plus or minus $g(E_1,E_2,\ldots,E_k,\ldots,E_n)$, since we may always go from 1 to the other by interchanging pairs of arguments. With this thought in mind, we see that the last sum of (V.1.5) can be written with $\pm g(E_1,E_2,\ldots,E_k,\ldots,E_n) = g(E_{i_1},E_{i_2},\ldots,E_{i_k},\ldots,E_{i_n})$ factored out to obtain

$$g(X_1,X_2,\ldots,X_k,\ldots,X_n) = g(E_1,E_2,\ldots,E_k,\ldots,E_n)\cdot \det_0 (X_1,X_2,\ldots,X_k,\ldots,X_n), \tag{a}$$

where $\det_0$ represents the "left-over" factor

$$\det_0 (X_1,\ldots,X_n) = \sum_{i_1=1}^{n} \cdots \sum_{i_n=1}^{n} \pm a_{1i_1}\cdots a_{ni_n}.$$

Whatever this function $\det_0$ is, it does not depend on g but depends only on the permutations of the arguments [see the defining sum of (V.1.5)]. This establishes the "moreover" part of our theorem, that is, existence of $\det_0$, and uniqueness. Just from observing (a), we see that $\det_0$ must be multilinear, antisymmetric, and $\det_0 (E_1,E_2,\ldots,E_n) = 1$. This establishes properties (1), (2), and (3) of Theorem V.1.5.

We pursue the question further, by showing now that $\det_0$ defined by the last line of (V.1.5) and (a), is the classical determinant (Definition IV.2.1).

Now the ordered set of subscripts $\{i_1,i_2,\ldots,i_k,\ldots,i_n\}$ (since they are distinct), is just the set of integers $\{1,2,\ldots,n\}$ in some scrambled order. That is, $\{i_1,i_2,\ldots,i_k,\ldots,i_n\}$ may be thought of as the ordered range of the permutation

$$i = \begin{pmatrix} 1 & 2\cdots k\cdots n \\ i_1 & i_2\cdots i_k\cdots i_n \end{pmatrix}.$$

Observe that Proposition V.1.4 tells us that

$$g(E_{i_1},E_{i_2},\ldots,E_{i_k},\ldots,E_{i_n}) = \operatorname{sgn} \begin{pmatrix} 1 & 2\cdots k\cdots n \\ i_1 & i_2\cdots i_k\cdots i_n \end{pmatrix} g(E_1,E_2,\ldots,E_k,\ldots,E_n) \tag{V.1.6}$$

for *any* antisymmetric function (multilinear or not). Effecting the substitution of (V.1.6) into the last equality of (V.1.5), and factoring out the common term $g(E_1,E_2,\ldots,E_k,\ldots,E_n)$, yields

$$g(X_1,X_2,\ldots,X_k,\ldots,X_n) = \Bigg[\underbrace{\sum_{i_1=1}^{n}\sum_{i_2=1}^{n}\cdots\sum_{i_k=1}^{n}\cdots\sum_{i_n=1}^{n} \operatorname{sgn}\begin{pmatrix} 1 & 2\cdots k\cdots n \\ i_1 & i_2\cdots i_k\cdots i_n \end{pmatrix}}_{\sum_i \operatorname{sgn}(i)} a_{1i_1}a_{2i_2}\cdots a_{ki_k}\cdots a_{ni_n}\Bigg] \times g(E_1,E_2,\ldots,E_k,\ldots,E_n). \tag{V.1.7}$$

But the summation above (inside the brackets), is our cumbersome old friend, the classical determinant function, which we denote here by $\det_0$. In fact, by comparing this sum (V.1.7) with (IV.2.1) of Definition IV.2.1, what we have here is $\det_0((a_{ij})_{ij})$, that is, the determinant of the $n \times n$ matrix whose kth *row* (not column) is the vector X_k. No matter. $\det_0(A) = \det_0(A^t)$ (Theorem IV.2.8) so that the sum above equally represents $\det_0((a_{ij})_{ji})$, the determinant of the $n \times n$ matrix whose kth *column* is the vector X_k.

Injecting this observation into (V.1.7) gives us

$$g(X_1,X_2,\ldots,X_k,\ldots,X_n) = \det_0(X_1,X_2,\ldots,X_k,\ldots,X_n)g(E_1,E_2,\ldots,E_k,\ldots,E_n).$$

If A is the $n \times n$ matrix $(a_{ji})_{ij}$, whose kth column is the n-vector X_k, then we may write the last equality as

$$g(A) = \det_0(A)g(I_n),$$

which was to be proved.

REMARK. In the proof of Theorem V.1.5, we attempted to appeal to 2 audiences simultaneously, namely, those who are familiar with the classical determinant function, and those who are not. The point of departure came with the totally unfriendly looking multiple sum at the end of (V.1.5). For those unfamiliar with the classical determinant, we said, "This sum does not depend on g; let us give it a simple name like $\det_0(X_1,X_2,\ldots,X_k,\ldots,X_n)$." Thus, we rewrote (V.I.5) as (**a**), from which properties (1), (2), and (3) followed almost by inspection. For those familiar with the classical determinant, we did not let the matter rest there. We continued to pounce upon the disagreeable multiple sum of (V.1.5), and forced it into a form which was familiar, namely, the summation form of Definition IV.2.1 which defines the classical determinant function.

And isn't that interesting! Who would have suspected that from just properties (1) and (2) of Definition V.1.2, the classical sum defining the determinant (IV.2.1) of Definition IV.2.1 would appear?

Now that we have functionals that obey (1) and (2) of Definition V.1.2 fairly well pinned down, it is time to evoke property (c). We will now see that Definition V.1.2 does, indeed, characterize the one and only function $\det_0$, which we met in Chapter IV.

Corollary V.1.6 If g is a functional satisfying all 3 properties of Definition V.1.2, then g can only equal $\det_0$, the classical determinant function of Definition IV.2.1.

Proof: From Theorem V.1.5, we have that

$$g(A) = \det_0(A) \cdot g(I_n), \quad \text{for all } n \times n \text{ matrices } A.$$

To invoke property (3) of Definition V.1.2, is to require that $g(I_n) = 1$. Thus, the equation above reads

$$g(A) = \det_0(A), \quad \text{for all } n \times n \text{ matrices } A.$$

The proof (however brief) is done.

Thanks to Corollary V.1.6, we are certain that the function det defined in Definition V.1.2 is *unique* (so that we may speak of *the* function "det," instead of *a* function "det"); moreover, the function det is precisely the *classical* determinant function of Chapter IV, which we have denoted as $\det_0$ in *this* chapter. That also settles the question of *existence*. Now that distinctions between the nonclassical function det, and the classical function $\det_0$ are unnecessary, we drop the zero subscript of $\det_0$ once and for all. We now have 2 ways of defining the *same* function, det, namely, Definition IV.2.1 and Definition V.1.2.

EXERCISES

V.1.3 Suppose f is some antisymmetric function sending ordered 7-tuples of vectors to the scalars. Suppose that $f(X_1,X_2,X_3,X_4,X_5,X_6,X_7) = \lambda_0$. Compute

(a) $f(X_1,X_2,X_7,X_4,X_5,X_6,X_3)$
(b) $f(X_2,X_1,X_4,X_3,X_6,X_5,X_7)$
(c) $f(X_7,X_6,X_5,X_4,X_3,X_2,X_1)$
(d) $f(X_i,X_i,X_i,X_i,X_i,X_i,X_i)$
(e) $f(X_7,X_1,X_2,X_3,X_4,X_5,X_7)$
(f) $f(X_3,X_1,X_2,X_4,X_7,X_5,X_6)$.

V.1.4 (A converse to Proposition V.1.3.) Let f be any multilinear functional sending ordered n-tuples of vectors $(X_1,X_1,\ldots,X_i,\ldots,X_n)$ to the scalars, that is, property (1) of Definition V.1.2 obtains. Suppose f has the further property that $f(X_1,X_2,\ldots,X_i,\ldots,X_n) = 0$ whenever any pair of the vector arguments, $X_1,X_2,\ldots,X_i,\ldots,X_n$, are identical. Prove that, necessarily, f is antisymmetric, that is, property (2) of Definition V.1.2 obtains. [*Hint:* Consider $f(X_1,\ldots,X_i + X_j,\ldots,X_i + X_j,\ldots,X_n)$,

$$\underset{i\text{th place}}{\uparrow} \qquad \underset{j\text{th place}}{\uparrow}$$

which has the vector $X_i + X_j$ in both the ith and jth places.]

V.1.5 (The effect of elementary column operations on the value of the determinant function of Definition V.1.2.) Let f be any antisymmetric, multilinear functional as per properties (1) and (2) of Definition V.1.2. Prove directly from Definition V.1.2 that, for any $n \times n$ matrix $A = [A_1,A_2,\ldots,A_k,\ldots,A_n]$ whose kth column is the n-vector A_k,

(a) $f(C_{ij}(A)) = -f(A)$,
that is,

$$f(A_1,\ldots,\underset{\substack{\uparrow\\ i\text{th}\\ \text{place}}}{A_i},\ldots,\underset{\substack{\uparrow\\ j\text{th}\\ \text{place}}}{A_j},\ldots,A_n) = -f(A_1,\ldots,\underset{\substack{\uparrow\\ i\text{th}\\ \text{place}}}{A_j},\ldots,\underset{\substack{\uparrow\\ j\text{th}\\ \text{place}}}{A_i},\ldots,A_n).$$

(b) $f(C_i(a)(A)) = a \cdot f(A)$ for all scalars a,
that is,

$$f(A_1,\ldots,aA_i,\ldots,A_n) = af(A_1,\ldots,A_n).$$

(c) $f(C_i{}^j a)) = f(A)$ for all scalars a,
that is,

$$f(A_1,\ldots,\underset{\substack{\uparrow\\ i\text{th}\\ \text{place}}}{A_i},\ldots,\underset{\substack{\uparrow\\ j\text{th}\\ \text{place}}}{a \cdot A_i + A_j},\ldots,A_n) = f(A_1,\ldots,\underset{\substack{\uparrow\\ i\text{th}\\ \text{place}}}{A_i},\ldots,\underset{\substack{\uparrow\\ j\text{th}\\ \text{place}}}{A_j},\ldots,A_n).$$

V.1.6 Suppose f is a functional sending ordered n-tuples of vectors (that is, $n \times n$ matrices) to scalars which is lucky enough to satisfy the product rule

$$f(A \cdot B) = f(A)f(B) \quad \text{for all } n \times n \text{ matrices } A, B.$$

That is, f is a multiplicative homomorphism from $\mathcal{M}_{n,n}$ to the scalars (see Definition II.6.15).

(a) Prove that $f(I_n) = 1$; or else $f(I_n) = 0$, which forces $f(A) = 0$ for *all* $n \times n$ matrices A.

(b) Let C_{ij} be the elementary $n \times n$ matrix obtained form I_n by interchanging the ith and jth columns (the same matrix results when the ith and jth rows of I_n are interchanged). Prove that $f(C_{ij})$ can only equal $+1$ or -1. (*Hint*: $C_{ij} \cdot C_{ij} = I_n$; use the product property.)

Note: If f is assumed to be multilinear [Definition V.1.2(1)] as well, then it can be shown that $f(C_{ij}) = -1$ [see (b)]. Thus,

$$\begin{aligned} f(A_1,\ldots,\underset{\substack{\uparrow\\ i\text{th}\\ \text{place}}}{A_j},\ldots,\underset{\substack{\uparrow\\ j\text{th}\\ \text{place}}}{A_i},\ldots,A_n) &= f(A \cdot C_{ij}) = f(A)f(C_{ij}) && \text{multiplicative property of } f \\ &= f(A)(-1) \\ &= -f(A_1,\ldots,A_i,\ldots,A_j,\ldots,A_n). \end{aligned}$$

That is, f would necessarily be antisymmetric [Definition V.1.2(2)]. Part (a) established the validity of Definition V.1.2(3) and property (1) of Definition V.1.2 is part of our hypothesis. Thus, *another* way of characterizing the one and only determinant function is to speak of a multilinear functional which obeys the product law $f(A \cdot B) = f(A) \cdot f(B)$ for all $n \times n$ matrices A and B, assuming $f(A)$ is not zero for all A.

V.1.7 Construct (define) a function f on ordered pairs of n-vectors (X_1,X_2), which is antisymmetric and *not* multilinear.

V.1.8 *$C_{ij}(a)$ leaves the determinant invariant.* Prove that the elementary column operation $C_i{}^j(a)$ (the addition to column j of a times column i) does not affect the det. That is, show from Definition V.1.1 that

$$\det(X_1,\ldots, X_i,\ldots, X_j,\ldots, X_n) = \det(X_1,\ldots, X_i,\ldots, aX_i + X_j,\ldots, X_n)$$

or $$\det(A) = \det(C_j{}^i\,(a)A)$$

for all $n \times n$ matrices A.

V.1.9 Show that the elementary function $C_i(a)$ has the effect of multiplying the determinant by the scalar a, that is, show that

$$\det(C_i(a) \cdot A) = a \cdot \det(A).$$

2. *Consequences of the Nonclassical Definition for det*

Exercises V.1.5, V.1.6, and V.1.8 indicate that the effects of elementary row and column operations (Theorem IV.2.10). are easily obtained from the nonclassical Definition V.1.2.

We will prove 2 results with this new summation-free, permutation-free definition. One result, the product theorem, we have seen as Theorem IV.4.1. The second result will be one we have not yet encountered, except for the statement at the beginning of this chapter.

Let us reprove the product theorem.

Theorem V.2.1 (Theorem IV.4.2) If A and B are $n \times n$ matrices, then

$$\det(A \cdot B) = \det(A)\det(B).$$

Proof: Let A be an $n \times n$ matrix. Let $B = [B_1, B_2, \ldots, B_k, \ldots, B_n]$ be an $n \times n$ matrix; the kth column of B is denoted B_k, $k = 1, 2, \ldots, n$.

What is the kth column of the $n \times n$ product $A \cdot B$? From the very method of matrix multiplication (Theorem II.6.13), the kth column of $A \cdot B$ is precisely the $n \times 1$ matrix $A \cdot B_k$. This is illustrated below.

$$\begin{pmatrix} \\ A \\ \\ \end{pmatrix} \cdot \left(\begin{array}{ccc|c|ccc} * & \cdots & * & b_{1k} & * & \cdots & * \\ \vdots & & \vdots & b_{2k} & \vdots & & \vdots \\ \vdots & & \vdots & \vdots & \vdots & & \vdots \\ * & \cdots & * & b_{nk} & * & \cdots & * \end{array}\right) \to \left(\begin{array}{ccc|c|ccc} * & \cdots & * & & * & \cdots & * \\ \vdots & & \vdots & A \cdot B_k & \vdots & & \vdots \\ * & \cdots & * & & * & \cdots & * \end{array}\right).$$

$$\uparrow \qquad\qquad\qquad\qquad\qquad \uparrow$$

$$B_k \qquad\qquad\qquad k\text{th column of } A \cdot B$$

Therefore, to find $\det(A \cdot B)$ is to find

$$\det(A \cdot B_1, A \cdot B_2, \ldots, A \cdot B_k, \ldots, A \cdot B_n),$$

considered as an antisymmetric, multilinear functional acting on the n ordered columns of the $n \times n$ matrix $A \cdot B$.

A trick! Let us regard the matrix A as arbitrary, but *fixed*. Let us regard the matrix B (or the columns $B_1, B_2, \ldots, B_k, \ldots, B_n$) as "variable."

Thus, det $(A \cdot B_1, A \cdot B_2, \ldots, A \cdot B_k, \ldots, A \cdot B_n)$ may be considered as a certain function g_A (which depends on this fixed A) whose variables are just the n column vectors $B_1, B_2, \ldots, B_k, \ldots, B_n$. That is to say, we *define* the function g_A by

$$g_A(B_1, B_2, \ldots, B_k, \ldots, B_n) = \det(A \cdot B_1, A \cdot B_2, \ldots, A \cdot B_k, \ldots, A \cdot B_n),$$

▶ $$= \det(A \cdot B) \text{ (equivalent notation for the above)} \tag{V.2.1}$$

for all $n \times 1$ column vectors $B_1, B_2, \ldots, B_k, \ldots, B_n$ (that is, for all $n \times n$ matrices B).

What has been achieved in (V.2.1)?

First, since det is multilinear, and since matrix multiplication is distributive $[A \cdot (B_k + C_k) = A \cdot B_k + A \cdot C_k$ for all $n \times n$ matrices A, all $n \times 1$ matrices B_k, C_k; see Exercises II.6.83 and II.6.84], the function g_A is necessarily multilinear, that is, linear in each of its n arguments (Exercise V.2.1).

Secondly, since det is antisymmetric, g_A is also antisymmetric, that is, $g_A(B_1, \ldots, B_i, \ldots, B_j, \ldots, B_n) = -g_A(B_1, \ldots, B_j, \ldots, B_i, \ldots, B_n)$ (Exercise V.2.2).

So far, then, g_A is a functional whose behavior is wholly in accord with properties (1) and (2) of Definition V.1.2. Such functions have been studied in Theorem V.1.5; they are "essentially" the determinant function multiplied by some scalar. To be precise, Theorem V.1.5 tells us that

$$\det(A \cdot B) = g_A(B_1, B_2, \ldots, B_k, \ldots, B_n) \qquad \text{definition of } g_A \text{ in (V.2.1)}$$

$$= g_A(I_n)\, \overbrace{\det(B_1, B_2, \ldots, B_k, \ldots, B_n)}^{\det(B)} \qquad \text{Theorem V.1.5} \tag{V.2.2}$$

But $g_A(I_n) = \det(A \cdot I_n) = \det(A)$. Substituting this into (V.2.2), reveals that

$$\det(A \cdot B) = \det(A) \det(B)$$

which ends the proof of the product theorem.

Our next (hitherto unencountered) result is that if a matrix P is of triangular block form

$$P = \begin{pmatrix} A & * & \cdots & * \\ 0 & B & \ddots & \vdots \\ \vdots & \vdots & \ddots & * \\ 0 & 0 & \cdots & K \end{pmatrix}, \text{ where } A, B, \ldots, K$$

are *square* matrices, *not* necessarily of the same size, then the determinant of the matrix is the product of the determinants of the diagonal blocks, that is, det $(P) =$ det $(A) \cdot$ det $(B) \cdots$ det (K).

In building toward this result, we restrict ourselves to the (apparently) special case where only 2 diagonal block matrices, A and B, appear. Of course this is no restriction at all, for any matrix of form P can be considered as consisting of only 2 diagonal block entries

$$P = \begin{pmatrix} \boxed{A} & * & \cdots & * \\ 0 & B & \cdots & * \\ \vdots & 0 & \ddots & \vdots \\ 0 & 0 & \cdots & K \end{pmatrix}, \qquad \underbrace{\begin{matrix} B & \cdots & * \\ 0 & \ddots & \vdots \\ 0 & \cdots & K \end{matrix}}_{B'}$$

and if the result is true, $\det(P) = \det(A) \cdot \det(B')$. But $\det(B')$ can be computed by the same theorem; it, too, is a triangular block matrix. Repetition of this judicious partitioning of P into matrices with only 2 diagonal block entries, gives us the general result.

The following fact will be of use.

Proposition V.2.2 If I_p denotes the $p \times p$ identity matrix, and if B is any $q \times q$ matrix, C is any $p \times q$ matrix, then

$$\det \begin{pmatrix} I_p & C \\ O & B \end{pmatrix} = \det(B);$$

the lower left-hand entry O denotes the $q \times p$ matrix whose every entry is zero.

Proof: We use the fact that the elementary column operations $C(a)$ leave the determinant unaltered (see Exercise V.1.8). In fact, we may take multiples of the first p columns of $\begin{pmatrix} I_p & C \\ O & B \end{pmatrix}$, and completely annihilate the matrix C, without altering the determinant. That is, by taking certain multiples of the column vector

$$i\text{th entry} \rightarrow \begin{pmatrix} 0 \\ \vdots \\ 0 \\ 1 \\ 0 \\ \vdots \\ 0 \end{pmatrix} \text{ which is just the } i\text{th column of } \begin{pmatrix} I_p & C \\ O & B \end{pmatrix}, \; i = 1, 2, \ldots, p,$$

and adding these multiples to the last q columns of our original matrix, we can conclude that

(a) $$\det \begin{pmatrix} I_p & C \\ O & B \end{pmatrix} = \det \begin{pmatrix} I_p & O \\ O & B \end{pmatrix}.$$

Let us now regard

$$\det \begin{pmatrix} I_p & O \\ O & B \end{pmatrix}$$

as a function of the last q columns, that is, of the columns of B. Thus,

(b) $$\det \begin{pmatrix} I_p & O \\ O & B \end{pmatrix} = g\,(B)$$

defines g as multilinear (since det is), and antisymmetric (since det is). From Theorem V.1.5 and Corollary V.1.6,

(c) $$\det \begin{pmatrix} I_p & O \\ O & B \end{pmatrix} = g(B) = \det\,(B) g(I_q).$$

From the defining Equation (a) for g,

$$g\,(I_q) = \det \begin{pmatrix} I_p & O \\ O & I_q \end{pmatrix} = \det\,(I_{p+q}) = 1,$$

so that (b) becomes

$$\det \begin{pmatrix} I_p & O \\ O & B \end{pmatrix} = g(B) = \det\,(B) \cdot 1.$$

Combine (a) and (c) to obtain the statement of Proposition V.2.2.

With this special case on ice (upper left-hand block matrix $= I_p$), we are prepared to prove the general Theorem V.2.3.

Theorem V.2.3 Suppose A is any $p \times p$ matrix, B is any $q \times q$ matrix. Then for the $(p+q) \times (p+q)$ triangular block matrix $\begin{pmatrix} A & C \\ O & B \end{pmatrix}$, we have

$$\det \begin{pmatrix} A & C \\ O & B \end{pmatrix} = \det\,(A) \det\,(B).$$

Proof: Let us take a hard look at the determinant $\det \begin{pmatrix} A & C \\ O & B \end{pmatrix}$. If we assume that the matrices B and C (that is, the last q columns) are fixed. Then $\det \begin{pmatrix} A & C \\ O & B \end{pmatrix}$ may be regarded as some function (depending on the fixed matrices B and C) of just the p columns $A_1, A_2, \ldots, A_p$, of the

$p \times p$ matrix A. That is, we may *define* $g_{B,C}(\cdot)$ by

$$\begin{aligned}\det\begin{pmatrix} A & C \\ O & B\end{pmatrix} &= g_{B,C}(A_1,A_2,\ldots,A_i,\ldots,A_p) \\ &= g_{B,C}(A) \quad \text{(notation for above)}\end{aligned}$$

for all $p \times p$ matrices A (for all $p \times 1$ columns $A_1, A_2,\ldots, A_p$).

But what sort of a function (of p variables) is $g_{B,C}$?

Because of the fact that det is multilinear (especially that det is linear in each of its first p arguments), so, too, $g_{B,C}$ is multilinear (that is, linear in each of its p arguments).

The property of antisymmetry, which obtains for the function det, is reflected in $g_{B,C}$; that is, $g_{B,C}$ is also antisymmetric.

Voila! $g_{B,C}$ is not just a pedestrian function of p variables; it is both *multilinear* and *antisymmetric*. Calling upon the services of Theorem V.1.5, which deals with such doubly endowed functions, we may honestly deduce that

$$g_{B,C}(A_1,A_2,\ldots,A_i,\ldots,A_p) = g_{B,C}(I_p)\det(A_1,A_2,\ldots,A_i,\ldots,A_p). \tag{V.2.4}$$

Now what is this scalar $g_{B,C}(I_p)$ which appears in (V.2.4)? From the defining Equation (V.2.3) for $g_{B,C}$, we have

$$g_{B,C}(I_p) = \det\begin{pmatrix} I_p & C \\ O & B\end{pmatrix}.$$

To review and conclude, then, we write the equalities

$$\begin{aligned}\det\begin{pmatrix} A & C \\ O & B\end{pmatrix} &= g_{B,C}(A_1,A_2,\ldots,A_i,\ldots,A_p) && \text{from (V.2.3)} \\ &= g_{B,C}(I_p)\,\underbrace{\det(A_1,A_2,\ldots,A_i,\ldots,A_p)}_{\det(A)} && \text{Theorem V.1.5} \\ &= \underbrace{\det\begin{pmatrix} I_p & C \\ O & B\end{pmatrix}}\ \det(A) && \text{from (V.2.3)} \\ &= \det(B)\cdot\det(A) && \text{Proposition V.2.2}\end{aligned}$$

which proves the theorem.

EXERCISES

V.2.1 Prove that the function g_A, defined by

$$g_A(X_1,\ldots,X_i,\ldots,X_n) = \det(A\cdot X_1,\ldots,A\cdot X_i,\ldots,A\cdot X_n)$$

for all $n \times 1$ column vectors, $X_1, X_2, \ldots, X_i, \ldots, X_n$, is multilinear (that is, linear in each of its n arguments) because the function det is multilinear. The $n \times n$ matrix A is fixed throughout.

V.2.2 With g_A defined in Exercise V.2.1, show that g_A is antisymmetric since the function det is antisymmetric.

V.2.3 Using the property of determinants which says that

$$\det\begin{pmatrix} a_{11} & * & \cdots & * \\ 0 & a_{22} * & \cdots & * \\ \vdots & \vdots & \ddots & \vdots \\ 0 & 0 & \cdots & a_{nn} \end{pmatrix} = a_{11} \cdot a_{22} \cdots a_{nn}$$

for triangular matrices, prove that $\det\begin{pmatrix} I_p & 0 \\ 0 & B \end{pmatrix} = \det(B)$ as follows: If $B = (B_1, B_2, \ldots, B_k, \ldots, B_q)$ is the $q \times q$ matrix whose kth column is B_k, then define $g_I(B_1, B_2, \ldots, B_k, \ldots, B_q) = g_I(B)$ by

$$g_I(B_1, B_2, \ldots, B_k, \ldots B_q) = \det\begin{pmatrix} I_p & 0 \\ 0 & B \end{pmatrix}.$$

(a) Show that g_I is multilinear (linear in each of its q arguments) since the function det is multilinear (linear in each of its $p + q$ arguments).

(b) Show that g_I is antisymmetric, since det is antisymmetric.

(c) Show that, as a result of (a) and (b), we may conclude that

$$\det\begin{pmatrix} I_p & 0 \\ 0 & B \end{pmatrix} = \det(B).$$

V.2.4 Compute the determinant of the following matrices:

(a) $\begin{pmatrix} 5 & 16 & 61 \\ 0 & 3 & 0 \\ 0 & 2 & 1 \end{pmatrix}$

(b) $\begin{pmatrix} 3 & 1 & 6 & 7 & 4 & 7 \\ 2 & 4 & 1 & 2 & 3 & 1 \\ 0 & 0 & 1 & 2 & 1 & 6 \\ 0 & 0 & 3 & 1 & 6 & 1 \\ 0 & 0 & 0 & 0 & 5 & 1 \\ 0 & 0 & 0 & 0 & 1 & 1 \end{pmatrix}$

(c) $\begin{pmatrix} 1 & 1 & 1 & 0 & 0 \\ 1 & -1 & 1 & 2 & 1 \\ 0 & 1 & 2 & 1 & 2 \\ 0 & 0 & 0 & 3 & 2 \\ 0 & 0 & 0 & 0 & 1 \end{pmatrix}$

(d) $\begin{pmatrix} 2 & 1 & 0 & 0 & 0 & 0 \\ 3 & 1 & 0 & 0 & 0 & 0 \\ 1 & 0 & 5 & 8 & 0 & 0 \\ 3 & 1 & 0 & 1 & 0 & 0 \\ 1 & 1 & 2 & 2 & 2 & 1 \\ 2 & 1 & 3 & 1 & 1 & 2 \end{pmatrix}$

(e) $\begin{pmatrix} 1 & 0 & 0 & 0 & 0 & 0 & 0 & 0 & 0 \\ 3 & a & b & 0 & 0 & 0 & 0 & 0 & 0 \\ 1 & 1 & 1 & 0 & 0 & 0 & 0 & 0 & 0 \\ 8 & 1 & 3 & a & 0 & 0 & 0 & 0 & 0 \\ 2 & 3 & 1 & 1 & b & 3 & 0 & 0 & 0 \\ 6 & 8 & 8 & 2 & 0 & 1 & 0 & 0 & 0 \\ 6 & 7 & 2 & 0 & 1 & 1 & 2 & 0 & 1 \\ 4 & 4 & 0 & 3 & 4 & 8 & a & 1 & 0 \\ 2 & 0 & 2 & 1 & 4 & 1 & 1 & b & 0 \end{pmatrix}$

V.2.5 For which x are the following matrices nonsingular?

(a) $\begin{pmatrix} x & 2 & 0 & 0 & 0 \\ 3 & 1 & 2 & 0 & 0 \\ 3 & 0 & 2 & 0 & 0 \\ 6 & 1 & 1 & 1 & x \\ 2 & 4 & 1 & 2 & 2 \end{pmatrix}$

(b) $\begin{pmatrix} x & 0 & 0 & 0 & 0 \\ x & 1 & 0 & 0 & 0 \\ 3 & x & 2 & 1 & 0 \\ 2x^3 & x^2 & 5 & 3 & 0 \\ 5 & 9x & 3 & 1 & 4 \end{pmatrix}$

(c) $\begin{pmatrix} x & 0 \\ 1 & a \end{pmatrix}$

(d) $\begin{pmatrix} x & a \\ b & c \end{pmatrix}$

(e) $\begin{pmatrix} x & a & 0 & 0 & 0 & 0 \\ b & c & 0 & 0 & 0 & 0 \\ 1 & 3 & x & a & 0 & 0 \\ 2 & 1 & b & c & 0 & 0 \\ 4 & 6 & 8 & 7 & x & a \\ 1 & 0 & 2 & 3 & b & c \end{pmatrix}$

(f) $\begin{pmatrix} 8 & 0 & 0 & 0 & 0 & 0 \\ 0 & 6 & 0 & 0 & 0 & 0 \\ 1 & 5 & 6 & 7 & 9 & 2 \\ 2 & 1 & 0 & 2 & 6 & 0 \\ 3 & 5 & 0 & 0 & 1 & 0 \\ x & 6 & 0 & 0 & 0 & 2 \end{pmatrix}$

V.2.6 Prove that the determinant of a block diagonal matrix, whose each diagonal block is either upper or lower triangular, is equal to the product of the diagonal entries, for example, if

$$P = \left(\begin{array}{cccc|cccc} a_{11} & a_{12} & \cdots & a_{1n_1} & * & & \cdots & * \\ 0 & a_{22} & \cdots & a_{2n_1} & * & & \cdots & * \\ \vdots & & & \vdots & \vdots & & & \vdots \\ 0 & & 0 & a_{n_1n_1} & * & & \cdots & * \\ \hline & & & & b_{11} & 0 & \cdots & 0 \\ & & 0 & & b_{21} & b_{22} & \cdots & 0 \\ & & & & \vdots & & \ddots & \\ & & & & b_{n_21} & b_{n_22} & \cdots & b_{n_2n_2} \end{array}\right),$$

$\det(P) = a_{11}a_{22} \cdots a_{n_1n_2}b_{11}b_{22} \cdots b_{n_2n_2}$.

V.2.7 (a) Compute directly

$$\det\begin{pmatrix} a & 3 & 4 & 5 & 7 \\ 0 & b & 1 & 1 & 8 \\ 0 & 0 & c & 2 & 1 \\ 0 & 0 & 0 & d & 0 \\ 0 & 0 & 0 & 5 & e \end{pmatrix}$$

(b) Compare the result of (a) with the result obtained from the procedure of Exercise V.2.6.

Chapter V Summary

Basic in this chapter's development is the notion of a multilinear functional (Definition V.1.1). It is shown that multilinear, antisymmetric functionals are all scalar multiples of one fixed universal, such functional (Theorem V.1.5). This single unique, multilinear antisymmetric functional det is called the determinant, and in fact, coincides with our classical determinant functional from Chapter IV. With the functional approach to determinants, we are able to derive the product theorem quite easily, without encountering permutations or elementary matrices (Theorem V.2.1). A determinantal property for block triangular matrices is also given (Theorem V.2.3), which did not appear in Chapter IV.

CHAPTER VI

Structure of Operators

We have only begun to study the properties of linear transformations $\mathbf{T} \in \mathscr{L}(V_n, V_n)$ in relation to properties of the $n \times n$ matrix $A = mx(\mathbf{T})_{\mathscr{X}\mathscr{X}}$ induced by $\mathbf{T}$ (Definition II.6.2). In fact, there are many matrices $mx(\mathbf{T})_{\mathscr{X}\mathscr{X}}$ which are induced by $\mathbf{T}$; indeed, a different square matrix $mx(\mathbf{T})_{\mathscr{X}\mathscr{X}}$ generally arises for each ordered basis $\mathscr{X} = \{X_1, X_2, \ldots, X_n\}$ chosen for V_n. How, then, does the *determinant* vary from matrix to matrix? Amazingly enough, no matter *which* $n \times n$ matrix $mx(\mathbf{T})_{\mathscr{X}\mathscr{X}}$ we select to represent the linear operator $\mathbf{T} \in \mathscr{L}(V_n, V_n)$, the determinant *remains the same* (see Theorem VI.1.5).

This phenomenon prompts us to ask what other "quantities" are the same (that is, *invariant*) from matrix to matrix? For example, the scalar in the upper left-hand entry certainly does *not* remain invariant as we change bases and compare $mx(\mathbf{T})_{\mathscr{X}\mathscr{X}}$ with $mx(\mathbf{T})_{\mathscr{Y}\mathscr{Y}}$. However, the *sum* of the diagonal entries *does* remain invariant from matrix to matrix (see Exercise VI.1.14).

Before these questions of invariance can be answered, we need to know what *all* the $m \times n$ representing matrices are for any given linear transformation $\mathbf{T} \in \mathscr{L}(V_n, V_m)$. The development of the following section proceeds with this end in view.

Note: When a linear transformation $\mathbf{T}$ sends a vector space V_n to *itself*, that is, when $\mathbf{T} : V_n \to V_n$, we call $\mathbf{T}$ a **linear operator** on $\mathbf{V}_n$.

1. Change of Bases

Of interest will be the linear operator $\mathbf{U}_{\mathscr{X}} : V_n \to V_n$ which "changes the name," or converts each basis vector X_i in $\mathscr{X} = \{X_1, X_2, \ldots, X_n\}$ to a *standard* basis vector E_i. We define this operator now in Definition VI.1.1.

Definition VI.1.1 Let $\mathscr{X} = \{X_1, X_2, \ldots, X_i, \ldots X_n\}$ be a basis for V_n and let $\mathscr{E} = \{E_1, E_2, \ldots, E_i, \ldots, E_n\}$ denote the standard basis for V_n. Then the linear operator $\mathbf{U}_{\mathscr{X}} \in \mathscr{L}(V_n, V_n)$ is defined by

$$\mathbf{U}_{\mathscr{X}}(X_i) = E_i \quad \text{for each } X_i \text{ in } \mathscr{X}.$$

Example Let

$$\mathscr{X} = \{X_1, X_2, X_3\} = \left\{\begin{pmatrix}1\\1\\0\end{pmatrix}, \begin{pmatrix}1\\-1\\0\end{pmatrix}, \begin{pmatrix}0\\0\\2\end{pmatrix}\right\}$$

be an ordered basis for V_3; as usual, the standard basis for V_3 is the ordered set

$$\mathscr{E} = \left\{\begin{pmatrix}1\\0\\0\end{pmatrix}, \begin{pmatrix}0\\1\\0\end{pmatrix}, \begin{pmatrix}0\\0\\1\end{pmatrix}\right\}.$$

Then, from Definition VI.1.1, $\mathbf{U}_{\mathscr{X}}$ is defined for the basis vectors X_1, X_2, X_3 in $\mathscr{X}$ as follows:

$$\mathbf{U}_{\mathscr{X}}(X_1) = U_{\mathscr{X}}\begin{pmatrix}1\\1\\0\end{pmatrix} = \begin{pmatrix}1\\0\\0\end{pmatrix} = E_1$$

$$\mathbf{U}_{\mathscr{X}}(X_2) = U_{\mathscr{X}}\begin{pmatrix}1\\-1\\0\end{pmatrix} = \begin{pmatrix}0\\1\\0\end{pmatrix} = E_2 \qquad \textbf{(a)}$$

$$\mathbf{U}_{\mathscr{X}}(X_3) = U_{\mathscr{X}}\begin{pmatrix}0\\0\\2\end{pmatrix} = \begin{pmatrix}0\\0\\1\end{pmatrix} = E_3.$$

But what is $\mathbf{U}_{\mathscr{X}}(X)$ where X is *arbitrary* and not necessarily one of the 3 vectors in the basis $\mathscr{X}$? We may easily answer this question using the linearity (Definition II.2.1) of $\mathbf{U}_{\mathscr{X}}$. In fact, let

$$X = \begin{pmatrix}x_1\\x_2\\x_3\end{pmatrix}$$

be an arbitrary vector of V_3. Let us find

$$\mathbf{U}_{\mathscr{X}}(X) = \mathbf{U}_{\mathscr{X}}\begin{pmatrix}x_1\\x_2\\x_3\end{pmatrix}.$$

Expanding X in terms of the basis vectors X_1, X_2, X_3, we observe that

$$X = \begin{pmatrix}x_1\\x_2\\x_3\end{pmatrix} = \frac{x_1 + x_2}{2}\underset{\substack{\uparrow\\X_1}}{\begin{pmatrix}1\\1\\0\end{pmatrix}} + \frac{x_1 - x_2}{2}\underset{\substack{\uparrow\\X_2}}{\begin{pmatrix}1\\-1\\0\end{pmatrix}} + \frac{x_3}{2}\underset{\substack{\uparrow\\X_3}}{\begin{pmatrix}0\\0\\2\end{pmatrix}},$$

which implies that

$$\mathbf{U}_{\mathscr{X}}(X) = \frac{x_1 + x_2}{2}\underset{\substack{\uparrow \\ \mathbf{U}_{\mathscr{X}}(X_1)}}{\begin{pmatrix}1\\0\\0\end{pmatrix}} + \frac{x_1 - x_2}{2}\underset{\substack{\uparrow \\ \mathbf{U}_{\mathscr{X}}(X_2)}}{\begin{pmatrix}0\\1\\0\end{pmatrix}} + \frac{x_3}{2}\underset{\substack{\uparrow \\ \mathbf{U}_{\mathscr{X}}(X_3)}}{\begin{pmatrix}0\\0\\1\end{pmatrix}} = \begin{pmatrix}(x_1 + x_2)/2\\(x_1 - x_2)/2\\x_3/2\end{pmatrix} \qquad [\text{see } (\mathbf{a})]$$

for any and every

$$X = \begin{pmatrix}x_1\\x_1\\x_3\end{pmatrix} \in V_3.$$

Just for fun, let's find $\mathbf{U}_{\mathscr{X}}^{-1}$. Once we know the action $\mathbf{U}_{\mathscr{X}}^{-1}$ on some basis, we can extend linearly, to find the action of $\mathbf{U}_{\mathscr{X}}^{-1}$ for *all* vectors in V_3. But **(a)** tells us what $\mathrm{U}_{\mathscr{X}}^{-1}$ is for 3 vectors E_1, E_2, E_3 in the standard, ordered basis for V_3, namely,

$$\mathbf{U}_{\mathscr{X}}^{-1}(E_1) = \begin{pmatrix}1\\1\\0\end{pmatrix}$$

$$\mathbf{U}_{\mathscr{X}}^{-1}(E_2) = \begin{pmatrix}1\\-1\\0\end{pmatrix} \qquad (\mathbf{b})$$

$$\mathbf{U}_{\mathscr{X}}^{-1}(E_3) = \begin{pmatrix}0\\0\\2\end{pmatrix}$$

Since any

$$X = \begin{pmatrix}x_1\\x_2\\x_3\end{pmatrix} = x_1E_1 + x_2E_2 + x_3E_3,$$

we see from (**b**) that

$$\mathrm{U}_{\mathscr{X}}^{-1}\begin{pmatrix}x_1\\x_2\\x_3\end{pmatrix} = x_1\underset{\substack{\uparrow \\ \mathrm{U}_{\mathscr{X}}^{-1}}}{\begin{pmatrix}1\\1\\0\end{pmatrix}} + x_2\underset{\substack{\uparrow \\ \mathrm{U}_{\mathscr{X}}^{-1}(E_2)}}{\begin{pmatrix}1\\-1\\0\end{pmatrix}} + x_3\underset{\substack{\uparrow \\ \mathrm{U}_{\mathscr{X}}^{-1}(E_3)}}{\begin{pmatrix}0\\0\\2\end{pmatrix}} = \begin{pmatrix}x_1 + x_2\\x_1 - x_2\\2x_3\end{pmatrix} \qquad (\mathbf{c})$$

for any and every

$$\begin{pmatrix}x_1\\x_2\\x_3\end{pmatrix} \in V_3.$$

With this information in hand, why not find the 3×3 matrix $mx(\mathbf{U}_{\mathscr{X}}^{-1})_{\mathscr{E}\mathscr{E}}$? This is easy. From **(b)** and from Definition II.6.2, we have

$$mx(\mathbf{U}_{\mathscr{X}}^{-1})_{\mathscr{E}\mathscr{E}} = \begin{pmatrix}1 & 1 & 0\\1 & -1 & 0\\0 & 0 & 2\end{pmatrix} = [X_1X_2X_3]. \qquad (\mathbf{d})$$

To check this, observe that Theorem II.2.3 tells us that any linear transformation $\mathbf{T}$ is effected as a left-hand multiplication by its matrix $mx(\mathbf{T})_{\mathscr{E}\mathscr{E}}$. And sure enough, for $\mathbf{T} = \mathbf{U}_{\mathscr{X}}^{-1}$,

$$mx(\mathbf{U}_{\mathscr{X}}^{-1})_{\mathscr{E}\mathscr{E}}\begin{pmatrix} x_1 \\ x_2 \\ x_3 \end{pmatrix} = \begin{pmatrix} 1 & 1 & 0 \\ 1 & -1 & 0 \\ 0 & 0 & 2 \end{pmatrix}\begin{pmatrix} x_1 \\ x_2 \\ x_3 \end{pmatrix} \qquad \text{from } \mathbf{(d)}$$

$$= \begin{pmatrix} x_1 + x_2 \\ x_1 - x_2 \\ 2x_3 \end{pmatrix}, \qquad \text{while}$$

$$\mathbf{U}_{\mathscr{X}}^{-1}\begin{pmatrix} x_1 \\ x_2 \\ x_3 \end{pmatrix} = \begin{pmatrix} x_1 + x_2 \\ x_1 - x_2 \\ 2x_3 \end{pmatrix} \qquad \text{from } \mathbf{(c)}.$$

There are some observations to be made about $\mathbf{U}_{\mathscr{X}}$.

First: $\mathbf{U}_{\mathscr{X}}$ has been defined only for the n-vectors $X_1, X_2, \ldots, X_n$. What is $\mathbf{U}_{\mathscr{X}}(Y)$, where Y is arbitrary? Linearity of $\mathbf{U}_{\mathscr{X}}$ answers this question. Since $\{X_1 X_2, \ldots, X_n\}$ is a *basis* for V_n, there are unique scalars $a_1, a_2, \ldots, a_n$ (depending on Y), such that $Y = a_1X_1 + a_2X_2 + \cdots + a_nX_n$ (Theorem I.6.8). Thus $\mathbf{U}_{\mathscr{X}}(Y)$ may be written

$$\begin{aligned} \mathbf{U}_{\mathscr{X}}(a_1X_1 + a_2X_2 + \cdots + a_nX_n) &= a_1\mathbf{U}_{\mathscr{X}}(X_1) + a_2\mathbf{U}_{\mathscr{X}}(X_2) + \cdots \\ &\quad + a_n\mathbf{U}_{\mathscr{X}}(X_n) \qquad \text{linearity of } \mathbf{U}_{\mathscr{X}} \\ &= a_1E_1 + a_2E_2 + \cdots + a_nE_n \qquad \text{definition of } \mathbf{U}_{\mathscr{X}} \\ &= \begin{pmatrix} a_1 \\ a_2 \\ \vdots \\ a_n \end{pmatrix} \qquad \text{definition of } E_i, \end{aligned} \tag{VI.1.1}$$

which describes the action of $U_{\mathscr{X}}$ for *any* vector Y in V_n.

Secondly: $\mathbf{U}_{\mathscr{X}}$ may be interpreted from (VI.1.1) as a "coefficient-lifting" operation. That is to say, $\mathbf{U}_{\mathscr{X}}$ lifts or extracts the n-ordered coefficients $a_1, a_2, \ldots, a_n$, from $Y = a_1X_1 + a_2X_2 + \cdots + a_nX_n$, the $\mathscr{X}$ representation for Y, and inserts them, intact, as an ordered column [see

(IV.1.1)]. And where have we seen this borrowing and columnizing of coefficients before? The very definition for $mx(\mathbf{T})_{\mathscr{X}\mathscr{Y}}$ (Definition II.6.2) uses this procedure to describe the various columns of the matrix. Recall that for basis $\mathscr{X} = \{X_1, X_2, \ldots, X_n\}$ of V_n and $\mathscr{Y} = \{Y_1, Y_2, \ldots, Y_m\}$ of V_m, $\mathbf{T}(X_i) = a_{1i}Y_1 + a_{2i}Y_2 + \cdots + a_{mi}Y_m$ was used to define the ith column vector of $mx(\mathbf{T})_{\mathscr{X}\mathscr{Y}}$ as

$$\begin{pmatrix} a_{1i} \\ a_{2i} \\ \vdots \\ a_{mi} \end{pmatrix},$$

that is the ith column of $mx(\mathbf{T})_{\mathscr{X}\mathscr{Y}}$ is $\mathbf{U}_{\mathscr{Y}}(\mathbf{T}(X_i))$, where $\mathbf{U}_{\mathscr{Y}} \in \mathscr{L}(V_m, V_m)$ is defined by $\mathbf{U}_{\mathscr{Y}}(Y_j) = E_j$ for all $j = 1, 2, \ldots, m$.

Thirdly: $\mathbf{U}_{\mathscr{X}} \in \mathscr{L}(V_n, V_n)$ is an *invertible* operator, that is, $\mathbf{U}_{\mathscr{X}}^{-1}$ exists and may be defined by

$$\mathbf{U}_{\mathscr{X}}^{-1}(E_i) = X_i, \qquad \text{for all } i = 1, 2, \ldots, n.$$

Again, this defines $\mathbf{U}_{\mathscr{X}}^{-1}$ only for basis elements $E_1, E_2, \ldots, E_n$, but for *linear* operators, this is sufficient (see Theorem II.3.1). Incidently, the definition of $\mathbf{U}_{\mathscr{X}}$ assures us that $\mathbf{U}_{\mathscr{X}}^{-1}$, its inverse, *does* exist; in fact, *any* linear operator in $\mathscr{L}(V_n, V_n)$, which sends 1 basis onto another (as $\mathbf{U}_{\mathscr{X}}$ sends $\mathscr{X}$ onto $\mathscr{E}$, the standard basis), is *automatically* invertible. Conversely, the invertible linear operators always send any and every basis to some (other) basis (Exercises VI.1.1 and VI.1.2).

Now we know that every linear operator $\mathbf{T}$ in $\mathscr{L}(V_n, V_m)$ can be realized as a left-hand multiplication by a certain $m \times n$ matrix A_{T}, that is, for any $X \in V_n$, $\mathbf{T}(X)$ (read "$\mathbf{T}$ of X") equals $A_{\mathrm{T}} \cdot X$ (read "matrix A_{T} *multiplied by* the $n \times 1$ matrix X"). In fact, A_{T} is none other than $mx(\mathbf{T})_{\mathscr{E}\mathscr{E}}$, the matrix of $\mathbf{T}$ relative to the standard ordered basis (see Theorem II.2.3 and Definition II.6.2).

The problem of characterizing the operator $\mathbf{U}_{\mathscr{X}}$ as a left-hand multiplication by a certain $n \times n$ matrix becomes particularly easy. We begin by developing such a characterization for $\mathbf{U}_{\mathscr{X}}^{-1}$, the inverse operator.

Proposition VI.1.2 Suppose $\mathscr{X} = \{X_1, X_2, \ldots, X_n\}$ is a basis for V_n. Then for all $Y \in V_n$,

$$\mathbf{U}_{\mathscr{X}}^{-1}(Y) = [X_1\ X_2\ \cdots\ X_n] \cdot Y,$$

where $[X_1\ X_2\ \cdots\ X_n]$ denotes the $n \times n$ matrix whose ith column is the $n \times 1$ matrix X_i, $i = 1, 2, \ldots, n$.

Moreover, the $n \times n$ matrix $[X_1\, X_2 \cdots X_n] = mx(\mathbf{U}_{\mathscr{X}}^{-1})_{\mathscr{E}\mathscr{E}}$, the matrix of $\mathbf{U}_{\mathscr{X}}^{-1}$ relative to the standard ordered basis $\mathscr{E}$.

Proof: The proof rests on 1 simple observation; namely, that in order to obtain the ith column of any $n \times n$ matrix A, one merely multiplies A on the right by E_i, the ith standard basis vector ($n \times 1$ matrix). In particular, for the $n \times n$ matrix $[X_1X_2 \cdots X_i \cdots X_n]$ whose ith column is

$$X_i = \begin{pmatrix} x_{1i} \\ x_{2i} \\ \vdots \\ x_{ni} \end{pmatrix},$$

we have

$$\underbrace{\begin{pmatrix} x_{11} & x_{12} & \cdots & x_{1i} & \cdots & x_{1n} \\ x_{21} & x_{22} & \cdots & x_{2i} & \cdots & x_{2n} \\ \vdots & \vdots & & \vdots & & \vdots \\ x_{n1} & x_{n2} & \cdots & x_{ni} & \cdots & x_{nn} \end{pmatrix}}_{[X_1\ X_2\ \cdots\ X_i\ \cdots\ X_n]} \cdot \underbrace{\begin{pmatrix} 0 \\ \vdots \\ 0 \\ 1 \\ 0 \\ \vdots \\ 0 \end{pmatrix}}_{E_i} = \underbrace{\begin{pmatrix} x_{1i} \\ x_{2i} \\ \vdots \\ x_{ni} \end{pmatrix}}_{X_i} \qquad \text{(VI.1.2)}$$

On the other hand, if we observe the effect of $\mathbf{U}_{\mathscr{X}}^{-1}$ on E_i, we have

$$\mathbf{U}_{\mathscr{X}}^{-1} \underbrace{\begin{pmatrix} 0 \\ \vdots \\ 0 \\ 1 \\ 0 \\ \vdots \\ 0 \end{pmatrix}}_{E_i} = \underbrace{\begin{pmatrix} x_{1i} \\ x_{2i} \\ \vdots \\ x_{ni} \end{pmatrix}}_{X_i} \qquad \text{(VI.1.3)}$$

since, by definition, $\mathbf{U}_{\mathscr{X}}(X_i) = E_i$ [hence, $\mathbf{U}_{\mathscr{X}}^{-1}(E_i) = X_i$].

Comparison of Equations (VI.1.2) and (VI.1.3) reveals that in *multiplying* E_i (considered as an $n \times 1$ matrix) on the left by $[X_1X_2 \cdots X_n]$, we obtain the *same* vector as in applying the operator $\mathbf{U}_{\mathscr{X}}^{-1}$ to E_i, namely,

$$[X_1X_2\cdots X_n]\cdot E_i = \mathbf{U}_{\mathscr{X}}^{-1}E_i), \quad \text{for all } E_i,\ i = 1, 2, \ldots, n.$$

But if a pair of linear operators (left-hand multiplication by a square matrix, and $\mathbf{U}_{\mathscr{X}}^{-1}$) agree on basis elements, they agree for *all* vectors Y. Thus, for *all* Y (not just for $Y = E_i$), we have

$$[X_1X_2\cdots X_n]\cdot Y = \mathbf{U}_{\mathscr{X}}^{-1}(Y),$$

proving the first part of the proposition.

The proof that $mx(\mathbf{U}_{\mathscr{X}}^{-1})_{\mathscr{E}\mathscr{E}} = [X_1X_2\cdots X_n]$ is immediate. The ith column of $mx(\mathbf{U}_{\mathscr{X}}^{-1})$ is, by Definition II.6.2 the ordered coefficients of the vector $\mathbf{U}_{\mathscr{X}}^{-1}(E_i)$ written in terms of the basis $\{E_1,E_2,\ldots,E_n\}$. But

$$\mathbf{U}_{\mathscr{X}}^{-1}(E_i) = X_i = \begin{pmatrix} x_{1i} \\ x_{2i} \\ \vdots \\ x_{ni} \end{pmatrix} = \underbrace{x_{1i}E_1 + x_{2i}E_2 + \cdots + x_{ni}E_n}_{i\text{th column of } mx(\mathbf{U}_{\mathscr{X}}^{-1})_{\mathscr{E}\mathscr{E}}}.$$

That is, the ith column of $mx(\mathbf{U}_{\mathscr{X}}^{-1})_{\mathscr{E}\mathscr{E}}$ is just the vector X_i itself. In other words,

$$mx(\mathbf{U}_{\mathscr{X}}^{-1})_{\mathscr{E}\mathscr{E}} = [X_1X_2\cdots X_n],$$

the $n \times n$ matrix of (VI.1.2). The proposition is proved.

Example Let

$$\mathscr{X} = \left\{\begin{pmatrix}1\\2\\0\end{pmatrix}, \begin{pmatrix}0\\1\\3\end{pmatrix}, \begin{pmatrix}1\\1\\1\end{pmatrix}\right\} = \{X_1,X_2,X_3\}$$

be a basis for V_3. Find $mx(\mathbf{U}_{\mathscr{X}})_{\mathscr{E}\mathscr{E}}$ and $mx(\mathbf{U}_{\mathscr{X}}^{-1})_{\mathscr{E}\mathscr{E}}$.

In this case, $\mathbf{U}_{\mathscr{X}}$ and $\mathbf{U}_{\mathscr{X}}^{-1}$ are defined by

$$\mathbf{U}_{\mathscr{X}}\begin{pmatrix}1\\2\\0\end{pmatrix} = \begin{pmatrix}1\\0\\0\end{pmatrix} \qquad \mathbf{U}_{\mathscr{X}}^{-1}\begin{pmatrix}1\\0\\0\end{pmatrix} = \begin{pmatrix}1\\2\\0\end{pmatrix}$$

$$\mathbf{U}_{\mathscr{X}}\begin{pmatrix}0\\1\\3\end{pmatrix} = \begin{pmatrix}0\\1\\0\end{pmatrix} \quad \text{and} \quad \mathbf{U}_{\mathscr{X}}^{-1}\begin{pmatrix}0\\1\\0\end{pmatrix} = \begin{pmatrix}0\\1\\3\end{pmatrix}$$

$$\mathbf{U}_{\mathscr{X}}\begin{pmatrix}1\\1\\1\end{pmatrix} = \begin{pmatrix}0\\0\\1\end{pmatrix} \qquad \mathbf{U}_{\mathscr{X}}^{-1}\begin{pmatrix}0\\0\\1\end{pmatrix} = \begin{pmatrix}1\\1\\1\end{pmatrix}.$$

Now Proposition VI.2.1 tells us that $\mathbf{U}_{\mathscr{X}}^{-1}$ of a vector Y is the same as the matrix product of

$$[X_1X_2X_3] = \begin{pmatrix} 1 & 0 & 1 \\ 2 & 1 & 1 \\ 0 & 3 & 1 \end{pmatrix}$$

and the vector Y. That is,

$$mx(\mathbf{U}_{\mathscr{X}}^{-1})_{\mathscr{E}\mathscr{E}} = \begin{pmatrix} 1 & 0 & 1 \\ 2 & 1 & 1 \\ 0 & 3 & 1 \end{pmatrix}.$$

To find $mx(\mathbf{U}_{\mathscr{X}})_{\mathscr{E}\mathscr{E}}$, the inverse of the matrix $mx(\mathbf{U}_{\mathscr{X}}^{-1})_{\mathscr{E}\mathscr{E}}$ above, we may employ elementary row or column operations, or by computing the adjoint matrix and dividing by the determinant of $mx(\mathbf{U}_{\mathscr{X}}^{-1})_{\mathscr{E}\mathscr{E}}$. In any case, $mx(\mathbf{U}_{\mathscr{X}})_{\mathscr{E}\mathscr{E}}$, the inverse of $mx(\mathbf{U}_{\mathscr{X}}^{-1})_{\mathscr{E}\mathscr{E}}$, computes to be

$$mx(\mathbf{U}_{\mathscr{X}})_{\mathscr{E}\mathscr{E}} = \tfrac{1}{4}\begin{pmatrix} -2 & 3 & -1 \\ -2 & 1 & 1 \\ 6 & -3 & 1 \end{pmatrix}.$$

Indeed, one may compute the actual matrix products to see that

$$\mathbf{U}_{\mathscr{X}}(X_1) = mx(\mathbf{U}_{\mathscr{X}})_{\mathscr{E}\mathscr{E}} \cdot X_1 = \tfrac{1}{4}\begin{pmatrix} -2 & 3 & -1 \\ -2 & 1 & 1 \\ 6 & -3 & 1 \end{pmatrix} \cdot \begin{pmatrix} 1 \\ 2 \\ 0 \end{pmatrix} = \begin{pmatrix} 1 \\ 0 \\ 0 \end{pmatrix} = E_1,$$

and, in general,

$$\mathbf{U}_{\mathscr{X}}(X_i) = mx(\mathbf{U}_{\mathscr{X}})_{\mathscr{E}\mathscr{E}} \cdot X_i = E_i \quad \text{for} \quad i = 1, 2, 3.$$

This completes another example which describes the name-changing operator $\mathbf{U}_{\mathscr{X}} : X_i \to E_i$, and its inverse $\mathbf{U}_{\mathscr{X}}^{-1} : E_i \to X_i$ in terms of left-hand multiplications by certain square matrices; namely, $mx(\mathbf{U}_{\mathscr{X}})_{\mathscr{E}\mathscr{E}}$ and $mx(\mathbf{U}_{\mathscr{X}}^{-1})_{\mathscr{E}\mathscr{E}}$, respectively.

EXERCISES

VI.1.1 Suppose $\mathbf{T} \in \mathscr{L}(V_n, V_n)$ and $\mathbf{T}$ sends 1 basis $\{X_1, X_2, \ldots, X_n\}$ for V_n onto another basis $\{Y_1, Y_2, \ldots, Y_n\}$ for V_n. That is, $\mathbf{T}(X_i) = Y_i$ for each $i = 1, 2, \ldots, n$. Prove that, necessarily, the linear operator $\mathbf{T}^{-1}$ exists in $\mathscr{L}(V_n, V_n)$.

VI.1.2 (Converse to Exercise VI.1.1.) Suppose that both $\mathbf{T}$ and $\mathbf{T}^{-1}$ exist in $\mathscr{L}(V_n, V_n)$. Prove that $\mathbf{T}$ sends any basis onto a basis. That is, if $\mathbf{T}$ is invertible, and if $\{X_1, X_2, \ldots, X_n\}$ is a basis for V_n, then $\{\mathbf{T}(X_1), \mathbf{T}(X_2), \ldots, \mathbf{T}(X_n)\}$ is also a basis for V_n.

VI.1.3 Prove that if $\mathbf{T} \in \mathscr{L}(V_n, V_n)$ is an invertible operator, then the induced $n \times n$ matrix $A = mx(\mathbf{T})_{\mathscr{X}\mathscr{X}}$ is invertible for any basis $\mathscr{X}$ of V_n. Show, in fact, that $mx(\mathbf{T}^{-1})_{\mathscr{X}\mathscr{X}} = [mx(\mathbf{T})_{\mathscr{X}\mathscr{X}}]^{-1}$. Conversely, if $A = mx(\mathbf{T})_{\mathscr{E}\mathscr{E}}$ is an invertible $n \times n$ matrix, then prove that the inducing linear operator, $\mathbf{T} \in \mathscr{L}(V_n, V_n)$, is invertible. (*Hint*: Use the definition of matrix product, Definition II.6.12, which defines the homomorphism mx).

VI.1.4 (See Example on page 371.) Given the basis $\mathscr{X} = \{X_1, X_2, \ldots, X_i, \ldots, X_n\}$ find the $n \times n$ matrices A and B [that is, $mx(\mathbf{U})_{\mathscr{E}\mathscr{E}}$ and $mx(\mathbf{U}^{-1})_{\mathscr{E}\mathscr{E}}$] such that $\mathbf{U}_{\mathscr{X}}(X) = A \cdot X$ and $\mathbf{U}(X)_{\mathscr{X}}^{-1} = B \cdot X$ for all n-vectors X in V_n.

(a) $\mathscr{X} = \left\{\begin{pmatrix} 0 \\ 1 \end{pmatrix}, \begin{pmatrix} 2 \\ 2 \end{pmatrix}\right\}$

(b) $\mathscr{X} = \left\{\begin{pmatrix}0\\1\\0\end{pmatrix}, \begin{pmatrix}1\\0\\0\end{pmatrix}, \begin{pmatrix}0\\1\\1\end{pmatrix}\right\}$

(c) $\mathscr{X} = \left\{\begin{pmatrix}2\\0\\1\end{pmatrix}, \begin{pmatrix}1\\2\\2\end{pmatrix}, \begin{pmatrix}1\\0\\0\end{pmatrix}\right\}$

(d) $\mathscr{X} = \left\{\begin{pmatrix}a\\0\\1\end{pmatrix}, \begin{pmatrix}b\\1\\0\end{pmatrix}, \begin{pmatrix}1\\0\\c\end{pmatrix}\right\}$ $a \cdot c \neq 1$

(e) $\mathscr{X} = \left\{\begin{pmatrix}0\\1\\2\\1\end{pmatrix}, \begin{pmatrix}1\\0\\1\\0\end{pmatrix}, \begin{pmatrix}2\\1\\1\\2\end{pmatrix}, \begin{pmatrix}1\\1\\3\\1\end{pmatrix}\right\}$

VI.1.5 By Definition VI.1.1, $\mathbf{U}_{\mathscr{X}}(X_i) = E_i$ and $\mathbf{U}_{\mathscr{Y}}(Y_j) = E_j$ for each $X_i \in \mathscr{X} = \{X_1, X_2, \ldots, X_n\}$ and each $Y_j \in \mathscr{Y} = \{Y_1, Y_2, \ldots, Y_n\}$. Hence, $\mathbf{U}_{\mathscr{X}}^{-1}(E_i) = X_i$ and $\mathbf{U}_{\mathscr{Y}}^{-1}(E_j) = Y_j$.

(a) Describe the linear operator $\mathbf{V} \in \mathscr{L}(V_n, V_n)$ which sends basis $\mathscr{X}$ onto basis $\mathscr{Y}$ [that is, $\mathbf{V}(X_i) = Y_i$ for each $i = 1, 2, \ldots, n$] in terms of (some of) the operators $\mathbf{U}_{\mathscr{X}}$, $\mathbf{U}_{\mathscr{Y}}$, $\mathbf{U}_{\mathscr{X}}^{-1}$, $\mathbf{U}_{\mathscr{Y}}^{-1}$.

(b) Is *every* invertible linear operator $\mathbf{T}$ a product of the form $\mathbf{T} = \mathbf{U}_{\mathscr{Y}} \circ \mathbf{U}_{\mathscr{X}}^{-1}$ (see Exercises VI.1.1 and VI.1.2)?

VI.1.6 (See Exercise VI.1.5.) Given the *pair* of ordered bases $\mathscr{X} = \{X_1, X_2, \ldots, X_n\}$ and $\mathscr{Y} = \{Y_1, Y_2, \ldots, Y_n\}$ for V_n, compute the $n \times n$ matrix A, such that $A \cdot X_i = Y_i$ for each $i = 1, 2, \ldots, n$.

(a) $\mathscr{X} = \left\{\begin{pmatrix}1\\2\end{pmatrix}, \begin{pmatrix}8\\1\end{pmatrix}\right\}$, $\quad \mathscr{Y} = \left\{\begin{pmatrix}1\\0\end{pmatrix}, \begin{pmatrix}0\\1\end{pmatrix}\right\}$

(b) $\mathscr{X} = \left\{\begin{pmatrix}1\\2\end{pmatrix}, \begin{pmatrix}8\\1\end{pmatrix}\right\}$, $\quad \mathscr{Y} = \left\{\begin{pmatrix}0\\1\end{pmatrix}, \begin{pmatrix}1\\0\end{pmatrix}\right\}$

(c) $\mathscr{X} = \left\{\begin{pmatrix}1\\0\\1\end{pmatrix}, \begin{pmatrix}2\\1\\0\end{pmatrix}, \begin{pmatrix}0\\1\\1\end{pmatrix}\right\}$, $\quad \mathscr{Y} = \left\{\begin{pmatrix}1\\0\\0\end{pmatrix}, \begin{pmatrix}0\\1\\0\end{pmatrix}, \begin{pmatrix}0\\0\\1\end{pmatrix}\right\}$

(d) $\mathscr{X} = \left\{\begin{pmatrix}1\\0\\1\end{pmatrix}, \begin{pmatrix}2\\1\\0\end{pmatrix}, \begin{pmatrix}0\\1\\1\end{pmatrix}\right\}$, $\quad \mathscr{Y} = \left\{\begin{pmatrix}0\\1\\0\end{pmatrix}, \begin{pmatrix}0\\0\\1\end{pmatrix}, \begin{pmatrix}1\\0\\0\end{pmatrix}\right\}$

(e) $\mathscr{X} = \left\{\begin{pmatrix}1\\0\\1\end{pmatrix}, \begin{pmatrix}2\\1\\0\end{pmatrix}, \begin{pmatrix}0\\1\\1\end{pmatrix}\right\}$, $\quad \mathscr{Y} = \left\{\begin{pmatrix}1\\0\\1\end{pmatrix}, \begin{pmatrix}2\\1\\0\end{pmatrix}, \begin{pmatrix}0\\1\\1\end{pmatrix}\right\}$

(f) $\mathscr{X} = \left\{\begin{pmatrix}1\\0\\1\end{pmatrix}, \begin{pmatrix}2\\1\\0\end{pmatrix}, \begin{pmatrix}0\\1\\1\end{pmatrix}\right\}$, $\quad \mathscr{Y} = \left\{\begin{pmatrix}0\\1\\1\end{pmatrix}, \begin{pmatrix}2\\1\\0\end{pmatrix}, \begin{pmatrix}1\\0\\1\end{pmatrix}\right\}$

(g) $\mathscr{X} = \left\{ \begin{pmatrix} 1 \\ 0 \\ 1 \end{pmatrix}, \begin{pmatrix} 2 \\ 1 \\ 0 \end{pmatrix}, \begin{pmatrix} 0 \\ 1 \\ 1 \end{pmatrix} \right\}$, $\quad \mathscr{Y} = \left\{ \begin{pmatrix} 2 \\ 2 \\ 2 \end{pmatrix}, \begin{pmatrix} 1 \\ 1 \\ 0 \end{pmatrix}, \begin{pmatrix} 0 \\ 2 \\ 1 \end{pmatrix} \right\}$

(h) $\mathscr{X} = \left\{ \begin{pmatrix} 0 \\ 1 \\ 0 \\ 1 \end{pmatrix}, \begin{pmatrix} 1 \\ 0 \\ 1 \\ 0 \end{pmatrix}, \begin{pmatrix} 2 \\ 1 \\ 0 \\ 0 \end{pmatrix}, \begin{pmatrix} 0 \\ 0 \\ 1 \\ 1 \end{pmatrix} \right\}$, $\quad \mathscr{Y} = \left\{ \begin{pmatrix} 2 \\ 0 \\ 0 \\ 0 \end{pmatrix}, \begin{pmatrix} 1 \\ 1 \\ 1 \\ 1 \end{pmatrix}, \begin{pmatrix} 0 \\ 0 \\ 2 \\ 1 \end{pmatrix}, \begin{pmatrix} 1 \\ 0 \\ 1 \\ 1 \end{pmatrix} \right\}$

We are close to a complete characterization of *all* the $m \times n$ matrices $mx(\mathbf{T})_{\mathscr{X}\mathscr{Y}}$, induced by the linear transformation $\mathbf{T} \in \mathscr{L}(V_n, V_m)$, relative to the (varying) bases $\mathscr{X}$ and $\mathscr{Y}$ for V_n and V_m, respectively. In fact, as we shall see in Theorem VI.1.4, once we know $mx(\mathbf{T})_{\mathscr{E}\mathscr{E}'}$ relative to the standard ordered bases $\mathscr{E}$ of V_n, and $\mathscr{E}'$ of V_m, then we may obtain any and all other matrices $mx(\mathbf{T})_{\mathscr{X}\mathscr{Y}}$ simply by multiplying $mx(\mathbf{T})_{\mathscr{E}\mathscr{E}'}$ on the left and the right by certain square matrices.

For convenience we introduce notation VI.1.3.

Notation VI.1.3 We shall use the symbol

$$\mathscr{E} = \{E_1, E_2, \ldots, E_n\}$$

to denote the n element standard basis for the concrete vector-space V_n, and

$$\mathscr{E}' = \{E_1', E_2', \ldots, E_m'\}$$

will denote the standard basis for V_m. Hence, $A_{\mathbf{T}} = mx(\mathbf{T})_{\mathscr{E}\mathscr{E}'}$ is the $m \times n$ matrix of $\mathbf{T} \in \mathscr{L}(V_n, V_m)$ induced by $\mathbf{T}$ relative to the pair of standard bases, $\mathscr{E}$ and $\mathscr{E}'$ (see Definition I.6.11).

With the notational distinction between standard bases for distinct concrete vector-spaces, we approach the theorem which characterizes *all* $m \times n$ matrices of $\mathbf{T}$ in terms of the single $m \times n$ matrix $mx(\mathbf{T})_{\mathscr{E}\mathscr{E}'}$.

Theorem VI.1.4 (Change of Basis Theorem)

For any $\mathbf{T} \in \mathscr{L}(V_n, V_m)$, and for any pair of bases $\mathscr{X} = \{X_1, X_2, \ldots, X_n\} \subset V_n$, $\mathscr{Y} = \{Y_1, Y_2, \ldots, Y_m\} \subset V_m$, we always have

$$mx(\mathbf{T})_{\mathscr{X}\mathscr{Y}} = mx(\mathbf{U}_{\mathscr{Y}})_{\mathscr{E}'\mathscr{E}'} \cdot mx(\mathbf{T})_{\mathscr{E}\mathscr{E}'} \cdot mx(\mathbf{U}_{\mathscr{X}}^{-1})_{\mathscr{E}\mathscr{E}}$$

where $\mathscr{E}$ and $\mathscr{E}'$ are the standard bases for V_n and V_m, respectively.

Proof: Let us set down the $m \times n$ $mx(\mathbf{T})_{\mathscr{X}\mathscr{Y}}$ in terms of the very definition which gave birth to the symbol $mx(\mathbf{T})_{\mathscr{X}\mathscr{Y}}$, namely, Definition

II.6.2. Now to say

$$mx(\mathbf{T})_{\mathscr{X}\mathscr{Y}} = \begin{pmatrix} a_{11} & a_{12} & \cdots & a_{1n} \\ a_{21} & a_{22} & \cdots & a_{2n} \\ \vdots & \vdots & & \vdots \\ a_{m1} & a_{m2} & \cdots & a_{mn} \end{pmatrix} = (a_{ij})_{ij} \quad \begin{matrix} i = 1, 2, \ldots, m \\ j = 1, 2, \ldots, n \end{matrix} \qquad \text{(VI.1.4)}$$

coordinates of $\mathbf{T}(X_1)$ $\mathbf{T}(X_2)$ $\mathbf{T}(X_n)$ in terms of $\mathscr{Y} = \{Y_1, Y_2, \ldots, Y_m\}$

means only that for each $j = 1, 2, \ldots, n$,

$$\overbrace{\mathbf{T}(X_j) = a_{1j}Y_1 + a_{2j}Y_2 + \cdots + a_{ij}Y_i + \cdots + a_{mj}Y_m.}^{j\text{th column of } mx(\mathbf{T})_{\mathscr{X}\mathscr{Y}}} \qquad \text{(VI.1.5)}$$

(see Definition II.6.2).

We now call into action the 2 linear "name-changing" transformations $\mathbf{U}_{\mathscr{X}} : V_n \to V_n$ and $\mathbf{U}_{\mathscr{Y}} : V_m \to V_m$ given by Definition VI.1.1. We do this so as to effect substitutes for X_j and $Y_1, Y_2, \ldots, Y_m$ in (VI.1.5). From Definition VI.1.1, we have

$$\mathbf{U}_{\mathscr{X}}(X_j) = E_j, \quad \text{and} \quad \mathbf{U}_{\mathscr{Y}}(Y_i) = E_i',$$

or equivalently,

$$X_j = \mathbf{U}_{\mathscr{X}}^{-1}(E_j), \quad \text{and} \quad Y_i = \mathbf{U}_{\mathscr{Y}}^{-1}(E_i'), \qquad \text{(VI.1.6)}$$

for $i = 1, 2, \ldots, m$, $j = 1, 2, \ldots, n$.
Substitution of (VI.1.6) into (VI.1.5) yields

$$\begin{aligned} \mathbf{T}(\mathbf{U}_{\mathscr{X}}^{-1}(E_j)) &= a_{1j}\mathbf{U}_{\mathscr{Y}}^{-1}(E_1') + a_{2j}\mathbf{U}_{\mathscr{Y}}^{-1}(E_2') + \cdots + a_{mj}\mathbf{U}_{\mathscr{Y}}^{-1}(E_m') \\ &= \mathbf{U}_{\mathscr{Y}}^{-1}(a_{1j}E_1' + a_{2j}E_2' + \cdots + a_{mj}E_m') \\ &\qquad \text{from linearity of } \mathbf{U}_{\mathscr{Y}}^{-1}. \qquad \text{(IV.1.7a)} \end{aligned}$$

We now apply the linear transformation $\mathbf{U}_{\mathscr{Y}}$ to both (left-hand) sides of (VI.1.7a) which cancels the $\mathbf{U}_{\mathscr{Y}}^{-1}$ and yields

$$(\mathbf{U}_{\mathscr{Y}} \circ \mathbf{T} \circ \mathbf{U}_{\mathscr{X}}^{-1})E_j = \overbrace{a_{1j}E_1' + a_{2j}E_2' + \cdots + a_{ij}E_i' + \cdots + a_{mj}E_m'}, \qquad \text{(VI.1.7b)}$$

for all $j = 1, 2, \ldots, n$.

Do you realize what (VI.1.7b) *means*? We can interpret (VI.1.7b) to say that we have found the $m \times n$ matrix of the *composite* linear transformation $\mathbf{U}_{\mathscr{Y}} \circ \mathbf{T} \circ U_{\mathscr{X}}^{-1}$, relative to the pair of (standard) bases, $\mathscr{E}$ and $\mathscr{E}'$.

A comparison of (VI.1.7b) with Definition II.6.2, should convince one of that. Remarkably, the *same* matrix coefficients arise in both (VI.1.7b), which defines $mx(\mathbf{U}_{\mathscr{Y}} \circ \mathbf{T} \circ \mathbf{U}_{\mathscr{X}}^{-1})_{\mathscr{E}\mathscr{E}'}$, and in (VI.1.5), which defines $mx(\mathbf{T})_{\mathscr{X}\mathscr{Y}}$. Thus (now watch this!),

$$\begin{aligned} mx(\mathbf{T})_{\mathscr{X}\mathscr{Y}} &= mx(\mathbf{U}_{\mathscr{Y}} \circ \mathbf{T} \circ \mathbf{U}_{\mathscr{X}}^{-1})_{\mathscr{E}\mathscr{E}'} && \text{compare (VI.1.5), (VI.1.7b)} \\ &= mx(\mathbf{U}_{\mathscr{Y}})_{\mathscr{E}'\mathscr{E}'} \cdot mx(\mathbf{T})_{\mathscr{E}\mathscr{E}'} \cdot mx(\mathbf{U}_{\mathscr{X}}^{-1})_{\mathscr{E}\mathscr{E}} && \text{Definition II.6.12,} \end{aligned}$$

which proves the theorem.

Our next example tackles the problem of finding $mx(\mathbf{T})_{\mathscr{X}\mathscr{Y}}$ in 2 ways: First, we apply Definition II.6.12 *directly* so as to obtain the coefficients which eventually fill the columns of $mx(\mathbf{T})_{\mathscr{X}\mathscr{Y}}$ (the old method). Secondly, we use Theorem VI.1.4, which involves the computation of the 3 square matrices induced by $\mathbf{U}_{\mathscr{Y}}$, $\mathbf{T}$, and $\mathbf{U}_{\mathscr{Y}}^{-1}$ relative to the standard bases $\mathscr{E}$, $\mathscr{E}'$ (the new method).

Example Given the pair of ordered bases

$$\mathscr{X} = \{X_1, X_2, X_3\} = \left\{\begin{pmatrix}1\\0\\2\end{pmatrix}\begin{pmatrix}1\\1\\0\end{pmatrix}\begin{pmatrix}0\\1\\0\end{pmatrix}\right\} \quad \text{and} \quad \mathscr{Y} = \{Y_1, Y_2\} = \left\{\begin{pmatrix}3\\0\end{pmatrix}\begin{pmatrix}2\\1\end{pmatrix}\right\}$$

and the linear operator $\mathbf{T} \in \mathscr{L}(V_3, V_2)$ defined by

$$\mathbf{T}\begin{pmatrix}x_1\\x_2\\x_3\end{pmatrix} = \begin{pmatrix}2x_1 - x_2\\x_2 + x_3\end{pmatrix} \quad \text{for all} \quad X = \begin{pmatrix}x_1\\x_2\\x_3\end{pmatrix} \in V_3. \tag{VI.1.8}$$

Find the 3×2 matrix, $mx(\mathbf{T})_{\mathscr{X}\mathscr{Y}}$.

▶ FIRST APPROACH. Using Definition II.6.2 directly, we are obliged to write each $\mathbf{T}(X_i)$ in terms of the Y_j's, that is, from (VI.1.8), we obtain

$$\begin{aligned} \mathbf{T}(X_1) &= \mathbf{T}\begin{pmatrix}1\\0\\2\end{pmatrix} = \begin{pmatrix}2\\2\end{pmatrix} = a_{11}\overset{Y_1}{\overset{\downarrow}{\begin{pmatrix}3\\0\end{pmatrix}}} + a_{21}\overset{Y_2}{\overset{\downarrow}{\begin{pmatrix}2\\1\end{pmatrix}}} \\ \mathbf{T}(X_2) &= \mathbf{T}\begin{pmatrix}1\\1\\0\end{pmatrix} = \begin{pmatrix}1\\1\end{pmatrix} = a_{12}\overset{Y_1}{\overset{\downarrow}{\begin{pmatrix}3\\0\end{pmatrix}}} + a_{22}\overset{Y_2}{\overset{\downarrow}{\begin{pmatrix}2\\1\end{pmatrix}}} \\ \mathbf{T}(X_3) &= \mathbf{T}\begin{pmatrix}0\\1\\0\end{pmatrix} = \begin{pmatrix}-1\\1\end{pmatrix} = a_{13}\overset{Y_1}{\overset{\downarrow}{\begin{pmatrix}3\\0\end{pmatrix}}} + a_{23}\overset{Y_2}{\overset{\downarrow}{\begin{pmatrix}2\\1\end{pmatrix}}}. \end{aligned} \tag{VI.1.9}$$

Once we are able to find the unknown scalars a_{ij} in (VI.1.9), we will have discovered the column matrices for $mx(\mathbf{T})_{\mathscr{X}\mathscr{Y}}$ (see Definition II.6.2). Now what does this involve? *Each* of the 3 vector equations in (VI.1.9) represents a system of 2 linear equations in the the 2 unknowns a_{1i} and a_{2i}, $i = 1, 2, 3$. For example, the first vector equation of (VI.1.9) can be written terms of the 2 equations

$$3 \cdot a_{11} + 2 \cdot a_{21} = 2$$

$$0 \cdot a_{11} + 1 \cdot a_{21} = 2$$

in the 2 unknowns a_{11} and a_{21}. In total, then, (VI.1.9) represents 6 linear equations in the 6 unknowns, a_{ij}, $i = 1, 2, j = 1, 2, 3$.

Generally, the task of computing the 6 solutions might take a considerable measure of time. In our example, however, there are enough zero entries of the basis vectors, so that the solutions are not difficult to come by (a true textbook example). Whichever method of solution is employed, (IV.1.9) can be rewritten in its final form

$$\begin{aligned}
& \qquad\qquad\qquad\qquad\quad Y_1 \qquad\qquad Y_2 \\
\mathbf{T}(X_1) &= \begin{pmatrix} 2 \\ 2 \end{pmatrix} = -\tfrac{2}{3} \cdot \begin{pmatrix} 3 \\ 0 \end{pmatrix} + 2 \cdot \begin{pmatrix} 2 \\ 1 \end{pmatrix} \\
\mathbf{T}(X_2) &= \begin{pmatrix} 1 \\ 1 \end{pmatrix} = -\tfrac{1}{3} \cdot \begin{pmatrix} 3 \\ 0 \end{pmatrix} + 1 \cdot \begin{pmatrix} 2 \\ 1 \end{pmatrix} \qquad \text{(VI.1.10)} \\
\mathbf{T}(X_3) &= \begin{pmatrix} -1 \\ 1 \end{pmatrix} = -1 \cdot \begin{pmatrix} 3 \\ 0 \end{pmatrix} + 1 \cdot \begin{pmatrix} 2 \\ 1 \end{pmatrix}.
\end{aligned}$$

Since each of the 3 equations gives us our coefficient couple which we use for the *columns* of $mx(\mathbf{T})_{\mathscr{X}\mathscr{Y}}$ (Definition II.6.2), we conclude from (VI.1.10) that

$$mx(\mathbf{T})_{\mathscr{X}\mathscr{Y}} = \begin{pmatrix} -\frac{2}{3} & -\frac{1}{3} & -1 \\ 2 & 1 & 1 \end{pmatrix}. \qquad \text{(VI.1.11)}$$

▶ SECOND APPROACH. Using Theorem VI.1.4, we see that our desired matrix, $mx(\mathbf{T})_{\mathscr{X}\mathscr{Y}}$, is simply the product of the 3 matrices $mx(\mathbf{U}_{\mathscr{Y}})_{\mathscr{E}'\mathscr{E}'}$, $mx(\mathbf{T})_{\mathscr{E}\mathscr{E}'}$, and $mx(\mathbf{U}_{\mathscr{X}}^{-1})_{\mathscr{E}\mathscr{E}}$. These 3 matrices are especially easy to find since only the standard ordered bases $\mathscr{E}$, $\mathscr{E}'$ need to be considered. Watch!

To find $mx(\mathbf{U}_{\mathscr{X}}^{-1})_{\mathscr{E}\mathscr{E}}$. Proposition IV.1.2 tells us to construct $mx(\mathbf{U}_{\mathscr{X}}^{-1})_{\mathscr{E}\mathscr{E}}$ by requiring the jth column to coincide with X_j; that is,

▶ $$mx(\mathbf{U}_{\mathscr{X}}^{-1})_{\mathscr{E}\mathscr{E}} = [X_1 X_2 X_3] = \begin{pmatrix} 1 & 1 & 0 \\ 0 & 1 & 1 \\ 2 & 0 & 0 \end{pmatrix}. \qquad \text{(VI.1.12)}$$

(Indeed, you may check that $mx(\mathbf{U}_{\mathscr{X}}^{-1})_{\mathscr{E}\mathscr{E}} \cdot E_i = X_i$, $i = 1, 2, 3$, which agrees with the action of the operator $\mathbf{U}_{\mathscr{X}}^{-1}$ on X_i.)

To find $mx(\mathbf{T})_{\mathscr{E}\mathscr{E}'}$. Easy. Observe that the entires of any n-tuple (vector) X in V_n are exactly the coefficients of X when expanded in terms of the standard basis $\mathscr{E}$. That is, from (VI.1.8), we have

$$\mathbf{T}(E_1) = \mathbf{T}\begin{pmatrix}1\\0\\0\end{pmatrix} = \begin{pmatrix}2\\0\end{pmatrix} = 2 \cdot E_1' + 0 \cdot E_2'$$

$$\mathbf{T}(E_2) = \mathbf{T}\begin{pmatrix}0\\1\\0\end{pmatrix} = \begin{pmatrix}-1\\1\end{pmatrix} = (-1) \cdot E_1' + 1 \cdot E_2' \tag{VI.1.13}$$

$$\mathbf{T}(E_3) = \mathbf{T}\begin{pmatrix}0\\0\\1\end{pmatrix} = \begin{pmatrix}0\\1\end{pmatrix} = 0 \cdot E_1' + 1 \cdot E_2'.$$

Because of the fact that we are working with the *standard* ordered bases $\mathscr{E}$, $\mathscr{E}'$, (VI.1.13) involves *no* further solution, as did the equations of (VI.1.9). Thus, from (VI.1.13) and Definition II.6.2,

$$\blacktriangleright \qquad mx(\mathbf{T})_{\mathscr{E}\mathscr{E}'} = \begin{pmatrix}2 & -1 & 0\\0 & 1 & 1\end{pmatrix}. \tag{VI.1.14}$$

To find $mx(\mathbf{U}_{\mathscr{Y}})_{\mathscr{E}'\mathscr{E}'}$. First we can write $mx(\mathbf{U}_{\mathscr{Y}}^{-1})_{\mathscr{E}'\mathscr{E}'}$ by inspection, due to Proposition VI.1.2. We eventually seek $mx(\mathbf{U}_{\mathscr{Y}})_{\mathscr{E}'\mathscr{E}'}$, the *inverse* of $mx(\mathbf{U}_{\mathscr{Y}}^{-1})_{\mathscr{E}'\mathscr{E}'}$. Thus

$$mx(\mathbf{U}_{\mathscr{Y}}^{-1})_{\mathscr{E}'\mathscr{E}'} = \begin{pmatrix}3 & 2\\0 & 1\end{pmatrix}.$$
$$\begin{matrix}\uparrow & \uparrow\\ Y_1 & Y_2\end{matrix}$$

The 2×2 inverse of this matrix is easily computed (by the adjoint method, for example; see Corollary IV.4.5). We obtain

$$\blacktriangleright \qquad mx(\mathbf{U}_{\mathscr{Y}}^{-1})_{\mathscr{E}'\mathscr{E}'}^{-1} = mx(\mathbf{U}_{\mathscr{Y}})_{\mathscr{E}'\mathscr{E}'} = \begin{pmatrix}\frac{1}{3} & -\frac{2}{3}\\0 & 1\end{pmatrix}. \tag{VI.1.15}$$

Our desired matrix, $mx(\mathbf{T})_{\mathscr{X}\mathscr{Y}}$, is equal to $mx(\mathbf{U}_{\mathscr{Y}})_{\mathscr{E}'\mathscr{E}'} \cdot mx(\mathbf{T})_{\mathscr{E}\mathscr{E}} \cdot mx(\mathbf{U}_{\mathscr{X}}^{-1})_{\mathscr{E}\mathscr{E}}$ (Theorem VI.1.4). Equations (VI.1.12), (VI.1.14), and (VI.1.15) tell us that we have all the information necessary to construct this product.

Thus,

$$mx(\mathbf{T})_{\mathscr{X}\mathscr{Y}} = \overbrace{\begin{pmatrix}\frac{1}{3} & -\frac{2}{3}\\ 0 & 1\end{pmatrix}}^{mx(\mathbf{U}_{\mathscr{Y}})_{\mathscr{E}'\mathscr{E}'}} \cdot \underbrace{\overbrace{\begin{pmatrix}2 & -1 & 0\\ 0 & 1 & 1\end{pmatrix}}^{mx(\mathbf{T})_{\mathscr{E}\mathscr{E}'}} \cdot \overbrace{\begin{pmatrix}1 & 1 & 0\\ 0 & 1 & 1\\ 2 & 0 & 0\end{pmatrix}}^{mx(\mathbf{U}_{\mathscr{X}}^{-1})_{\mathscr{E}\mathscr{E}}}}$$

$$= \begin{pmatrix}\frac{1}{3} & -\frac{2}{3}\\ 0 & 1\end{pmatrix} \cdot \begin{pmatrix}2 & 1 & -1\\ 2 & 1 & 1\end{pmatrix}$$

$$= \begin{pmatrix}-\frac{2}{3} & -\frac{1}{3} & -1\\ 2 & 1 & 1\end{pmatrix},$$

which is in agreement with (VI.1.11).

A very easy consequence of Theorem VI.1.4, is the fact that for $\mathbf{T} \in \mathscr{L}(V_n, V_n)$, the determinant for $mx(\mathbf{T})_{\mathscr{X}\mathscr{X}}$ (notice that only *1* ordered basis $\mathscr{X}$ appears here) is the *same* for any and all ordered bases $\mathscr{X}$. This is the content of our next result, where $m = n$, and $\mathscr{E} = \mathscr{E}'$.

Theorem VI.1.5 Let $\mathscr{X}$ and $\mathscr{Y}$ be any pair of ordered bases for V_n. Let $\mathbf{T} \in \mathscr{L}(V_n, V_n)$ be a linear operator on V_n, whose $n \times n$ matrices relative to the bases $\mathscr{X}$ and $\mathscr{Y}$ are $mx(\mathbf{T})_{\mathscr{X}\mathscr{X}}$ and $mx(\mathbf{T})_{\mathscr{Y}\mathscr{Y}}$, respectively. Then $\det(mx(\mathbf{T})_{\mathscr{X}\mathscr{X}}) = \det(mx(\mathbf{T})_{\mathscr{Y}\mathscr{Y}})$.

Proof: Theorem VI.1.4 tells us that (for $\mathscr{E} = \mathscr{E}'$, since $m = n$),

$$mx(\mathbf{T})_{\mathscr{X}\mathscr{X}} = mx(\mathbf{U}_{\mathscr{X}})_{\mathscr{E}\mathscr{E}} \cdot mx(\mathbf{T})_{\mathscr{E}\mathscr{E}} \cdot mx(\mathbf{U}_{\mathscr{X}}^{-1})_{\mathscr{E}\mathscr{E}}.$$

Applying the determinant to both sides of this equation, and using the product theorem for determinants (Theorem IV.4.1, Theorem V.2.1), reveals that

$$\begin{aligned}
\det[mx(\mathbf{T})_{\mathscr{X}\mathscr{X}}] &= \det[mx(\mathbf{U}_{\mathscr{X}})_{\mathscr{E}\mathscr{E}} \cdot mx(\mathbf{T})_{\mathscr{E}\mathscr{E}} \cdot mx(\mathbf{U}_{\mathscr{X}}^{-1})_{\mathscr{E}\mathscr{E}}]\\
&= \det[mx(\mathbf{U}_{\mathscr{X}})_{\mathscr{E}\mathscr{E}}) \cdot \det(mx(\mathbf{T})_{\mathscr{E}\mathscr{E}} \cdot mx(\mathbf{U}_{\mathscr{X}}^{-1})_{\mathscr{E}\mathscr{E}}] && \text{Theorem IV.4.1,}\\
&= \det[mx(\mathbf{T})_{\mathscr{E}\mathscr{E}} \cdot mx(\mathbf{U}_{\mathscr{X}}^{-1})_{\mathscr{E}\mathscr{E}}) \det(mx(\mathbf{U}_{\mathscr{X}})_{\mathscr{E}\mathscr{E}}]\\
&= \det[mx(\mathbf{T})_{\mathscr{E}\mathscr{E}} \cdot mx(\mathbf{U}_{\mathscr{X}}^{-1})_{\mathscr{E}\mathscr{E}} \cdot mx(\mathbf{U}_{\mathscr{X}})_{\mathscr{E}\mathscr{E}}] && \text{Theorem IV.4.1,}\\
&= \det[mx(\mathbf{T})_{\mathscr{E}\mathscr{E}} \cdot mx(\overbrace{\mathbf{U}_{\mathscr{X}}^{-1} \circ \mathbf{U}_{\mathscr{X}}}^{I_n})_{\mathscr{E}\mathscr{E}}] && \text{Definition II.6.12}\\
&= \det[mx(\mathbf{T})_{\mathscr{E}\mathscr{E}}].
\end{aligned}$$

Similarly, $\det[mx(\mathbf{T})_{\mathscr{Y}\mathscr{Y}}] = \det[mx(\mathbf{T})_{\mathscr{E}\mathscr{E}}]$, so that the determinant is invariant under basis change. The proof is complete.

Theorem VI.1.5 tells us that any and all ordered bases $\mathscr{X}$ of V_n lead to the same value for the determinant of the $n \times n$ matrix $mx(\mathbf{T})_{\mathscr{X}\mathscr{X}}$. This allows us to define the determinant of a *linear operator* **T** as follows:

Definition VI.1.6 Let $\mathbf{T} \in \mathscr{L}(V_n, V_n)$. Then det (**T**), the **determinant of the linear operator T**, is defined by

$$\det(\mathbf{T}) = \det[mx(\mathbf{T})_{\mathscr{X}\mathscr{X}}],$$

where $\mathscr{X}$ is any ordered basis for V_n.

EXERCISES

In Exercises VI.1.7 through VI.1.10, a linear operator $\mathbf{T} \in \mathscr{L}(V_n, V_m)$ is given along with a pair $\mathscr{X}$, $\mathscr{Y}$, of ordered basis for V_n and V_m. In each case, find the $m \times n$ matrix, $mx(\mathbf{T})_{\mathscr{X}\mathscr{Y}}$.

VI.1.7 $\mathbf{T}\begin{pmatrix} x_1 \\ x_2 \end{pmatrix} = \begin{pmatrix} x_1 \\ 0 \\ 2x_1 - x_2 \end{pmatrix}$ for all $\begin{pmatrix} x_1 \\ x_2 \end{pmatrix} \in V_1$.

(a) Choose $\mathscr{X} = \left\{\begin{pmatrix} 1 \\ 0 \end{pmatrix}, \begin{pmatrix} 0 \\ 1 \end{pmatrix}\right\}$, $\mathscr{Y} = \left\{\begin{pmatrix} 1 \\ 0 \\ 0 \end{pmatrix}, \begin{pmatrix} 0 \\ 1 \\ 0 \end{pmatrix}, \begin{pmatrix} 0 \\ 0 \\ 1 \end{pmatrix}\right\}$, the standard ordered bases.

(b) Choose $\mathscr{X} = \left\{\begin{pmatrix} 1 \\ 1 \end{pmatrix}, \begin{pmatrix} 2 \\ 0 \end{pmatrix}\right\}$, $\mathscr{Y} = \left\{\begin{pmatrix} 1 \\ 1 \\ 1 \end{pmatrix}, \begin{pmatrix} 1 \\ 0 \\ 0 \end{pmatrix}, \begin{pmatrix} 1 \\ -1 \\ 0 \end{pmatrix}\right\}$.

(c) Choose $\mathscr{X} = \left\{\begin{pmatrix} 2 \\ 1 \end{pmatrix}, \begin{pmatrix} 1 \\ 2 \end{pmatrix}\right\}$, $\mathscr{Y} = \left\{\begin{pmatrix} 3 \\ 0 \\ 0 \end{pmatrix}, \begin{pmatrix} 1 \\ 0 \\ 1 \end{pmatrix}, \begin{pmatrix} 0 \\ 1 \\ 0 \end{pmatrix}\right\}$.

(d) Choose $\mathscr{X} = \left\{\begin{pmatrix} 0 \\ 1 \end{pmatrix}, \begin{pmatrix} 2 \\ 2 \end{pmatrix}\right\}$, $\mathscr{Y} = \left\{\begin{pmatrix} 3 \\ 1 \\ 1 \end{pmatrix}, \begin{pmatrix} 1 \\ 0 \\ 1 \end{pmatrix}, \begin{pmatrix} 2 \\ 0 \\ 0 \end{pmatrix}\right\}$.

VI.1.8 $\mathbf{T}\begin{pmatrix} x_1 \\ x_2 \\ x_3 \end{pmatrix} = \begin{pmatrix} x_1 \\ x_2 \\ x_3 \end{pmatrix}$ for all $\begin{pmatrix} x_1 \\ x_2 \\ x_3 \end{pmatrix} \in V_3$;

that is, **T** is the identity operator on V_3.

(a) Choose $\mathscr{X} = \mathscr{Y} = \{E_1, E_2, E_3\}$, the standard ordered basis.

(b) Choose $\mathscr{X} = \mathscr{Y} = \{X_1, X_2, X_3\}$ to be a pair of arbitrary, but *equal* bases.

(c) Choose $\mathscr{X} = \left\{\begin{pmatrix} 0 \\ 1 \\ 0 \end{pmatrix}, \begin{pmatrix} 2 \\ 1 \\ 1 \end{pmatrix}, \begin{pmatrix} 0 \\ 1 \\ 1 \end{pmatrix}\right\}$, $\mathscr{Y} = \left\{\begin{pmatrix} 2 \\ 1 \\ 0 \end{pmatrix}, \begin{pmatrix} 0 \\ 1 \\ 1 \end{pmatrix}, \begin{pmatrix} 3 \\ 3 \\ 1 \end{pmatrix}\right\}$.

VI.1.9 $\mathbf{T}\begin{pmatrix} x_1 \\ x_2 \\ x_3 \end{pmatrix} = \begin{pmatrix} 2x_1 - x_2 + x_3 \\ x_2 \\ x_1 - x_2 - x_3 \end{pmatrix}$ for all $\begin{pmatrix} x_1 \\ x_2 \\ x_3 \end{pmatrix} \in V_3$.

(a) Choose $\mathscr{X} = \mathscr{Y} = \{E_1, E_2, E_3\}$ the standard ordered basis for V_3.

(b) Choose $\mathscr{X} = \mathscr{Y} = \left\{\begin{pmatrix}1\\2\\0\end{pmatrix}, \begin{pmatrix}3\\0\\1\end{pmatrix}, \begin{pmatrix}1\\1\\1\end{pmatrix}\right\}$.

(c) Choose $\mathscr{X} = \mathscr{Y} = \left\{\begin{pmatrix}3\\1\\0\end{pmatrix}, \begin{pmatrix}1\\2\\0\end{pmatrix}, \begin{pmatrix}1\\2\\3\end{pmatrix}\right\}$.

(d) Choose $\mathscr{X} = \left\{\begin{pmatrix}1\\4\\0\end{pmatrix}, \begin{pmatrix}2\\1\\1\end{pmatrix}, \begin{pmatrix}1\\1\\1\end{pmatrix}\right\}$, $\mathscr{Y} = \left\{\begin{pmatrix}0\\1\\0\end{pmatrix}, \begin{pmatrix}2\\3\\1\end{pmatrix}, \begin{pmatrix}1\\4\\1\end{pmatrix}\right\}$.

(e) Choose $\mathscr{X} = \left\{\begin{pmatrix}2\\0\\0\end{pmatrix}, \begin{pmatrix}5\\1\\3\end{pmatrix}, \begin{pmatrix}1\\1\\2\end{pmatrix}\right\}$, $\mathscr{Y} = \left\{\begin{pmatrix}1\\0\\1\end{pmatrix}, \begin{pmatrix}4\\1\\1\end{pmatrix}, \begin{pmatrix}2\\2\\0\end{pmatrix}\right\}$.

VI.1.10 $\mathbf{T}\begin{pmatrix}x_1\\x_2\\x_3\\x_4\end{pmatrix} = \begin{pmatrix}x_1 + 2x_2 - 3x_3 + x_4\\ x_1 - x_2 \\ 3x_2 - 3x_3 + x_4\\ x_1 + x_2 + x_3 + x_4\end{pmatrix}$ for all $\begin{pmatrix}x_1\\x_2\\x_3\\x_4\end{pmatrix} \in V_4$.

(a) Choose $\mathscr{X} = \mathscr{Y} = \{E_1, E_2, E_3, E_4\}$, the standard ordered basis for V_4.

(b) Choose $\mathscr{X} = \mathscr{Y} = \left\{\begin{pmatrix}1\\0\\1\\1\end{pmatrix}, \begin{pmatrix}1\\2\\0\\1\end{pmatrix}, \begin{pmatrix}0\\3\\1\\1\end{pmatrix}, \begin{pmatrix}0\\1\\2\\0\end{pmatrix}\right\}$.

(c) Choose $\mathscr{X} = \left\{\begin{pmatrix}1\\0\\0\\0\end{pmatrix}, \begin{pmatrix}0\\2\\1\\1\end{pmatrix}, \begin{pmatrix}0\\1\\4\\0\end{pmatrix}, \begin{pmatrix}0\\1\\1\\1\end{pmatrix}\right\}$, $\mathscr{Y} = \left\{\begin{pmatrix}1\\2\\0\\0\end{pmatrix}, \begin{pmatrix}2\\1\\0\\0\end{pmatrix}, \begin{pmatrix}1\\0\\1\\0\end{pmatrix}, \begin{pmatrix}0\\1\\0\\1\end{pmatrix}\right\}$.

In Exercises VI.1.11 through VI.1.13, you are asked 2 questions:

First, you are to find det (**T**), the determinant of the linear operator **T** (see Definition VI.1.6 and Theorem VI.1.5) by computing the determinant of one or more matrices induced by **T**.

Secondly, you are asked to find the sum of the diagonal elements of the indicated matrices (induced by **T**); this sum is called the *trace* of the matrix.

VI.1.11 Find the determinant and the trace (sum of the main diagonal elements) for the 2×2 matrix $mx(\mathbf{T})_{\mathscr{X}\mathscr{X}}$, where $\mathbf{T}\begin{pmatrix}x_1\\x_2\end{pmatrix} = \begin{pmatrix}2x_1 - x_2\\ x_1 + x_2\end{pmatrix}$, for

(a) $\mathscr{X} = \left\{\begin{pmatrix}0\\1\end{pmatrix}, \begin{pmatrix}2\\0\end{pmatrix}\right\}$.

(b) $\mathscr{X} = \left\{\begin{pmatrix}1\\-1\end{pmatrix}, \begin{pmatrix}0\\3\end{pmatrix}\right\}$.

(c) $\mathscr{X} = \left\{\begin{pmatrix}0\\2\end{pmatrix}, \begin{pmatrix}-1\\-1\end{pmatrix}\right\}$.

VI.1.12 Find the determinant and the trace for the 3×3 matrix
(a) of Exercise VI.1.9(a)
(b) of Exercise VI.1.9(b)
(c) of Exercise VI.1.9(c)
(d) of Exercise VI.1.9(d)
(e) of Exercise VI.1.9(e).

VI.1.13 Find the determinant and the trace for the 4×4 matrix
(a) of Exercise VI.1.10(a)
(b) of Exercise VI.1.10(b)
(c) of Exercise VI.1.10(c).

If all went well you will have noticed that the trace of $mx(\mathbf{T})$ (like the determinant), is invariant; that is, unaltered by basis change. That is, the same value for the trace should have emerged for the matrices of Exercise VI.1.8(a), (b), of Exercises VI.1.9(a), (b), (c), and of Exercises VI.1.10(a), (b). Exercise VI.1.14 asks you to prove that this is no accident.

VI.1.14 Let $A = (a_{ij})_{ij}$ be an $n \times n$ matrix. Define tr (A), the trace of A, by tr $(A) = \sum_{i=1}^{n} a_{ii}$, the sum of the main diagonal entries of A.
(a) Prove that for all $n \times n$ (square) matrices A and B, tr $(A \cdot B) =$ tr $(B \cdot A)$.
(b) (See Theorem VI.1.4.) Prove that for all ordered bases $\mathscr{X}$ for V_n, tr $(mx(\mathbf{T})_{\mathscr{X}\mathscr{X}}) =$ tr $(mx(\mathbf{T})_{\mathscr{E}\mathscr{E}})$, where $\mathscr{E}$ is the standard ordered basis for V_n. Hence, tr $(mx(\mathbf{T})_{\mathscr{X}\mathscr{X}}) =$ tr $(mx(\mathbf{T})_{\mathscr{Y}\mathscr{Y}})$ for all ordered bases $\mathscr{X}$ and $\mathscr{Y}$ of V_n [of course, $\mathbf{T} \in \mathscr{L}(V_n, V_n)$].

VI.1.15 Prove that if f is a function sending $n \times n$ matrices to scalars (such as the determinant and trace functions), and if f has the property that $f(A \cdot B) = f(B \cdot A)$ for all $n \times n$ matrices A and B (as is the case for the determinant and trace functions), then f is basis-invariant in the sense that $f(mx(\mathbf{T})_{\mathscr{X}\mathscr{X}}) = f(mx(\mathbf{T})_{\mathscr{E}\mathscr{E}})$ for all ordered bases $\mathscr{X}$ of V_n; hence, $f(mx(\mathbf{T})_{\mathscr{X}\mathscr{X}}) = f(mx(\mathbf{T})_{\mathscr{Y}\mathscr{Y}})$ for all pairs of ordered bases $\mathscr{X}$ and $\mathscr{Y}$, of V_n, where $\mathbf{T} \in \mathscr{L}(V_n, V_n)$.

VI.1.16 Prove that $\mathbf{T}$ is an invertible operator in $\mathscr{L}(V_n, V_n)$ if and only if, for *all* pairs of ordered bases, $\mathscr{X}$ and $\mathscr{Y}$, of V_n, det $(mx(\mathbf{T})_{\mathscr{X}\mathscr{Y}}) \neq 0$.

VI.1.17 Can an $n \times n$ invertible matrix A exist such that tr $(A) = 0$?

VI.1.18 Prove that the linear operator (not a matrix!) $\mathbf{T} \in \mathscr{L}(V_n, V_n)$ has an inverse if and only if det $(\mathbf{T}) \neq 0$.

2. *Eigenvalues and Eigenvectors*

We consider a linear operator $\mathbf{T} \in \mathscr{L}(V_n, V_n)$ and try to understand its structure by seeking those vectors of V_n (if any) which are "stretched," "contracted," or "reversed in direction" by $\mathbf{T}$. These, roughly speaking, are the **eigenvectors of T.** To be a bit more specific, suppose there exists a nonzero vector $X_0 \in V_n$ such that $\mathbf{T}(X_0) = \lambda_0 X_0$ for some scalar λ_0, which depends on X_0 and $\mathbf{T}$. X_0 is called an **eigenvector of T**. If $\lambda_0 = 3$, say, we may think of $\mathbf{T}$ "stretching" X_0 to 3 times its original length; if $\lambda_0 = \frac{1}{2}$, say, then $\mathbf{T}$ "contracts" X_0 to one half its original size; if $\lambda_0 = -2$, say, we may think of $\mathbf{T}$ as reversing the direction of X_0 and then stretching X_0 to twice its original length.

We present a formal definition now.

Definition VI.2.1 Let $\mathbf{T} \in \mathscr{L}(V_n, V_n)$ be a linear operator on V_n. If there exists a nonzero vector $X \in V_n$ such that

$$\mathbf{T}(X) = \lambda X \quad \text{for some scalar } \lambda \text{ depending on } X,$$

then X is called an **eigenvector of T (for T),** and the scalar λ is called the **eigenvalue of T (for T)** corresponding to, or associated with, the eigenvector X. Equivalent terms used for "eigenvector," are **proper vector, latent vector,** and **characteristic vector.** Similarly, **proper value, latent value,** and **characteristic value** are synonyms for "eigenvalue."

Once our definition is presented, we are naturally led to the questions: Do thing such as eigenvectors (and corresponding eigenvalues) always exist for arbitrary linear operators $\mathbf{T} \in \mathscr{L}(V_n, V_n)$? If they do exist, is there a reasonable means for finding them? How do they help in understanding properties of $\mathbf{T}$? For an answer to the first question, we offer some examples.

Example We define $\mathbf{T} \in \mathscr{L}(V_3, V_3)$ by

$$\mathbf{T}\begin{pmatrix} x_1 \\ x_2 \\ x_3 \end{pmatrix} = \begin{pmatrix} 2x_1 \\ -3x_2 \\ 0 \end{pmatrix} \quad \text{for all } X = \begin{pmatrix} x_1 \\ x_2 \\ x_3 \end{pmatrix} \in V_3.$$

We observe that

$$\mathbf{T}\begin{pmatrix} 1 \\ 0 \\ 0 \end{pmatrix} = 2\begin{pmatrix} 1 \\ 0 \\ 0 \end{pmatrix}, \quad \mathbf{T}\begin{pmatrix} 0 \\ 1 \\ 0 \end{pmatrix} = -3\begin{pmatrix} 0 \\ 1 \\ 0 \end{pmatrix}, \quad \text{and } \mathbf{T}\begin{pmatrix} 0 \\ 0 \\ 1 \end{pmatrix} = 0\begin{pmatrix} 0 \\ 0 \\ 1 \end{pmatrix}.$$

Thus,

$$\begin{pmatrix} 1 \\ 0 \\ 0 \end{pmatrix}, \begin{pmatrix} 0 \\ 1 \\ 0 \end{pmatrix}, \text{and } \begin{pmatrix} 0 \\ 0 \\ 1 \end{pmatrix}$$

are all eigen*vectors* for $\mathbf{T}$ with corresponding eigen*values* 2, -3, and 0, respectively.

We shall be able to see later that the 3 vectors given above are *all* the eigenvectors of $\mathbf{T}$ up to nonzero scalar multiples, that is, if X is an eigenvector of $\mathbf{T}$, then any nonzero scalar multiple $\alpha\mathbf{X}$ is also an eigenvector of $\mathbf{T}$ (Exercise VI.2.2). In this sense, the eigenvectors are not unique; however, the subspace of all the eigenvectors corresponding to a certain eigenvalue *is* unique (Theorem VI.2.2). Moreover, it is easy to see that any eigenvector X can give rise to 1 and *only* 1 associated eigenvalue λ (Exercise VI.2.3).

Example Put $\mathbf{T}(X) = 3 \cdot X$ for all $X \in V_n$. In this case, *every* nonzero vector X is an eigenvector with corresponding real eigenvalue 3.

Example Define $\mathbf{T} \in \mathscr{L}(V_2, V_2)$ by

$$\mathbf{T}\begin{pmatrix} x_1 \\ x_2 \end{pmatrix} = \begin{pmatrix} 0 & -1 \\ 1 & 0 \end{pmatrix} \cdot \begin{pmatrix} x_1 \\ x_2 \end{pmatrix} = \begin{pmatrix} -x_2 \\ x_1 \end{pmatrix} \quad \text{for all} \begin{pmatrix} x_1 \\ x_2 \end{pmatrix} \in V_2.$$

For $\mathbf{T}$ defined above, there are no eigenvectors X so long as we restrict ourselves to 2-vectors X whose each entry is a *real* number, In fact, let us try to find an $X \in V_2$, and a *real* number λ such that $\mathbf{T}(X) = \lambda X$. That is, we seek $\begin{pmatrix} x_1 \\ x_2 \end{pmatrix}$ such that $\mathbf{T}\begin{pmatrix} x_1 \\ x_2 \end{pmatrix} = \begin{pmatrix} -x_2 \\ x_1 \end{pmatrix}$ also equals $\lambda \begin{pmatrix} x_1 \\ x_2 \end{pmatrix}$. This leads to the vector equation $\begin{pmatrix} -x_2 \\ x_1 \end{pmatrix} = \begin{pmatrix} \lambda x_1 \\ \lambda x_2 \end{pmatrix}$ which, in turn, leads to the 2 linear equations

$$\begin{aligned} -x_2 &= \lambda x_1 \\ x_1 &= \lambda x_2, \end{aligned}$$

or, equivalently,

$$\begin{aligned} \lambda x_1 + x_2 &= 0 \\ -x_1 + \lambda x_2 &= 0, \end{aligned}$$

a system of 2 equations in 2 unknowns, x_1 and x_2. Now Theorem III.2.4 tells us that the dimension of the solution space is $2 - r$, where r is the rank of the 2×2 coefficient matrix $\begin{pmatrix} \lambda & 1 \\ -1 & \lambda \end{pmatrix}$. We desire a *non*zero solution for this system, so that in our case, the rank r must not equal 2; that is, $r < 2$. But this is equivalent to saying that the coefficient matrix $\begin{pmatrix} \lambda & 1 \\ -1 & \lambda \end{pmatrix}$ is not invertible (Theorem III.3.5), which is equivalent to saying that the determinant of $\begin{pmatrix} \lambda & 1 \\ -1 & \lambda \end{pmatrix}$ is zero (Theorem IV.2.9). Thus, a nonzero $X = \begin{pmatrix} x_1 \\ x_2 \end{pmatrix}$ exists for the vector equation $\mathbf{T}\begin{pmatrix} x_1 \\ x_2 \end{pmatrix} = \lambda \begin{pmatrix} x_1 \\ x_2 \end{pmatrix}$, for some λ if and only if $\det \begin{pmatrix} \lambda & 1 \\ -1 & \lambda \end{pmatrix} = 0$; that is, if and only if $\lambda^2 + 1 = 0$.

Clearly, if λ is restricted to the real numbers so that λ^2 is consequently non-negative, then $\lambda^2 + 1$ can *never* be zero. Therefore, *no* eigenvector $X = \begin{pmatrix} x_1 \\ x_2 \end{pmatrix}$ with real-number entries can exist either. If, on the other hand, *complex* numbers are allowed, then a (complex) value for λ can always be found; in our case, $\lambda^2 + 1 = 0$ so that $\lambda = i$ or $\lambda = -i$; i denotes the complex number with the property that $i^2 = -1$. Observe that

$$\mathbf{T}\begin{pmatrix} i \\ 1 \end{pmatrix} = \begin{pmatrix} 0 & -1 \\ 1 & 0 \end{pmatrix} \cdot \begin{pmatrix} i \\ 1 \end{pmatrix} = \begin{pmatrix} -1 \\ i \end{pmatrix} = i \begin{pmatrix} i \\ 1 \end{pmatrix},$$

so that $\begin{pmatrix} i \\ 1 \end{pmatrix}$ or any scalar multiple thereof, is an eigenvector for $\mathbf{T}$ in $V_2(\mathbf{C})$, the vector-space of all 2-vectors with *complex* number entries, with corresponding complex eigenvalue $\lambda = -i$.

REMARK. This example points out a situation (for the first time in this book) where real numbers prove inadequate. In asking for solutions for the equation $\lambda^2 + 1 = 0$, we are *forced* to step outside the real numbers and search the *complex* numbers to obtain the desired solutions. All that has been said in this book prior to this chapter is valid when the scalars are considered either as complex numbers or as real numbers (the real numbers may be thought of as a subset of the set of complex numbers).

▶ Thus, we may interpret V_n as either $V_n(\mathbf{C})$, the vector-space of ordered n-vectors with entries from the set of complex numbers $\mathbf{C}$, or as $V_n(\mathbf{R})$, the vector space of ordered n-vectors with entries from the set of real numbers R.

▶ Similarly, $\mathscr{M}_{m,n}$ may be interpreted as either $\mathscr{M}_{m,n}(\mathbf{C})$ or as $\mathscr{M}_{m,n}(\mathbf{R})$, the space of all $m \times n$ matrices with complex or real number entries, respectively.

Appendix B presents a sketch of some properties of complex numbers.

Prior to our examples, we glibly made some statements about eigenvectors and eigenvalues, which we now support with the following theorem.

Theorem VI.2.2 Let $\mathbf{T} \in \mathscr{L}(V_n, V_n)$ and let λ_0 be an eigenvalue for $\mathbf{T}$. Then the set $\mathscr{M}_{\lambda_0}$ of all eigenvectors of V_n associated with the eigenvalue λ_0, along with the zero O, forms a nonempty subspace of V_n.

Proof: Our hypothesis tells us that λ_0 exists as an eigenvalue for $\mathbf{T}$. Thus, from the very definition of eigenvalue (Definition VI.2.1), we are assured that a nonzero vector $X \in V_n$ exists such that $\mathbf{T}(X) = \lambda_0 X$. So? Well, this means that the set $\mathscr{M}_{\lambda_0}$ contains at least one nonzero vector, which guarantees that if $\mathscr{M}_{\lambda_0}$ is a subspace of V_n, it is a nontrivial (that is, a nonzero) subspace.

Now ordinarily, to show that the set $\mathscr{M}_{\lambda_0} = \{\text{all nonzero eigenvectors}\} \cup \{O\}$ is, in fact, a vector-space in its own right, we would have to verify tediously each of the 10 properties of Definition I.5.2, which defines a vector-space. But $\mathscr{M}_{\lambda_0}$ happens to be a subset of a vector space (V_n to be exact). Thus we call upon Theorem I.7.2, which says that we need only verify 2 of the 10 properties; the other 8 will follow automatically. The 2 properties which imply the other 8 are the properties of closure of $\mathscr{M}_{\lambda_0}$ under addition, and closure under scalar multiplication (properties A.1 and B.1, respectively, of Definition I.5.2).

▶ VERIFICATION OF A.1 (Definition I.5.2). Let X and Y be vectors in $\mathscr{M}_{\lambda_0}$. If at least 1 of them, X say, is O, then surely $X + Y = O + Y = Y$ still belongs to $\mathscr{M}_{\lambda_0}$. Hence, let us assume that both X and Y are nonzero eigenvectors in $\mathscr{M}_{\lambda_0}$. This means that

$$\mathbf{T}(X) = \lambda_0 X \quad \text{and} \quad \mathbf{T}(Y) = \lambda_0 Y. \tag{VI.2.1}$$

Is the sum $X + Y$ *still* an eigenvector for **T** relative to the eigenvalue λ_0? We answer this question by observing that

$$\begin{aligned} \mathbf{T}(X+Y) &= \mathbf{T}(X) + \mathbf{T}(Y) && \text{linearity of } \mathbf{T} \\ &\quad\ \ \downarrow \qquad\quad \downarrow \\ &= \lambda_0 X \ + \lambda_0 Y && \text{from (VI.2.1)} \\ &= \lambda_0(X + Y) && \text{axiom B.5, Definition I.5.2, which obtains in } V_n. \end{aligned} \tag{VI.2.2}$$

The sequence (VI.2.2) of equalities shows that the effect of **T** on the vector $X + Y$, is simply to multiply *it* by the scalar λ_0, that is, $X + Y$ is an eigenvector of **T** corresponding to the eigenvalue λ_0. We have shown then, that $X + Y \in \mathscr{M}_{\lambda_0}$ whenever $X \in \mathscr{M}_{\lambda_0}$ and $Y \in \mathscr{M}_{\lambda_0}$. The closure axiom for addition, A.1, is now proven valid relative to $\mathscr{M}_{\lambda_0}$.

▶ VERIFICATION OF B.1. Exercise VI.2.2.

We now have that the set $\mathscr{M}_{\lambda_0}$ which includes O and all eigenvectors of **T** associated with the eigenvalue λ_0, forms a bonafide vector (sub)space of V_n.

The time has come to present a concrete method of obtaining eigenvalues along with the associated eigenvectors for any given linear operator **T**. We shall use the fact that any linear operator $\mathbf{T} \in \mathscr{L}(V_n, V_n)$ can be effected by multiplying the vectors of V_n (that is, the $n \times 1$ matrices) on the left by the $n \times n$ matrix $mx(\mathbf{T})_{\mathscr{E}\mathscr{E}}$, the matrix of **T** relative to the standard ordered basis for **T** (Theorem II.2.3 and Definition II.6.2). For the purpose of abbreviating our symbols, we offer this remark.

REMARK. Recall that the symbol A_{T} denotes the $n \times n$ matrix $mx(\mathbf{T})_{\mathscr{E}\mathscr{E}}$, where **T** is a linear operator on V_n (Notation VI.1.3). Thus, Theorem II.2.3 allows us to write, for any $X \in V_n$,

$$\mathbf{T}(X)(\mathbf{T} \text{ of } X) = A_{\mathrm{T}} \cdot X \ (A_{\mathrm{T}} \text{ multiplied by } X) \tag{VI.2.3}$$

We present a very simple rule for finding all the eigenvalues for an arbitrary linear operator $\mathbf{T} \in \mathscr{L}(V_n, V_n)$.

Theorem VI.2.3 Let $\mathbf{T} \in \mathscr{L}(V_n, V_n)$ and put $A_{\mathrm{T}} = mx(\mathbf{T})_{\mathscr{E}\mathscr{E}}$. Then the scalar λ is an eigenvalue for **T** is and only if $\det(A_{\mathrm{T}} - \lambda I_n) = 0$. In other words, the eigenvalues for **T** are exactly those scalars λ (real or complex) for which $\det(A_{\mathrm{T}} - \lambda I_n)$ equals zero.

Proof: λ is an eigenvalue for **T** if and only if

$$\mathbf{T}(X) = \lambda X \quad \text{for some nonzero } X \in V_n \quad \text{(Definition VI.2.1).}$$

This statement is true if and only if, for some nonzero $X \in V_n$,

$$\Leftrightarrow A_{\mathrm{T}} \cdot X = \lambda I_n \cdot X \qquad \text{(since } I_n \cdot X = X\text{)}$$

$$\Leftrightarrow (A_{\mathrm{T}} \cdot X) - (\lambda I_n \cdot X) = O$$

$$\Leftrightarrow \underset{\substack{\uparrow \\ n \times n \\ \text{matrix}}}{(A_{\mathrm{T}} - \lambda I_n)} \cdot \underset{\substack{\uparrow \\ n \times 1 \\ \text{matrix}}}{X} = O \qquad \text{(distributivity of matrix multiplication; see Exercises II.6.83 and II.6.84)}$$

This last equality is very revealing. It says that for some *nonzero* X in V_n, we have $(A_{\mathrm{T}} - \lambda I_n) \cdot X = O$. This is equivalent to the statement that the $n \times n$ matrix $(A_{\mathrm{T}} - \lambda I_n)$ must have rank r less than and *not equal to* n (Theorem III.2.4 or Proposition III.2.8), which is equivalent to saying that the inverse $(A_{\mathrm{T}} - \lambda I_n)^{-1}$ does *not exist* (Theorem III.3.5). To say that the inverse of the matrix $(A_{\mathrm{T}} - \lambda I_n)$ does not exist is just to say that $\det (A_{\mathrm{T}} - \lambda I_n) = 0$ (Theorem IV.2.9) and the proof is done.

The following example illustrates the previous theorem.

Example Given

$$\mathbf{T}\begin{pmatrix} x_1 \\ x_2 \\ x_3 \end{pmatrix} = \begin{pmatrix} x_1 + 2x_3 \\ x_2 + x_3 \\ x_1 + 2x_2 + x_3 \end{pmatrix}$$

for all

$$X = \begin{pmatrix} x_1 \\ x_2 \\ x_3 \end{pmatrix} \in V_3.$$

Since

$$\mathbf{T}\begin{pmatrix} 1 \\ 0 \\ 0 \end{pmatrix} = \begin{pmatrix} 1 \\ 0 \\ 1 \end{pmatrix}, \; T\begin{pmatrix} 0 \\ 1 \\ 0 \end{pmatrix} = \begin{pmatrix} 0 \\ 1 \\ 2 \end{pmatrix} \quad \text{and} \quad \mathbf{T}\begin{pmatrix} 0 \\ 0 \\ 1 \end{pmatrix} = \begin{pmatrix} 2 \\ 1 \\ 1 \end{pmatrix},$$

we have that

$$mx(\mathbf{T})_{\mathscr{E}\mathscr{E}} = A_{\mathrm{T}} = \begin{pmatrix} 1 & 0 & 2 \\ 0 & 1 & 1 \\ 1 & 2 & 1 \end{pmatrix};$$

equivalently,

$$\mathbf{T}\begin{pmatrix} x_1 \\ x_2 \\ x_3 \end{pmatrix} = \begin{pmatrix} 1 & 0 & 2 \\ 0 & 1 & 1 \\ 1 & 2 & 1 \end{pmatrix} \cdot \begin{pmatrix} x_1 \\ x_2 \\ x_3 \end{pmatrix} \quad \text{for all } X = X_2 \in V_3$$

[see (VI.2.3)]. To find the eigenvalues for $\mathbf{T}$ is to find precisely those scalars λ for which det $(A_T - \lambda I_3)$ equals zero (Theorem VI.2.3). But

$$\begin{aligned}\det(A_T - \lambda I_3) &= \det\left(\begin{pmatrix}1 & 0 & 2\\ 0 & 1 & 1\\ 1 & 2 & 1\end{pmatrix} - \begin{pmatrix}\lambda & 0 & 0\\ 0 & \lambda & 0\\ 0 & 0 & \lambda\end{pmatrix}\right)\\ &= \det\left(\begin{pmatrix}1-\lambda & 0 & 2\\ 0 & 1-\lambda & 1\\ 1 & 2 & 1-\lambda\end{pmatrix}\right),\end{aligned} \tag{VI.2.4}$$

which, when expanded by the first column (Theorem IV.3.4),

$$\begin{aligned}&= (1-\lambda)\det\begin{pmatrix}1-\lambda & 1\\ 2 & 1-\lambda\end{pmatrix} - 0\det\begin{pmatrix}0 & 2\\ 2 & 1-\lambda\end{pmatrix} + 1\det\begin{pmatrix}0 & 2\\ 1-\lambda & 1\end{pmatrix}\\ &= (1-\lambda)[(1-\lambda)^2 - 2] - 0 + (1\times 2)(1-\lambda)\\ &= (1-\lambda)[(1-\lambda)^2 - 2 - 2] \quad [\text{since } (1-\lambda) \text{ factors out of both terms}]\\ &= (1-\lambda)[(1-\lambda)^2 - 4].\end{aligned}$$

Now in the last line of (VI.2.4), we could proceed to multiply the factors out, to obtain that for all scalars λ, $\det(A_T - \lambda I_3) = -\lambda^3 + 3\lambda^2 - 7\lambda - 3$, a polynomial of degree 3. But this does not allow a convenient form for finding the roots, namely those particular scalars λ, for which $-\lambda^3 + 3\lambda^2 - 7\lambda - 3 = 0$. Returning to (VI.2.4), then, we have

$$\det(A_T - \lambda I_3) = (1-\lambda)[(1-\lambda)^2 - 4]$$

for our particular choice of $\mathbf{T} \in \mathscr{L}(V_3, V_3)$. It is easy to see exactly when this product of scalars, $(1-\lambda)[(1-\lambda)^2 - 4]$ equals zero; when and only when at least 1 of the 2 factors, $(1-\lambda)$ or $(1-\lambda)^2 - 4$, equals zero. Thus,

$$\det(A_T - \lambda I_3) = (1-\lambda)[(1-\lambda)^2 - 4] = 0$$

if and only if

(a) $$(1-\lambda) = 0,$$

or

(b) $$(1-\lambda)^2 - 4 = 0.$$

From (a), we see that $\lambda = 1$; from (b) we conclude that $(1-\lambda)^2 = 4$, or that $(1-\lambda) = 2$ or else $(1-\lambda) = -2$. This can occur only for $\lambda = -1$ or $\lambda = 3$, respectively. The problem of finding all the eigenvalues is solved. The only scalars λ for which $\det(A_T - \lambda I_3) = (1-\lambda)[(1-\lambda)^2 - 4] = 0$, are the scalars $\lambda_1 = 1$ [from (a)], $\lambda_2 = -1$, and $\lambda_3 = 3$ [from (b)].

Intoxicated with success, we shall continue our efforts. We tackle the problem of finding the *eigenvectors* which correspond to eigenvalues 1, -1, and 3.

$\lambda_1 = 1$: For which

$$X = \begin{pmatrix}x_1\\ x_2\\ x_3\end{pmatrix}$$

does $\mathbf{T}(X) = 1 \cdot X$? Now

$$\mathbf{T}\begin{pmatrix} x_1 \\ x_2 \\ x_3 \end{pmatrix} = \begin{pmatrix} x_1 + 2x_3 \\ x_2 + x_3 \\ x_1 + 2x_2 + x_3 \end{pmatrix} = 1 \cdot \begin{pmatrix} x_1 \\ x_2 \\ x_3 \end{pmatrix} \tag{VI.2.5}$$

leads to the homogeneous system of equations

$$\begin{aligned} 0x_1 + 0x_2 + 2x_3 &= 0 \\ 0x_1 + 0x_2 + 1x_3 &= 0 \\ 1x_1 + 2x_2 + 0x_3 &= 0 \end{aligned} \tag{VI.2.6}$$

when corresponding entries of the last 2 vectors of (VI.2.5) are equated. It is to be observed that (VI.2.6) is nothing more than the homogeneous system of equations $(A_{\mathbf{T}} - \lambda I_3) \cdot X = 0$, where λ is chosen to be 1. In general (VI.2.6) would be solved by the methods of Gaussian Elimination, Chapter I, or the methods of Chapter III, but here, the system can be solved by inspection. The first 2 equations of (VI.2.6) (which are really the same since the first is twice the second), reveal at once that $x_3 = 0$. The third equation tells us that $x_1 + 2x_2 = 0$, or that x_1 is always -2 times x_2. Thus, we have complete liberty to choose x_2 at will since only the *ratio* of x_2 to x_1 is restricted by the third equation of (VI.2.6). Put $x_2 = 1$, say, so that x_1 is forced to equal -2. We conclude that *an* eigenvector X_1 of $\mathbf{T}$ corresponding to the eigenvalue $\lambda_1 = 1$ is the vector

$$X_1 = \begin{pmatrix} -2 \\ 1 \\ 0 \end{pmatrix}.$$

The typical eigenvector for $\mathbf{T}$ corresponding to the eigenvalue $\lambda_1 = 1$, is generally a scalar multiple of this vector X_1 (see Theorem VI.2.4). The student may easily test the authenticity of a suspected eigenvector X by applying $\mathbf{T}$ to it, and observing whether or not the equation $\mathbf{T}(X) = \lambda_1 X$ obtains for the particular eigenvalue $\lambda_1 = 1$. In our case, we have that for

$$X_1 = \begin{pmatrix} -2 \\ 1 \\ 0 \end{pmatrix},$$

$$\mathbf{T}(X_1) = \mathbf{T}\begin{pmatrix} -2 \\ 1 \\ 0 \end{pmatrix} = 1\begin{pmatrix} -2 \\ 1 \\ 0 \end{pmatrix} = 1 \cdot X_1;$$

X_1 is therefore a bonafide eigenvector corresponding to the eigenvalue $\lambda_1 = 1$. For thc cases $\lambda_2 = -1$, $\lambda_3 = 3$, see Exercise VI.2.5(f).

Example We define $T \in \mathscr{L}(V_4, V_4)$ by

$$\mathbf{T}\begin{pmatrix} x_1 \\ x_2 \\ x_3 \\ x_4 \end{pmatrix} = \begin{pmatrix} 2x_1 + x_3 - x_4 \\ 2x_2 + x_3 \\ x_3 - x_4 \\ x_3 + x_4 \end{pmatrix} = \underbrace{\begin{pmatrix} 2 & 0 & 1 & -1 \\ 0 & 2 & 1 & 0 \\ 0 & 0 & 1 & -1 \\ 0 & 0 & 1 & 1 \end{pmatrix}}_{A_{\mathbf{T}} = mx(\mathbf{T})_{\mathscr{E}}} \cdot \begin{pmatrix} x_3 \\ x_2 \\ x_3 \\ x_4 \end{pmatrix} \tag{VI.2.7}$$

for all $X \in V_4$. For this linear operator $\mathbf{T} \in \mathscr{L}(V_4, V_4)$ we must distinguish between $V_4 = V_4(\mathbf{R})$, the 4-dimensional vector-space of 4-tuples whose entries are restricted to *real* numbers, and $V_4 = V_4(\mathbf{C})$, the 4-dimensional vector-space of 4-tuples whose entiries are *complex* numbers. As we shall see, if $V_4 = V_4(\mathbf{R})$, then only 2 linearly independent eigenvectors [in $V_4(\mathbf{R})$] can be found for **T**. On the other hand, with the richer vector-space $V_4 = V_4(\mathbf{C})$ at our disposal, 4 linearly independent eiegnvectors [in $V_4(\mathbf{C})$] can be found for **T**. Note that the number of linearly independent vectors is mentioned, as it is the *subspace* corresponding to each eigenvalue that we wish to characterize (Theorem VI.2.2). We present the details now.

Consider det $(A_{\mathrm{T}} - \lambda I_4)$ for the (real or complex) scalar λ.

$$\det (A_{\mathrm{T}} - \lambda I_4) = \det \begin{pmatrix} 2-\lambda & 0 & 1 & -1 \\ 0 & 2-\lambda & 1 & 0 \\ 0 & 0 & 1-\lambda & -1 \\ 0 & 0 & 1 & 1-\lambda \end{pmatrix} \tag{VI.2.8}$$

from (VI.2.7)

$$= (2-\lambda)^2 [(1-\lambda)^2 + 1].$$

Now the (polynomial) expression $(2-\lambda)^2[(1-\lambda)^2 + 1]$ of (VI.2.8) equals zero when

(a) $$(2-\lambda)^2 = 0,$$

or when

(b) $$[(1-\lambda)^2 + 1] = 0.$$

Equation (a) is satisfied by the 2 (nondistinct) *real* scalars $\lambda_1 = 2$, $\lambda_2 = 2$. Equation (b) is satisfied by the 2 *complex* scalars $\lambda_3 = 1 + i$, $\lambda_4 = 1 - i$. At exactly this point, we are led to the necessity of distinguishing between **T** as an operator on *real* vectors of $V_4(\mathbf{R})$, or on the *complex* vectors of $V_4(\mathbf{C})$. Since det $(A_{\mathrm{T}} - \lambda I_4)$ is always a fourth-degree polynomial in the scalar variable λ, there will *always* be 4 (not necessarily distinct) *complex* values of λ for which det $(A_{\mathrm{T}} - \lambda I_4)$ equals zero; that is, there exist 4 complex roots. If the roots are all real, and if the matrix A_{T} has real entries, then **T** has corresponding eigenvectors in $V_4(\mathbf{R})$. If, however, all roots of the polynomial det $(A_{\mathrm{T}} - \lambda I_4)$ are *not* real (as in our example) we *cannot* be sure that it is always possible to find eigenvectors in $V_4(\mathbf{R})$ for the complex roots.

To return to our example, we find eigenvectors corresponding to the eigenvalues $\lambda_1 = 2$, $\lambda_2 = 2$, $\lambda_3 = 1 + i$, $\lambda_4 = 1 - i$.

▶ $\lambda_1 = \lambda_2 = 2$: Setting $\mathbf{T}(X) = 2 \cdot X$ for $X \in V_4$, we obtain from (VI.2.7)

$$\mathbf{T}\begin{pmatrix} x_1 \\ x_2 \\ x_3 \\ x_4 \end{pmatrix} = \begin{pmatrix} 2x_1 + x_3 - x_4 \\ 2x_2 + x_3 \\ x_3 - x_4 \\ x_3 + x_4 \end{pmatrix} = \begin{pmatrix} 2x_1 \\ 2x_2 \\ 2x_3 \\ 2x_4 \end{pmatrix}.$$

Matching corresponding coefficients of the last 2 vectors above, we obtain the homogeneous system

$$\begin{aligned} x_3 - x_4 &= 0 \\ x_3 &= 0 \\ -x_3 - x_4 &= 0 \\ x_3 - x_4 &= 0. \end{aligned}$$

The second equation tells us that $x_3 = 0$; hence (from the first equation, say), $x_4 = 0$. Notice that x_1 and x_2 are *totally unrestricted*. We conclude, then, that the eigenvectors for **T** corresponding to the eigenvalue $\lambda = 2$ are all of the form

$$X = \begin{pmatrix} x_1 \\ x_2 \\ 0 \\ 0 \end{pmatrix}.$$

This set of vectors forms a subspace of V_4, spanned by the 2 linearly independent vectors

$$\begin{pmatrix} 1 \\ 0 \\ 0 \\ 0 \end{pmatrix} \text{ and } \begin{pmatrix} 0 \\ 1 \\ 0 \\ 0 \end{pmatrix}.$$

Of course, any *other* basis would suffice to characterize $\mathscr{M}_{\lambda_1}$, the subspace of eigenvectors of **T** associated with the eigenvalue $\lambda_1 = 2$; for example,

$$\begin{pmatrix} 1 \\ 1 \\ 0 \\ 0 \end{pmatrix} \text{ and } \begin{pmatrix} 1 \\ -1 \\ 0 \\ 0 \end{pmatrix}.$$

▶ $\lambda_3 = 1 + i$: If we write

$$\mathbf{T}\begin{pmatrix} x_1 \\ x_2 \\ x_3 \\ x_4 \end{pmatrix} = \begin{pmatrix} 2x_1 + x_3 - x_4 \\ 2x_2 + x_3 \\ x_3 - x_4 \\ x_3 + x_4 \end{pmatrix} = \begin{pmatrix} (1+i)x_1 \\ (1+i)x_2 \\ (1+i)x_3 \\ (1+i)x_4 \end{pmatrix},$$

then, after equating corresponding coefficients of the vectors above, we are led to the homogeneous system of equations

$$\begin{aligned} (1-i)x_1 \qquad\qquad + x_3 - x_4 &= 0 \\ (1-i)x_2 + x_3 \qquad &= 0 \\ -ix_3 - x_4 &= 0 \\ x_3 - ix_4 &= 0. \end{aligned} \tag{VI.2.9}$$

This system may be solved, using the techniques of Chapter III where the coefficient matrix is reduced to a row-echelon form which yields an equivalent system [with the same solutions as the original system (VI.2.9)]. Thus, we would have, for example,

$$\underbrace{\begin{pmatrix} (1-i) & 0 & 1 & -1 \\ 0 & (1-i) & 1 & 0 \\ 0 & 0 & -i & -1 \\ 0 & 0 & 1 & -i \end{pmatrix}}_{\substack{\text{coefficient matrix} \\ \text{of (VI.2.9)}}} \xrightarrow{R_3{}^4(-i)} \underbrace{\begin{pmatrix} (1-i) & 0 & i & -1 \\ 0 & (1-i) & 1 & 0 \\ 0 & 0 & -i & -1 \\ 0 & 0 & 0 & 0 \end{pmatrix}}_{\substack{\text{coefficient matrix of} \\ \text{equivalent system}}} \tag{VI.2.10}$$

which shows that the coefficient matrix of (VI.2.9) is of rank 3, since the resulting matrix of (VI.2.10) has exactly 3 nonzero rows. Thus, the solution space for (VI.2.9) is of dimension $4-3=1$ (Theorem III.2.4). The "equivalent" system resulting from (VI.2.10) is exactly the original system (VI.2.9) except for the deletion of the fourth equation. We are at liberty to assign an arbitrary value to x_4: Since the solution space is 1-dimensional, this should fix x_1, x_2, and x_3. Let us choose $x_4=1$. Then the equations of (VI.2.9) force us to conclude that $x_1=1$, $x_2=\frac{1}{2}(1-i)$, and $x_3=i$. Thus X_3, an eigenvalue of $\mathbf{T}$ corresponding to the eigenvalue $\lambda_3=(1+i)$, is

$$X_3=\begin{pmatrix}1\\ \frac{1}{2}(1-i)\\ i\\ 1\end{pmatrix}.$$

Since the solution space for (VI.2.9) is 1-dimensional, any other eigenvector corresponding to $\lambda_3=(1+i)$ is a scalar multiple of this vector (see also Theorem VI.2.2). To eliminate the fraction $\frac{1}{2}$ in X_3, one might represent the eigenvector as

$$X_3'=2X_3=\begin{pmatrix}2\\ 1-i\\ 2i\\ 2\end{pmatrix}.$$

The student should verify for himself that X_3 (and X_3', too) performs as it ought to, that is, that $\mathbf{T}(X_3)=(1+i)X_3$ [and that $\mathbf{T}(X_3')=(1+i)X_3'$]. This small task is left as Exercise VI.2.6.

▶ $\lambda_4=(1-i)$: The problem of finding an $X_4\in V_n(\mathbf{C})$ such that $\mathbf{T}(X)=(1-i)X$ is left to the student (Exercise VI.2.7).

We observe that we have selected a very particular $n\times n$ matrix A_{T} to represent the linear operator $\mathbf{T}\in\mathscr{L}(V_n,V_n)$ in Theorem VI.2.4. Thus, in order to find the eigenvalues of the *operator* $\mathbf{T}$, we construct the $n\times n$ *matrix* $A_{\mathrm{T}}=mx(\mathbf{T})_{\mathscr{E}\mathscr{E}}$ and then compute the n roots of the nth degree polynomial $\det(A_{\mathrm{T}}-\lambda I_n)$. The natural question to pose at this point is the following: Will any *other* $n\times n$ matrix $A=mx(\mathbf{T})_{\mathscr{Y}\mathscr{Y}}$ give us the eigenvalues for the linear operator $\mathbf{T}$ as does A_{T}? The answer is, that the polynomial $\det(A_{\mathrm{T}}-\lambda I_n)$ is precisely the same polynomial as $\det(A-\lambda I_n)$, where $A=mx(\mathbf{T})_{\mathscr{Y}\mathscr{Y}}$ is *any* $n\times n$ matrix $\mathbf{T}$, relative to the arbitrary ordered basis $\mathscr{Y}=\{Y_1,Y_2,\ldots,Y_n\}$ for V_n. This fact is the assertion of the following theorem which tells us that the polynomial $\det(mx(\mathbf{T})_{\mathscr{Y}\mathscr{Y}}-\lambda I_n)$ is unaltered by change of basis $\mathscr{Y}$, and depends, therefore, *only* upon the linear operator $\mathbf{T}\in\mathscr{L}(V_n,V_n)$.

Theorem VI.2.4 For any ordered basis $\mathscr{Y}=\{Y_1,Y_2,\ldots,Y_n\}$ of V_n, $\det(mx(\mathbf{T})_{\mathscr{Y}\mathscr{Y}}-\lambda I_n)=\det(mx(\mathbf{T})_{\mathscr{E}\mathscr{E}}-\lambda I_n)$.

Proof: The following will be of importance here:

(a) $$\begin{aligned} mx(\mathbf{T})_{\mathscr{Y}\mathscr{Y}}&=mx(\mathbf{U}_{\mathscr{Y}})_{\mathscr{E}\mathscr{E}}\cdot mx(\mathbf{T})_{\mathscr{E}\mathscr{E}}\cdot mx(\mathbf{U}_{\mathscr{Y}}^{-1})_{\mathscr{E}\mathscr{E}} \qquad \text{(Theorem VI.1.4)}\\ &=U\cdot mx\mathbf{T}_{\mathscr{E}\mathscr{E}}\cdot U^{-1}\end{aligned}$$

where U is a shorthand symbol representing $mx(\mathbf{U}_{\mathscr{Y}})_{\mathscr{E}\mathscr{E}}$. Now let us consider the polynomial

$$\begin{aligned}
\det(mx(\mathbf{T})_{\mathscr{Y}\mathscr{Y}} - \lambda I_n) &= \det(U \cdot mx(\mathbf{T})_{\mathscr{E}\mathscr{E}} \cdot U^{-1} - \lambda I_n) && \text{[from (a)]} \\
&= \det(U \cdot mx(\mathbf{T})_{\mathscr{E}\mathscr{E}} \cdot U^{-1} - \lambda U \cdot U^{-1}) \\
&= \det(U \cdot (mx(\mathbf{T})_{\mathscr{E}\mathscr{E}} - \lambda I_n) \cdot U^{-1}) \\
& && \text{(distributivity of matrix multiplication)} \\
&= \det(U)\det(mx(\mathbf{T})_{\mathscr{E}\mathscr{E}} - \lambda I_n)\det(U^{-1}) \\
& && \text{(Theorem IV.4.1)} \\
&= \underbrace{\det(U)\det(U^{-1})}_{\downarrow}\det(mx(\mathbf{T})_{\mathscr{E}\mathscr{E}} - \lambda I_n) \\
&= \det(U \cdot U^{-1})\det(mx(\mathbf{T})_{\mathscr{E}\mathscr{E}} - \lambda I_n) \\
& && \text{(Theorem IV.4.1)} \\
&= \overset{1}{\det(I_n)}\det(mx(\mathbf{T})_{\mathscr{E}\mathscr{E}} - \lambda I_n) \\
&= \det(mx(\mathbf{T})_{\mathscr{E}\mathscr{E}} - \lambda I_n).
\end{aligned}$$

The theorem is proved.

As promised, we have shown that $\det(mx(\mathbf{T})_{\mathscr{Y}\mathscr{Y}} - \lambda I_n)$ depends *only* on the linear operator $\mathbf{T} \in \mathscr{L}(V_n, V_n)$, and *not* on the ordered basis $\mathscr{Y}$. For this reason, we are able to define $\det(mx(\mathbf{T})_{\mathscr{Y}\mathscr{Y}} - \lambda I_n)$ as a polynomial depending *only* on $\mathbf{T}$—the presence of the ordered basis $\mathscr{Y}$ in the symbol is, in a sense, an illusion, as a *change* of basis in no way affects the polynomial $\det(mx(\mathbf{T})_{\mathscr{Y}\mathscr{Y}} - \lambda I_n)$. The definition follows.

Definition VI.2.5 Let $\mathbf{T} \in \mathscr{L}(V_n, V_n)$. Then the nth degree polynomial

$$p_{\mathbf{T}}(\lambda) = \det(mx(\mathbf{T})_{\mathscr{Y}\mathscr{Y}} - \lambda \mathbf{1}_n),$$

is called the **characteristic polynomial of the linear operator T**, where $\mathscr{Y}$ is any ordered basis of V_n.

Similarly, if A is any $n \times n$ matrix in the algebra $\mathscr{M}_{n,n}$, then the nth degree polynomial

$$p_A(\lambda) = \det(A - \lambda I_n)$$

is called the **characteristic polynomial of the matrix** A.

Example Given $\mathbf{T} \in \mathscr{L}(V_2, V_2)$ defined by $\mathbf{T}\begin{pmatrix} x_1 \\ x_2 \end{pmatrix} = \begin{pmatrix} 2x_1 - ix_2 \\ x_1 - x_2 \end{pmatrix}$ for all $\begin{pmatrix} x_1 \\ x_2 \end{pmatrix} \in V_2(\mathbf{C})$, we shall compute $p_{\mathbf{T}}(\lambda) = \det(mx(\mathbf{T})_{\mathscr{Y}\mathscr{Y}} - \lambda \mathbf{1}_n)$ in 2 cases, namely, where $\mathscr{Y} = \left\{\begin{pmatrix} 1 \\ 0 \end{pmatrix}, \begin{pmatrix} 0 \\ 1 \end{pmatrix}\right\}$, and where $\mathscr{Y} = \left\{\begin{pmatrix} 1 \\ i \end{pmatrix}, \begin{pmatrix} 2 \\ 1 \end{pmatrix}\right\}$. As usual, $i^2 = -1$.

CASE 1. $\mathscr{Y} = \left\{\begin{pmatrix} 1 \\ 0 \end{pmatrix}, \begin{pmatrix} 0 \\ 1 \end{pmatrix}\right\} = \{E_1, E_2\}$.

Relative to the standard ordered basis $\mathscr{Y} = \mathscr{E}$, we have

$$mx(\mathbf{T})_{\mathscr{Y}\mathscr{Y}} = \begin{pmatrix} 2 & -i \\ 1 & -1 \end{pmatrix},$$

so that

$$\begin{aligned} \det\,(mx(\mathbf{T})_{\mathscr{Y}\mathscr{Y}} - \lambda I_2) &= \det\left(\begin{pmatrix} 2 & -i \\ 1 & -1 \end{pmatrix} - \begin{pmatrix} \lambda & 0 \\ 0 & \lambda \end{pmatrix}\right) \\ &= \det\begin{pmatrix} 2-\lambda & -i \\ 1 & -(1+\lambda) \end{pmatrix} \\ &= -(1+\lambda)(2-\lambda) + i \\ &= -[-\lambda^2 + \lambda + 2] + i \\ &= \lambda^2 - \lambda - 2 + i. \end{aligned} \qquad \text{(VI.2.11)}$$

CASE 2. $\mathscr{Y} = \left\{\begin{pmatrix} 1 \\ i \end{pmatrix}, \begin{pmatrix} 2 \\ 1 \end{pmatrix}\right\} = \{Y_1, Y_2\}.$

To construct $mx(\mathbf{T})_{\mathscr{Y}\mathscr{Y}}$, we may use the availability of $mx(\mathbf{T})_{\mathscr{E}\mathscr{E}}$ above, along with Theorem VI.1.4, which says that

$$mx(\mathbf{T})_{\mathscr{Y}\mathscr{Y}} = mx(\mathbf{U}_{\mathscr{Y}})_{\mathscr{E}\mathscr{E}} \cdot mx(\mathbf{T})_{\mathscr{E}\mathscr{E}} \cdot mx(\mathbf{U}_{\mathscr{Y}}^{-1})_{\mathscr{E}\mathscr{E}}. \qquad \text{(VI.2.12)}$$

Since $\mathbf{U}_{\mathscr{Y}}^{-1}(E_i) = Y_i$, $i = 1, 2$, we have

$$mx(\mathbf{U}_{\mathscr{Y}}^{-1})_{\mathscr{E}\mathscr{E}} = \begin{pmatrix} 1 & 2 \\ i & 1 \end{pmatrix} \quad \text{(Proposition VI.1.2)} \qquad \text{(VI.2.13)}$$

so that in computing $mx(\mathbf{U}_{\mathscr{Y}})_{\mathscr{E}\mathscr{E}}$, the inverse of (IV.2.13), we obtain

$$mx(\mathbf{U}_{\mathscr{Y}})_{\mathscr{E}\mathscr{E}} = \frac{1}{(1-2i)}\begin{pmatrix} 1 & -2 \\ -i & 1 \end{pmatrix}. \qquad \text{(VI.2.14)}$$

Substituting (VI.2.13) and (VI.2.14) in (VI.2.12) gives us

$$\begin{aligned} mx(\mathbf{T})_{\mathscr{Y}\mathscr{Y}} &= \frac{1}{(1-2i)}\begin{pmatrix} 1 & -2 \\ -i & 1 \end{pmatrix} \cdot \begin{pmatrix} 2 & -i \\ 1 & -1 \end{pmatrix} \cdot \begin{pmatrix} 1 & 2 \\ i & 1 \end{pmatrix} \\ &= \frac{1}{(1-2i)}\begin{pmatrix} 1+2i & 2-i \\ 1-4i & -4i \end{pmatrix} \\ &= \begin{pmatrix} \frac{1}{5}(-3+4i) & \frac{1}{5}(4+3i) \\ \frac{1}{5}(9-2i) & \frac{1}{5}(8-4i) \end{pmatrix}. \end{aligned} \qquad \text{(VI.2.15)}$$

The student may verify the validity of (VI.2.15) by checking whether the equations

$$\mathbf{T}(Y_1) = \tfrac{1}{5}(-3+4i)Y_i + \tfrac{1}{5}(9-2i)Y_2$$

and

$$\mathbf{T}(Y_2) = \tfrac{1}{5}(4+3i)Y_1 + \tfrac{1}{5}(8-4i)Y_2$$

as Definition II.5.1 and II.6.2 require. Recall that in our example,

$$Y_1 = \begin{pmatrix}1\\i\end{pmatrix},\ Y_2 = \begin{pmatrix}2\\1\end{pmatrix},\ \text{and } T\begin{pmatrix}x_1\\x_2\end{pmatrix} = \begin{pmatrix}2x_1 - ix_2\\x_1 - x_2\end{pmatrix} \text{ for all } \begin{pmatrix}x_1\\x_2\end{pmatrix} \in V_2(\mathbf{C}).$$

Finally, we are able to write

$$\det\,(mx(\mathbf{T})_{\mathscr{Y}\mathscr{Y}} - \lambda \mathbf{I}_2) = \det\left(\begin{pmatrix}a_{11} & a_{12}\\a_{21} & a_{22}\end{pmatrix} - \begin{pmatrix}\lambda & 0\\0 & \lambda\end{pmatrix}\right)$$

where $a_{11} = \frac{1}{5}(-3 + 4i)$, $a_{12} = \frac{1}{5}(4 + 3i)$, $a_{21} = \frac{1}{5}(9 - 2i)$, $a_{22} = \frac{1}{5}(8 - 4i)$ [see (VI.2.15)]. Thus,

$$\begin{aligned}\det\,(mx(\mathbf{T})_{\mathscr{Y}\mathscr{Y}} - \lambda I_2) &= \det\begin{pmatrix}a_{11} - \lambda & a_{12}\\a_{21} & a_{22} - \lambda\end{pmatrix}\\&= (a_{11} - \lambda)(a_{22} - \lambda) - a_{12}a_{21}\\&= \lambda^2 - (a_{11} + a_{22})\lambda + a_{11}a_{22} - a_{12}a_{21}.\end{aligned}$$

if we replace the a_{ij} terms by their respective values [see (IV.2.15)], we obtain

$$\det\,(mx(\mathbf{T})_{\mathscr{Y}\mathscr{Y}} - \lambda I_2) = \lambda^2 - (1)\lambda - 2 + i$$

which agrees with (VI.2.11).

EXERCISES

VI.2.1 Given $\mathbf{T} \in \mathscr{L}(V_2, V_2)$ defined by

$$\mathbf{T}\begin{pmatrix}x_1\\x_2\end{pmatrix} = \begin{pmatrix}0\\--x_2\end{pmatrix} \text{ for all } X = \begin{pmatrix}x_1\\x_2\end{pmatrix} \text{ in } V_2.$$

(a) Show that $X_0\begin{pmatrix}0\\1\end{pmatrix}$ is an eigenvector for $\mathbf{T}$ corresponding to the eigenvalue $\lambda_0 = -1$.

(b) Show that for any nonzero scalar α, αX_0 is again an eigenvector corresponding to the eigenvalue $\lambda_0 = -1$.

VI.2.2 Suppose $\mathbf{T} \in \mathscr{L}(V_n, V_n)$ and that X is a nonzero eigenvector for $\mathbf{T}$ corresponding to the eigenvalue λ, that is, $\mathbf{T}(X) = \lambda X$. Prove that for any and every nonzero scalar α, the vector αX is *also* an eigenvector for $\mathbf{T}$ corresponding to the eigenvalue λ. How does this result establish axioms B.1 and A.4 of Definition I.5.2, relative to $\mathscr{M}_{\lambda_0}$? (See proof of Theorem VI.2.2.)

VI.2.3 (a) Prove that any (nonzero) eigenvector X of a linear operator $\mathbf{T} \in \mathscr{L}(V_n, V_n)$, corresponds to 1 and *only* 1 eigenvalue λ.

(b) Is it true, conversely, that an eigenvalue always corresponds to 1 and only 1 eigenvector?

(c) If X and Y are eigenvectors corresponding to distinct eigenvalues λ and μ, respectively, then show that X and Y are necessarily linearly independent.

VI.2.4 Find the eigenvectors and eigenvalues for the following linear operators in $\mathscr{L}(V_n(\mathbf{R}), V_n(\mathbf{R}))$.

(a) $\mathbf{T}\begin{pmatrix} x_1 \\ x_2 \end{pmatrix} = \begin{pmatrix} 2x_1 \\ 3x_2 \end{pmatrix}$ for all $\begin{pmatrix} x_1 \\ x_2 \end{pmatrix}$ in $V_2(\mathbf{R})$.

(b) $\mathbf{T}\begin{pmatrix} x_1 \\ x_2 \end{pmatrix} = \begin{pmatrix} x_2 \\ x_1 \end{pmatrix}$ for all $\begin{pmatrix} x_1 \\ x_2 \end{pmatrix} \in V_2(\mathbf{R})$.

(c) $\mathbf{T}\begin{pmatrix} x_1 \\ x_2 \\ x_3 \end{pmatrix} = \begin{pmatrix} 0 \\ 3x_2 \\ -x_3 \end{pmatrix}$ for all $\begin{pmatrix} x_1 \\ x_2 \\ x_3 \end{pmatrix} \in V_3(\mathbf{R})$.

(d) $\mathbf{T}(X) = 0$ for all $X \in V_n(\mathbf{R})$.

(e) $\mathbf{T}\begin{pmatrix} x_1 \\ x_2 \end{pmatrix} = \begin{pmatrix} x_1 \\ 2x_1 + 3x_2 \end{pmatrix}$ for all $\begin{pmatrix} x_1 \\ x_2 \end{pmatrix} \in V_2(\mathbf{R})$.

(f) $\mathbf{T}\begin{pmatrix} x_1 \\ x_2 \\ x_3 \end{pmatrix} = \begin{pmatrix} -x_1 + 2x_3 \\ 2x_2 + x_3 \\ x_3 \end{pmatrix}$ for all $\begin{pmatrix} x_1 \\ x_2 \\ x_3 \end{pmatrix} \in V_3(\mathbf{R})$.

(g) $\mathbf{T}\begin{pmatrix} x_1 \\ x_2 \\ x_3 \end{pmatrix} = \begin{pmatrix} x_2 \\ x_3 \\ 0 \end{pmatrix}$ for all $\begin{pmatrix} x_2 \\ x_2 \\ x_3 \end{pmatrix} \in V_3(\mathbf{R})$.

VI.2.5 Find the eigenvectors and (possibly complex) eigenvalues for the following linear operators in $\mathscr{L}(V_n(\mathbf{C}), V_n(\mathbf{C}))$.

(a) $\mathbf{T}\begin{pmatrix} x_1 \\ x_2 \end{pmatrix} = \begin{pmatrix} ix_1 \\ x_2 \end{pmatrix}$ for all $\begin{pmatrix} x_1 \\ x_2 \end{pmatrix} \in V_2(\mathbf{C})$.

(b) $\mathbf{T}\begin{pmatrix} x_1 \\ x_2 \end{pmatrix} = \begin{pmatrix} -x_2 \\ x_1 \end{pmatrix}$ for all $\begin{pmatrix} x_1 \\ x_2 \end{pmatrix} \in V_2(\mathbf{C})$.

(c) $\mathbf{T}\begin{pmatrix} x_1 \\ x_2 \\ x_3 \end{pmatrix} = \begin{pmatrix} x_2 \\ x_3 \\ x_1 \end{pmatrix}$ for all $\begin{pmatrix} x_1 \\ x_2 \\ x_3 \end{pmatrix} \in V_3(\mathbf{C})$.

(d) $\mathbf{T}\begin{pmatrix} x_1 \\ x_2 \end{pmatrix} = \begin{pmatrix} x_1 - x_2 \\ x_1 + x_2 \end{pmatrix}$ for all $\begin{pmatrix} x_1 \\ x_2 \end{pmatrix} \in V_2(\mathbf{C})$.

(e) $\mathbf{T}\begin{pmatrix} x_1 \\ x_2 \\ x_3 \end{pmatrix} = \begin{pmatrix} 2x_1 \\ ix_1 + 2x_2 \\ x_2 + ix_3 \end{pmatrix}$ for all $\begin{pmatrix} x_1 \\ x_2 \\ x_3 \end{pmatrix} \in V_3(\mathbf{C})$.

(f) $\mathbf{T} \in \mathscr{L}(V_3, V_3)$ defined by (VI.2.5).

VI.2.6 For $\mathbf{T} \in \mathscr{L}(V_4(\mathbf{C}), V_4(\mathbf{C}))$ defined by (VI.2.7,) show that

$$\begin{pmatrix} 1 \\ \frac{1}{2}(1 - i) \\ i \\ 1 \end{pmatrix}$$

is an eigenvector for $\mathbf{T}$ corresponding to the eigenvalue $\lambda = 1 + i$.

VI.2.7 For $\mathbf{T} \in \mathscr{L}(V_4(\mathbf{C}), V_4(\mathbf{C}))$ defined by (VI.2.7), find an eigenvector for $\mathbf{T}$ corresponding to the eigenvalue $\lambda = 1 - i$.

VI.2.8 Recall that an $n \times n$ matrix is called upper (respectively lower) triangular if all entries below (respectively above) the main diagonal are zero. Suppose for $\mathbf{T} \in \mathscr{L}(V_n, V_n)$, the $n \times n$ matrix $A_{\mathbf{T}} = mx(\mathbf{T})_{\mathscr{E}\mathscr{E}}$ is either upper triangular or lower triangular. Prove that the eigenvalues for $\mathbf{T}$ are just the n main diagonal entries of $A_{\mathbf{T}}$. Moreover, at least 1 of the eigenvectors of $\mathbf{T}$ is a standard basis vector of V_n.

VI.2.9 Suppose $\mathbf{T}$ and $\mathbf{U}$ belong to $\mathscr{L}(V_n, V_n)$ and that $\mathbf{T}$ and $\mathbf{U}$ commute with each other, that is, $\mathbf{T} \circ \mathbf{U} = \mathbf{U} \circ \mathbf{T}$. Prove that if $X \in V_n$ is an eigenvector for $\mathbf{T}$, corresponding to the eigenvalue λ, then $\mathbf{U}(X)$ is also an eigenvector for $\mathbf{T}$ corresponding to the eigenvalue λ. In other words, you are to show that $\mathbf{U}$ sends any vector of the subspace $\mathscr{M}_\lambda$ for $\mathbf{T}$, into another vector of $\mathscr{M}_\lambda$. That is, if $\mathscr{M}_\lambda$ is the set of all X such that $\mathbf{T}(X) = \lambda X$ (with O included), then $\mathbf{U}$ leaves $\mathscr{M}_\lambda$ invariant as long as $\mathbf{U}$ commutes with $\mathbf{T}$.

VI.2.10 Suppose λ_0 is an eigenvalue for $\mathbf{T} \in \mathscr{L}(V_n, V_n)$ corresponding to some eigenvector X_0 in V_n.

(a) Prove that $(\lambda_0)^2$ is an eigenvalue of $\mathbf{T}^2 = \mathbf{T} \circ \mathbf{T}$ corresponding to the (same) eigenvector X_0.

(b) Prove that $(\lambda_0)^3$ is an eigenvalue of $\mathbf{T}^3$ corresponding to the eigenvector X_0.

(c) If $\mathbf{T} \in \mathscr{L}(V_n, V_n)$ and if $\mathbf{T}^k = \mathbf{0}$, the zero operator, for some positive integer k, then show that the zero scalar is the only possible eigenvalue for $\mathbf{T}$.

VI.2.11 For the following $n \times n$ matrices, find the characteristic polynomial. Compare the constant term of the characteristic polynomial to the determinant of each matrix.

(a) $A = \begin{pmatrix} 1 & 2 \\ 2 & 1 \end{pmatrix}.$

(b) $B = \begin{pmatrix} 1 & 0 \\ 2 & 1 \end{pmatrix} \cdot \begin{pmatrix} 1 & 2 \\ 2 & 1 \end{pmatrix} \cdot \begin{pmatrix} 1 & 0 \\ 2 & 1 \end{pmatrix}^{-1}.$

(c) $B = \begin{pmatrix} 1 & 0 \\ 2 & 1 \end{pmatrix} \cdot \begin{pmatrix} 1 & 2 \\ 2 & 1 \end{pmatrix} \cdot \begin{pmatrix} 1 & 0 \\ -2 & 1 \end{pmatrix}.$

(d) $C = \begin{pmatrix} 1 & 2 \\ 2 & 4 \end{pmatrix}.$

(e) $A \cdot C$ and $C \cdot A$ [see (a) and (d)].

(f) $D = \begin{pmatrix} 1 & 0 & -3 \\ 0 & 1 & 3 \\ 1 & 2 & 4 \end{pmatrix}.$

(g) $E = \begin{pmatrix} 0 & 1 & 0 \\ 2 & 2 & 2 \\ 1 & 3 & 1 \end{pmatrix}.$

(h) $D \cdot E \cdot D^{-1}$.

(i) $D \cdot E$ and $E \cdot D$.

VI.2.12 (See Exercise VI.2.11.) Show that for any $n \times n$ matrix A, the constant term a_0 of the characteristic polynomial $p_A(\lambda) = (-\lambda)^n + a_{n-1}\lambda^{n-1} + \cdots + a_2\lambda^2 + a_1\lambda + a_0$, is equal to det (A), the determinant of A.

VI.2.13 Show that a linear operator **T** (and hence its induced $n \times n$ matrix A_{T}) is invertible if and only if the scalar 0 is *not* an eigenvalue of **T**.

3. *Diagonalizable Operators*

Let us compare the 2 linear operators, **T** and **U**, in $\mathscr{L}(V_2, V_2)$ defined by

$$\mathbf{T}\begin{pmatrix} x_1 \\ x_2 \end{pmatrix} = \begin{pmatrix} x_1 \\ ax_1 + x_2 \end{pmatrix} = \underbrace{\begin{pmatrix} 1 & 0 \\ a & 1 \end{pmatrix}}_{A_{\mathrm{T}}} \cdot \begin{pmatrix} x_1 \\ x_2 \end{pmatrix} \quad \text{for all } \begin{pmatrix} x_1 \\ x_2 \end{pmatrix} \in V_2,\ a \neq 0,$$

and

$$\mathbf{U}\begin{pmatrix} x_1 \\ x_2 \end{pmatrix} = \begin{pmatrix} x_1 \\ x_2 \end{pmatrix} = \underbrace{\begin{pmatrix} 1 & 0 \\ 0 & 1 \end{pmatrix}}_{A_{\mathrm{U}}} \cdot \begin{pmatrix} x_1 \\ x_2 \end{pmatrix} \quad \text{for all } \begin{pmatrix} x_1 \\ x_2 \end{pmatrix} \in V_2.$$

We observe that the characteristic polynomials for each operator p_{T} and p_{U} (or $p_{A_{\mathrm{T}}}$ and $p_{A_{\mathrm{U}}}$) are identical (see Definition VI.2.5), that is, $p_{\mathrm{T}}(\lambda) = p_{\mathrm{U}}(\lambda) = (1 - \lambda)^2$ for all scalars λ. The (nondistinct) pair of roots for this common polynomial are the scalars 1 and 1. Thus, 1 is the only eigenvalue possible for both **T** and **U** (Theorem VI.2.3).

How do the respective eigen*vectors* for **T** and **U** compare? At this point, we discern a distinct difference between the personalities of **T** and **U**. In a sense, **U** has "more" eigenvectors than **T**. More exactly, the subspace $\mathscr{M}^{\mathrm{U}}$ of V_2, which is generated (spanned) by the eigenvectors of **U** is 2-dimensional (and hence, $\mathscr{M}^{\mathrm{U}}$ is *all* of V_2), while the subspace $\mathscr{M}^{\mathrm{T}}$ of V_2, generated by the eigenvectors of **T**, is only 1-dimensional (see Theorem VI.2.2). In fact, from the defining equation for **T**, we see that $\mathbf{T}(X) = 1 \cdot X$ only for $X = \begin{pmatrix} x_1 \\ x_2 \end{pmatrix}$ where $x_1 = 0$; that is, any eigenvector for **T** is merely a scalar multiple of the vector $\begin{pmatrix} 0 \\ 1 \end{pmatrix}$, so that the subspace $\mathscr{M}^{\mathrm{T}}$ generated by $\begin{pmatrix} 0 \\ 1 \end{pmatrix}$ has $\left\{\begin{pmatrix} 0 \\ 1 \end{pmatrix}\right\}$ as its basis. Thus, $\mathscr{M}^{\mathrm{T}}$ is only 1-dimensional. On the other hand, $\mathbf{U}(X) = 1 \cdot X$ for any and all $X \in V_2$. $\mathscr{M}^{\mathrm{U}}$, the subspace generated by the nonzero eigenvectors of **U**, is spanned by *any* 2 (nonzero) linearly independent vectors. Thus, $\mathscr{M}^{\mathrm{U}}$ is 2-dimensional. Thus, we have an example of an operator **U**, the identity operator on V_2, whose eigenvectors are sufficiently abundant that we may form a *basis* from them. By contrast, we have an operator **T** whose eigenvectors all lie in a 1-dimensional space; the eigenvectors of **T** will never constitute an authentic basis for V_2.

QUESTION 1. Which linear operators in $\mathscr{L}(V_n, V_n)$ have sufficiently many eigenvectors (not all necessarily corresponding to the *same* eigenvalue) so that a *basis* for V_n can be selected from them? In other words, when does a linear operator on V_n account for n linearly independent eigenvectors? (Not *all* operators are so generously endowed as the operator $\mathbf{T} \in \mathscr{L}(V_2, V_2)$ above illustrates.)

QUESTION 2. What is so wonderful about linear operators on V_n which *have* as many as n linearly independent eigenvectors?

In answer to Question 2, our interest in operators **T** on V_n, whose eigenvectors contain a basis for V_n, is due to the following (yet to be proven) fact: $mx(\mathbf{T})_{\mathscr{X}\mathscr{X}}$, the $n \times n$ matrix of **T** relative to the ordered basis $\mathscr{X} = \{X_1, X_2, \ldots, X_n\}$, is a *diagonal* matrix (all entries off the main diagonal are zero) if and only if the basis $\mathscr{X}$ consists entirely of eigenvectors of **T** (Theorem VI.3.2). To say that $\mathscr{X}$ is a basis of eigenvectors of **T**, is to say, then, that **T** has n linearly independent eigenvectors. Question 1 asks: Which operators **T** on V_n, do have n linearly independent eigenvectors (and hence, have a diagonal matrix representation)? A *partial* answer is those $\mathbf{T} \in \mathscr{L}(V_n, V_n)$ whose n eigenvalues [roots of the nth degree characteristic polynomial $p_{\mathbf{T}}(\cdot)$] are *distinct* from one another (Theorem VI.3.4).

Before we prove that eigenvector bases always induce diagonal matrices, and vice versa, we ought to say precisely what we mean by diagonalization relative to a linear operator.

Definition VI.3.1 A linear operator $\mathbf{T} \in \mathscr{L}(V_n, V_n)$ is called **diagonalizable**, or **diagonalable** if and only if relative to at least 1 ordered basis $\mathscr{X} = \{X_1, X_2, \ldots, X_n\}$ for V_n, the $n \times n$ matrix induced by **T** relative to $\mathscr{X}$ is a diagonal matrix, that is,

$$mx(\mathbf{T})_{\mathscr{X}\mathscr{X}} = \text{diag}\,(\lambda_1, \lambda_2, \ldots, \lambda_n) = \begin{pmatrix} \lambda_1 & 0 & \cdots & 0 \\ 0 & \lambda_2 & & \vdots \\ \vdots & & \ddots & \\ 0 & \cdots & & \lambda_n \end{pmatrix}$$

for at least one ordered basis $\mathscr{X}$.

Similarly, an $n \times n$ matrix A is called **diagonalizable**, or **diagonable** if and only if for at least 1 invertible $n \times n$ matrix S, $S^{-1} \cdot A \cdot S$ is a diagonal matrix, that is,

$$S^{-1} \cdot A \cdot S = \begin{pmatrix} \lambda_1 & 0 & \cdots & 0 \\ 0 & \lambda_2 & \cdots & 0 \\ \vdots & & \ddots & \vdots \\ 0 & 0 & \cdots & \lambda_n \end{pmatrix} = \text{diag}\,(\lambda_1,\lambda_2,\ldots\lambda_n)$$

for certain scalars $\lambda_1, \lambda_2, \ldots, \lambda_n$.

We state the theorem which says that the only time $mx(\mathbf{T})_{\mathscr{X}\mathscr{X}}$ is a diagonal matrix is when the n linearly independent vectors of the basis $\mathscr{X}$ are all eigenvectors for $\mathbf{T}$. Conversely, if $\mathscr{X}$ is a basis of eigenvectors, then $mx(\mathbf{T})_{\mathscr{X}\mathscr{X}}$ is necessarily diagonal.

Theorem VI.3.2 Let $\mathbf{T} \in \mathscr{L}(V_n,V_n)$. Then $mx(\mathbf{T})_{\mathscr{X}\mathscr{X}}$, the $n \times n$ matrix of $\mathbf{T}$ relative to the ordered basis $\mathscr{X} = \{X,X,\ldots,X\}$, is a diagonal matrix (that is, $\mathbf{T}$ is diagonalizable) $\Leftrightarrow$ each vector X_i of the basis $\mathscr{X}$ is an eigenvector for $\mathbf{T}$. Moreover, the diagonal entries, $\lambda_1, \lambda_2, \ldots, \lambda_n$, of $mx(\mathbf{T})_{\mathscr{X}\mathscr{X}} = \text{diag}\,(\lambda_1,\lambda_2,\ldots,\lambda_n)$ are exactly the (not necessarily distinct) eigenvalues of $\mathbf{T}$ corresponding to the eigenvectors $X_1, X_2,\ldots,X_n$, respectively; that is, $\mathbf{T}(X_i) = \lambda_i X_i$ for each $i = 1, 2,\ldots, n$.

Proof: In consulting Definition II.6.2, we see that relative to the ordered basis $\mathscr{X} = \{X_1,X_2,\ldots,X_n\}$, $mx(\mathbf{T})_{\mathscr{X}\mathscr{X}} = (a_{ij})_{ij}$ if and only if

$$\begin{aligned} \mathbf{T}(X_1) &= a_{11}X_1 + a_{21}X_2 + \cdots + a_{n1}X_n \\ \mathbf{T}(X_2) &= a_{12}X_1 + a_{22}X_2 + \cdots + a_{n2}X_n \\ &\vdots \\ \mathbf{T}(X_n) &= a_{1n}X_1 + a_{2n}X_2 + \cdots + a_{nn}X_n. \end{aligned} \tag{VI.3.1}$$

Now let us ask ourselves the question: When is $mx(\mathbf{T})_{\mathscr{X}\mathscr{X}}$ a diagonal matrix, and when is it not? In terms of the action of $\mathbf{T}$ upon the basis vectors $X_1, X_2,\ldots, X_n$, we see from (VI.3.1) that

$$mx(\mathbf{T})_{\mathscr{X}\mathscr{X}} \text{ is a diagonal matrix} \Leftrightarrow \mathbf{T}(X_i) = a_{ii}X_i \quad \text{for each } i = 1, 2,\ldots, n. \tag{VI.3.2}$$

We can be even more specific than (VI.3.2). We can [from (VI.3.1)] state exactly *which* diagonal matrix is involved; to wit,

$$mx(\mathbf{T})_{\mathscr{X}\mathscr{X}} = \text{diag}\,(a_{11},a_{22},\ldots,a_{nn}) \Leftrightarrow \mathbf{T}(X_i) = a_{ii}X_i, \quad \text{for each } i = 1, 2,\ldots, n. \tag{VI.3.3}$$

But the right-hand side of (VI.3.3) says that each basis vector, X_i, happens to be an eigenvector of $\mathbf{T}$ corresponding to the eigenvalue a_{ii} (Definition

VI.2.1). If we set $a_{ii} = \lambda_i$ for each $i = 1, 2, \ldots, n$, then we have the theorem proved in the form given. That is, we have shown that $mx(\mathbf{T})_{\mathscr{X}\mathscr{X}} =$ diag $(\lambda_1, \lambda_2, \ldots, \lambda_n)$ if and only if the ordered basis $\mathscr{X} = \{X_1, X_2, \ldots, X_n\}$, happens to be a basis of *eigen*vectors for $\mathbf{T}$, where each X_i corresponds to the eigenvalue λ_i appearing on the main diagonal of $mx(\mathbf{T})_{\mathscr{X}\mathscr{X}}$.

Example The identity operator $\mathbf{I}_n \in \mathscr{L}(V_n, V_n)$ defined by $\mathbf{I}_n(X) = X$ for all $X \in V_n$ is diagonalizable. In fact, *every* X in V_n is an eigenvector corresponding to the eigenvalue 1, and for any bases $\mathscr{X}$, $mx(\mathbf{I}_n)_{\mathscr{X}\mathscr{X}} =$ diag $(1, 1, \ldots, 1)$.

Example Define $\mathbf{T} \in \mathscr{L}(V_3, V_3)$ by

$$\mathbf{T}\begin{pmatrix} x_1 \\ x_2 \\ x_3 \end{pmatrix} = \underbrace{\begin{pmatrix} 3 & -1 & 0 \\ -1 & 3 & 0 \\ 0 & 0 & 3 \end{pmatrix}}_{A_{\mathbf{T}}} \begin{pmatrix} x_1 \\ x_2 \\ x_3 \end{pmatrix} \begin{pmatrix} 3x_1 - x_2 \\ -x_1 + 3x_2 \\ 3x_3 \end{pmatrix}$$

for all

$$\begin{pmatrix} x_1 \\ x_2 \\ x_3 \end{pmatrix} \in V_3.$$

Now $\mathbf{T}$ has 3 distinct eigenvalues (roots of the characteristic polynomial $p_{\mathbf{T}}(\lambda) = (3 - \lambda)[(3 - \lambda)^2 - 1]$—see Definition VI.2.5 and Theorem VI.3.3) $\lambda_1 = 2$, $\lambda_2 = 3$, and $\lambda_3 = 4$. Eigenvectors corresponding to these eigenvalues are, respectively,

$$X_1 = \begin{pmatrix} 1 \\ 1 \\ 0 \end{pmatrix}, X_2 = \begin{pmatrix} 0 \\ 0 \\ 1 \end{pmatrix}, \text{ and } X_3 = \begin{pmatrix} 1 \\ -1 \\ 0 \end{pmatrix}; \tag{VI.3.4}$$

that is,

$$\mathbf{T}(X_i) = \lambda_i X_i, \; i = 1, 2, 3.$$

Of course, any nonzero scalar multiple of these eigenvectors is still an eigenvector (Theorem VI.2.2), so that we do not have uniqueness of eigenvectors. Observe that the ordered set of eigenvectors $\mathscr{X} = \{X_1, X_2, X_3\}$, is actually a *basis* for V_n, since they form a linearly independent set (check this). Relative to this basis $\mathscr{X} = \{X_1, X_2, X_3\}$, let us construct $mx(\mathbf{T})_{\mathscr{X}\mathscr{X}}$. Comparing (VI.3.4) with Definition II.6.2, we are forced to conclude,

$$mx(\mathbf{T})_{\mathscr{X}\mathscr{X}} = \begin{pmatrix} 2 & 0 & 0 \\ 0 & 3 & 0 \\ 0 & 0 & 4 \end{pmatrix} = \text{diag } (2, 3, 4).$$

Thus, $\mathbf{T}$ is diagonalizable and, in accordance with Theorem VI.3.2, the basis $\mathscr{X}$ which induces the diagonal form, is a basis of eigenvectors. Observe, too, that the eigenvalues of $\mathbf{T}$ appear as the diagonal entries.

We provide a test which determines when a linear operator on V_n has as many as n linearly independent eigenvectors. A sufficient (but not necessary) condition is that the n eigenvalues of $\mathbf{T} \in \mathscr{L}(V_n, V_n)$ are distinct (see example above). The following theorem shows that k distinct eigenvalues account for at least as many linear independent corresponding eigenvectors. A complete *basis* of eigenvectors is forced, therefore, when k is set equal to n (Corollary VI.3.4).

Theorem VI.3.3 Suppose $\mathbf{T} \in \mathscr{L}(V_n, V_n)$ and that $X_1, X_2, \ldots, X_k$ are distinct eigenvectors of $\mathbf{T}$ corresponding to the eigenvalues $\lambda_1, \lambda_2, \ldots, \lambda_k$, respectively. If the eigenvalues are *distinct* ($\lambda_i \neq \lambda_j$ when $i \neq j$), then the set of eigenvectors $\{X_1, X_2, \ldots, X_k\}$ is necessarily linearly independent.

Proof: Our approach is actually a thinly disguised proof by induction (see Appendix A). We shall see the following:

▶ STEP A. Under the assumption of k *distinct* eigenvalues $\lambda_1, \lambda_2, \ldots, \lambda_k$, we shall show that if we assume further, that the k-element set of eigenvectors, $\{X_1, X_2, \ldots, X_k\}$, is *linearly dependent*, then removal of a certain one of the vectors, X_k, say, always leaves a $(k-1)$ element set, $\{X_1, X_2, \ldots, X_{k-1}\}$, which is, again, *linearly dependent*. That is, we shall show that if the k eigenvalues are distinct, then (renumbering if necessary) linear dependence of the original set implies linear dependence of a "deleted" set, that is,

$$\{X_1, X_2, \ldots, X_{k-1}, X_k\}\text{LD} \Rightarrow \{X_1, X_2, \ldots, X_{k-1}\}\text{LD}.$$

▶ STEP B. We observe from Theorem I.6.6, that any 1-element set $\{X\}$, where $X \neq 0$, is always linearly independent. Therefore, for the special case where $X = X_i$ is a (necessarily nonzero) eigenvector for $\mathbf{T}$, the 1-element set $\{X_i\}$ is always linearly independent.

Step A in conjunction with Step B leads us to the unavoidable conclusion that the original set of eigenvectors $\{X_1, X_2, \ldots, X_k\}$ must have been linearly independent to begin with, not linearly dependent; this will prove our theorem. Why?

Let us actually take the bold step of assuming that $\{X_1, X_2, \ldots, X_k\}$ is linearly dependent. Step A says that by throwing away a certain well-chosen vector, X_k say, our depleted set is linearly dependent also. Repeat this procedure; that is, let us throw away another strategic vector, X_{k-1} say, from the linearly dependent set $\{X_1, X_2, \ldots, X_{k-1}\}$ to obtain a "smaller" linearly dependent set $\{X_1, X_2, \ldots, X_{k-2}\}$ with only $k-2$ eigenvectors. Step A tells us that we can *always* produce a "smaller" linearly dependent set, provided the original set of eigenvectors is linearly dependent. If we persist in disposing of eigenvectors, we eventually are left with a linearly dependent set with only 1 eigenvector. But, in contradiction to this, Step B says that single eigenvectors always form a linearly *independent* set $\{X_i\}$.

A contradiction has materialized. And whose fault is it? The fault lies with anyone who assumes what we have assumed, namely:

(1) The eigenvalues $\{\lambda_1,\lambda_2,\ldots,\lambda_k\}$ are pairwise distinct, and

(2) The set of corresponding eigenvectors $\{X_1,X_2,\ldots,X_k\}$ is linearly dependent.

Since assumptions (1) and (2) *together* lead to the contraction that a certain set of eigenvectors forms a linearly dependent (via Step A), and that all 1-element sets of eigenvectors are linearly independent (via Step B), *one* of these assumptions may not stand. Assumption (2) must fall since we wish to analyze operators with distinct eigenvalues, and our theorem is proven—once Step A and Step B are proven. And this is our next task.

▶ *Proof of Step A:* If the set of nonzero eigenvectors $\{X_1,X_2,\ldots,X_k\}$ is linearly dependent, then 1 of the vectors can be written as a linear combination of the other $k-1$ vectors (Definition I.6.3). By renumbering if necessary, let us say that X_k is a linear combination of the $k-1$ remaining eigenvectors $X_1, X_2,\ldots, X_{k-1}$. That is to say, we can find scalars, $\alpha_1, \alpha_2,\ldots, \alpha_{k-1}$, not all equal to zero (although *some* of the α_i's could equal zero), such that

$$X_k = \alpha_1 X_1 + \alpha_2 X_2 + \cdots + \alpha_{k-1} X_{k-1}. \tag{VI.3.5}$$

Now we use the fact that each X_i is an eigenvector for $\mathbf{T}$, that is, $\mathbf{T}(X_i) = \lambda_i X_i$ for each $i = 1, 2,\ldots, n$. If we apply $\mathbf{T}$ to both sides of (VI.3.5), and replace $\mathbf{T}(X_i)$ by $\lambda_i X_i$ for each i (as our last equation allows us to do), we obtain

$$\lambda_k X_k = \alpha_1\lambda_1 X_1 + \alpha_2\lambda_2 X_2 + \cdots + \alpha_{k-1}\lambda_{k-1} X_{k-1}. \tag{VI.3.6}$$

Now if we simply multiply both sides of (VI.3.5) through by the scalar λ_k, we obtain

$$\lambda_k X_k = \alpha_1\lambda_k X_1 + \alpha_2\lambda_k X_2 + \cdots + \alpha_{k-1}\lambda_k X_{k-1}. \tag{VI.3.7}$$

We now subtract corresponding sides of (VI.3.7) from (VI.3.6) to obtain

$$O = \alpha_1(\lambda_1 - \lambda_k)X_1 + \alpha_2(\lambda_2 - \lambda_k)X_2 + \cdots + \alpha_i(\lambda_i - \lambda_k)X_i + \cdots + \alpha_{k-1}(\lambda_{k-1} - \lambda_k)X_{k-1}. \tag{VI.3.8}$$

To say that (VI.3.8) is valid, is to say that the "smaller" set of $k-1$ eigenvectors $\{X_1,X_2,\ldots,X_{k-1}\}$ is linearly dependent (Theorem I.6.5) since, indeed, not all the scalar coefficients in (VI.3.8) equal zero. To see this, we observe that at least 1 of the α_i's is nonzero [see (VI.3.5)] since X_k is a nonzero vector (as are all eigenvectors). Secondly, by assuming distinct eigenvalues, $\lambda_i - \lambda_k$ is never zero for any and all $i = 1, 2,\ldots, k-1$. Thus, the product, $\alpha_k(\lambda_i - \lambda_k)$, is a nonzero scalar for at least 1

$i \in \{1, 2, \ldots, k-1\}$. This means that at least 1 of the scalar coefficients in (VI.6.8) is nonzero which, from Theorem I.6.5, establishes the linear *dependence* of the set $\{X_1, X_2, \ldots, X_{k-1}\}$.

▶ *Proof of Step B:* It is immediate that 1-element sets $\{X\}$, where X is a nonzero vector (eigenvector or not), are always linearly independent. This follows directly from Theorem II.6.6.

As our preview of strategy indicated at the start of this proof, we have a contradiction: Step A says that if we start with a LD set of eigenvectors, we can continue to throw away vectors and yet preserve the LD property—even down to the case where only 1 eigenvector remains. But single eigenvectors form LI sets (Step B), which is a contradiction if ever there was one. The assumption of linear dependence on the eigenvectors must be dropped. But to be convinced that distinct eigenvalues imply linearly independent eigenvectors, is to have proved the theorem.

Recall that we were seeking a condition on the operator $\mathbf{T} \in \mathscr{L}(V_n, V_n)$ which would suffice (but would not be necessary) in order that $\mathbf{T}$ have n linearly independent eigenvectors which, in turn, would mean that $\mathbf{T}$ is diagonalizable. (Theorem VI.3.2.) From Theorem VI.3.3, we have Corollary VI.3.4.

Corollary VI.3.4 If $\mathbf{T} \in \mathscr{L}(V_n, V_n)$ has exactly n *distinct* eigenvalues $\lambda_1, \lambda_2, \ldots. \lambda_n$, then $\mathbf{T}$ has n linearly independent eigenvectors, $X_1, X_2, \ldots, X_n$, in V_n. Relative to this ordered basis $\mathscr{X} = \{X_1, X_2, \ldots, X_n\}$ of eigenvectors, where $\mathbf{T}(X_i) = \lambda_i X_i$, $mx(\mathbf{T})_{\mathscr{X}\mathscr{X}} = \text{diag}(\lambda_1, \lambda_2, \ldots, \lambda_n)$.

Proof: The first statement of this theorem is exactly the statement of our previous result, Theorem VI.3.3, where $k = n$. The second statement is a repeat performance on the part of Theorem VI.3.2.

There we have it. We now know that if a linear operator on V_n has n distinct eigenvalues (that is, the nth degree characteristic polynomial $p_{\mathrm{T}}(\lambda) = \det(A_{\mathrm{T}} - \lambda I_n)$ has n distinct roots; see Theorem VI.2.3), then relative to an ordered basis $\mathscr{X} = \{X_1, X_2, \ldots, X_n\}$ of eigenvectors, $mx(\mathbf{T})_{\mathscr{X}\mathscr{X}}$ is a diagonal $n \times n$ matrix. Corollary VI.3.4 gives us good reason to regard with some kindness those linear operators whose characteristic polynomials issue forth n *distinct* (possibly complex) roots.

The following result, which ties our previous results together, uses Theorem VI.1.4, which prescribes a method for constructing $mx(\mathbf{T})_{\mathscr{X}\mathscr{Y}}$ ($\mathscr{X}$ and $\mathscr{Y}$ need not be equal, nor even in the same vector-space).

Theorem VI.3.5 For $\mathbf{T} \in \mathscr{L}(V_n, V_n)$, where $A_{\mathrm{T}} = mx(\mathbf{T})_{\mathscr{E}\mathscr{E}}$, the matrix of $\mathbf{T}$ relative to the standard ordered basis $\mathscr{E}$, the following conditions are equivalent to each other:

(1) **T** is diagonalizable
(2) **T** has n linearly independent eigenvectors X_i, where $\mathbf{T}(X_i) = \lambda_i X_i$, $i = 1, 2, \ldots, n$
(3) diag $(\lambda_1, \lambda_2, \ldots, \lambda_n) = S^{-1} \cdot A_{\mathrm{T}} \cdot S$, where $S = [X_1 X_2 \cdots X_n]$ is the invertible $n \times n$ matrix whose ith column agrees with the vector X_i, which is, moreover, an eigenvector for **T**. In short, $mx(\mathbf{T})_{\mathscr{E}\mathscr{E}}$ is diagonalizable.

Proof: The equivalence of (1) and (2) has already been established in Theorem VI.3.2.

We now show that (1) is equivalent to (3). From Proposition VI.1.2, we see that the $n \times n$ matrix $S = [X_1 X_2 \cdots X_n]$ is exactly $mx(\mathbf{U}_{\mathscr{X}})^{-1}_{\mathscr{E}\mathscr{E}}$, where $\mathbf{U}_{\mathscr{X}} \in \mathscr{L}(V_n, V_n)$ is defined by $\mathbf{U}_{\mathscr{X}}(X_i) = E_i$. Thus, $S^{-1} = mx(\mathbf{U}_{\mathscr{X}})_{\mathscr{E}\mathscr{E}}$ and, consequently,

$$S^{-1} \cdot A_{\mathrm{T}} \cdot S = mx(\mathbf{U}_{\mathscr{X}})_{\mathscr{E}\mathscr{E}} \cdot mx(\mathbf{T})_{\mathscr{E}\mathscr{E}} \cdot mx(\mathbf{U}_{\mathscr{X}})^{-1}_{\mathscr{E}\mathscr{E}}.$$

But from Theorem VI.1.4, the right-hand side of this equation is equal to $mx(\mathbf{T})_{\mathscr{X}\mathscr{X}}$, the matrix of **T** relative to the ordered basis $\mathscr{X} = \{X_1, X_2, \ldots, X_n\}$. That is,

$$S^{-1} \cdot A_{\mathrm{T}} \cdot S = mx(\mathbf{T})_{\mathscr{X}\mathscr{X}}. \tag{VI.3.9}$$

With this information in hand, we prove first that (3) $\Rightarrow$ (1), and then that (1) $\Rightarrow$ (3).

(3) $\Rightarrow$ (1): If $S^{-1} \cdot A \cdot S = \text{diag}\,(\lambda_1, \lambda_2, \ldots, \lambda_n)$, then, from (VI.3.9) we are saying that

$$mx(\mathbf{T})_{\mathscr{X}\mathscr{X}} = \text{diag}\,(\lambda_1, \lambda_2, \ldots, \lambda_n)$$

which says that **T** is diagonalizable (Definition VI.3.1). Thus, (3) $\Rightarrow$ (1).

(1) $\Rightarrow$ (3): If **T** is diagonalizable, then, for some ordered basis $\mathscr{X} = \{X_1, X_2, \ldots, X_n\}$,

$$mx(\mathbf{T})_{\mathscr{X}\mathscr{X}} = \text{diag}\,(\lambda_1, \lambda_2, \ldots, \lambda_n).$$

From Theorem VI.1.4, again, $mx(\mathbf{T})_{\mathscr{X}\mathscr{X}} = mx(\mathbf{U}_{\mathscr{X}})_{\mathscr{E}\mathscr{E}} \cdot A_{\mathrm{T}} \cdot mx(\mathbf{U}_{\mathscr{X}})^{-1}_{\mathscr{E}\mathscr{E}}$, so that the above equation becomes

$$mx(\mathbf{U}_{\mathscr{X}})_{\mathscr{E}\mathscr{E}} \cdot A_{\mathrm{T}} \cdot mx(\mathbf{U}_{\mathscr{X}})^{-1}_{\mathscr{E}\mathscr{E}} = \text{diag}\,(\lambda_1, \lambda_2, \ldots, \lambda_n).$$

Simply set $S = mx(\mathbf{U}_{\mathscr{X}})^{-1}_{\mathscr{E}\mathscr{E}}$, and we have justified the existence of a matrix S such that

$$S^{-1} \cdot A_{\mathrm{T}} \cdot S = \text{diag}\,(\lambda_1, \lambda_2, \ldots, \lambda_n),$$

provided **T** is diagonalizable. That is, (1) $\Rightarrow$ (3). Our proof is done.

Example Find the 3×3 matrix S such that $S^{-1} \cdot A_{\mathbf{T}} \cdot S$ is a diagonal matrix for $\mathbf{T} \in \mathscr{L}(V_n, V_n)$ defined in the previous example, that is,

$$\mathbf{T}\begin{pmatrix} x_1 \\ x_2 \\ x_3 \end{pmatrix} = \underbrace{\begin{pmatrix} 3 & -1 & 0 \\ -1 & 3 & 0 \\ 0 & 0 & 3 \end{pmatrix}}_{A_{\mathbf{T}}} \cdot \begin{pmatrix} x_1 \\ x_2 \\ x_3 \end{pmatrix} = \begin{pmatrix} 3x_1 - x_2 \\ -x_1 + 3x_2 \\ 3x_3 \end{pmatrix}$$

for all

$$\begin{pmatrix} x_1 \\ x_2 \\ x_3 \end{pmatrix} \in V_3.$$

Since one ordering of our 3 linearly independent eigenvectors would be, X_1, X_2, X_3, we may, by virtue of Theorem VI.3.5 (3), construct $S = [X_1 X_2 X_3]$ to be

$$S = \begin{pmatrix} 1 & 0 & 1 \\ 1 & 0 & -1 \\ 0 & 1 & 0 \end{pmatrix}.$$

We compute the inverse matrix S^{-1}, and find

$$S^{-1} = \tfrac{1}{2}\begin{pmatrix} 1 & 1 & 0 \\ 0 & 0 & 2 \\ 1 & -1 & 0 \end{pmatrix}.$$

Indeed, we may verify that

$$\begin{aligned} S^{-1} \cdot A_{\mathbf{T}} \cdot S &= \tfrac{1}{2}\begin{pmatrix} 1 & 1 & 0 \\ 0 & 0 & 2 \\ 1 & -1 & 0 \end{pmatrix} \cdot \begin{pmatrix} 3 & -1 & 0 \\ -1 & 3 & 0 \\ 0 & 0 & 3 \end{pmatrix} \cdot \begin{pmatrix} 1 & 0 & 1 \\ 1 & 0 & -1 \\ 0 & 1 & 0 \end{pmatrix} \\ &= \begin{pmatrix} 2 & 0 & 0 \\ 0 & 3 & 0 \\ 0 & 0 & 4 \end{pmatrix}, \end{aligned}$$

which is the 3×3 diagonal matrix diag (2,3,4) whose main diagonal entries are the eigenvalues of $\mathbf{T}$.

Example Diagonalize the 2×2 matrix $A = \begin{pmatrix} i & -(1+2i) \\ 0 & 2 \end{pmatrix}$, that is, find a 2×2 matrix S such that $S^{-1} \cdot A \cdot S$ is a diagonal matrix.

We proceed exactly as in the previous example, treating A as the matrix of an operator in $\mathscr{L}(V_2(\mathbf{C}), V_2(\mathbf{C}))$ relative to the standard ordered basis $\mathscr{E}$.

▶ FIRST STEP: COMPUTE EIGENVALUES. In our case, A is a triangular matrix, so that the 2 eigenvalues, i and 2, already appear along the main diagonal (see Exercise VI.2.8). To see this directly, observe that $p_A(\cdot)$, the characteristic polynomial of A, is defined by

$$p_A(\lambda) = \det(A - \lambda I_2) = \det\begin{pmatrix} i - \lambda & -(1+2i) \\ 0 & 2 - \lambda \end{pmatrix} = (i - \lambda)(2 - \lambda),$$

whose 2 roots (eigenvalues) are $\lambda_1 = i$, and $\lambda_2 = 2$.

▶ SECOND STEP: FIND CORRESPONDING EIGENVECTORS. The matrix equation $A \cdot X = \lambda_1 X = i \cdot X$ is expressed as

$$\begin{pmatrix} i & -(1+2i) \\ 0 & 2 \end{pmatrix}\begin{pmatrix} x_1 \\ x_2 \end{pmatrix} = i\begin{pmatrix} x_1 \\ x_2 \end{pmatrix},$$

which leads to the pair of equations

(a) $ix_1 - (1+2i)x_2 = ix_1,$

and

(b) $2x_2 = ix_2.$

From Equation (b) above, we must set $x_2 = 0$. Hence, x_1 may be chosen at will, so let us say x_1 equals 1. Thus, an eigenvector corresponding to the eigenvalue $\lambda_1 = i$, is $X_1 = \begin{pmatrix} 1 \\ 0 \end{pmatrix}$.

Similarly, writing $A \cdot X = \lambda_2 X = 2 \cdot X$, leads to the pair of equations

(c) $ix_1 - (1+2i)x_2 = 2x_1$

(d) $2x_2 = 2x_2.$

No information is derived from Equation (d), but Equation (c) tells us that $(-2+i)x_1 = (1+2i)x_2$. In other words, the ratio

$$\frac{x_1}{x_2} = \frac{(1+2i)}{(-2+i)} = \frac{-i(i-2)}{(-2+i)} = -i.$$

Thus, if we set $x_2 = 1$, then x_1 must equal $-i$, that is, an eigenvector corresponding to the eigenvalue $\lambda_2 = 2$, is the vector $X_2 = \begin{pmatrix} -i \\ 1 \end{pmatrix}$.

▶ THIRD STEP: CONSTRUCT $S = [X_1 X_2]$ AND S^{-1}.

$S = [X_1 X_2] = \begin{pmatrix} 1 & -i \\ 0 & 1 \end{pmatrix}$ [see Theorem VI.3.5(3)]. In computing the inverse matrix, we have $S^{-1} = \begin{pmatrix} 1 & i \\ 0 & 1 \end{pmatrix}$.

▶ FOURTH STEP: CHECK THAT $S^{-1} \cdot A \cdot S$ IS DIAGONAL.

$$\begin{pmatrix} 1 & i \\ 0 & 1 \end{pmatrix} \cdot \begin{pmatrix} i & -(1+2i) \\ 0 & 2 \end{pmatrix} \cdot \begin{pmatrix} 1 & -i \\ 0 & 1 \end{pmatrix} = \begin{pmatrix} i & 0 \\ 0 & 2 \end{pmatrix} = \text{diag}\,(i,2)$$

the diagonal matrix of eigenvalues of A.

REMARK. The 4 steps enumerated in this example offer a convenient procedure to follow in diagonalizing any $n \times n$ matrix.

EXERCISES

VI.3.1 From Exercises VI.2.8, we have that a triangular $n \times n$ matrix (which equals $mx(\mathbf{T})_{\mathscr{E}\mathscr{E}}$ for some $\mathbf{T} \in \mathscr{L}(V_n, V_n)$) has its n eigenvalues along the main diagonal. Let $A = \begin{pmatrix} a & b \\ 0 & c \end{pmatrix}$ be such a matrix, where a, b, and c are (real or complex) scalars.

(a) If $A = mx(\mathbf{T})_{\mathscr{E}\mathscr{E}}$ relative to the standard ordered basis $\mathscr{E}$, describe

$\mathbf{T} \in \mathscr{L}(V_2, V_2)$, that is, describe $\mathbf{T}\begin{pmatrix} x_1 \\ x_2 \end{pmatrix}$ for all $\begin{pmatrix} x_1 \\ x_2 \end{pmatrix} \in V_2$.

(b) Assume $a = c$. Prove that $\mathbf{T}$ is diagonalizable if and only if $b = 0$. Thus, every nonzero vector is an eigenvector for $\mathbf{T}$ in this case.

(c) Assume $a \neq c$. Prove that $\mathbf{T}$ is diagonalizable regardless of the value of b. Find the eigenvectors of $\mathbf{T}$.

VI.3.2 Given

$$A_{\mathbf{T}} = mx(\mathbf{T})_{\mathscr{E}\mathscr{E}} = \begin{pmatrix} 0 & a & b \\ 0 & 0 & c \\ 0 & 0 & 0 \end{pmatrix},$$

where a, b, and c are all nonzero scalars.

(a) Find the eigenvalues of $\mathbf{T}$.

(b) Prove that any eigenvector

$$\begin{pmatrix} x_1 \\ x_2 \\ x_3 \end{pmatrix}$$

for $\mathbf{T}$ must have its third entry equal to zero, that is, show $x_3 = 0$.

(c) Can there exist 3 linearly independent vectors in V_3 (eigenvectors or not) such that each vector has, for its third coordinate, the scalar zero?

(d) Prove from (b) and (c) that $\mathbf{T}$ is not diagonalizable, without finding any eigenvectors for $\mathbf{T}$ explicitly.

(e) Find the eigenvectors for $\mathbf{T}$, and the dimension of the subspace of V_3 spanned by them.

VI.3.3 Prove that no operator on V_n can ever have more than n distinct eigenvalues.

VI.3.4 Diagonalize the following matrices, noting that each matrix is understood to be of the form $mx(\mathbf{T})_{\mathscr{E}\mathscr{E}}$, the matrix of a linear operator $\mathbf{T} \in \mathscr{L}(V_n, V_n)$ relative to the standard ordered basis $\mathscr{E}$. Thus, you are to find a matrix S such that $S^{-1} \cdot A \cdot S$ is a diagonal matrix, where A represents the original matrix.

(a) $\begin{pmatrix} -1 & -2 \\ 4 & 5 \end{pmatrix}$ (c) $\begin{pmatrix} 4i & 6i \\ -2i & -2i \end{pmatrix}$

(b) $\begin{pmatrix} 9 & 20 \\ -4 & -9 \end{pmatrix}$ (d) $\begin{pmatrix} -2+2i & 1-i \\ 2-2i & -1+i \end{pmatrix}$

(e) $\begin{pmatrix} 0 & 0 & 0 \\ 2 & 1 & 0 \\ 0 & 0 & -1 \end{pmatrix}$ (h) $\begin{pmatrix} 3 & a & 0 & 0 \\ 0 & 1 & 0 & 0 \\ 0 & 0 & 2 & 0 \\ 0 & 0 & b & 0 \end{pmatrix}$

(f) $\begin{pmatrix} 1 & 0 & 2 \\ 0 & 2 & 1 \\ 0 & 0 & 3 \end{pmatrix}$ (i) $\begin{pmatrix} 2 & 1 & 0 \\ 0 & 1 & 1 \\ 0 & 0 & -1 \end{pmatrix}$

(g) $\begin{pmatrix} 3 & 1 & 0 & 0 \\ 0 & 1 & 0 & 0 \\ 0 & 0 & 2 & 0 \\ 0 & 0 & 1 & 0 \end{pmatrix}$ (j) $\begin{pmatrix} 2 & a & b \\ 0 & 1 & c \\ 0 & 0 & -1 \end{pmatrix}$

VI.3.5 Let $\{i_1,i_2,\ldots,i_n\}$ represent the integers $\{1,2,\ldots,n\}$, given in some order. Thus, if $X_2, X_2, \ldots, X_n$ are linearly independent eigenvectors for some operator $\mathbf{T} \in \mathscr{L}(V_n,V_n)$, then $\mathscr{X}_1 = \{X_1,X_2,\ldots,X_n\}$ and $\mathscr{X}_2 = \{X_{i1},X_{i2},\ldots,X_{in}\}$ represent 2 orderings of this eigenvector basis for V_n. Prove that a reordering of the basis simply induces a corresponding reordering on the diagonal elements of $mx(\mathbf{T})_{\mathscr{X}_1\mathscr{X}_1}$. That is, show that if $mx(\mathbf{T})_{\mathscr{X}_1\mathscr{X}_1} =$ diag $(\lambda_1,\lambda_2,\ldots,\lambda_n)$, then

$$mx(\mathbf{T})_{\mathscr{X}_2\mathscr{X}_2} = \text{diag}\ (\lambda_{i_1},\lambda_{i_2},\ldots,\lambda_{i_n}).$$

VI.3.6 Show that if the $n \times n$ matrix A is diagonalizable, then $A + \alpha I_n$ is also diagonalizable, where α is *any* fixed scalar. Use this result to prove its own converse. That is, show that if $A + \alpha I_n$ is diagonalizable, then A is necessarily diagonalizable.

VI.3.7 We know from Exercise VI.3.1, part (b), that the matrix $A = \begin{pmatrix} 1 & 0 \\ a & 1 \end{pmatrix}$ is not diagonalizable for any and every nonzero scalar a. Prove that the properties of diagonalizability and nondiagonalizability are not preserved under addition. Specifically:

(a) Prove, by example, that the sum of 2 nondiagonalizable matrices can yield a diagonalizable matrix.

(b) Prove, by example, that the sum of 2 diagonalizable matrices can result in a nondiagonalizable matrix.

VI.3.8 (Hamilton-Cayley Theorem for diagonalizable matrices.)

(a) Suppose $A =$ diag $(a_1,a_2,\ldots,a_n)$, and $B =$ diag $(b_1,b_2,\ldots,b_n)$, are $n \times n$ diagonal matrices. Prove that for this special kind of matrix the product $A \cdot B$ can be computed by multiplying the n^2 corresponding entries of each matrix.

More exactly, prove that

$$A \cdot B = B \cdot A \text{ diag } (a_1b_1, a_1b_2,\ldots,a_nb_n).$$

(b) If $A =$ diag $(a_1,a_2,\ldots,a_n)$, then prove that $A^k =$ diag $(a_1{}^k,a_2{}^k,\ldots,a_n{}^k)$, for any positive integer k.

(c) Suppose $p(\lambda) = c_k\lambda^k + c_{k-1}\lambda^{k-1} + \cdots + c_1\lambda + c_0$ is a polynomial defined for each scalar λ by the $k + 1$ fixed scalars $c_k,c_{k-1},\ldots, c_1, c_0$. If $A =$ diag $(a_1,a_2,\ldots,a_n)$, then prove that

(1) $c_kA^k + c_{k-1}A^{k-1} + \cdots + c_1A + c_0I_n =$ diag $(p(a_1), p(a_2),\ldots,p(a_n))$.

Roughly speaking, in order to evaluate a polynomial $p(\cdot)$, whose argument, or "input" element, is a diagonal matrix A instead of a scalar λ [that is, in order to evaluate the left-hand side of (**1**)], one simply evaluates the polynomial for each main diagonal entry a_i [that is, compute the right-hand side of (**1**)].

(d) Suppose A and B are $n \times n$ matrices such that for some $n \times n$ matrix S, $A = S^{-1} \cdot B \cdot S$. Prove that for all positive integers k, $A^k = S^{-1} \cdot B^k \cdot S$. (Hence, A and B have the same characteristic polynomial; see Theorem VI.2.4 and Definition VI.2.5.) As a consequence of this, show that for fixed scalars $c_k, c_{k-1}, \ldots, c_1, c_0$,

$$\textbf{(2)}\quad c_kA^k + c_{k-1}A^{k-1} + \cdots + c_1A + c_0I_n = S^{-1} \cdot (c_kB^k + c_{k-1}B^{k-1} + \cdots + c_1B + c_0I_n) \cdot S.$$

(e) Assume now, that B is a diagonaliazble $n \times n$ matrix with eigenvalues $\lambda_1, \lambda_2, \ldots, \lambda_n$. Then, from Theorem VI.3.5, there exists an $n \times n$ matrix S such that

$$A = \text{diag}\,(\lambda_1,\lambda_2,\ldots,\lambda_n) = S^{-1} \cdot B \cdot S.$$

If $p(\lambda) = c_n\lambda^n + c_{n-1}\lambda^{n-1} + \cdots + c_1\lambda + c_0$ represents the characteristic polynomial of B (or of $A = S^{-1} \cdot B \cdot S$), evaluated for any scalar λ, then prove that

$$c_nA^n + c_{n-1}A^{n-1} + \cdots + c_1A + c_0I_n = O, \text{ the } n \times n \text{ zero } \textit{matrix}.$$

As a consequence of this equation, and of (**2**), part (d), prove that the original matrix B, is also a "matrix root" of its own characteristic polynomial $p(\cdot)$, that is, prove that necessarily,

$$c_nB^n + c_{n-1}B^{k-1} + \cdots + c_1B + c_0I_n = O,$$

the zero *matrix*.

Note: In Exercise VI.3.8, we have shown that every diagonalizable $n \times n$ matrix B is a "matrix root" of its own characteristic polynomial. The Hamilton-Cayley Theorem yields the same conclusion for all $n \times n$ matrices, as opposed to diagonalizable ones only.

VI.3.9 Prove directly that pairs of eigenvectors X_1, X_2 corresponding to pairs of *distinct* eigenvalues λ_1, λ_2, $\lambda_1 \neq \lambda_2$, form a linearly independent set $\{X_1,X_2\}$. *Hint*: Write

(a) $O = \alpha_1X_1 + \alpha_2X_2$.

In order to show that necessarily $\alpha_1 = \alpha_2 = 0$, perform 2 operations: (1) Apply $\mathbf{T}$ to both sides of (a), bearing in mind that X_1, X_2, are eigenvectors for $\mathbf{T}$ ($\mathbf{T}(X_i) = \lambda_iX_i$, $i = 1,2$). (2) Multiply both sides of (a) through by 1 of the eigenvalues, λ_1, say. Subtract the result of (1) from that of (2) to obtain the fact that the set $\{X_1,X_2\}$ is linearly independent.

An immediate corollary to Theorem VI.3.5 is a spectral decomposition for a square matrix A (or for a linear operator $\mathbf{T}$ on V_n) whose n eigenvalues are distinct. This spectral decomposition will allow us to effect matrix computations which might otherwise prove tedious and uninspiring. Let us present this spectral theorem now.

Theorem VI.3.6 **(A spectral theorem for square matrices.)** Let A be an $n \times n$ matrix with n distinct eigenvalues $\lambda_1, \lambda_2, \ldots, \lambda_n$. Then A can always be represented in the form

$$A = \lambda_1 P_1 + \lambda_2 P_2 + \cdots + \lambda_n P_n , \tag{VI.3.10}$$

where the $n \times n$ matrices $P_1, P_2, \ldots, P_n$ have the following properties

(a) $P_1{}^2 = P_i$ for all $i = 1, 2, \ldots, n$ (idempotence)

(b) $P_i \cdot P_j = O$ whenever $i \neq j$

(c) $P_1 + P_2 + \cdots + P_n = I_n$ (decomposition of I_n)

the set of n idempotents $\{P_1, P_2, \ldots, P_n\}$ is called a **spectral decomposition of the identity induced by A.**

Proof: Consider the linear operator $\mathbf{T} : V_n \to V_n$ defined by $\mathbf{T}(X) = A \cdot X$, that is, A times column-vector X for all column-vectors $X \in V_n$. Then A becomes $A_{\mathbf{T}} = mx(\mathbf{T})_{\mathscr{E}\mathscr{E}}$, the matrix of $\mathbf{T}$ relative to the standard ordered basis $\mathscr{E}$ (see Theorem II.2.3 and Definition II.6.2). Now let $\mathscr{X} = \{X_1, X_2, \ldots, X_n\} \subset V_n$ be the ordered basis of eigenvectors of $\mathbf{T}$ corresponding to the respective distinct eigenvalues $\lambda_1, \lambda_2, \ldots, \lambda_n$ (Corollary VI.3.4 assures the existence of this basis of eigenvectors). Thus, $\mathbf{T}(X_i) = \lambda_i X_i$, or $A_i X_i = \lambda_i X_i$, all $i = 1, 2, \ldots, n$. An easy reading of Theorem VI.3.5 allows us to present the following matrix equation:

$$\underbrace{\begin{pmatrix} \lambda_1 & & & 0 \\ & \lambda_2 & & \\ & & \ddots & \\ 0 & & & \lambda_n \end{pmatrix}}_{\operatorname{diag}(\lambda_1, \lambda_2, \ldots, \lambda_n)} = S^{-1} \cdot A_{\mathrm{T}} \cdot S \tag{VI.3.11}$$

where $S = [X_1 X_2 \cdots X_n]$ is the $n \times n$ matrix whose ith column is the ith eigenvector X_i.

The proof now rests in carving up $\operatorname{diag}(\lambda_1, \lambda_2, \ldots \lambda_n)$ of (VI.3.11) in a suitable way. In fact, let us write

$$\begin{pmatrix} \lambda_1 & & & 0 \\ & \lambda_2 & & \\ & & \ddots & \\ 0 & & & \lambda_n \end{pmatrix} = \lambda_1 E_{11} + \lambda_2 E_{22} + \cdots + \lambda_i E_{ii} + \cdots + \lambda_n E_{nn} \tag{VI.3.12}$$

$$\text{where } E_{ii} = \begin{pmatrix} 0 & & & \overset{i\text{th column}}{\downarrow} & & & 0 \\ & \ddots & & & & & \\ & & 0 & & & & \\ & & & 1 & & & \\ & & & & 0 & & \\ & & & & & \ddots & \\ 0 & & & & & & 0 \end{pmatrix} \leftarrow i\text{th row}$$

the $n \times n$ matrix with zeros everywhere except for a lone 1 on the main diagonal in the ith row and ith column. If we but equate the right-hand sides of (VI.3.11) and (VI.3.12), we obtain

$$S^{-1} \cdot A_{\mathrm{T}} \cdot S = \lambda_1 E_{11} + \lambda_2 E_{22} + \cdots + \lambda_i E_{ii} + \cdots + \lambda_n E_{nn}. \quad \text{(VI.3.3)}$$

Now with (VI.3.13) in hand, we are on the threshold of a spectral decomposition for our original matrix $A = A_{\mathrm{T}}$. We solve for A_{T} in (VI.3.13) by multiplying through on the left by S [thereby cancelling S^{-1} in (VI.3.13)] and on the right by S^{-1} [cancelling S in (VI.3.13)] and obtain the matrix equation

$$\begin{aligned} A = A_{\mathrm{T}} &= S \cdot (\lambda_1 E_{11} + \lambda_2 E_{22} + \cdots + \lambda_i E_{ii} + \cdots + \lambda_n E_{nn}) \cdot S^{-1} \\ &= (\lambda_1 S \cdot E_{11} + \lambda_2 S \cdot E_{22} + \cdots + \lambda_i S \cdot E_{ii} + \cdots + \lambda_n S \cdot E_{nn}) \cdot S \\ &\qquad \text{(distribute } S \text{ through on the left)} \\ &= \lambda_1 \underbrace{(S \cdot E_{11} \cdot S^{-1})} + \lambda_2 \underbrace{(S \cdot E_{22} \cdot S^{-1})} + \cdots + \lambda_n \underbrace{(S \cdot E_{nn} \cdot S^{-1})} \\ &\qquad \text{(distribute } S^{-1} \text{ through on the right)} \\ &= \lambda_1 P_1 + \lambda_2 P_2 + \cdots + \lambda_i P_i + \cdots + \lambda_n P_n \\ &\qquad \text{(setting } P_i = S \cdot E_{ii} \cdot S^{-1}\text{).} \end{aligned} \quad \text{(VI.3.14)}$$

We claim that (VI.3.14) leads us to the desired result (VI.3.10). All we need do now is verify that properties (a), (b), and (c) for our Theorem VI.3.6 hold relative to the n matrices $\{P_1, P_2, \ldots, P_n\}$ defined in (VI.3.14), where each $P_i = S \cdot E_{ii} \cdot S^{-1}$ for $S = [X_1 X_2 \cdots X_n]$, the $n \times n$ matrix whose columns are eigenvectors. But this is easy. In fact, it is routine to check properties (a), (b), and (c) for the special set of matrices $\{E_{11}, E_{22}, \ldots, E_{nn}\}$. This automatically forces (a), (b), and (c) to hold for any other set of the form $\{S \cdot E_{11} \cdot S^{-1}, S \cdot E_{22} \cdot S^{-1}, \ldots, S \cdot E_{nn} \cdot S^{-1}\}$ for *any* invertible $n \times n$ matrix S (see Exercise VI.3.10). This ends the proof.

• REMARK. Why is the spectral theorem (Theorem VI.3.6) so wonderful? Well, suppose for example, we have for some 3×3 matrix A,

$$A = \lambda_1 P_1 + \lambda_2 P_2 + \lambda_3 P_3 \quad \text{(VI.3.15)}$$

where, as per Theorem VI.3.6, $P_i^2 = P_i$, $P_i \cdot P_j = 0$ for $i \neq j$, and $P_1 + P_2 + P_3 = I_3$. Then we can *easily* compute A^2, A^3, A^n $6A^2 + 3A^4$ or *any* polynomial in A. In particular, A^2 becomes

$$\begin{aligned} A \cdot A &= (\lambda_1 P_1 + \lambda_2 P_2 + \lambda_3 P_3) \cdot (\lambda_1 P_1 + \lambda_2 P_2 + \lambda_3 P_3) \text{ from (VI.3.15)} \\ &= \lambda_1{}^2 P_1{}^2 + \lambda_2{}^2 P_2{}^2 + \lambda_3{}^2 P_3{}^2 \\ &\quad + \lambda_1\lambda_2 \underbrace{P_1 P_2}_{O} + \lambda_1\lambda_3 \underbrace{P_1 P_3}_{O} + \lambda_2\lambda_1 \underbrace{P_2 P_1}_{O} \\ &\quad + \lambda_2\lambda_3 \underbrace{P_2 P_3}_{O} + \lambda_3\lambda_1 \underbrace{P_3 P_1}_{O} + \lambda_3\lambda_2 \underbrace{P_3 P_2}_{O}. \end{aligned} \tag{VI.3.16}$$

This is an easy computation? It is because in (VI.3.16) *all* cross terms $P_i \cdot P_j$, $i \neq j$, equal the zero matrix. Also, all the square terms $P_i{}^2$ return to the form P_i (idempotence). We reread (VI.3.16) now to say

$$A^2 = \lambda_1{}^2 P_1 + \lambda_2{}^2 P_2 + \lambda_3{}^2 P_3. \tag{VI.3.17}$$

Thus, in order to *square* a matrix A, one simply squares each (eigenvalue) scalar which appears in the spectral decomposition (VI.3.10) for A; an easy process? By induction, we may prove the following corollary to our spectral theorem.

Corollary VI.3.7 Let A be an $n \times n$ matrix with spectral representation

$$A = \lambda_1 P_1 + \lambda_2 P_2 + \cdots + \lambda_n P_n,$$

where $P_i{}^2 = P_i$, $P_i \cdot P_j = 0$ for $i \neq j$ and $P_1 + P_2 + \cdots + P_n = I_n$. Let $f(x) = a_0 + a_1 x + a_2 x^2 + \cdots + a_k x^k$ be a kth degree polynomial, where each a_i is a complex number. Then $\mathbf{f}(A)$, the polynomial in A, defined by

$$\mathbf{f}(A) = a_0 I + a_1 A + a_2 A^2 + \cdots + a_k A^k$$

computes to be

$$\mathbf{f}(A) = f(\lambda_1) P_1 + f(\lambda_2) P_2 + \cdots + f(\lambda_n) P_n,$$

that is, to obtain the **polynomial function of A**, merely apply to polynomial to each of the scalars $\lambda_1, \lambda_2, \ldots, \lambda_n$ which appear in the spectral (VI.3.10) form for A.

Moreover, if each $\lambda_i \neq 0$, then A^{-1} exists and is of the form

$$A^{-1} = \lambda_1^{-1} P_1 + \lambda_2^{-2} P_2 + \cdots + \lambda_n^{-1} P_n.$$

Proof: Exercise VI.3.10.

Example Suppose by some means (which we fully describe in the next example) we discover that for

$$A = \begin{pmatrix} 3 & 0 & 0 \\ 0 & 4 & -2 \\ 0 & -2 & 4 \end{pmatrix},$$

we have

$$A = \begin{pmatrix} 3 & 0 & 0 \\ 0 & 4 & -2 \\ 0 & -2 & 4 \end{pmatrix} = \underset{\lambda_1}{3} \underbrace{\begin{pmatrix} 1 & 0 & 0 \\ 0 & 0 & 0 \\ 0 & 0 & 0 \end{pmatrix}}_{P_1} + \underset{\lambda_2}{2} \underbrace{\begin{pmatrix} 0 & 0 & 0 \\ 0 & \frac{1}{2} & \frac{1}{2} \\ 0 & \frac{1}{2} & \frac{1}{2} \end{pmatrix}}_{P_2} + \underset{\lambda_3}{6} \underbrace{\begin{pmatrix} 0 & 0 & 0 \\ 0 & \frac{1}{2} & -\frac{1}{2} \\ 0 & -\frac{1}{2} & \frac{1}{2} \end{pmatrix}}_{P_3}. \quad \text{(VI.3.18)}$$

It is easy to check that $P_i^2 = P_i$, $P_i \cdot P_j = 0$, $i \neq j$, and $P_1 + P_2 + P_3 = I_3$. According to our Corollary VI.3.7, then, for the polynomials $f_1(x) = x^{12}$ and $f_2(x) = 5x^2 + 10x$, we easily compute the corresponding matrix functions $\mathbf{f}_1(A)$ and $\mathbf{f}_2(A)$ as follows:

(a) $$\mathbf{f}_1(A) = A^{12} = 3^{12}\begin{pmatrix} 1 & 0 & 0 \\ 0 & 0 & 0 \\ 0 & 0 & 0 \end{pmatrix} + 2^{12}\begin{pmatrix} 0 & 0 & 0 \\ 0 & \frac{1}{2} & \frac{1}{2} \\ 0 & \frac{1}{2} & \frac{1}{2} \end{pmatrix} + 6^{12}\begin{pmatrix} 0 & 0 & 0 \\ 0 & \frac{1}{2} & -\frac{1}{2} \\ 0 & -\frac{1}{2} & \frac{1}{2} \end{pmatrix},$$

(b) $$\underbrace{5A^2 + 10A}_{\mathbf{f}_2(A)} = \underbrace{(5 \cdot 3^2 + 10 \cdot 3)}_{f_2(3)}P_1 + \underbrace{(5 \cdot 2^2 + 10 \cdot 2)}_{f_2(2)}P_2 + \underbrace{(5 \cdot 6^2 + 10 \cdot 6)}_{f_2(6)}P_3$$

$$= 75\begin{pmatrix} 1 & 0 & 0 \\ 0 & 0 & 0 \\ 0 & 0 & 0 \end{pmatrix} + 40\begin{pmatrix} 0 & 0 & 0 \\ 0 & \frac{1}{2} & \frac{1}{2} \\ 0 & \frac{1}{2} & \frac{1}{2} \end{pmatrix} + 240\begin{pmatrix} 0 & 0 & 0 \\ 0 & \frac{1}{2} & -\frac{1}{2} \\ 0 & -\frac{1}{2} & \frac{1}{2} \end{pmatrix}$$

$$= \begin{pmatrix} 75 & 0 & 0 \\ 0 & 140 & -100 \\ 0 & -100 & 140 \end{pmatrix}.$$

One may also verify (b) directly by computing A^2 from (VI.3.18), multiplying by 5, and adding $10A$ to the result.

Now Corollary VI.3.7 tells us that A^{-1} can easily be found from the spectral decomposition (VI.3.18) by simply inverting each of the nonzero λ_i's, that is,

(c) $$A^{-1} = 3^{-1}P_1 + 2^{-1}P_2 + 6^{-1}P_3$$

$$= \frac{1}{3}\begin{pmatrix} 1 & 0 & 0 \\ 0 & 0 & 0 \\ 0 & 0 & 0 \end{pmatrix} + \frac{1}{2}\begin{pmatrix} 0 & 0 & 0 \\ 0 & \frac{1}{2} & \frac{1}{2} \\ 0 & \frac{1}{2} & \frac{1}{2} \end{pmatrix} + \frac{1}{6}\begin{pmatrix} 0 & 0 & 0 \\ 0 & \frac{1}{2} & -\frac{1}{2} \\ 0 & -\frac{1}{2} & \frac{1}{2} \end{pmatrix}$$

$$= \begin{pmatrix} \frac{1}{3} & 0 & 0 \\ 0 & \frac{1}{3} & \frac{1}{6} \\ 0 & \frac{1}{6} & \frac{1}{3} \end{pmatrix}.$$

Sure enough, A^{-1} [from (c)], when multiplied by A [from (VI.3.18)] gives us $A \cdot A^{-1} = A^{-1} \cdot A = I_3$.

We offer a set of rules by which actual spectral decompositions for matrices with distinct eigenvalues can be found.

Rules VI.3.8 Computing spectral decompositions for $n \times n$ matrices with distinct eigenvalues $\lambda_1, \lambda_2, \ldots, \lambda_n$.

▶ FIRST: Compute the eigenvalues for A by finding the roots for the characteristic polynomial, that is, find those λ_i such that $p_A(\lambda_i) = \det(A - \lambda_i I_n) = 0$.

▶ SECOND: Find the n linearly independent eigenvectors $X_1, X_2, \ldots, X_n$, corresponding to the respective eigenvalues $\lambda_1, \lambda_2, \ldots, \lambda_n$, that is, find $X_i \in V_n$ where $A \cdot X_i = \lambda_i X_i$.

▶ THIRD: Construct the $n \times n$ matrix $S = [X_1 X_2 \cdots X_n]$ whose ith column is the (column) eigenvector X_i corresponding to eigenvalue λ_i. Now construct its $n \times n$ inverse S^{-1}.

▶ FOURTH: Write

$$A = \lambda_1 \underbrace{(S \cdot E_{11} \cdot S^{-1})}_{P_1} + \lambda_2 \underbrace{(S \cdot E_{22} \cdot S^{-1})}_{P_2} + \cdots + \lambda_n \underbrace{(S \cdot E_{nn} \cdot S^{-1})}_{P_n},$$

where $E_{ii} = \text{diag}\,(0, \ldots, 0, \underset{\substack{\uparrow \\ i\text{th place}}}{1}, 0, \ldots, 0)$ for each $i = 1, 2, \ldots, n$.

Example Let $A = \begin{pmatrix} 1 & 1 \\ 0 & 2 \end{pmatrix}$.

We follow the 4-step approach of Rule VI.3.8.

▶ FIRST: (Eigenvalues)

$$p_n(\lambda) = \det\,(A - \lambda I_2) = \det \begin{pmatrix} 1-\lambda & 1 \\ 0 & 2-\lambda \end{pmatrix} = (1-\lambda)(2-\lambda) = 0$$

only for $\lambda_1 = 1$ and $\lambda_2 = 2$. Our eigenvalues $\lambda_1 = 1$, $\lambda_2 = 2$, are distinct; a sufficient condition for us to proceed.

▶ SECOND: (Eigenvectors) For all $X = \begin{pmatrix} x_1 \\ x_2 \end{pmatrix} \in V_2$, we have

$$A \cdot X = \begin{pmatrix} 1 & 1 \\ 0 & 2 \end{pmatrix} \cdot \begin{pmatrix} x_1 \\ x_2 \end{pmatrix} = \begin{pmatrix} x_1 + x_2 \\ 2x_2 \end{pmatrix} = \lambda \begin{pmatrix} x_1 \\ x_2 \end{pmatrix}. \tag{VI.3.19}$$

If we set $\lambda = \lambda_1 = 1$, then $x_2 = 0$ from (VI.3.19) and x_1 is unrestricted; choose $x_1 = 1$, say, so that

$$A \cdot \begin{pmatrix} 1 \\ 0 \end{pmatrix} = \underset{\substack{\uparrow \\ \lambda_1}}{1} \cdot \underset{\substack{\uparrow \\ X_1}}{\begin{pmatrix} 1 \\ 0 \end{pmatrix}} = 1 \cdot X_1.$$

Set $\lambda = \lambda_2 = 2$ in [(VI.3.19)] to obtain no restriction on x_2, while $x_1 + x_2 = 2x_1$, that is, $x_1 = x_2$. Choose $x_1 = x_2 = 1$. Sure enough,

$$A \cdot \begin{pmatrix} 1 \\ 1 \end{pmatrix} = \underset{\substack{\uparrow \\ \lambda_2}}{2} \underset{\substack{\uparrow \\ X_2}}{\begin{pmatrix} 1 \\ 1 \end{pmatrix}}, \text{ that is, } X_2 = \begin{pmatrix} 1 \\ 1 \end{pmatrix}.$$

▶ THIRD: (Construct S, S^{-1}.)

$$S = [X_1 X_2] = \begin{pmatrix} 1 & 1 \\ 0 & 1 \end{pmatrix}; \text{ therefore } S^{-1} = \begin{pmatrix} 1 & -1 \\ 0 & 1 \end{pmatrix}.$$

▶ FOURTH: (Spectral decomposition.)

To write $A = \lambda_1 S \cdot E_{11} \cdot S^{-1} + \lambda_2 S \cdot E_{22} \cdot S^{-1}$, is to write

$$A = \underset{\lambda_1}{\underset{\uparrow}{1}} \cdot \underbrace{\begin{pmatrix} 1 & 1 \\ 0 & 1 \end{pmatrix}}_{S} \cdot \underbrace{\begin{pmatrix} 1 & 0 \\ 0 & 0 \end{pmatrix}}_{E_{11}} \cdot \underbrace{\begin{pmatrix} 1 & -1 \\ 0 & 1 \end{pmatrix}}_{S^{-1}} + \underset{\lambda_2}{\underset{\uparrow}{2}} \cdot \underbrace{\begin{pmatrix} 1 & 1 \\ 0 & 1 \end{pmatrix}}_{S} \cdot \underbrace{\begin{pmatrix} 0 & 0 \\ 0 & 1 \end{pmatrix}}_{E_{22}} \cdot \underbrace{\begin{pmatrix} 1 & -1 \\ 0 & 1 \end{pmatrix}}_{S^{-1}} \qquad \text{(VI.3.20)}$$

$$= 1 \cdot \underbrace{\begin{pmatrix} 1 & -1 \\ 0 & 0 \end{pmatrix}}_{P_1} + 2 \cdot \underbrace{\begin{pmatrix} 0 & 1 \\ 0 & 1 \end{pmatrix}}_{P_2}.$$

Notice that we have many checks on our computation, namely, in computing $1 \cdot P_1 + 2 \cdot P_2$ from the last equation, we find that, indeed, $1 \cdot P_1 + 2 \cdot P_2 = \begin{pmatrix} 1 & 1 \\ 0 & 2 \end{pmatrix}$, the very matrix A we started with. Moreover, with $P_1 = \begin{pmatrix} 1 & -1 \\ 0 & 0 \end{pmatrix}$, $P_2 = \begin{pmatrix} 0 & 1 \\ 0 & 1 \end{pmatrix}$, we see that $P_i^2 = P_i$, $i = 1, 2$, $P_1 \cdot P_2 = P_2 \cdot P_1 = \begin{pmatrix} 0 & 0 \\ 0 & 0 \end{pmatrix}$, and $P_1 + P_2 = \begin{pmatrix} 1 & 0 \\ 0 & 1 \end{pmatrix}$, just as a spectral decomposition of idempotents ought to. From (VI.3.20) we may employ Corollary VI.3.7 to effect some (now) easy computations on $A = \begin{pmatrix} 1 & 1 \\ 0 & 2 \end{pmatrix}$. For example,

(a) $$\begin{pmatrix} 1 & 1 \\ 0 & 2 \end{pmatrix}^5 = 1^5 P_1 + 2^5 P = 1\begin{pmatrix} 1 & -1 \\ 0 & 0 \end{pmatrix} + 32\begin{pmatrix} 0 & 1 \\ 0 & 1 \end{pmatrix} = \begin{pmatrix} 1 & 31 \\ 0 & 32 \end{pmatrix}.$$

(b) $$\begin{pmatrix} 1 & 1 \\ 0 & 2 \end{pmatrix}^{-1} = 1^{-1} P_1 + 2^{-1} P_2 = 1\begin{pmatrix} 1 & -1 \\ 0 & 0 \end{pmatrix} + \frac{1}{2}\begin{pmatrix} 0 & 1 \\ 0 & 1 \end{pmatrix} = \begin{pmatrix} 1 & -\frac{1}{2} \\ 0 & \frac{1}{2} \end{pmatrix}.$$

(c) $A^7 - 2A^5$. Note $A^7 = 1^7 P_1 + 2^7 P_2$

$$= 1\begin{pmatrix} 1 & -1 \\ 0 & 0 \end{pmatrix} + 128\begin{pmatrix} 0 & 1 \\ 0 & 1 \end{pmatrix} = \begin{pmatrix} 1 & 127 \\ 0 & 128 \end{pmatrix}$$

$A^5 = \begin{pmatrix} 1 & 31 \\ 0 & 32 \end{pmatrix}$ from (a) above, so $A^7 - 2A^5$

$$= \begin{pmatrix} 1 & 127 \\ 0 & 128 \end{pmatrix} - 2\begin{pmatrix} 1 & 31 \\ 0 & 32 \end{pmatrix} = \begin{pmatrix} 1 & 127 \\ 0 & 128 \end{pmatrix} - \begin{pmatrix} 2 & 62 \\ 0 & 64 \end{pmatrix} = \begin{pmatrix} -1 & 65 \\ 0 & 64 \end{pmatrix}.$$

(d) We can find 4 *square roots* for A, that is, matrices B where $B \cdot B = A$. In fact,

$$B_1 = \sqrt{1}\, P_1 + \sqrt{2}\, P_2, \qquad B_2 = 1 P_1 - \sqrt{2}\, P_1$$

$$B_3 = -1 P_1 + \sqrt{2}\, P_2, \qquad B_4 = -1 P_1 - \sqrt{2}\, P_2$$

are all square roots of A. For example,

$$B_3 = \begin{pmatrix} -1 & 1 + \sqrt{2} \\ 0 & \sqrt{2} \end{pmatrix} \quad \text{is such that } B_3 \cdot B_3 = \begin{pmatrix} 1 & 1 \\ 0 & 2 \end{pmatrix}.$$

EXERCISES

VI.3.10 (a) Prove that the $n \times n$ matrices $\{E_{11}, E_{22}, \ldots, E_{nn}\}$ form a spectral decomposition of the identity, where for each i,

$$E_{ii} = \begin{pmatrix} 0 & & & & 0 \\ & \ddots & & & \\ & & 1 & & \\ & & & \ddots & \\ 0 & & & & 0 \end{pmatrix}$$

has zeros everywhere except for a 1 in the iith place. That is, show conditions (a), (b), and (c) of Theorems VI.3.6 obtain for this set.

(b) If S is any invertible $n \times n$ matrix, prove that the set

$$\{A \cdot E_{11} \cdot S^{-1}, S \cdot E_{22} \cdot S^{-1}, \ldots, S \cdot E_{nn} \cdot S^{-1}\}$$

is also a spectral family of matrices, that is, fulfill conditions (a), (b), and (c) of Theorem VI.3.6.

VI.3.11 Given the polynomial functions $f_1(x) = x^2$, $f_2(x) = x^5$, $f_3(x) = 2x^2 - x^5$, compute $\mathbf{f}_1(A)$, $\mathbf{f}_2(A)$, and $\mathbf{f}_3(A)$ where A is the matrix of Exercise

(a) VI.3.4(a) (e) VI.3.4(e) (h) VI.3.4(h)
(b) VI.3.4(b) (f) VI.3.4(f) (i) VI.3.4(i)
(c) VI.3.4(c) (g) VI.3.4(g) (j) VI.3.4(j).
(d) VI.3.4(d)

VI.3.12 Find the four square roots of each of the matrices of Exercise VI.3.4(a) and VI.3.4(b) [see Example above, part (d)].

4. The Inner Product

For the purpose of this discussion, we shall consider a 2-vector $X = \begin{pmatrix} x_1 \\ x_2 \end{pmatrix}$ to be the "name" of the geometric point in the plane, whose distance from the vertical axis is x_1, and whose distance from the horizontal axis is x_2. For the time being, we *require* the scalar entries, x_1 and x_2, to be *real* numbers. That is, the 2-dimensional vector-space we consider, is $V_2(\mathbf{R})$, the space of 2-vectors with *real* entries. In Figure 6.1, x_1 is positive and x_2 is negative.

The purpose of identifying 2-vectors with points (or "arrows") in the plane, is that we are able to take advantage of the geometry of the plane in discussing such notions as *length* of a 2-vector, and the *angle* between a pair of 2-vectors. In developing the concepts of angles between vectors and length, we shall use the following results from trigonometry:

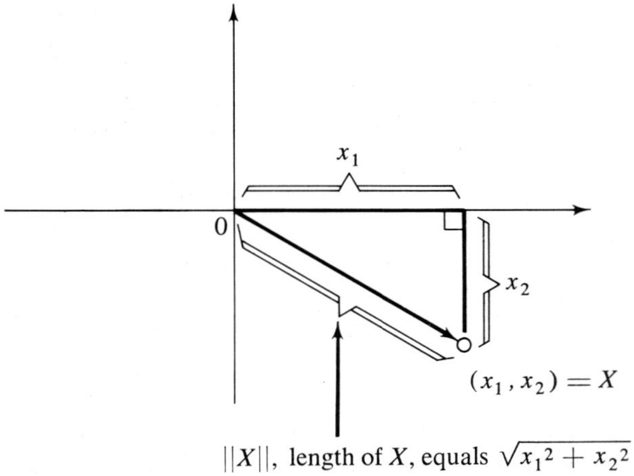

Figure 6.1 $X = \begin{pmatrix} x_1 \\ x_2 \end{pmatrix}$ ***identified with the point*** (x_1, x_2).

(1) $\cos(\theta_n - \theta_1) = \cos(\theta_1)\cos(\theta_2) + \sin(\theta_1)\sin(\theta_2)$ for all angles θ_1, θ_2, where $\sin(\cdot)$ and $\cos(\cdot)$ are the sine and cosine functions, respectively.

(2) (Pythagorean theorem.) If a right triangle has 2 legs whose lengths are a and b, and whose hypotenuse length is c, then necessarily, $a^2 + b^2 = c^2$.

Let us see what the cosine of $\theta_2 - \theta_1$, the angle between vectors X and Y in Figure 6.2, computes to be in terms of the real vectors X and Y. Note that θ_1 is the angle formed by the vector X and the horizontal axis; θ_2 is the angle formed by the vector Y and the horizontal axis. (See Figure 6.3.)

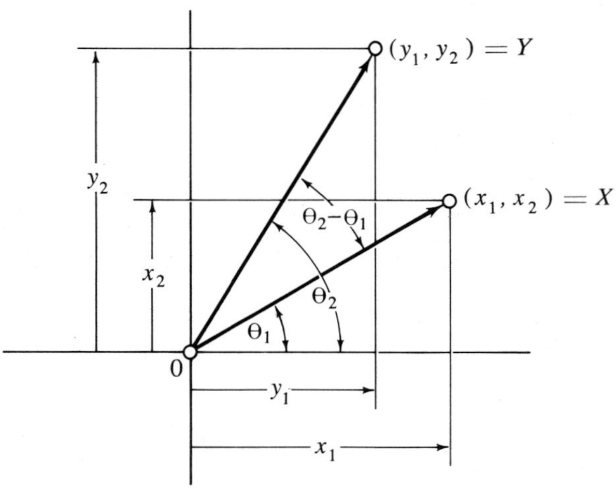

Figure 6.2

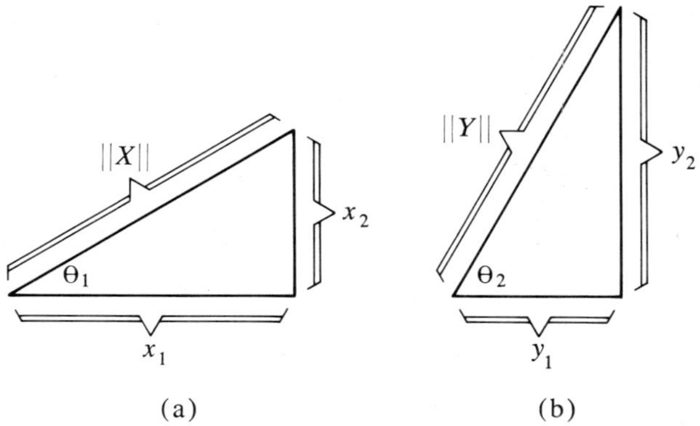

Figure 6.3

Thus, from the definition of the trigonometric functions sin (·) and cos (·),

$$\cos \theta_1 = \frac{x_1}{\| X \|}, \ \cos \theta_2 = \frac{y_1}{\| Y \|}, \ \sin \theta_1 = \frac{x_2}{\| X \|}, \ \sin \theta_2 = \frac{y_2}{\| Y \|}$$

where $\| X \|$ and $\| Y \|$ represent the lengths of the vectors X and Y, respectively. Substituting these values into (1) yields an expression for the cosine of the included angle $\theta_2 - \theta_1$, namely,

$$\cos(\theta_2 - \theta_1) = \frac{x_1 y_1}{\| X \| \cdot \| Y \|} + \frac{x_2 y_2}{\| X \| \cdot \| Y \|}.$$

From this equality, we can derive the following: If $X = \begin{pmatrix} x_1 \\ x_2 \end{pmatrix}$ and $Y = \begin{pmatrix} y_1 \\ y_2 \end{pmatrix} \in V_2(\mathbf{R})$, then

$$\| X \| \, \| Y \| \cos(\theta) = x_1 y_1 + x_2 y_2, \qquad \text{(VI.4.1)}$$

where $\| X \|$ and $\| Y \|$ represent the lengths of the real vectors X and Y, respectively, and θ is the included angle, that is, the angle formed by the vectors X and Y.

Notice the distinct flavor of each side of Equation (VI.4.1). The left-hand side, $\| X \| \, \| Y \| \cos(\theta)$, is a *geometric* statement about lengths and the included *angle*; the right-hand side, $x_1 y_1 + x_2 y_2$, is nongeometric in nature, and is purely *analytic*, that is, the right-hand side is a statement involving the real number entries (coordinates) of the 2-vectors X and Y. Now it is the *right*-hand side of (VI.4.1) that can easily be extended to real spaces of higher dimension; the left-hand side cannot be extended, as our 3-dimensional world fails us when we try to conceive of an angle formed by a pair of real vectors in spaces of higher dimension. Also, the analytic right-hand side of (VI.4.1) lends itself to a natural extension to the *complex* case where the vectors X and Y enjoy complex entries, that is, $X, Y \in V_n(\mathbf{C})$. We present, then, a definition which extends (and does not contradict) the geometrically motivated expression of (VI.4.1).

Definition VI.4.1 Given the (real or complex) n-vectors

$$X = \begin{pmatrix} x_1 \\ x_2 \\ \vdots \\ x_n \end{pmatrix}, \; Y = \begin{pmatrix} y_1 \\ y_2 \\ \vdots \\ y_n \end{pmatrix}.$$

Then $\langle X, Y \rangle$, the **inner product** of X and Y, is defined by

(a) $$\langle X,Y\rangle = x_1\bar{y}_1 + x_2\bar{y}_2 + \cdots + x_n\bar{y}_n = \sum_{i=1}^{n} x_i\bar{y}_i.$$

As usual, $\bar{y}_i$ is the complex conjugate of the complex number y_i. The *length* of a vector X, $\|X\|$, is defined by

(b) $$\|X\| = \langle X,Y\rangle^{1/2}.$$

The *cosine* of the angle formed by vectors X and $Y \in V_n(\mathbf{R})$ is defined to be

(c) $$\cos(\theta) = \frac{\langle X,Y\rangle}{\|X\| \cdot \|Y\|}$$

whenever $\|X\| \neq 0$, and $\|Y\| \neq 0$.

Example Suppose

$$X = \begin{pmatrix} 2+i \\ 3 \\ 1+i \end{pmatrix} \quad \text{and} \quad Y = \begin{pmatrix} i \\ -2i \\ 1-i \end{pmatrix}.$$

Then according to Definition VI.4.1,

$$\begin{aligned} \langle X,Y\rangle &= (2+i)(\bar{i}) + 3(\overline{-2i}) + (1+i)(\overline{1-i}) \\ &= (2+i)(-i) + 3(+2i) + (1+i)(1+i) \\ &= -2i - i^2 + 6i + 1 + i^2 + 2i \\ &= 1 + 6i. \end{aligned}$$

We have used i to denote the square root of minus 1, that is, $i^2 = -1$, and $\overline{a+ib}$, the conjugate of $a+ib$, a, b real, is the complex number $a - ib$. To continue our example,

$$\begin{aligned} \langle Y,X\rangle &= i(\overline{2+i}) + (-2i)(\bar{3}) + (1-i)(\overline{1+i}) \\ &= i(2-i) + (-2i)(3) + (1-i)(1-i) \\ &= 2i - i^2 - 6i + 1 + i^2 - 2i \\ &= 1 - 6i. \end{aligned}$$

Observe that in this special case, $\langle X,Y\rangle = \overline{\langle Y,X\rangle}$, the complex conjugate of the complex number $\langle Y,X\rangle$.

$$\begin{aligned}\langle X,X\rangle &= (2+i)\overline{(2+i)} + 3(\bar{3}) + (1+i)\overline{(1+i)}\\ &= (2+i)(2-i) + 3\cdot 3 + (1+i)(1-i)\\ &= 5+9+2\\ &= 16.\end{aligned}$$

Therefore, $\|X\| = \sqrt{\langle X,X\rangle} = 4$. Similarly, $\langle Y,Y\rangle = 7$.

We note that in Definition VI.4.1, Equation (a), we have the natural extension of Display (VI.4.1) to spaces of arbitrary finite dimension, and to the complex cese as well. Equation (c) is also inspired by (VI.4.1) where we have "solved for cos (θ)" in the real case. Observe, also, that for vectors in $V_n(\mathbf{R})$, Equation (b) in conjunction with Equation (a) is simply a restatement of the Pythagorean theorem when $n=2$.

Example Consider the pair of real 2-vectors X, Y and their graphical representations in Figure 6.4.

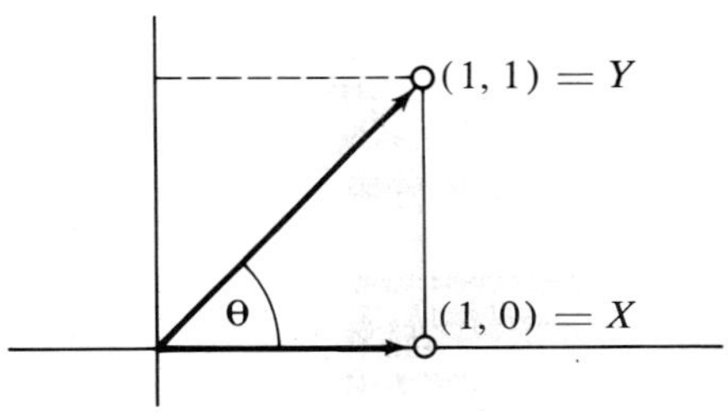

Figure 6.4

Here, X is a vector forming 1 side of a square whose sides are of length 1; Y forms a diagonal of this unit square. Now since $X = \begin{pmatrix}1\\0\end{pmatrix}$, and $Y = \begin{pmatrix}1\\1\end{pmatrix}$, we have from (a) of Definition VI.4.1 that

$$\langle X,Y\rangle = x_1y_1 + x_2y_2 = 1\cdot 1 + 0\cdot 1 = 1.$$

We use (b) to compute the lengths

$$\|X\| = \sqrt{\langle X,X\rangle} = \sqrt{x_1^2 + x_2^2} = \sqrt{1^2+0^2} = 1$$

$$\|Y\| = \sqrt{\langle Y,Y\rangle} = \sqrt{y_1^2 + y_2^2} = \sqrt{1^2+1^2} = \sqrt{2}$$

which is easy to see from the diagrams, also. From (c) we have

$$\cos(\theta) = \frac{\langle X,Y\rangle}{\|X\|\cdot\|Y\|} = \frac{1}{1\cdot\sqrt{2}} = \frac{1}{\sqrt{2}}\left(\frac{\sqrt{2}}{\sqrt{2}}\right) = \frac{\sqrt{2}}{2}.$$

But $\sqrt{2}/2$ is the cosine of 45 degrees, and from the diagram, we see that 45 degrees is, indeed, the angle formed by vectors X and Y.

Now this example revealed no information that couldn't have been obtained directly from the diagram above. But this is only because the dimension involved (namely 2) is small enough to permit a diagram, or picture. For higher dimensions, no such pictures can help us in analyzing properties of vectors, lengths and angles; we *must* appeal to the analytic interpretations of our geometric notations as given as in Definition VI.4.1.

A striking example of such an analytic extension of a geometric notion appears now, as we extend to *any* dimension, the notion of one vector being perpendicular to another. For example, we shall know what is meant for a 17-dimensional vector to be perpendicular to another 17-dimensional vector, even though no "picture" can be drawn to represent this perpendicularity. And that is quite a feat. The definition follows

Definition VI.4.2 Let X and Y be vectors of $V_n(\mathbf{R})$ [respectively, $V_n(\mathbf{C})$]. Then X is said to be **perpendicular** to Y (respectively, X is **orthogonal** to Y), symbolized $\mathbf{X} \perp \mathbf{Y}$, or $\mathbf{Y} \perp \mathbf{X}$, if and only if $\langle X,Y\rangle = 0$.

Notice that this notion of perpendicularity for real vectors in $V_n(\mathbf{R})$ is wholly consistent with the geometric notion of perpendicularity for the special 2-dimensional space. In (VI.4.1), or from Definition VI.4.1(c), we see that $\|X\| \, \|Y\| \cos(\theta) = \langle X,Y\rangle$, equals zero (for nonzero vectors X and Y) if and only if $\cos(\theta) = 0$. That is to say, for dimension 2, $\langle X,Y\rangle$ equals zero if and only if $\theta = 90$ degrees, or $\theta = 270$ degrees, which says that X and Y are, indeed, perpendicular in the intuitive sense. Definition VI.4.2 does extend the notion of perpendicularity to complex vectors and to all finite dimensions.

Eventually (Theorem VI.4.4), we shall show that any set $\mathscr{X} = \{X_1, X_2, \ldots, X_k\}$ of **pairwise perpendicular** vectors or **pairwise orthogonal** vectors (that is, for every pair of distinct vectors X_i, $X_j \in \mathscr{X}$, we have $X_i \perp X_j$), is automatically a *linearly independent* set of vectors in V_n. Before this, however, we shall do well to isolate some of the basic properties of the inner product.

Theorem VI.4.3 Properties of the inner product function $\langle\cdot,\cdot\rangle$.

(a) $\langle X,X\rangle > 0$ for all nonzero vectors $X \in V_n$. Moreover, $\langle X,X\rangle = 0$ if and only if $X = O \in V_n$.

(b) $\langle X,Y\rangle = \overline{\langle Y,X\rangle}$ for all X, $Y \in V_n$.

(c) $\langle \alpha X + \beta Y, Z\rangle = \alpha\langle X,Z\rangle + \beta\langle Y,Z\rangle$, for all X, Y, $Z \in V_n$ and all scalars α and β, that is, the inner product is linear in the first argument.

(d) $\langle X, \alpha Y + \beta Z\rangle = \bar{\alpha}\langle X,Y\rangle + \bar{\beta}\langle X,Z\rangle$, for all X, Y, $Z \in V_n$ and all scalars α and β, that is, the inner product is conjugate linear in the second argument.

Proof: The proofs of (a) through (d) follow immediately from Definition VI.4.1, the definition of the inner product, and are left as exercises.

REMARK. In Theorem VI.4.3(a), we are told that if $X \neq O$, then $\langle X,X \rangle$ is always positive. This is in agreement with our intuitive notion that "length" should always be positive [see Definition VI.4.1(b)]. Property (b) of Theorem VI.4.3 says that when we interchange the 2 arguments, or variables X and Y, we effect a complex conjugation of the value of the inner product. Of course, if X and Y have only real entries, we always have $\langle X,Y \rangle = \langle Y,X \rangle$. Conditions (c) and (d) together tell us that the inner product (considered as a function of 2 ordered variables X and Y) is a functional sending pairs of vectors to scalars. For the real case ($X, Y \in V_n(\mathbf{R})$), the inner product is 2-linear; see Definition V.1.1.

We come to Theorem VI.4.4, which says, roughly, that pairwise perpendicularity (in the real case) or pairwise orthogonality (in the complex case), always implies linear independence.

Theorem VI.4.4 Let $\mathscr{X} = \{X_1, X_2, \ldots, X_k\}$ be a set of k distinct nonzero vectors of V_n such that for each pair of vectors, X_i and X_j in $\mathscr{X}$, $X_i \perp X_j$, that is, $\langle X_i, X_j \rangle = 0$. Then the set $\mathscr{X}$ is linearly independent.

Proof: We are given that the set $\mathscr{X} = \{X_1, X_2, \ldots, X_n\}$ is pairwise perpendicular or orthogonal. How can we establish linear independence as a necessary consequence? According to Theorem I.6.6, the ordered set $\mathscr{X} = \{X_1, X_2, \ldots, X_k\}$ is linearly independent if and only if $O = \alpha_1 X_1 + \alpha_2 X_2 + \cdots + \alpha_k X_k$ *only* for the trivial case where each and every scalar α_i equals zero. This is our test, then. We shall set $O = \alpha_1 X_1 + \alpha_2 X_2 + \cdots + \alpha_k X_k$ and if the *only* scalars $\alpha_1, \alpha_2, \ldots, \alpha_k$ which suffice to make the equality obtain are the zero scalars, then $\mathscr{X} = \{X_1, X_2, \ldots, X_k\}$ can only be linearly, independent. Accordingly, we assume

$$O = \alpha_1 X_1 + \alpha_2 X_2 + \cdots + \alpha_i X_i + \cdots + \alpha_k X_k; \qquad \text{(VI.4.2)}$$

we are to show that for each and every $i = 1, 2, \ldots, k$, α_i necessarily equals zero. We choose i arbitrarily, $i = 1, 2, \ldots, k$ and fix it from here on. Thus,

$$\begin{aligned} 0 &= \langle O, X_i \rangle && \text{from Definition VI.4.1} \\ &= \langle \alpha_1 X_1 + \alpha_2 X_2 + \cdots + \alpha_i X_i + \cdots + \alpha_k X_k, \ X_i \rangle && \text{from (VI.4.2)} \\ &= \alpha_1 \overset{0}{\langle X_1, X_i \rangle} + \alpha_2 \overset{0}{\langle X_2, X_i \rangle} + \cdots + \alpha_i \langle X_i, X_i \rangle + \cdots + \alpha_k \overset{0}{\langle X_k, X_i \rangle} \\ & && \text{from Theorem VI.4.3(c).} \end{aligned}$$

Now our hypothesis of pairwise perpendicularity (or orthogonality) forces us to conclude that each of the above inner product terms is zero—*except* for the ith term $\langle X_i,X_i\rangle$ which cannot be zero since X_i is not the zero vector [see Theorem VI.4.3(a)]. Our equality above then reads

$$0 = \alpha_i\langle X_i,X_i\rangle$$

which can occur only if the scalar $\alpha_i = 0$. Now i was chosen arbitrarily, so we must conclude that for *all* i, $i = 1, 2, \ldots, k$, $\alpha_i = 0$. That is, $O = \alpha_1 X_1 + \alpha_2 X_2 + \cdots + \alpha_i X_i + \cdots + \alpha_k X_k$ *only* for the trivial case $\alpha_1 = \alpha_2 = \cdots = \alpha_k = 0$; there is no other choice possible for the α_i's, and the set $\mathscr{X} = \{X_1,X_2,\ldots,X_k\}$ is therefore linearly independent. The theorem is done.

EXERCISES

VI.4.1 Given the vectors

$$X = \begin{pmatrix} 0 \\ 2 \\ 2 \end{pmatrix}, \qquad Y = \begin{pmatrix} 2i \\ -1 \\ 1 \end{pmatrix}, \qquad Z = \begin{pmatrix} -i \\ 2 \\ 0 \end{pmatrix},$$

compute

(a) $\langle X,X\rangle$, $\quad \|Y\|$, $\quad \langle Z,Z\rangle$, $\quad \|Z\|$

(b) $\langle X,Y\rangle$, $\quad \langle Y,Z\rangle$, $\quad \langle Z,Y\rangle$, $\quad \langle X,Z\rangle$

(c) $\langle 3iX, -2Y\rangle$, $\quad \langle 3Y, -2Z\rangle$, $\quad \langle 3Y,4iY\rangle$

(d) $\langle 2X + 3iY,Z\rangle$, $\quad 2\langle X,Y\rangle + 3i\langle Y,Z\rangle$

(e) $\langle 2Y,4X - 2Z\rangle$, $\quad 2\cdot 4\langle Y,X\rangle - 2\cdot 2\langle Y,Z\rangle$.

VI.4.2 Given the vectors

$$X = \begin{pmatrix} 2 \\ 1 \\ 0 \\ 1 \end{pmatrix}, \qquad Y = \begin{pmatrix} 0 \\ i \\ 2 \\ -1 \end{pmatrix}, \qquad Z = \begin{pmatrix} 2 \\ 1 \\ i \\ 1 \end{pmatrix}$$

in V_4, compute the expressions described in parts (a) through (e) of Exercise VI.4.1.

VI.4.3 Given the typical vectors $X = \begin{pmatrix} x_1 \\ x_2 \end{pmatrix}$, $Y = \begin{pmatrix} y_1 \\ y_2 \end{pmatrix}$, and $Z = \begin{pmatrix} z_1 \\ z_2 \end{pmatrix} \in V_2$, along with the arbitrary scalars α and β. Use Definition VI.4.1 directly to verify each of the 4 properties (a) through (d) of Theorem VI.4.3.

VI.4.4 (See Exercise VI.4.3.) Use Definition VI.4.1 to prove the validity of each of the properties (a) through (d) of Theorem VI.4.3 for arbitrary vectors X, Y, and Z in V_n.

VI.4.5 Show that in Theorem VI.4.3, property (d) does not have to be proved independently of the other properties; in fact, property (d) is a direct consequence of properties (b) and (c).

VI.4.6 Consider X and Y to be (possibly complex) 2-vectors. Theorem VI.4.3 [properties (b) and (c)], presents the inner product $\langle\cdot,\cdot\rangle$ as a 2-linear functional sending the ordered pairs of 2-vectors (X,Y) to the scalar $\langle X,Y\rangle$. Now, Definition V.1.2 also defines the determinant functional as a certain 2-linear functional sending the ordered pair of 2-vectors (X,Y) to the scalar $\det(X,Y)$.

For which vectors X, $Y \in V_2$, is it possible that $\langle X,Y\rangle = \det(X,Y)$?

VI.4.7 In the statement of Theorem VI.4.4, k *distinct nonzero* vectors are chosen for study.

(a) Is the adjective "nonzero" necessary in the hypothesis. Why or why not?

(b) Is the term "distinct" necessary in the hypothesis, or is this automatically implied by the condition of pairwise orthogonality?

VI.4.8 Given fixed vectors X, Y and $Z \in V_n$.

(a) Prove that $\langle X+\alpha Y, Z\rangle = \langle X,Z\rangle$ for all scalars α if and only if $Y \perp Z$.

(b) Is statement (a) above true when the phrase "for all scalars α" is replaced by "for at least one nonzero scalar α"?

VI.4.9 If X and $Y \in V_n$, then prove that

(a) the value of $\langle \alpha X, Y\rangle$ is the same for all scalars α if and only if $X \perp Y$.

(b) Is statement (a) above true when the phrase "for all scalars α" is replaced with "for some pair of scalars $\alpha = \alpha_1$ and $\alpha = \alpha_2$"?

VI.4.10 What is the maximum number of pairwise perpendicular or orthogonal vectors in any subset $\{X_1, X_2, \ldots, X_k\}$ of V_n?

VI.4.11 Prove that in V_n, the zero vector O is the only vector which is perpendicular to every other vector of V_n; that is, if $\langle Y, X_0\rangle = 0$ for *every* nonzero $Y \in V_n$, then necessarily, $X_0 = O$.

VI.4.12 Prove that for all scalars α and all $X \in V_n$, $\|\alpha X\| = |\alpha|\,\|X\|$, where $|\alpha|$ is the absolute value (modulus) of the complex number α.

VI.4.13 Given arbitrary nonzero vectors X and $Y \in V_n$, we define the vector

$$Z \in V_n \text{ by } Z = X - \frac{\langle X,Y\rangle}{\langle Y,Y\rangle} Y.$$

(a) Prove that $Z \perp Y$.

(b) Prove that the 3 vectors X, $\dfrac{\langle X,Y\rangle}{\langle Y,Y\rangle} Y$, and Z obey the law of Pythagoras in the sense that $\left\|\dfrac{\langle X,Y\rangle}{\langle Y,Y\rangle}\cdot Y\right\|^2 + \|Z\|^2 = \|X\|^2$.

(c) **(Cauchy-Schwarz inequality.)** From the inequality $0 \le \|Z\|^2$, prove that $|\langle X,Y\rangle|^2 \le \langle X,X\rangle\langle Y,Y\rangle$. When does exact equality hold?

VI.4.14 In this exercise, you are given $n-1$ vectors $X_1, X_2, \ldots, X_{n-1}$ of V_n. If possible, find at least 1 vector X_n of length 1 which is perpendicular to or orthogonal to each X_i, $i = 1, 2, \ldots, n-1$.

(a) $X_1 = \begin{pmatrix} 2 \\ 0 \end{pmatrix}$

(b) $X_1 = \begin{pmatrix} 2 \\ 0 \\ 1 \end{pmatrix}$, $X_2 = \begin{pmatrix} 0 \\ 1 \\ 0 \end{pmatrix}$

(c) $X_1 = \begin{pmatrix} 2 \\ 1 \\ 0 \end{pmatrix}, \quad X_2 = \begin{pmatrix} 1 \\ 2 \\ 0 \end{pmatrix}$

(d) $X_1 = \begin{pmatrix} 1 \\ 1 \\ 1 \end{pmatrix}, \quad X_2 = \begin{pmatrix} 1 \\ 1 \\ 1 \end{pmatrix}$

(e) $X_1 = \begin{pmatrix} 1 \\ 1 \\ 0 \end{pmatrix}, \quad X_2 = \begin{pmatrix} 1 \\ -1 \\ 0 \end{pmatrix}$

(f) $X_1 = \begin{pmatrix} 1 \\ 0 \\ 0 \\ 0 \end{pmatrix}, X_2 = \begin{pmatrix} 0 \\ 1 \\ 0 \\ 0 \end{pmatrix}, X_3 = \begin{pmatrix} 0 \\ 0 \\ i \\ 0 \end{pmatrix}$

(g) $X_1 = \begin{pmatrix} 2 \\ 1 \\ 0 \\ 0 \end{pmatrix}, X_2 = \begin{pmatrix} 1 \\ 2 \\ 0 \\ 0 \end{pmatrix}, X_3 = \begin{pmatrix} 1 \\ i \\ 0 \\ 0 \end{pmatrix}$

(h) $X_1 = \begin{pmatrix} 2 \\ 1 \\ 0 \\ 0 \end{pmatrix}, X_2 = \begin{pmatrix} 1 \\ 2 \\ 0 \\ 0 \end{pmatrix}, X_3 = \begin{pmatrix} 2 \\ 2 \\ 1 \\ 0 \end{pmatrix}$

(i) $X_1 = \begin{pmatrix} 0 \\ 0 \\ 0 \\ 1 \end{pmatrix}, X_2 = \begin{pmatrix} 2 \\ 0 \\ 1 \\ 0 \end{pmatrix}, X_3 = \begin{pmatrix} 1 \\ i \\ 1 \\ 1 \end{pmatrix}$

VI.4.15 Let M be a subspace of V_n (Definition I.7.1). Prove that $M^\perp$, the set of all vectors in V_n which are perpendicular or orthogonal to each and every vector of M, is also a sub*space* of V_n; that is, the one and only test for a vector X to be a qualified member of $M^\perp$ is that $\langle X,Y \rangle = 0$ for all $Y \in M$, and this set $M^\perp$ is a sub(vector) space of V_n.

VI.4.16 Suppose the set $\mathscr{X} = \{X_1, X_2, \ldots, X_n\} \subset V_n$ is an orthonormal basis for V_n. The term **orthonormal** means that $\mathscr{X}$ is a set of pairwise perpendicular or orthogonal vectors (hence, $\mathscr{X}$ is a linearly independent set of n vectors. $\mathscr{X}$ is therefore a *basis*). Pairwise perpendicularity or orthogonality accounts for the prefix "ortho." Moreover, each X_i in $\mathscr{X}$ is of length 1, which accounts for the "normal" in "orthonormal." What we have said, in brief, is that for $X_i, X_j \in V_n$,

$$\langle X_i, X_j \rangle = 0 \quad \text{if } i = j,$$

and

$$\langle X_i, X_j \rangle = 1 \quad \text{if } i = j.$$

Since $\mathscr{X}$ is a basis, we know that any $Y \in V_n$ has a representation as $Y = \alpha_1 X_1 + \alpha_2 X_2 + \cdots + \alpha_n Y_n$, for *unique* scalars $\alpha_1, \alpha_2, \ldots, \alpha_n$. (Theorem I.6.8.) Prove that these unique scalars are necessarily of the form $\alpha_i =$

$\langle Y, X_i \rangle$ for each $i = 1, 2, \ldots, n$; that is, prove that

$$Y = \langle Y, X_1 \rangle X_1 + \langle Y, X_2 \rangle X_2 + \cdots + \langle Y, X_n \rangle X_n.$$

Hint: Write $Y = \alpha_1 X_1 + \alpha_2 X_2 + \cdots + \alpha_n X_n$ and isolate α_i by considering $\langle Y, X_i \rangle$; use Theorem VI.4.3(c).

Definition VI.4.5 Let $A = (a_{ij})_{ij}$, $i = 1, 2, \ldots, m$, $j = 1, 2, \ldots, n$, be an $m \times n$ matrix. Then A^*, the **conjugate transpose** of the matrix A is the $n \times m$ matrix defined by $A^* = (\bar{a}_{ij})_{ji}$. That is, the jith entry of A^* is the scalar $\bar{a}_{ij}$, the complex conjugate of a_{ij}. Of course, if each a_{ij} is a real number, $\bar{a}_{ij} = a_{ij}$ and $A^* = A^t$, the transpose of A.

Example If

$$A = \begin{pmatrix} 1 & i & -1 \\ 2+i & 0 & 0 \end{pmatrix},$$

then

$$A^* = \begin{pmatrix} 1 & 2-i \\ -i & 0 \\ -1 & 0 \end{pmatrix}.$$

Our interest in conjugate transpose matrices stems from the relation between A and A^* induced through the inner product. In fact, we see how the conjugate transpose matrix becomes a natural object of study in Theorem VI.4.6.

Theorem VI.4.6 Let $A = (a_{ij})$, $i = 1, 2, \ldots, m$, $j = 1, 2, \ldots, n$ be an $m \times n$ matrix with complex entries a_{ij}. Then for all (column) vectors $X \in V_n$ and $Y \in V_m$,

$$\langle A \cdot X, Y \rangle = \langle X, A^* \cdot Y \rangle \tag{VI.4.3}$$

where $A^* = (\bar{a}_{ij})_{ji}$ is the $n \times m$ conjugate transpose of A.

Note: The left-hand side of (VI.4.3) represents the inner product of the m-vectors, $A \cdot X$ and Y; the right-hand side denotes the inner product of n-vectors X and $A^* \cdot Y$.

Proof: There is little more to the proof of (VI.4.3) than a mere verification by computation in accordance with Definition VI.4.1. In fact, set

$$X = \begin{pmatrix} x_1 \\ x_2 \\ \vdots \\ x_n \end{pmatrix}, \quad Y = \begin{pmatrix} y_1 \\ y_2 \\ \vdots \\ y_m \end{pmatrix}.$$

The time-worn rule of matrix multiplication tells us that for $A = (a_{ij})_{ij}$

and $A^* = (\bar{a}_{ij})_{ji}$,

$$\underbrace{\begin{pmatrix} a_{11} & a_{12} & \cdots & a_{1n} \\ a_{21} & a_{22} & \cdots & a_{2n} \\ \vdots & \vdots & & \vdots \\ a_{m1} & a_{m2} & \cdots & a_{mn} \end{pmatrix}}_{A} \cdot \underbrace{\begin{pmatrix} x_1 \\ x_2 \\ \vdots \\ x_n \end{pmatrix}}_{X} = \begin{pmatrix} \sum_k a_{1k}x_k \\ \sum_k a_{2k}x_k \\ \vdots \\ \sum_{k=1}^{n} a_{ik}x_k \\ \vdots \\ \sum_k a_{mk}x_k \end{pmatrix} \begin{matrix} \leftarrow \text{first entry of } A \cdot X \\ \leftarrow \text{second entry of } A \cdot X \\ \\ \leftarrow i\text{th entry of } A \cdot X \\ \\ \leftarrow m\text{th entry of } A \cdot X, \end{matrix}$$

$$A \cdot X \in V_m \qquad \text{(VI.4.4a)}$$

and

$$\underbrace{\begin{pmatrix} \bar{a}_{11} & \bar{a}_{21} & \cdots & \bar{a}_{m1} \\ \bar{a}_{12} & \bar{a}_{22} & \cdots & \bar{a}_{m2} \\ \vdots & \vdots & & \vdots \\ a_{1n} & a_{2n} & \cdots & a_{mn} \end{pmatrix}}_{A^*} \cdot \underbrace{\begin{pmatrix} y_1 \\ y_2 \\ \vdots \\ y_m \end{pmatrix}}_{Y} = \begin{pmatrix} \sum_i \bar{a}_{i1}y_i \\ \sum_i \bar{a}_{i2}y_i \\ \vdots \\ \sum_{i=1}^{m} \bar{a}_{ik}y_i \\ \vdots \\ \sum_i \bar{a}_{in}y_i \end{pmatrix} \begin{matrix} \leftarrow \text{first entry of } A^* \cdot Y \\ \leftarrow \text{second entry of } A^* \cdot Y \\ \\ \leftarrow k\text{th entry of } A^* \cdot Y \\ \\ \leftarrow n\text{th entry of } A^* \cdot Y. \end{matrix}$$

$$A^* \cdot Y \in V_n \qquad \text{(VI.4.4b)}$$

Roughly speaking, Definition VI.4.1 tells us to compute $\langle X, Y\rangle$ by "matching" ith entries of each vector, multiplying them together (not to forget conjugating the entries of Y), and adding these products. Thus,

$$\langle A \cdot X, Y\rangle = \sum_{i=1}^{m} \Big(\underbrace{\sum_{k=1}^{n} a_{ik}x_k}_{\substack{i\text{th entry of } A \cdot X;\\ \text{see (VI.4.4a)}}} \Big) \underbrace{\overline{y_i}}_{i\text{th entry of } Y} \qquad \text{(Definition VI.4.1)} \qquad \text{(VI.4.5)}$$

and

$$\langle X, A^*Y\rangle = \sum_{k=1}^{n} \underbrace{x_k}_{k\text{th entry of } X} \underbrace{\overline{\Big(\sum_{i=1}^{m} \bar{a}_{ik}y_i \Big)}}_{\substack{k\text{th entry of } A^*Y;\\ \text{see (VI.4.4b)}}} \qquad \text{(Definition VI.4.1)}$$

$$= \sum_{k=1}^{n} x_k \Big(\sum_{i=1}^{m} a_{ik}\overline{y_i} \Big) \qquad \text{effect of conjugation on } \sum_i \bar{a}_{ik}y_i \qquad \text{(VI.4.6)}$$

$$= \sum_{i=1}^{m} \sum_{k=1}^{n} a_{ik}x_k\bar{y}_i \qquad \text{interchanging order of summation.}$$

Compare (VI.4.5) with (VI.4.6); observing equality is to establish (VI.4.3) and hence, proof of our theorem.

As a matter of taste, we may take the *matrix* result of Theorem VI.4.6 and convert it to a result concerning (abstract) *linear transformations* $\mathbf{T} \in \mathscr{L}(V_n, V_m)$. We use Theorem II.2.3 which says that any linear transformation $\mathbf{T}$ is affected only by a left-hand matrix multiplication, that is, $\mathbf{T}(X)$ ($\mathbf{T}$ *of* X) always equals $A \cdot X$ (A *times* X), where $A = mx(\mathbf{T})_{\mathscr{E}\mathscr{E}'}$, the $m \times n$ matrix of $\mathbf{T}$ relative to the standard ordered bases $\mathscr{E}$ and $\mathscr{E}'$ of V_n and V_m, respectively. Theorem VI.4.6 restated in terms of linear transformations, then becomes Corollary VI.4.7.

Corollary VI.4.7 Let $\mathbf{T} : V_n \to V_m$ be any linear transformation sending V_n to V_m. Then there exists a unique linear transformation $\mathbf{T}^* : V_m \to V_n$ such that for all $x \in V_n$ and all $Y \in V_m$,

$$\langle \mathbf{T}(X), Y \rangle = \langle X, \mathbf{T}^*(Y) \rangle. \tag{VI.4.7}$$

Moreover, if $\mathbf{T}(X) = A \cdot X$ for the $m \times n$ matrix A, then $\mathbf{T}^*(Y) = A^* \cdot Y$ for all $Y \in V_m$.

Proof: This one is really easy. Since $\mathbf{T}(X)$, $\mathbf{T}$ of X, can be thought of as a matrix product,

$\mathbf{T}(X) = A_{\mathrm{T}} \cdot X$ for the $m \times n$ matrix $A_{\mathrm{T}} = mx(\mathbf{T})_{\mathscr{E}\mathscr{E}}$. (Theorem II.2.3)

$$\begin{aligned} \langle \mathbf{T}(X), \quad Y \rangle &= \langle A_{\mathrm{T}} \cdot X, \quad Y \rangle && \text{Theorem II.2.3} \\ &= \langle X, A_{\mathrm{T}}^* \cdot Y \rangle && \text{Theorem VI.4.6} \\ &= \langle X, \mathbf{T}^*(Y) \rangle && \text{define } \mathbf{T}^*(Y) = A_{\mathrm{T}}^* \cdot Y. \end{aligned}$$

In the last step, we *defined* $\mathbf{T}^* : V_m \to V_n$ to be the left-hand multiplication by the $n \times m$ conjugate transpose matrix A_{T}^*, of A_{T}, that is, $\mathbf{T}^*(Y) = A_{\mathrm{T}}^* \cdot Y$ for all $Y \in V_m$. Certainly $\mathbf{T}^*$ does send m-vectors to n-vectors and is, moreover, *linear* (Definition II.2.1), that is, $\mathbf{T}^* \in \mathscr{L}(V_m, V_n)$ as promised, and is constructed in accordance with our "moreover" phrase in the statement of Corollary VI.4.7.

This last result, which saw us construct a linear transformation $\mathbf{T}^* \in \mathscr{L}(V_m, V_n)$ form $\mathbf{T} \in \mathscr{L}(V_n, V_m)$, motivates the following definition.

Definition VI.4.8 Let $\mathbf{T} \in \mathscr{L}(V_n, V_m)$. Then the linear transformation $\mathbf{T}^* \in \mathscr{L}(V_m, V_n)$, which depends on $\mathbf{T}$, with the property that

$$\langle \mathbf{T}(X), Y \rangle = \langle X, \mathbf{T}^*(Y) \rangle$$

for all $X \in V_n$, $Y \in V_m$, is called the **adjoint** of $\mathbf{T}$.

REMARK. Thanks to Corollary VI.4.7, we know that such adjoint transformations always can be found [since $\mathbf{T}^*(Y)$ is just $A_{\mathrm{T}}^* \cdot Y$, where $\mathbf{T}(X) = A_{\mathrm{T}} \cdot X$]; thus, we are ethically justified in presenting Definition VI.4.8 since we have no fears about nonexistence of such $\mathbf{T}^*$'s.

Also, the adjoint $\mathbf{T}^*$ of a linear *transformation* $\mathbf{T}$ is not to be confused with adj (A), the *matrix* adjoint of the *matrix* A (Definition IV.4.3).

EXERCISES

In Exercises VI.4.17 through VI.4.20, you are given a rectangular $m \times n$ matrix A. In each case,

(a) Construct the $n \times m$ conjugate transpose matrix A^*.

(b) Construct the m-vector $A \cdot X$ and the n-vector $A^* \cdot Y$, where X and Y are arbitrary vectors in V_n and V_m, respectively.

(c) Use Definition VI.4.1 to compute separately, the quantities $\langle A \cdot X, Y \rangle$ and $\langle X, \quad A^*Y \rangle$.

VI.4.17 $$A = \begin{pmatrix} 1 & 2 & -1 \\ 0 & 3 & 1 \end{pmatrix}.$$

VI.4.18 $$A = \begin{pmatrix} i & 1 \\ 2 & 0 \\ i & -1 \end{pmatrix}.$$

VI.4.19 $$A = \begin{pmatrix} 3 & 0 & 0 \\ 0 & 3 & 0 \\ 0 & 0 & 0 \end{pmatrix}.$$

VI.4.20 $A = (1 \quad i \quad 0 \quad -2)$, a 1×4 matrix.

VI.4.21 Let $A = \text{diag }(a,a,\ldots,a) = aI_n$ be the $n \times n$ diagonal matrix whose diagonal entries are each equal to a. If we define $\mathbf{T} \in \mathscr{L}(V_n,V_n)$ by $\mathbf{T}(X) = A \cdot X$, then describe $\mathbf{T}^*(X)$ for arbitrary $X \in V_n$.

VI.4.22 Suppose $\mathbf{T} \in \mathscr{L}(V_3,V_3)$ is that linear operator which annihilates the first and third entries of any concrete 3-vector $X \in V_3$, that is,

$$\mathbf{T}\begin{pmatrix} x_1 \\ x_2 \\ x_3 \end{pmatrix} = \begin{pmatrix} 0 \\ x_2 \\ 0 \end{pmatrix} \quad \text{for all } \begin{pmatrix} x_1 \\ x_2 \\ x_3 \end{pmatrix} \in V_3.$$

(a) Use Definition VI.4.8 directly to show that $\mathbf{T} = \mathbf{T}^*$ ($\mathbf{T}$ equals its own adjoint).

(b) Find $A_{\mathbf{T}} = mx(\mathbf{T})_{\mathscr{E}\mathscr{E}}$ so that $\mathbf{T}(X) = A_{\mathbf{T}} \cdot X$. Show that $A_{\mathbf{T}} = A_{\mathbf{T}}^*$, an alternate proof that $\mathbf{T} = \mathbf{T}^*$ (compare Corollary VI.4.7).

(c) Find ker($\mathbf{T}$) the null space of $\mathbf{T}$ [the set of all X such that $\mathbf{T}(X) = O$] and find its dimension.

(d) Find a basis for rg($\mathbf{T}$), the range of $\mathbf{T}$ [the set of all "output" vectors of the form $\mathbf{T}(X)$], and compute its dimension (see Theorem II.3.6).

(e) Show that every vector in rg($\mathbf{T}$) is orthogonal to every vector in ker($\mathbf{T}$), that is, show that if $X \in \text{rg}(\mathbf{T})$, and if $Y \in \ker(\mathbf{T})$, then $\langle X,Y \rangle = 0$.

(f) Is it true that if $\langle X,Y \rangle = 0$, then *necessarily* 1 of the vectors, X or Y, belongs to rg($\mathbf{T}$), while the other belongs to ker($\mathbf{T}$)?

(g) If $\langle X,Y \rangle = 0$, and $X \in \text{rg}(\mathbf{T})$, does $Y \in \ker(\mathbf{T})$?

VI.4.23 Given $\mathbf{T} \in \mathscr{L}(V_2,V_2)$ defined by

$$\mathbf{T}\begin{pmatrix} x_1 \\ x_2 \end{pmatrix} = \begin{pmatrix} 2x_1 \\ 3x_2 \end{pmatrix} \quad \text{for all } \begin{pmatrix} x_1 \\ x_2 \end{pmatrix} \in V_2.$$

Answer the 7 questions, (a) through (g) of Exercise VI.4.22 relative to the operator **T** defined above.

VI.4.24 Given **T** and $\mathbf{U} \in \mathscr{L}(V_2, V_2)$ defined by:

$$\mathbf{T}\begin{pmatrix} x_1 \\ x_2 \end{pmatrix} = \begin{pmatrix} 2x_1 - x_2 \\ x_1 \end{pmatrix}, \quad \mathbf{U}\begin{pmatrix} x_1 \\ x_2 \end{pmatrix} = \begin{pmatrix} 3x_1 - 6x_2 \\ -x_1 + x_2 \end{pmatrix}$$

for all $\begin{pmatrix} x_1 \\ x_2 \end{pmatrix} \in V_2$. Relative to the standard ordered basis $\mathscr{E} = \{E_1, E_2\}$ for V_2, find the matrices of the following operators:

(a) **T** and **T***
(b) **U** and **U***
(c) $\mathbf{T} \circ (\mathbf{U}^*)$
(d) $(\mathbf{U} \circ \mathbf{T}^*)^*$
(e) $\mathbf{T}^* \circ \mathbf{U} \circ \mathbf{T}$
(f) $(\mathbf{T}^* \circ \mathbf{U}^* \circ \mathbf{T})^*$.

VI.4.25 Given the linear operators, **T** and **U**, defined in Exercise VI.4.24. For any given $X = \begin{pmatrix} x_1 \\ x_2 \end{pmatrix} \in V_n$, find

(a) $\mathbf{T}^*(X)$ and $\mathbf{U}^*(X)$
(b) $\mathbf{T} \circ (\mathbf{U}^*)(X)$
(c) $\mathbf{T}^* \circ \mathbf{U} \circ \mathbf{T}(X)$.

VI.4.26 Given **T** and $\mathbf{U} \in \mathscr{L}(V_3, V_3)$ defined by

$$\mathbf{T}\begin{pmatrix} x_1 \\ x_2 \\ x_3 \end{pmatrix} = \begin{pmatrix} x_1 \\ x_2 \\ 0 \end{pmatrix}, \text{ and } \mathbf{U}\begin{pmatrix} x_1 \\ x_2 \\ x_3 \end{pmatrix} = \begin{pmatrix} ix_2 \\ x_1 - x_3 \\ 0 \end{pmatrix} \text{ for all } \begin{pmatrix} x_1 \\ x_2 \\ x_3 \end{pmatrix} \in V_3.$$

Relative to the standard ordered basis $\mathscr{E} = \{E_1, E_2, E_3\}$ for V_3, find the matrices of the following operators:

(a) **T** and **T***
(b) **U** and **U***
(c) $\mathbf{U}^* \circ \mathbf{T} \circ \mathbf{U}$
(d) $\mathbf{U} \circ \mathbf{U}^*$
(e) $\mathbf{U}^* \circ \mathbf{U}$
(f) $\mathbf{T}^5$.

VI.4.27 With the linear operators, **T** and **U**, defined in Exercise VI.4.26, and for any

$$X = \begin{pmatrix} X_1 \\ x_2 \\ x_3 \end{pmatrix} \in V_3,$$

find

(a) $\mathbf{T}^*(X)$ and $\mathbf{U}^*(X)$
(b) $\mathbf{U} \circ \mathbf{T} \circ \mathbf{U}(X)$
(c) $\mathbf{U} \circ \mathbf{U}^*(X)$
(d) $\mathbf{U}^* \circ \mathbf{U}(X)$
(e) $\mathbf{T}^5(X)$.

VI.4.28 With the linear transformations **T** given below, find $mx\,(\mathbf{T})_{\mathscr{E}\mathscr{E}'}$, $mx\,(\mathbf{T})^*{}_{\mathscr{E}'\mathscr{E}}$ and $\mathbf{T}^*(X)$ for all $X \in \operatorname{dom}(\mathbf{T})$.

(a) $$\mathbf{T}\begin{pmatrix} x_1 \\ x_2 \end{pmatrix} = \begin{pmatrix} x_1 + x_2 \\ x_1 - x_2 \\ 3x_3 \end{pmatrix} \quad \text{for all } \begin{pmatrix} x_1 \\ x_2 \end{pmatrix} \in V_2.$$

(b) $$\mathbf{T}\begin{pmatrix} x_1 \\ x_2 \\ x_3 \\ x_4 \end{pmatrix} = \begin{pmatrix} 3ix_1 \\ x_2 - ix_3 \\ x_2 - ix_3 \end{pmatrix} \quad \text{for all } \begin{pmatrix} x_1 \\ x_3 \\ x_3 \\ x_4 \end{pmatrix} \in V_4.$$

(c) $\mathbf{T}(x) = \begin{pmatrix} 2x \\ 3ix \\ -x \end{pmatrix}$ for all $x \in V_1$.

(d) $\mathbf{T}\begin{pmatrix} x_1 \\ x_2 \\ x_3 \end{pmatrix} = (x_1 + x_2 + x_3)$ for all $\begin{pmatrix} x_1 \\ x_2 \\ x_3 \end{pmatrix} \in V_3$.

5. Unitary and Orthogonal Transformations

Definition VI.5.1 Let U be an $n \times n$ matrix. Then U is called a **unitary matrix** (respectively **orthogonal matrix**) if and only if for all $X, Y \in V_n(\mathbf{C})$ [respectively for all $X, Y \in V_n(\mathbf{R})$],

$$\langle U \cdot X, \quad U \cdot Y\rangle = \langle X,Y\rangle.$$

The linear operator $\mathbf{T}_U \in \mathscr{L}(V_n, V_n)$ defined by $\mathbf{T}_U(X) = U \cdot X$ (so that $U = mx(\mathbf{T})_{\mathscr{E}\mathscr{E}}$) is called a **unitary** or **orthogonal operator** according as $V_n = V_n(\mathbf{C})$, or $V_n = V_n(\mathbf{R})$, respectively. Equivalently, then, $\mathbf{T}$ is **unitary** (respectively **orthogonal**) if and only if

$$\langle \mathbf{T}(X), \quad \mathbf{T}(Y)\rangle = \langle X, Y\rangle$$

for all X, $Y \in V_n(\mathbf{C})$ [respectively, for all X, $Y \in V_n(\mathbf{R})$].

REMARK. By this time, the reader should be at home with the device used in Definition VI.5.1, namely, passing from a matrix A to the linear transformation $\mathbf{T}(X) = A \cdot X$ (see Theorem II.2.3), and conversely, passing from linear operators $\mathbf{T}$ to their concrete representation matrices $A_{\mathbf{T}} = \mathrm{m}x(\mathbf{T})_{\mathscr{E}\mathscr{E}}$ [so that $\mathbf{T}(X) = A_{\mathbf{T}} \cdot X$].

In a certain sense, the unitary or orthogonal transformations are those operators on V_n that deform the vector-space V_n as little as possible. We know that such remarkable operators do exist; the identity is one such. To fix ideas, let us consider an orthogonal operator $\mathbf{T}$ on $V_2(R)$, which we may picture as the set of vectors in the plane. If we set $X = Y$ in Definition VI.5.1, we obtain, for all $X \in V_2(\mathbf{R})$, the equation

$$\langle \mathbf{T}(X),\mathbf{T}(X)\rangle = \langle X,X\rangle,$$

or

$$\|T(X)\| = \|X\| \quad \text{[Definition VI.4.1(b)]}$$

which says that the linear orthogonal operator $\mathbf{T}$ preserves the *length* of every vector in $V_2(R)$, that is, $\|X\|$, the length of X, is always equal to $\|\mathbf{T}(X)\|$, the length of its image under the action of $\mathbf{T}$. A consequence of this, is that no vector X can get sent to $O \in V_2(\mathbf{R})$, unless $X = O$ to begin with. Thus, the orthogonal operator $\mathbf{T}$ is always invertible since $\ker(\mathbf{T}) = \{O\}$ (see Theorem II.3.9).

We observe another geometric characteristic of any orthogonal operator $\mathbf{T}$ on the plane, namely, $\mathbf{T}$ preserves the *angle* between pairs of vectors (the angle between vectors X and Y is not easy to visualize for *complex* 2-vectors). That is to say, if θ is the angle formed by X and Y, then θ is also the angle formed by the image vectors, $\mathbf{T}(X)$ and $\mathbf{T}(Y)$. To see this, recall that in the real case, $\langle X,Y\rangle = \|X\| \cdot \|Y\| \cdot \cos(\theta_{XY})$, where $\theta_{X,Y}$ is the angle formed by X and Y. That is,

$$\begin{aligned}
\cos(\theta_{X,Y}) &= \langle X,Y\rangle / \|X\| \cdot \|Y\| \\
&= \langle \mathbf{T}(X),\mathbf{T}(Y)\rangle / \|X\| \cdot \|Y\| \quad \text{Definition VI.5.1} \\
&= \langle \mathbf{T}(X),\mathbf{T}(Y)\rangle / \|\mathbf{T}(X)\| \cdot \|\mathbf{T}(Y)\| \quad \text{since } \mathbf{T} \text{ preserves length} \\
&= \cos(\theta_{\mathrm{T}(X),\mathrm{T}(Y)}).
\end{aligned}$$

Thus, $\theta_{\mathrm{T}(X),\mathrm{T}(Y)}$, the angle formed by $\mathbf{T}(X)$ and $\mathbf{T}(Y)$, is equal to $\theta_{X,Y}$, the angle formed by X and Y. Figure 6.5 illustrates the length and angle-preserving property of orthogonal operators.

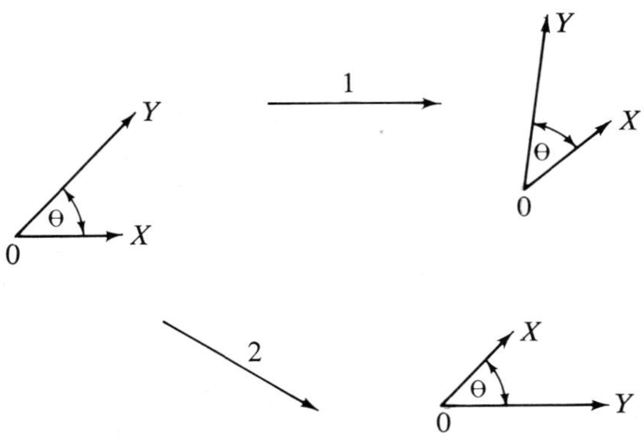

Figure 6.5

We now present properties of unitary and orthogonal operators $\mathbf{T}$, along with parallel results concerning their matrices A_{T} [where $A_{\mathrm{T}} = mx(\mathbf{T})_{\mathcal{E}\mathcal{E}}$, which implies $\mathbf{T}(X) = A_{\mathrm{T}} \cdot X$]. We present Theorem VI.5.2.

Theorem VI.5.2 Let $\mathbf{T}$ be a unitary operator on $V_n(\mathbf{C})$ [respectively orthogonal operator on $V_n(\mathbf{R})$]. Let $mx(\mathbf{T})_{\mathcal{E}\mathcal{E}} = A_{\mathrm{T}}$ be the $n \times n$ unitary (respectively orthogonal) matrix induced by $\mathbf{T}$, so that $\mathbf{T}(X) = A_{\mathrm{T}} \cdot X$, for all X. Then

(a) $\mathbf{T}$ is an invertible operator and $A_{\mathbf{T}}$ is an invertible matrix. Moreover, $\mathbf{T}^{-1} = \mathbf{T}^*$, the adjoint of $\mathbf{T}$, and $A_{\mathbf{T}}^{-1} = A_{\mathbf{T}}^*$, the conjugate transpose of $A_{\mathbf{T}}$.

(b) The set of the n rows and the set of the n columns of the unitary (respectively orthogonal) matrix $A_{\mathbf{T}}$, are both *orthonormal* sets of row and column vectors, respectively. Conversely, if an $n \times n$ matrix A consists of n columns which form an orthonormal set of column vectors, then that matrix is necessarily a unitary, or orthogonal, matrix. Hence the linear transformation $\mathbf{T}(X) = A \cdot X$ is a unitary or orthogonal transformation.

Proof: ***Condition*** (*a*). In our preliminary discussion of orthogonal operators on the plane, we established invertibility. In general, we have the same result; we establish invertibility by showing that the *only* vector X which is sent to O, is the zero vector itself. From Theorem II.3.9, this will establish invertibility. Accordingly, suppose $\mathbf{T}(X) = O$. Then $\|\mathbf{T}(X)\|$, the length of $\mathbf{T}(X)$ is zero. In any case, $\mathbf{T}$ preserves length ($\|\mathbf{T}(X)\| = \|X\|$) which is seen by setting $X = Y$ in Definition VI.5.1. Thus $0 = \|\mathbf{T}(X)\| = \|X\|$. Thus, $X = O$, since the only vector X with zero length, is the zero-vector $X = O$ [Theorem VI.4.3(a)]. To see that $\mathbf{T}^{-1} = \mathbf{T}^*$, we observe that for all $X, Y \in V_n$,

$$
\begin{aligned}
\langle \mathbf{T}^* \circ \mathbf{T}(X), Y\rangle &= \langle \mathbf{T}^*(\mathbf{T}(X)), Y\rangle && \text{Definition II.1.5} \\
&= \langle \mathbf{T}(X), \mathbf{T}^{**}(Y)\rangle && \text{Definition VI.4.8} \\
&= \langle \mathbf{T}(X), \mathbf{T}(Y)\rangle && \text{since } \mathbf{T}^{**} = \mathbf{T}\text{, see} \\
& && \text{Exercise VI.5.1} \qquad \text{(VI.5.1)} \\
&= \langle X, Y\rangle && \text{Definition VI.5.1} \\
&= \langle \mathbf{I}(X), Y\rangle && \mathbf{I} \text{ is the identity operator.}
\end{aligned}
$$

From (VI.5.1), we have that for all $X, Y \in V_n$, $\langle \mathbf{T}^* \circ \mathbf{T}(X), Y\rangle = \langle \mathbf{I}(X), Y\rangle$; equivalently,

$$\langle ((\mathbf{T}^* \circ \mathbf{T}) - \mathbf{I})X, Y\rangle = 0 \quad \text{for all } X, Y. \qquad \text{(VI.5.2)}$$

Using an old trick again, we argue as follows:

(1) $(\mathbf{T}^* \circ \mathbf{T} - \mathbf{I})(X)$ is orthogonal to all Y, from (VI.5.2).

(2) Hence $(\mathbf{T}^* \circ \mathbf{T} - \mathbf{I})X = O$ since only O is orthogonal to *all* vectors.

(3) Hence, $\mathbf{T}^* \circ \mathbf{T}(X) = X$ for all X, that is, $\mathbf{T}^* \circ \mathbf{T} = \mathbf{I}$ so that $\mathbf{T}^*$ is a left (hence, 2-sided) inverse for $\mathbf{T}$. That is, $\mathbf{T}^* = \mathbf{T}^{-1}$, as promised.

We may use the multiplicative property of the function $mx(\cdot)_{\mathscr{E}\mathscr{E}}$ (Definition II.6.12) to show a corresponding result for the unitary (or orthogonal) matrix $A_{\mathrm{T}} = mx(\mathbf{T})_{\mathscr{E}\mathscr{E}}$. That is, we have

$$\begin{aligned} A_{\mathrm{T}}^* \cdot A_{\mathrm{T}} &= mx(\mathbf{T}^*)_{\mathscr{E}\mathscr{E}} \cdot mx(\mathbf{T})_{\mathscr{E}\mathscr{E}} && mx(\mathbf{T}) = A_{\mathrm{T}} \Rightarrow mx(\mathbf{T}^*) = A_{\mathrm{T}}^*, \text{ Corollary VI.4.7} \\ &= mx(\mathbf{T}^* \circ \mathbf{T})_{\mathscr{E}\mathscr{E}} && \text{Definition II.6.12} \\ &= mx(\mathbf{I})_{\mathscr{E}\mathscr{E}} && \text{from (c)} \\ &= I_n && \text{the identity matrix.} \end{aligned} \tag{VI.5.3}$$

From (VI.5.3), we have that A_{T}^* is a left (hence, 2-sided) inverse for A_{T}, that is, $A_{\mathrm{T}}^* = A_{\mathrm{T}}^{-1}$ and $\mathbf{T}^* = \mathbf{T}^{-1}$ for unitary (orthogonal) $\mathbf{T} \in \mathscr{L}(V_n, V_n)$. Part (a) is proven.

Condition (*b*). Let $\mathbf{T}$ be unitary (or orthogonal) on V_n, and let $A_{\mathrm{T}} = mx(\mathbf{T})_{\mathscr{E}\mathscr{E}}$ be its induced unitary (or orthogonal) matrix, so that $\mathbf{T}(X) = A_{\mathrm{T}} \cdot X$. Now there is no easier way to isolate the *i*th column of a matrix A than to multiply A on the right by E_i, the *i*th standard basis vector of V_n. That is, if

$$A_{\mathrm{T}} = [A_1 A_2 \cdots A_i \cdots A_n],$$

where A_i is the *i*th column of A_{T}, then (VI.5.4)

$$A_i = A_{\mathrm{I}} \cdot E_i \quad \text{for each } i = 1, 2, \ldots, n.$$

Accordingly, for i, j (not necessarily distinct) $= 1, 2, \ldots, n$,

$$\begin{aligned} \langle A_i, A_j \rangle &= \langle A_{\mathrm{T}} \cdot E_i, \quad A_{\mathrm{T}} \cdot E_j \rangle && \text{from (VI.5.4)} \\ &= \langle \mathbf{T}(E_i), \quad \mathbf{T}(E_j) \rangle && \text{Theorem II.2.3} \\ &= \langle E_i, E_j \rangle && \text{Definition VI.5.1} \\ &= \begin{Bmatrix} 0 & \text{if} & i \neq j \\ 1 & \text{if} & i = j \end{Bmatrix} && \{E_1, \ldots, E_i, \ldots, E_n\} \text{ is orthonormal.} \end{aligned} \tag{VI.5.5}$$

Therefore, from (VI.5.5), we see that the *columns* of A_{T} form an o.n. set. To see that the rows of A_{T} are also o.n., we use the fact that if A_{T} is orthogonal, so is A_{T}^t, the transpose of A_{T} (Exercises VI.5.3 and VI.5.4); rows of A_{T} then become columns of A_{T}^t and our result above, applies. To see that an o.n. set of columns is *sufficient* to induce the unitary property on a matrix A, is left as Exercise VI.5.6. So much for the proof of this theorem.

Example We consider the orthogonal operator on the plane [a pictorial model for $V_2(\mathbf{R})$] which rotates every vector counter-clockwise through a fixed angle θ. It is not obvious that such a transformation, call it **Rot**, should be linear, but we shall establish this fact. That is, we shall show that if we first add $X + Y$, then rotate the sum, we obtain the same result as when we rotate X and Y separately, then add

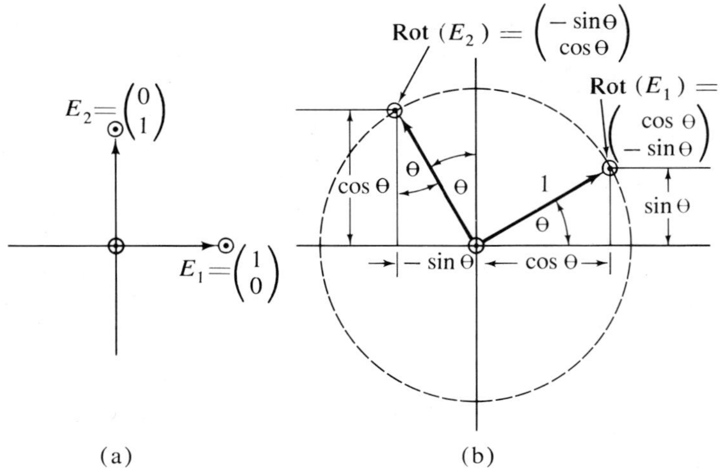

Figure 6.6

these results, that is, $\mathbf{Rot}(X + Y) = \mathbf{Rot}(X) + \mathbf{Rot}(Y)$. Similarly, we must verify that $\alpha\,\mathbf{Rot}(X) = \mathbf{Rot}(\alpha X)$ for all real scalars α, and all $X \in V_2(\mathbf{R})$.

As Figure 6.6 indicates, a rotation of $E_1 = \begin{pmatrix} 1 \\ 0 \end{pmatrix}$ through the angle θ, results in the unit-vector whose x-coordinate is $\cos(\theta)$, and whose y-coordinate is $\sin(\theta)$, that is, $\mathbf{Rot}(E_1) = \begin{pmatrix} \cos(\theta) \\ \sin(\theta) \end{pmatrix}$.

Similarly

$$\mathbf{Rot}(E_2) = \begin{pmatrix} -\sin(\theta) \\ \cos(\theta) \end{pmatrix}.$$

To put it another way,

$$\mathbf{Rot}(E_1) = \begin{pmatrix} \cos(\theta) \\ \sin(\theta) \end{pmatrix} = \cos(\theta)E_1 + \sin(\theta)E_2$$

$$\mathbf{Rot}(E_2) = \begin{pmatrix} -\sin(\theta) \\ \cos(\theta) \end{pmatrix} = -\sin(\theta)E_1 + \cos(\theta)E_2.$$

From Definition II.6.2, we have

$$mx(\mathbf{Rot})_{\mathscr{E}\mathscr{E}} = A_{\text{Rot}} = \begin{pmatrix} \cos(\theta) & -\sin(\theta) \\ \sin(\theta) & \cos(\theta) \end{pmatrix}. \qquad \text{(VI.5.6)}$$

Geometrically, the inverse operation of **Rot**, is that rotation **Rot′** in the opposite direction to θ, that is, clockwise. Thus, $\mathbf{Rot} \circ \mathbf{Rot}' = \mathbf{I}$, the identity operator. Indeed, it may be verified that

$$\begin{aligned} A_{\text{Rot}}^{-1} &= A_{\text{Rot}}^{*} && \text{Theorem VI.2.6(b)} \\ &= \begin{pmatrix} \cos(\theta) & \sin(\theta) \\ -\sin(\theta) & \cos(\theta) \end{pmatrix} \\ &= \begin{pmatrix} \cos(-\theta) & -\sin(-\theta) \\ \sin(-\theta) & \cos(-\theta) \end{pmatrix} && \text{since } \cos(\theta) = \cos(-\theta) \\ & && \text{and } \sin(-\theta) = -\sin(\theta) \end{aligned}$$

is exactly of the form (VI.5.6) where $-\theta$ has replaced θ, that is, A_{Rot}^{-1}, the inverse matrix of A_{Rot}, is truly a rotation through an angle of $-\theta$, which returns each vector to its original position, after A_{Rot} is effected. It is easy to verify by direct matrix computation that $A_{\text{Rot}} \cdot A_{\text{Rot}}^{-1} = \mathbf{1}_2$, and that A_{Rot} is a bona-fide orthogonal matrix, that is, that

$$\langle A_{\text{Rot}} \cdot X, A_{\text{Rot}} \cdot Y\rangle = \langle X,Y\rangle \quad \text{for all } X, Y \in V_2(\mathbf{R}).$$

From the geometric definition of the operator **Rot**, one can see that **Rot** has *no* eigenvectors unless $\theta = 180$ degrees or $\theta = 0$ degrees, that is, unless **Rot** $= -$**I**, or Rot $=$ **I**. That is, no vector is simply stretched or contracted by **Rot**, since, after all, a rotation is being effected. Computing the characteristic polynomial of **Rot** (of A_{Rot}) will offer an algebraic (as opposed to geometric) proof of this fact, inasmuch as this polynomial will have no real roots (that is, no real eigenvalues) and hence, **Rot** has no corresponding eigenvectors [see Exercise VI.5.7(d)], unless $\theta = 0°$ or $\theta = 180$ degrees.

EXERCISES

VI.5.1 Prove that for any linear transformation $\mathbf{T}: V_n \to V_m$ (it is not necessary that $m = n$), $\mathbf{T}^{**} = \mathbf{T}$. (*Hint*: Use Corollary VI.4.7 and reduce the question to one concerning conjugate transpose matrices.)

VI.5.2 Let $A = (a_{ij})_{ij}$ be an $m \times n$ matrix, and let X and Y be concrete column vectors of V_n and V_m, respectively. Use Definition VI.4.1 directly to prove

(a) $\overline{\langle A \cdot X, Y\rangle} = \langle \overline{A} \cdot \overline{X}, \overline{Y}\rangle$,

where the left-hand side of (a) denotes the complex conjugate of the (complex) scalar $\langle A \cdot X,Y\rangle$, and the matrices $\overline{A}$, $\overline{X}$, and $\overline{Y}$ in the right-hand side of (a) are obtained from A, X and Y by replacing each entry by its complex conjugate [that is, if $A = (a_{ij})_{ij}$, then $\overline{A} = (\overline{a}_{ij})_{ij}$].

VI.5.3 (See Exercise VI.5.2.) Let $A = (a_{ij})_{ij}$ be a unitary (respectively orthogonal) $n \times n$ matrix. Show that necessarily,

(a) $\overline{A} = (\overline{a}_{ij})_{ij}$, the conjugate of A, is also unitary (respectively orthogonal)

(b) $A^t = (a_{ji})_{ij}$, the transpose of A is also unitary (respectively orthogonal)

Hint: Use the fact that A^*, the conjugate transpose of A can be represented as $A^* = (\overline{A})^t = (\overline{A^t})$; also, $\overline{\overline{A}} = A$ and $(A^t)^t = A$.

VI.5.4 (See Exercise VI.5.3.) Prove that any unitary or orthogonal matrix A has an o.n. set of *row* vectors.

VI.5.5 [Any linear operator which sends the standard basis $\mathscr{E}$, to an orthonormal basis $\mathscr{X}$, is necessarily unitary (orthogonal).] Given $\mathbf{U} \in \mathscr{L}(V_n,V_n)$, with the property that $\langle \mathbf{U}(E_i), \mathbf{U}(E_j)\rangle = \langle E_i,E_j\rangle$, that is, the set

$$\mathscr{X} = \{\mathbf{U}(E_1),\mathbf{U}(E_2),\ldots,\mathbf{U}(E_n)\}$$

is a set of unit vectors which are pairwise orthogonal (perpendicular), as is the set of standard basis vectors $\mathscr{E} = \{E_1,E_2,\ldots,E_n\}$. Prove that $\mathbf{U}$ is a unitary (orthogonal) operator on V_n. That is, for arbitrary $X = \sum_{i=1}^n x_1E_i$, and $Y = \sum_{j=1}^n y_jE_j$, show that $\langle \mathbf{U}(X),\mathbf{U}(Y)\rangle = \langle X,Y\rangle$.

VI.5.6 Suppose $A_{\mathbf{U}}$ is the $n \times n$ matrix $mx(\mathbf{U})_{\mathscr{E}\mathscr{E}}$, for some $\mathbf{U} \in \mathscr{L}(V_n, V_n)$. Suppose the n columns of $A_{\mathbf{U}}$ form an orthonormal set of n-vectors. Prove that this is sufficient evidence for concluding that $\mathbf{U}$ is a unitary (orthogonal) operator on V_n—see Exercise VI.5.5; also see the proof of Theorem VI.6.2.(c).

VI.5.7 Let A_{Rot} be the 2×2 orthogonal matrix defined in the example following the proof of Theorem VI.5.2 (page 435).
Verify
(a) $A_{\text{Rot}} \cdot A^t_{\text{Rot}} = A^t_{\text{Rot}} \cdot A_{\text{Rot}} = I_2$.
(b) If $\theta = 120$ degrees or $\frac{2}{3}\pi$ radians, then $A^3_{\text{Rot}} = I_2$.
(c) If $\theta = 90$ degrees or $\pi/2$ radians, then $A^4_{\text{Rot}} = I_2$.
(d) $p_A(\cdot)$, the characteristic polynomial of A_{Rot}, has real roots if and only if $\sin(\theta) = 0$, that is, if and only if $\theta = 0$, or $\theta = 180$ degrees, which is to say, if and only if $A_{\text{Rot}} = I_2$ or $A_{\text{Rot}} = -I_2$.
(e) For all X, Y, $\in V_2(\mathbf{R})$, for all θ, $\langle A_{\text{Rot}} \cdot X, A_{\text{Rot}} \cdot Y\rangle = \langle X, Y\rangle$.

VI.5.8 Which of the following matrices are unitary (or orthogonal)?

(a) $\begin{pmatrix} 1 & 0 \\ 0 & -1 \end{pmatrix}$

(b) $\begin{pmatrix} 0 & 1 \\ -1 & 0 \end{pmatrix}$

(c) $\begin{pmatrix} x & -y \\ y & x \end{pmatrix}$, $x^2 + y^2 = 1$

(d) $\begin{pmatrix} \frac{1}{2} & 0 & \frac{\sqrt{3}}{2} \\ 0 & 1 & 0 \\ \frac{\sqrt{3}}{2} & 0 & -\frac{1}{2} \end{pmatrix}$

VI.5.9 Find the values (if possible) of the unknown variables which appear in the following matrices, so that they become unitary or orthogonal.

(a) $\begin{pmatrix} 1 & x \\ 0 & y \end{pmatrix}$ (c) $\begin{pmatrix} \frac{\sqrt{3}}{2} & x & 0 \\ 0 & 0 & z \\ \frac{1}{2} & y & 0 \end{pmatrix}$

(b) $\begin{pmatrix} x & u \\ y & v \end{pmatrix}$

VI.5.10 Let $\mathbf{U}$ and $\mathbf{V}$ be the (linear)? functions sending $V_2(\mathbf{R})$ to itself, defined geometrically in Figure 6.7 as follows:
$\mathbf{U}$ reflects each vector about the x-axis
$\mathbf{V}$ reflects each vector about the y-axis.

(a) Find $\mathbf{U}\begin{pmatrix} x_1 \\ x_2 \end{pmatrix}$ and $\mathbf{V}\begin{pmatrix} x_1 \\ x_2 \end{pmatrix}$ for each $\begin{pmatrix} x_1 \\ x_2 \end{pmatrix} \in V_2(\mathbf{R})$.

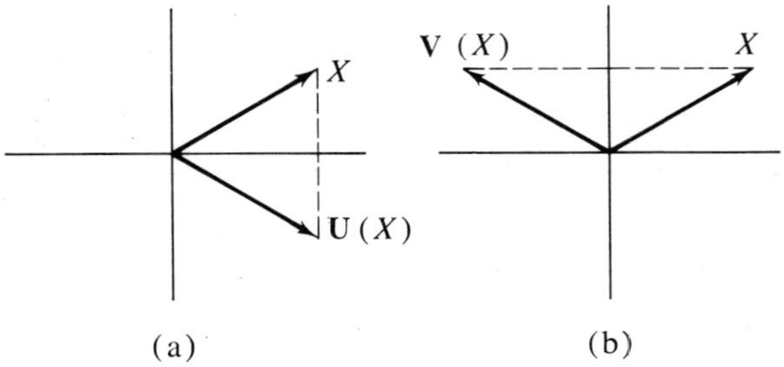

Figure 6.7

(b) Prove that the functions **U** and **V** are linear.

(c) Find $A_{\mathbf{U}}$ and $A_{\mathbf{V}}$, the 2×2 matrices of **U** and **V**, respectively, relative to the standard ordered basis $\mathscr{E} = \{E_1, E_2\}$.

(d) Are the operators **U** and **V** orthogonal operators?

(e) Show that $\mathbf{U} \circ \mathbf{U} = \mathbf{V} \circ \mathbf{V} = \mathbf{I}$, and by direct computation, that

$$A_{\mathbf{U}} \cdot A_{\mathbf{U}} = A_{\mathbf{V}} \cdot A_{\mathbf{V}} = I_2.$$

(f) Find all eigenvalues and corresponding eigenvectors in $V_2(\mathbf{R})$ for both operators **U** and **V**.

VI.5.11 Prove that for any unitary (respectively, orthogonal) operator **U** on V_n, all eigenvalues are necessarily scalars of length one, that is,

$$\mathbf{U}(X) = \lambda X \to |\lambda| = 1 \text{ [respectively, } \mathbf{U}(X) = \lambda X \Rightarrow \lambda = \pm 1].$$

6. Self-adjoint Operators and Their Matrices

We shall interest ourselves now in those square matrices A which are equal to their conjugate transpose A^* (that is, $\overline{A}^t$); translating into the language of operators, we are concerned with those $\mathbf{T} \in \mathscr{L}(V_n, V_n)$ with the property that $\mathbf{T} = \mathbf{T}^*$.

Definition VI.6.1 Let $A = (a_{ij})_{ij}$ be an $n \times n$ matrix. Then A is said to be **hermitian** if and only if $A = A^*$ (that is, if and only if $a_{ij} = \bar{a}_{ji}$ for all i, j). A is called **symmetric** if and only if $A = A^t$ (that is, if and only if $a_{ij} = a_{ji}$ for all i,j), where A^t is the transpose of A.

Similarly, a linear operator $\mathbf{T} \in \mathscr{L}(V_n, V_n)$ is called **self-adjoint** if and only if $\mathbf{T} = \mathbf{T}^*$, where $\mathbf{T}^*$ is the adjoint of $\mathbf{T}$.

It is easy to characterize those $\mathbf{T} \in \mathscr{L}(V_n, V_n)$ for which $\mathbf{T} = \mathbf{T}^*$. We present Proposition VI.6.2.

Proposition VI.6.2 (Matrix characterization of self-adjoint operators.) Let $\mathbf{T} \in \mathscr{L}(V_n, V_n)$. Then $\mathbf{T} = \mathbf{T}^*$ (**T** is a self-adjoint *operator*) if and only if $A_{\mathbf{T}} = A_{\mathbf{T}}^*$ ($A_{\mathbf{T}}$ is a hermitian *matrix*), where $A_{\mathbf{T}} = mx(\mathbf{T})_{\mathscr{E}\mathscr{E}}$, that is, $\mathbf{T}(X) = A_{\mathbf{T}} \cdot X$ for all $X \in V_n$.

Proof: From Corollary VI.4.7, we see that for any $\mathbf{T} \in \mathscr{L}(V_n, V_n)$, $\mathbf{T}^*(X) = A_\mathbf{T}^* \cdot X$ for all $X \in V_n$, where $\mathbf{T}(X) = A_\mathbf{T} \cdot X$ defines the $n \times n$ matrix $A_\mathbf{T}$. Surely, then, $\mathbf{T}^* = \mathbf{T}$ if and only if $A_\mathbf{T} = A_\mathbf{T}^*$, that is, if and only if $A_\mathbf{T}$ is a hermitian matrix.

Example Let $A = \begin{pmatrix} 2 & i \\ -i & 0 \end{pmatrix}$. Then $A = A^*$, and hence, the linear operator $\mathbf{T}$ on V_2 defined by A

$$\mathbf{T}\begin{pmatrix} x_1 \\ x_2 \end{pmatrix} = A \cdot \begin{pmatrix} x_1 \\ x_2 \end{pmatrix} = \begin{pmatrix} 2 & i \\ -i & 0 \end{pmatrix} \cdot \begin{pmatrix} x_1 \\ x_2 \end{pmatrix} = \begin{pmatrix} 2x_1 + ix_2 \\ -ix_1 \end{pmatrix} \quad \text{for all} \begin{pmatrix} x_1 \\ x_2 \end{pmatrix} \in V_2$$

is a *self-adjoint* operator. Indeed, actual computation would reveal that

$$\langle \mathbf{T}(X), Y \rangle = \langle X, \mathbf{T}(Y) \rangle$$

for all X, $Y \in V_2$ (see Definition VI.4.1).

Let us focus our attention upon those self-adjoint linear operators on V_n which have n *not necessarily distinct* eigenvalues $\lambda_1, \lambda_2, \ldots, \lambda_n$. We shall discover that each eigenvalue λ_i is necessarily a *real* number [even for $V_n = V_n(\mathbf{C})$], and that there exist n correspondingly linearly independent eigenvectors X_1, X_2, $\ldots$, X_n (compare Theorem VI.3.3) which may be chosen to be pairwise *orthogonal* (*perpendicular*, for the real case).

It comes as no surprise that any linear operator $\mathbf{T} \in \mathscr{L}(V_n(\mathbf{C}), V_n(\mathbf{C}))$ has n not necessarily distinct (and generally complex) eigen values $\lambda_1, \lambda_2, \ldots, \lambda_n$. Why? Because λ_i is an eigenvalue for $\mathbf{T}$ if and only if λ_i is also a root of the nth degree characteristic polynomial (with complex coefficients), $\det(A_\mathbf{T} - \lambda \mathbf{I}_n)$ (Theorem VI.2.3). Every such nth degree polynomial has n, not necessarily distinct, complex roots (Fundamental Theorem of Algebra); in other words, every $\mathbf{T} \in \mathscr{L}(V_n(\mathbf{C}), V_n(\mathbf{C}))$ has n, not necessarily distinct, complex eigenvalues. For each *distinct* eigenvalue, we are assured of at least 1 eigenvector in $V_n(\mathbf{C})$; this follows, because if λ_i is an eigenvalue then $\det(A_\mathbf{T} - \lambda_i \mathbf{I}_n) = 0$. That is, the $n \times n$ matrix $A_\mathbf{T} - \lambda_i I_n$ is not invertible (Theorem IV.2.9). Thus, we may assert that r, the rank of $A_\mathbf{T} - \lambda_i I_n$, is less than and not equal to n (Theorem III.3.5). Thus, a *nonzero* n-vector $X \in V_n(\mathbf{C})$ exists such that

$$(A_\mathbf{T} - \lambda_i I_n) \cdot X = O$$

(Theorem III.2.4). To write this equation in equivalent form, we have: If λ_i is a root of the characteristic polynomial $\det\,(A_\mathbf{T} - \lambda I_n)$, then there exists a nonzero $X \in V_n(\mathbf{C})$ such that

$$A_\mathbf{T} \cdot X = \lambda_i X;$$

that is, X is an eigenvector for $\mathbf{T}$ (see Theorem VI.2.3). Thus, we have at least as many eigenvectors as we have distinct eigenvalues.

Theorem VI.6.3 Let $\mathbf{T} = \mathbf{T}^* \in \mathscr{L}(V_n, V_n)$ be a self-adjoint operator on V_n which has the k *distinct* eigenvalues $\lambda_1, \lambda_2, \ldots, \lambda_k$ (necessarily, $1 \leq k \leq n$) corresponding to the k (nonzero) eigenvectors $X_1, X_2, \ldots, X_k$, respectively; that is, $\mathbf{T}(X_i) = \lambda_i X_i$ (equivalently, $A_\mathbf{T} \cdot X_i = \lambda_i X_i$) for each $i = 1, 2, \ldots, k$. Then λ_i is always a real scalar ($\lambda_i = \bar{\lambda}_i$) for each $i = 1, 2, \ldots, k$, and moreover, the set of eigenvectors $\{X_1, X_2, \ldots, X_k\}$ is pairwise *orthogonal* (pairwise *perpendicular*, for the real case).

Proof: With $\mathbf{T} = \mathbf{T}^*$, where $\mathbf{T}(X_i) = \lambda_i X_i$, we have

$$\begin{aligned}
\lambda_i\langle X_i,X_i\rangle &= \langle \lambda_i X_i,X_i\rangle && \text{Theorem VI.4.3(c) (set } Y = O) \\
&= \langle \mathbf{T}(X_i),X_i\rangle && \text{Definition of eigenvector } X_i \\
&= \langle X_i,\mathbf{T}^*(X_i)\rangle && \text{Definition VI.4.8} \\
&= \langle X_i,\mathbf{T}(X_i)\rangle && \text{Hypothesis that } \mathbf{T} = \mathbf{T}^* \\
&= \langle X_i,\lambda_i X_i\rangle && \text{Definition of eigenvector } X_i \\
&= \bar{\lambda}_i\langle X_i,X_i\rangle && \text{Theorem VI.4.3(d) (set } Z = O).
\end{aligned}$$

This string of equalities tells us that for any eigenvector X_i of $\mathbf{T}$, with corresponding eigenvalue λ_i,

$$(\lambda_i - \bar{\lambda}_i)\langle X_i,X_i\rangle = 0.$$

Since X_i is an eigenvector, and is therefore nonzero, $\|X_i\|^2 = \langle X_i,X_i\rangle \neq 0$. In order for the last equality to hold, then, $\lambda_i - \bar{\lambda}_i = 0$, that is, λ_i is a real scalar, which proves the first part of Theorem VI.6.3.

We wish to show, now, that the set of eigenvectors $\{X_1,X_2,\ldots,X_k\}$ is pairwise orthogonal (perpendicular), that is, for all i, j, $i \neq j$, we must prove that $\langle X_i,X_j\rangle = 0$. Accordingly, let us consider the following chain of equalities:

$$\begin{aligned}
\lambda_i\langle X_i,X_j\rangle &= \langle \lambda_i X_i,X_j\rangle && \text{Theorem VI.4.3} \\
&= \langle \mathbf{T}(X_i),X_j\rangle && \text{Definition of eigenvector } X_i \\
&= \langle X_i,\mathbf{T}^*(X_j)\rangle && \text{Definition VI.4.8} \\
&= \langle X_i,\lambda_j X_j\rangle && \text{Definition of eigenvector } X_j \text{ for } \mathbf{T} = \mathbf{T}^* \\
&= \lambda_j\langle X_i,X_j\rangle && \text{note that } \bar{\lambda}_j = \lambda_j.
\end{aligned}$$

From these equalities, we conclude that

$$\text{for all } i, j, \ i \neq j, \quad (\lambda_i - \lambda_j)\langle X_i,X_j\rangle = 0. \tag{VI.6.1}$$

Now $(\lambda_i - \lambda_j)$ cannot possibly equal zero since, by hypothesis, the eigenvalues $\lambda_1, \lambda_2, \ldots, \lambda_k$, are *distinct* from one another. (This is the first and only time we use the hypothesis of distinctness in our proof.) Since (VI.6.1) holds true, and since $(\lambda_i - \lambda_j) \neq 0$, the burden of being zero falls entirely upon the scalar factor $\langle X_i, X_j \rangle$. That is, $\langle X_i, X_j \rangle = 0$ necessarily, and our proof is done.

REMARK. There should be some unsettlement in the mind of the reader in light of the hypotheses of Theorem VI.6.3. In that theorem, we chose a self-adjoint operator $\mathbf{T}$ with the extraordinary good fortune of having $k \leq n$ *distinct* eigenvalues! The natural question to ask is: How can we be sure that a self-adjoint operator $\mathbf{T}$ on the real vector-space $V_n(\mathbf{R})$, has any eigenvalues at all? (Of course, every operator on *complex* space $V_n(\mathbf{C})$ will have n, not necessarily distinct, eigenvalues since the characteristic nth degree polynomial always has n complex roots). In answer to this question, we present the following existence theorem for the *real* case.

Proposition VI.6.4 Let $\mathbf{T}$ be a self-adjoint operator on $V_n(\mathbf{R})$. Then $\mathbf{T}$ has n, not necessarily distinct, real eigenvalues, that is, the characteristic polynomial of $\mathbf{T}$ has n real roots.

Proof: Let

$$p_{\mathrm{T}}(\lambda) = (-\lambda)^n + a_{n-1}\lambda^{n-1} + \cdots + a_1\lambda + a_0 = \det (A_{\mathrm{T}} - \lambda I_n)$$

be the nth degree characteristic polynomial of $\mathbf{T}$, whose coefficients $a_{n-1}, a_{n-2}, \ldots, a_1, a_0$, are real scalars since all entries of the matrix A_{T} are real. From the Fundamental Theorem of Algebra, $p_{\mathrm{T}}(\cdot)$ enjoys n *complex* (but, hopefully, real) roots $\lambda_1, \lambda_2, \ldots, \lambda_n$, and may be factored in the form

$$p_{\mathrm{T}}(\lambda) = (\lambda_1 - \lambda)(\lambda_2 - \lambda)\cdots(\lambda_n - \lambda) \qquad \text{for all real } \lambda. \qquad \text{(VI.6.2)}$$

Now we brace ourselves for a good trick. We define an extended operator, $\tilde{\mathbf{T}}$, on the *complex* space $V_n(\mathbf{C})$ by defining

$$\tilde{\mathbf{T}}(X) = A_{\mathrm{T}} \cdot X \quad \text{for all } X \in V_n(\mathbf{C}) \text{ [not just for } X \in V_n(\mathbf{R})],$$

where $A_{\mathrm{T}} = mx(\mathbf{T})_{\mathscr{E}\mathscr{E}}$ is the $n \times n$ *real*-entried matrix induced by the *real* operator $\mathbf{T}$ on $V_n(\mathbf{R})$, relative to the standard ordered basis $\mathscr{E}$. $\tilde{\mathbf{T}}$ is an *extension* of $\mathbf{T}$ in the sense that $\tilde{\mathbf{T}}(X) = \mathbf{T}(X)$ whenever the entries of X are real, that is, whenever $X \in V_n(\mathbf{R}) \subset V_n(\mathbf{C})$. We observe that the coefficients of $p_{\tilde{\mathrm{T}}}(\cdot)$, the characteristic polynomial of $\tilde{\mathbf{T}}$ are *equal* to those of $p_{\mathrm{T}}(\cdot)$, the characteristic polynomial of $\mathbf{T}$. This follows, since the *complex* operator $\tilde{\mathbf{T}}$, and the *real* operation $\mathbf{T}$, both induce the *same* $n \times n$ matrix relative to the standard ordered basis $\mathscr{E}$, that is, $A_{\tilde{\mathrm{T}}} = A_{\mathrm{T}}$, and hence, for real λ at least, $p_{\tilde{\mathrm{T}}}(\lambda) = \det (A_{\tilde{\mathrm{T}}} - \lambda I_n) = \det (A_{\mathrm{T}} = \lambda I_n) = p_{\mathrm{T}}(\lambda)$. It is also true that in extending $\mathbf{T}$ to $\tilde{\mathbf{T}}$, we retain the property of self-adjointness, that is, $\mathbf{T} = \mathbf{T}^* \Rightarrow \tilde{\mathbf{T}} = \tilde{\mathbf{T}}^*$ (Exercise VI.6.5). Finally, the n complex roots, λ_1,

$\lambda_2, \ldots, \lambda_n$, of $p_{\mathbf{T}}(\cdot)$, which do exist (from the Fundamental Theorem of Algebra), are *real* since $\tilde{\mathbf{T}}$ is self-adjoint (Theorem VI.6.3). That is, the n roots of $p_{\mathbf{T}}(\cdot)$, the characteristic polynomial of $\mathbf{T}$ which appears in (VI.6.2), are all real. The existence of the n real eigenvalues (roots) for $\mathbf{T}$ is proven.

Our attention now rests on the study of certain self-adjoint operators—called orthogonal projections—which will prove to be the building blocks for any and all self-adjoint operators.

Definition VI.6.5 A linear operator $\mathbf{P} \in \mathscr{L}(V_n, V_n)$ is called an **idempotent** if and only if $\mathbf{P}^2 = \mathbf{P}$. Any self-adjoint idempotent, that is, $P^2 = P = P^*$, is called an **orthogonal projection**.

REMARK. The term orthogonal is used in Definition VI.6.5, since for idempotent operators, self-adjointness ($\mathbf{P} = \mathbf{P}^*$) is equivalent to the fact that the kernel of $\mathbf{P}$ is orthogonal to the range of $\mathbf{P}$. See Exercise VI.6.10(b) for one direction of this equivalence, namely, for any operator $\mathbf{T}$, idempotent or not, $\mathbf{T} = \mathbf{T}^* \Rightarrow \text{rg}\,(\mathbf{T}) \perp \ker\,(\mathbf{T})$. Exercise VI.6.10(a) also asks you to develop the converse for idempotent operators, that is, $\mathbf{T} = \mathbf{T}^2$, and $\text{rg}\,(\mathbf{T}) \perp \ker\,(\mathbf{T}) \Rightarrow \mathbf{T} = \mathbf{T}^*$.

Example We define the operator $\mathbf{P}$ on $V_2(\mathbf{R})$ geometrically by considering a vector $X \in V_2(\mathbf{R})$, as a point or arrow in the plane; $E_1 = \begin{pmatrix}1\\0\end{pmatrix}$ and $E_2 = \begin{pmatrix}0\\1\end{pmatrix}$ are identified with the unit vectors (1,0) and (0,1), which lie on the x-axis and y-axis, respectively. Roughly speaking, $\mathbf{P}$ projects each vector X onto the straight line L, which lies in the first and third quadrants, passes through (0,0), and makes an angle of 45° with the x-axis and the y-axis (see Figure 6.8). In Figure 6.8 we may intuitively check the properties of idempotence ($\mathbf{P}^2 = \mathbf{P}$), and orthogonality in the sense that ker ($\mathbf{P}$), those vectors which project onto $O = (0,0)$, lie on the line which is *perpendicular* to $L = \text{rg}\,(\mathbf{P})$. What may not be clear, is that this function $\mathbf{P}$, sending $V_2(\mathbf{R})$ onto L,

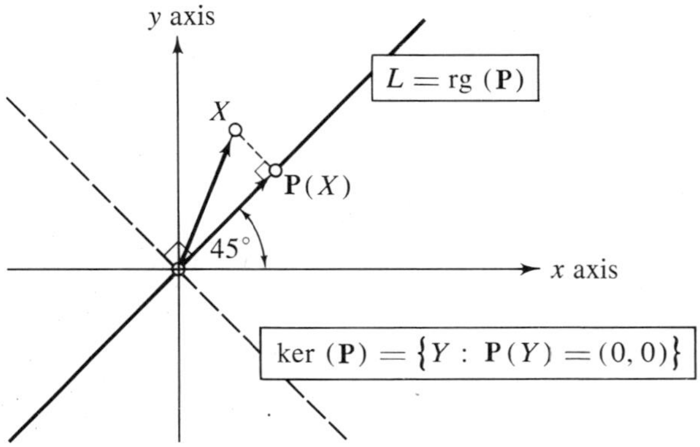

Figure 6.8

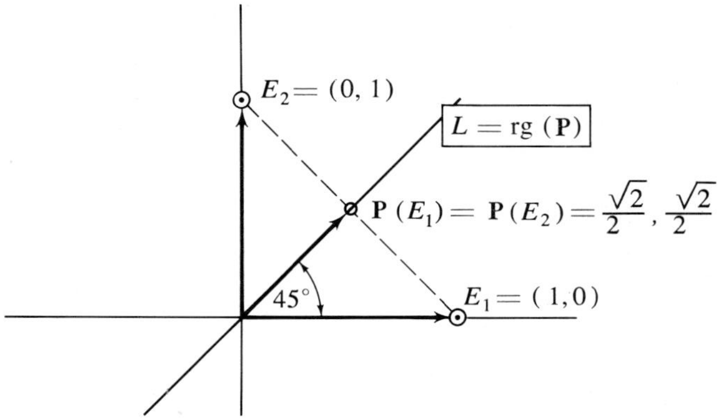

Figure 6.9

is *linear*. We observe that in this case, **P** projects both basis vectors, E_1 and E_2, onto the same image vector on the line L that is, $P(E_1) = P(E_3) = \left(\frac{\sqrt{2}}{2}, \frac{\sqrt{2}}{2}\right)$, which we identify with the 2-vector $\begin{pmatrix} \sqrt{2}/2 \\ \sqrt{2}/2 \end{pmatrix} \in V_2(\mathbf{R})$ (see Figure 6.9). Assuming that we are satisfied with the linearity of **P**, we are free to construct its 2×2 matrix $A_{\mathbf{P}} = mx(\mathbf{P})_{\mathscr{E}\mathscr{E}}$. From Definition II.5.1, or Definition II.6.2, we have

$$A_{\mathbf{P}} = mx(\mathbf{P})_{\mathscr{E}\mathscr{E}} = \begin{pmatrix} \frac{\sqrt{2}}{2} & \frac{\sqrt{2}}{2} \\ \\ \frac{\sqrt{2}}{2} & \frac{\sqrt{2}}{2} \end{pmatrix} = \frac{\sqrt{2}}{2}\begin{pmatrix} 1 & 1 \\ 1 & 1 \end{pmatrix}.$$

Once the matrix has been constructed, we are able to test **P** for self-adjointness. Since $A_{\mathbf{P}} = \bar{A}^t_P = A_{\mathbf{P}}{}^t$, **P** is, indeed, self-adjoint (Proposition VI.6.2). It is easy to verify that $\mathbf{P}^2 = \mathbf{P}$, since the corresponding property obtains for the induced matrix $A_{\mathbf{P}}$, that is,

$$(A_{\mathbf{P}})^2 = \frac{\sqrt{2}}{2}\begin{pmatrix} 1 & 1 \\ 1 & 1 \end{pmatrix} \cdot \frac{\sqrt{2}}{2}\begin{pmatrix} 1 & 1 \\ 1 & 1 \end{pmatrix} = \frac{\sqrt{2}}{2}\begin{pmatrix} 1 & 1 \\ 1 & 1 \end{pmatrix} = A_{\mathbf{P}}.$$

Using this matrix representation for **P**, we can find $\mathbf{P}(X)$ for any $X = V_2(\mathbf{R})$, namely, for all $X = \begin{pmatrix} x_1 \\ x_2 \end{pmatrix} \in V_2(\mathbf{R})$,

$$\begin{aligned} \mathbf{P}(X) &= A_{\mathbf{P}} \cdot X && \text{Theorem II.2.3} \\ &= \frac{\sqrt{2}}{2}\begin{pmatrix} 1 & 1 \\ 1 & 1 \end{pmatrix} \cdot \begin{pmatrix} x_1 \\ x_2 \end{pmatrix} \\ &= \begin{pmatrix} \frac{\sqrt{2}}{2}(x_1 + x_2) \\ \\ \frac{\sqrt{2}}{2}(x_1 + x_2) \end{pmatrix}, \end{aligned}$$

a vector whose first and second coordinates agree. But the 45 degree line L is *exactly* the set of points (x,y) in the plane whose first and second coordinates agree, that is, $L = \{(x,y) : x = y\}$, which shows that the range of $\mathbf{P}$ truly lies on L.

For convenience, we shall at first consider the special case of self-adjoint operators which enjoy the maximum number of *distinct* eigenvalues, that is, we shall assume n distinct eigenvalues for $\mathbf{T}$. We will now show that projections which project V_n onto a ray (projections with 1-dimensional range as in our previous example), generate *all* such operators. This is a spectral theorem, and we offer it now.

Theorem VI.6.6 (A spectral theorem for self-adjoint operators: case of distinct eigenvalues.) Let $\mathbf{T} = \mathbf{T}^* \in \mathscr{L}(V_n, V_n)$, where $\mathbf{T}$ has n distinct eigenvalues $\alpha_1, \alpha_2, \ldots, \alpha_n$, corresponding to the pairwise orthogonal eigenvectors $X_1, X_2, \ldots, X_n$, respectively. Then

(1) $$\mathbf{T} = \alpha_1 \mathbf{P}_{X_1} + \alpha_2 \mathbf{P}_{X_2} + \cdots + \alpha_n \mathbf{P}_{X_n},$$

where each $\mathbf{P}_{X_i}$ is the orthogonal projection whose range is sp $[X_i]$. Moreover,

(2) $$\mathbf{P}_{X_1} + \mathbf{P}_{X_2} + \cdots + \mathbf{P}_{X_n} = \mathbf{I},$$

the identity operator, and

(3) $$\mathbf{P}_{X_i} \circ \mathbf{P}_{X_j} = \mathbf{P}_{X_j} \circ \mathbf{P}_{X_i} = \mathbf{0} \quad \text{for } i \neq j.$$

Proof: As in Theorem VI.3.6, we cast our problem in a matrix setting. Accordingly, but $A_{\mathrm{T}} = mx(\mathbf{T})_{\mathscr{E}\mathscr{E}}$ so that automatically, $\mathbf{T}(X) = A_{\mathrm{T}} \cdot X$ for all $X \in V_n$ (Theorem II.2.3). Now $\mathbf{T}$ has n distinct eigenvalues α_1, $\alpha_2, \ldots, \alpha_n$, and employing the matrix spectral theorem (Theorem III.3.6), we may write

$$A_{\mathrm{T}} = \alpha_1 P_1 + \alpha_2 P_2 + \cdots + \alpha_i P_i + \cdots + \alpha_n P_n, \tag{VI.6.3}$$

where for each i, $P_i = S \cdot E_{ii} \cdot S^{-1}$; recall, $S = [X_1 X_2 \cdots X_n]$ is the $n \times n$ matrix whose ith column is the eigenvector X_i of length one, and $E_{ii} =$ $\text{diag}\,(0,\ldots,0,\underset{\substack{\uparrow\\ i\text{th place}}}{1},0,\ldots,0)$. From (VI.6.3) along with Theorem VI.3.6, we are almost finished with the proof. All that remains is to verify that each P_i of (VI.6.3) is in actuality the orthogonal projection onto sp $[X_i]$ the subspace spanned by eigenvector X_i.

Now we have from (VI.6.3), that $P_i = S \cdot E_{ii} \cdot S^{-1}$. Aha! But S is the matrix whose n columns (unit-length eigenvectors) are pairwise *orthogonal* (Theorem VI.6.3). But this means exactly that S is a unitary matrix [Theorem VI.5.2(b)], so that $S^{-1} = S^*$ [Theorem VI.5.2(a)]. From this we have

$$\begin{aligned} P_i^* &= (S \cdot E_{ii} \cdot S^{-1})^* && \text{Definition of } P_i, \text{ see (VI.6.3)} \\ &= (S \cdot E_{ii} \cdot S^*)^* && \text{since unitary } S^{-1} = S^* \\ &= (S^{**} \cdot E_{ii}^* \cdot S^*) && \text{since } (A \cdot B)^* = B^* \cdot A^* \text{ for all matrices } A, B \\ &= S \cdot E_{ii} \cdot S^* && S^{**} = S,\ E_{ii}^* = E_{ii}. \\ &= P_i. \end{aligned} \tag{VI.6.4}$$

Thus, $P_i = P_i^*$ is a hermitian matrix. Also,

$$\begin{aligned} P_i^2 &= (S \cdot E_{ii} \cdot \underbrace{S^{-1}) \cdot (S}_{I_n} \cdot E_{ii} \cdot S^{-1}) \\ &= S \cdot E_{ii} \cdot E_{ii} \cdot S^{-1} \\ &= S \cdot E_{ii} \cdot S^{-1} \quad \text{since } E_{ii}^2 = E_{ii} \\ &= P_i. \end{aligned} \tag{VI.6.5}$$

In short, $P_i = P_i^*$ [from (IV.6.4)] and $P_i = P_i^2$ [from (VI.6.5)], so that, indeed, P_i is (the matrix of) an orthogonal projection (Definition VI.6.5).

Once the student completes Exercise VI.6.12, which establishes that for the operator $\mathbf{P}_{X_i}$ defined by $\mathbf{P}_{X_i}(X) = P_i \cdot X$, that rg $(\mathbf{P}_{X_i}) = \text{sp}\,[X_i]$, then we obtain our spectral decomposition for $\mathbf{T}$ in the form

$$\mathbf{T} = \alpha_1\mathbf{P}_{X_1} + \alpha_2\mathbf{P}_{X_2} + \cdots + \alpha_i\mathbf{P}_{X_i} + \cdots + \alpha_n\mathbf{P}_{X_n};$$

equivalently, for the matrix $A_\mathbf{T} = mx(\mathbf{T})_{\mathcal{E}\mathcal{E}}$,

$$A_\mathbf{T} = \alpha_1 P_1 + \alpha_2 P_2 + \cdots + \alpha_i P_i + \cdots + \alpha_n P_n,$$

where $P_i = S \cdot E_{ii} \cdot S^{-1}$.

Example We consider the self-adjoint operator $\mathbf{T}$, defined on $V_3(\mathbf{C})$ by

$$\mathbf{T}\begin{pmatrix} x_1 \\ x_2 \\ x_3 \end{pmatrix} = \begin{pmatrix} -ix_2 \\ ix_1 \\ 3x_3 \end{pmatrix} = \underbrace{\begin{pmatrix} 0 & -i & 0 \\ i & 0 & 0 \\ 0 & 0 & 3 \end{pmatrix}}_{A_\mathbf{T}} \cdot \begin{pmatrix} x_1 \\ x_2 \\ x_3 \end{pmatrix}, \tag{VI.6.6}$$

for all

$$\begin{pmatrix} x_1 \\ x_2 \\ x_3 \end{pmatrix} \in V_3(\mathbf{C}).$$

We can see that $A_\mathbf{T} = mx(\mathbf{T})_{\mathcal{E}\mathcal{E}}$ is hermitian ($A_\mathbf{T} = \overline{A}_\mathbf{T}$), so that we are assured that $\mathbf{T} = \mathbf{T}^*$ (Proposition VI.6.2). Of course, from the defining Equation (VI.6.6), one may also verify directly that $\langle \mathbf{T}(X), Y \rangle = \langle X, \mathbf{T}(Y) \rangle$ for all $X, Y \in V_3(\mathbf{C})$.

EIGENVALUES

We unearth the (necessarily real) eigenvalues for **T** of our example by computing roots of the characteristic polynomial $p_T(\cdot)$ (Definition VI.2.1, Theorem VI.2.3). Thus, for any complex scalar λ, we have

$$\begin{aligned} p_T(\lambda) &= \det (A_T - \lambda I_3) \\ &= \det \begin{pmatrix} -\lambda & -i & 0 \\ i & -\lambda & 0 \\ 0 & 0 & 3-\lambda \end{pmatrix} = (\lambda^2 - 1)(3-\lambda) \\ &= (\lambda - 1)(\lambda + 1)(3 - \lambda). \end{aligned}$$

Clearly, the scalars which $p_T(\cdot)$ sends to zero, are the real numbers $\alpha_1 = 1$, $\alpha_2 = -1$ and $\alpha_3 = 3$, that is, $p_T(1) = p_T(-1) = p_T(3) = 0$, and these are the only scalars, real or complex, which are roots of $p_T(\cdot)$.

EIGENVECTORS

Using (VI.6.6) which, after all, defined **T**, we eventually find that

$$\mathbf{T}(X_1) = \alpha_1 X_1 = 1 \cdot X_1 \qquad \text{for (any scalar multiple of) } X_1 = \begin{pmatrix} i/\sqrt{2} \\ -1/\sqrt{2} \\ 0 \end{pmatrix}$$

$$\mathbf{T}(X_2) = \alpha_2 X_2 = -1 \cdot X_2 \qquad \text{for } X_2 = \begin{pmatrix} 1/\sqrt{2} \\ -i/\sqrt{2} \\ 0 \end{pmatrix} \tag{VI.6.7}$$

$$\mathbf{T}(X_3) = \alpha_3 X_3 = 3 \cdot X_3 \qquad \text{for } X_3 = \begin{pmatrix} 0 \\ 0 \\ 1 \end{pmatrix}.$$

It should be noted that the scalar $1/\sqrt{2}$ appears in the entries of X_1 and X_2 so as to obtain a vector of *unit* length in both cases. In agreement with Theorem VI.6.3, we see that the set of eigenvectors $\{X_1, X_2, X_3\}$ defined in (VI.6.7), is pairwise orthogonal. In fact, we have

$$\langle X_i, X_j \rangle = \begin{pmatrix} 1 & \text{if} & i = j \\ 0 & \text{if} & i \neq j \end{pmatrix}. \tag{VI.6.8}$$

The routine verification of (VI.6.8) is an exercise.

DIAGONALIZATION OF A_T

We construct the unitary matrix $S=[X_1X_2X_3]$. Thus, we have

$$S=\begin{pmatrix} i/\sqrt{2} & 1/\sqrt{2} & 0 \\ -1/\sqrt{2} & -i/\sqrt{2} & 0 \\ 0 & 0 & 1 \end{pmatrix}, \quad S^{-1}=\bar{S}^t=\begin{pmatrix} -i/\sqrt{2} & -1/\sqrt{2} & 0 \\ 1/\sqrt{2} & i/\sqrt{2} & 0 \\ 0 & 0 & 1 \end{pmatrix},$$

and so, as Theorem VI.3.6 predicts, we have,

$$\bar{S}^t \cdot A_T \cdot S = \text{diag}\,(1,-1,3) = 1 \cdot E_{11} + (-1) \cdot E_{22} + 3 \cdot E_{33},$$

that is,

$$\underbrace{\begin{pmatrix} -i/\sqrt{2} & -1/\sqrt{2} & 0 \\ 1/\sqrt{2} & i/\sqrt{2} & 0 \\ 0 & 0 & 0 \end{pmatrix}}_{S^{-1}=\bar{S}^t} \cdot \underbrace{\begin{pmatrix} 0 & -i & 0 \\ i & 0 & 0 \\ 0 & 0 & 3 \end{pmatrix}}_{A_T} \cdot \underbrace{\begin{pmatrix} i/\sqrt{2} & 1/\sqrt{2} & 0 \\ -1/\sqrt{2} & -i/\sqrt{2} & 0 \\ 0 & 0 & 0 \end{pmatrix}}_{S}$$

$$= \underbrace{\begin{pmatrix} 1 & 0 & 0 \\ 0 & -1 & 0 \\ 0 & 0 & 3 \end{pmatrix}}_{\text{diag}(1,-1,3)}. \tag{VI.6.9}$$

SPECTRAL DECOMPOSITION OF A_T

We isolate the matrix A_T from (VI.6.9) by multiplying through on the left by S, and on the right by $\bar{S}^t=S^{-1}$. Thus we have

$$\begin{aligned} A_T=\begin{pmatrix} 0 & i & 0 \\ i & 0 & 0 \\ 0 & 0 & 3 \end{pmatrix} &= S \cdot (1 \cdot E_{11} + (-1) \cdot E_{22} + 3 \cdot E_{33})\bar{S}^t \\ &= 1 \cdot S \cdot E_{11} \cdot \bar{S}^t + (-1) \cdot S \cdot E_{22} \cdot \bar{S}^t + 3 \cdot E_{33} \cdot \bar{S}^t. \end{aligned} \tag{VI.6.10}$$

We compute the hermitian idempotent matrices $P_{X_i}=S \cdot E_{ii} \cdot \bar{S}^t$, $i=1, 2, 3$, to obtain

$$\begin{aligned} S \cdot E_{11} \cdot \bar{S}^t = P_{X_1} &= \begin{pmatrix} 1/2 & -i/2 & 0 \\ i/2 & 1/2 & 0 \\ 0 & 0 & 0 \end{pmatrix} \\ S \cdot E_{22} \cdot \bar{S}^t = P_{X_2} &= \begin{pmatrix} 1/2 & i/2 & 0 \\ -i/2 & 1/2 & 0 \\ 0 & 0 & 0 \end{pmatrix} \\ S \cdot E_{33} \cdot \bar{S}^t = P_{X_3} &= \begin{pmatrix} 0 & 0 & 0 \\ 0 & 0 & 0 \\ 0 & 0 & 1 \end{pmatrix}. \end{aligned} \tag{VI.6.11}$$

From (VI.6.11), we see that (VI.6.10) becomes

$$A_T = \begin{pmatrix} 0 & -i & 0 \\ i & 0 & 0 \\ 0 & 0 & 3 \end{pmatrix} = \overset{\alpha_1}{\overset{\downarrow}{1}} \cdot \underbrace{\begin{pmatrix} 1/2 & -i/2 & 0 \\ i/2 & 1/2 & 0 \\ 0 & 0 & 0 \end{pmatrix}}_{P_{X_1}} + \overset{\alpha_2}{\overset{\downarrow}{(-1)}} \cdot \underbrace{\begin{pmatrix} \frac{1}{2} & i/2 & 0 \\ -i/2 & 1/2 & 0 \\ & 0 & 0 \end{pmatrix}}_{P_{X_2}}$$

$$+\overset{\alpha_3}{\overset{\downarrow}{3}} \cdot \underbrace{\begin{pmatrix} 0 & 0 & 0 \\ 0 & 0 & 0 \\ 0 & 0 & 1 \end{pmatrix}}_{P_{X_3}} \qquad \text{(VI.6.12)}$$

which is the spectral decomposition for A_T. It is for the student to verify that P_{X_i}, $i = 1, 2, 3$ of (VI.6.11), is an idempotent, hermitian matrix, and that $P_{X_i} \cdot P_{X_j} = 0$ if $i \neq j$. It is also to be verified that the image vector $P_{X_i} \cdot X$, for any and all $X \in V_3(\mathbf{C})$, is always a scalar multiple of the eigenvector X_i, $i = 1, 2, 3$ (Exercise VI.6.12).

APPLICATION OF THE SPECTRAL DECOMPOSITION

With the spectral decomposition (VI.6.12) at hand, we are easily able to compute powers $A_T{}^n$ of A_T, and even the inverse A_T^{-1}. In using Corollary VI.3.7, we have

$$A_T{}^8 = \alpha_1{}^8 P_{X_1} + \alpha_2{}^8 P_{X_2} + \alpha_3{}^8 P_{X_3}$$

$$= (1)^8 \cdot \begin{pmatrix} 1/2 & -i/2 & 0 \\ i/2 & 1/2 & 0 \\ 0 & 0 & 0 \end{pmatrix} + (-1)^8 \begin{pmatrix} 1/2 & i/2 & 0 \\ -i/2 & 1/2 & 0 \\ 0 & 0 & 0 \end{pmatrix} + (3)^8 \begin{pmatrix} 0 & 0 & 0 \\ 0 & 0 & 0 \\ 0 & 0 & 1 \end{pmatrix}$$

$$= \begin{pmatrix} 1 & 0 & 0 \\ 0 & 1 & 0 \\ 0 & 0 & 3^8 \end{pmatrix}.$$

Similarly, if we set $B = \alpha_1^{-1} P_{X_1} + \alpha_2^{-1} P_{X_2} + \alpha_3^{-1} P_{X_3}$ ($\alpha_i \neq 0$), we see that $A_T \cdot B = B \cdot A_T = I_3$ since each P_{X_i} is idempotent, and cross-terms, $P_{X_i} \cdot P_{X_j}$, vanish. That is, $B = (A_T)^{-1}$. For this example, $\alpha_1 = 1$, $\alpha_2 = -1$, $\alpha_3 = 3$, so that

$$(A_T)^{-1} = (1)^{-1} \cdot \begin{pmatrix} 1/2 & -i/2 & 0 \\ i/2 & 1/2 & 0 \\ 0 & 0 & 0 \end{pmatrix} + (-1)^{-1} \cdot \begin{pmatrix} 1/2 & i/2 & 0 \\ -i/2 & 1/2 & 0 \\ 0 & 0 & 0 \end{pmatrix}$$

$$= + (3)^{-1} \begin{pmatrix} 0 & 0 & 0 \\ 0 & 0 & 0 \\ 0 & 0 & 1 \end{pmatrix} = \begin{pmatrix} 0 & -i & 0 \\ i & 0 & 0 \\ 0 & 0 & 1/2 \end{pmatrix}.$$

EXERCISES

VI.6.1 Show that if $\mathbf{T} \in \mathscr{L}(V_n, V_n)$, and if $\mathbf{T} = \mathbf{T}^*$, then $A_\mathbf{T} = mx(\mathbf{T})_{\mathscr{E}\mathscr{E}}$ must have real scalars along the main diagonal. (Do not assume that $A_\mathbf{T}$ is already a diagonal matrix.)

VI.6.2 Given $\mathbf{T} \in \mathscr{L}(V_3, V_3)$ defined by

$$\mathbf{T}\begin{pmatrix} x_1 \\ x_2 \\ x_3 \end{pmatrix} = \begin{pmatrix} ix_1 \\ 2x_2 \\ 0 \end{pmatrix} \quad \text{for all } \begin{pmatrix} x_1 \\ x_2 \\ x_3 \end{pmatrix} \in V_3.$$

(a) Find $A_\mathbf{T}$ and $A_{\mathbf{T}^*}$, the matrices of $\mathbf{T}$ and $\mathbf{T}^*$, respectively, relative to the standard ordered basis $\mathscr{E}$.

(b) Prove that $\mathbf{T} \circ \mathbf{T}^* = \mathbf{T}^* \circ \mathbf{T}$, and yet $\mathbf{T}^* \neq \mathbf{T}$.

(c) Find a basis for rg $(\mathbf{T})$, the range of $\mathbf{T}$, and a basis for ker $(\mathbf{T})$, the kernel of $\mathbf{T}$.

(d) Prove that rg $(\mathbf{T}) \perp$ ker $(\mathbf{T})$, by direct computation.

VI.6.3 Given the operator $\mathbf{T}$ on $V_2(\mathbf{C})$ defined by

$$\mathbf{T}\begin{pmatrix} x_1 \\ x_2 \end{pmatrix} = \begin{pmatrix} 2x_1 + ax_2 \\ -2ix_1 + 2x_2 \end{pmatrix} \quad \text{for all } \begin{pmatrix} x_1 \\ x_2 \end{pmatrix} \in V_2(\mathbf{C}).$$

(a) Find $A_\mathbf{T} = mx(\mathbf{T})_{\mathscr{E}\mathscr{E}}$.

(b) For which scalar a (if any) is $\mathbf{T}$ self-adjoint?

(c) Find a basis for rg $(\mathbf{T})$ and for ker $(\mathbf{T})$, and show that rg $(\mathbf{T}) \perp$ ker$(\mathbf{T})$, when a is chosen so that $\mathbf{T} = \mathbf{T}^*$.

VI.6.4 Show that rg $(\mathbf{T}) \perp$ ker $(\mathbf{T})$ for the self-adjoint operator on $V_3(\mathbf{R})$ defined by

$$\mathbf{T}\begin{pmatrix} x_1 \\ x_2 \\ x_3 \end{pmatrix} = \begin{pmatrix} x_1 - x_3 \\ 2x_2 \\ -x_1 + x_3 \end{pmatrix} \quad \text{for all } \begin{pmatrix} x_1 \\ x_2 \\ x_3 \end{pmatrix} \in V_3(\mathbf{R}).$$

Find a set of pairwise perpendicular eigenvectors of $\mathbf{T}$.

VI.6.5 [Extending a self-adjoint operator on $V_n(\mathbf{R})$ to one to $V_n(\mathbf{C})$.] Given $\mathbf{T} \in \mathscr{L}(V_n(\mathbf{R}), V_n(\mathbf{R}))$. Thus, for the $n \times n$ matrix $A_\mathbf{T} = mx(\mathbf{T})_{\mathscr{E}\mathscr{E}}$, we have $\mathbf{T}(X) = A_\mathbf{T} \cdot X$, for all *real* vectors ($n \times 1$ matrices) $X \in V_n(\mathbf{R})$ (Theorem II.2.3). Using the same matrix $A_\mathbf{T}$, which has only real entries, we define the (extended) operator $\tilde{\mathbf{T}}$ on $V_n(\mathbf{C})$ by $\tilde{\mathbf{T}}(X) = A_\mathbf{T} \cdot X$ for all *complex* vectors ($n \times 1$ matrices) $X \in V_n(\mathbf{C})$.

(a) Prove that $\tilde{\mathbf{T}}$ extends $\mathbf{T}$ in the sense that $\tilde{\mathbf{T}}(X) = \mathbf{T}(X)$ whenever X is a real entried, concrete vector of $V_n(\mathbf{R})$.

(b) Prove that if $\mathbf{T} = \mathbf{T}^*$, then $\tilde{\mathbf{T}} = \tilde{\mathbf{T}}^*$. (*Hint*: See Proposition VI.6.2.)

VI.6.6 (Characterizing the vectors in the null space of an idempotent, $\mathbf{P} = \mathbf{P}^2$.) Let $\mathbf{P} = \mathbf{P}^2 \in \mathscr{L}(V_n, V_n)$. Prove the following:

(a) If $Z \in V_n$ is of the form $Z = X - \mathbf{P}(X)$ for some $X \in V_n$, then $\mathbf{P}(Z) = O$.

(b) Conversely, if $\mathbf{P}(Z) = O$, then $Z = X - \mathbf{P}(X)$ for some $X \in V_n$. [*Hint*: Find this certain X for which $Z = X - \mathbf{P}(X)$.]

From (a) and (b), we conclude that ker $(\mathbf{P})$, the kernel of $\mathbf{P}$, is exactly the set of all vectors of the form $X - \mathbf{P}(Z)$.

VI.6.7 Let $\mathbf{P} = \mathbf{P}^2 \in \mathscr{L}(V_n, V_n)$. Prove that for any $X \in V_n$, X may be written as the sum of 2 vectors, Z_1 and Z_2, one of which belongs to rg $(\mathbf{P})$, the range of $\mathbf{P}$, and the other belongs to ker $(\mathbf{P})$, the null space of $\mathbf{T}$. That is to say, for any $X \in V_n$, $X = Z_1 + Z_2$, where $Z_1 \in \text{rg}(\mathbf{P})$, and $Z_2 \in \ker(\mathbf{P})$. [*Hint*: rg $(\mathbf{P})$ is, by definition, the set of all vectors of the form $\mathbf{P}(Y)$, $Y \in V_n$; ker $(\mathbf{P})$ is characterized in Exercise VI.6.6.]

VI.6.8 Let $\mathbf{P}$ be an idempotent operator on V_n. Prove that if $\mathbf{P}(X + Y) = Y$ for some X, $Y \in V_n$, then $\mathbf{P}(X) = O$.

VI.6.9 (Uniqueness of the decomposition of Exercise VI.6.7.) Given

$$\mathbf{P} = \mathbf{P}^2 \in \mathscr{L}(V_n, V_n)$$

(a) Suppose that the vector O decomposes as follows:

$$O = Z_1 + Z_2,$$

where $Z_1 \in \text{rg}(\mathbf{P})$, and $Z_2 \in \ker(\mathbf{P})$. Prove that necessarily, $Z_1 = O$, and $Z_2 = O$ [*Hint*: $Z_1 \in \text{rg}(\mathbf{P})$ must be of the form $Z_1 = \mathbf{P}(X)$ for some $X \in V_n$; $Z_2 \in \ker(\mathbf{P})$ must (from Exercise VI.6.6)] be of the form $Z_2 = Y - \mathbf{P}(Y)$ for some $Y \in V_n$. Exercise VI.6.8 may be used to show that $Z_1 = O$, from which it follows that $Z_2 = O$.

(b) Suppose X has 2 decompositions

$$X = Z_1 + Z_2 = W_1 + W_2,$$

where Z_1, $W_1 \in \text{rg}(\mathbf{P})$, and Z_2, $W_2 \in \ker(\mathbf{P})$. Prove that, necessarily, $Z_1 = W_1$, and $Z_2 = W_2$, so that this decomposition, which exists from Exercise VI.6.7, is also unique.

VI.6.10 (a) Given the idempotent operator $\mathbf{P} = \mathbf{P}^2$ on V_n with the property that rg $(\mathbf{P}) \perp \ker(\mathbf{P})$, that is, for every vector $\mathbf{P}(X) \in \text{rg}(\mathbf{P})$, and for every $Y - \mathbf{P}(Y) \in \ker(\mathbf{P})$(see Exercise VI.6.6), we have $\langle \mathbf{P}(X), Y - \mathbf{P}(Y)\rangle = 0$. Prove that $\mathbf{P}$ is necessarily self-adjoint, that is, for all X, $Y \in V_n$, $\langle \mathbf{P}(X), Y\rangle = \langle X, \mathbf{P}(Y)\rangle$. [*Hint*: Consider $\langle \mathbf{P}(X), Y\rangle$ and write $Y = \mathbf{P}Y + (Y - \mathbf{P}(Y))$; consider $\langle X, \mathbf{P}(Y)\rangle$ and write $X = \mathbf{P}X + (X - \mathbf{P}(X))$.]

(b) Conversely, show if $\mathbf{T}^* = \mathbf{T}$ (it is *not* assumed that $\mathbf{T}^2 = \mathbf{T}$), then rg $(\mathbf{T}) \perp \ker(\mathbf{T})$.

VI.6.11 Prove that for any idempotent (not necessarily self-adjoint) operator $\mathbf{P} = \mathbf{P}^2$ on V_n, the only possible eigenvalues are the real scalars 0 or 1.

VI.6.12 (The projection matrix $P_{X_i} = [X_1 X_2 \cdots X_n] \cdot E_{ii} \cdot [X_1 X_2 \cdots X_n]^{-1}$ has for its range, the 1-dimensional subspace sp$[X_i]$.) In the proof of the spectral theorem, Theorem VI.6.6, the matrix P_{X_i} was defined by

$$P_{X_i} = S \cdot E_{ii} \cdot S^{-1} = S \cdot E_{ii} \cdot \bar{S}^t.$$

where $S = [X_1 X_2 \cdots X_n]$ is the unitary (orthogonal) matrix whose column vectors X_1, X_2, $\ldots$, X_n, are of length one, and pairwise orthogonal (perpendicular). Recall the fact that for any $n \times n$ matrix A, $A \cdot E_i = i$th column vector of A, where E_i is the ith standard basis vector of V_n.

(a) Prove that $S \cdot E_i = X_i$, and consequently, $S^{-1} \cdot X_i = \bar{S}^t \cdot X_i = E_i$ for each $i = 1, 2, \ldots, n$.

(b) Prove that $E_{ii} \cdot E_j = O$ if $i \neq j$, and $E_{ii} \cdot E_i = E_i$.

(c) For $P_{X_i} = S \cdot E_{ii} \cdot S^{-1}$, prove that $P_{X_i} \cdot X_j = O \in V_n$ if $i \neq j$, and $P_{X_i} \cdot X_i = X_i \in V_n$.

(d) Since the orthonormal set $\{X_1, X_2, \ldots, X_n\}$ is a basis for V_n, for *every* $Y \in V_n$, there exists a unique set of scalars $\alpha_1, \alpha_2, \ldots, \alpha_n$ such that $Y = \alpha_1 X_1 + \alpha_2 X_2 + \cdots + \alpha_n X_n$ (Theorem I.6.8). Prove that for this typical vector Y of V_n, $P_{X_i} \cdot Y = \alpha_i X_i$, that is, the range of the operator $\mathbf{P}_{X_i}$, which is left-hand multiplication by P_{X_i}, consists of all scalar multiples of X_i.

(e) Prove that $P_{X_i} \cdot P_{Xj} = O$ whenever $i \neq j$.

VI.6.13 (A characterization of projections $\mathbf{P}$ with 1-dimensional range $\mathrm{sp}[X_0]$.) Let X_0 be a unit vector of V_n. We define the function $\mathbf{P}_{X_0}(\cdot) = \langle \cdot, X_0 \rangle X_0$, sending V_n to itself, by

$$\mathbf{P}_{X_0}(Y) = \langle Y, X_0 \rangle X_0 \qquad \text{for all } Y \in V_n$$

(X_0 is a *fixed* unit vector of V_n).

(a) Prove that $\mathbf{P}_{X_0}(\cdot)$ is a linear function from V_n to itself.

(b) Prove that $\mathbf{P}_{X_0}(\cdot)$ is a projection, that is,

$$\mathbf{P}_{X_0} \circ \mathbf{P}_{X_0} = \mathbf{P}_{X_0}, \text{ and } \mathbf{P}^*_{X_0} = \mathbf{P}_{X_0}.$$

VI.6.14 (Spectral discomposition of hermitian (symmetric) matrices.) For the given square matrix A, you are asked to:

(1) Find the eigenvalues $\alpha_1, \alpha_2, \ldots, \alpha_n$, and corresponding eigenvectors $X_1, X_2, \ldots, X_n$.

(2) Find the unitary matrices, $S = [X_1 X_2 \cdots X_n]$ and S^{-1}, which diagonalize A, that is, find S such that $\mathrm{diag}(\alpha_1, \alpha_2, \ldots, \alpha_n) = S^{-1} \cdot A \cdot S$.

(3) Find the spectral decomposition for A, that is, find projection matrices $P_{X_i} = S \cdot E_{ii} \cdot S^{-1}$ and verify directly that $A = \alpha_1 P_{X_1} + \alpha_2 P_{X_2} + \cdots + \alpha_n P_{X_n}$, and that $P_{X_i}{}^2 = P_{X_i} = \overline{(P_{X_i})^t}$, $P_{X_i} \cdot P_{X_j} = O$ for $i \neq j$. Also verify that $P_{X_1} + P_{X_2} + \ldots + P_{X_n} = I_n$.

(4) Find A^{-1}, using this spectral decomposition, provided the eigenvalues $\alpha_1, \alpha_2, \ldots, \alpha_n$ are each nonzero.

(5) Find A^5, using the spectral decomposition.

(a) $A = \begin{pmatrix} -1 & 2 \\ 2 & 2 \end{pmatrix}$

(b) $A = \begin{pmatrix} -1 & -2 \\ -2 & 2 \end{pmatrix}$

(c) $A = \begin{pmatrix} 4 & 2 \\ 2 & 1 \end{pmatrix}$

(d) $A = \begin{pmatrix} 0 & 4 & 0 \\ 4 & 0 & 0 \\ 0 & 0 & 1 \end{pmatrix}$

(e) $A = \begin{pmatrix} 0 & 3 & 0 & 0 \\ 3 & 0 & 0 & 0 \\ 0 & 0 & 0 & 1 \\ 0 & 0 & 1 & 0 \end{pmatrix}$

(f) $A = \begin{pmatrix} x_1{}^2 & x_1x_2 \\ x_1x_2 & x_2{}^2 \end{pmatrix}$, $x_1{}^2 + x_2{}^2 = 1$, $x_1, x_2 \in \mathbf{R}$

(g) $A = \begin{pmatrix} 1 & 1 & 1 \\ 1 & 1 & 1 \\ 1 & 1 & 1 \end{pmatrix}$

[*Hint*: In (g), the 3 eigenvalues are not distinct. However, a set of 3 orthonormal eigenvectors can be found, after which, the diagonalization of A can proceed.]

(h) $A = \begin{pmatrix} a+b & 0 & (a-b) \\ 0 & 0 & 0 \\ (a-b) & 0 & a+b \end{pmatrix}$, $\quad a, b \in \mathbf{R}$. $a \neq b$.

VI.6.15 Suppose $\mathbf{P}$ and $\mathbf{Q} \in \mathscr{L}(V_n, V_n)$ are self-adjoint, that is, $\mathbf{P} = \mathbf{P}^*$, $\mathbf{Q} = \mathbf{Q}^*$. Prove that $\mathbf{P} \circ \mathbf{Q} = \mathbf{0}$, the zero operator on V_n, if and only if rg ($\mathbf{P}$) $\perp$ rg ($\mathbf{Q}$), that is, if and only if any vector of rg ($\mathbf{P}$) is orthogonal (perpendicular) to any vector of rg ($\mathbf{Q}$).

We now are prepared to abandon the assumption of *distinctness* for the eigenvalues $\alpha_1, \alpha_2, \ldots, \alpha_n$ of the self-adjoint $\mathbf{T} = \mathbf{T}^* \in \mathscr{L}(V_n, V_n)$ and prove a spectral decomposition like Theorem VI.6.6 in this most general case.

What was so wonderful about distinctness of eigenvalues anyway? In short, distinctness of eigenvalues assured a basis full of eigenvectors, and hence diagonalizability of $\mathbf{T}$, from which our spectral analysis followed (see Corollary VI.3.4).

Under the assumption of only k distinct eigenvectors, $1 \leq k \leq n$, we are guaranteed (at first glance) a set of only k pairwise orthogonal (hence linearly independent) eigenvectors $\{X_1, X_2, \ldots, X_k\}$ of $\mathbf{T}$ (see Theorem VI.4.4).

If we could *always* be sure that sufficiently many eigenvectors existed for $\mathbf{T}$ so as to form a *basis* $\{X_1, X_2, \ldots, X_k, \ldots, X_n\}$ for V_n; then we could proceed to develop a spectral theorem for general self-adjoint linear operators $\mathbf{T} = \mathbf{T}^*$ on V_n by essentially mimicking the proof of Theorem VI.6.6 from (VI.6.2) on.

At this, the end of the book, it seems appropriate to provide Theorem VI.6.7.

Theorem VI.6.7 For any $\mathbf{T} = \mathbf{T}^* \in \mathscr{L}(V_n, V_n)$, $\mathbf{T}$ always has n pairwise orthogonal eigenvectors $\{X_1, X_2, \ldots, X_n\}$ which form a basis for V_n. Moreover, if α_i is an eigenvalue of $\mathbf{T}$ which appears p_i times in the list of eigenvalues $\{\alpha_1, \alpha_2, \ldots, \alpha_i, \ldots, \alpha_n\}$, then there are exactly p_i eigenvectors of $\mathbf{T}$ corresponding to α_i. That is to say, if α_i appears p_i times, $1 \leq p_i \leq n$, then the subspace of eigenvectors M_{α_i} corresponding to α_i is of dimension p_i.

Before proving the theorem, we present an example.

Example Let $\mathbf{T} \in \mathscr{L}(V_3, V_3)$ be defined by

$$\text{(a)} \quad \mathbf{T}\begin{pmatrix} x_1 \\ x_2 \\ x_3 \end{pmatrix} = \begin{pmatrix} 2x_1 \\ 2x_2 \\ 3x_3 \end{pmatrix} \quad \text{for all } \begin{pmatrix} x_1 \\ x_2 \\ x_3 \end{pmatrix} \in V_3.$$

Then

$$A_{\mathbf{T}} = mx(\mathbf{T})_{\mathscr{E}\mathscr{E}} = \begin{pmatrix} 2 & 0 & 0 \\ 0 & 2 & 0 \\ 0 & 0 & 3 \end{pmatrix},$$

which is hermitian ($A_{\mathbf{T}} = A_{\mathbf{T}}{}^{t}$). Thus, $\mathbf{T}$ is self-adjoint (Proposition VI.6.2). Observe that $p_{\mathbf{T}}(\lambda) = \det(A_{\mathbf{T}} - \lambda I_3) = (2-\lambda)^2(3-\lambda)$ which has as its 3 roots

$$\text{(b)} \quad \alpha_1 = 2, \qquad \alpha_2 = 2, \qquad \alpha_3 = 3.$$

According to our theorem, we ought to be able to find 2 pairwise orthogonal eigenvectors for the eigenvalue 2, which appears *twice* in (b) and 1 eigenvector, perpendicular to the other two, for the eigenvalue 3, which appears *once* in (b).

▶ Eigenvalue $\alpha_1 = \alpha_2 = 2$.

For which eigenvectors $X = \begin{pmatrix} x_1 \\ x_2 \\ x_3 \end{pmatrix}$ do we have

$$\text{(c)} \qquad \mathbf{T}\begin{pmatrix} x_1 \\ x_2 \\ x_3 \end{pmatrix} = 2\begin{pmatrix} x_1 \\ x_2 \\ x_3 \end{pmatrix}?$$

That is, for which $X = \begin{pmatrix} x_1 \\ x_2 \\ x_3 \end{pmatrix}$ is it true that

$$\begin{pmatrix} 2x_1 \\ 2x_2 \\ 3x_3 \end{pmatrix} = \begin{pmatrix} 2x_1 \\ 2x_2 \\ 2x_3 \end{pmatrix} \quad \text{(a) substituted into (c).}$$

This equation implies that necessarily, $x_3 = 0$. Hence, (c) obtains for any and all $X \in V_3$ whose third coordinate x_3 equals zero. Thus,

$$X = \begin{pmatrix} x_1 \\ x_2 \\ 0 \end{pmatrix}$$

is the *typical* eigenvector for $\mathbf{T}$ corresponding to the eigenvalue 2; x_1 and x_2 are 2 unrestricted variables (or "degrees of freedom"). The subspace of V_3, of *all* such X

$$\begin{pmatrix} x_1 \\ x_2 \\ 0 \end{pmatrix}$$

has for a *basis*, *many* pairs of orthogonal (hence, linearly independent) vectors. For example,

$$\left\{\underset{\substack{\uparrow\\ X_1}}{\begin{pmatrix}1\\0\\0\end{pmatrix}},\underset{\substack{\uparrow\\ X_2}}{\begin{pmatrix}0\\1\\0\end{pmatrix}}\right\},\quad \text{since}\begin{pmatrix}x_1\\x_2\\0\end{pmatrix}=x_1\begin{pmatrix}1\\0\\0\end{pmatrix}+x_2\begin{pmatrix}0\\1\\0\end{pmatrix},$$

$$=x_1X_1+x_2X_2,$$

or

$$\left\{\underset{\substack{\uparrow\\ X_1'}}{\begin{pmatrix}1\\-1\\0\end{pmatrix}},\underset{\substack{\uparrow\\ X_2'}}{\begin{pmatrix}1\\1\\0\end{pmatrix}}\right\},\quad \text{since}\begin{pmatrix}x_1\\x_2\\0\end{pmatrix}=\frac{(x_1-x_2)}{2}\begin{pmatrix}1\\-1\\0\end{pmatrix}+\frac{(x_1+x_2)}{2}\begin{pmatrix}1\\1\\0\end{pmatrix}$$

$$=\frac{(x_1-x_2)}{2}\cdot X_1'+\frac{(x_1+x_2)}{2}\cdot X_2'.$$

▶ Eigenvalue $\alpha_3 = 3$.
From (a) we conclude that $\mathbf{T}(X) = 3X$ only for

$$X=\begin{pmatrix}0\\0\\x_3\end{pmatrix}=x_3\begin{pmatrix}0\\0\\1\end{pmatrix}.$$

That is, the subspace of eigenvectors for the eigenvalue 3 is 1-dimensional with the single-element basis,

$$\left\{\begin{pmatrix}0\\0\\1\end{pmatrix}\right\},$$

or any nonzero scalar multiple of

$$X_3=\begin{pmatrix}0\\0\\1\end{pmatrix}.$$

Thus, we have

$$\begin{array}{cccc} & \alpha_1=2, & \alpha_2=2, & \alpha_3=3\\ & \downarrow & \downarrow & \downarrow\\ \mathscr{X}= & \left\{X_1=\begin{pmatrix}1\\0\\0\end{pmatrix},\right. & X_2=\begin{pmatrix}0\\1\\0\end{pmatrix}, & \left.X_3=\begin{pmatrix}0\\0\\1\end{pmatrix}\right\},\\ & \downarrow & \downarrow & \downarrow\\ \mathscr{X}'= & \left\{X_1'=\begin{pmatrix}1\\-1\\0\end{pmatrix},\right. & X_2'=\begin{pmatrix}1\\1\\0\end{pmatrix}, & \left.X_3'=\begin{pmatrix}0\\0\\3\end{pmatrix}\right\},\end{array}$$

where $\mathscr{X}$ and $\mathscr{X}'$ are just 2 sets of eigenvector bases fulfilling the conclusions of the theorem, namely, they are pairwise orthogonal, and as many eigenvectors appear in $\mathscr{X}$ or $\mathscr{X}'$ as the corresponding eigenvalues.

Proof of Theoren VI.6.7: We present this proof in very broad strokes. That is, we outline the strategy, leaving details as exercises (with hints) for the student.

Let $\alpha_1, \alpha_2, \ldots, \alpha_k$ represent a listing of the k distinct (necessarily real) eigenvalues for $\mathbf{T}$. Thus, each α_i, $i = 1, 2, \ldots, k$, accounts for at least 1 eigenvector X_i; existence is assured by Theorem VI.2.3. Moreover, these k eigenvectors $X_1, X_2, \ldots, X_k$ are pairwise orthogonal by Theorem VI.6.3. But how do we dig up as many linearly independent eigenvectors, as there are appearances of α_i? If, as in Theorem VI.6.6, *all* eigenvalues were distinct, we would then have all our n eigenvectors, forming a basis. Here we have only k, $k \leq n$, such eigenvectors. Recall (see Exercise VI.4.15) that for any subspace M, $M^\perp$, the set of *all* vectors perpendicular to M is also a subspace, that is, $M^\perp = \{X\colon$ whenever $Y_M \in M$, we always have $\langle X, Y_M \rangle = 0\}$. Moreover, we can be sure that if the dimension of M, dim $(M) < n$, (that is, $M \notin V_n$), then $M^\perp$ is nontrivial. In fact

▶ (1) *If* dim $(M) = p$, *then* dim $(M^\perp) = n - p$.

Hence, if $M \neq V_n$, $M^\perp \neq \{0\}$ (Exercise VI.6.17). We shall use the following fact:

▶ (2) *If* $\mathbf{T} = \mathbf{T}^*$, *and if* M *is a subspace of eigenvectors for* $\mathbf{T}$ [*so that* $\mathbf{T}(M) \subset M$], *then* $\mathbf{T}$ *sends* $M^\perp$ *to itself* $(\mathbf{T}(M^\perp) \subset M^\perp)$. *Moreover* $\mathbf{T}$ *is self-adjoint on* $M^\perp$.

Now let M be the k-dimensional subspace spanned by the eigenvectors $X_1, X_2, \ldots, X_k$. With the invariance of $M^\perp$ under $\mathbf{T}$ described by (2), we consider the operator $\mathbf{T}$ *restricted* to the $n-k$ dimensional subspace $M^\perp$, denoted $\mathbf{T}/_{M^\perp}$; that is, treat $n-k$-dimensional $M^\perp$ not as a subspace of anything, but as a vector-space in its own right, so that $\mathbf{T}/_{M^\perp} \in \mathscr{L}(M^\perp, M^\perp)$. We now reason that we are starting afresh with $\mathbf{T}/_{M^\perp} \in \mathscr{L}(M^\perp, M^\perp)$ so that $\mathbf{T}$ has at least 1 eigenvector, call it X_{k+1}, in $M^\perp$. But to which eigen*value* does X_{k+1} correspond? The eigenvalue α_{k+1} of $\mathbf{T}/_{M^\perp}$ must already be in our original list $\{\alpha_1, \alpha_2, \ldots, \alpha_k\}$, since $\mathbf{T}$ can have no different eigenvectors (hence no different eigenvalues, either) from $\mathbf{T}/_{M^\perp}$. Thus, α_{k+1} is one of the repetitions of some α_i, $1 \leq i \leq k$. We conclude from this discussion, that if $\mathbf{T} = \mathbf{T}^*$, and $\alpha_1, \alpha_2, \ldots, \alpha_k$ are eigenvalues for $\mathbf{T}$, then we can always go to $M^\perp$, the subspace orthogonal to sp $[X_1, X_2, \ldots, X_k]$, where $\mathbf{T}(X_i) = \alpha_i X_i$ $i = 1, 2, \ldots, k$, and find *another* eigenvector X_{k+1}, corresponding to one of the repetitions of the eigenvalues $\alpha_1, \alpha_2, \ldots, \alpha_k$. Since $X_{k+1} \in \text{sp}\,[X_1, X_2, \ldots, X_k]^\perp$, the augmented set of eigenvectors $\{X_1, X_2, \ldots, X_k, X_{k+1}\}$ is pairwise orthogonal.

The proof ends by successive repetition of this argument. Always find another eigenvector in $\text{sp}\,[X_1,X_2,\ldots,X_k,\ldots,X_\ell]^\perp$ corresponding to a repeated eigenvalue, until $\text{sp}\,[X_1,X_2,\ldots,X_k,\ldots,X_\ell]^\perp = \{0\}$, that is, until $\ell = n$, whereupon a *basis* of pairwise orthogonal eigenvectors materializes.

EXERCISES

VI.6.16 (Gram-Schmidt orthogonalization)

(a) Suppose $\{X_1,X_2\}$ is a linearly independent set. Define X_2' by

$$X_2' = X_2 - \frac{\langle X_2,X_1\rangle}{\langle X_1,X_1\rangle} X_1.$$

(1) Show that $X_1 \perp X_2'$, that is, $\langle X_1,X_2'\rangle = 0$,

and

(2) $\text{sp}\,[X_1,X_2] = \text{sp}\,[X_1,X_2']$.

(b) Suppose $\{X_1,X_2,X_3\}$ is a linearly independent set. Define X_2' and X_3' by

$$X_2' = X_2 - \frac{\langle X_2,X_1\rangle}{\langle X_1,X_1\rangle} X_1,$$

$$X_3' = X_3 - \frac{\langle X_3,X_2'\rangle}{\langle X_2',X_2'\rangle} X_2' - \frac{\langle X_3,X_1\rangle}{\langle X_1,X_1\rangle} X_1.$$

(1) Show that the set $\{X_1,X_2',X_3'\}$ is pairwise orthogonal, that is, $\langle X_1,X_2'\rangle = \langle X_1,X_3'\rangle = \langle X_2',X_3'\rangle = 0$.

(2) Show that $\text{sp}\,[X_1,X_2] = \text{sp}\,[X_1,X_2']$, and $\text{sp}\,[X_1,X_2,X_3] = \text{sp}\,[X_1,X_2',X_3']$.

(c) Inductive definition: Given a linearly independent set $\{X_1,X_2,\ldots,X_k\}$. Show that this set can be orthogonalized as follows (Gram-Schmidt orthogonalization process):

Define $X_1 = X_1'$, and X_2' and X_3' as in part (b). In general, define X_p' in terms of your previously generated *pairwise orthogonal* vectors $X_1, X_2',\ldots, X_{p-1}'$, by

$$X_p' = X_p - \sum_{i=1}^{p-1} \frac{\langle X_p,X_i'\rangle}{\langle X_i',X_i'\rangle} X_i'.$$

(1) Show that $X_p' \perp X_i'$ for all $i = 1, 2,\ldots, p-1$.

(2) Show that for all $i = 1, 2,\ldots, k$, $\text{sp}\,[X_1,X_2,\ldots,X_i] = \text{sp}\,[X_1',X_2',\ldots,X_i']$.

(d) Given the following linearly independent sets $\{X_1,X_2,\ldots,X_k\}$, find the Gram-Schmidt orthogonalization $\{X_1',X_2',\ldots,X_k'\}$ given in part (c).

(a) $\left\{\begin{pmatrix}1\\0\\1\end{pmatrix}, \begin{pmatrix}0\\1\\1\end{pmatrix}, \begin{pmatrix}1\\2\\0\end{pmatrix}\right\}$

(b) $\left\{\begin{pmatrix}1\\1\\1\end{pmatrix},\begin{pmatrix}1\\1\\2\end{pmatrix},\begin{pmatrix}0\\1\\0\end{pmatrix}\right\}$

(c) $\left\{\begin{pmatrix}1\\1\\0\\0\end{pmatrix},\begin{pmatrix}1\\0\\0\\0\end{pmatrix}\right\}$

(d) $\left\{\begin{pmatrix}1\\0\\0\\0\end{pmatrix},\begin{pmatrix}0\\1\\3\\0\end{pmatrix},\begin{pmatrix}0\\3\\1\\0\end{pmatrix}\right\}$

(e) $\left\{\begin{pmatrix}1\\0\end{pmatrix},\begin{pmatrix}a\\b\end{pmatrix}\right\}$ $b \neq 0$

VI.6.17 Show that if $\dim(M) = p$, $M \subset V_n$, then $\dim(M^\perp) = n - p$, where $M^\perp = \{X\colon \text{whenever } Y \in M,\ \langle X,Y\rangle = 0\}$.
Hint: Let $\{X_1,X_2,\ldots,X_p\}$ be a basis for M ($\text{sp}\,[X_1,X_2,\ldots,X_p] = M$), and extend this basis to $\{X_1,X_2,\ldots,X_p,Y_1,Y_2,\ldots,Y_{n-p}\}$ for all of V_n (see Theorems I.6.14 and I.7.5). Use the Gram-Schmidt orthogonalization process to construct from this basis, the basis

$$\{X_1',X_2',\ldots,X_p',Y_1',Y_2'\ldots,Y_{n-p}'\}$$

which is pairwise orthogonal, and where $M = \text{sp}\,[X_1,X_2,\ldots,X_p] = \text{sp}\,[X_1',X_2',\ldots,X_p']$. Conclude that $\{Y_1',Y_2',\ldots,Y_{n-p}'\}$ is a basis for $M^\perp$ and therefore $M^\perp$ has dimension $n - p$.

VI.6.18 Prove that if $\mathbf{T} = \mathbf{T}^* \in \mathscr{L}(V_n,V_n)$, and if for some subspace M, spanned by eigenvectors we have $\mathbf{T}(X) \in M$ whenever $X \in M$ ($\mathbf{T}(M) \subset M$), then necessarily, $\mathbf{T}(X) \in M^\perp$ whenever $X \in M^\perp$ ($\mathbf{T}(M^\perp) \subset M^\perp$). *Hint:* Let $\{X_1,X_2,\ldots,X_k\}$ be a basis of eigenvectors for M. Consider $Y \in M^\perp$ and $\langle\mathbf{T}(X_i),Y\rangle$ and $\langle X_i,\mathbf{T}^*(Y)\rangle$.

VI.6.19 Prove that if $\mathbf{T} = \mathbf{T}^*$, then $mx(\mathbf{T})_{\mathscr{X}\mathscr{X}}$ is of form $\text{diag}\,(\alpha_1,\alpha_2,\ldots,\alpha_n)$ relative to the basis $\mathscr{X} = \{X_1,X_2,\ldots,X_n\}$ of pairwise orthogonal eigenvectors for $\mathbf{T}$. (As before, $\alpha_1,\alpha_2,\ldots,\alpha_n$, denote the not necessarily distinct eigenvalues of $\mathbf{T}$). *Hint*: Proof of Theorem VI.6.6; especially (VI.6.2).

VI.6.20 Show that if $\alpha_1, \alpha_2,\ldots, \alpha_k$ are the *distinct* eigenvalues of $\mathbf{T}$, where each α_i appears p_i times, then $p_1 + p_2 + \cdots + p_k = n$, and

$$mx(\mathbf{T})_{\mathscr{X}\mathscr{X}} = \alpha_1 F_1 + \alpha_2 F_2 + \cdots + \alpha_k F_k,$$

where F_1 is the $n \times n$ matrix with zeros everywhere except for p_1 1's down the first p_1 main diagonal entries; F_2 has zeros everywhere except for p_2 1's along the next p_2 main diagonal entries, and so on (see Exercise VI.6.19).

VI.6.21 Show that $F_i^2 = F_i = \bar{F}_i^t$ for each $i = 1, 2,\ldots, k$ (see Exercise VI.6.20).

VI.6.22 If $X = [X_1X_2\cdots X_n]$ is the unitary matrix whose ith column is the eigenvector X_i for $\mathbf{T} = \mathbf{T}^* \in \mathscr{L}(V_n,V_n)$, then show that $S \cdot F_i \cdot \bar{S}^t$ is always a projection where $\dim\,(\text{rg}\,(S \cdot F_i \cdot \bar{S}^t)) = p_i$.

VI.6.23 **Spectral Theorem.** Prove that if $\mathbf{T} = \mathbf{T}^* \in \mathscr{L}(V_n, V_n)$ with (real) distinct eigenvalues $\alpha_1, \alpha_2, \ldots, \alpha_k$, where each α_i appears p_i times as a root of the characteristic polynomial $p_{\mathrm{T}}(\lambda) = \det(A_{\mathrm{T}} - \lambda I_n)$, then

$$\mathbf{T} = \alpha_1 \mathbf{P}_2 + \alpha_2 \mathbf{P}_2 + \cdots + \alpha_k \mathbf{P}_k$$

where $\mathbf{P}_i$ is a projection with p_i-dimensional range, $\mathbf{P}_i \circ \mathbf{P}_j = \mathbf{0}$ whenever $i \neq j$, and $\mathbf{P}_1 + \mathbf{P}_2 + \cdots + \mathbf{P}_k = \mathbf{1}$, the identity operator in $\mathscr{L}(V_n, V_n)$. *Hint*: See proof of Theorem VI.6.6.

VI.6.24 (See Exercise VI.6.23.) If $\mathbf{T} = \mathbf{T}^* \in \mathscr{L}(V_n, V_n)$, where *none* of the distinct eigenvalues $\alpha_1, \alpha_2, \ldots, \alpha_k$ is zero, then show that

$$\mathbf{T}^{-1} = \alpha^{-1} \mathbf{P}_1 + \alpha^{-1} \mathbf{P}_2 + \cdots + \alpha_k{}^{-1} \mathbf{P}_k.$$

VI.6.25 For the following hermitian matrices A, use the spectral theorem to compute A^2, $A,^5$ and A^{-1} if defined:

(a) $A = \begin{pmatrix} 2 & 0 \\ 0 & 2 \end{pmatrix}$

(b) $A = \begin{pmatrix} 5 & i & 0 \\ -i & 5 & 0 \\ 0 & 0 & 4 \end{pmatrix}$

(c) $A = \begin{pmatrix} 1 & 0 & 2 \\ 0 & 0 & i \\ 2 & -i & 3 \end{pmatrix}$

Chapter VI Summary

The purpose of Section 1 is to show that *all* matrices for a given linear transformation can be generated from just 1 matrix $A_{\mathbf{T}}$. In fact, by multiplying $A_{\mathbf{T}}$ on the left and on the right by certain invertible matrices, we create any and every matrix $mx(\mathbf{T})_{\mathscr{X}\mathscr{Y}}$ that $\mathbf{T}$ is capable of inducing (Theorem VI.1.4). A consequence of this is that the determinant does not vary for all the square matrices $mx(\mathbf{T})_{\mathscr{X}\mathscr{X}}$, even though the basis $\mathscr{X}$ may vary (Theorem VI.1.5). Section 2 introduces eigenvalues and eigenvectors, and Theorem VI.2.3 relates eigenvalues to determinants. In Section 3, diagonalizable operators are characterized as those operators having sufficiently many eigenvectors to form a basis for the space (Theorem VI.3.2). The powerful spectral theorem (Theorem VI.3.6) appears for *matrices* with distinct eigenvalues. Section 4 introduces the inner product. After discussion of adjoints and unitary operators (Section 5), the ideas of the chapter are all brought to bear on the self-adjoint operators in the final section (Section 6). For example, when $\mathbf{T} = \mathbf{T}^*$, it is shown that eigenvectors corresponding to distinct eigenvalues are necessarily orthogonal (Theorem VI.6.3). Once projections are introduced (Definition VI.6.5), it is shown that every self-adjoint operator (at least if all eigenvalues are distinct) decomposes as a certain sum of orthogonal projections (Theorem VI.6.6). This is the spectral theorem for *operators* which allows for a transparent means for computing powers of self-adjoint linear operators, and even inverses, when inverses exist. The section concludes with the more general spectral decomposition theorem (without requiring *distinct* eigenvalues) whose proof is only sketched since, in large measure, it duplicates that of Theorem VI.6.6.

Chart of Theorems and Definitions

We list below all theorems, corollaries, propositions, definitions, and notations along with their page locations and a sketch of their content. The abbreviations used are as follows: **Th.** = Theorem, **Cor.** = Corollary, **Prop.** = Proposition, **Def.** = Definition, and **Not.** = Notation.

Chapter I. Vector Spaces and Applications

1. *Addition and Scalar Multiplication of Vectors*

Def. I.1.1. column vector, row vector, V_n, $V_n(\mathbf{R})$, $V_n(\mathbf{C})$.

Def. I.1.2. $X + Y$.

Def. I.1.3. αX.

Def. I.1.4. $X = Y$.

2. *Products of Vectors*

Def. I.2.1. $X \cdot Y = \sum x_i y_i$.

Def. I.2.2. $(x_{ij})_{ij}$, $m \times n$ matrix.

Def. I.2.3. $(x_{ij})_{ij} \cdot (y_{pq})_{pq} = \left(\sum_k x_{ik} y_{kj}\right)_{ij}$.

3. *Simultaneous Linear Equations*

4. *Linear Programming*

Prop. I.4.1. The set of all vectors $U = \lambda_1 X + \lambda_2 Y, \lambda_1 + \lambda_2 = 1, 1 \geq \lambda_1, \lambda_2 \geq 0$, describes all points between X and Y.

Prop. I.4.2. $au + bv > e$ (respectively, $<e$), defines the set of all (u,v) in a half plane whose boundary is the set of (u,v) satisfying $au + bv = e$.

Prop. I.4.3. Functionals attain their maxima at endpoints of a line segment. If the maximum occurs at both endpoints, the functional is constant on the entire line segment.

Def. I.4.4. Finite convex polygon.

Th. I.4.5. Functionals on a finite convex polygon attain their maximum (minimum) at a vertex.

5. *Abstract Vector Spaces*

Def. I.5.1. $X + Y + Z$ represents either $(X + Y) + Z$ or $X + (Y + Z)$.

Def. I.5.2. The 10 defining properties of a vector space—
A.1 + closure, A.2 + associativity, A.3 + commutivity, A.4 exsitence of a zero, A.5 existence of an additive inverse $-x$ for each x.
B.1 scalar multiplication closure, B.2 $\alpha(\beta X) = (\alpha\beta)X$, B.3 $\alpha(X + Y) = \alpha X + \alpha Y$, B.4 $(\alpha + \beta)X = \alpha X + \beta X$, B.5 $1 \cdot X = X$.

6. *Bases and Dimension*

Def. I.6.1. Linear combination of ordered vectors.
Def. I.6.2. A set spans another set.
Def. I.6.3. $\{X_1,X_2,\ldots,X_k\}$ LD $\Leftrightarrow$ one X_i is a linear combination of the others.
Def. I.6.4. $\{X_1,X_2,\ldots,X_k\}$ LI $\Leftrightarrow \{X_1,X_2,\ldots,X_k\}$ not LD.
Th. I.6.5. $\{X_1,X_2,\ldots,X_k\}$ LD $\Leftrightarrow$ there is a nontrivial means for generating zero as a linear combination of the X_i's.
Th. I.6.6. $\{X_1,X_2,\ldots,X_k\}$ LI $\Leftrightarrow$ the *only* way to generate zero with the X_i's as $O = \alpha_1 X_1 + \alpha_2 X_2 - \cdots + \alpha_k X_k$, is the trivial way, that is, to set $\alpha_1 = \alpha_2 = \cdots = \alpha_k = 0$.
Def. I.6.7. $\mathscr{X} = \{X_1,X_2,\ldots,X_k\}$ is a basis for a vector space $\Leftrightarrow$ (1) $\mathscr{X}$ spans, and (2) $\mathscr{X}$ is LI.
Th. I.6.8. $\mathscr{X} = \{X_1,X_2,\ldots,X_k\}$ is a basis for a vector-space $\Leftrightarrow \mathscr{X}$ spans uniquely.
Th. I.6.9. Every spanning set has at least as many elements as every other LI set.
Cor. I.6.10. Every basis has the same number of elements as every other basis of the space.
Def. I.6.11. Standard basis $\{E_i\}$ for V_n.
Th. I.6.12. Every basis of V_n has n elements.
Def. I.6.13. Dimension = number of elements in any basis.
Th. I.6.14. Every LI subset of V_n extends to a basis.

7. *Subspaces*

Def. I.7.1. M a subspace of $V_n \Leftrightarrow M \subset V_n$ and M is a vector space.
Th. I.7.2. M a subspace of $V_n \Leftrightarrow M$ is closed under addition and scalar multiplication.
Def. I.7.3. sp $[X_1,X_2,\ldots,X_n]$.
Prop. I.7.4. sp $[X_1,X_2,\ldots,X_n]$ is always a subspace of V where $X_1, X_2, \ldots, X_n \in V$.
Th. I.7.5. Subspaces of V_n all have dimension $\leq n$.
Def. I.7.6. $M + N$ (M, N subspaces of V_n).
Th. I.7.7. $M + N$ is always a subspace of V_n if M and N are subspaces of V_n.
Th. I.7.8. $M \cap N$ is always a subspace of V_n if M and N are subspaces of V_n.
Th. I.7.9. $\dim (M + N) = \dim (M) + \dim (N) - \dim (M \cap N)$ whenever M, N are subspaces of V_n.
Def. I.7.10. $M \oplus N = M + N$ where $M \cap N = \{O\}$.
Def. I.7.11. $\bigoplus_{i=1}^{k} M_i$.

Chapter II. Linear Transformations and Their Matrices

1. *Functions*

Def. II.1.1. Function $f : X \to Y$, domain, range, $f(\cdot)$.
Not. II.1.2. $(x,y) \approx (x,f(x))$.
Def. II.1.3. Impact set = any set containing rg (f).
Def. II.1.4. Into, onto, to, for $f : X \to Y$.

Def. II.1.5. $g \circ f$, composition of functions.
Def. II.1.6. One-to-one function.
Prop. II.1.7. f is 1-1 $\Rightarrow f^{-1}$ is a function, and is 1-1.
Def. II.1.8. f^{-1}, the inverse of a function f.

2. ***Linear Transformations***
Def. II.2.1. $\mathbf{T} : V_n \to V_m$ linear $\Leftrightarrow$ (II.2.1) $\mathbf{T}(\alpha X) = \alpha\mathbf{T}(X)$, and (II.2.2) $\mathbf{T}(X + Y) = \mathbf{T}(X) + \mathbf{T}(Y)$.
Prop. II.2.2. $\mathbf{T} : V_n \to V_m$ linear $\Leftrightarrow \mathbf{T}(\alpha X + \beta Y) = \alpha\mathbf{T}(X) + \beta\mathbf{T}(Y)$.
Th. II.2.3. $\mathbf{T} : V_n \to V_m$ linear $\Leftrightarrow \mathbf{T}(X) = A \cdot X$ for some $m \times n$ matrix A.

3. ***Rank and Nullity of a Linear Transformation***
Prop. II.3.1. $\mathbf{T} : V_n \to V_m$ linear $\Rightarrow$ range ($\mathbf{T}$) is a subspace of V_m.
Def. II.3.2. $\mathbf{T}$ onto, into.
Def. II.3.3. dom ($\mathbf{T}$), rg ($\mathbf{T}$), null space of ($\mathbf{T}$) = ker ($\mathbf{T}$) = $\mathbf{T}^{-1}(O)$.
Prop. II.3.4. $\mathbf{T} : V_n \to V_m$ linear $\Rightarrow$ ker ($\mathbf{T}$) is a subspace of V_n.
Prop. II.3.5. $O \in$ ker ($\mathbf{T}$); thus, ker ($\mathbf{T}$) $\neq \varnothing$.
Th. II.3.6. dim (dom $\mathbf{T}$) = dim (rg $\mathbf{T}$) + dim (ker $\mathbf{T}$).
Cor. II.3.7. Linear transformations maintain or reduce dimension.
Def. II.3.8. Rank ($\mathbf{T}$) = dim (rg $\mathbf{T}$), Nullity of $\mathbf{T}$ = dim (ker $\mathbf{T}$).
Th. II.3.9. $\mathbf{T} : V_n \to V_m$ linear. Then $\mathbf{T}$ is 1-1 $\Leftrightarrow$ ker ($\mathbf{T}$) = $\{O\}$.
Cor. II.3.10. $\mathbf{T} : V_n \to V_m$ linear. Then $\mathbf{T}$ is 1-1 $\Leftrightarrow$ dim (rg $\mathbf{T}$) = n.
Th. II.3.11. $\mathbf{T} : V_n \to V_m$ linear. If $\mathbf{T}^{-1}$ exists, then $\mathbf{T}^{-1}$ is always a linear transformation.

4. ***Algebraic Properties of Linear Transformations***
Def. II.4.1. $\mathbf{T}_1 + \mathbf{T}_2$.
Th. II.4.2. $\mathbf{T}_1 + \mathbf{T}_2$ is always linear if $\mathbf{T}_1$, $\mathbf{T}_2$ are.
Def. II.4.3. $\alpha\mathbf{T}$.
Th. II.4.4. $\alpha\mathbf{T}$ is always linear if $\mathbf{T}$ is.
Def. II.4.5. $\mathscr{L}(V_n, V_m)$.
Th. II.4.6. $\mathscr{L}(V_n, V_m)$ is a vector space.
Th. II.4.7. $\mathbf{U}$, $\mathbf{T}$ linear $\Rightarrow \mathbf{U} \circ \mathbf{T}$ is linear if $\mathbf{U} \circ \mathbf{T}$ is defined.
Th. II.4.8. $\mathbf{T} \circ (\mathbf{U} + \mathbf{V}) = \mathbf{T} \circ \mathbf{U} + \mathbf{T} \circ \mathbf{V}$, $(\mathbf{T} + \mathbf{U}) \circ \mathbf{V} = \mathbf{T} \circ \mathbf{V} + \mathbf{U} \circ \mathbf{V}$ for all linear $\mathbf{T}$, $\mathbf{U}$, $\mathbf{V}$, where the products are defined.
Def. II.4.9. $\mathbf{1} \circ \mathbf{T} = \mathbf{T} \circ \mathbf{1} = \mathbf{T}$ for all $\mathbf{T}$, defines identity $\mathbf{1} \in \mathscr{L}(V_n, V_n)$.
Def. II.4.10. Group, vector space, ring, algebra.

5. ***Matrices of Linear Transformations (the 2 × 2 case)***
Def. II.5.1. $mx(\mathbf{T})_{\mathscr{X}\mathscr{Y}}$ for $\mathbf{T} \in \mathscr{L}(V_2, V_2)$.
Th. II.5.2. $mx(\cdot)_{\mathscr{X}\mathscr{Y}} : \mathscr{L}(V_2, V_2) \to 2 \times 2$ matrices is 1-1, onto.

6. ***Matrices of Linear Transformations (the m × n case)***
Def. II.6.1. sigma notation $\sum$ for sums.
Def. II.6.2. $mx(\mathbf{T})_{\mathscr{X}\mathscr{Y}}$ for $\mathbf{T} \in \mathscr{L}(V_n, V_m)$.
Not. II.6.3. $\mathscr{M}_{m,n}$ = the set of all $m \times n$ matrices.
Th. II.6.4. $mx(\cdot)_{\mathscr{X}\mathscr{Y}} : \mathscr{L}(V_n, V_m) \to \mathscr{M}_{m,n}$ is a 1-1, onto function.
Def. II.6.5. (addition on $\mathscr{M}_{m,n}$) $mx\mathbf{T}_{\mathscr{X}\mathscr{Y}} + mx(\mathbf{U})_{\mathscr{X}\mathscr{Y}}$ defined to be $mx(\mathbf{T} + \mathbf{U})_{\mathscr{X}\mathscr{Y}}$.
Not. II.6.6. $(a_{ij})_{ij}$, $i = 1, 2, \ldots, m$, $j = 1, 2, \ldots, n$, is the $m \times n$ matrix whose ijth entry is a_{ij}.

Th. II.6.7. $(a_{ij})_{ij} + (b_{ij})_{ij} = (a_{ij} + b_{ij})_{ij}$.

Th. II.6.8. $\mathcal{M}_{m,n}$ is a group relative to addition.

Def. II.6.9. (Scalar multiplication on $\mathcal{M}_{m,n}$) $\alpha \cdot mx(\mathbf{T})_{\mathcal{X}\mathcal{Y}}$ defined to be $mx(\alpha\mathbf{T})_{\mathcal{X}\mathcal{Y}}$.

Th. II.6.10. $\alpha(a_{ij})_{ij} = (\alpha a_{ij})_{ij}$.

Th. II.6.11. $\mathcal{M}_{m,n}$ is a vector space.

Def. II.6.12. (Pairwise multiplication for matrices $mx(\mathbf{T})_{\mathcal{X}\mathcal{X}}$, where $\mathbf{T} \circ \mathbf{U}$ is defined.

Th. II.6.13. $(a_{ik})_{ik}(b_{lj})_{lj} = (\sum_{k=1}^{p} a_{ik} \cdot b_{kl})_{ij}$, $\quad i = 1, 2, \ldots, m$; $k, l = 1, 2, \ldots, p$; $j = 1, 2, \ldots, n$.

Th. II.6.14. $\mathcal{M}_{n,n}$ is an algebra.

Def. II.6.15. Group homomorphism and isomorphism, vector-space homomorphism and isomorphism, algebra homomorphism and isomorphism.

Th. II.6.16. (Fundamental Isomorphism Theorem) $\mathcal{L}(V_n, V_m)$ and $\mathcal{M}_{m,n}$ are vector space isomorphic under the isomorphism $mx(\cdot)_{\mathcal{X}\mathcal{Y}}$. If $m = n$, $\mathcal{X} = \mathcal{Y}$, then $\mathcal{L}(V_n, V_n)$ and $\mathcal{M}_{n,n}$ are algebra isomorphic, where $mx(\cdot)_{\mathcal{X}\mathcal{X}}$ is an algebra isomorphism.

Chapter III. Systems of Linear Equations

1. *Row- and Column-echelon Matrices*

Def. III.1.1. System of equations $A \cdot X = Y_0$. Where A is the $m \times n$ coefficient matrix. $X_0 \in V_n$ is a *solution* if $A \cdot X_0 = Y_0$. $A : Y$ is the augmented matrix of the system.

Def. III.1.2. $A = [A_1 A_2 \cdots A_n] = \begin{bmatrix} A^1 \\ A^2 \\ \cdot \\ \cdot \\ \cdot \\ A^m \end{bmatrix}$,

an $m \times n$ matrix. The elementary operations R_{ij}, $R_i(a)$, $R_i{}^j(a)$ C_{ij}, $C_i(a)$, $C_i{}^j(a)$ are defined.

Def. III.1.3. Row-echelon and column-echelon matrix. Echelon rank of a row (respectively, column) echelon matrix equals the number of nonzero rows (respectively columns).

Th. III.1.4. Every matrix is reducible to row (respectively column) echelon form via elementary row (respectively column) operations.

2. *Systems of Equations Revisited*

Def. III.2.1. $A \cdot X = O$ is a *homogeneous* system.

Th. III.2.2. Row space (respectively column space) of a matrix is unaltered by application of elementary row (respectively column) operations.

Cor. III.2.3. Row rank (respectively column rank) of a matrix A is unaltered by application of elementary row (respectively column) operations, where row rank = dim (row space), column rank = dim (column space) of A.

Th. III.2.4. $A \cdot X = O_m$ for A an $m \times n$ matrix $\Rightarrow \dim \{X : A \cdot X = O\} = n -$ row rank.

Th. III.2.5. For any matrix A, row rank = column rank.

Def. III.2.6. Rank of matrix A.

Th. III.2.7. If $A \cdot X = Y$, then $X = X_0 + Z$ where X_0 is a fixed particular solution (that is, $A \cdot X_0 = Y$), and Z is a null solution (that is, $A \cdot Z = O$).

Prop. III.2.8. $A \cdot X = Y$ has unique vector solution X, where A is an $m \times n$ matrix, $\Leftrightarrow$ rank $(A) = n$. Hence $m \geq n$.

3. ***Elementary Matrices and Invertibility***

Th. III.3.1. Elementary row (respectively column) operation on matrix A can always be effected by left- (respectively right-) hand multiplications on A by certain (elementary) matrices.

Def. III.3.2. Elementary matrix.

Def. III.3.3. A^{-1} is a (two-sided) inverse for $n \times n$ matrix $A \Leftrightarrow A^{-1} \cdot A = A \cdot A^{-1} = I_n$.

Th. III.3.4. A is reducible to echelon form I_n if and only if rank $(A) = n$, where A is an $n \times n$ matrix.

Th. III.3.5. A^{-1} exists for $A \Leftrightarrow$ rank $(A) = n \Leftrightarrow$ every 1-sided inverse for A is a 2-sided inverse.

Th. III.3.6. $A =$ product of elementary matrices $\Leftrightarrow A^{-1}$ exists, where A is an $n \times n$ matrix.

Chapter IV. Determinants: A Classical Approach

1. ***Permutations***

Def. IV.1.1. σ is a permutation $\in S_n \Leftrightarrow \sigma\{1,2,\ldots,n\} \xrightarrow[\text{onto}]{1-1} \{1,2,\ldots,n\}$.

Not. IV.1.2. $\sigma = \begin{pmatrix} 1 & 2 & \cdots & n \\ \sigma(1) & \sigma(2) & \cdots & \sigma(n) \end{pmatrix}$.

Def. IV.1.3. *Inversion* induced by σ, *even* and *odd* permutation.

Def. IV.1.4. Transposition.

Th. IV.1.5. Transpositions reverse parity.

Cor. IV.1.6. Transpositions are odd permutations.

2. ***Definition of the Determinant***

Def. IV.2.1. det $((a_{ij})_{ij}) = \sum_\sigma \operatorname{sgn}(\sigma) a_{1\sigma(1)} a_{2\sigma(2)} \cdots a_{n\sigma(n)}$.

Th. IV.2.2. det $R_{ij}(A) = -$det (A).

Cor. IV.2.3. If 2 rows of A agree, then det $(A) = 0$.

Th. IV.2.4. det $(R_i(a)A) = a$ det (A).

Th. IV.2.5. det $(R_i{}^j(a)A) =$ det (A).

Cor. IV.2.6. det $R_{ij} =$ det $(R_{ij}) = -1$, det $R_i(a) =$ det $(R_i(a)^t) = a$, det $(R_i{}^j(a)) =$ det $(R_i{}^j(a)^t) = 1$, for elementary matrices R_{ij}, $R_i(a)$, $R_i{}^j(a)$.

Cor. IV.2.7. If F is an elementary matrix, then det $(F \cdot A) =$ det $(F) \cdot$ det (A), for all square matrices A.

Th. IV.2.8. det $(A) =$ det (A^t) for all $n \times n$ matrices A.

Th. IV.2.9. det $(A) = 0 \Leftrightarrow$ rank $(A) < n \Leftrightarrow A^{-1}$ does not exist.

Th. IV.2.10. Statement of Theorems 2.2, 2.4, 2.5, and Corollary 2.3 with "rows" replaced by "columns."

3. ***Expansion by Minors***

Not. IV.3.1. $A(i,j)$ the $(n-1) \times (n-1)$ matrix obtained from $n \times n$ A, by deleting row i, column j; det $(A(i,j))$ is a *minor* of A.

Th. IV.3.2. (Expansion by first row) det $\sum_{k=1}^{n} (-1)^{k+1} a_{1k}$ det $(A(1,k))$.

Th. IV.3.3. (Expansion by ith row) $\det(A) = \sum_{k=1}^{n} (-1)^{i+k} a_{ik} \det(A(i,k))$, i fixed.

Th. IV.3.4. (Expansion by jth column) $\det(A) = \sum_{i=1}^{n} (-1)^{i+j} \det(A(i,j))$, j fixed.

4. *The Product Theorem and Applications of det (·)*

Th. IV.4.1. For all pairs of $n \times n$ matrices, A, B, $\det(A \cdot B) = \det(A) \cdot \det(B)$.

Cor. IV.4.2. $\det(A^{-1}) = (\det(A))^{-1}$.

Def. IV.4.3. $\operatorname{adj}(A) = ((-1)^{i+j} \det A(j,i))_{ij}$.

Th. IV.4.4. For all square A, $A \cdot \operatorname{adj}(A) = (\operatorname{adj}(A)) \cdot A = (\det(A)) \cdot I_n$.

Cor. IV.4.5. A^{-1} exists $\Rightarrow A^{-1} = (\operatorname{adj}(A))/\det(A)$.

Th. IV.4.6. Cramer's rule for n equations, n unknowns.

Chapter V. Determinants: A Nonclassical Approach

1. *Definition of det (·)*

Def. V.1.1. Multilinear functional.

Def. V.1.2. det (·) is a multilinear, antisymmetric functional which takes on the value 1 on the identity matrix I_n.

Prop. V.1.3. g antisymmetric $\Rightarrow g(\ldots, X, \ldots, X, \ldots) = 0$.

Prop. V.1.4. $g(E_{i_1}, E_{i_2}, \ldots, E_{i_n}) = \operatorname{sgn}(i) g(E_1, E_2, \ldots, E_n)$ for any g antisymmetric.

Th. V.1.5. Any multilinear, antisymmetric g is a multiple of the determinant, that is, $g(A) = \det(A) \cdot g(I_n)$.

Cor. V.1.6. Uniqueness of the determinant function from Def. V.1.2.

2. *Consequences of the Nonclassical Definition for det*

Th. V.2.1. $\det(A \cdot B) = \det(A) \cdot \det(B)$.

Prop. V.2.2. $\det\begin{pmatrix} I & C \\ 0 & B \end{pmatrix} = \det(B)$, I, B square.

Th. V.2.3. $\det\begin{pmatrix} A & C \\ 0 & B \end{pmatrix} = \det(A)\det(B)$, A, B square.

Chapter VI. Structure of Operators

1. *Change of Bases*

Def. VI.1.1. $\mathbf{U}_{\mathscr{X}}(X_i) = E_i$, $i = 1, 2, \ldots, n$.

Prop. VI.1.2. $\mathbf{U}_{\mathscr{X}}^{-1}(Y) = [X_1 X_2 \cdots X_n] \cdot Y$, or $mx(\mathbf{U}_{\mathscr{X}}^{-1})_{\mathscr{E}\mathscr{E}} = [X_1 X_2 \cdots X_n]$.

Not.VI.1.3. $A_{\mathbf{T}}$ denotes $mx(\mathbf{T})_{\mathscr{E}\mathscr{E}'}$ for $\mathbf{T} \in \mathscr{L}(V_n, V_m)$.

Th.VI.1.4. (Change of basis theorem) $mx(\mathbf{T})_{\mathscr{X}\mathscr{Y}} = mx(\mathbf{U}_{\mathscr{Y}})_{\mathscr{E}'\mathscr{E}'} mx(\mathbf{T})_{\mathscr{E}\mathscr{E}'} mx(\mathbf{U}_{\mathscr{X}}^{-1})_{\mathscr{E}\mathscr{E}}$.

Th. VI.1.5. For $\mathbf{T} \in \mathscr{L}(V_n, V_n)$, $\det(mx(\mathbf{T})_{\mathscr{X}\mathscr{X}}) = \det(mx(\mathbf{T})_{\mathscr{Y}\mathscr{Y}})$, for all $\mathscr{X}, \mathscr{Y} \subset V_n$.

Def. VI.1.6. $\det(\mathbf{T}) = \det mx(\mathbf{T})_{\mathscr{X}\mathscr{X}}$ for any $\mathscr{X} \subset V_n$.

2. *Eigenvalues and Eigenvectors*

Def. VI.2.1. $\mathbf{T}(X) = \lambda X$. λ an eigenvalue, $X \neq O$, X an eigenvector.

Th. VI.2.2. The set of all eigenvectors corresponding to a fixed eigenvalue of $\mathbf{T} \in \mathscr{L}(V_n, V_n)$, is always a subspace of V_n once O is adjoined.

Th. VI.2.3. λ an eigenvalue for $\mathbf{T} \Leftrightarrow \det(A_{\mathbf{T}} - \lambda I_n) = 0$.

Th. VI.2.4. $\det(mx(\mathbf{T})_{\mathscr{E}\mathscr{E}} - \lambda I_n) = \det(mx(\mathbf{T})_{\mathscr{X}\mathscr{X}} - \lambda I_n)$ for all bases $\mathscr{X}$.

Def. VI.2.5. $p_{\mathbf{T}}(\lambda) = \det(mx(\mathbf{T})_{\mathscr{E}\mathscr{E}} - \lambda I_n)$.

3. *Diagonalizable Operators*

Def. VI.3.1. $\mathbf{T} \in \mathscr{L}(V_n,V_n)$ *diagonalizable* means $mx(\mathbf{T})_{\mathscr{X}\mathscr{X}}$ is a diagonal matrix for some basis $\mathscr{X}$. Matrix A is *diagonalizable* means $S^{-1} \cdot A \cdot S$ is a diagonal matrix for some matrix S.

Th. VI.3.2. $mx(\mathbf{T})_{\mathscr{X}\mathscr{X}}$ is diagonal $\Leftrightarrow \mathscr{X}$ is a basis of eigenvectors.

Th. VI.3.3. Single eigenvectors corresponding to distinct eigenvalues form linearly independent sets.

Cor. VI.3.4. $\mathbf{T} \in \mathscr{L}(V_n,V_n)$ has n distinct eigenvalues $\Rightarrow \mathbf{T}$ has n linearly independent eigenvectors.

Th. VI.3.5. $\mathbf{T} \in \mathscr{L}(V_n,V_n)$. $\mathbf{T}$ is diag $\Leftrightarrow \mathbf{T}$ has n *LI* eigenvectors

$$\{X_1,X_2,\ldots,X_n\} \Leftrightarrow \text{diag}\,(\lambda_1,\lambda_2,\ldots,\lambda_n) = S^{-1} \cdot A_{\mathrm{T}} \cdot S$$

where λ_i's are eigenvalues, $S = [X_1X_2\cdots X_n]$, and $A_{\mathrm{T}} = mx(\mathbf{T})_{\mathscr{E}\mathscr{E}}$.

Th. VI.3.6. (a spectral theorem). A an $n \times n$ matrix with distinct eigenvalues $\lambda_1, \lambda_2, \ldots, \lambda_n$. Then $A = \sum \lambda_i P_i$, where $P_i = S \cdot E_{ii} \cdot S^{-1}$, so that $\sum P_i = I_n$, $P_i{}^2 = P_i$, $P_i \cdot P_j = O$ for $i \neq j$.

Cor. VI.3.7. If f is a polynomial, then $\mathbf{f}(A) = \sum f(\lambda_i)P_i$ where A is $n \times n$ with distinct eigenvalues $\lambda_i, \lambda_2,\ldots,\lambda_n$. If, moreover, $\lambda_i \neq 0$ for all i, then $A^{-1} = \sum \lambda_i{}^{-1}P_i$.

Rule VI.3.8. To find spectral form for matrix A: (1) Find eigenvalues λ_i: (2) Find eigenvectors X_i: (3) Construct matrices $S = [X_1X_2\cdots X_n]$ and S^{-1}: (4) Write $A = \sum \lambda_i S \cdot E_{ii} \cdot S^{-1}$.

4. *The Inner Product*

Def. VI.4.1. $\langle X,Y\rangle = \sum x_i\overline{y_i}$.

Def. VI.4.2. $X \perp Y \Leftrightarrow \langle X,Y\rangle = 0$, orthogonality (perpendicularity).

Th. VI.4.3. (a) $\langle X,X\rangle \geq 0$; (b) $\langle X,Y\rangle = \overline{\langle Y,X\rangle}$; (c) $\langle \alpha X + \beta Y,Z\rangle = \alpha\langle X,Z\rangle + \beta\langle Y,Z\rangle$; (d) $\langle X,\alpha Y + \beta Z\rangle = \bar{\alpha}\langle X,Y\rangle + \bar{\beta}\langle X,Z\rangle$.

Th. VI.4.4. $\{X_1,X_2,\ldots,X_n\}$ pairwise orthogonal $\Rightarrow$ linearly independence.

Def. VI.4.5. A an $m \times n$ matrix. $A^* = \overline{A}^t$, the conjugate transpose of A.

Th. VI.4.6. $\langle A \cdot X, Y\rangle = \langle X, A^* \cdot Y\rangle$ for all $X \in V_n$, $Y \in V_m$, where A is an $m \times n$ matrix.

Cor. VI.4.7. If $A = mx(\mathbf{T})_{\mathscr{E}\mathscr{E}'}$ (so that $\mathbf{T}(X) = A \cdot X$), then for all $X \in V_n$, $Y \in V_m$, $\langle \mathbf{T}(X),Y\rangle = \langle X, \mathbf{T}^*(Y)\rangle$, where $\mathbf{T}^* \in \mathscr{L}(V_m,V_n)$ is defined by $T^*(Y) = A^* \cdot Y$ for all $Y \in V_m$.

Def. VI.4.8. $\mathbf{T}^*$, the *adjoint* of $\mathbf{T} \in \mathscr{L}(V_n,V_n)$.

5. *Unitary and Orthogonal Transformations*

Def. VI.5.1. $\mathbf{T} \in \mathscr{L}(V_n,V_n)$ is a unitary (orthogonal) operator means

$$\langle \mathbf{T}(X), \mathbf{T}(Y)\rangle = \langle X, Y\rangle,$$

all X, $Y \in V_n$. U is a unitary (orthogonal) matrix means $\langle U \cdot X, U \cdot Y\rangle = \langle X,Y\rangle$, all X, $Y \in V_n$.

Th. VI.5.2. $\mathbf{T}$ is unitary (orthogonal) on V_n. Then (a) $\mathbf{T}^{-1} = \mathbf{T}^*$. [Hence $(A_{\mathrm{T}})^{-1} = (A_{\mathrm{T}})^*$ where A_{T} is the unitary matrix $A_{\mathrm{T}} = mx(\mathbf{T})_{\mathscr{E}\mathscr{E}}$.] (b) A_{T} has an o.n. set for its row and for its columns.

6. *Self-adjoint Operators and Their Matrices*

Def. VI.6.1. $\mathbf{T} \in \mathscr{L}(V_n,V_n) \cdot \mathbf{T} = \mathbf{T}^*$ means $\mathbf{T}$ is self-adjoint. A an $n \times n$ matrix. $A = A^t$ means A is symmetric; $A = A^*$ means A is hermitian.

Prop. VI.6.2. $\mathbf{T} = \mathbf{T}^* \Leftrightarrow A_{\mathbf{T}} = (A_{\mathbf{T}})^*$.

Th. VI.6.3. $\mathbf{T} = \mathbf{T}^* \Rightarrow$ eigenvalues of $\mathbf{T}$ are real, and distinct eigenvalues correspond to orthogonal eigenvectors.

Prop. VI.6.4. $\mathbf{T} = \mathbf{T}^*$ on $V_n(\mathbf{R})$. $\mathbf{T}$ still has n real eigenvalues.

Def. VI.6.5. $\mathbf{P} \in \mathscr{L}(V_n, V_n)$. $\mathbf{P}$ an orthogonal projection means $\mathbf{P} = \mathbf{P}^2 = \mathbf{P}^*$.

Th. VI.6.6. (A spectral theorem for s.a. operators.) If $\mathbf{T} = \mathbf{T}^*$, $\mathbf{T}$ has distinct (real) eigenvalues $\alpha_1, \alpha_2, \ldots, \alpha_n$, then $\mathbf{T} = \sum \alpha_i \mathbf{P}_{X_i}$, where $\mathbf{P}_{X_i}$ is the orthogonal projection with one-dimensional range $\text{sp}\,[X_i]$; $\{X_1, X_2, \ldots, X_n\}$ is an o.n. set of eigenvectors for $\mathbf{T}$.

Th. VI.6.7. (A spectral theorem for s.a. operators.) Same as Th. VI.6.6, but without the hypothesis of distinctness on the eigenvalues $\alpha_1, \alpha_2, \ldots, \alpha_n$.

APPENDIX A

Mathematical Induction (An Informal Approach)

We shall be considering an infinite list of sentences or statements $P(n)$, each of which depends upon the non-negative integer n.

For example, for each non-negative integer n, $P(n)$ might be the sentence "The integer n is even," and we symbolize this by writing

$$P(n) \equiv \text{The integer } n \text{ is even.}$$

If we start with the integer $n = 0$, then we have the following infinite list of sentences:

$$\begin{aligned}
P(0) &\equiv \text{The integer 0 is even.}\\
P(1) &\equiv \text{The integer 1 is even.}\\
P(2) &\equiv \text{The integer 2 is even.}\\
P(3) &\equiv \text{The integer 3 is even.}\\
&\vdots\\
P(n) &\equiv \text{The integer } n \text{ is even.}\\
&\vdots
\end{aligned} \tag{A.1}$$

We raise the question of "how many" of these sentences in the list are *true*. In this particular case, we see that "half are true and half are not true, or valid." That is, $P(n)$ is *true* when 2 is a factor of n (n is even), and $P(n)$ is *not true* when 2 is not a factor of n (n is odd).

WHAT MATHEMATICAL INDUCTION DOES

The purpose of the technique of mathematical induction is to establish the truth, or validity, of an infinite list of consecutive sentences with one fell swoop. With lists such as (A.1), this is not possible, since the consecutive sentences are not, in fact, valid (although every other sentence *is* valid). We shall shortly present a list of sentences, $P(1)$, $P(2)$, $P(3), \ldots,$ $P(n), \ldots,$ where the successive sentences *are* indeed valid, and we shall prove the validity by means of the technique of mathematical induction. It behooves us, then, to ask, "Exactly what *is* this remarkable technique?"

THE ANATOMY OF MATHEMATICAL INDUCTION

The process of mathematical induction proceeds in 2 parts.

PART 1. Prove that at least 1 of the sentences $P(n)$ is true; that is, select a particular n, such as $n = 1$, and show that $P(1)$ is true.
PART 2. Prove that truth is contagious in the following sense: For any and all n, whenever $P(n)$ is *assumed* to be true, then a necessary consequence of that *assumption* on $P(n)$, is that the *next* sentence of the list, $P(n + 1)$, is also true. That is, Part 2 consists in establishing the *implication*

$$P(n) \text{ true} \Rightarrow P(n+1) \text{ true} \quad \text{for all } n.$$

REMARK. Part 2, in and of itself, does not *establish* the truth of any single sentence of our infinite list of sentences; rather, it establishes a logical relationship between adjacent, or successive, *pairs* of sentences. The relationship established in Part 2 can be crudely summarized as, "If, by some stroke of good fortune, $P(n)$ is true, then an unavoidable consequence is the fact that $P(n + 1)$ is true." Now Part 1, in *conjunction with* Part 2, establishes that all sentences from $P(1)$ on, must be true. That is to say, we have:

▶ $P(1)$ is true (from Part 1).

▶ If $P(1)$ is true (which it is, from Part 1), then the next sentence, $P(1 + 1) = P(2)$ is true (the implication of Part 2).

▶ If $P(2)$ is true (which it is, from previous line), then the next sentence, $P(2 + 1) = P(3)$ is true (the implication of Part 2).

▶ If $P(3)$ is true (which it is, from previous line), then the next sentence, $P(3 + 1) = P(4)$ is true (the implication of Part 2).

By continuing in this manner, we establish the truth, or validity, of every successive sentence in the list *starting* with the first, or earliest, verified sentence of Part 1. For this reason, Part 1 is usually attempted by verifying the earliest possible sentence [usually $P(1)$] on the list, since this is the springboard from which the truth of subsequent sentences will flow, due to Part 2. In (A.2), we offer a diagram of the principle of mathematical induction.

LIST OF SENTENCES

	$P(1)$	
	$P(2)$	← Truth of $P(2)$, the springboard sentence, established independently (Part 1).
Sentences,	$P(3)$	
from $P(2)$	$P(4)$	
onward,	$P(5)$	
are true	$P(6)$	
	$\vdots$	(A.2)
	$P(n)$	For arbitrary n, establish the contagious nature of truth between pairs of sentences,
	$P(n+1)$	that is, *assume* $P(n)$ true, and show that truth of $P(n+1)$ must follow (Part 2).
	$\vdots$	

Example For $P(n)$, we shall take the sentence, "The sum of consecutive integers, whose first term is 1, and whose last term is n, is equal to $n(n+1)/2$." We abbreviate this verbiage by writing

$$P(n) \equiv [1 + 2 + \cdots + n = n(n+1)/2].$$

The statement within the square brackets is truly a sentence; the sum

$$\text{“}1 + 2 + \cdots + n\text{”}$$

is the subject of the sentence, the symbol "$=$" is only a shorthand for the *predicate verb* "equals," "is," or "is equal to," and the quantity "$n(n+1)/2$" is the *predicate nominative*. Recalling that n is the last term of the sum in $P(n)$, our infinite list of sentences may be diagrammed as follows:

$$
\begin{array}{lr}
P(1) & 1 = \dfrac{1(1+1)}{2} \\[2ex]
P(2) & 1 + 2 = \dfrac{2(2+1)}{2} \\[2ex]
P(3) & 1 + 2 + 3 = \dfrac{3(3+1)}{2} \\[2ex]
\vdots & \vdots \\[2ex]
P(n) & 1 + 2 + 3 + \cdots + n = \dfrac{n(n+1)}{2} \\[2ex]
P(n+1) & 1 + 2 + 3 + \cdots + n + (n+1) = \dfrac{(n+1)((n+1)+1)}{2} \\[2ex]
\vdots & \vdots \qquad \vdots
\end{array}
\tag{A.3}
$$

▶ *Effecting Part 1*. This is easy. $P(1)$ is exactly the sentence "1 equals $\dfrac{1(1+1)}{2}$," which is true once you are convinced that $1 = 1$. In fact, it is must as easy to check that $P(2)$, the sentence "$1 + 2$ equals $\frac{2(2+1)}{2}$," is also true, but we try to directly verify the earliest sentence on our list, if possible. In this case, we can verify the very earliest, or first sentence, $P(1)$.

▶ *Effecting Part 2.* We now play a game of pretense in that we pretend that $P(n)$ is true for some arbitrary but fixed n. [Of course, from Part 1, we know that this is not a totally vain assumption inasmuch as $P(n)$ is really true when $n = 1$, that is, $P(1)$ is our springboard sentence]. We show that the truth of the subsequent sentence, $P(n + 1)$, must necessarily follow if $P(n)$ is true. Let us do just that. Assume (pretend) that for some fixed n, $P(n)$ is valid, that is,

$$1 + 2 + \cdots + n \quad \text{really } \textit{does} \text{ equal the quantity } \frac{n(n+1)}{2}. \tag{A.4}$$

Question: What does $1 + 2 + \cdots + n + (n + 1)$ equal? Well,

$$\begin{aligned}
1 + 2 + \cdots + n + (n+1) &= \underbrace{(1 + 2 + \cdots + n)}_{\text{from (A.4)}} + \underbrace{(n+1)} \\
&= \frac{n(n+1)}{2} + (n+1) \\
&= \frac{n(n+1)}{2} + \frac{2(n+1)}{2} \\
&= \frac{(n+1)(n+2)}{2} \qquad \text{factor } (n+1) \text{ from each term above; combine terms.} \\
&= \frac{(n+1)((n+1)+1)}{2}.
\end{aligned} \tag{A.5}$$

Thus, if we read (A.5) from beginning to end, we have the valid equality

$$\left[1 + 2 + \cdots + n + (n+1) = \frac{(n+1)((n+1)+1)}{2}\right] \equiv P(n+1) \qquad \text{[see (A.3)]}$$

as a consequence of the (presumed) truth of $P(n)$ [see (A.4)]. In other words

$$P(n) \text{ true, forces } P(n+1) \text{ to be true.}$$

Part 2 is complete.

Thus, since $P(1)$ *is* true (Part 1) (we need not pretend or assume its truth), $P(2)$ is automatically true from the implication of Part 2. Since $P(2)$ is true, Part 2 says that, unavoidably, $P(3)$ is true, and so on. We conclude that, in truth,

$$1 + 2 + \cdots + n \text{ is } \textit{always} \text{ equal to } \frac{n(n+1)}{2},$$

which gives us a neat formula for the sum of the first n integers expressed purely in terms of the last integer n.

Example Let $\mathbf{T}: V_n \to V_m$ be a linear function sending vector space V_n to V_m. Thus, for any *pair* of vectors $X_1, X_2 \in V_n$, we have

$$\mathbf{T}(X_1 + X_2) = \mathbf{T}(X_1) + \mathbf{T}(X_2);$$

see Definition II.2.1, Equation (II.2.2). We wish to show, by induction, that this property extends to any p variables $X_1, X_2, \ldots, X_p$, that is, we wish to prove

For all positive integers q, $\mathbf{T}(X_1 + X_2 + \cdots + X_q) = \mathbf{T}(X_1) + \mathbf{T}(X_2) + \cdots + \mathbf{T}(X_q)$;

see (II.2.3). Note that we have replaced the symbol n by the symbol q in order to stress the underlying *substance* of the argument, as opposed to the shallow *form* of the argument. In both cases, we are symbolizing a positive integer. What, then, shall be our list of sentences $P(1), P(2), \ldots, P(q), P(q+1), \ldots$? The displayed equation, that sentence which we hope to prove valid for all p, is the typical sentence desired, that is, set

$$P(q) \equiv [\mathbf{T}(X_1 + X_2 + \cdots + X_q) = \mathbf{T}(X_1) + \mathbf{T}(X_2) + \cdots + \mathbf{T}(X_q)].$$

Thus, a diagram of our infinite list would be

$$\begin{array}{lll}
P(1) & \mathbf{T}(X_1) & = \mathbf{T}(X_1) \\
P(2) & \mathbf{T}(X_1 + X_2) & = \mathbf{T}(X_1) + \mathbf{T}(X_2) \\
P(3) & \mathbf{T}(X_1 + X_2 + X_3) & = \mathbf{T}(X_1) + \mathbf{T}(X_2) + \mathbf{T}(X_3) \\
\vdots & & \vdots \\
P(q) & \mathbf{T}(X_1 + X_2 + \cdots + X_q) & = \mathbf{T}(X_1) + \mathbf{T}(X_2) + \cdots + \mathbf{T}(X_q) \\
P(q+1) & \mathbf{T}(X_1 + X_2 + \cdots + X_q + X_{q+1}) & = \mathbf{T}(X_1) + \mathbf{T}(X_2) + \cdots + \mathbf{T}(X_q) + \mathbf{T}(X_{q+1}) \\
\vdots & & \vdots
\end{array}$$

Part 1. Obviously $P(1)$, $\mathbf{T}(X_1) = \mathbf{T}(X_1)$, is true, and from the definition of additivity, $P(2)$, $\mathbf{T}(X_1 + X_2) = \mathbf{T}(X_1) + \mathbf{T}(X_2)$ is also true. $P(1)$ [or $P(2)$] can serve as our springboard sentence, that is, our earliest true sentence on the list.

Part 2. Assume for some positive integer q,

$$\mathbf{T}(X_1 + X_2 + \cdots + X_q) = \mathbf{T}(X_1) + \mathbf{T}(X_2) + \cdots + \mathbf{T}(X_q), \tag{A.6}$$

that is, assume, $P(q)$ is really true for some q. What can be said, then, about $\mathbf{T}(X_1 + X_2 + \cdots + X_q + X_{q+1})$?

$$\begin{aligned}
\mathbf{T}(X_1 + X_2 + \cdots + X_q + X_{q+1}) &= \mathbf{T}((X_1 + X_2 + \cdots + X_2) + (X_{q+1})) \\
&\qquad \text{since } \mathbf{T} \text{ is additive on the } \textit{pair} \\
&\qquad \text{of vectors } (X_1 + X_2 + \cdots + X_q), \\
&\qquad \text{and } (X_{q+1}), \text{ we have} \\
&= \underbrace{\mathbf{T}(X_1 + X_2 + \cdots + X_q)}_{\text{from (A.6)}} + \underbrace{\mathbf{T}(X_{q+1})} \\
&= \overbrace{\mathbf{T}(X_1) + \mathbf{T}(X_2) + \cdots + \mathbf{T}(X_q)} + \overbrace{\mathbf{T}(X_{q+1})}.
\end{aligned} \tag{A.7}$$

Reading (A.7) from top to bottom, reveals the truth of the valid equality.

$$[\mathbf{T}(X_1 + X_2 + \cdots + X_q + X_{q+1}) = \mathbf{T}(X_1) + \mathbf{T}(X_2) + \cdots + \mathbf{T}(X_q) + \mathbf{T}(X_{q+1})] \equiv P(q+1),$$

provided $P(q)$ is valid, that is,

$P(q)$ true, forces $P(q+1)$ to be true.

Since $P(1)$ *is* true, so then is $P(2)$ true, and so on. Thus, $P(q)$ is true for all integers $q \geq 1$; that is, if it is additive relative to 2 variables, then it is additive relative to any integer number of variables.

EXERCISES

Prove the validity of the following statements by using mathematical induction.

A.1 $1^2 + 2^2 + 3^2 + \cdots + n^2 = \dfrac{n(n+1)(2n+1)}{6}.$

A.2 $2 + 4 + 6 + \cdots + 2n = n(n+1).$

A.3 $2^2 + 4^2 + 6^2 + \cdots + (2n)^2 = \frac{2}{3}\, n(n+1)(2n+1).$

A.4 $1 + 3 + 5 + \cdots + (2n-1) = n^2.$

A.5 $1^2 + 3^2 + 5^2 + \cdots + (2n-1)^2 = \dfrac{n}{3}(2n-1)(2n+1).$

A.6 $1^3 + 2^3 + 3^3 + \cdots + n^3 = \dfrac{n^2(n+1)^2}{4}.$

A.7 $1 + 4 + 7 + \cdots + (3n-2) = \dfrac{n}{2}(3n-1).$

A.8 $4 + 8 + 12 + \cdots + 4n = 2n(n+1).$

A.9 $5 + 10 + 15 + \cdots + 5n = \frac{5}{2}\, n(n+1).$

A.10 $1 + x + x^2 + \cdots + x^n = \dfrac{1 - x^{n+1}}{1 - x}$ for all $x \neq 1$.

A.11 Let f be a function sending non-negative integers to real numbers. Furthermore, f has the following properties.

(a) $f(0) = 1$.

(b) For all integers k, $f(k) \cdot f(1) = f(k+1)$.
Prove that for all $n = 0, 1, 2, \ldots,$ $f(k) \cdot f(n) = f(k+n)$.

A.12 Let x be any scalar. We *define*

(a) $x^0 = 1$.

(b) For all integers k, x^{k+1} is to equal $x^k \cdot x^1$.
Prove that for all $n = 0, 1, 2, \ldots,$ $x^k \cdot x^n = x^{k+n}$. [See Exercise A.11, and choose $f(k) = x^k$. Otherwise, a direct proof by induction is feasible.]

A.13 Given the distributive property for a ring R where

$$x \cdot (y_1 + y_2) = x \cdot y_1 + x \cdot y_2 \qquad \text{for all } x, y_1, y_2, \in R.$$

Prove that for all $x, y_1, y_2, \ldots, y_n \in R$,

$$x \cdot (y_1 + y_2 + \cdots + y_n) = x \cdot y_1 + x \cdot y_2 + \cdots + x \cdot y_n.$$

A.14 Prove that $x^n - y^n$ is divisible by $x - y$ for each positive integer n (that is $x - y$ is a factor of $x^n - y^n$ for each n). *Hint:* Verify the equality

$$x^{n+1} - y^{n+1} = x^n(x - y) + y(x^n - y^n)$$

for all x, y, and for all positive integers n.

A.15 Prove that $x^{2n} - y^{2n}$ is divisible by $x + y$ for all positive integers n. *Hint:* Verify the equality

$$x^{2(n+1)} - y^{2(n+1)} = (x^{2n} - y^{2n})x^2 + (x^2 - y^2)y^{2n}$$

for all x, y, and for all positive integers n.

A.16 Find the error in the following "proof" that every girl is beautiful.
We take for our sentence $P(n)$, the following:

$P(n) \equiv$ "If a set consisting of exactly n girls, has at least 1 beautiful girl as an element, then *all* n girls of the set are beautiful."

Notice that $P(n)$ is a sentence which is, in itself, an implication. It says, roughly, that beauty is contagious, for if one girl of a set is beautiful, then *all* the girls become infected with the beauty bug. To put it another way, $P(n)$ says that there are no sets of girls of mixed composition, that is, we never have a set of girls where some girls are beautiful, and some girls are not. Either all girls of any n-element set are beautiful, or else they are all not-so-beautiful—this is the essence of the statement $P(n)$.

We have now, to decide the *truth* of $P(n)$, the statement which says that the beauty of 1 girl in a set of n girls, forces (or implies) the beauty of all the other girls.

• *Part 1.* $P(1)$ is the statement, "If a set consisting of exactly one girl has at least 1 beautiful girl as an element, then *all* the girls of the set are beautiful." $P(1)$ is obviously true, for if 1 girl of a 1-element set is beautiful, then, inescapably, the whole (1-element) set consists of beautiful girl(s).

• *Part 2.* Let us pretend that $P(n)$ is *true* for some fixed n, that is, whenever we have any n-element set of girls, the beauty of 1 always implies the beauty of all. (We know this is possible for $n = 1$, at least.) We are obliged to show that $P(n + 1)$ is true as a consequence. Thus, let us consider a set of $n + 1$ girls, and let us presume that at least 1 of them is a living doll, that is, is beautiful. Does the truth of $P(n)$ force all $n + 1$ girls to be beautiful? We will see.

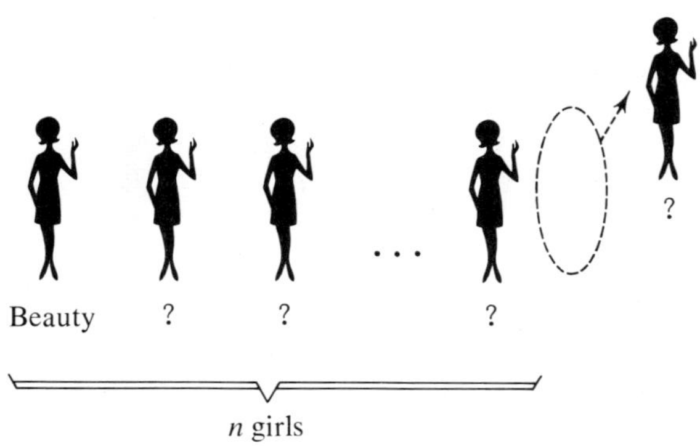

Figure A.1

Consider an arbitrary set of $n+1$ girls in our classroom, at least 1 of whom is beautiful, the other n girls being of (as yet) unknown description. We send 1 of these "other" unclassified girls out into the hall, leaving behind a set of n girls, one of whom is beautiful (see Figure A.1). From the assumed truth of $P(n)$, the presence of this 1 beautiful girl is sufficient to induce beauty in all the other $n-1$ girls in the classroom. Thus we have n beautiful girls in the classroom, and a girl in the hall which is either (1) beautiful, or (2) nonbeautiful.

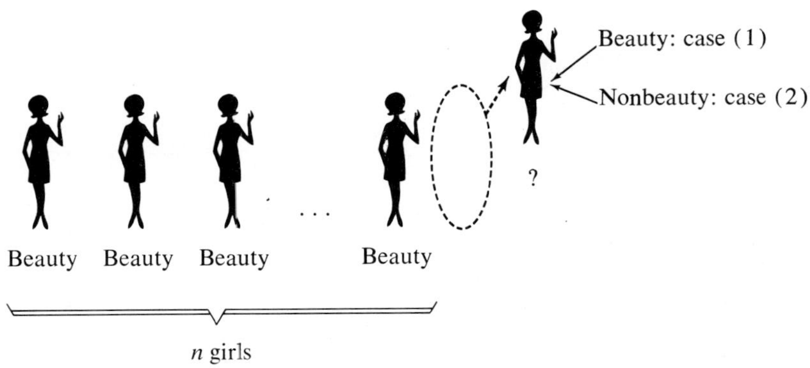

Figure A.2

If case (1) occurs, where the girl in the hall *is* beautiful, then we are done. We merely invite her back into the classroom, and we have a set of $n+1$ beautiful girls, induced by the beauty of (at least) 1 of the remaining girls, that is, $P(n) \Rightarrow P(n+1)$ true.

If case (2) occurs, where the girl in the hall is a nonbeauty, then we simply ask her to re-enter the classroom and change places with some (necessarily beautiful) girl in the classroom (see Figure A.2). Now we have 1 beautiful girl in the hall, and $n-1$ beautiful girls in the classroom, alongside our (nonbeauty) hallway girl. Invoking the truth of $P(n)$, which disallows mixed compositions of sets of girls with n elements, we conclude that the hallway girl must contract the beauty virus, that is,

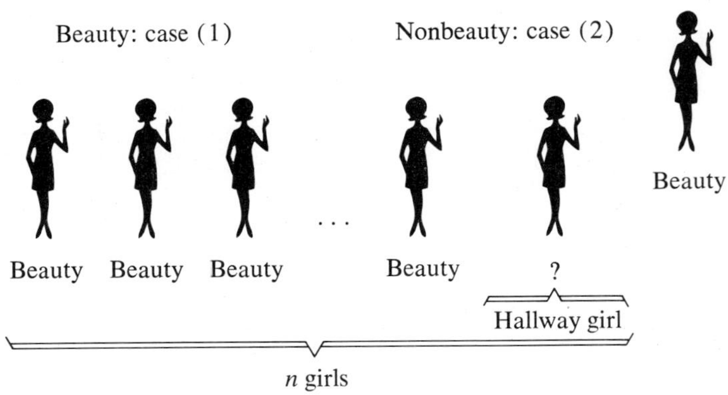

Figure A.3

the beauty of at least 1 other girl in the classroom, must, by the truth of $P(n)$, force even the nonbeauty hallway girl to be beautiful. We now have n beautiful girls in the classroom, and once we recall our replacement girl from the hall (who was beautiful; see Figure A.3) and we have our set of $n + 1$ beautiful girls. Once again, $P(n)$ true $\Rightarrow P(n + 1)$ true. Therefore, all girls are beautiful.

APPENDIX B

Complex Numbers

1. *Why Complex Numbers?*

Certain equations that arise need "solutions"—and sometimes these solutions prove to be objects, or numbers, which may seem strange or new to us.

For example, consider the equation

$$x + 3 = 8. \tag{B.1}$$

To find a "solution" to Equation (B.1) is to find at least 1 object which when substituted for x, yields a valid statement of fact. As it turns out, 1 and only 1 object will do, namely, the number 5. Thus, if "x" in (B.1) is replaced by "5," we obtain the statement

$$5 + 3 = 8,$$

which is a valid statement.

Now consider the equation

$$x + 3 = 3. \tag{B.2}$$

In this case, replace x by the "number" zero, denoted 0, and we have the valid statement

$$0 + 3 = 3.$$

Equation (B.2) is mentioned because the number zero was not readily accepted into the set of integers when it was first discovered (or invented). That is, 0 seemed to be, somehow, different; it seemed, intuitively, almost not to exist.

Now consider the equation

$$x + 3 = 1. \tag{B.3}$$

What is x? Is it a number which is "negative"? Indeed, the negative numbers were not only snubbed, their very *existence* was doubted at first. In the sense that negative numbers introduce no contradictions into our number system, they do indeed exist. Thus, replace x in (B.3) by -2, the negative of 2, and we obtain the valid statement

$$-2+3=1.$$

Thus, in Equations (B.1), (B.2), and (B.3), we have need of the *integers* $Z=\{0,\pm 1,\pm 2,\ldots\}$. In a similar fashion, the equation

$$5x+3=4 \tag{B.4}$$

requires for its solution *none* of the numbers we have encountered so far, that is, no number of the form $\{0,\pm 1,\pm 2,\ldots\}$ will yield a valid statement when it is substituted for x in (B.4). More and different numbers are needed? Yes. Numbers p/q which are quotients of integers p and q are needed. For Equation (B.4), the rational number $\frac{1}{5}$ is the solution.

Yet, for all this, "numbers" outside the rationals are still necessary. Consider the equation

$$x^2-2=0. \tag{B.5}$$

There are 2 solutions to this one, and neither of them is *rational*. The solutions, usually denoted $\sqrt{2}$ and $-\sqrt{2}$ cannot be more exactly expressed than to say, "$\sqrt{2}$ is that number, which when multiplied by itself, yields the integer 2. No, $\sqrt{2}$ is not an integer . . . nor is it a quotient of integers. It is something else. It is described only by what it looks like after it is multiplied by itself." We beg the question of existence, and state only that the *irrational* numbers are given a *reason* for existence because of equations such as (B.5). Thus, we have encountered integers, negative numbers, and rationals and irrationals, the total collection of which is called "the set of real numbers."

And we are not done, for the set of real numbers proves *inadequate*. Consider, then, the equation

$$x^2+1=0. \tag{B.6}$$

For a solution, we require an object, call it i, whose square is the number -1! That is to say, $i^2=-1$. Well, no one ever heard of any *real* number with such a property (that is, its square equals -1), so we must go *outside* the real numbers for a solution. The tendency here is to be prejudiced against this entity i, as people of the olden days were against *negative* numbers. Nevertheless, we introduce the special symbol i to be that "number" or "object" with the property that $i^2=-1$. i is not a *real* number, so what kind of number shall it be called? Out of exasperation more than anything else, i is dubbed an **imaginary number**. Thus. any number, such as $3 + 4 \cdot i$, involving i, is called a **complex number**.

And we are done; for the set of complex numbers is finally adequate. In fact we have the following theorem, which we present without proof.

Theorem (Fundamental Theorem of Algebra) *For the equation*

$$a_0 + a_1x + a_2x^2 + \cdots + a_nx^n = 0, \quad a_n \neq 0, n \geq 1, \tag{B.7}$$

where each a_i is a complex number, there is at least 1 *complex number z, which, when substituted for the symbol x, yields a valid statement; that is, there exists a complex number solution for* (B.7).

This theorem should be a relief, for no further progression of equations such as (B.1), (B.2), (B.3), (B.4), (B.5), and (B.6), can force us to invent any other numbers outside the complex numbers, for solutions.

2. The Complex Numbers

We present a definition of complex numbers and then define the operations of addition and multiplication between pairs of complex numbers.

Definition B.1 A **complex number** z is any element of the form

$$z = a + ib \quad \text{(equivalently, } a + bi\text{)},$$

where a and b are real numbers, and $i \cdot i = i^2 = -1$. Two complex numbers $z_1 = a_1 + ib_1$, and $z_2 = a_2 + ib_2$ are said to be **equal** (written $z_1 = z_2$) if and only if $a_1 = a_2$ and $b_1 = b_2$.

REMARK. The symbol "+" in Definition B.1 is different from the plus used in an expression such as

$$4 + 5$$

in the following sense: With $4 + 5$, we may telescope or contract the expression with the shorter expression "9." With $a + ib$, there is no such telescoping of the terms. We may think of the (noncontractable) number $a + ib$ as "a along with ib."

We see, then, that any complex number is fully described by means of its 2 component real parts, a and b. Since the *ordered pair* of *real* numbers a and b describes the complex number $a + ib$, we introduce an alternate notation for $a + ib$, or its equivalent, $a + bi$.

Notation B.1 The symbol (a,b) shall represent the complex number $z = a + ib$, that is,

$$(a,b) = a + ib = a + bi$$

where a and b are real numbers. The first real entry, a, of (a,b) is called the **real part** of $(a,b) = a + ib$, and the second real entry, b, is called the **imaginary part** of $(a,b) = a + ib$. The set of *all* complex numbers $(a,b) = a + ib$, is denoted by **C**.

Let us now induce a pairwise addition on the set of complex numbers. This will be done in the most natural and (hopefully) obvious way. For example, if $a_1 + ib_1$, and $a_2 + ib_2$ are 2 complex numbers, then which complex number (of form $a_3 + ib_3$) shall we assign to the sum $(a_1 + ib_1) + (a_2 + ib_2)$? It seems natural to combine (that is, add) the real parts and the imaginary parts, respectively. Thus, for addition in **C**, we have Definition B.2.

Definition B.2 Let $a_1 + ib_1 = (a_1,b_2)$ and $a_2 + ib_2 = (a_2,b_2)$ be any 2 complex numbers. Then $(a_1 + ib_1) + (a_2 + ib_2)$ is defined by

$$(a_1 + ib_1) + (a_2 + ib_2) = (a_1 + a_2) + i(b_1 + b_2).$$

Using our alternate notation to say exactly the same thing,

$$(a_1,b_1) + (a_2,b_2) = (a_1 + a_2, b_1 + b_2).$$

Example Find the sum of the complex numbers $3 + i(-2) = (3,-2)$, and $1 + i4 = (1,4)$. Definition B.2 says that $(3 + i(-2)) + (1 + i4) = (3 + 1) + i(-2 + 4) = 4 + i2$. Using the ordered pair notation, we form the complex sum by adding corresponding entries, that is,

$$(3,-2) + (1,4) = (3 + 1, -2 + 4) = (4,2).$$

The virtue of the original notation $a + ib$ for complex numbers becomes apparent once we attempt to define a *multiplication* between pairs of complex numbers [the ordered pair notation (a,b) has a virtue or two which will be described later]. Thus, to define $(a_1 + ib_1) \cdot (a_2 + ib_2)$, we pretend that the symbol "$+i$" in the symbol "$a + ib$" acts like the familiar $+$ between real numbers. This means that we have a distributive law working in our behalf, so that

$$\begin{aligned}(a_1 + ib_1) \cdot (a_2 + ib_2) &= (a_1 + ib_1) \cdot a_2 + (a_1 + ib_1) \cdot ib_2 \\ &\qquad \text{(distributing from the left)} \\ &= a_1a_2 + ib_1a_2 + a_1ib_2 + ib_2 \cdot ib_2 \\ &\qquad \text{(distributing on the right)} \\ &= a_1a_2 + i^2b_1b_2 + ib_1a_2 + a_1ib_2 \\ &= (a_1a_2 - b_1b_2) + i(a_1b_2 + a_2b_1). \\ &\qquad \text{(since } i^2 = -1\text{)}.\end{aligned}$$

This is the basis for Definition B.3.

Definition B.3 Let $a_1 + ib_1 = (a_1,b_1)$ and $a_2 + ib_2 = (a_2,b_2)$ be any 2 complex numbers. Then $(a_1 + ib_1) \cdot (a_2 + ib_2)$ is defined by

$$(a_1 + ib_1) \cdot (a_2 + ib_2) = (a_1a_2 - b_1b_2) + i(a_1b_2 + a_2b_1).$$

Using the equivalent ordered pair notation, we say the same thing in a (possibly) less intuitive manner by,

$$(a_1,b_1) \cdot (a_2,b_2) = (a_1a_2 - b_1b_2,\ a_1b_2 + a_2b_1).$$

Example Let us multiply the complex numbers (Is the *order* of the factors important?) $2 + i6$, and $3 + i(-1)$. Thus, from Definition B.3,

$$\begin{aligned}(2 + i6) \cdot (3 + i(-1)) &= (2 \cdot 3 - 6 \cdot (-1)) + i(2 \cdot (-1) + 6 \cdot 3), \\ &= (6 + 6) + i(-2 + 18) = 12 + i16.\end{aligned}$$

Using the ordered pair notation, the above equations become

$$\begin{aligned}(2,6) \cdot (3,-1) &+ (2 \cdot 3 - 6 \cdot (-1), 2 \cdot (-1) + 6 \cdot 3) \\ &= (6 + 6, -2 + 18) = (12,16).\end{aligned}$$

We can take no pride in effecting an addition (Definition B.2) and a multiplication (Definition B.3) on **C**, unless they are "good" operations. That is, if z_1, z_2, and z_3 represent arbitrary complex numbers, then we should wish to know the answers to questions such as:

(a) Does $z_1 + z_2 = z_2 + z_1$, and $z_1 \cdot z_2 = z_2 \cdot z_1$ always? (commutivity)

(b) Is the equation $z_1 . (z_2 + z_3) = z_1 \cdot z_2 + z_1 \cdot z_3$ always true? (distributivity)

(c) Do "feeble" elements exist? That is, is there some "zero," z^* such that $z^* + z = z$ for all z? Is there some "unity" z^{**} such that $z^{**} \cdot z = z$ for all z?

(d) Do inverses exist relative to the previously described "feeble" elements z^* and z^{**} (assuming *their* existence)? That is, for every $z \in \mathbf{C}$, is there a z' and a z'' such that

$$z + z' = z^*,$$

and

$$z \cdot z'' = z^{**}?$$

The questions are left as exercises. The following notation will prove convenient.

Notation B.2 The symbol $a - ib$ (equivalently, $a - bi$) shall stand for the complex number $a + i(-b) = (a,-b)$ (b can be either positive or negative).

If $z = a + ib = (a,b)$, then $\bar{z}$, the *complex conjugate* of z, is defined by

$$\bar{z} = a - ib = (a,-b).$$

EXERCISES

B.1 Find the sums, $z_1 + z_2$, of the following complex numbers, z_1 and z_2:

(a) $z_1 = 6 - i2,\ z_2 = 6 + i2$
(b) $z_1 = 4 + 6i,\ z_2 = -2 - 3i$
(c) $z_1 = 0 + 2i,\ z_2 = 2 + 0i$
(d) $z_1 = 0 + 0i,\ z_2 = a + bi$
(e) $z_1 = -a + bi,\ z_2 = a - bi$
(f) $z_1 = a + 2ai,\ z_2 = a + 3ai$

B.2 Find the products, $z_1 \cdot z_2$, of the following complex numbers, z_1 and z_2:

(a) $z_1 = 1 + 1i,\ z_2 = 2 + 3i$
(b) $z_1 = -1 - 2i,\ z_2 = 3 + 4i$
(c) $z_1 = \frac{\sqrt{3}}{2} - \frac{1}{2}i,\ z_2 = \frac{\sqrt{3}}{2} - \frac{1}{2}i$
(d) $z_1 = a + ai,\ z_2 = b + bi$
(e) $z_1 = 0 + 3i,\ z_2 = 0 - 3i$
(f) $z_1 = 2 - 2i,\ z_2 = 3 + 4i$

B.3 Find the neutral complex number, $z_0 = a + ib$, which is "feeble" under addition, that is, find $z_0 \in \mathbf{C}$ which has the property that $z_0 + z = z + z_0$ for any and every $z \in \mathbf{C}$.

B.4 (Commutivity of $+$ and $\cdot$) Prove that for any pair of complex numbers, z_1 and z_2, $z_1 + z_2 = z_2 + z_1$, and $z_1 \cdot z_2 = z_2 \cdot z_1$.

B.5 (Distributivity) Prove that for any triple of complex numbers z_1, z_2, z_3, we have $z_1 \cdot (z_2 + z_3) = z_1 \cdot z_2 + z_1 \cdot z_3$.

B.6 Find the neutral complex number, $z_1 = a + ib$, which is "feeble" under multiplication, that is, find $z_1 \in \mathbf{C}$ which has the propety that $z_1 \cdot z = z \cdot z_1$ for any and every $z \in \mathbf{C}$.

B.7 Let $z = a + ib$ be a complex number so that $\bar{z} = a - ib$ is its complex conjugate. Prove that

(a) $z + \bar{z} = 2a + i0$
(b) $z - \bar{z} = 0 + 2bi$
(c) $z \cdot \bar{z} = (a^2 + b^2) + i \cdot 0$
(d) $\bar{\bar{z}} = z$
(c) $\overline{z_1 + z_2} = \bar{z}_1 + \bar{z}_2$
(f) $\overline{z_1 \cdot z_2} = \bar{z}_1 \cdot \bar{z}_2$

B.8 Show that $z \cdot z = z$ *only* for $z = (0,0)$ or $z = (1,0)$.

3. *Geometric Interpretation*

Which complex numbers $(a,b) = a + ib$ correspond to the familiar *real* numbers? Once this correspondence is established, we may think of the real numbers as being imbedded in (or a special case of) the complex numbers.

QUESTION. Which complex number $(a,b) = a + ib$ corresponds to (is actually) the real number 1? That is, we are to find $(a,b) = a + ib$ such that $1 = (a,b)$.

ANSWER. Of necessity, $1 = (1,0)$. To see this, 1 is the (feeble) multiplicative identity for the reals. If we were to discover $1 = (a,b)$, then (a,b) must be multiplicatively feeble on complex numbers. That is, we require

$$1 \cdot (c,d) = (c,d) \quad \text{for all complex } (c,d). \tag{B.8}$$

Now put $1 = (a_0,b_0)$ and let's see why $a_0 = 1$, and $b_0 = 0$. Thus,

$$1 \cdot (c,d) = \overbrace{(a_0,b_0)}^{1} \cdot (c,d) = (c,d)$$

means

$$(a_0c - b_0d, b_0c + a_0d) = (c,d) \quad \text{Definition B.3.}$$

This last equation says that for all real c, d,

(a) $a_0c - b_0d = c$, and

(b) $b_0c + a_0d = d$ (see Definition B.1).

Set $c = 1$, $d = 0$ in (a) and (b) to obtain $a_0 = 1$ and $b_0 = 0$. Thus, if 1 corresponds to any complex number at all, then 1 corresponds to the complex number $(1,0) = 1 + 0 \cdot i$ (see Exercise B.6).
That is,

$$1 \to (1,0) = 1 + 0 \cdot i.$$

By our rules of addition, $2 = 1 + 1 \to (1,0) + (1,0) = (2,0)$. By induction, we have that for any positive integer n,

$$n \to (n,0) = n + 0 \cdot i.$$

We may argue similarly to extend the above correspondence to rational numbers, since

$$1 = \underbrace{\frac{1}{p} + \frac{1}{p} + \cdots + \frac{1}{p}}_{p \text{ terms}}, \quad \text{and}$$

$$(1,0) = \underbrace{\left(\frac{1}{p}, 0\right) + \left(\frac{1}{p}, 0\right) + \cdots + \left(\frac{1}{p}, 0\right)}_{p \text{ terms}},$$

we have the correspondence

$$\frac{1}{p} \to \left(\frac{1}{p}, 0\right).$$

Finally, for *all* real numbers a, we have the correspondence

$$a \to (a,0) = a + 0 \cdot i,$$

which we emphasize by writing

$$a = (a, 0) = a + 0 \cdot i,$$

for all real numbers a.

Due to the above discussion, which equates real numbers a with complex numbers $a + 0 \cdot i$, we have Notation B.3.

Notation B.3 We will write a in place of the symbol $a + 0 \cdot i = (a, 0)$, and ib for the symbol $0 + ib = (0, b)$, that is,

(a) $$a = (a, 0),$$

(b) $$ib = (0, b).$$

REMARK. We note that (a) has been fully justified by our previous discussion. Part (b) must be justified by showing that when $b = 1$, the equation $i = (0,1)$ holds true. $i = (0,1)$ has the property it is supposed to have, namely, in both cases, the square is -1. To see this, note that the left-hand side of part (b), when squared, yields $i^2 = -1$, which is true from the definition of i. Squaring the right-hand side yields $(0,1)^2 = (-1,0)$ (Definition B.3). But from part (a), $(-1,0) = -1$. Thus, Part (b) is valid at least when $b = 1$.

Let us use the fact that complex numbers $a + ib = (a,b)$ are, in fact, *ordered pairs* of real numbers. We have already introduced a graphical model for such ordered pairs, namely, the plane. Thus, in "picture form," the set of complex numbers **C** looks exactly like the set of 2-vectors $V_2(\mathbf{R})$ with real coordinates *except* that **C** has a multiplication defined, while $V_2(\mathbf{R})$ does not. (See Figure B.1.) By our previous discussion, we see that the real numbers are just the complex numbers (a,b) whose imaginary part (second entry) b, is equal to zero. That is, all points on the *horizontal axis* are the real numbers, imbedded into the set of complex numbers (the plane). Hence the horizontal axis is called the *real axis* [and is not called the x-axis, as in the case where the plane represents $V_2(\mathbf{R})$]. Since the

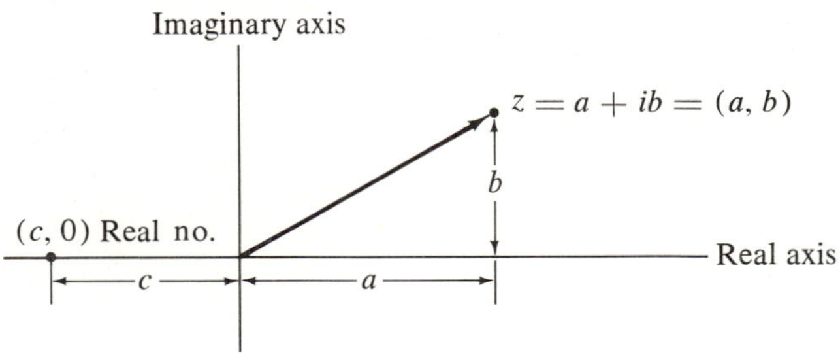

Figure B.1

"pure imaginary" complex numbers $0 + ib = (0,b)$, whose real part (first entry) is zero, are determined as points lying on the vertical axis, that axis is called the *imaginary axis*.

With this picture in mind, we see that the operation of addition (Definition B.2) exactly coincides with the coordinatewise addition defined for 2-vectors in $V_2(\mathbf{R})$ (Definition I.1.2). Thus, the parallelogram model for vector addition obtains for *both* $V_2(\mathbf{R})$ and for $\mathbf{C}$. A convenient graphical representation for multiplication of complex numbers must await the development of de Moivre's formula.

We are ready for the notion of *division* by complex numbers. Our objective is to describe that complex number z^{-1} (necessarily of the form $z^{-1} = a + ib$) for each nonzero $z \in \mathbf{C}$, with the property that $z^{-1} \cdot z = z \cdot z^{-1} = 1 = (1,0)$. It is easy to see that at least for nonzero *real* numbers $z = (a,0) = a$, the multiplicative inverse $z^{-1} = (a^{-1},0)$, where $aa^{-1} = a^{-1}a = 1$, that is, a^{-1} is the (real number) multiplicative inverse for a. To test this, imbed the number a into the complex numbers $a \to (a,0)$. We see that even under complex multiplication,

$$(a^{-1},0) \cdot (a,0) = (a,0) \cdot (a^{-1},0) = (1,0) = 1.$$

That is, $(a^{-1},0)$ *is* the multiplicative inverse of $(a,0)$; symbolically, $(a^{-1},0) = (a,0)^{-1}$.

We do not know yet that inverses for *arbitrary* nonzero complex numbers $z = (a,b)$ exist. Even so, we define the notion of inverse formally with Definition B.4.

Definition B.4 Let $z = (a,b)$ be a complex number. The complex number $(a,b)^{-1} = z^{-1}$, the **multiplicative inverse** of z, is that number (c,d) (which depends on z) such that

$$(a,b) \cdot (c,d) = (c,d) \cdot (a,b) = (1,0) = 1.$$

That is,

$$(a,b) \cdot (a,b)^{-1} = (a,b)^{-1} \cdot (a,b) = (1,0) = 1.$$

An equivalent *notation* for $(a,b)^{-1} = (a + ib)^{-1}$ is $(a + ib)^{-1} = \dfrac{1}{a + ib}$. If $w = (a',b') = a' + ib'$ is any other complex number, then the product of $a' + ib'$ with $(a + ib)^{-1}$ is denoted by

$$(a' + ib') \cdot (a + ib)^{-1} = (\text{notation}) \, \frac{(a' + ib')}{(a + ib)}.$$

If r is a real number, then we have

$$\frac{a + ib}{r} = \frac{a}{r} + i\frac{b}{r},$$

since $(r,0)^{-1} = (r^{-1},0)$.

REMARK. This definition sets $(a+ib)^{-1} = \dfrac{1}{a+ib}$; the latter symbol $\dfrac{1}{a+ib}$ does not apparently represent a complex number since any complex number must be an ordered pair (c,d). That such an ordered pair $(c,d) = c + id$ does exist for each nonzero $z = (a,b)$, will be shown presently. First, we introduce the notion of length, or modulus of a complex number z.

Definition B.5 Suppose $z = (a,b) = a + ib \in C$. Then $|z|$, the non-negative **modulus (length)** of the complex number z, is defined by

$$|z|^2 = a^2 + b^2.$$

Equivalently,

$$|z| = \sqrt{a^2 + b^2}.$$

REMARK. As it turns out, the notion of modulus or length can be intimately tried to the idea of complex conjugate (see Notation B.2). It is an easy exercise [see Exercise B.7(c)] to show that for all $z = a + ib$,

$$z \cdot \bar{z} = \bar{z} \cdot z = a^2 + b^2.$$

In light of Definition B.5, this just means

$$\text{(a)} \qquad z \cdot \bar{z} = |z|^2 \quad \text{for all } z \in \mathbf{C},$$

a simple observation which shall prove to be very useful. In fact, we use the relation (a) to establish Theorem B.1.

Theorem B.1 *Suppose $z = (a,b) = a + ib$, where $z \neq 0$ (that is, $a^2 + b^2 > 0$). Then z^{-1}, the complex multiplicative inverse of z is described by*

$$\begin{aligned} &\quad\text{(a)} \qquad \text{(b)} \qquad\qquad \text{(c)} \qquad\quad \text{(d)} \\ z^{-1} &= (a,b)^{-1} = \frac{1}{a+ib} = \frac{\bar{z}}{|z|^2} \\ &= \frac{a - ib}{a^2 + b^2} \\ &= \left(\frac{a}{a^2+b^2}\right) - i\left(\frac{b}{a^2+b^2}\right). \end{aligned}$$

Proof: We must note that the terms marked (a), (b), and (c) above are only different notations for the inverse of z. The 3 specific symbols which appear under the (d), represent that ordered pair which pretends to be the inverse z^{-1} of z. And so, we *test* the alleged z^{-1} by multiplying against

z to see whether or not the number $1 = (1,0)$ arises. Accordingly,

$$(a+ib)\cdot\left(\frac{a-ib}{a^2+b^2}\right) = (a+ib)\cdot\left(\frac{a}{a^2+b^2} - i\frac{b}{a^2+b^2}\right)$$

$$= \frac{a^2}{a^2+b^2} + i\frac{ba}{a^2+b^2} - i\frac{ab}{a^2+b^2} - i^2\frac{b^2}{a^2+b^2}$$

$$= \frac{a^2}{a^2+b^2} + \frac{b^2}{a^2+b^2}$$

$$= \frac{a^2+b^2}{a^2+b^2} = 1.$$

Thus, the complex number $\dfrac{a-ib}{a^2+b^2}$ that is, $\dfrac{\bar{z}}{|z|^2}$ equals z^{-1}. A brief *symbolic* derivation of this result follows:

$$z^{-1} = \frac{1}{z} = \frac{1}{z}\left(\frac{\bar{z}}{\bar{z}}\right) = \frac{\bar{z}}{z\cdot\bar{z}} = \frac{\bar{z}}{|z|^2}.$$

(In the middle step, $\frac{\bar{z}}{\bar{z}}$ is struck through and marked 1.)

In any case, we must agree that z^{-1}, the multiplicative inverse of z is just the complex number $\dfrac{\bar{z}}{|z|^2}$.

Example Find $(2+3i)^{-1}$. Thus we write

$$(2+3i)^{-1} = \frac{1}{2+3i} = \frac{1}{(2+3i)}\cdot\frac{2-3i}{(2-3i)}$$

so that in multiplying the denominator $2+3i$ by the complex conjugate, $2-3i$, we obtain the *real* number $2^2+3^2 = |2+3i|^2$. Continuing the above equality, we have

$$(2+3i)^{-1} = \frac{2-3i}{(2+3i)(2-3i)} = \frac{(2-3i)}{2^2+9^2} = \frac{2}{13} - \frac{3}{13}i$$

which is a complex number of the form $a+bi$, where $a = \dfrac{2}{13}$ and $b = \dfrac{-3}{13}$. A check reveals that

$$(2+3i)\left(\frac{2}{13} - \frac{3}{13}i\right) = \frac{4}{13} + \frac{6i}{13} - \frac{6i}{13} - \frac{9i^2}{13} = \frac{4}{13} + \frac{9}{13} = 1,$$

that is,

$$\left(\frac{2}{13} - \frac{3}{13}i\right) = (2+3i)^{-1}, \quad \text{or} \quad \left(\frac{2}{13} - \frac{3}{13}i\right)^{-1} = (2+3i).$$

A restatement of Theorem B.1 is the following:

Every nonzero complex number $z = a + ib \in \mathbf{C}$ *has a multiplicative inverse* $z^{-1} \in \mathbf{C}$.

Since $z^{-1} = \dfrac{\bar{z}}{|z|^2}$, existence is assured whenever the denominator $|z|^2$ is not zero.

That is, an inverse z^{-1} exists for $z = a + ib$ whenever $|z|^2 = z \cdot \overline{Z} = a^2 + b^2 \neq 0$. But the sum of 2 non-negative real numbers, a^2 and b^2, is zero *only* when both a and $b = 0$, that is, only when $z = 0$. Whenever $z \neq 0$, we have $|z|^2 = a^2 + b^2 \neq 0$, so that $z^{-1} = \dfrac{\bar{z}}{|z|^2}$ does exist (see Exercise B.20).

Example Compute $\dfrac{(1-3i)}{(6+2i)}$. Again, to bring the denominator into positive real-number form, multiply both numerator and denominator by the complex conjugate $6 - 2i$. Thus

$$\begin{aligned}(1-3i)\cdot(6+2i)^{-1} = \frac{(1-3i)}{(6+2i)} &= \frac{(1-3i)}{6+2i}\cdot\frac{(6-2i)}{(6-2i)} \\ &= \frac{(1-3i)(6-2i)}{(6+2i)(6-2i)} \\ &= \frac{6-18i-2i+6i^2}{36+12i-12i-4i^2} \\ &= \frac{-20i}{40} = -\frac{1}{2}i = \left(0, -\frac{1}{2}\right).\end{aligned}$$

Check: We should have $(1 - 3i) = (-\frac{1}{2}i)(6 + 2i)$, which is easily seen to be true.

EXERCISES

B.9 Represent the sums of Exercise B.1 as points (or vectors) in the plane.

B.10 Represent the products of Exercise B.2 as points (or vectors) in the plane.

B.11 Describe the position of the complex number $z = (a,b) = a + ib$ relative to the position of its complex conjugate $\bar{z} = (a, -b) = a - ib$ when considered as points (or vectors) in the plane.

B.12 Show that $\overline{z^{-1}} = \bar{z}^{-1}$. Symbolically, show that $\overline{\left(\dfrac{1}{z}\right)} = \dfrac{1}{(\bar{z})}$. *Hint:* Use the fact that $z^{-1} \cdot z = 1$ and see Exercise B.7(f).

B.13 Prove that $z = \bar{z}$ if and only if z is real.

B.14 Use Definition B.3 to find $(0,b)^2$ for any real number b. How does the result compare to the number $(ib)^2 = i^2b^2 = -1b^2 = -b^2$? Is the real number $-b^2$ the same as the complex number $(-b^2,0)$?

B.15 The identification of real numbers a with the complex number $(a,0)$ was shown by equating the complex multiplicative identity $(1,0)$ to the real number multiplicative identity 1. Is it still true, that for all real a,b, $(a,0) \cdot (b,0) = (ab,0)$?

B.16 We have seen that in order to multiply a 2-vector (x_1,x_2) by a scalar λ, we multiply each entry by λ, that is, $\lambda(x_1,x_2) = (\lambda x_1,\lambda x_2)$. Does the same result hold if the ordered pair (x_1,x_2) is regarded as the complex number $x_1 + ix_2$ and the real scalar λ is identified with the complex number $(\lambda,0)$? That is, do we have $\lambda \cdot (x_1,x_2) = (\lambda,0) \cdot (x_1,x_2) = (\lambda x_1,\lambda x_2)$ for all real λ?

B.17 Find the complex multiplicative inverses for:

(a) $3 + 3i$ (d) $4 - 5i$ (g) $\sqrt{2} - i\sqrt{3}$

(b) $3 - 38$ (e) $(9,1)$ (h) i

(c) 6 (f) $-7 + 2i$ (i) $2 - 10i$.

B.18 Find the complex number represented by:

(a) $\dfrac{-1+2i}{1-2i}$ (d) $\dfrac{3-4i}{2i}$

(b) $\dfrac{7-i}{5}$ (e) $\dfrac{-6+7i}{-6-7i}$

(c) $\dfrac{6+2i}{6-2i}$ (f) $\dfrac{-6+7i}{-6+7i}$

B.19 Show that $\left|\dfrac{\bar{z}}{z}\right| = 1$ for all $z \neq 0$. (*Hint:* $|w|^2 = w\bar{w}$ for any complex number w.)

B.20 Show that $|z| = 0$ if and only if $z = 0$.

B.21 Show that the nonzero complex number $\dfrac{z}{|z|}$ always has modulus one, that is, $\left|\dfrac{z}{|z|}\right| = 1$ for all nonzero $z \in \mathbf{C}$.

4. Polar Form for Complex Numbers

The following discussion appeals to the graphical representation of complex numbers as points or vectors in the plane. Properties of complex numbers are then expressed in trigonometric terms. (See Figure B.2.) For example, we may define the *length* of the complex number $z = a + ib$ to be the length of the hypotenuse of the triangle in the above figure. Pythagorus has already told us how to find this number:

$$r = \text{length } z = \sqrt{a^2 + b^2}.$$

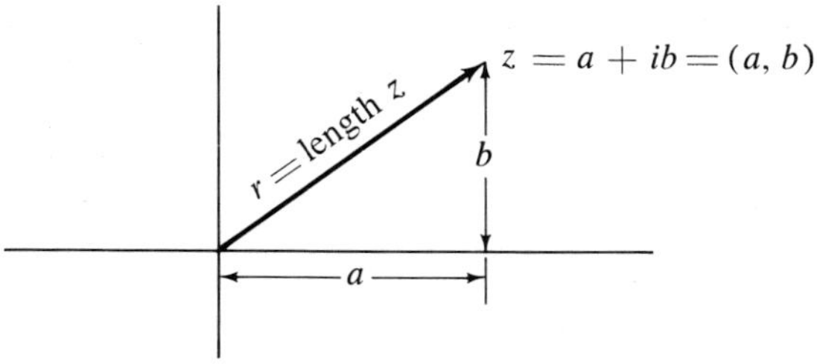

Figure B.2

But this is exactly what we have *already* called the modulus of z in Definition B.5. Thus, for $z = a + ib$,

$$|z|^2 = z \cdot \bar{z} = a^2 + b^2$$

can be thought of as a measure of the *length* of z only because we have a graphical model for z as a vector in the plane. Now it becomes *intuitively* clear that $z = 0$ [the origin (0,0) if and only if $|z|$, the length of z, is zero (see Exercise B.20)].

Consider trigonometric decomposition of z in Figure B.3:

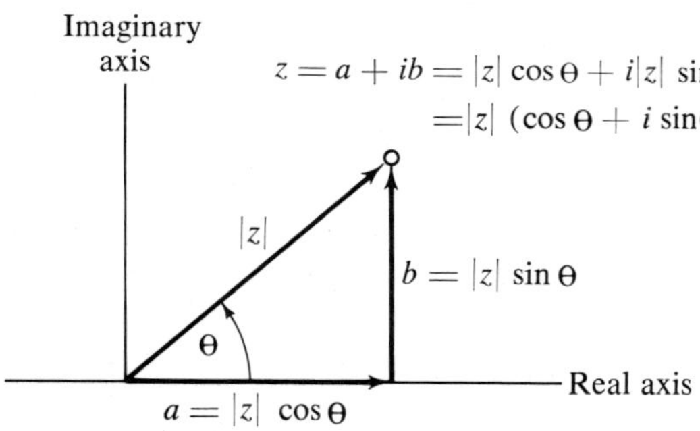

Figure B.3

A fact of trigonometry is that the length of legs of the right triangle can always be expressed in terms of $|z|$, the lenth of the hypotenuse, and θ, one of the interior angles. Specifically, we have

$$\begin{aligned} a \text{ (the horizontal side)} &= |z| \cos(\theta) \\ b \text{ (the vertical side)} &= |z| \sin(\theta). \end{aligned}$$

With these equations at hand, we are forced to conclude that for any complex number $z = (a,b) = a + ib$,

$$\begin{aligned} z = a + ib &= |z| \cos(\theta) + i|z| \sin(\theta) \\ &= |z|[\cos(\theta) + i \sin(\theta)]. \end{aligned} \tag{B.9}$$

We see that any complex number z which accounts for an angle θ with the horizontal axis, has the form $z = |z|\ (\cos(\theta) + i\sin(\theta))$. The complex number $(\cos(\theta) + i\sin(\theta)) = (\cos(\theta), \sin(\theta))$ which appears deserves special attention. Hence, Notation B.4.

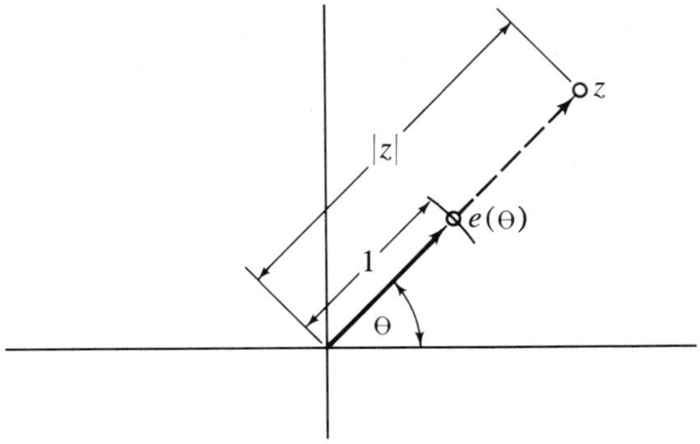

Figure B.4

Notation B.4 For any angle θ, the symbol $e(\theta)$ is defined by

$$e(\theta) = \cos(\theta) + i\sin(\theta) = (\cos(\theta), \sin(\theta)).$$

It is easy to see that $e(\theta)$, is a unit vector (that is, has length $|e(\theta)|$ equal to *one*), and that, like z, it forms an angle of θ with the horizontal axis. (See Figure B.4.) Intuitively, it makes sense to think of constructing z by choosing the unit vector $e(\theta)$ which "points in the same direction as z" (accounts for the same angle θ, as does z), and then, by multiplying (stretching) that unit vector by $|z|$, the length of z. Thanks to (B.9) and Notation B.4, we have Definition B.6.

Definition B.6 For any complex number $z = a + ib = (a,b)$, the representation

$$z = |z|e(\theta) = |z|(\cos(\theta) + i\sin(\theta))$$

is called the **polar form** (representation) of z. In solving $z = |z|e(\theta)$ for $e(\theta)$, we obtain

$$e(\theta) = \frac{z}{|z|}.$$

This agrees with our intuition in that any vector (complex number), when divided by its length, results in a *unit* vector pointing in the *same* direction (see Exercise B.21).

Example Express the complex numbers $z_1 = \sqrt{3} + i1$ and $z_2 = 3 - 4i$ in polar form.

In order to write $z_i = |z_i|e(\theta_i)$ $i = 1, 2$, we must find (1) $|z_i|$, the length of z_i, and (2) $e(\theta_i)$, the unit vector which enjoys the same angle of direction θ_i, as does z_i. Part (1) is easy. Since for any $z = a + ib$, $|z| = \sqrt{a^2 + b^2}$ (Definition B.5), we see that

$$\text{(a)} \quad |z_1| = |\sqrt{3} + i1| = \sqrt{(\sqrt{3})^2 + 1^2} = \sqrt{3 + 1} = \sqrt{4} = 2,$$

and

$$\text{(b)} \quad |z_2| = |3 - 4i| = \sqrt{3^2 + (-4)^2} = \sqrt{9 + 16} = \sqrt{25} = 5.$$

As we have mentioned above, any nonzero complex number z, divided by its modulus $|z|$, is a complex number of modulus, or length 1 (see Exercise B.21). Thus, we may write

$$z_1 = \frac{|z_1|}{|z_1|} z_1 = \underbrace{|z_1|}_{\text{modulus of } z_1} \underbrace{\frac{z_1}{|z_1|}}_{\text{complex number of modulus one}}$$

so that $\dfrac{z_1}{|z_1|}$ is that unit vector (complex number) whose length is one, and whose direction is the same as that of z_1 since $\dfrac{z_1}{|z_1|}$ is just a positive scalar multiple of z_1. The above equation represents z_1 in its polar form $z_1 = |z_1|e(\theta)$ where $e(\theta) = \dfrac{z_1}{|z_1|}$. Thus,

$$z_1 = |z_1| \frac{z_1}{|z_1|} = 2\left(\frac{\sqrt{3} + i1}{2}\right) = \underset{|z_1|}{\underbrace{2}} \underbrace{\left(\frac{\sqrt{3}}{2} + \frac{1}{2}i\right)}_{e(\theta_2)},$$

and

$$z_2 = |z_2| \left(\frac{z_2}{|z_2|}\right) = 5\left(\frac{3 - 4i}{5}\right) = \underset{|z_2|}{\underbrace{5}} \underbrace{\left(\frac{2}{5} - \frac{4}{5}i\right)}_{e(\theta_2)}.$$

If any doubts remain as to the modulus of $e(\theta_i)$, $i = 1, 2$, we need only check

$$|e(\theta_1)| = \left|\frac{\sqrt{3}}{2} + \frac{1}{2}i\right| = \sqrt{\frac{3}{4} + \frac{1}{4}} = \sqrt{1} = 1,$$

and

$$|e(\theta_2)| = \left|\frac{3}{5} - \frac{4}{5}i\right| = \sqrt{\frac{9}{25} + \frac{16}{25}} = \sqrt{\frac{25}{25}} = 1.$$

Why do we concentrate on the polar representation of complex numbers? For $z \in \mathbf{C}$ expressed in its original (cartesian) form $z = (a,b) = a + ib$, we must agree that problems of addition and subtraction are almost trivial. In fact, addition is effected by coordinate addition. Multiplication and division, however, are another matter. Happily, the polar form almost trivializes this type of problem. For example, we shall soon be able to find, without much effort, all the complex cube roots, say, for the complex number $1 + i$.

Let z_1 and z_2 be any 2 nonzero complex numbers. Then

$$\begin{aligned} z_1 \cdot z_2 &= |z_1|\left(\frac{z_1}{|z_1|}\right) \cdot |z_2|\left(\frac{z_2}{|z_2|}\right) \\ &= |z_1|\, e(\theta_1) \cdot |z_2|\, e(\theta_2) \\ &= |z_1|\, |z_2|\, e(\theta_1)\, e(\theta_2) \\ &= |z_1 z_2|\, e(\theta_1)\, e(\theta_2) \qquad \text{Exercise B.7(f).} \end{aligned}$$

With this last equation, it behoves us to evaluate the complex product $e(\theta_1)e(\theta_2)$ of the unit complex numbers, $e(\theta_1)$ and $e(\theta_2)$. Amazingly, the product of $e(\theta_1)$ with $e(\theta_2)$ is also a unit complex number $e(\theta_3)$. Moreover, θ_3, the angle of the product is the *sum* of the angles θ_1 and θ_2. That is, Theorem B.2.

Theorem B.2 *Let* $e(\theta_1) = \cos(\theta_1) + i \sin(\theta_1)$, *and* $e(\theta_2) = \cos(\theta_2) + i \sin(\theta_2)$ *be 2 unit vectors (complex numbers). Then*

$$e(\theta_1)\cdot e(\theta_2) = e(\theta_2)\cdot e(\theta_1) = e(\theta_1 + \theta_2).$$

Proof: We need only to set down the following pair of trigonometric identities:

(a) $$\sin(\theta_1 + \theta_2) = \sin(\theta_2)\cos(\theta_n) + \cos(\theta_1)\sin(\theta_2)$$

(b) $$\cos(\theta_1 + \theta_2) = \cos(\theta_2)\cos(\theta_n) - \sin(\theta_1)\sin(\theta_2).$$

Thus,

$$\begin{aligned} e(\theta_1)e(\theta_2) &= (\cos(\theta_1) + i\sin(\theta_1))\cdot(\cos(\theta_2) + i\sin(\theta_2)) \\ &= \cos(\theta_1)\cos(\theta_2) + i^2 \sin(\theta_1)\sin(\theta_2) + i\sin(\theta_1)\cos(\theta_2) \\ &\qquad + \cos(\theta_1) i \sin(\theta_2) \\ &= (\cos(\theta_1)\cos(\theta_2) - \sin(\theta_1)\sin(\theta_2)) + i(\sin(\theta_1)\cos(\theta_2) \\ &\qquad \text{from (b)} \qquad\qquad \text{from (a)} \quad + \cos(\theta_1)\sin(\theta_2)) \\ &= \cos(\theta_1 + \theta_2) + i\sin(\theta_1 + \theta_2) \\ &= e(\theta_1 + \theta_2). \qquad \text{(Notation B.4).} \end{aligned}$$

This completes the proof.

An easy corollary of this theorem describes an intuitive way in which complex multiplication proceeds. It may be described as follows: In order to obtain the product of the complex numbers z_1 and z_2, construct the complex number (in polar form), whose length is $|z_1 \cdot z_2| = |z_1| \cdot |z_2|$, the product of the lengths of z_1, and z_2, and whose angle with the real (horizontal) axis is $\theta_1 + \theta_2$, the sum of the angles θ_1 and θ_2 described by z_1 and z_2, respectively. More formally, we have Corollary B.1.

Corollary B.1 *For* $z_1 = |z_1|e(\theta_1)$, *and* $z_2 = |z_2|e(\theta_2)$. The polar form of the product $z_1 z_2$ is given by

$$z_1 \cdot z_2 = |z_1 \cdot z_2| e(\theta_1 + \theta_2).$$

Proof: In partial repetition of the discussion previous to Theorem B.2, we have

$$\begin{aligned} z_1 \cdot z_2 &= |z_1| e(\theta_1) \cdot |z_2| e(\theta_{n2}) \\ &= |z_1| \cdot |z_2| \cdot e(\theta_1) e(\theta_2) \\ &= |z_1 \cdot z_2| e(\theta_1) \cdot e(\theta_2) \qquad \text{Exercise B.7(f)} \\ &= |z_1 \cdot z_2| e(\theta_1 + \theta_2) \qquad \text{Theorem B.2,} \end{aligned}$$

and the corollary is established.

Choose $z_1 = \sqrt{3} + i$, and $z_2 = -2\sqrt{3} + 2i$. (See Figure B.5.) Regarding

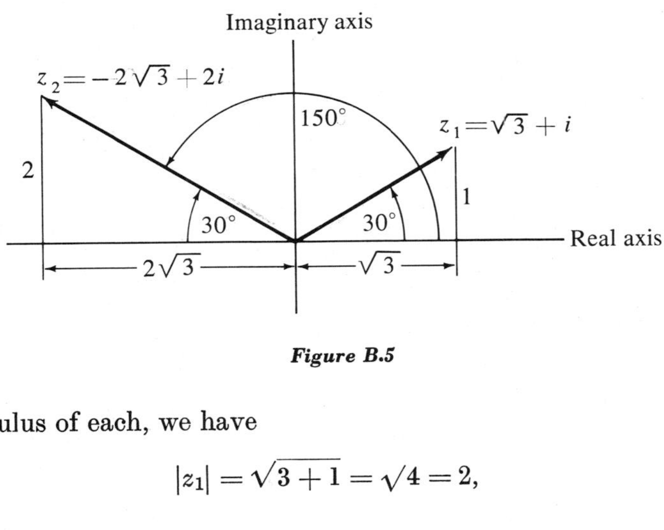

Figure B.5

the modulus of each, we have

$$|z_1| = \sqrt{3 + 1} = \sqrt{4} = 2,$$

and

$$|z_2| = \sqrt{4 \cdot 3 + 4} = \sqrt{16} = 4.$$

The polar terms for z_1 and z_2 are then

$$z_1 = |z_1| \frac{z_1}{|z_1|} = 2\left(\frac{\sqrt{3} + i}{2}\right) = 2\left(\frac{\sqrt{3}}{2} + \frac{1}{2}i\right) + 2e(\theta_i),$$

and

$$z_2 = |z_2| \frac{z_2}{|z_2|} = 4\left(\frac{-2\sqrt{3} + 2i}{4}\right) = 4\left(\frac{-\sqrt{3}}{2} + \frac{1}{2}i\right) = 4e(\theta_2).$$

In this convenient textbook example, we have chosen

$$e(\theta_1) = \left(\frac{\sqrt{3}}{2} + \frac{1}{2}i\right) = \cos(30°) + i \sin(30°)$$

and

$$e(\theta_2) = \left(-\frac{\sqrt{3}}{2} + \frac{1}{2}i\right) = \cos(150°) + i \sin(150°).$$

That is, θ_1 and θ_2 are the familiar (albeit unlikely) angles 30° ($\pi/6$ radians) and 150° ($5/6\pi$ radians), respectively. If Corollary B.1 maintains, then the *length* or *modulus* of $z_1 \cdot z_2$ is the product of the lengths, that is,

$$|z_1 \cdot z_2| = |z_1| \cdot |z_2| = 2 \cdot 4 = 8.$$

Also, the angle θ_3 of $z_1 \cdot z_2$ is the *sum* of the angles θ_1 and θ_2, that is, $\theta_3 = \theta_1 + \theta_2 = 30° + 150° = 180°$. That is,

$$\begin{aligned} z_1 \cdot z_2 &= 2e(30°) \cdot 4e(150°) \\ &= 8e(30° + 150°) = 8e(180°). \end{aligned}$$

But what *is* the complex number $z_1 \cdot z_2 = 8e(180°)$? It is nothing more than a complex number of modulus 8, forming an angle of $180° = \pi$ radians, that is, in Figure B.6 $8e(180°)$ lies right on the real axis, and is in the

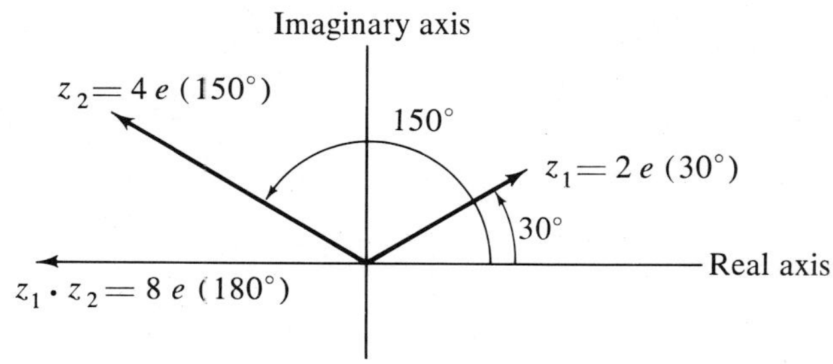

Figure B.6

negative section, so that $8e(180°)$ is actually the real number -8.

Let us check this by our old-fashioned, bare hands multiplication. Thus,

$$\begin{aligned} z_1 \cdot z_2 &= (\sqrt{3} + i)(-2\sqrt{3} + 2i) \\ &= -2\sqrt{3} \cdot \sqrt{3} + i^2 \cdot 2 - 2i\sqrt{3} + 2i\sqrt{3} \\ &= -(2 \cdot 3) - 2 \\ &= -6 - 2 = -8, \end{aligned}$$

which is exactly the result predicted by the polar form of multiplication.

EXERCISES

B.22 Represent the given complex numbers in polar form.

(a) $3 + 3i$ (e) $3 + 12i$

(b) $3 - 3i$ (f) $\sqrt{3} - i$

(c) $-3 + 3i$ (g) $a + 2i$

(d) $-3 - 3i$ (h) $6i$

B.23 Find the following products (see Theorem B.2).

(a) $6(\cos 15° + i \sin 15°) \cdot 2(\cos (30°) + i \sin (30°))$

(b) $3(\cos (10°) + i \sin (10°)) \cdot 4(\cos 80° + i \sin 80°)$

(c) $[2(\cos 36° + i \sin 36°)]^5$

(d) $[3(\cos 45° + i \sin 45°)]^4 \cdot [4(\cos 5° + i \sin 5°)]^2$

(e) $5(\cos (-20°) + i \sin(-20°)) \cdot 6(\cos 35° + i \sin 35°)$.

B.24 Show that a complex number $z = |z|e(\theta) = |z|(\cos (\theta) + i \sin (\theta))$ is real if and only if $\theta = k \cdot 180° = k \cdot \pi$ radians, where $k = 0, \pm 1, \pm 2, \ldots$.

B.25 Using the fact that for all θ, $\cos (\theta) = \cos (-\theta)$ and $-\sin (\theta) = \sin (-\theta)$, prove that $\overline{e(\theta)}$, the conjugate of $e(\theta)$, equals $e(-\theta)$. Prove the more general fact that if $z = |z|e(\theta)$, then $\bar{z} = |z|e(-\theta)$.

B.26 **(De Moivre)** Prove that for all $n = 1, 2, 3, \ldots,$

$$e(\theta)^n = e(n\theta),$$

that is,

$$(\cos (\theta) + i \sin (\theta))^n = \cos (n\theta) + i \sin (n\theta).$$

Hint: See Theorem B.2; use mathematical induction on n.

B.27 (See Exercise B.26.) Find the following powers:

(a) $(1 - i)^5$ (d) $(1 - \sqrt{3}\, i)^7$

(b) $\left(\frac{\sqrt{3}}{2} + \frac{1}{2} i\right)^{10}$ (e) $(e + i)^3$

(c) $(5 + 5i)^{12}$ (f) $(2\sqrt{3} - 2i)^9$

B.28 Prove that the polar form for the number 1 is $1 = e(\theta)$, where $\theta = k \cdot 360° = k \cdot 2\pi$ radians, $k = 0, \pm 1, \pm 2, \ldots$. Note that the angle θ is not unique since 0° is the same physical angle as $\pm 360°$, $\pm 720°$, and so on.

B.29 Prove that $e(\theta)^{-1} = e(-\theta)$, that is, show that

$$e(\theta) \cdot e(-\theta) = 1.$$

B.30 **(Extension of De Moivre's Theorem)** Suppose θ is an angle such that $n\theta$ is some integer multiple of 360°. That is to say, suppose

$$n\theta = \underbrace{\theta + \theta + \cdots + \theta}_{n \text{ terms}} = k \cdot 360° \text{ for some fixed } k = 0, \pm 1, \pm 2, \ldots.$$

(a) Prove that $e(n\theta) = 1$ (see Exercise B.28).

(b) Prove that $e(\theta)^n = 1$ also (see Theorem B.2).

Thus, $e(\theta) = 1^{1/n}$ is an nth root of *unity*.

B.31 (Division with polar form.) Suppose z is a nonzero complex number (that is, $|z| \neq 0$). Prove that if $z = |z|\, e(\theta)$, then $z^{-1} = |z|^{-1} e(-\theta)$. In other words, verify the equation

$$\underbrace{|z| \cdot e(\theta)}_{z} \cdot |z|^{-1} e(-\theta) = 1$$

which will establish that the factor $|z|^{-1}e(-\theta) = z^{-1}$ (see Exercise B.29).

B.32 Use the result of Exercise B.29 to find the polar form for z^{-1} for the given complex numbers z. By multiplication, verify the equation $z \cdot z^{-1} = 1$ in each case.

(a) $z = 3(\cos(16°) + i \sin(6°))$

(b) $z = 4(\cos(45°) + i \sin(45°))$

(c) $z = \frac{1}{2}(\cos(90°) + i \sin(90°))$

(d) $z = \frac{1}{2}(\cos(180°) + i \sin(180°))$

(e) $z = 5\left(\frac{1}{2} + \frac{i}{2}\sqrt{3}\right)$

(f) $z = 2 - 2i$

(g) $z = \sqrt{3} - i$

(h) $z = 4 - 4i$

We tackle the problem of ***n*th roots of complex numbers**. That is, given $z_0 \in \mathbf{C}$, we are to find all $z \in \mathbf{C}$ such that $z^n = z^3$. We proceed by example.

Problem. Find all z such that $z^3 = 1 + i$. Such z (if any exist) will be called *cube roots* of $1 + i$, and are denoted by the symbol $z = (1 + i)^{1/3}$. First, we put $1 + i$ into polar form. Observe that $|1 + i| = \sqrt{1^2 + 1^2} = \sqrt{2}$, we have

$$1 + i = \sqrt{2}\,\frac{(1+i)}{\sqrt{2}} + \sqrt{2}\left(\frac{1}{\sqrt{2}} + i\,\frac{1}{\sqrt{2}}\right) = \sqrt{2}(\cos(45°) + i \sin(45°))$$
$$= \sqrt{2}\, e(45°).$$

From the polar form, it is clear that $z = (\sqrt{2})^{1/3}\, e\left(\frac{45°}{3}\right) = (\sqrt{2})^{1/3}\, e(15°)$ is a cube root of $1 + i$, for

$$\begin{aligned} z^3 &= (\sqrt{2})^{1/3}\, e(15°) \cdot (\sqrt{2})^{1/3}\, e(15°) \cdot (\sqrt{2})^{1/3}\, e(15°) \\ &= (\sqrt{2})^{1/3}(\sqrt{2})^{1/3}(\sqrt{2})^{1/3}\, e(15°)\, e(15°)\, e(15°) \\ &= \sqrt{2}[e(15° + 15° + 15°)] \text{ Theorem B.2} \\ &= \sqrt{2}[e(45°)] = 1 + i. \end{aligned}$$

In general, if $z = r \cdot [e(\theta)]$ is a cube root of $\sqrt{2}[e(45°)]$, we must have $z^3 = \sqrt{2}[e(45°)]$, or

$$\begin{aligned} z^3 &= r[e(\theta)] \cdot r[e(\theta)] \cdot r[e(\theta)] \\ &= r^3[e(\theta)e(\theta)e(\theta)] \\ &= r^3[e(3\theta)] = \sqrt{2}\,[e(45°)]. \end{aligned}$$

This implies that $r^3 = \sqrt{2}$, or that

$$r = (\sqrt{2})^{1/3},$$

the cube root of $\sqrt{2}$, the length of $1 + i$. Thus any cube root z is of the form

$$z = (\sqrt{2})^{1/3}[e(\theta)].$$

But now θ must be an angle such that

$$3\theta = 45°.$$

But 45° "looks like" 360° + 45°, or 720° + 45°, so that

$$3\theta = 45° + k \cdot 360°, \; k = 0, 1, 2, \ldots.$$

Summary

In order to find $z = (1 + i)^{1/3} = (\sqrt{2} \cdot e(45°))^{1/3}$, we must have $|z| = (\sqrt{2})^{1/3}$, so that $z = (\sqrt{2})^{1/3} \cdot e(\theta)$, where $3\theta = 45°$. We write this last equation in apparently distinct forms as

(1) $3\theta = 45°$

(2) $3\theta = 45° + 360°$

(3) $3\theta = 45° + 720°$

(4) $3\theta = 45° + 1080°$

$\vdots \quad \vdots$

The respective solutions to these equations are

(1) $\theta = 15°$

(2) $\theta = 15° + 120°$

(3) $\theta = 15° + 240°$

(4) $\theta = 15° + 360°$ — same angle as 15° from (1).

Thus 3 *distinct* angles $\theta_1 = 15°$, $\theta_2 = 15° + 120° = 135°$, and $\theta_3 = 15° + 240° = 255°$ arise with the property that $3\theta_i = 45°(+k360°)$ $i = 1, 2, 3$. Moreover, these angles are *evenly* spaced around the circle in that they differ by 120° (see Figure B.7). This can only mean that we have unearthed 3

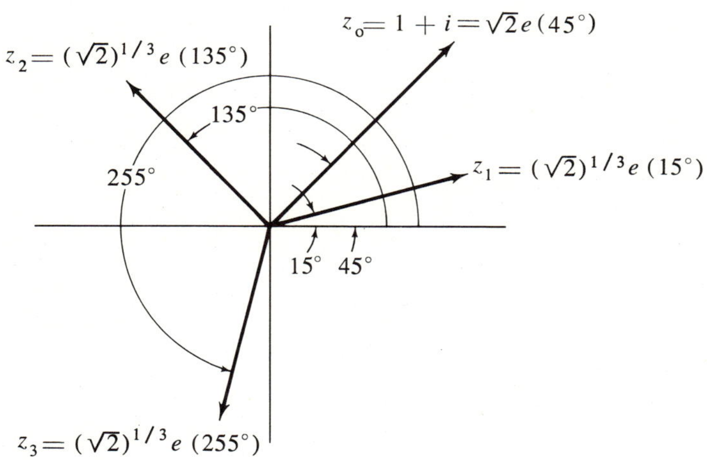

Figure B.7

cube roots z_1, z_2, and z_3 for $1 + i$, namely,

$$z_1 = (\sqrt{2})^{1/3}e(15°) = (\sqrt{2})^{1/3}(\cos 15° + i \sin 15°)$$

$$z_2 = (\sqrt{2})^{1/3}e(135°) = (\sqrt{2}^{1/3}(\cos 135° + i \sin 135°)$$

$$z_3 = (\sqrt{2})^{1/3}e(255°) = (\sqrt{2})^{1/3}(\cos 255° + i \sin 255°)$$

Cube roots $z_0 = 1 + i$

B.33 Find the following roots and graph them:

(a) $(i)^{1/4}$

(b) $(-1)^{1/4}$

(c) $(1)^{1/4}$

(d) $[4\sqrt{2}(1 + i)]^{1/3}$

(e) $(i)^{1/2}$

(f) $(-\sqrt{3} + i)^{1/3}$

(g) $[8(1 + \sqrt{3}\, i)]^{1/4}$

(h) $16^{1/4}$

(*i*) $(-1) + i^{1/3}$

(j) $\left(\frac{\sqrt{3}}{2} - \frac{1}{2}i\right)^{1/5}$

B.34 Show that the nth roots of unity (all $z \in \mathbf{C}$ such that $z^n = 1$ are the n complex numbers of modulus one (that is, on the unit circle) which are equally spaced $\left(\frac{360°}{n}\right)$ apart. The real number $z = 1$ (whose angle is 0°) is always an nth root of unity; hence, the angles of the other $n - 1$ roots are

$$\frac{360°}{n}, \quad \frac{2}{n} \cdot 360°, \quad \frac{3}{n} \cdot 360°, \ldots, \frac{n-1}{n} 360°.$$

Index

Z